An Introduction to Neural and Electronic Networks

Second Edition

This is a volume in
NEURAL NETWORKS: FOUNDATIONS TO APPLICATIONS

Edited by Steven F. Zornetzer, Joel Davis, Clifford Lau, and Thomas McKenna
Office of Naval Research
Arlington, Virginia

NEURAL NETWORKS: FOUNDATIONS TO APPLICATIONS

Steven F. Zornetzer, Joel Davis, Clifford Lau,
and Thomas McKenna, editors

Thomas McKenna, Joel Davis, and Steven Zornetzer
Single Neuron Computation

Randall D. Beer, Roy E. Ritzman, and Thomas McKenna
Biological Neural Networks in Neuroethology and Robotics

Ronald J. MacGregor
Theoretical Mechanics of Biological Neural Networks

Vasant Honavar and Leonard Uhr
Artificial Intelligence and Neural Networks

An Introduction to Neural and Electronic Networks

Second Edition

EDITED BY

Steven F. Zornetzer, Joel L. Davis, Clifford Lau, and Thomas McKenna

Office of Naval Research, Arlington, Virginia

ACADEMIC PRESS

San Diego New York Boston London Sydney Tokyo Toronto

Cover artwork (paperback edition only): The cover is a three-dimensional representation of cellular activity in a network model of olfactory cortex constructed by James M. Bower, Matthew Wilson, and Alex Protopapas (described in Chapter 1). The image was rendered using the GENESIS simulation system in collaboration with Tom Defanti and Jason Leigh at the Electronic Visualization Laboratory of the University of Illinois at Chicago.

This book is printed on acid-free paper. ♾

Academic Press, Inc.
A Division of Harcourt Brace & Company
525 B Street, Suite 1900, San Diego, California 92101-4495

United Kingdom Edition published by
Academic Press Limited
24-28 Oval Road, London NW1 7DX

Library of Congress Cataloging-in-Publication Data

An introduction to neural and electronic networks / edited by Steven F. Zornetzer ... {et al.]. -- 2nd ed.
p. cm. -- (Neural networks, foundations to applications series)
Includes bibliographical references and indexes.
ISBN 0-12-781882-0 (case), -- ISBN 0-12-781883-9 (paper)
1. Neural networks (computer science) 2. Electronic circuits.
I. Zornetzer, Steven F. II. Series; Neural networks, foundations to applications.
QA76.87.I5895 1994
006.3--dc20 94-20815
CIP

PRINTED IN THE UNITED STATES OF AMERICA
94 95 96 97 98 99 EB 9 8 7 6 5 4 3 2 1

Contents

Contributors

Numbers in parentheses indicate the pages on which the authors' contributions begin.

Lex A. Akers (359), Center for Solid State Electronics Research, Arizona State University, Tempe, Arizona 85287

Charles H. Anderson (45), Department of Anatomy and Neurobiology, Washington University School of Medicine, St. Louis, Missouri 63110

Randall D. Beer (165), Department of Computer Engineering and Science, and Department of Biology, Case Western Reserve University, Cleveland, Ohio 44106

David Blackman (141), Department of Biochemistry and Biophysics, University of Pennsylvania, Philadelphia, Pennsylvania 19104

James M. Bower (3), Division of Biology, Computation and Neural Systems Program, California Institute of Technology, Pasadena, California 91125

Gail A. Carpenter (465), Center for Adaptive Systems, and Department of Cognitive and Neural Systems, Boston University, Boston, Massachusetts 02215

Hillel J. Chiel (165), Department of Biology, and Department of Neuroscience, Case Western Reserve University, Cleveland, Ohio 44106

Leon N Cooper (229), Department of Physics, and Center for Neural Science, Brown University, Providence, Rhode Island 02912

Robert Coultrip (31), Center for the Neurobiology of Learning and Memory, University of California at Irvine, Irvine, California 92717

Rodney Douglas (277), Medical Research Council Anatomical Neuropharmacology Unit, Department of Pharmacology, University of Oxford, Oxford OX1 3TH, United Kingdom

Rolf Eckmiller (99), Division of Biocybernetics, Department of Biophysics, Heinrich-Heine University, Düsseldorf, D-4000 Düsseldorf 1, Germany

Gerald M. Edelman (205), The Neurosciences Institute, and The Rockefeller University, New York, New York 10021

Federico Faggin (297), Synaptics Incorporated, San Jose, California 95134

David K. Ferry (359), Center for Solid State Electronics Research, Arizona State University, Tempe, Arizona 85287

Leif H. Finkel (205), The Neurosciences Institute, and The Rockefeller University, New York, New York 10021

Walter J. Freeman (119), Department of Molecular and Cell Biology, University of California at Berkeley, Berkeley, California 94720

Roy Furman (141), Department of Neurology, University of Pennsylvania, Philadelphia, Pennsylvania 19104

Mark A. Gluck (77, 91), Center for Molecular and Behavioral Neurosciences, Rutgers University, Newark, New Jersey 07102

Richard Granger (31), Center for the Neurobiology of Learning and Memory, University of California at Irvine, Irvine, California 92717

Robert O. Grondin (359), Center for Solid State Electronics Research, Arizona State University, Tempe, Arizona 85287

Stephen Grossberg (465), Center for Adaptive Systems, and Department of Cognitive and Neural Systems, Boston University, Boston, Massachusetts 02215

Norberto M. Grzywacz (405), The Smith-Kettlewell Eye Research Institute, San Francisco, California 94115

Dan Hammerstrom (335), Adaptive Solutions, Inc., Beaverton, Oregon 97006

Karl Kilborn (31), Center for the Neurobiology of Learning and Memory, University of California at Irvine, Irvine, California 92717

Christof Koch (315), Computation and Neural Systems Program, 216-76 Division of Biology, California Institute of Technology, Pasadena, California 91125

John Lazzaro (185), Department of Computer Science, California Institute of Technology, Pasadena, California 91125

Gary Lynch (31), Center for the Neurobiology of Learning and Memory, University of California at Irvine, Irvine, California 92717

Misha Mahowald (277), Medical Research Council Anatomical Neuropharmacology Unit, Department of Pharmacology, University of Oxford, Oxford OX1 3TH, United Kingdom

Carver Mead (185, 297), Department of Computer Science, California Institute of Technology, Pasadena, California 91125

Paul Mueller (141), Department of Biochemistry and Biophysics, University of Pennsylvania, Philadelphia, Pennsylvania 19104

Catherine E. Myers (77, 91), Center for Molecular and Behavioral Neurosciences, Rutgers University, Newark, New Jersey 07102

Richard Myers (31), Center for the Neurobiology of Learning and Memory, University of California at Irvine, Irvine, California 92717

Tomaso Poggio (405), Center for Biological and Computational Learning, and Artificial Intelligence Laboratory, Massachusetts Institute of Technology, Cambridge Massachusetts 02139

George N. Reeke, Jr. (205), The Neurosciences Institute, and The Rockefeller University, New York, New York 10021

Douglas L. Reilly (229), Nestor Incorporated, Providence, Rhode Island 02906

Roy E. Ritzmann (165), Department of Biology and Department of Neuroscience, Case Western Reserve University, Cleveland, Ohio 44106

David E. Rumelhart (431), Department of Psychology, Stanford University, Stanford, California 94305

Terrence J. Sejnowski (391), Department of Biophysics, Johns Hopkins University, Baltimore, Maryland 21218

Koji Shimoide (119), Department of Molecular and Cell Biology, University of California at Berkeley, Berkeley, California 94720

Patric K. Stanton (391), Department of Biophysics, Johns Hopkins University, Baltimore, Maryland 21218

Richard F. Thompson (91), Neural Information and Behavioral Sciences (NIBS), University of Southern California, Los Angeles, California 90089

David C. Van Essen (45), Department of Anatomy and Neurobiology, Washington University School of Medicine, St. Louis, Missouri 63110

Christoph von der Malsburg (447), Institut für Neuroinformatik, Ruhr-Universität Bochum, Germany, and Computer Science Department and Section for Neurobiology, University of Southern California, Los Angeles, California 90089

Eric Whelpley (31), Center for the Neurobiology of Learning and Memory, University of California at Irvine, Irvine, California 92717

Jim Whitson (31), Center for the Neurobiology of Learning and Memory, University of California at Irvine, Irvine, California 92717

Bernard Widrow (251), Department of Electrical Engineering, Stanford University, Stanford, California 94305

Rodney Winter (251), Department of Electrical Engineering, Stanford University, Stanford, California 94305

Preface to the Second Edition

In the five years that have passed since we presented our view of the neural network world, great progress has been made. We have reorganized this edition to reflect that progress. The first chapter, by Bower, provides a good example of changes in the field. The discovery of oscillations in the activity of the central nervous system led to a rush of "top-down" approaches to the modeling effort. However, Bower's work shows that attention to structural details (the "bottom-up" approach) may yield important new insights into network design.

The U.C. Irvine group (Granger *et al.*) continues to demonstrate how simulation of brain networks with anatomically specified architectures and physiological function can yield novel algorithms. The authors present a new sequence-recognition network, based on the hippocampus, that is capable of storing 50 million sequences with an extremely low rate of error.

Van Essen and Anderson have updated their chapter to emphasize the need for dynamic control of information in the brain, showing the significant importance of network models in what was originally a largely descriptive enterprise. Advances in cortical microcircuitry have led to a cascaded network architecture that is useful for understanding information processing in the visual system.

In a new chapter, Gluck and Myers present a neurocomputational theory of stimulus representation and learning that is based on hippocampal function. Their model accurately predicts a wide range of learning and generalization behaviors in both intact and lesioned animals. Then, in Chapter 5, Gluck and Myers join with Thompson to present new findings arising from their computational model of the cerebellum. They also offer the beginnings of a network approach to link motor-reflex (cerebellar) conditioning and higher-order (hippocampal) functions.

Nonlinear concepts in neural information processing is a much stronger theme in this edition. In a new chapter, Freeman and Shimoide utilize the nonlinear approach in examining EEG data as the basis for optimizing parameters in models used to simulate brain function. Freeman's model is taken from the vertebrate olfactory system, in which the number of neurons ranges upward to 100 million, but the degrees of freedom needed for modeling are far fewer because of neuronal cooperativity.

In a new chapter in our Emulated and Simulated Systems section, Beer, Ritzmann, and Chiel describe a design for motor control in an autonomous robot based on the nervous system of the cockroach. Their work suggests computational neuroethology as a new approach to robotics.

The Electronic Networks section also has a new addition. Douglas and Mahowald describe their first step toward providing a complete set of circuit modules for constructing analogs of individual neurons in complementary metal-oxide semiconductor (CMOS), very large scale integrated (VLSI) technology. Their approach is based on a "physically motivated" description of the neuron derived from neurophysiological data.

In updating his chapter on vision chips, Koch shows how the resistive network ideas he described in the first edition have led toward object-based vision chips that are now commercially produced. Further, Koch compares analog and digital systems and speculates about the future of these systems.

Hammerstrom also provides an update of his work with digital VLSI architecture. His early model has been physically implemented (CNAPS architecture) in two chips, one with 64 PNs and the other with 16 PNs. Hammerstrom discusses the implementation of various biological and artificial neural networks on his chip and offers a number of current real-world applications.

The chapter by Akers *et al.* predicts an "explosive growth" of neural network systems and applications in the 1990s. As the authors point out, the great challenge lies in developing synthetic neural chips with on-chip learning. Their solution is a design that implements an unsupervised learning algorithm as an integral part of the neural circuit, with a processing node that acts as a preprocessing stage in order to reduce the dimension of its inputs while generating efficient representations for additional layers of processing elements.

Grzywacz and Poggio consider the neuronal computations associated with motion analysis. The computation of motion by neuronal circuitry is presented by first developing computational models and then considering the properties of the neural machinery. Grzywacz and Poggio's argument for biological fidelity in model building echoes a theme central to this volume.

In an elegant update of his neuronal and computational models, von der Malsburg describes neural network self-organization and ontogenesis, and significantly expands his first-edition discussion of orientation domains. His treatment of intracortical receptive fields and intrinsic cortical activity patterns elevates the discussion of orientation selectivity to new levels.

Carpenter and Grossberg provide an exciting and completely revised closing chapter for this second edition. The rapid and sophisticated evolution of their Adaptive Resonance Theory (ART), as described in the first edition, is described here in their most recent formulation of ARTMAP. The generality, robustness, utility, comparative performance data, and computational power of their new ARTMAP architectures will fascinate and stimulate the reader.

Preface to the First Edition

Biological nervous systems have evolved as a result of millions of years of nature's research and development to culminate in the ultimate computing entity—the mammalian brain. Understanding the brain remains one of the greatest scientific challenges, and one likely to evade full comprehension for the foreseeable future.

In the past two decades the amount of information known about biological nervous systems has expanded explosively. At the same time dramatic advances in silicon chip technology and conventional computer science also occurred. These concurrent events, coupled with the growing awareness of the limits of von Neumann computing in dealing with ill-posed and "fuzzy" real-world computational problems have set the stage for what is likely to be a scientific revolution, having dramatic and wide-spread influences throughout science, technology, and accordingly, society.

Neural networks and neurocomputing represent radical departures from conventional approaches to digital computers, both architecturally and algorithmically. Neurally inspired and biologically constrained processing elements interconnected in ways that permit self-organization and reorganization have already been shown to have the capacity to do useful complex computations that in some cases outperform similar computations made by conventional computers.

The progress made thus far represents the first steps of a nascent scientific endeavor, the full potential of which can only be imagined at this time. If this endeavor is to fully succeed, some essential ingredients must be present (and have been present) from the outset. Perhaps most important among these ingredients is the open exchange and enthusiastic collaboration of scientists and engineers from diverse fields. Principal contributions to the budding efforts thus far have come from physicists, electronic engineers, psychologists, mathematicians, neurobiologists, and computer scientists. If the effort is to expand and achieve its potential, active and sustained interdisciplinary exchange and cooperation is essential.

What is it we would like neural networks to compute? What are the aspects of biological intelligence we would like to emulate in a new generation of computational devices? The answers to these questions are central driving forces pushing the basic research into an arena filled with applications and technological opportunities. Optimally, we would like neural nets to learn spontaneously as a result of experience; to seek optimal solution paths in a given problem; to process accurately inexact, ambiguous, or fuzzy data that provide only approximate matches to what has already been stored in memory; to perform hypothesis testing and probability estimation, and to adjust dynamically implicit rules used in processing information when presented with new input data.

Such capabilities, if achieved efficiently, should provide a foundation for special function chips and neurocomputers that have the ability to: perform real-time pattern recognition at speeds at least as fast as the rates at which new patterns can be entered; provide real-time control in multiple constraint environments; retrieve original memories from fragments of the original; recognize distorted patterns, remember relational associations when the relationship may not be obvious; learn solutions to new problems in real-time based on demonstrated solutions to similar problems; and recall memories even if individual processing elements fail so that performance degrades gracefully and not catastrophically.

Achieving these capabilities will require a close coupling of new information derived from biological nervous systems with advances in interconnect technology, computer design, optical information processing, algorithm development, and adaptive signal processing. Central to the enterprise, certainly in the long view, will be the successful extraction of relevant principles, constraints, forcing functions, and architectures from biological nervous systems. Ultimately, it is these abilities we seek to emulate in artificial neural systems. Clearly we will never duplicate a real nervous system in all of its mystery and complexity. It is not, and probably should not be, a goal of the neural network research enterprise to do so. Rather, by asking the appropriate research questions about biological neural computation, questions not typically central in the minds of most traditional neurobiologists, new insights into principles of biological computation will emerge. Such insights should help us develop new approaches to machine-based parallel processing, signal extraction from noise, data sharing/merging/expansion and reduction, multiple uses of inhibition in circuit design, and more.

Section I. Neurobiological Substrates

It is, of course, not news to anyone interested in the field of neural networks that one of the original presumptions of the pioneers in this enterprise was the possibility of making computational systems based, to a greater or lesser extent, on biological nervous systems. In fact, the brain remains the sole existence proof for many of the aspirations of neural network investigators. The authors in this volume's first section, "Neurobiological Substrates," have approached neural networks from a biological, or, more specifically, a neuroscience point of view. It may surprise readers who approach neural networks from a mathematical or engineering background that the computational strategies adopted by the authors in this section are somewhat unique within the rapidly growing neuroscience discipline, where a "biomedical" rather than "engineering" model has represented the research driver. A brief examination of the contributions from various laboratories will reveal that increasing numbers of neuroscientists have begun to describe the phenomena they study by using simulations, models, and algorithmic techniques in order to more clearly understand the complex, interactive structure they study. The recent adoption of this methodology allows for increased communications between neuroscientists and their scientific brethren in other disciplines; an interaction to be facilitated, we hope, by this book.

Dr. James Bower's chapter suggests a reverse engineering approach to understanding the nervous system. He has chosen to base his model on a portion of the olfactory (smell) system known as the piriform cortex. He has done this for a number of reasons, including its precisely defined input and output, its regular cellular architecture, and, most importantly, the great deal of specific information regarding cellular connectivity within its structure. The ability to profitably incorporate an extensive neuroscience database is probably a prerequisite for the success of this type of modeling effort, since the initial data collection usually requires an effort measured in many man years.

After building his neural network (i.e., piriform cortex) using simulation tools based on a hypercube-class parallel supercomputer, Bower describes how olfaction can be used as a test system for computational theory. More specifically, he has taken the experimental observation that piriform cortex exhibits naturally occurring oscillations or frequencies that can be correlated with events in an animal's history, to suggest that this time-dependent signal processing may be an indication of stimulus classification. Oscillatory phenomena seem to be a feature of all biological networks. These rhythms were examined in detail by early neuroscience pioneers but later became less fashionable. More recently, because of renewed interest in computational neuroscience, the concept of a coherent neural network oscillation composed of oscillations of cellular subsets has enjoyed new interest among scientists interested in both perceptual and cognitive processes. Dr. Bower's models, closely tied to multi-neuron recording experiments in behaving animals, provides a real-world test bed to examine hypotheses suggested by the piriform simulation.

The chapter by Drs. Lynch and Granger and colleagues both contrasts and complements the previous chapter. This collaboration describes work on biologically based neural networks used for active perceptual processing. Once again the choice is the olfactory system, for many of the same reasons suggested earlier. The Lynch-Granger model also incorporates empirically derived biological principles into successively higher levels of their model, which are organized into three categories of feature dimensions: architecture, performance rules, and learning rules. The learning rules are based on the characteristics of long-term potentiation (LTP), which has been defined as "a persistent increase in synaptic strength produced by brief periods of high frequency stimulation." If piriform cortex provides a wealth of architectural data upon which to base a neural network model, LTP yields a similar abundance of data regarding the necessary and/or sufficient conditions required to produce this learning-like phenomenon. Strength and timing (frequency) of the signal are both crucially important in producing this learning phenomenon, or, as the neuroscientists like to say, "plasticity." LTP is of further interest because recent pharmacological experiments have provided neurochemical and neuroanatomical bases for changing synaptic strength. Readers interested in pursuing this approach in more detail are encouraged to examine the LTP work described in the neuroscience literature.

The authors of the first two chapters seem to agree that "categorization" is an important property of cortical networks, and furthermore, a property of neural systems that proceeds in an unsupervised fashion. Lynch and Granger have placed more emphasis on the interaction that can arise with successive samplings of the environment. The biologically based network simulations employed by these authors have suggested computational reasons for why animals typically take three samples of an odorant ("sniffs") before making perceptual judgments. The model also describes a solution to a computationally difficult but biologically "easy" problem, i.e., the "hidden odor" task of detecting and recognizing weak familiar stimuli in the presence of other, much stronger stimuli.

In contrast to the choice of the olfactory system as a model system, Drs. Van Essen and Anderson have emphasized a more traditional research focus, for those interested in machine intelligence. Although the olfactory system has become relatively less efficient during the final steps of evolution leading to primates, vision has become the most highly developed of the senses. The central anatomy of higher cortical processing within the visual system lacks the precision of the cellular architecture often found in other portions of the central nervous system. Nonetheless, the visual system's central relevance to the performance of useful tasks by an intelligent system requires an understanding of its function. Van Essen and Anderson's chapter charts visual system information sensing and analysis processes from light receptors on the retina to complex processors of inferotemporal cortex responsible for identifying complex patterns (e.g., faces).

At the retinal level, the visual system is divided, morphologically and functionally, into two parallel systems. Low to moderate temporal frequencies and luminance information is carried via the P (parvocellular) system and luminance at lower spatial resolution, (higher temporal rates) carried via the M (magnocellular) system. This distinction can be followed through a number of subcortical anatomical structures to the initial cortical regions responsible for visual processing. This so called "primary visual cortex" was first described in the 1960s by David Hubel and Torsten Wiesel in a series of classic papers that describe how neurophysiological analyses can be used to understand a biological information processing system.

Dr. Van Essen and his colleagues have described more than thirty distinct visual information processing areas in the mammalian cortex. Many of these remain to be described in greater functional and architectural detail from research in the years to come. This chapter describes some of the visual processing strategies suggested by a laminar analysis of cells in these cortical areas in addition to the cascaded network provided by the areas themselves.

Professor Edmund Rolls provides an interesting bridge in this series of chapters because he begins with a brief review of his own work in olfaction and vision before moving on to his major topic—the hippocampus. The hippocampus is part of a phylogenetically very old portion of the brain called the "limbic system." During the last twenty-five years, neuroscientists have spent a great deal of effort in studying the hippocampus for two important reasons. First, it contains a well organized anatomically distinct microstructure that seems, like piriform cortex and cerebellum, to lend itself to circuit analysis. Second, clinical observations suggest that patients with bilateral surgical removal of the hippocampus and surrounding brain matter cannot form new memories, although they are often capable of recalling detailed events of their youth. This and other findings clearly suggest a role for this structure in memory formation.

Professor Rolls and his colleagues have carefully incorporated anatomical data regarding connectivity probabilities among hippocampal cellular layers to suggest a model of computational theory based on an autoassociation (or autocorrelation) matrix memory proposed for the hippocampus. One striking feature of the hippocampus is the presence of strong recurrent collaterals (particularly in one layer known as CA3). This feature may fulfill a requirement for the creation of "autoassociative," content-addressable memories so prominent in the parallel distributed models of memory suggested by Rumelhart, McClelland, Hinton, and others.

The emphasis in the last two chapters of this section shifts slightly to the output side of biological activity—motor systems and movement. Drs. Thompson, Gluck, and Myers have adopted classical conditioning, a set of learning rules very different from long-term potentiation. In their model classical conditioning, an experimental paradigm first developed by Pavlov, a neutral stimulus (the conditioned stimulus—a tone in Thompson's study) is paired with another stimulus that evokes a response (the unconditioned stimulus—an air-puff to the eye, and unconditioned stimulus-eyeblink, respectively). When learning occurs, the conditioned stimulus acquires the ability to produce a response similar to the response produced previously by the unconditioned stimulus. Thompson's contribution has been to trace the neural circuitry responsible for this learning and memory phenomenon and to begin to understand the synaptic changes or "weights" that allow it to happen. Once again the task is made more tractable by the fact that the neuroanatomy of the cerebellum is well described with inputs, outputs, and some connective features computationally defined well enough to begin the modeling process. One important feature of the collaborative effort described in this chapter is the demonstration that the mathematical formulation of cerebellar classical conditioning is essentially identical to the Least Mean Square rule. Accordingly, these experimental data might be used as a link between neurobiological, psychological, and engineering approaches to neural network studies.

Professor Eckmiller's chapter provides a very useful bridge between our authors whose neural network approach has been most strongly influenced by neurobiological substrates and those contributors whose interest is more strongly motivated by the emulation and simulation of those systems. As with the previous chapter, Dr. Eckmiller chooses to use the cerebellum (and oculomotor nuclei, which control eye movement) as his neural substrate.

The general concept behind Dr. Eckmiller's schema is that intelligent robots (and primates) receive information via specific sensory modules in "spatio-temporal trajectories." This information can be processed by either biological or nonbiological self-organized, "trajectory-handling neural nets."

Eckmiller's networks are in the form of a neural triangular lattice that contains identical nodes ("neurons" with analog features) connected only to their six immediate neighbors. Although the architecture itself is unbiological, the synaptic connectivity portion of the model contains biologically derived constraints, such as excitation, inhibition, and presynaptic facilitation, working simultaneously to produce motor programs that drive the robotic output.

Section II. Emulated and Simulated Systems

The emulated and simulated systems portion of this book describes efforts by six groups to integrate biological data into a functional system that is task oriented. These investigators have begun with a problem and have sought the solution in a neural network framework. The exact simulation or mapping of the neural system is irrelevant. The emphasis is on the extraction of neurobiological principles to provide direction and clues for alternative solutions.

The first two chapters of this section describe the application of a strategy very similar to the solution of two dissimilar problems. Professor Mueller and his colleagues describe a multiple-layer visual perception system that mimics activity at various retinal layers, intermediate relay structures (lateral geniculate), and cerebral cortex (areas 17 and 18). The system also differs from biology, e.g., center-surround retinal organization proceeds without lateral inhibition. In fact, inhibition seems to be a mechanism that is quite often omitted from neural network simulations, although biological systems would fail to function without it. Dr. Mueller's system employs Gabor functions and the production of a series of sequentially applied matched filters. It does not require iterative computations, back propagation, gradient descent or other learning procedures of this type.

The second emulation effort describes a model of auditory localization based on biological systems that use extensive time-domain processing to create a representation of auditory space. The time dimension of these networks makes use of a topology and learning structure that is inspired by signal processing in avian auditory systems and is quite different from traditional multi-layer artificial neural networks. Professor Mead and his colleagues have chosen to model a biological system with well known passive localization capabilities—the barn owl. The integrated circuit model of the owl's time-coding pathway models both the structure and function of that pathway. In fact, most subcircuits of Mead's chip have an anatomical correlate. This chapter demonstrates the importance of the new silicon technology in the production of devices that can be seen as small but significant approximations of the nervous system. This approach also suggests a future role for neural selected regions of implementation as a test bed for auditory neuroscience, thus making the usual netware-software-hardware sequence a loop, rather than a serial process.

Chapters 9 and 10 share a common interest in motor or robotics systems. Professor Arbib's primary concerns are the generation of stereotyped rhythmic activity and, secondly, the control of that activity. Rhythmic patterned activity has been studied in widely diverse species but with special emphasis on invertebrates. The comparatively simple nervous systems of these animals provides the investigator with clearer access to the neural network, often termed a central pattern generator. In addition to the obvious explanation that stereotyped rhythmic activity is produced by a "pacemaker" cell; ring networks and reciprocal inhibition have also been demonstrated to produce these motor behaviors.

Arbib's model of movement control is closely tied to the generation of eye movements, or saccades. There are many types of saccades and, therefore, many types of oculomotor control systems, which are usually operating together. Systems approaches have probably had their greatest impact on neuroscience in this area, and it would seem likely that efforts based on neural network models may provide a descriptive framework to understand the complex interactions required for visual pursuit and fixation of gaze by a system in motion.

Professor Edelman and his colleagues have also chosen to base their system on known neurobiological properties. Although the Rockefeller group incorporates many details of anatomy, physiology, and synaptic plasticity, they place their emphasis not on the activities of the single neuron, or even on the small groups of cells in a microcircuit, but rather on massively large (for modeling, if not biological conditions) collections of highly interconnected units. This approach (synthetic neural modeling) requires massive computer support to implement the simulation as "Darwin III." This neural network is "unsupervised," yet manages to learn categorization problems and respond to novel stimuli in either a positive or negative manner. The robotic arm also learns to reach and touch by first moving randomly, coming into contact with various objects, feeling them neutral or noxious (bumpy versus smooth), and eventually learning to swat the noxious object away. A caveat is perhaps appropriate at this juncture. Readers who require a mathematical or algorithmic description of Edelman's neural model will be disappointed. More details can be found in Professor Edelman's book, *Neural Darwinism*, but the details that would allow other investigators to exactly reproduce Darwin III have not been published.

Chapter 11 provides an excellent review of the algorithmic substrates of neural network models with a special emphasis on learning rules. Professor Cooper's RCE (Reilly, Cooper, Elbaum) neural network has three layers, low internal connectivity, feedforward architecture, and high-density memory storage. These properties allow the development of software applications that can be run in real time. A number of RCE networks can be combined with a controller model that directs the training of the individual RCE network processor. The resultant system has been applied to a number of real-world problems with great success.

The last chapter in this section is contributed by a neural networks pioneer. Professor Widrow reminds us that a highly simplified type of neural network, the adaptive filter, has enjoyed great commercial success in telecommunications for a number of years. In this sense, the first widespread adoption of neural network technology has already taken place. Professor Widrow produced adaptive elements in his Stanford laboratory before current microcircuit technology became possible by using electrochemical "batteries" to store charges that later became known as weights in neural network units. These MADELINE algorithms are now leading to neural network implementations using very large scale (VLSI) technology, as described in the next section.

Section III. Electronic Networks

The Electronic Networks section describes the two major technologies for implementing the neural network models presented in the previous chapters. There are many reasons for putting neural networks on hardware. First, there is the inevitable interplay between science and technology. Science discovers the laws and principles of nature, and technology puts them to good use. Often in science, research is not complete until a device can be made to demonstrate the principles. For example, a physicist would not be content with paper design of a new transistor until he could fabricate and demonstrate it. So it is in neural networks research; in order to validate the mathematical models, it is necessary to build hardware models to demonstrate similar functions. Although some of the neural network models can be validated by computer simulations, something is irrecoverably lost in a digital simulation for neural processing that is inherently analog. Furthermore, simulation of the most simple neural network can take hours on a supercomputer. Hardware is the only alternative to speeding up the computation.

There are many different technologies for implementing artificial neural networks, including silicon, gallium arsenide, integrated optoelectronics, and optics. Even on silicon there are possibilities of charge coupled devices (CCD), polysilicon floating gates (FAMOS), and metal nitride oxide semiconductor (MNOS) technologies for charge storage. The implementations can range from full analog circuits, full digital, or hybrids. In Chapter 13, Faggin and Mead explore the key issues and alternatives of neural network implementation using silicon-based very large scale integration (VLSI) technology. They explicitly look to neurobiology for inspiration and consider analog computation as the most appropriate paradigm. Although they strive for biological fidelity, they are not blinded by the constraints VLSI technology places on the implementation architecture. They suggest that the well-known interconnect limitations of VLSI technology can be addressed by techniques such as time division multiplexing. Based on energy calculations, they argue that the brain has a Factor 10^4 advantage over the most advanced digital VLSI technology. Conventional digital computing architectures simply cannot match the processing power of the central nervous system. Thus, radically different computing architectures that are biologically inspired must be devised to meet the high-throughput real-time processing needs. The main point they make is that we have something to learn from neurobiology.

In a similar vein in Chapter 14, Koch's neural network architecture is also biologically inspired but with particular application to the early vision problem. The specific architecture he presents is the hexagonal resistive grid network in a fashion after the lateral inhibition circuits in the mammalian retina. This is an excellent example of exploiting the physics of VLSI devices in a nearest-neighbor interconnect architecture to emulate the retina. The behavior of the resistive grid, in the limit, is governed by the diffusion equation, and is equivalent to convolution by a Gaussian function. These techniques are well known in image processing, but now it is actually implemented in hardware using VLSI technology. The advantage of this technology over digital simulations, both in terms of power consumption and throughput rate, is clear.

Moving from the sensory system, Bailey, Hammerstrom, Mates, and Rudnick in Chapter 15 consider the problems of implementing a silicon association cortex. They present some of the architectural techniques for the various tradeoffs that are inevitable when one tries to build a device emulating a structure as complex as the cortex. The technique they employ is as complex as graph theory and mapping between the network and the graph. Their perspective is that of a computer architect who, when faced with the problems of designing a high performance parallel processor system, must trade off processor complexity with interconnection/communication complexity. The single most important issue in network communication is locality. For locally connected architectures, the network can be efficiently mapped to silicon. The problem is to avoid long (or global) wires. They suggest that one solution is augmented broadcast hierarchy, whereby local communication is by means of broadcast and global communication is by point-to-point connections. With these techniques, it might be possible to eventually emulate an association cortex.

Taking a different tack, Akers, Ferry, and Grondin in Chapter 16 approach the artificial neural network implementation problem from the perspective of a VLSI designer. Their goal is not to emulate biological systems but rather to design high performance, fault tolerant associative memories. Neural networks turn out to be ideally suited to their purpose. They describe in some detail the history of VLSI technology, leading to the need for unconventional and fault tolerant architectures. Of particular interest is the connection (isomorphism) they make between neural networks and cellular automata. Again the interconnection limitations of VLSI technology figure prominently in their design. They suggest a local-limited-interconnect architecture of a multilayer perceptron network.

To ease the interconnectivity problem, people have looked to alternative technologies. Psaltis, Gu, and Brady, in Chapter 17, use holographic techniques for massively interconnected neural networks. In optics, photons can safely cross paths without interference. This is an advantage for massively interconnected networks such as associative memories and multilayer perceptron networks. Holograms (both planar and volumetric) offer a technology that would significantly reduce the complexity of the hardware that is needed to perform the massive interconnections in neural networks. Full N^2 interconnection can be easily realized using planar hologram, and up to N^3 interconnections can be realized with volume hologram. This would be very impressive if interconnections were all that is needed in a neural network. Unfortunately the real computation is still done at the neural processing element, which is much more amenable to electronic implementation. Furthermore, any optical neural network would have to ultimately be interfaced to a computer, which is almost always electronic. Thus, integrated optoelectronic technology seem to offer the best alternative, whereby electrons are used for the neural processing element and photons are used for the interconnections. However, that technology is still years away.

Anderson, in Chapter 18, presents a wholly optical implementation of a competitive and cooperative neural network. Here he strives to imbed the mathematical models of neural networks into the physics of optical systems, far removed from neurobiology. The competitive/cooperative dynamics of his optical system is strikingly similar to those in neural network models. The photorefractive material can be used in a manner very much like the memory in neural networks. The optical implementation involves the interplay between saturable gain and saturable loss in a feedback circuit to maintain the lock-in of a strong (desired) image. This is the essence of a competitive/cooperative network. Anderson's perspective is that of a physicist who sees the similarity between nonlinear optical circuits and conventional neural networks. That similarity manifests itself in the same language, the mathematics of the network, which is the subject of the next section.

Section IV. Computational and Mathematical Considerations

This section describes the computational models of a variety of networks of interconnected neurons. In an effort to explain brain function, these models can be broadly divided into two classes: phenomenological models, which explain behavior purely on the basis of stimulus–response paradigm; and physiological models, which explain behavior on the basis of the underlying neural structures and circuitry. Psychologists have traditionally used the former (or top-down approach) to explain human behavior, whereas neurophysiologists have taken the latter (or bottom-up approach) to explain brain function. Both groups have contributed significantly to recent advances in understanding brain function. Computer scientists in the 1960s and 1970s borrowed many concepts from experimental psychology and developed the field of artificial intelligence (AI). Some success in putting "expertise" into conventional computers has been attributed to the success of AI. However, many intractable signal processing and pattern recognition problems remain. Recently researchers have turned to neural network models and massively parallel computer architectures to provide the necessary speed increase in throughput rate and in algorithmic intelligence. This section presents the authors' views on the kind of computation that is going on in the brain. It must be realized that these are simplified models of a very complex organ and that these models may or may not reflect real brain computation. As to the validity of these computational models, only time and more research will tell. Notwithstanding, these computational models are useful in their own right to motivating us to look to neurobiology for radically different computing structures.

In Chapter 19, Sejnowski and Stanton present some convincing arguments about learning and plasticity in the central nervous system and show that simplified models, in spite of their limitation, can help to understand synaptic plasticity in the mammalian hippocampus. They look to neurobiology and long-term potentiation (LTP) for evidence in support of the validity of the Hebbian synaptic modification rule. They also present some evidence for associative long-term depression (LTD), and suggest a theoretical covariance model of associative memory for the hippocampus.

In Chapter 20, Grzywacz and Poggio concentrate their neuronal computation model at the retina and particularly at the computation of motion. This deceptively simple task, done by humans and animals routinely and automatically in real time, is actually quite difficult to model. They review for us the spatial and temporal properties of the excitary and inhibitory pathways and present some evidence in favor of the shunting model. Many models and algorithms for the computation of motion require the calculation of the "optical flow." Since there are many definitions of optical flow, the definition of choice should be dependent on the properties of the biophysical mechanisms at the retina. Although the shunting inhibition model is a good candidate, it is certainly not fully validated. Other mechanisms for computing motion in the visual cortex, such as the shifter circuit of Van Essen and Anderson (see Chapter 3), are also possible.

Rumelhart's model of brain-style computation (Chapter 21) is at a higher level of abstraction, taking the neuron as the basic processing unit and stripping away the underlying biochemical and ionic complexity. The neuron is simply considered as a processor with multiple input connections, multiple output connections, and an activation function. A general purpose brain-like computing device would consist of a large number of such units interconnected into a network. Their goal is not to truly understand the architecture and function of the cortex, but rather to understand the kinds of problems that can most naturally be solved by such a network. This kind of network comes in many forms and has been referred to as a multilayer perceptron network, artificial neural network, parallel distributed processing (PDP) system, or neurocomputer. Whether or not these PDP networks mimic the cortex, one thing is very clear—these networks have proven to be extremely robust in terms of application to pattern recognition and optimization problems. Since these brain-style networks are not programmed like a conventional computer, they must be trained by supervised learning. The training (or learning) results in finding a set of connection strengths that allow the network to carry out the desired computation. The learning procedure, which is a generalization of the ADELINE least mean square (LMS) minimization algorithm, has come to be known as back propagation. In other words, the output error is back propagated to adjust the connection weights so as to minimize the output error. The principal advantage of PDP network over the perceptron and ADELINE of the 1960s is the existence of the hidden layer, which turns out to be indispensable for solving pattern recognition problems that are not separable by hyperplanes.

Moving from a relatively fixed network that requires supervised training, von der Malsburg, in Chapter 22, discusses network self-organization, with the help of examples taken from the ontogenesis of the visual cortex. In other words, he addresses the question of how brain circuitry is organized. He suggests that two types of variables are relevant to network self-organization: signals and interconnections. Although competition and cooperation exist at the local level, self-organization is the emergence of a globally ordered state which ensures optimal mutual consistency of all local rules. Ontogenetic developments of fibers from the retina to the tectum, of ocularity domains, and of orientation domains are used as examples in this self-organization.

Moving away from neurobiology, Feldman, Fanty, Goddard, and Lynne, in Chapter 23, describe computation with massively parallel computational network. Their perspective is that of a computer scientist who is trying to design, realize, and analyze these computational structures, known as connectionist network, to solve hard problems, particularly in AI. A connectionist network consists of many simple computational units, each of which is connected to all the other units. Much of their effort involves using such a connectionist network in AI application. They describe in some detail their Rochester Connectionist Simulator, as well as simulations on the SUN3/260 and the BBN Butterfly multiprocessor computer. This chapter is a good example of how an emerging new field (neural networks) can effectively merge with an established field (AI) to the mutual benefit of both disciplines.

A book on neural networks would certainly not be complete without a chapter from Grossberg and Carpenter. In Chapter 24, they describe the process by which humans learn recognition codes in real-time through the process of self-organization and present in some detail their adaptive resonance theory (ART) model. The driving force behind the ART model is the stability–plasticity dilemma. In other words, a system has to be stable and yet remain plastic and adaptive to changing stimulus input. The ART architecture is a neural network that self-organizes for stable recognition codes in real-time. The advantage of ART model is that it is self-organizing and therefore does not require any supervised learning and training. Additionally, ART architecture can recognize novel input patterns instantly without resorting to slow correlation and memory search processes. An ART network will learn old and new patterns up to its maximum capacity. Although ART is motivated by the desire to build an intelligent computational system for solving real-world pattern recognition problems, it is also inspired by decades of research in understanding the biological and psychological processes of brain and behavior. Regardless of whether adaptive resonance is really the way the brain stores and processes information, this research has certainly contributed significantly to brain theory and to the recent advances in neural networks.

Acknowledgments

The editors wish to acknowledge the many individuals who have contributed in various ways to this book. We acknowledge those scientists who have the insight and courage to break new intellectual ground and cross traditional disciplinary boundaries. We acknowledge our publisher, who early on saw wisdom in doing a book in a new and unorthodox area. We acknowledge the Office of Naval Research, which served as a catalyst in helping to stimulate the early phases of neural network research. We acknowledge Donna L. Bruggeman, without whose secretarial and editing efforts this book might not be.

I

Neurobiological Substrates

1

Reverse Engineering the Nervous System: An *In Vivo, In Vitro*, and *In Computo* Approach to Understanding the Mammalian Olfactory System

James M. Bower

INTRODUCTION

In the first edition of this chapter, I stated our hope and expectation that unanticipated functional insights would result from efforts to model realistically the nervous system. As such, a distinction was drawn between "bottom-up" modeling efforts, the initial principal intent of which was to capture the structure of the nervous system, and more "top-down" approaches in which modeling serves primarily as a means to demonstrate the feasibility of preconceived functional ideas. The clearly stated assumption was that the expected close association of the structure of the brain to its function would put bottom-up models in a better position to reveal fundamentally new ideas concerning brain function.

Although our early efforts to model the oscillations seen in olfactory cortex certainly had suggested new functional ideas (Wilson & Bower, 1988), at the time this chapter was originally written, it was still unclear how valuable or important these ideas might be. As documented in this revised chapter, there is now ample evidence that attention to structural detail yields important new functional insights. For example, prompted by experimental results in visual cortex (Eckhorn et al., 1988; Gray, Konig, Engel, & Singer, 1989; Gray & Singer, 1989; Kreiter & Singer, 1992), the last several years have generated many top-down interpretive models of the oscillatory behavior of cerebral cortical networks (Sporns, Gally, Reeke, & Edelman, 1989; Schuster & Wagner, 1990a,b; Schillen & Konig, 1991; Grannan, Kleinfeld, & Sompolinsky, 1993; Neibur, Koch, & Rosin, 1993). Inspired by a previously suggested solution for the abstract problem of feature binding in machine vision (von der Malsburg & Schneider, 1986), cortical oscillations have been proposed as a solution to this supposed problem in biological vision (Buhmann & von der Malsburg, 1991; Sporns, Tononi, & Edelman, 1991). Furthermore, extensions have been made to visual attention (Neibur et al., 1993) and even to the mechanism of consciousness (Crick & Koch, 1990; Crick, 1994). On the other hand, as described in the first version of this chapter, our bottom-up modeling of cerebral cortical activity patterns suggested that similar oscillations in olfactory cortex might be more related to the flow of information within cortical networks rather than a code for the information itself (Wilson & Bower, 1991, 1992). The models also generated specific predictions concerning the nature of the oscillations that have been shown to be largely correct, resulting in a reinterpretation of the significance of oscillations (Gray, 1994). Thus, in our view, the explosion of top-down modeling generated by the observation of oscillations in visual cortex clearly demonstrates the contrasting risks and benefits of top-down and bottom-up approaches. As a result, the revised version of this chapter much more strongly asserts the importance of letting neuronal structure, rather than more abstract considerations, lead the way to neuronal function.

MODELING IN NEUROBIOLOGY: TOP-DOWN VERSUS BOTTOM-UP

Neurobiologists interested in how mammalian brains work are faced with trying to understand the function of networks composed of many millions or billions of complex neurons, making, in total, billions or trillions of equally complex connections. As experimental neurobiology has become more and more sophisticated, it is increasingly clear that detailed physiological and anatomical descriptions of these systems will not be enough to infer how neural circuits function. Instead, experimental neurobiology will increasingly require the context provided by modeling both to direct experimental questions and to interpret experimental results (Bower, 1992).

Although modeling efforts inevitably will play an increasingly important role in neurobiology, traditional theoretical approaches in neuroscience and behavior have primarily emphasized the construction of abstract top-down models in which the detailed anatomical and physiological structures of nervous systems themselves are taken into consideration only to a relatively minor degree if at all (Marr, 1982). It is the thesis of this chapter, however, that such top-down approaches are much less likely to result in an understanding of how the brain works than modeling efforts that place primary emphasis on first capturing the structure of the brain. Top-down models, by definition, are essentially demonstrations of the feasibility of preconceived ideas. For this reason, if they include any neural detail, it is usually only those components that fit the proposed theory. Bottom-up models, on the other hand, are intended to capitalize on the likely close relationship between the structure of the nervous system and its function. As such, I believe that these models are in a much better position to uncover new computational principles.

Of course, there are a number of well-known criticisms of modeling efforts of this type. Perhaps the most common is that realistic simulations only substitute one problem (a complex network in an animal) for another (a complex network in a computer; cf. Churchland & Sejnowski 1988). In our experience, however, this criticism does not adequately take into account the fact that the construction of realistic models is really a process rather than an end in itself. As a comparison of this chapter and its previous version will document, realistic modeling can support the natural evolution of functional ideas, as new physiological data are obtained and new models are built. In contrast, top-down modeling efforts are almost always completed once the property in question has been demonstrated (cf. Lisberger & Sejnowski, 1992).

Criticisms of realistic modeling are often also based on a more fundamental assumption, or hope, that the specific details of the nervous system might not matter. As David Marr proposed (Marr, 1982), in this view, any particular biological neuron or network should be thought of as just one implementation of a more general computational algorithm. Following a physics model, understanding the abstracted algorithm is thought to represent a more general form of understanding than can be provided by considering the details of the system itself. However, it is not clear that the "technology of understanding" in the simple systems typically selected for study by physicists will readily apply to exceedingly complex systems like brains, or even devices that do the kinds of things brains do. It is fairly clear, for example, that "closed form" analytic solutions will not be developed any time soon for models that include neurobiological details, and thus for the brain itself.

The more serious question often at the heart of the top-down, bottom-up debate is whether a particular neural system is just one of several possible implementations of an algorithm to solve a particular sensory or motor problem, or instead represents a much more limited solution set, or even, in some sense, an "optimal" solution. The more a particular system is "the best" possible solution rather than just one of many equivalents, the more attention will need to be paid to its structural details. The fact that in each case in which precise measurements of the performance of the nervous system have been possible biology has always been shown to be operating near physical limits, lends support to the proposal that we should take brain structure very seriously. Thus, for example, the retina detects single photons (Nakatani, Tamura, & Yau, 1991), whereas the auditory system operates with sensitivities just above Brownian noise (Denk, Webb, & Hudspeth, 1986). Furthermore, there are numerous instances in which completely unrelated animals have evolved essentially identical neural circuits. For example, the retina of a primate and that of an octopus are virtually identical, despite the fact that both have independently evolved many millions of components.

Another argument in favor of paying precise attention to the actual structure of the nervous system has to do with the complexity of the computational problems being solved. The more difficult these problems are, the closer attention should probably be paid to the particularities of existing solutions. In this regard, the history of abstract theoretical studies of animal behavior, as well as the supposedly brain-related efforts in artificial intelligence and neural networks, are all rife with underestimates of the magnitude of the problems being solved by real neural circuits. The increasing complexity of neural networks and "connectionist" architectures, despite the predisposition of practitioners in these fields toward simplicity, suggests that the extreme complexity of nervous systems might reflect something more fundamental about the complexities of the problems being solved. Hard problems may simply require complicated solutions not easily reduced to general algorithmic principles. It may be that if the brain could perform the same functions with less neurons, it would. In the absence of clear evidence that this is not the case, it seems reasonable to err on the side of the biologically conservative and assume that a detailed consideration of the structure of the nervous system will be essential for progress.

Although this chapter does take a hard stand for a bottom-up approach, it is not intended to negate the usefulness of more abstract models. In some sense, of course, all models are abstractions. However, the distinction made here is that top-down models usually abstract known biological features, like the complexities of single cells, to explore particular ideas. Even we generate these types of models once experiments and our more detailed models have suggested new functional ideas (cf. Hasselmo, Anderson, & Bower, 1992). In these instances, abstract models can play several important roles including providing a more thorough understanding of parameter space (Hasselmo et al., 1992); "translating" functional ideas into a language better understood by engineers, mathematicians, and physicists (Hasselmo, 1993; 1994); and, in some cases, promoting the transfer of new ideas to artificial computing devices. Accordingly, abstract modeling can be very useful if the clear direction of information transfer is from the bottom up and is based on the initial construction of a structurally realistic model. It is also much easier to make a realistic model abstract than it is to make an abstract model realistic.

Abstract modeling intended to study more abstract problems can also provide useful contexts for realistic modeling efforts. Thus, in our own work, abstract models of auto-associative memory have provided a valuable context for our initial studies of the olfactory system. However, when these abstract models are promoted as actual representatives of the nervous system, without specific reference to the detailed structure of its networks and neurons, I believe the opportunity to learn from that neural structure is lost. Again, this is the value of bottom-up modeling, to provide a mechanism for the structure of the nervous system to reveal things not known before, and which may not have been discovered otherwise.

STRUCTURAL MODELS OF REAL NEURAL NETWORKS

The following sections describe our use of structurally realistic models to reverse engineer the mammalian olfactory system (Wilson & Bower, 1988, 1989, 1992; Nelson, Furmanski, & Bower, 1989; Hasselmo & Bower, 1993; Bhalla & Bower, 1993). As will be clear from the text to follow, in our hands, these models are used interdependently with ongoing experimental investigations. In fact, it is the iterative use of structural models and experimental work that provides the power of the approach. For this reason, we believe that the growth of this type of modeling in computational neuroscience will be driven by experimentalists, like ourselves, interested in how the structures we study might work (Bower, 1992). The next sections define what is meant here by *structural modeling*, and how we have used these models to begin to understand the mammalian olfactory system.

Definition

Structurally realistic simulations are computer-based implementations of models whose first objective is to capture what is known about the anatomical structure and physiological characteristics of the neural system of interest. Models of this sort are not constructed to demonstrate how neural-like structures could theoretically perform some preconceived function (cf. Granger, Ambros-Ingerson, Henry, & Lynch, 1988), instead they are intended to serve as experimental tools for revealing new functional ideas. For this reason, our models are first built to replicate *functionally neutral* cellular or network properties. For example, typically, our single-cell models are first tuned to replicate neuronal responses to artificially applied current and voltage clamps (Bhalla & Bower, 1993; DeSchutter & Bower, 1994a) and then parameters are frozen while the effects of more natural synaptic inputs are explored (Bhalla & Bower, 1993; DeSchutter & Bower, 1994b,c). As described here, our network models are tuned to generate physiological responses to artificially applied shocks before more "natural" patterns of activation are applied.

Although the distinction between models that establish structural realism to reveal function and models that employ various degrees of structural realism to promote functional ideas can, at times, become blurred (cf. Granger et al., 1988; Ambros-Ingerson, Granger, & Lynch, 1990), on closer inspection, it is usually obvious when a model has been designed specifically to demonstrate a particular idea. From our modeling philosophy, it naturally follows that the ultimate question that any modeler must be prepared to answer is What do you know now that you did not know before you constructed your model? If the only answer is That I was correct, then, in our view, the real power of modeling has not been realized.

Practical Advantages

Independent of any advantage that structurally realistic simulations might have in discovering neural function per se, this modeling approach has several practical advantages related to model construction. First, and very importantly, the process of building such a simulation forces the modeler to quantify exactly what is known about a particular circuit. In our experience, it is often the case that early stages of building a model of this kind point out more vividly what is not known about a network than what is. In this way, structural models lead to a much better understanding of the limits of our knowledge, at the same time naturally suggesting which additional experiments are necessary to improve that knowledge.

Structural models also have numerous advantages simply from the point of view of building and manipulating a simulation. First, biological realism allows known neuroanatomical and neurophysiological data to be used as constraints for model parameters. This is an important feature for models of complex systems which otherwise can posit an almost limitless number of components and interactions. Second, the relationship between the parameters in a structural model and real biological measurements limits the parameter space of the models which must be explored (Bhalla

& Bower, 1993). Because most of the time spent working with a model involves determining the dependence of its activity on model parameters, this greatly increases the chances that modeling will yield something interesting. Finally, by generating biologically relevant outputs, structural modeling results are readily comparable to data from actual experiments, making predictions testable. Testability is one of the most serious problems facing all models of complex systems. Thus, modelers should not only be held to the standard, What do you know now that you did not know before? but also How can you determine that it is true? On both counts, biologically realistic models have a distinct advantage.

Modeling Steps

Ultimately, the power of this approach is in the simulation-based interplay between the structure of a network, the experimental determination of model parameters, the evolving, structurally guided description of the network's functional organization, and the experimental testing of modeled results. The success of these interactions is dependent on including, from the outset, sufficient structural detail in the simulation to be able to replicate well-characterized physiological responses.

We have found that it is most useful to start with "nonphysiological responses" that nevertheless reflect the overall organization of the structure in question. Thus, for example, in the following network modeling effort with piriform cortex, the initial objective was to replicate cortical responses obtained experimentally by artificially shocking the input pathway to the cortex (Wilson & Bower, 1992). Replicating responses from these direct electrical shocks turned out to be a good initial test of the model because such massively evoked cortical activity was less dependent on more detailed network properties not yet included in the simulation. Furthermore, cortical responses to shock stimuli are quite invariant in form (Ketchum & Haberly, 1993), which reduces the complexity of tuning model parameters. Finally, as modelers, we believed that the activity induced in the cortex by an electrical shock was unlikely to have any particular functional significance. Accordingly, at the outset of the modeling, our ideas about the functional organization of piriform cortex did not interfere with constructing the model. As described here, however, even this stage of modeling generated unexpected functional ideas.

In our experience, a second important feature of this initial stage of model building involves simulating results from as many different recording methods as possible. Our simulations, for example, were designed from the start to generate intracellular potentials from simulated neurons, extracellular spike train activity, and extracellular bulk electrical responses such as evoked potentials and EEGs (Figure 4). In general, the more types of responses the model generates, the more rigorously it can be tested, and also the more accurately the parameter space can be explored (Bhalla & Bower, 1993). This effort at the beginning also ensures that the results of later simulations can be tested by using diverse types of real experimental data.

Once the model has been tuned on presumably "function neutral" measures and is capable of replicating multiple types of responses, the modeler has more confidence that the model itself includes enough of the essential features to start to explore function. However, as will be described, new functionally interesting features of neural structure often become apparent even in the process of model tuning (Wilson & Bower, 1992). This repeated experience is one basis for the claim that paying attention to the details increases the chance that the nervous system itself will provide clues to the function of its components.

REVERSE ENGINEERING: PIRIFORM CORTEX

Over the last several years, we have been developing a general purpose neural network simulation system (called GENESIS) to support the construction of structural simulations of real neural networks (Wilson, Bhalla, Uhley, & Bower 1989; Bower & Hale, 1991). Currently, our own modeling efforts using GENESIS extend to several neural systems including the neurons and circuitry of the olfactory bulb (Bhalla, Wilson, & Bower, 1988; Bhalla & Bower, 1993), the inferior olive (Lee & Bower, 1988), the hippocampus (DeSchutter & Bower, 1993); the cerebellum (Jaeger, DeSchutter, & Bower, 1993; DeSchutter & Bower, 1994a,b,c), and the central pattern generators (Ryckebusch, Bower, & Mead, 1989). In the interest of space and clarity, this chapter considers only selected results from our ongoing efforts to model olfactory (piriform) cortex (Wilson, Bower, & Haberly, 1986; Bower, Nelson, Wilson, Fox, & Furmanski, 1989; Wilson & Bower, 1988, 1989, 1992; Hasselmo & Bower, 1992; Hasselmo et al., 1992; Vanier & Bower, 1993; Protopapas & Bower, 1994).

The General Structure of Piriform Cortex

The model that will be described is based on the known structure of the piriform cortex. Thus, all the principal parameters in the model, as well as its anatomical structure, are constrained, as far as possible, by experimentally measured values (Wilson & Bower, 1992). Accordingly, to understand the significance of the simulation results, it will be necessary first to describe briefly the essential structure of this cortex. Unfortunately, space does not allow a detailed description of the structure of the real network or the specific implementation of the model itself. Readers are referred to the review by Haberly (1985) for additional information on piriform cortex and to several papers published in collaboration with Matthew Wilson and Michael Hasselmo for simulation details (Wilson & Bower, 1988, 1989, 1992; Hasselmo et al., 1992; Hasselmo & Bower, 1993).

Piriform cortex is the primary region of cerebral cortex devoted exclusively to processing olfactory information (Haberly & Price, 1978; Haberly, 1985). It receives a direct projection from the olfactory bulb which itself receives input directly from olfactory receptors in the olfactory epithelium (Figure 1). Its output is to forebrain structures, such as the entorhinal cortex and through the entorhinal cortex to the hippocampus, frontal cortex, and thalamus. We are interested in studying this cortical structure because of its presumed role in the classification of olfactory stimuli (Tanabe, Lino, & Takagi 1975; Haberly, 1985; Bower, 1991a,b,c, 1993), its close proximity to the sensory periphery (Figure 1), its relatively simple structure in comparison with other cerebral cortical areas (Haberly, 1985; Bower & Haberly, 1986), and its close association with structures like the hippocampus, which are believed to be directly involved in such functions as memory acquisition and storage (Haberly, 1985; McNaughton & Morris, 1987).

The input to the piriform cortex from the olfactory bulb is delivered via a fiber bundle known as the *lateral olfactory tract* (LOT; Figure 2A). This fiber tract appears to make sparse, nontopographic, excitatory connections with cortical neurons across the extent of the cortex (Devor, 1976), but importantly, the influence of this afferent input is much greater in the rostral than caudal cortex (Figure 2A; Haberly & Shepard, 1973; Schwob & Price, 1978). It should also be noted that this primary afferent input spreads across the surface of the piriform cortex (Figure 2A), whereas the more common neocortical arrangement has afferents entering vertically (e.g., geniculo-cortical input; Shepherd, 1979).

In addition to the input connections from the olfactory bulb, there is also an extensive set of connections between the neurons intrinsic to the cortex (Figure 2B). For example, the association fiber system arises from the principal cortical cells, the pyramidal cells, and makes sparse, distributed excitatory connections with other cells of this type throughout the cortex (Figure 2B; Haberly & Bower, 1984; Haberly & Presto, 1986). Each of these excitatory fiber systems has a characteristic conduction velocity and dendritic termination zone, and has a different amount of net influence, depending on the cortical region in question (Figure 2B; Haberly & Shepherd, 1973). Pyramidal cells also make excitatory connections with nearby feedforward and feedback inhibitory interneurons (Figure 2C and D). In the model, these inhibitory interneurons get input from and make reciprocal connec-

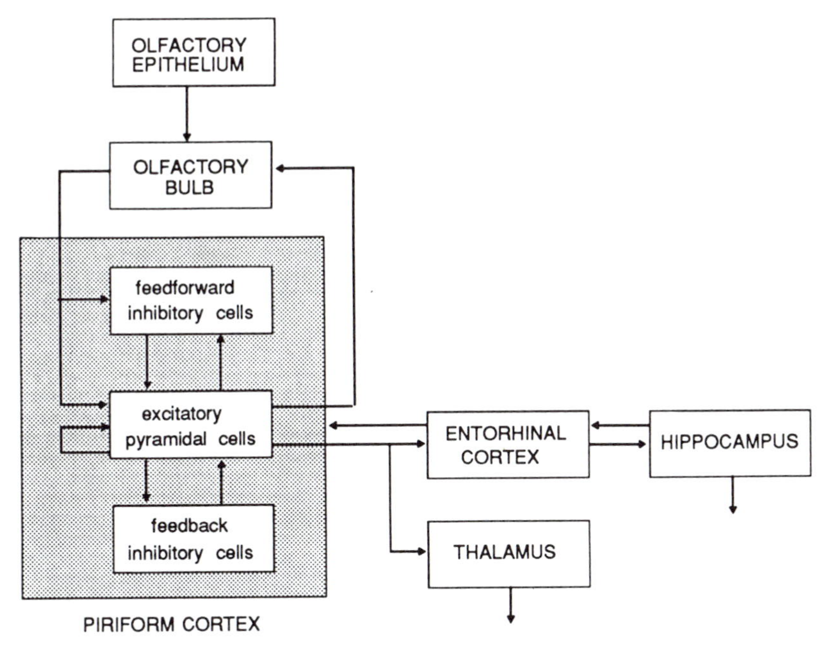

FIGURE 1 Several of the pathways providing input to and output from the piriform cortex. The gray shaded box indicates the principal cellular components of the model of piriform cortex discussed in the chapter. (Figure modified from Wilson & Bower, 1989.)

tions with nearby pyramidal cells. The feedforward neuron appears to get a predominant input directly from the lateral olfactory tract (LOT; Figure 2C and D). Experimental evidence indicates that these two types of inhibitory neurons are responsible for two different types of inhibition on pyramidal cells (Figure 2D; Tseng & Haberly, 1987; Sato, Mori, Tazawa, & Takagi, 1982). The feedforward inhibitory cells are believed to be responsible for a long latency, long duration hyperpolarization of pyramidal cells, whereas the feedback inhibitory interneurons are responsible for a fast acting, short duration current shunting type IPSP. Figure 3 summarizes the influences on a single pyramidal cell in the model of the cortex.

Modeling Cortical Oscillations

The first stage in our modeling effort involved an attempt to replicate, in simulation, cortical responses to several different types of afferent stimuli. The results that follow demonstrate that the simulation, as described, is capable of replicating a large number of response patterns without a change in the basic parameters describing its cell types, their dynamics, or their interconnectivity. This fact makes us more confident in the overall structure of the model.

Electrical Shock of the LOT

One of the ways in which experimentalists have traditionally probed the structure of piriform cortex is with direct electrical shock stimulation of the lateral olfactory tract (Haberly, 1973; Haberly & Bower, 1984; Bower & Haberly, 1986; Freeman, 1968a,b). The response of the cortex to this type of stimulation is interesting because different spatial and temporal patterns of surface-evoked potentials are generated depending on the strength of the shock presented (Figure 4;

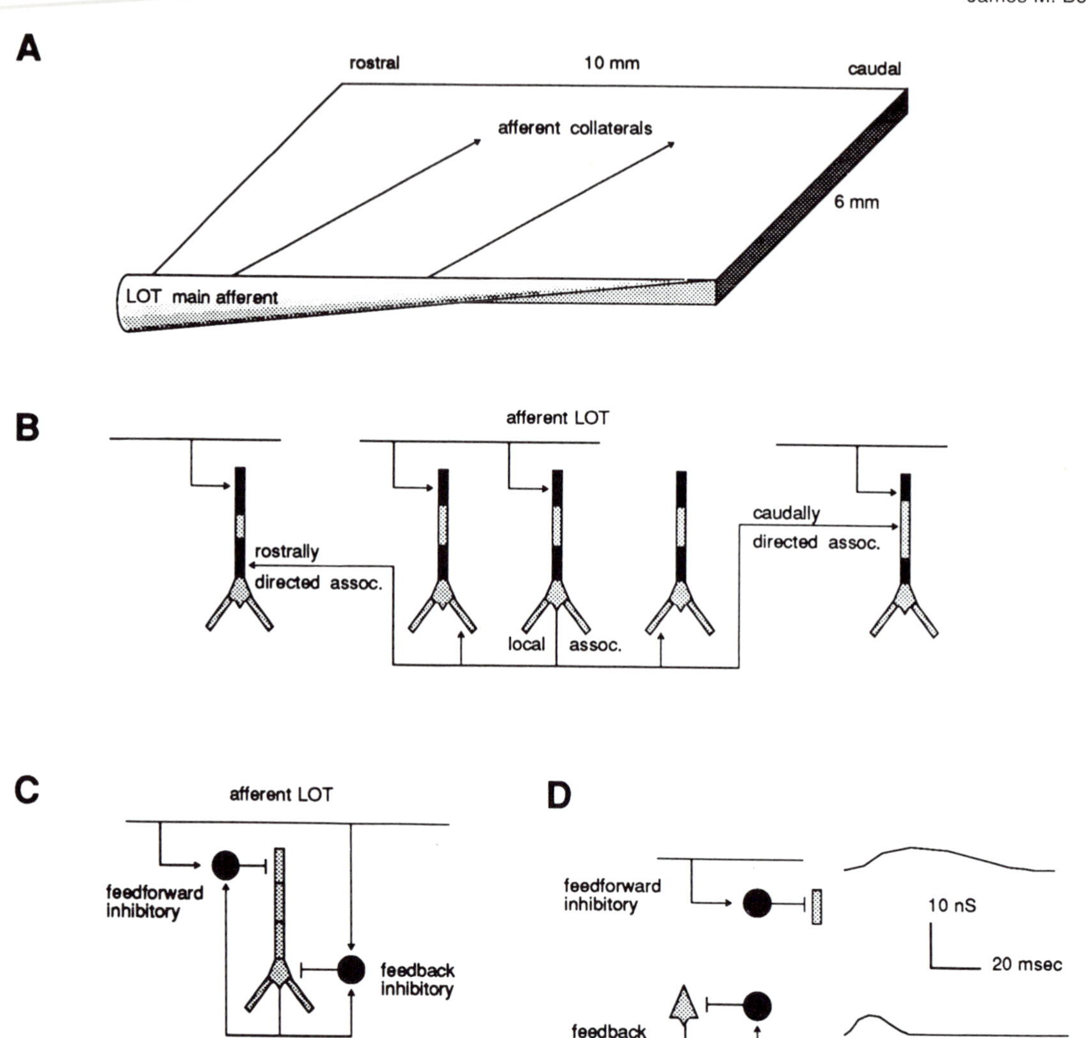

FIGURE 2 Schematic representation of the structural features of piriform cortex which have served as the basis for the simulations discussed in the text. A indicates the spatial pattern of projection of lateral olfactory tract (LOT) axons into and across the cortex. It also indicates the decrease in influence of this afferent input as the LOT courses toward the caudal end of the cortex. B shows the excitatory connections made by afferent and pyramidal cell axons within the cortex and simulation. Note that pyramidal cell "association" connections are both local and distant. The differently shaded regions of the figure indicate the relative contributions of the different fiber systems at different rostro-caudal positions in the cortex. C indicates the basic pattern of interconnections between pyramidal cells and inhibitory interneurons. The two classes of inhibitory neurons modeled are shown. D shows the different temporal properties of the conductances induced by the two classes of inhibitory neurons as discussed in the text. (Figure modified from Wilson & Bower, 1989.)

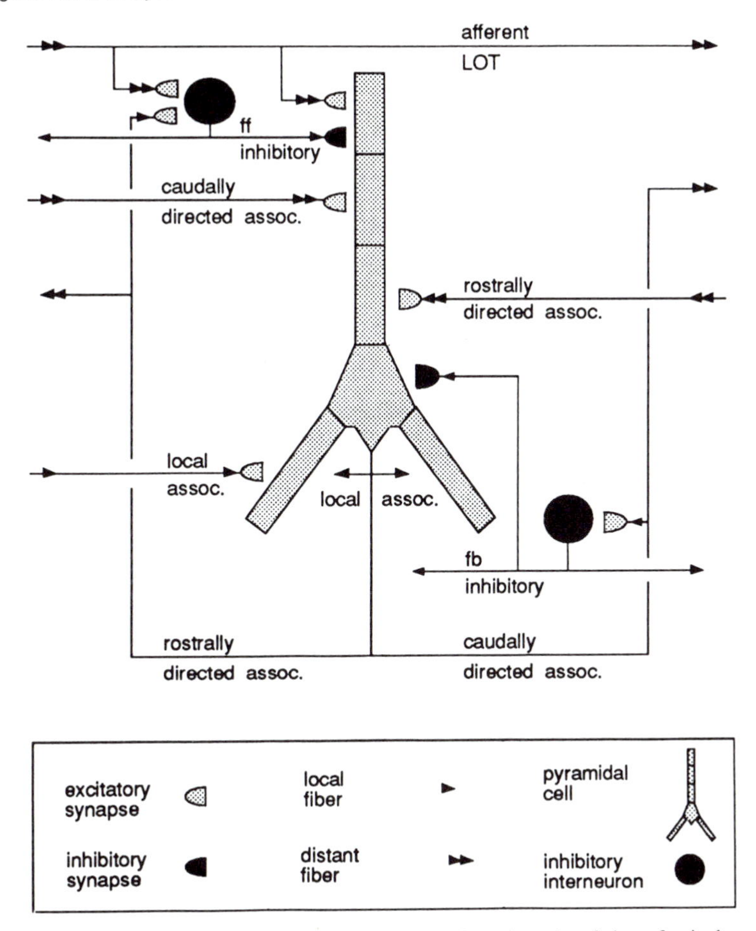

FIGURE 3 Summary of the structure of the model from the point of view of a single pyramidal cell and one each of the associated inhibitory interneurons (see text and key for explanation). (Figure modified from Wilson & Bower, 1988.)

Freeman, 1962, 1968b). Specifically, when the LOT is stimulated with high shock strengths, short duration, biphasic-evoked potentials are recorded (Figure 4A). On the other hand, weak shocks to the same LOT locus evoke multiphase prolonged oscillatory surface field potentials (Figure 4B). Note that this stimulus strength dependence is somewhat counterintuitive, in that a weak shock generates more sustained activity than a strong shock.

The simulation of the cortex replicates quite well the shock strength dependence of cortical-evoked potentials (Figure 4A and B). Figure 5 shows graphically the sequence of simulated events that generates the different evoked potential profiles shown in Figure 4. The specific sequence of events for a large shock strength is as follows: During the first few milliseconds after shocking the LOT, afferent activity sweeps into the cortex, depolarizing the rostral pyramidal cells it first encounters (Figure 5A). Action potentials initiated in the axons of these pyramidal cells as a consequence of afferent activation spread excitation locally to the feedback inhibitory interneurons which, in turn,

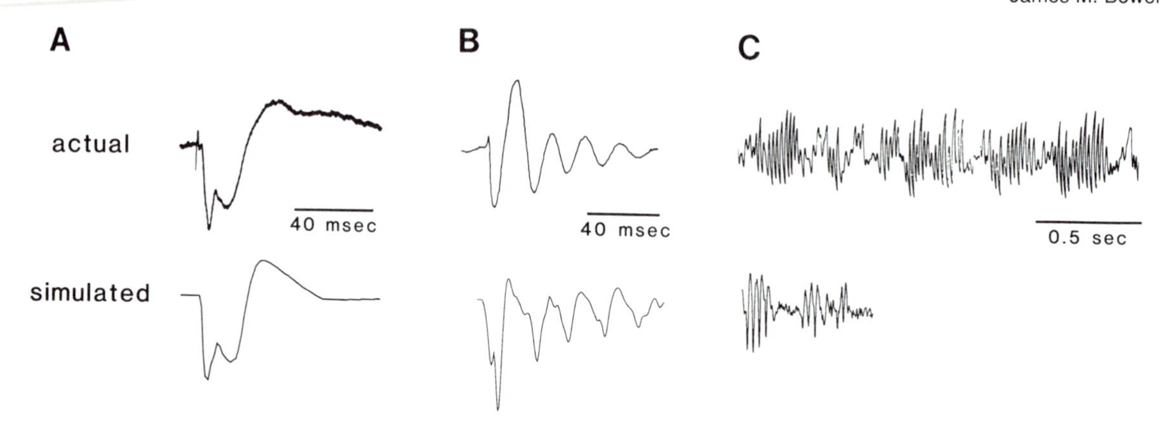

FIGURE 4 Comparison of actual physiological records with those generated by our simulation. In A and B, field potentials evoked by large amplitude shocks of the LOT (A) are compared to potentials evoked by weak shocks (B). In C, EEG recordings from awake behaving animals are compared to simulated results with continuously presented random input to the model. Real field potentials in A are taken from Haberly, 1973; those in B are taken from Freeman, 1968a. EEG recordings shown in C are taken from Bressler, 1984.

reciprocally inhibit nearby pyramidal cells. As previously mentioned, these feedback interneurons set up a short latency, relatively short duration current shunting type inhibition in these pyramidal cells (Figure 2D). As time proceeds in simulation, the substantial amount of afferent activity induced by the large LOT shock continues to spread, caudally depolarizing pyramidal cells in the rest of the cortex (Figure 5B). At the same time, the action potentials of pyramidal cells activated by the initial afferent volley also spread toward caudal cortex via the association fiber system. However, with large shock strengths, afferent input to caudal cells is large enough to depolarize these cells to threshold even before the more slowly conducting association fiber activity reaches them (Figure 5B). Of course these caudal cells, once active, in turn spread excitation back toward rostral cortex (Figure 5C). It is here that a critical point in the behavior of the simulation is seen. Because rostral cells are still under the influence of the current shunting inhibition set up by the initial wave of activation, the rostrally directed activity fails to reactivate rostral neurons (Figure 5D). As a result, the evoked potential response of the cortex to a large stimulus has a simple biphasic waveform representing one cycle of activation (Figure 4).

Figure 5 documents that the pattern of simulated cortical activity resulting from stimulation with a weaker shock input to the LOT is quite different. Initially, the response of the cortex to weak and strong shock conditions looks somewhat similar (Figure 5A). In both, the afferent volley excites rostral pyramidal cells which, in turn, activate local inhibitory neurons and spread activity caudally along the association fiber system. A critical difference is seen, however, as activity sweeps into caudal cortex (Figure 5B, C, and D). Because the influence of the LOT is, in general, much less in caudal than in rostral cortex (Figure 2A and B; Haberly, 1973; Schwob & Price, 1978), and weaker shock strengths activate fewer afferent axons, afferent excitation depolarizes cells to threshold only in rostral regions of the cortex. As a consequence, with weak shock strengths, spike activity in simulated caudal pyramidal cells turns out to be principally dependent on depolarization resulting from the caudally directed association fiber system rather than the LOT. Because the rostral to caudal association fiber system has approximately half the conduction velocity of the LOT fibers (Haberly, 1973), with weak shocks, spiking in caudal pyramidal cells is delayed when compared to larger shocks (Figure 5C and D). It is this change in

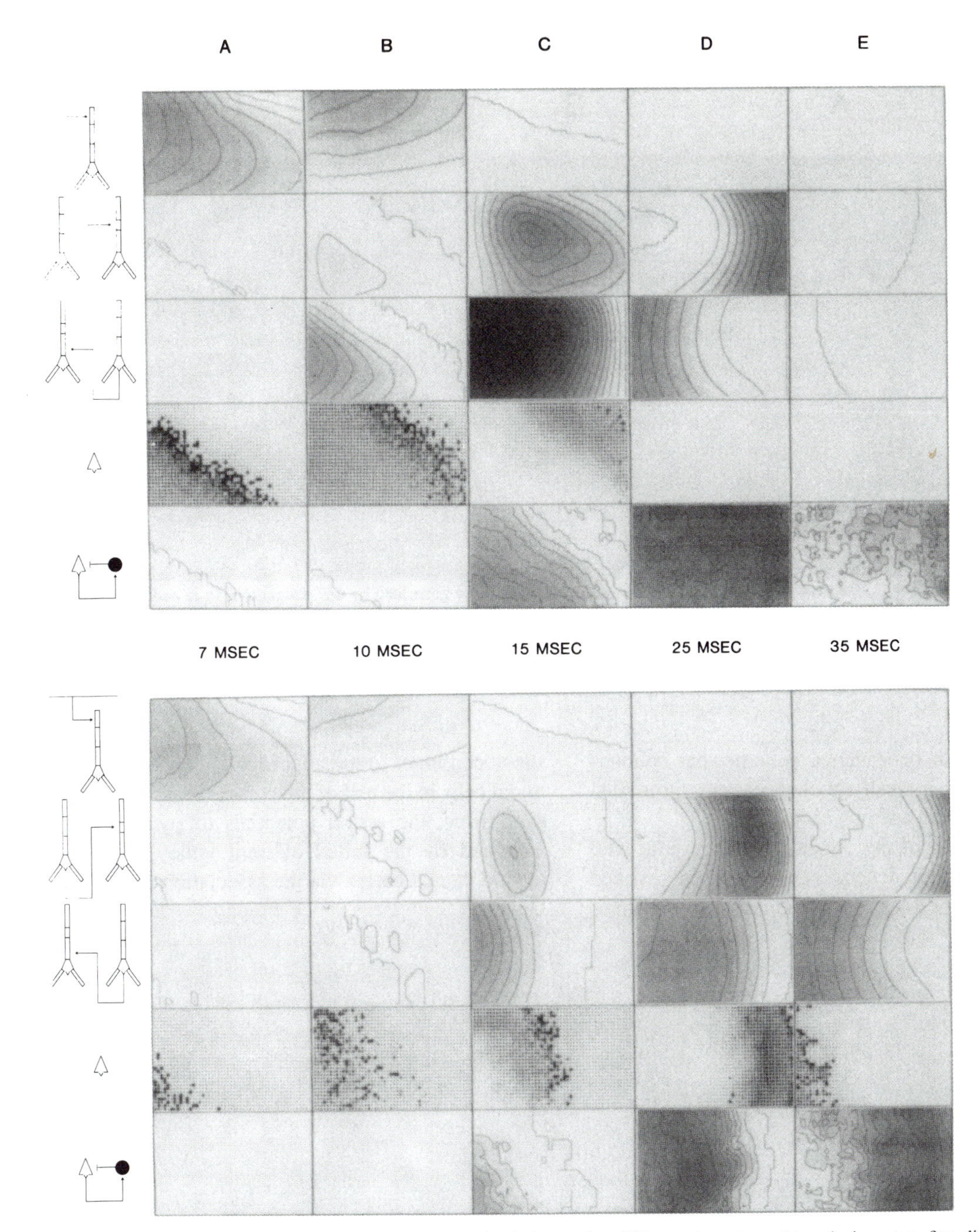

FIGURE 5 The effects of weak and strong shock strengths in simulation at five different time steps. At each time step, five different features of the simulation are compared for both the high (*upper panels*) and low (*bottom panels*) shock strength conditions. Each plot represents the activity at each time step across the full two-dimensional extent of the cortex. Rostral is to the left and caudal to the right. The icons at the left of the figure indicate the feature being displayed in each row of plots. The first row indicates the conductance changes due to the afferent input to the cortex. The next two rows represent conductance changes due to the influence of the rostral to caudal and caudal to rostral association fiber systems, respectively. The third row displays the level of depolarization of the pyramidal cell somas; the size of each box indicates the level of depolarization. Finally, the last row shows conductance changes in local pyramidal cells due to the feedback inhibitory interneurons. As discussed in the text, the principal feature of interest in this figure is the reactivation found in rostral cortex at T = 35 msec in response to a weak shock strength. (See text for explanation.)

the fiber system's influence on the eventual activation of caudal cortex and the resulting delays that are principally responsible for the simulation's differential response to shocks of different strength. This is not because of the later firing of caudal cells, per se, but because this, in turn, delays the influence of caudal cells on rostral cells. As can be seen in Figure 5E, the influence of caudal cortex in these stimulus conditions arrives in rostral cortex after the inhibitory activity induced by the initial afferent input has run its course locally. This leaves rostral pyramidal cells susceptible to reactivation. With a low enough LOT shock strength, enough rostral cells spike on reactivation to initiate another rostral to caudal wave of activity and the cycle repeats. In both the model and the real cortex, as the amplitude of the LOT shock is decreased, the number of these reverberating cortical cycles increases (Ketchum & Haberly, 1993; Freeman, 1962, 1968b).

Clearly, there are numerous ways that one could imagine generating the evoked potentials that have been described. In fact, more abstract modeling efforts have generated the same results based on a purely local mechanism of disinhibition (Freeman, 1975). Although the current simulations are entirely compatible with a local effect contributing to these patterns of activity, the simulation results have provided an additional mechanism that could support the observed behavior. Also, if correct, the simulation has revealed a relationship between cortical components that was not appreciated before. Specifically, the results show that this behavior of the simulation depends on an interaction between the time it takes information to travel across the full extent of the cortex and the time constants of local inhibitory neurons. This was our first indication that network dynamics in this cortex might be the result of a kind of tuning between local and global network features, which is an important component of several of the functional ideas discussed later. However, for the moment, two additional points should be made: (1) the global and local relationships seen in the model would not have been apparent had the simulation not included structural details; and (2) being structural, the simulation makes specific and testable predictions. For example, it makes the clear prediction that the relative timing of induced activity in rostral and caudal pyramidal cells should depend on the strength of the LOT shock.

Natural Oscillatory Behavior

Up to this point, the discussion has centered on the simulation of cortical responses induced by artificially administered shocks to the LOT. This work suggests that a kind of tuning between local and global features of the network might exist. The important question to be considered now is the relevance of this tuning to cortical responses induced by more natural patterns of input. To consider this question, it will first be necessary to describe the characteristic patterns of activity that are found in the olfactory cortex of an animal actively sniffing its environment.

Much of what is known about patterns of cortical activity in awake behaving animals comes from the behavioral studies of Walter Freeman and his colleagues (Freeman, 1970, 1975; Freeman & Schneider, 1982). Their work has shown that piriform cortex EEGs are characterized by a repeated pattern of oscillatory bursts occurring at a frequency of 7–10 Hz (Figure 4). They have demonstrated, furthermore, that these bursts tend to be correlated both with the inspiration cycle of the animal and with oscillations in the olfactory bulb (Bressler, 1984; Freeman & Schneider, 1982). Thus, a single burst is likely to occur in association with a single sniff. The possible significance of this coupling will be considered, but for now, it is important to point out that the principal frequency component found within one of these bursts is in the region of 40 Hz. The reader should note that the average principal frequency of the cortical response to weak shock strengths is also of this same magnitude (Figure 4). It is also worth mentioning that studies in other regions of cerebral cortex (Eckhorn et al., 1988; Gray et al., 1989) have indicated the presence of oscillations at the same frequency. Later sections consider these results explicitly.

The model just described has also been used to study the possible origins of the naturally occurring patterns of cortical oscillation. First, the model was presented with low levels of phasic afferent input of the sort expected to be sent to the cortex from the bulb (Bressler, 1984; Freeman & Schneider, 1982; Ketchum & Haberly, 1993). In this case, the input given the cortex consisted of high frequency bursts of activity mimicking the bulb's oscillating pattern occurring at a 7–10 Hz rate corresponding to the sniff cycle. Perhaps, not surprisingly, the simulation in these conditions replicates the 7–10 Hz component of the EEG (Figure 4C). It also, however, replicates the higher frequency components. Even more interesting is the result that the simulated cortex generates both oscillatory frequencies when presented with low levels of continuous, nonphasic afferent input (Figure 4C).

The fact that this simulation is capable of replicating both principal frequency components of the EEG given either phasic *or* tonic input suggests that these oscillatory patterns may be intrinsic properties of the circuitry of this structure. Freeman reached the same conclusion as a result of experiments in which the LOT was cut and low levels of stimulation were given to the cortical side of the cut tract (Freeman, 1968b). Thus, experimental and simulation results both support the idea that this cortex oscillates intrinsically at frequencies appropriate to the phasic patterns of afferent activity that it would naturally receive during the active sniffing cycle of the animal. Interestingly enough, our efforts to model the olfactory bulb suggest that it, too, oscillates intrinsically at behaviorally relevant frequencies (Bhalla et al., 1988). Thus, these complex mammalian neural structures may share with much smaller oscillating invertebrate central pattern generating networks (Ryckebusch et al., 1989; Selverston, 1985), the property of intrinsic oscillations that are modulated or entrained but not driven by afferent activity. It seems likely that this could become a common property across all neural networks that oscillate while they compute.

As previously discussed, the suggestion from the simulation that the cortex naturally oscillates at the behaviorally correct frequencies is consistent with experimental results (Freeman, 1968b). However, analysis of how the simulation generates these oscillations provides a new structural explanation for this behavior. This analysis again reveals a principal role for the time constants found in the influences of inhibitory neurons. Specifically, as in the case of the LOT shocks, the high frequency component of the EEG is patterned by the time constants of local feedback inhibition which effectively reduces the excitability of pyramidal cells for periods of, on average, 25–35 msec. In contrast, the lower frequency EEG bursting pattern seems to be under the influence of the other type of inhibitory neuron being modeled. During the higher frequency bursting activity, the feedforward component of the input to this neuron results in a slow buildup of long latency, long duration hyperpolarization on nearby pyramidal cells (Figure 6). At some point, the strength of this influence becomes large enough to suppress all cortical activity even in the presence of continuing afferent input (Figure 6). The length of this suppression, in simulation, is related to the duration of the effects of this inhibitory neuron on the pyramidal cell dendrites. When this influence subsides, the pyramidal cells are free to repolarize and the cycle repeats itself. In this way, the model not only replicates known cortical behavior, but also suggests different roles for the two types of inhibitory influences that have been demonstrated to exist in this cortex.

Functional Interpretations

Relationship between Single Cell Properties and Network Dynamics The final sections of the first version of this chapter considered the possible functional significance of these patterns of cortical oscillations. The text pointed out that afferent activation of the model generated a stereotyped spatial and temporal pattern of activity in the superficial dendrites of pyramidal cells throughout the cortex (cf. Figure 5). Afferent input (from the LOT) results in the depolarization of the distal most region of the apical dendrites of these cells, followed by a more proximal dendritic depolarization from the caudally directed association fiber system, with the most proximal part of the apical dendrite then being depolarized by synapses of the rostrally directed association fiber system. This pattern of activation, also observed experimentally (Ketchum &

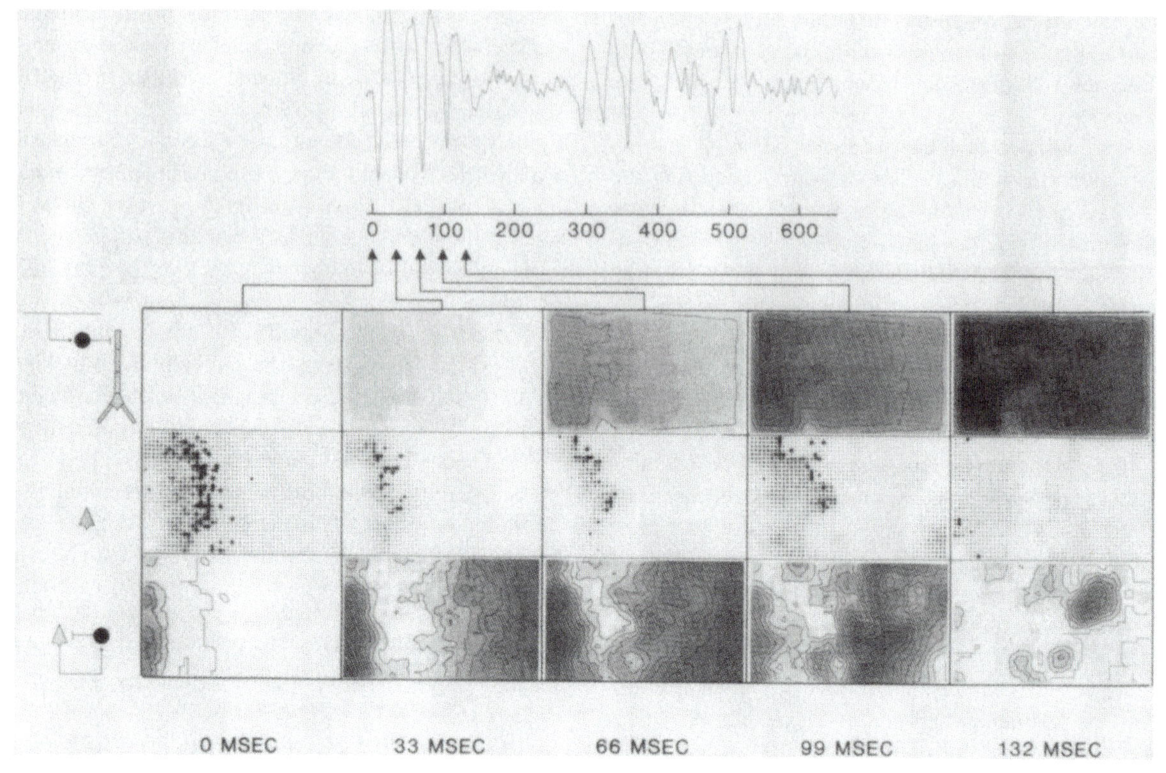

FIGURE 6 The patterns of pyramidal cell activity found in simulation at the beginning of each cycle of oscillation in the simulated EEG. The arrows and lines indicate which column of plots is associated with which cycle of the EEG oscillation. The icons again indicate the feature being displayed in each row of plots. The top and bottom rows indicate the conductance changes in the local pyramidal cells due to the feedforward and feedback neurons, respectively. The middle row shows the depolarization of the pyramidal cells across the cortex at each time indicated. Note the changing pattern of single neuron activity associated with each cycle of the EEG. The text describes the mechanism the simulation suggests may be in part responsible for this behavior.

Haberly, 1993), suggested that it was reasonable to assume that proper functioning of pyramidal cells (e.g., for synaptic learning) might require that their apical dendrites be activated with particular spatial temporal patterns. Accordingly, we proposed that the network might be designed to ensure that these spatial/temporal patterns occur (Bower, 1991a). Single pyramidal cell modeling experiments in our laboratory further suggest that this is the case (Protopapas & Bower, 1994).

A second, simulation-based speculation concerned the relative timing of activity in rostral and caudal pyramidal cells. In particular, it was pointed out that the 25–35 msec principal interval in the higher frequency oscillation of the EEG is also approximately the length of time it takes afferent-induced activity to spread from rostral to caudal and back to rostral cortex (Figure 4). Given that the olfactory bulb is also oscillating at the same frequencies, it seemed reasonable to suggest that rostral cortex repeatedly mixes raw bulbar input with feedback information from the rest of the cortex. On the basis of this observation of the model, and the fact that multiple such sweeps of activity take place following each sniff (Komisaruk 1970; Macrides, Eichenbaum, & Forbes, 1982), we proposed that the 40 Hz oscillations might reflect an iterative mechanism responsible for the convergence of the cortex on a particular odor recognition. By ensuring that network activity evoked in a previous sweep of cortical firing is superimposed onto afferent information arriving from the bulb, the network might facilitate its convergence to a stable spatial response pattern over the course of one sniff. In this way, the simulations suggest that the higher frequency oscillations in the EEG could reflect the regulation of an iterative process, operating at a fundamental interval of approximately 25 msec. More generally, the modeling results are consistent with the idea that the cortex is not performing a passive integration of sensory information over the course of sensory sampling periods (sniffs), but is dynamically altering the cortical conditions under which incoming sensory information is processed (Bower, 1991a).

Significance of 40 Hz Oscillations in Cerebral Cortex The common thread in each of our speculations concerning the functional significance of cortical oscillations was that the oscillations themselves serve to ensure "the relative timing of synaptic activities in different regions of the cortex." In particular, we previously stated that "the cortex may be tuned to ensure that specific patterns of neural activity are generated repeatedly by afferent input." This interpretation resulted from a detailed analysis of the way in which the model generated periodic behavior, not from some higher order proposal for network function.

The presence of 40 Hz oscillations has been demonstrated in several other cerebral cortical regions (Eckhorn et al., 1988; Gray et al., 1989). Perhaps the report that has captured the most attention is the finding that the 40 Hz oscillatory properties of groups of neurons in primary visual cortex appear to reflect specific properties of visual stimuli (Eckhorn et al., 1988; Gray et al., 1989). Given previous abstract modeling efforts in vision (von der Malsburg & Schneider, 1986), these results have led to a very different set of speculations on the functional role of cortical oscillations. In particular, it has been proposed that these oscillatory properties serve as a kind of code for a range of higher order information processing functions from figure ground separation (Buhmann & von der Malsburg, 1991; Sporns et al., 1991), to visual attention (Neibur et al., 1993), and even to general consciousness (Crick & Koch, 1990). These speculations have also produced a growing number of top-down models intent on describing the conditions under which such oscillations might occur (Sporns et al., 1989; Schuster & Wagner, 1990a,b; Schillen & Konig, 1991; Grannan et al., 1993), as well as neural network implementations (Andreou & Edwards, In press).

To shed light on the apparent conflict between our model-based view that the dynamical properties of cortical networks serve to coordinate the processing of information but do not themselves code that information, and the top-down interpretation of the experimental data in visual cortex that the oscillations themselves code information about the stimulus, we modified the model of piriform cortex to reflect basic structural features of neocortical network organization (Wilson & Bower, 1990, 1991). In particular, afferent projections were reorganized to be vertically parallel rather than horizontally serial, and the extent and strength of horizontal connections was reduced. These three changes being made, the model easily replicated the original Gray and Singer (1989) stimulus-induced results in visual cortex (Figure 7).

The results of our structural simulations of neocortical circuits first demonstrated that the same mechanisms that generate oscillatory properties in piriform cortex could also reproduce the specific features of the oscillations seen in neocortex (Wilson & Bower, 1991). Furthermore, observing model behavior led to the conclusion that the dynamical properties of neocortical networks are more likely to subserve the coordination of network activity than to represent stimulus-related information. We proposed that, unlike piriform cortex cells, which may all be equally involved in ongoing associative memory functions, different subsets of visual cortical cells might need to be "in synch" with each other, depending on the nature of the visual system. Our model demonstrated a mechanism for automatically linking neuronal activities

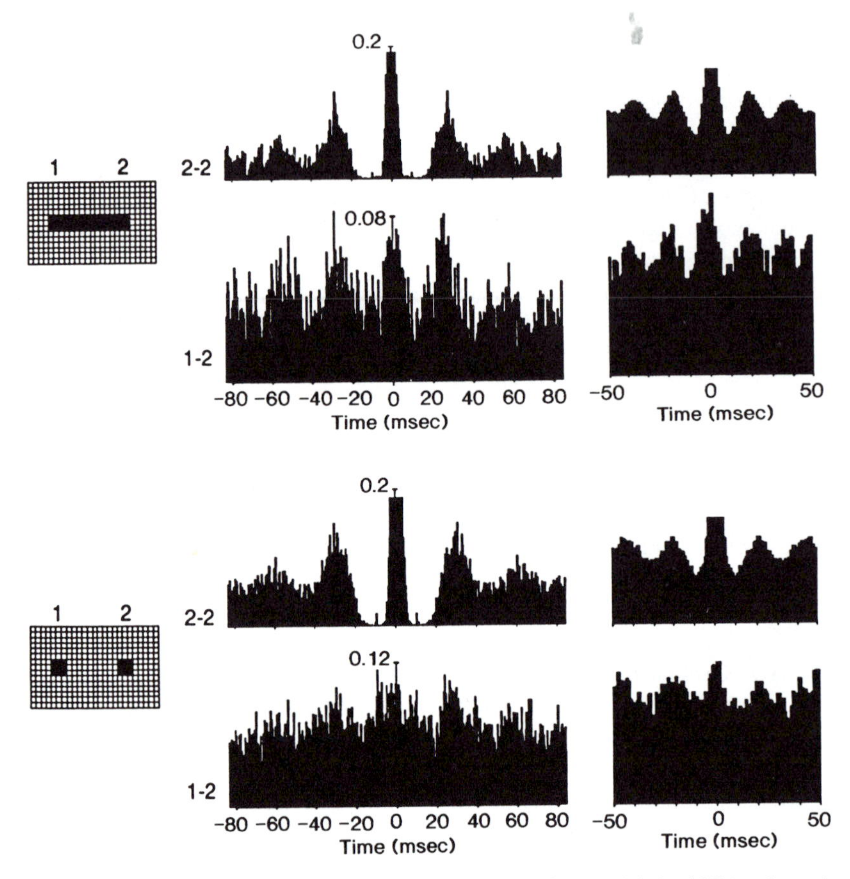

FIGURE 7 A comparison of the auto- and cross-correlations from modeled (middle) and actual (right; modified from Gray et al., 1989) visual cortex. The left column shows a diagram of the model with the stimulus region shaded in each case. The numbers indicate the location of the recording sites referred to in the auto- (2-2) and cross- (1-2) correlations. The correlations generated by presentation of a continuous and broken bar stimulus are shown in the upper and lower panels, respectively. (Figure used with permission from Wilson & Bower, 1990.)

related to a common input, suggesting, again, that cortical oscillations reflect the communication structure underlying the computation and not the computation itself.

Beyond suggesting a different interpretation for the significance of cortical oscillations, our model also leads to several specific related predictions concerning the structure of the oscillations. First, our model demonstrated that intracortical pathways with finite conduction velocities could establish and maintain zero-phase oscillations. At the time our model was generated, the prevailing view was that some structure outside the cortex itself was necessary to establish a zero-phase relationship. However, it is now generally accepted that intrinsic connections are responsible for the oscillations.

Second, we predicted that the oscillations in visual cortex should be quite variable in their occurrence from trial to trial, as well as in phase and frequency (Wilson & Bower, 1991). At the time, most abstract modelers were intent on constructing models that immediately generated these oscillations and sustained the oscillations throughout the entire stimulus presentation. Given the assumption that these oscillations served to bind stimulus features together, this is a reasonable modeling objective. In fact, we pointed out that the trial-to-trial and within-trial variability we predicted makes it unlikely that the oscillations themselves could serve as a stable source of information about the stimulus, or provide a reliable mechanism for more abstract functions like feature binding and/or attention. At the time of our modeling, the analysis of the real visual cortical data was based on averaging techniques making it difficult to determine the time to onset or persistence of the oscillatory patterns (Eckhorn et al., 1988; Gray et al., 1989). However, subsequent trial-by-trial analysis by two separate groups has shown that this prediction of our model was essentially correct: correlations in the oscillations come and go (Schuster, 1991; Gray, Koenig, Engel, & Singer, 1992). Nevertheless, top down models of oscillations generally continue to ignore this issue (Sporns et al., 1989; Schuster & Wagner, 1990a,b; Schillen & Konig, 1991; Grannan et al., 1993).

Finally, our models led to the speculation that the neurons within the cortex that should show the highest degree of oscillation should be the locally projecting inhibitory interneurons we suggested were responsible for regulating the overall dynamics of these networks. Thus, we predicted that any neurons with intrinsic high-frequency oscillatory properties should be inhibitory. Similarly, extracellularly recorded cells with strong oscillatory tendencies should also be inhibitory. Obviously, if the neurons with the strongest oscillatory and zero-phase properties were local projecting inhibitory neurons, it would substantially complicate the theory that these oscillations represent higher order features of the stimulus. For example, it is hard to imagine how feature binding between cortical regions could be dependent on neurons that do not project outside of their own cortical locations. Data concerning this final prediction, has not yet been published, however preliminary intracellular labeling results from the laboratory of Charles Gray suggests that we may be correct (Gray, personal communication).

We believe that each of our modeling predictions pose serious problems for the proposed "top down" functional roles of cortical oscillations. Nevertheless, the fact that our first two predictions have been demonstrated to be essentially correct has had little effect on the abstract modelers of this phenomenon. These speculations have even made it into the popular press (Crick 1994). In my opinion, this is one of the more substantial dangers of top-down approaches to understanding neuronal function: that a particular idea rapidly takes on a life of its own, separate from the nervous system, based on an intellectual appeal which, more times than not, is essentially its simplicity.

Modeling Cortical Function

This chapter, so far, has only dealt with the model-based mechanisms for replicating the oscillatory structure of cerebral cortex in general and olfactory cortex in particular. Initially, we viewed the ability of the model to replicate this behavior as simply a first necessary step toward studying the network's structure–function relationships. Although the discovery of 40 Hz oscillations in visual cortex has meant that this initial work has been, perhaps, more functionally relevant than we initially thought, we still regard these efforts as primarily laying the groundwork for more functional modeling efforts. The following sections describe our more recent efforts in this direction.

Associative Memory and Olfactory Cortex

Olfactory cortex is clearly important for olfactory object recognition. For example, lesions of piriform cortex, or its afferent input, significantly impair the capacity to perform olfactory discrimination tasks (Slotnick & Berman, 1980; Staubli, Schottler, & Nejat-Bina, 1987). Piriform cortex is also the largest region of olfactory cortex (Haberly, 1985), and occupies a central position between the sensory periphery and multimodal cortical structures such as the entorhinal cortex and hippocampus (Price, 1985) believed to be fundamental to general memory function (Squire, 1986a,b, 1987).

Although the olfactory cortex is clearly involved in object recognition, it is yet unclear in what way it contributes. However, we believe that there are several operational reasons to suspect that the olfactory cortex may implement an autoassociative memory (Bower 1991a,b,c, 1993). First, the olfactory cortex receives patterns of afferent input generated by diverse blends of molecules that are sometimes significant only in their relative concentrations (Epple et al., 1989) and abstract autoassociative memory models are particularly adept at generating stable outputs to such complex input patterns (Palm, 1980; Kohonen, 1984). Second, single bulbar neurons respond to a wide variety of odors, undoubtedly resulting in a highly distributed representation of olfactory sensory information (Mathews, 1972; Tanabe et al., 1975). Such a distributed representation is quite appropriate for subsequent processing by an autoassociative memory (Palm, 1980; Kohonen 1984). Experimental studies in our laboratory, in which many and single mitral cells have been recorded simultaneously in awake behaving rats (Bhalla et al., 1988; Bhalla & Bower, 1991), have also demonstrated considerable variability in mitral cell responses to single odors. Autoassociative memories are capable of recognizing an input even in the presence of such variability (Kohonen, 1984; Hopfield, 1982, 1984). Finally, particular objects are likely to give off changing blends of molecules and are still recognized as the same object (Bower, 1991a). Again, abstract autoassociative memory models can reconstruct original output patterns based on less-than-complete stimuli (Kohonen, 1984; Hopfield, 1982, 1984).

Modeling Autoassociative Function

Although there appear to be similarities between the computational task of the olfactory system and autoassociative memories, the primary basis for our speculation that the olfactory cortex is a form of autoassociative network is its structural similarity to more abstract autoassociative networks (Figure 8; Haberly, 1985; Haberly & Bower, 1984; Bower, 1991b). In particular, the extensive and diffuse distribution of both afferent projections from the olfactory bulb and the intrinsic connections within the cortex itself (Luskin & Price, 1983) are highly reminiscent of more abstract

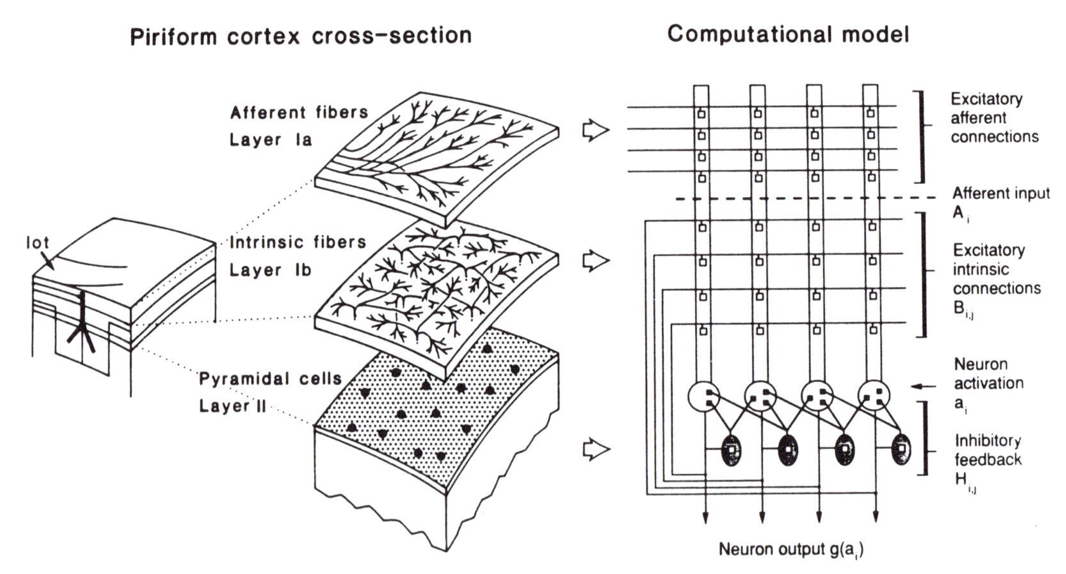

FIGURE 8 An overview of piriform cortex anatomical structure and computational modeling of the piriform cortex associative memory function. (Figure used with permission from Hasselmo & Bower, 1993.)

autoassociative memory models (Palm, 1980; Kohonen, 1984). Having constructed a network model with these features that was also capable of replicating the basic dynamical properties of piriform cortex, we began using a combination of experimental and modeling techniques to explore the possibility that this cortex actually implements an autoassociative memory (Wilson & Bower, 1988; Hasselmo, Wilson, Anderson, & Bower, 1990; Hasselmo & Bower, 1992; Hasselmo et al., 1992). To do this, a Hebb-type correlation learning rule (Hebb, 1949) was introduced into the network to govern activity-dependent changes in the synaptic strengths of modeled connections (Wilson & Bower, 1988). The model was then provided with input intended to represent loosely the activity of single neurons in the olfactory bulb and the resulting cortical activity was evaluated.

The results of these simulations demonstrated that the model was capable of basic autoassociative function (Figure 9; Wilson & Bower, 1988). More than that, however, the model predicted that the ability to generate stable patterns of neuronal activity, even with variable inputs, would be optimal when Hebbian synaptic modification was located primarily in the excitatory synapses of the association fiber system (Bower, 1991b, 1993; Hasselmo et al., 1990). In contrast, Hebbian synaptic modification in synapses associated with afferent projections from the olfactory bulb actually interfered with memory performance (Hasselmo et al., 1990). In contrast, the more top-down modeling of Granger and colleagues (Granger et al., 1988; Ambros-Ingerson et al., 1990) proposed that the principle site of synaptic modification was located in the afferent fiber system. In fact, their model made very little use of the association fiber system, which our model suggests is essential for the function of the network.

At the time our simulations were performed, no detailed information was yet available on the specific location of synaptic modification in piriform cortex, although one *in vivo* study using field potentials suggested that long-term potentiation (LTP) might occur in afferent synapses (Roman, Staubli, & Lynch, 1987). Motivated by our modeling results, we subsequently used a brain slice preparation to demonstrate that synaptic potentials evoked by association fiber stimulation, not afferents, show clear and consistent short-term potentiation (Hasselmo & Bower, 1990). Furthermore, Kanter and Haberly (1990) simultaneously showed a significantly greater level of long-term potentiation in association fiber synapses than in afferent fiber synapses, just as the model had predicted. This relative lack of modifiability in afferent projections is also consistent with the balance of previously published data demonstrating a lack of both short-term and substantial longer-term potentiation in afferent synapses (Racine & Milgram, 1983; Racine, Milgram, & Hafner, 1983; Stripling, Patneau, & Gramlich, 1988). Thus a specific prediction made by the model, that the largest amount of activity-dependent synaptic modification should take place in the association fiber system, was shown to be consistent with subsequent experimental results.

Neuromodulators and the Function of Piriform Cortex

The model's prediction of the experimentally supported result that synaptic modification should be primarily present in the association fiber system rather than in the afferent system, specifically focused our attention on the possible functional significance of differences between these two major populations of excitatory synapses. In particular, we elected to use experimental techniques to study the possible differential effects of the neuromodulator acetylcholine on these two populations of synapses. The piriform cortex receives extensive cholinergic innervation from a specific region of the basal forebrain, the horizontal limb of the diagonal band of Broca (Wenk, Meyer, & Bigl, 1977; Macrides, Davis, Youngs, Nadi, & Margolis, 1981), and odor recognition and memory performance have been demonstrated to be influenced by this neuromodulator (Hunter & Murray, 1989; Serby et al., 1989; Koss, Weiffenbach, Haxby, & Friedland 1988; Doty, Reyes, & Gregor, 1987; Coyle & Nicoll, 1984).

Alerted by the modeling results to pay particular attention to any differential effects of acetylcholine in

A

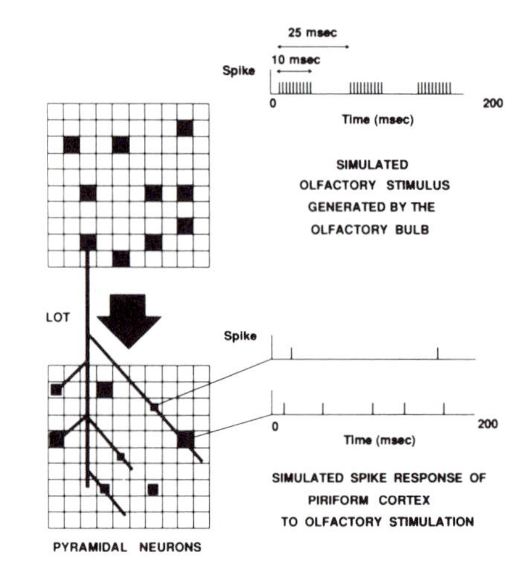

B

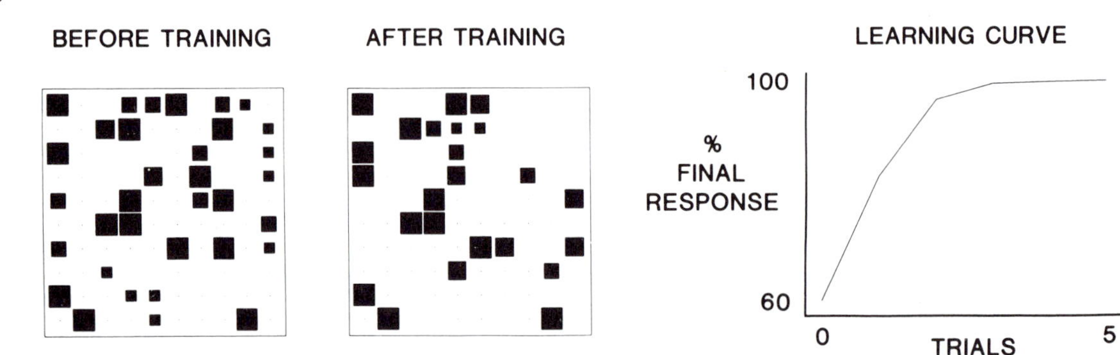

FIGURE 9 Results from simulations in which the memory properties of the piriform cortex model were explicitly studied. A indicates with circles those synapses undergoing activity-dependent changes in connection strengths in these simulations (see text and Wilson & Bower, 1988). As shown by the simulated intracellular records attached to the diagrammed electrodes, changes in the afferent synapses occurred only as a consequence of paired stimuli and were not permanent. All other circled synapses were capable of long-term Hebbian modifications, as discussed in the text. The diagram at the top right illustrates several features of these simulations. First, the random projection pattern of one of ten active bulbar neurons (100 total in the simulation) to cortical pyramidal cells is shown. The stereotyped pattern of activity in the active bulbar neurons used in these simulations is also shown. The cortical section of this diagram demonstrates the display convention for cortical activity in this and the next figure. As shown, the size of the black box overlying the position of each pyramidal cell in the two-dimensional simulated cortical array corresponds to the total number of action potentials the cell produced during 200 msecs of bulbar activity. B demonstrates the response of the simulated cortex to activity in a random set of 10 bulbar neurons. On the far left is shown the response of the cortex to the first presentation of the stimulus, that is, effectively before training. The middle diagram indicates the final stable pattern of activity induced in the cortex by this particular active set of bulbar neurons in the presence of synapse modification. The graph on the far right indicates the changes in the pattern of activity over the five trials necessary for the simulated cortex to converge on a fixed final response pattern. One trial, in this case, corresponds to one 200-msec period of stimulation as shown in A above. (Figure modified from Wilson & Bower, 1988.)

afferent and association fiber synapses, we were quite surprised to find that cholinergic agonists (e.g., carbachol) specifically suppressed the association fiber system, but not the afferent fiber system (Hasselmo & Bower, 1992). This presynaptic effect suppressed synaptic potentials more than 50% at concentrations less than 5 mM, and by more than 80% at 100 mM. In contrast, carbachol reduced the height of afferent fiber synaptic potentials by less than 12%, even at a concentration of 500 mM. We also demonstrated, as in previous studies (Woody & Gruen, 1993) that carbachol produced conductance changes in pyramidal cells, resulting in greater cell excitability (Hasselmo & Bower, 1992).

Modeling the Effects of Acetylcholine

Although we were pleased to find that acetylcholine produced differential synaptic effects, its suppressive effect on the association fiber system appeared initially to be somewhat paradoxical, as it is just these synapses that our models suggested should be most sensitive to synaptic modification during learning (Wilson & Bower, 1988; Hasselmo et al., 1990). Yet, the principal synaptic effect of cholinergic agents appeared to be the selective suppression of these very synapses.

To resolve this conflict, we incorporated cholinergic-like effects into an autoassociative memory model based on the structure of olfactory cortex (Hasselmo et al., 1992). By using this model we were able to show that the cholinergic effects we had seen should actually enhance the ability of the network to store input patterns and regenerate them later (Hasselmo et al., 1992). This enhanced performance was related to the fact that autoassociative networks all suffer from a tendency for multiple, unrelated neuronal patterns to become linked together as more input patterns are stored. This is especially the case if the input patterns share common components, as is very likely to be true for many olfactory stimuli (Bower, 1991c). This interference is, in fact, related to the mechanism that allows autoassociative networks to generate stable outputs given variable inputs. After the first pattern is stored, activation of the second pattern causes a partial "recall" of the first pattern. If the second pattern is a degraded version of the first pattern, this is a desirable effect. However, if the second pattern is a new pattern to be stored, the "recall" of the first pattern will have the effect of making the network store not only the new pattern but a mixture of the two patterns. The long-term effect of this process is a catastrophic drop in the network's storage capacity (Hasselmo et al., 1992; Hasselmo, 1993, 1994). By generally suppressing the strength of intrinsic fiber synapses while storing a new pattern, effectively what happens is that only those synapses strongly activated become part of the new pattern. Weakly activated synapses, which are the main agents responsible for unwanted pattern linkage, never become active enough to contribute in the presence of generalized suppression.

Our modeling also demonstrated that the benefits of the suppression of association fiber connections were enhanced by the increased postsynaptic excitability that we have also shown is induced by cholinergic agents (Hasselmo & Bower, 1993). Elevated postsynaptic excitability to afferent input increases the speed of memory storage, counteracting a general slowing effect resulting from the suppression of intrinsic synaptic connections. Although it has previously been suggested (but not modeled) that cholinergic increases in cell excitability might speed learning (Singer, 1987), our simulations demonstrate that this change alone makes learning more unstable and less efficient (Hasselmo et al., 1990, 1992). However, when all the experimentally demonstrated effects of cholinergic agents are included in the model, the net effect is a significant improvement in memory function (Hasselmo & Bower, 1993).

A NEW THEORY OF NEUROMODULATOR EFFECTS ON CORTICAL FUNCTION

Neuromodulators and Cortical "Learning State"

As just described, the interaction of modeling with experimental studies has resulted in the generation of a new theory of cortical memory function (Hasselmo & Bower, 1992). Before constructing our cortical models, we did not know what specific role the association fiber system, or acetylcholine, might play in network function. However, our modeling and experimental work have now suggested that neuromodulators may be responsible for modulating the "pattern storage state" of cortical networks. Our models have led us to propose that during periods in which novel stimulus patterns are being stored, acetylcholine levels should be high so as to reduce the possible contamination of old patterns with new ones (Hasselmo et al., 1992; Hasselmo, 1993). During periods in which previously stored patterns are being presented, the levels of acetylcholine should be low so as to allow the pattern-completion capabilities of the intrinsic synaptic connections to operate unfettered. Thus, if we are correct, the memory dynamics of the olfactory cortex may be under the direct control of neuromodulatory agents and their multiple physiological actions. This result was not "built into" our models, but emerged from them.

Testability

Earlier in this chapter, the claim was made that one reason for constructing realistic models capable of producing realistic outputs was that this made the generation of testable predictions much easier. In the current case, the model makes specific predictions concerning the pattern of EEG activity that would be expected during the presence of high amounts of acetylcholine. As described earlier, our previous model-based analysis of the oscillatory properties of olfactory cortex suggested a significant role for the association fiber system in cortical oscillations (Wilson & Bower, 1992). It is therefore not surprising that our cortical models suggest that these oscillatory properties change as a result of the suppressive effects of norepinephrine or acetylcholine on these connections. We have predicted that decreasing the strength of associational input will result in a reduction in 40 Hz oscillatory behavior, enhancing theta (7–10 Hz) components. Although the increased excitability of pyramidal cells will act to counteract somewhat this loss of oscillatory activity. Nevertheless, we do expect subtle changes in the oscillatory responses of the network that can be tested for experimentally.

In addition to these physiological predictions, examination of the oscillatory structure of the network in the presence of neuromodulators has also highlighted the role that might be played by the temporal dynamics of the olfactory cortex and its cells in memory function (Bower, 1991a). In particular, the ability of the network model to generate a stable output pattern given a previously presented input pattern appears to be dependent on an alternation in the influence of afferent and association fiber systems. In the first phase of memory recall, the memory behavior of the model is optimal if network activity is dominated by the pattern of afferent activity. This serves to establish the global response properties of the network to a given activation pattern. The second phase of memory recall allows the modified synaptic weights of the association fiber system to influence final network behavior. This later phase is critical for the stability of output patterns, as well as for functions such as pattern completion. In the model, this alternation results naturally from the structure of the piriform cortex network (Wilson & Bower, 1990, 1992). The relation between network behavior and the integration properties of single pyramidal cells is currently being explored in our laboratory by using both models and experiments (Protopapas & Bower, 1994).

Significance for Abstract Autoassociative Memory Models

This book is intended as an introduction to neural networks broadly defined. Although the sole focus of our research is to understand how the olfactory system works, our efforts may also have relevance to more abstract autoassociative memory modeling and theory (Bower, 1993; Hasselmo, 1993, 1994). Specifically, the problem of memory interference caused by runaway intrinsic activity also limits the capacity of more abstract autoassociative memory models, especially when overlapping (or nonorthogonal) input patterns are being stored. In some abstract memory models, the approach taken to overcome this limitation involves basing memory storage on a precomputed clean outer product of the input pattern (Hopfield, 1982). In these cases, the strengths of the connections which give the best overall performance are calculated beforehand, avoiding the problem of dynamically and independently establishing the correct weights. In models that do learn synaptic weights, modelers often externally clamp the activity of the units within the network to the value of the pattern to be recognized (Kohonen, 1984, Ackley, Hinton, & Sejnowski, 1985). This results in the application of Hebbian synaptic modification only to the connections between units directly activated by the input pattern. Accordingly, these models overcome memory interference difficulties through direct and specific external intervention. Hasselmo has shown mathematically that the more subtle and graded suppression of association connections seen in olfactory cortex actually improves the performance of these types of networks (Hasselmo, 1993). However, years of previous, more abstract investigations of autoassociative memories did not come up with this combination of memory storage-enhancing modifications. In my opinion, this again argues for an approach to understanding neural computation from the biology first.

Acknowledgments

I would like to acknowledge Matthew Wilson, who is principally responsible for the physiological simulation results described in this chapter and who composed most of the figures. I would also like to acknowledge Michael Hasselmo, who was largely responsible for both the experimental and modeling studies of the effects of acetylcholine on olfactory cortex. This work is supported by the NSF (EET-8700064) and the ONR (Contract N00014-88-K-0513).

References

Ackley, D. H., Hinton, G. E., & Sejnowski, T. J. (1985). A learning algorithm for Boltzmann machines. *Cognitive Science*, **9**, 147–169.

Ambros-Ingerson, J., Granger, R., & Lynch, G. (1990) . Simulation of paleocortex performs hierarchical clustering. *Science*, **247**, 1344–1348.

Amit, D. J. (1989) *Modeling brain function: The world of attractor neural networks.* New York: Cambridge University Press.

Andreou, A. G., & Edwards, T. G. (1994). VLSI phase locking architectures for feature linking in multiple target tracking systems. *Neural Information Processing Systems 1993*, pp. 866–873.

Bhalla, U. S., & Bower, J. M. (1993). Exploring parameter space in detailed single neuron models: Simulations of the mitral and granule cells of the olfactory bulb. *Journal of Neurophysiology*, **69**, 1948–1965.

Bhalla, U., & Bower, J. M. (1991). Multiple single unit recording from olfactory bulb of awake behaving rats. *Soc. Neurosci. Abstr.* **17**, 636.

Bhalla, U. S., Wilson, M. A., & Bower, J. M. (1988). Integration of computer simulations and multiunit recording in the rat olfactory system. *Society for Neuroscience Abstracts*, **14**, 1188.

Bower, J. M. (1991a). Relations between the dynamical properties of single cells and their networks in piriform (olfactory) cortex In J. McKenna, J. Davis, & S. Zornetzer (Eds.), *Single neuron computation.* New York: Academic Press, pp. 437–462.

Bower, J. M. (1991b). Associative memory in a biological network: Structural simulations of the olfactory cerebral cortex. In V. Miltunovic & S. Antognetti (Eds.), *Neural networks* (Vol. II). New Jersey: Prentice Hall.

Bower, J. M. (1991c). Piriform cortex and olfactory object recognition. In J. Davis & H. Eichenbaum (Eds.), *Olfaction as a model system for computational neuroscience.* Cambridge, MA: MIT Press.

Bower, J. M. (1992). Modeling the nervous system. *Trends in Neurosciences*, **15**, 411–412.

Bower, J. M. (1993). The modulation of learning state in a biological associative memory: An *in vitro*, *in vivo*, and *in computo* study of object recognition in mammalian olfactory cortex. *Artificial Intelligence Review*, **7**, 261–269.

Bower, J. M., & Haberly, L. B. (1986). Facilitating and nonfacilitating synapses on pyramidal cells: A correlation between physiology and morphology. *Proceedings of the National Academy of Sciences of the United States of America*, **83**, 1115–1119.

Bower, J. M., & Hale, J. (1991). Exploring neural circuits on graphics workstations. *Scientific Computing and Automation*, **7**, 35–46.

Bower, J. M., Nelson, M. E., Wilson, M. A., Fox, G. C., & Furmanski, W. (1989). Piriform (olfactory) cortex model on the hypercube. In G. Fox (Ed.), *Proceedings of 3rd conference on hypercube concurrent computers & applications.* New York: ACM.

Bressler, S. L. (1984). Spatial organization of EEGs from olfactory bulb and cortex. *Electroencephalography and Clinical Neurophysiology*, **57**, 270–276.

Buhmann, J., & von der Malsburg, C. (1991). Sensory segmentation by neural oscillators. In Proceedings of the international conference on neural networks (Vol II), pp. 603–607.

Churchland, P. S. & Sejnowski, T. J. (1988). Perspectives on cognitive neuroscience. *Science*, **242**, 741–745.

Coyle, A. E., & Nicoll, R. A. (1984). Characterization of a slow cholinergic postsynaptic potential recorded *in vitro* from rat hippocampal pyramidal cells. *Journal of Physiology*, (*London*), **365**, 47P.

Crick, F. (1994). *The astonishing hypothesis*. New York: Scribner's.

Crick, F., & Koch, C. (1990). Towards a neurobiological theory of consciousness. *Seminars in the Neurosciences*, **2**, 263–275.

Denk, W., Webb, W. W., Hudspeth, A. J. (1986). Optical measurement of the Brownian-motion spectrum of hairbundles in the transducing hair-cells of the frog auditory-system. *Biophysical Journal*, **49**, 21–31.

DeSchutter, E., & Bower, J. M. (1993). Sensitivity of synaptic plasticity to the CA^{2+} permeability of NMDA-Channels: A model of long-term potentiation in hippocampal neurons. *Neural Computation*, **5**, 681–694.

DeSchutter, E., & Bower, J. M. (1994a). An active membrane model of the cerebellar Purkinje cell: I. Simulation of current clamps in slice. *Journal of Neurophysiology*, **71**, 375–400.

DeSchutter, E., & Bower, J. M. (1994b). An active membrane model of the cerebellar Purkinje cell: II. Simulation of synaptic responses. *Journal of Neurophysiology*, **71**, 401–419.

DeSchutter, E., & Bower, J. M. (in press). Responses of cerebellar Purkinje cells are independent of the dendritic location of granule cell synaptic inputs. *Proceedings of the National Academy of Sciences of the United States of America*, **91**, 4736–4740.

Devor, M. (1976). Fiber trajectories of olfactory bulb efferents in the hamster. *Journal of Comparative Neurology*, **166**, 31–48.

Doty, R. L., Reyes, P. F., & Gregor, T. (1987). Presence of both odor and detection deficits in Alzheimer's disease. *Brain Research Bulletin*, **18**, 597.

Eckhorn, R., Bauer, R., Jordon, W., Brosch, M., Kruse, W., Munk, M., & Reitboeck, H. J. (1988). Coherent oscillations: A mechanisms of feature linking in ght visual cortex? *Biological Cybernetics*, **60**, 121–130.

Epple, G., Belcher, A., Greenfield, K. L., Kuderling, I., Nordstrom, K., & Smith, A. B. III (1989). Scent mixtures as social signals in two primate species: *Saguinus fuscicollis* and *Saguinus o. oedipus*. In D. G. Laing, W. S. Cain, R. L. McBride, & B. W. Ache (Eds.), *Perception of complex smells and tastes*. New York: Academic Press.

Freeman, W. J. (1962). Alterations in prepyriform evoked potential in relation to stimulus intensity. *Experimental Neurology*, **6**, 70–84.

Freeman, W. J. (1968a). Relation between unit activity and evoked potentials in prepiriform cortex of cats. *Journal of Neurophysiology*, **31**, 337–348.

Freeman, W. J. (1968b). Effects of surgical isolation and tetanization of prepyriform cortex in cats. *Journal of Neurophysiology*, **31**, 349–357.

Freeman, W. J. (1970). Amplitude and excitability changes of prepyriform cortex related to work performance in cats. *Journal of Biomedical Systems*, **1**, 3–29.

Freeman, W. J. (1975). Mass action in the nervous system New York: Academic Press.

Freeman, W. J., & Schneider, W. (1982). Changes in spatial patterns of rabbit olfactory EEG with conditioning to odors. *Psycophysiology*, **19**, 44–56.

Granger, R., Ambros-Ingerson, J., Anton, P. S., Whitson, J., & Lynch, G. (1990). Computational action and interaction of brain circuits. In S. Zornetzer, J. Davis, and C. Lau (Eds.), *An introduction to neural and electronic networks*. New York: Academic Press.

Granger, R., Ambros-Ingerson, J., Henry, H., & Lynch, G. (1988). Partitioning of sensory data by a cortical network. In D. Z. Anderson (Ed.), *Neural information processing systems*. New York: American Institute of Physics.

Grannan, E. R., Kleinfeld, D., & Sompolinsky, H. (1993). Stimulus-dependent synchronization of neuronal assemblies. *Neural Computation*, **5**, 550–569.

Gray, C. M. (1994). The significance of oscillations in cerebral cortical networks: A reinterpretation. *Journal of Computational Neuroscience*, **1**, 11–38.

Gray, C. M., Konig, P., Engel, A. K., & Singer, W. (1989). Oscillatory responses in cat visual cortex exhibit intercolumnar synchronization which reflects global stimulus properties. *Nature*, **338**, 334–337.

Gray, C. M., Konig, P., Engel, A. K., & Singer, W. (1992). Synchronization of oscillatory neuronal responses in cat visual cortex: Temporal properties. *Visual Neuroscience*, **8**, 337–347.

Gray, C. M., & Singer, W. (1989). Stimulus-specific neuronal oscillations in orientation columns of cat visual cortex. *Proceedings of the National Academy of Sciences of the United States of America*, **86**, 1698–1702.

Haberly, L. B. (1973). Unitary analysis of opossum prepyriform cortex. *Journal of Neurophysiology*, **36**, 762–774.

Haberly, L. B. (1985). Neuronal circuitry in olfactory cortex: Anatomy and functional implications. *Chemical Senses*, **10**, 219–238.

Haberly, L. B., & Bower, J. M. (1984). Analysis of association fiber system in piriform cortex with intracellular recording and staining techniques. *Journal of Neurophysiology*, **51**, 90–112.

Haberly, L. B., & Presto, S. (1986). Ultrastructural analysis of synaptic relationships of intracellular stained pyramidal cell axons in piriform cortex. *Journal of Comparative Neurology*, **248**, 464–474.

Haberly, L. B., & Price, J. L. (1978). Association and commissural fiber systems of the olfactory cortex of the rat: I. Systems originating in the piriform cortex and adjacent areas. *Journal of Comparative Neurology*, **178**, 711–740.

Haberly, L. B., & Shepard, G. M. (1973). Current density analysis of opossum prepyriform cortex. *Journal of Neurophysiology*, **36**, 789–802.

Hasselmo, M. E. (1993). Acetylcholine and learning in a cortical associative memory. *Neural Computation*, **5**, 32–44.

Hasselmo, M. E. (1994). Runaway synaptic modification in models of cortex: Implications for Alzheimer's disease. *Neural Networks*, **7**, 13–40.

Hasselmo, M. E., Anderson, B. P., & Bower, J. M. (1992). Cholinergic modulation of cortical associative memory function. *Journal of Neurophysiology*, **67**, 1230–1246.

Hasselmo, M. E., & Bower, J. M. (1990). Afferent and association fiber differences in short-term potentiation in piriform (olfactory) cortex. *Journal of Neurophysiology*, **64**, 179–190.

Hasselmo, M. E., & Bower, J. M. (1992). Cholinergic suppression specific to intrinsic not afferent fiber synapses in rat piriform (olfactory) cortex. *Journal of Neurophysiology*, **67**, 1222–1229.

Hasselmo, M. E., & Bower, J. M. (1993). Acetylcholine and memory. *Trends in Neurosciences*, **16**, 218–222.

Hasselmo, M. E., Wilson, M. A., Anderson, B., & Bower, J. M. (1990). Associative function in piriform (olfactory) cortex: Computational modeling and neuropharmacology. In The brain, Cold Spring Harbor symposium on quantitative biology, Volume LV. Cold Spring Harbor, New York: Cold Spring Harbor Press, pp. 599–610.

Hebb, D. O. (1949). The organization of behavior. New York: Wiley.

Hopfield, J. J. (1982). Neural networks and physical systems with emergent collective computational abilities. *Proceedings of the National Academy of Sciences of the United States of America*, **79**, 2554–2558.

Hopfield, J. J. (1984). Neurons with graded response have collective computational properties like those of 2-state neurons. *Proceedings of the National Academy of Sciences of the United States of America*, **81**, 3088–3092.

Hunter, A. J., & Murray, T. K. (1989). Cholinergic mechanisms in a simple test of olfactory learning in the rat. *Psychopharmacology*, **99**, 270–273.

Jaeger, D., De Schutter, E., & Bower, J. M. (1993). Prolonged activation following brief synaptic input in the cerebellar Purkinje cell: Intracellular recording and compartmental modeling. In J. M. Bower & F. Eeckman, (Eds.), *Computation and neural systems 1992*, Boston: Kluwer Press.

Kanter, E. D., & Haberly, L. B. (1990). NMDA-dependent induction of long-term potentiation in afferent and association fiber systems of piriform cortex in vitro. *Brain Research*, **525**, 175–179.

Ketchum, K. L., & Haberly, L. B. (1993). Membrane currents evoked by afferent fiber stimulation in rat piriform cortex. I. Current source density analysis. *Journal of Neurophysiology*, **69**(1), 248–260.

Kohonen, T. (1984). *Self-organization and auto-associative memory*. Berlin: Springer-Verlag.

Komisaruk, B. R. (1970). Synchrony between limbic system theta activity and rhythmical behavior in rats. *Journal of Comparative and Physiological Psychology*, **70**, 482–492.

Koss, E., Weiffenbach, J. M., Haxby, J. V., & Friedland, R. P. (1988). Olfactory detection and identification performance are dissociated in early Alzheimer's disease. *Neurology*, **38**, 1228.

Kreiter, A. K., & Singer, W. (1992). Oscillatory neuronal responses in the visual-cortex of the awake macaque monkey. *European Journal of Neuroscience*, **4**, 369–375.

Lee, M., & Bower, J. M. (1988). A structural simulation of the inferior olivary nucleus. *Society for Neuroscience Abstracts*, **14**, 184.

Lisberger, S. G., & Sejnowski, T. J. (1992). Motor learning in a recurrent network model based on the vestibuloocular reflex. *Nature*, **360**, 159–161.

Luskin, M. B., & Price, J. L. (1983). The topographic organization of associational fibers of the olfactory system in the rat, including centrifugal fibers to the olfactory bulb. *Journal of Comparative Neurology*, **216**, 264–291.

Macrides, F., Davis, B. J., Youngs, W. M., Nadi, N. S., & Margolis, F. L. (1981). Cholinergic and catecholaminergic afferents to the olfactory bulb in the hamster: A neuroanatomical, biochemical and histochemical investigation. *Journal of Comparative Neurology*, **203**, 495–514.

Macrides, F., Eichenbaum, H. B., & Forbes, W. B. (1982). Temporal relationship between sniffing and the limbic theta rhythm during odor discrimination reversal learning. *Journal of Neuroscience*, **2**, 1705–1717.

Marr, D. (1982). Vision San Francisco: W. H. Freeman.

Mathews, D. F. (1972). Response patterns of single units in the olfactory bulb of the rat to odor. *Brain Res.* **47**, 389.

McNaughton, B. L., & Morris, R. G. (1987). Hippocampal synaptic enhancement and information storage within a distributed memory system. *Trends in NeuroSciences*, **10**, 408–415.

Nakatani, K., Tamura, T., & Yau, K. W. (1991). Light adaptation in retinal rods of the rabbit and two other nonprimate mammals. *Journal of General Physiology*, **97**, 413–435.

Neibur, E., Koch, C., & Rosin, C. (1993). An oscillation-based model of the neuronal basis of attention. *Vision Research*, **33**, 2789–2802.

Nelson, M., Furmanski, W., & Bower, J. M. (1989). Simulating neural networks on parallel computers. In C. Koch & I. Segev (Eds.), Methods in neuronal modeling: From synapses to networks. Cambridge, MA: MIT Press.

Palm, G. (1980). On auto-associative memory. *Biological Cybernetics*, **36**, 19–31.

Price, J. L. (1985). Beyond the primary olfactory cortex: Olfactory related areas in the neocortex, thalamus and hypothalamus. *Chemical Senses*, **10**, 239–258.

Protopapas, A., & Bower, J. M. (1994). Sensitivity in the response of piriform cortex pyramidal cells to fluctuations in synaptic timing. In F. Eeckman (Ed.), *Computation and Neural Systems, 1993*. Boston: Kluwer Press.

Racine, R. J., & Milgram, M. W. (1983). Short-term potentiation phenomena in the rat limbic forebrain. *Brain Research*, **260**, 201.

Racine, R. J., Milgram, M. W., & Hafner, S. (1983). Long-term potentiation phenomena in the rat limbic forebrain. *Brain Research*, **260**, 217.

Roman, F., Staubli, U., & Lynch, G. (1987). Evidence for potentiation in a cortical network during learning. *Brain Research*, **418**, 221–226.

Ryckebusch, S., Bower, J. M., & Mead, C. (1989). Modeling small biological oscillating networks in VLSI. In D. Touretzky (Ed.), *Advances in neural network information processing systems*. San Mateo,CA: Morgan Kaufmann.

Sato, M., Mori, K., Tazawa, Y., & Takagi, S. F. (1982). Two types of postsynaptic inhibition in pyriform cortex of the rabbit: fast and slow inhibitory postsynaptic potentials. *Journal of Neurophysiology*, **48**, 1142–1156.

Schillen, T. B., & Konig, P. (1991). Stimulus-dependent assembly formation of oscillatory responses: I Synchronization. *Neural Computation*, **3**, 155–166.

Schuster, H. G. (1991). Nonlinear dynamics and neural oscillations. In H. G. Schuster (Ed.), *Nonlinear dynamics and neuronal networks: Proceedings of the 63rd W. E. Heraues seminar, Friedrichsdorf, 1990*. New York: VCH.

Schuster, H. G., & Wagner, P. (1990a). A model for neuronal oscillation in the visual cortex: I. Mean-field theory and the derivation of the phase equations. *Biological Cybernetics*, **64**, 77–82.

Schuster, H. G., & Wagner, P. (1990b). A model for neuronal oscillation in the visual cortex: II. Phase description of the feature dependent synchronization. *Biological Cybernetics*, **64**, 83–85.

Schwob, J. E., & Price, J. L. (1978). The cortical projection of the olfactory bulb: Development in fetal and neonatal rats correlated with quantitative variation in adult rats. *Brain Research*, **151**, 169–174.

Selverston, A. I. (1985). *Model neural networks and behavior*. New York: Plenum Press.

Serby, M., Flicker, C., Rypma, B., Weber, S., Rotrosen, J. P., & Ferris, S. H. (1989). Scopolamine and olfactory function. *Biological Psychiatry*, **28**, 79–82.

Shepard, G. (1979). The synaptic organization of the brain. New York: Oxford University Press.

Singer, W. (1987). Activity-dependent self-organization of synaptic connections as a substrate of learning. In J. P. Changeux & M. Konishi (Eds.), *The neural and molecular bases of learning*. New York: Wiley.

Slotnick, B. M., & Berman, E. J. (1980). Transection of the lateral olfactory tract does not produce anosmia. *Brain Research Bulletin*, **5**, 141.

Sporns, O., Gally, J. A., Reeke, G. A., & Edelman, G. M. (1989). Reentrant signaling among simulated neuronal groups leads to coherency in their oscillatory activity. *Proceedings of National Academy of Sciences of the United States of America*, **86**, 7265–7269.

Sporns, O., Tononi, G., & Edelman, G. M. (1991). Modeling perceptual grouping and figure–ground segregation by means of active reentrant connections. *Proceedings of the National Academy of Sciences*, (USA), **88**, 129–133.

Squire, L. R. (1986a). Mechanisms of memory. *Science*, **232**, 1612–1619.

Squire, L. R. (1986b). Memory: Brain systems and behavior. *Trends in Neurosciences*, **11**, 170–175.

Squire, L. R. (1987). *Memory and brain*. New York: Oxford University Press.

Staubli, U., Schottler, F., & Nejat-Bina, D. (1987). Role of dorsomedial thalamic nucleus and piriform cortex in processing olfactory information. *Behavior and Brain Research*, **25**, 117.

Stripling, J. S., Patneau, D. K., & Gramlich, C. A. (1988). Selective long-term potentiation in the pyriform cortex. *Brain Research*, **441**, 281.

Tanabe, T., Lino, M., & Takagi, S. F. (1975). Discrimination of odors in olfactory bulb, pyriform-amygdaloid areas and orbitofrontal cortex of the monkey. *Journal of Neurophysiology*, **38**, 1284–1296.

Tseng, G-F., & Haberly, L. B. (1987). Characterization of synaptically mediated fast and slow inhibitory processes in piriform cortex in an in vitro slice preparation. *Journal of Neurophysiology*, **59**, 1352–1376.

Vanier, M. C., & Bower, J. M. (1993). Differential effects of norepinephrine on synaptic transmission in layers IA and IB of the rat olfactory cortex. In J. M. Bower & F. Eeckman (Eds.), *Computation and neural systems 1992*. Boston: Kluwer Press.

von der Malsburg, C., & Schneider, W. (1986). A neural cocktail party processor. *Biological Cybernetics*, **54**, 29–40.

Wenk, H., Meyer, U., & Bigl, V. (1977). Centrifugal cholinergic connections in the olfactory system of rats. *Neuroscience*, **2**, 797–800.

Wilson, M., Bhalla, U., Uhley, J., & Bower, J. M. (1989). GENESIS: A system for simulating neural networks. In D. Touretzky (Ed.), *Advances in neural network information processing Systems*. San Mateo, CA: Morgan Kaufmann.

Wilson, M., & Bower, J. M. (1988). A computer simulation of olfactory cortex with functional implications for storage and retrieval of olfactory information. In D. Anderson (Ed.), *Neural information processing systems*. New York: American Institute of Physics.

Wilson, M., & Bower, J. M. (1989). The simulation of large-scale neuronal networks. In C. Koch & I. Segev (Eds.), *Methods in neuronal modeling: From synapses to networks*. Cambridge, MA: MIT Press.

Wilson, M. A., & Bower, J. M. (1990). Computer simulation of oscillatory behavior in cerebral cortical networks. In D. Touretzky (Ed.), *Advances in neural information processing systems*. (Vol. 2). San Mateo, CA: Morgan Kaufmann.

Wilson, M. A., & Bower, J. M. (1991). A computer simulation of oscillatory behavior in primary visual cerebral cortex. *Neural Computation*, **3**, –509.

Wilson, M., & Bower, J. M. (1992). Simulating cerebral cortical networks: Oscillations and temporal interactions in a computer simulation of piriform (olfactory) cortex. *Journal of Neurophysiology*, **67**, 981–995.

Wilson, M., Bower, J. M., & Haberly, L. B. (1986). Computer simulation of piriform cortex. *Society for Neuroscience Abstracts*, **12**, 1358.

Woody, C. D., & Gruen, E. (1993). Cholinergic and glutamatergic effects on neocortical neurons may support rate as well as development of conditioning. In A. C. Cuello (Ed.), *Progress in brain research*, (Vol. 98). New York: Elsevier Science Publishers.

2

Brains, and Their Applications

Richard Granger, Jim Whitson, Eric Whelpley, Robert Coultrip, Richard Myers, Karl Kilborn, and Gary Lynch

INTRODUCTION

Computational analysis of the physiology and anatomy of specific brain circuits has led to (1) new and testable hypotheses of the function of these structures, and (2) derivation of novel computational circuit designs based on these hypotheses. Distinct brain circuits vary widely in their architecture, physiology, and plasticity characteristics, and simulation studies have indicated quite different algorithms embodied in different networks. Somewhat less expected has been the identification of algorithms that emerge only from interactions among multiple circuits, without arising from the individual constituents. We briefly review (1) the anatomy, physiology, and synaptic plasticity rules for four distinct circuits in the olfactory–hippocampal pathway, (2) distinct emergent computations arising from each, and (3) interactions among these structures via feedforward and feedback pathways connecting them, yielding novel algorithms absent from the constituent structures themselves.

Biological systems face information processing problem domains significantly more complex than those that have been successfully addressed by artificial systems, and exhibit both a range of sensitivity and generalization capabilities that exceed those of engineered designs. These biological systems also successfully address the significant design problem of scaling. Within the mammals alone, brain size scales by roughly four orders of magnitude (from milligrams to kilograms) and exhibits two particularly impressive scaling characteristics: (1) the mammalian brain retains many common elements that must operate over this entire range of sizes, from small to large organisms, and yet (2) new brain structures, which confer added intellectual power, are added with size, and these new structures must interact seamlessly with the existing circuitry. The absence of these capabilities in extant information processing systems has been a significant impediment to the construction of systems able to interact effectively with real-world environmental stimuli.

The results described here are based on a research program consisting of analysis and simulation of brain networks with anatomically specified architectures and physiological function, yielding novel algorithms applicable to complex stimuli, multiple inputs, and time-varying signals, and exhibiting large capacities and linear scaling (Ambros-Ingerson, Granger, & Lynch, 1990). We have identified a collection of common, vertically integrated designs, from synapses to cells to local circuits, incorporated in different forms in many different networks (Antón, Lynch, & Granger, 1991; Antón, Granger, & Lynch, 1992; Coultrip & Granger, 1993; Coultrip, Granger, & Lynch, 1992). These designs represent elemental tools that are reused repeatedly in the design of a range of network architectures, giving rise to differential action in different networks (Lynch & Granger, 1991, 1992). These studies have focused predominantly thus far on networks comprising the olfactory–hippocampal pathway, chosen both for the relative anatomical simplicity of its constituents and its status as evolutionary precursor for neocortex, which grows with evolution to account for most of the human brain. This review describes results for four of these constituent networks: the olfactory bulb, olfactory cortex, hippocampal field CA3, and field CA1. Interactions among these fields are described in light of their different functions and the different feedforward and feedback pathways connecting them.

SYNAPTIC LEARNING IN THE BRAIN

Synaptic long-term potentiation (LTP) is the only known brain plasticity mechanism possessing the three major characteristics of long-lasting memory: (1) it is rapidly induced (seconds; Gustafsson & Wigstrom, 1990), (2) it lasts undecremented with time (weeks; Staubli & Lynch, 1987), (3) it has a huge capacity (synapse specificity; Larson & Lynch, 1986). In addition, it goes without saying that memory must somehow be induced by brain activity patterns that occur during learning, and LTP is best induced by precisely such physiological patterns (Larson, Wong, & Lynch, 1986) via known cellular mechanisms (Mott & Lewis, 1991). Moreover, pharmacological agents that block LTP block learning (Morris et al., 1986; Staubli, Thibault, DiLorenzo, & Lynch, 1989; del Cerro et al., 1992) and it has been shown that novel pharmacological agents that enhance LTP (Xiao et al., 1991) enhance learning (Staubli et al., 1994; Granger et al., 1993).

Artificial neural network architectures make use of learning rules based on an array of mathematical starting points, but few if any use the actual physiological induction and expression rules of the known mechanism, LTP. In virtually all extant network learning rules, two cells are connected via a synapse or weight, and changes in the weight alter the communication between the cells. In the simplest of these "correlational" learning rules, correlated activity levels of the two cells leads to a strengthening of the synapse, as in Hebb's (1949) postulate. It is widely assumed and often stated that many networks use this rule, but surprisingly, outside of this chapter, there exist no networks in the reported literature using this rule: Oddly enough, true Hebbian learning in artificial neural networks is virtually unknown. The commonly employed variant is one in which not only does correlated activity lead to synaptic strengthening, but in addition, uncorrelated activity leads to synaptic weakening, a mechanism not proposed by Hebb and one for which the biological evidence is still controversial. Published findings have included nonspecific (heterosynaptic) depression (decrease of those synapses whose inputs are silent during postsynaptic activation; Bindman, Christofi, Murphy, & Nowicky, 1991), reversal of previously induced LTP (Barrionuevo, Schottler, & Lynch, 1980; Staubli et al., 1989; Larson, Xiao, & Lynch, 1993), and synapse-specific decrement (Stanton & Sejnowski, 1989; Stevens, 1990; Dudek & Bear, 1992; Mulkey & Malenka, 1992)—yet, synapse-specific decrement in adult telencephelon, lasting without decay over the time course of memory (e.g., weeks), has yet to be demonstrated.

The minutiae of physiological LTP induction and expression rules may introduce computational constraints arising solely from biological necessities, without contributing to the computational power of the resulting LTP-based learning. For instance, if it turns out that LTP has no long-term decremental counterpart (LTD), then synapses only increase; these increases proceed by relatively fixed-size steps (Larson et al., 1986), and only up to a fixed maximum strength or ceiling. It is possible that these restrictions exist for reasons pertaining only to the biological needs of the system and are extraneous, or even an impediment, to the computations performed in neural circuitry. Alternatively, these detailed physiological properties may yield novel learning rules that confer useful computational abilities to networks that use them, and indeed this is what has been found in many reported instances (e.g., Granger, Ambros-Ingerson, & Lynch, 1989; Ambros-Ingerson et al., 1990; Granger, Whitson, Larson, & Lynch, 1994, in press). We review here a series of examples of specific computational utility, identified to arise directly from novel learning and performance rules based on the detailed induction and expression characteristics of LTP.

THE OLFACTORY–HIPPOCAMPAL PATHWAY

Figure 1 schematically illustrates the components of the hippocampus and its interfaces with cortex. Synaptic LTP has been demonstrated in every link in each of these components, and we take it as axiomatic that learning and memory in these structures are mediated by such synaptic change. The specific characteristics of LTP vary from circuit to circuit; different functionality arises from these different learning rules, as embedded in their different network architectures.

Olfactory Bulb

Anatomy

The olfactory bulb receives massively convergent input from approximately 50 million peripheral

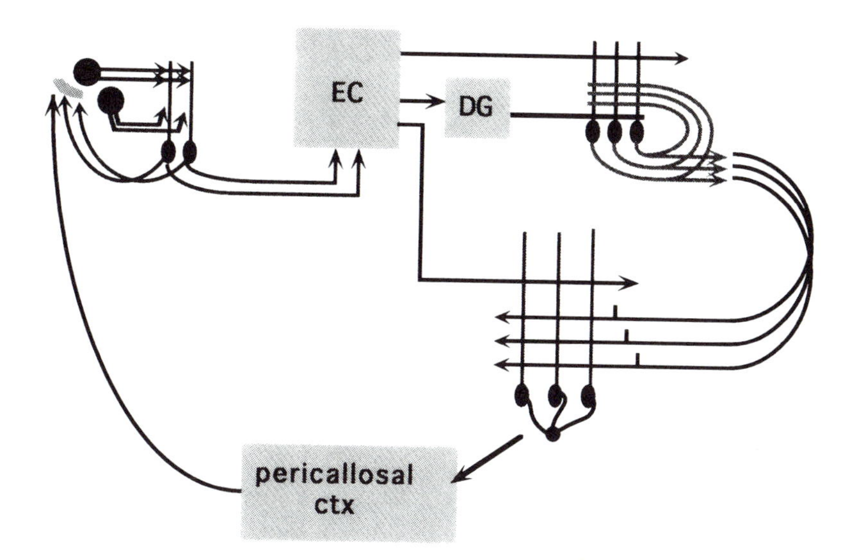

FIGURE 1 Hippocampus: its constituent circuitry and interfaces with cortex. Cortical circuitry (including olfactory paleocortex and polysensory association cortices) project to entorhinal cortex (EC), which in turn innervates all three constituents of hippocampus: dentate gyrus (DG), field CA3, and field CA1. The DG, in turn, generates the mossy fiber projection to CA3, and CA3 provides the Shaffer/collateral projection to CA1. CA1 projects out to deep layers of entorhinal cortex, to subiculum, and to pericallosal cortex, all of which in turn project back to association cortex.

olfactory receptors (in rabbit) onto the apical dendrites of roughly 175,000 primary excitatory mitral/tufted (MT) cells in the bulb. Olfactory receptor input (via the olfactory nerve) contacts bulb cells in about 1900 synaptic bundles, termed *glomeruli*, which contain the apical dendritic branches of roughly 92 MT cells per glomerulus. Each MT cell in higher animals typically sends a dendrite to only a single glomerulus, resulting in an architecture in which bulb cells are organized into groups of about 92 primary cells according to their glomerular membership. Mitral/tufted cells also send out long, laterally directed dendrites generating cross-glomerular interaction. GABAergic granule cells make unusual two-way dendro-dendritic contacts with MT cell dendrites, through which granule cells inhibit MT cells and MT cells excite granule cells. There is evidence that these nonaxonal contacts exhibit graded activity rather than the all-or-none activity of axons. There are roughly 3200 granule cells for every glomerular group of MT cells.

Modeling

Synaptic interactions among the multiple inputs to MT and granule cells were calculated via nonlinear summation of individual postsynaptic potentials (PSPs) resulting from the synaptic interactions among olfactory nerve axons, MT excitatory neurons, periglomerular cells, and granule cells. Physiological simulation capturing these interactive effects is achieved via iterative PSP functions derived from a lower-level, lumped-circuit representation of a cell (Antón et al., 1991), which uses the superposition theorem on the resistor-capacitor (RC) circuit representation of dendrites to obtain the voltage contribution of each individual synaptic activation. The result is a RC circuit for PSP transients combining all synaptic conductances and membrane conductance over time, and incorporating driving force and active and passive components of the voltage. These PSP functions can be summed linearly because of the use of the linear superposition theorem in the equation's derivation.

Results

The frequency of receptor cell firing is related to the intensity of chemical stimulation of the receptors, and therefore may carry information about the intensity of an olfactory stimulus. There is a gradient of MT cell excitability across distinct subpopulations of these cells, with the most superficial cells (superficial tufteds) most excitable, followed by middle tufteds, deep tufteds, and mitrals (Schneider & Scott, 1983; Mori, 1987). A given level of input activation (olfactory nerve axon frequency) will thus activate a given subset of MT cells within a glomerular patch; increasing input activation will activate those same cells plus others, as the higher excitability thresholds of deeper MT cells are reached and exceeded. Antón et al. (1991) found that the number of responding cells in an olfactory bulb simulation increased monotonically and nonlinearly with olfactory nerve stimulation frequency.

Piriform Cortex

Anatomy

Figure 2 schematically illustrates the major anatomical features of the primary olfactory (pyriform) cortex and its primary subcortical input structure, the olfactory bulb. A key architectural feature is the presence of a loop consisting of (1) excitatory feedforward fibers from bulb mitral cells to cortical layer II/III cells, and (2) an inhibitory feedback projection from cortex to bulb. In detail, the primary bulb excitatory cells, MT cells, project monosynaptically to layer I of olfactory (piriform) cortex where they synapse sparsely and nontopographically onto apical dendrites of layer II and III cortical cells (Price, 1973). Layer II cells emit collateral axons which flow predominantly caudally, contacting more caudal layer II and III cells and forming the primary input to lateral entorhinal cortex (Zimmer, 1971; Wyss, 1981). The axons of layer III cells in caudal piriform cortex flow predominantly in the opposite direction, mostly passing via layer III back to the bulb to synapse not on the excitatory MT cells, but rather on bulb granule cells, which in turn, inhibit MT

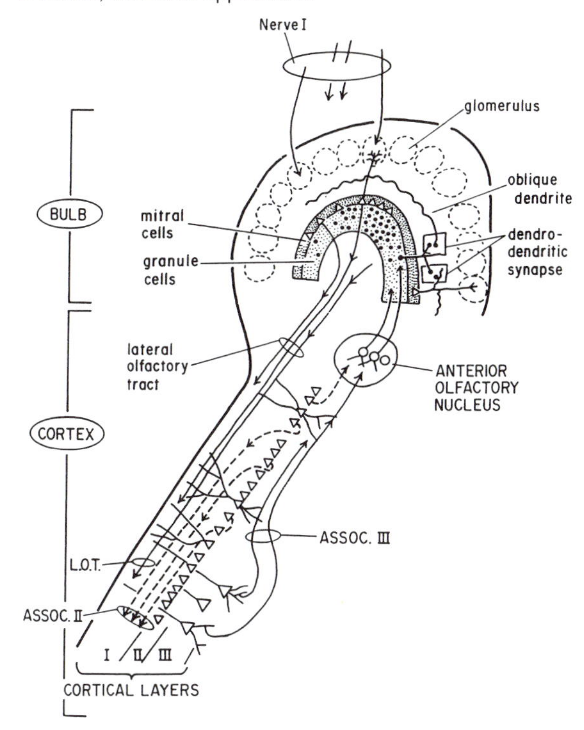

FIGURE 2 Olfactory system anatomy: Receptor cells in the nasal epithelium form a topography-preserving projection via the first cranial nerve to the excitatory mitral cells in the olfactory bulb; these cells in turn project sparsely and nontopographically to layer I of olfactory cortex, contacting apical dendrites of layer II and III cells. Cortical cells project back, both directly and via the anterior olfactory nucleus, to the inhibitory granule cells of the bulb, which dendrodendritically inhibit mitral cells.

cells (Mori, 1987). The other major architectural detail for purposes of modeling is that of excitatory–inhibitory interaction in local circuits in cortex. Excitatory cortical cells outnumber inhibitory cells by roughly two orders of magnitude; inhibitory cell axons contact excitatory cells relatively densely within a small radius and have no long axons, unlike excitatory neurons. Thus the inhibitory cells roughly "tile" the cortical surface, each inhibitory cell having a region of influence covering dozens to hundreds of excitatory cells. The resulting cortical "patches" can act as relatively coherent local circuits. Physiologically, inhibition is long, lasting hundreds of milliseconds, whereas excitation is brief (approximately 10 msec). In simulations of such local circuit patches, the excitatory neurons jointly innervate and receive feedback from a common inhibitory interneuron. The excitatory cell receiving the most input activation during a given activity cycle is the first to reach its spike threshold. Spiking excites the inhibitory cell, which in turn prevents other excitatory cells from responding. The result is the natural generation of a simple competitive or winner-take-all mechanism, allowing only the most strongly activated cell in a patch to respond with spiking activity. Analysis has shown that this simple mechanism closely approximates an optimal winner-take-all mechanism (Coultrip et al., 1992).

Synchronous Cyclic Activity

Normal, unrestrained animals actively exploring an environment exhibit synchronous EEG activity at roughly the 4–8 Hz theta rhythm in hippocampus (Hill, 1978). In an olfactory learning task, the entire olfactory–hippocampal pathway becomes synchronized to the theta rhythm (Komisaruk, 1970; Macrides, 1975; Eichenbaum, Kuperstein, Fagan, & Nagode, 1987), and the latency of peaks of this theta wave in each of the constituent circuits is roughly what would be expected from synaptic transmission delays between the circuits (Macrides, 1975).

It has been demonstrated that this rhythm is the optimal stimulation pattern for the induction of LTP: When afferents to a target are activated in brief (4-pulse) high-frequency (100 Hz) bursts separated by 200 msec (i.e., the 5 Hz theta rhythm), LTP is robustly induced, *in vitro* or *in vivo*, without the need for GABA blockers, voltage clamps, or physiologically implausible long (seconds) high-frequency bursts of activity (Larson & Lynch, 1986; Larson et al., 1986). The biophysical mechanism underlying the optimality of the 200-msec interval is understood: the initial afferent burst activates both excitatory and inhibitory responses in the target cell. These excitatory and inhibitory postsynaptic potentials (EPSPs and IPSPs) have characteristic different time courses, EPSPs lasting 10–20 msec and IPSPs lasting 100–300 msec. At

200 msec after this first burst, the initial IPSP has roughly returned to baseline, but the inhibitory synapse has become refractory (Mott & Lewis, 1991). Thus, the second burst activates predominantly only excitatory currents. These, therefore, summate with each other during the brief burst, enabling them to surpass the threshold of the voltage-sensitive NMDA receptor channel (Collingridge, Kehl, & McLennan, 1983), and allowing it to open and pass Ca^{2+}, which is the chemical messenger triggering the cascade of events inducing LTP (Lynch, Larson, Kelso, Barrionuevo, & Schottler, 1983).

Emergent Computation

LTP in olfactory cortex (Roman, Staubli, & Lynch, 1987; Kanter & Haberly, 1990; Jung, Larson, & Lynch, 1990) is induced via stimulation patterns (brief bursts at the theta rhythm) matching those found during learning in behaving animals (Komisaruk, 1970; Otto, Eichenbaum, Weiner, & Wible, 1991). This argues for investigations of cortical memory function based on this synchronized operating mode together with the induction and expression rules for LTP. Learning in this model results in the initial cortical responses becoming nearly identical to sufficiently similar inputs from the bulb, that is, the common statistical operation of clustering (Granger et al., 1989). Clustering readily arises from correlational rules in networks, and has been found by many researchers (von der Malsburg, 1973; Grossberg, 1976; Rumelhart & Zipser, 1985). Essentially, the resulting cortical response corresponds to general families or clusters of inputs; a given response signals membership of the input in a given cluster (e.g., "fruit" odors vs. "meat" odors vs. "floral" odors), thereby coarsely partitioning the input space.

Olfactory Bulb–Cortex Interaction

Hypotheses about and computational analysis of the olfactory system incorporating these features of LTP induction and expression have led to findings that suggest not just the encoding of sensory cues, but also organization of the resulting memories into structures not typically seen in neural network models. Implementation of a repetitive sampling feature meant to represent the cyclic sniffing behavior of mammals (Komisaruk, 1970) produced a system that exhibited successively finer-grained encodings of learned cues over sampling cycles. Each sampling cycle includes feedforward activity from bulb to cortex, followed by feedback from cortex to bulb. Because this feedback activates long-lasting inhibition in the bulb, the next sample of the input arrives against an inhibitory background in bulb, effectively masking part of the input and thereby resulting in different activity patterns in bulb and in cortex. Thus, resampling a fixed cue generates different cortical responses with each new sampling cycle.

The feedback from cortex to the bulb granule cell layer selectively inhibits those portions of the bulb response giving rise to the cortical firing pattern; this inhibition in the bulb lasts for hundreds of milliseconds (Nicoll, 1969). Resampling then causes new bulb and cortex activity against the background of this long-lasting inhibition. The resulting cortical response corresponds to odor components not shared across category members; that is, the first sample responses to a set of flowers will all be identical, signifying that these odors are all members of a single category. Subsequent samples correspond to the differences among different flowers, thereby effectively distinguishing among subcategories of floral odors. Thus learning via LTP in the model generates a multilevel hierarchical memory that uncovers statistical relationships inherent in collections of learned cues, and, during retrieval, sequentially traverses this hierarchical recognition memory. Moreover, the resulting algorithm exhibits time and space complexity rivaling that of the plethora of standard hierarchical clustering algorithms in the literature. The space complexity of the algorithm is $O(nN)$, where N is the dimensionality of the input and n the number of inputs to be learned; that is, the space costs are linear in the number and dimension of inputs. The three time costs of the algorithm for each input vector are (1) summation of target cell inputs, (2) computation of "winners" of the lateral inhibitory competition in local circuit patches, and (3) weight modifications or learning. Because of the inherent parallelism of the algorithm, (1) is $O(\log N)$, (2) is $O(\log n)$, and (3) is constant; the network rapidly converges during training, and to process a collection of n elements is $O(n(\log n)(\log nN))$ in parallel (Ambros-Ingerson et al., 1990).

Hippocampal Field CA3

Anatomical and Physiological Properties

Field CA3 of the hippocampus receives its primary inputs from dentate gyrus granule cell mossy fibers and perforant path axons from entorhinal cortex. In addition, CA3 pyramidal cells receive a relatively dense local excitatory feedback projection via recurrent collaterals of their own axons, thus creating an unusually dense recurrent excitatory loop of a sort that might be thought to support dynamical activity.

It has been shown that the potentiation produced by high frequency stimulation of the mossy fibers is not LTP (Staubli, Larson, & Lynch, 1990; Zalutsky & Nicoll, 1990). Unlike LTP, mossy fiber potentiation (MFP) changes the (presynaptic) frequency facilitation characteristics of the synapses, does not require NMDA receptor stimulation for induction, and is very likely not a postsynaptic effect. If MFP is purely presynaptic, requiring no associative postsynaptic component, it likely occurs at every synapse along an axon that is driven to fire at high frequency. In contrast, perforant path synapses on CA3 cells have been shown to exhibit normal LTP (Berger, 1984).

Results of Modeling

Simulations of field CA3 have incorporated physiologically realistic time summation via linear superposition of activation of a lumped-circuit cell model that can accurately combine multiple excitatory and inhibitory synaptic currents overlapping in time (Antón et al., 1992). Currents modeled include fast excitatory (AMPA), slow excitatory (NMDA), fast inhibitory (GABAa), slow inhibitory (GABAb) frequency facilitation and long-lasting, calcium-activated potassium currents (AHP). Typical simulations included 300 pyramidal cells and 120 inhibitory interneurons equally divided between GABAa and GABAb. Overall connectivity of the model was sparse (less than 0.1 probability of contact between afferents and targets), but local neighborhoods were densely connected via recurrent collaterals (probability of contact 0.6 within local neighborhood). It was found that very brief (30 msec) simulated input stimulation to the perforant path initiated dynamical activity which continued for long periods of time, up to many hundreds of milliseconds. After LTP of the perforant path—pyramidal cell synapses—responses to the same brief input stimulation lasted significantly longer, often for up to many seconds (thousands of milliseconds). Moreover, after LTP, the continuing response contains repetitive subsequences which reliably reoccur at regular intervals for the duration of the spiking activity after a brief stimulus. In addition, the subsequences are specific to particular initial stimuli, such that a signal-specific sequence recurs in response to a given brief stimulus. It is suggested that this maintained pattern provides a reverberating signal to a stimulus–response sequence that remains after the stimulus is gone and while the response continues. Arrival of the next extrinsic input would, in this scheme, collapse the existing pattern and replace it with transient cell firing or a new pattern (Lynch & Granger, 1992).

Hippocampal Field CA1

Activity Over Time

Field CA1 of the hippocampus receives afferent contacts from entorhinal cortex and from the adjacent hippocampal field (CA3). Both input systems make sparse, nontopographic contact. Excitatory–inhibitory local circuit interaction in field CA1 resembles that in cortex, such that lateral inhibition effectively causes excitatory cells to compete to respond to inputs. The temporal characteristics of inputs to CA1 from cortex and from CA3 raise the question of the behavior of such local circuits to inputs spread over time; as a result, the instantaneous competitive system has been extended to lateral inhibitory activity arriving over time, with the result that earliest-arriving inputs select a subset of best-responding cells, subsequent arrivers select best responders from that subset, and so on, "honing" the target cells down to the few best responders to the sequence of input (Granger et al., submitted).

Non-Hebbian LTP

As described, synchronous bursts of activity arriving at sequentially adjacent peaks of the theta rhythm carrier envelope are optimal for inducing LTP in synapses on pyramidal cells in field CA1 (Larson et al., 1986), for reasons that have become increasingly well understood (Mott & Lewis, 1991). The unlikelihood of complete synchrony of afferents during behaviorally relevant physiological activity raises the question of rules characterizing LTP induced by asynchronously arriving afferent stimuli within the envelope of single peaks of the theta rhythm. A "Hebbian" coactivity rule would predict that as asynchronous afferents arrive, increased depolarization of the target neuron over the staggered arrival times will cause later inputs to be strengthened more than earlier inputs. However, experiments using three small inputs stimulated in a staggered sequence over 70 msec showed that the earliest-arriving afferents potentiate their synapses the most, with subsequently arriving afferents causing successively less potentiation (Larson & Lynch, 1989; see Figure 3A).

Computational evaluation of this non-Hebbian, temporal LTP induction rule raises the question of how the resultant learning might be expressed during subsequent performance. Analysis with biophysical simulations using the SPICE program (Figure 3B) predicts that sequentially arriving inputs with different synaptic potencies will evoke the greatest depolarization in a target cell when they are activated in the order of their strengths (most potentiated input first). Given the order-dependence of the LTP rule, this means that a trained cell reacts most strongly to the same temporal sequence that it was trained on. Thus potentiation of an input sequence "codes" a cell to recognize that sequence subsequently, causing the cell to act as a form of "sequence detector." The predicted effect occurs robustly across a wide range of simulated conditions, one of which is illustrated in Figure 3C.

Emergent Computation of Field CA1

These physiological characteristics of LTP in field CA1, incorporated into a network of simplified cells derived from the SPICE and local–circuit models described earlier, enabled investigation of the computa-

FIGURE 3 Order-dependent potentiation: (A) excitatory synaptic afferents to a target activate both AMPA (A) and NMDA (N) receptors. In the figure, S1 and S2 carry brief bursts of high-frequency activity. Afferent S1 is activated 70 msec before S2. Initially, S1 activity only opens AMPA receptor channels; as this activity depolarizes the target, the voltage-sensitive NMDA receptor channels become activated. Subsequently, S2 activity arrives, opening its AMPA channels. The resulting continued depolarization maintains the open NMDA channel at S1, causing further potentiation at the S1 site (retrograde facilitation), whereas the depolarization at S1 shunts some current from S2, somewhat lessening its potentiation (anterograde suppression). The result is that S1 becomes more potentiated than S2 because of the combined retrograde facilitation and anterograde suppression (Larson & Lynch, 1989). NMDA receptors themselves remain relatively unpotentiated, whereas AMPA receptors express significant potentiation (Muller & Lynch, 1988; Muller et al., 1988; Kauer et al., 1988). (B) SPICE simulations of cell depolarization via sequential activation of differentially potentiated afferents. S1 is strongly potentiated and S2 weakly potentiated, as in (A); then, S1 and S2 are activated both in the order in which they were potentiated (S1–S2) and in the opposite order (S2–S1), and cell depolarization is measured. The simulated depolarization is 21% larger in response to the sequence S1–S2 than to the sequence S2–S1, predicting preferential expression of LTP in response to the learned sequence (Granger et al., submitted). (C) Test of the prediction from the SPICE simulation. S1 was strongly potentiated and S2 left unpotentiated, and extracellular field potentials were measured in response to S1–S2 versus S2–S1. The response to the sequence S1–S2 was 33% larger than the response to the reverse sequence S2–S1 (Adapted from Granger et al., 1994).

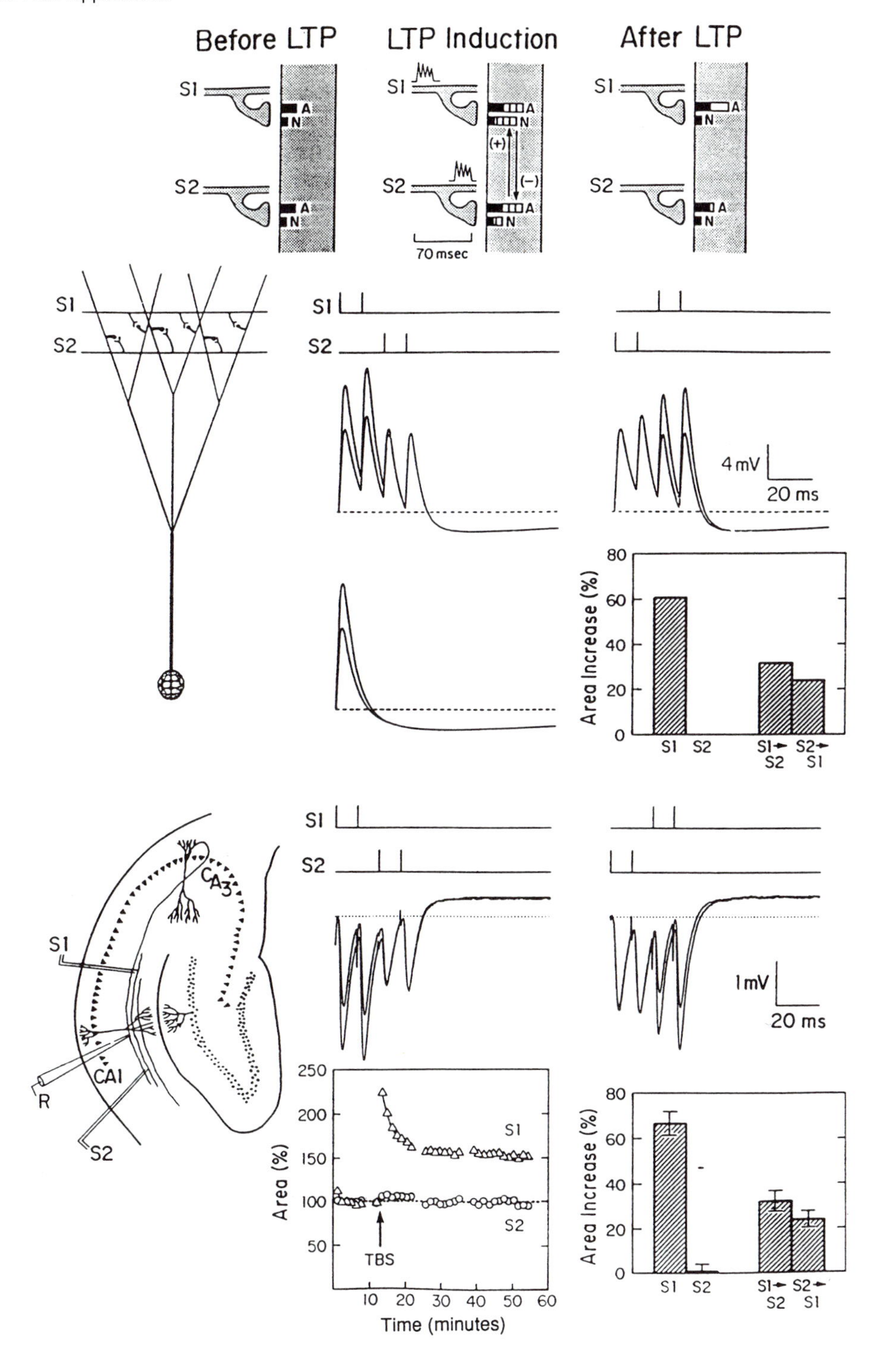

Before LTP
LTP Induction
After LTP
S1
S2
A
N
(+)
(−)
70 msec
4 mV
20 ms
1mV
Area Increase (%)
S1 S2
S1→ S2
S2→ S1
CA3
CA1
R
TBS
Area (%)
Time (minutes)

tional properties of this set of non-Hebbian LTP learning and performance rules. Intuitively, the resulting learning algorithm processes a temporal input sequence by performing order-dependent potentiation of the synapses of those cells that survive lateral inhibition in the competitive patches. The synapses of surviving cells are potentiated to a level commensurate with the order in which their inputs were activated: Contact with the first-arriving input is potentiated the most and contact with the last-arriving input is potentiated the least.

The performance algorithm preferentially activates only those cells whose synapses are appropriately potentiated at each input step. If any patch in the network contains no cells that survive this "honing" process through the set of temporal input steps, the input is rejected (not recognized); if each patch contains a surviving cell, the input is accepted (recognized). The resulting accept/reject or match/mismatch response to any recognized input X will be unique within a specified error tolerance (Granger et al., 1994 (in press)).

During performance, voltage summation uses potentiated weights, whereas during learning, summation is performed using only naive weights, implementing the distinction between changes to AMPA versus NMDA receptor conductance with LTP induction (Muller, Joly, & Lynch, 1988; Kauer, Malenka, & Nicoll, 1988). In other words, during episodes of learning (induction), the NMDA receptor-mediated voltage dominates cell responses and therefore predominantly determines which of the target cells will "win" the competition and be potentiated. Yet, these NMDA receptor-mediated voltages change little or not at all as a result of prior learning episodes; the changes are, rather, predominantly to the AMPA component. One computational consequence of this is notable: In contrast to the great majority of learning algorithms used in artificial neural networks, prior learning via LTP has little or no effect on subsequent LTP, preventing the formation of "attractor" cells (i.e., cells that tend to respond to any of a number of inputs similar to those previously trained).

The resulting extreme selectivity of cells, in combination with the order-dependency of the LTP-based learning and performance rules, confers unusually large capacity to the network. The capacity of the network can be cast in terms of its errors as a function of the number of sequences stored in a network of a given size. Two types of errors can be distinguished: errors of collision, in which a particular set of cells respond to more than one temporal string or "word" during training, and errors of commission, in which a target cell responds at testing (performance) time to a string on which it has not been trained. Theoretical analysis of the values of these error rates was derived by Granger et al. (submitted), who showed that these rates remained extremely low even with heavy "loading" of the network, that is, training of large numbers of sequences into a relatively small network. It was found, for example, that a network of 1000 cells could be trained to recognize 10,000 sequences of length 10, with an error rate of 0.0001. It was also shown that the capacity of the network scaled extremely well; that is, larger networks could learn comparably larger numbers of sequences, in contrast with many neural network approaches in which the network must be exponentially larger than the number of items it is to learn. In particular, given a network consisting of M competitive local circuits of C cells apiece, each with A synapses, the number n of sequences of length S that can be learned without exceeding error rate E is

$$n = \frac{C \log(1 - E^{1/SM})}{\log(1 - (1/A))} \tag{1}$$

Figure 4 is a contour graph of the relationship among the number of patches in the network, the number of cells in each patch, and the number of sequences that can be stored without exceeding a fixed recognition error of $p(\text{commission}) = 0.001$.

If cells are assumed to receive 10,000 synaptic contacts on their dendrites, then the point at the upper right corner of the contour graph corresponds to a 100,000-cell network (100 patches of 1000 cells each) storing approximately 5×10^7 (50 million) random sequences of length 10, with a recognition error rate of 0.001. It is worth noting that 100,000 cells with 10,000 synaptic contacts apiece constitutes a relatively small network in real terms (less than the size of one field CA1 in a

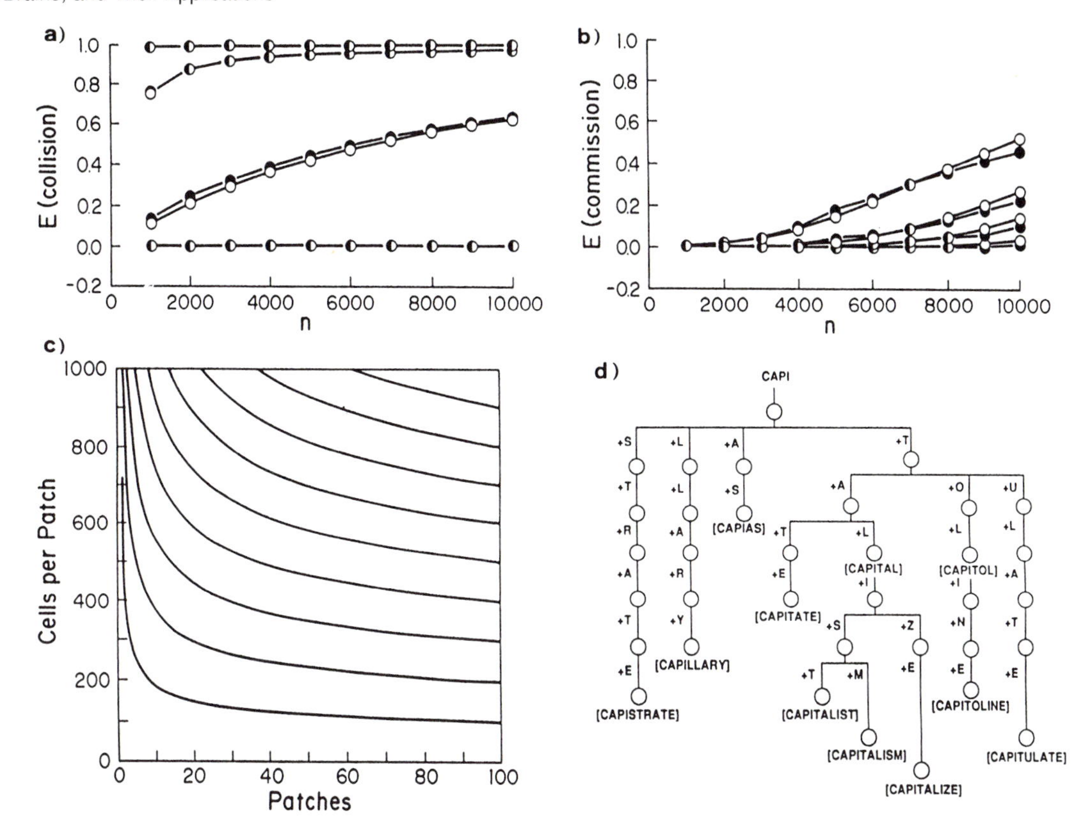

FIGURE 4 Capacity and scaling properties of the sequence-recognition network. Contour lines indicate the number n of sequences of length 10 that can be learned without exceeding a fixed error probability $E = 0.001$ for cells with 10,000 synapses each. The abcissa gives the number of patches in the network, and the ordinate denotes number of cells per patch. Each contour curve corresponds to a specific number n of learned sequences. The lowest curve denotes 5 million learned sequences. Following the curve shows that this many sequences can be learned by a network of 5 patches (x-axis) and about 250 cells per patch (y-axis), or by a network of 40 patches of 125 cells per patch, and so on. The contour curves increase by 5 million per curve; the highest curve shown corresponds to 45 million learned sequences (accomplished via 60 patches of 1000 cells each ranging to 100 patches of 900 cells each). The largest network for which theoretical values are plotted corresponds to the point at the upper-right corner of the contour graph: This network consists of 100,000 cells (100 patches of 1000 cells each) and has a capacity of 50 million sequences, with error rate $E \leq 0.001$.

rat), and yet 50 million learned sequences is equivalent to one novel sequence learned every 10 sec for 8 hours each day for 50 years, with a 0.001 recognition error rate. This probably exceeds the extent of the rat's capacity by orders of magnitude. Allowing for possible additional errors of transmission, contact, and noise, it still provides a scheme capable of accounting for the dramatic capacity of our own memory systems. Standing (1973) found that subjects exposed to 10,000 pictures for only seconds apiece nonetheless exhibited 90% recognition rates for those stimuli when tested weeks later. Neural network models do not exhibit the capability for this kind of rapid learning, long retention, and large capacity; rather, they typically require extensive training and cannot store many items without using very large networks. Surprisingly, what might have been thought to be excessively low-level biological detail of LTP induction and expression physiology gives rise to networks that directly address the problematic question of how large memories can be implemented in brain circuitry.

Acknowledgments

This work was supported in part by the Office of Naval Research under Grants N00014-89-J-1255 and N00014-92-J-1625.

References

Ambros-Ingerson, J., Granger, R., & Lynch, G. (1990). Simulation of paleocortex performs hierarchical clustering. *Science*, **247**, 1344–1348.

Antón, P. S., Lynch, G., & Granger, R. (1991). Computation of frequency-to-spatial transform by olfactory bulb glomeruli. *Biological Cybernetics*, **65**, 407–414.

Antón, P. S., Granger, R., & Lynch, G. (1992). Temporal information processing in synapses, cells, and circuits. In T. McKenna, J. Davis, & S. Zornetzer (Eds.), *Single neuron computation*. New York: Academic Press.

Arai, A., & Lynch, G. (1992). Factors regulating the magnitude of long-term potentiation induced by theta pattern stimulation. *Brain Research*, **598**, 173–184.

Barrionuevo, G., Schottler, F., & Lynch, G. (1980). The effects of repetitive low frequency stimulation on control and potentiated synaptic responses in the hippocampus. *Life Sciences*, **27**, 2385–2391.

Berger, T. (1984). Long-term potentiation of hippocampal synaptic transmission affects rate of behavioral learning. *Science*, **224**, 627–630.

Bindman, L., Christofi, G., Murphy, K., & Nowicky, A. (1991). In T. Stone (Ed.), *Aspects of synaptic transmission* (pp. 3–25). London: Taylor & Francis.

Collingridge, G. L., Kehl, S. L., & McLennan, H. (1983). Excitatory amino acids in synaptic transmission in the Schaffer collateral-commissural pathway of the rat hippocampus. *Journal of Physiology*, **334**, 33–46.

Coultrip, R., & Granger, R. (1993). LTP learning rules in sparse networks approximate Bayes classifiers via Parzen's method. *Neural Networks*.

Coultrip, R., Granger, R., & Lynch, G. (1992). A cortical model of winner-take-all competition via lateral inhibition. *Neural Networks*, **5**, 47–54.

del Cerro, S., Jung, M., & Lynch, G. (1992). Benzodiazepines block long-term potentiation in rat hippocampal and piriform cortex slices. *Neuroscience*, **49**, 1–6.

Dudek, S., & Bear, M. (1992). Homosynaptic long-term depression in area ca1 of hippocampus and the effects of NMDA receptor blockade. *Proceedings of the National Academy of Sciences of the United States of America*, **89**, 4363–4367.

Eichenbaum, H., Kuperstein, M., Fagan, M., & Nagode, J. (1987). Cue-sampling and goal approach correlates of hippocampal unit activity in rats performing an odor-discrimination task. *Journal of Neuroscience*, **7**, 716–732.

Granger, R., Ambros-Ingerson, J., and Lynch, G. (1989). Derivation of encoding characteristics of layer II cerebral cortex. *Journal of Cognitive Neuroscience*, **1**(1), 61–87.

Granger, R., Staubli, U., Davis, M., Perez, Y., Nilsson, L., Rogers, G., & Lynch, G. (1993). A drug that facilitates glutamatergic transmission reduces exploratory activity and improves performance in a learning dependent task. *Synapse*, **15**, 326–329.

Granger, R., Whitson, J., Larson, J., & Lynch, G. (1994) Non-Hebbian properties of LTP enable high-capacity encoding of temporal sequences. Proceedings of the National Academy of Sciences, 91: 10104–10108.

Grossberg, S. (1976). Adaptive pattern classification and universal recoding: I. Parallel development and coding of neural feature detectors. *Biological Cybernetics*, **23**, 121–134.

Gustafsson, B., & Wigstrom, H. (1990). Long-term potentiation in the ca1 region: Its induction and early temporal development. *Progress in Brain Research*, **83**, 223–232.

Hebb, D. O. (1949). The Organization of behavior. New York: Wiley.

Hill, A. (1978). First occurrence of hippocampal spatial firing in a new environment. *Experimental Neurology*, **62**, 282–297.

Jung, M., Larson, J., & Lynch, G. (1990). Long-term potentiation of monosynaptic EPSPs in rat piriform cortex *in vitro*. *Synapse*, **6**, 279–293.

Kanter, E. D., & Haberly, L. B. (1990). NMDA-dependent induction of long-term potentiation in afferent and association fiber systems of piriform cortex *in vitro*. *Brain Research*, **525**, 175–179.

Kauer, J. A., Malenka, R., & Nicoll, R. (1988). A persistent postsynaptic modification mediates long-term potentiation in the hippocampus. *Neuron*, **1**, 911–917.

Komisaruk, B. R. (1970). Synchrony between limbic system theta activity and rhythmical behavior in rats. *Journal of Comparative Physiology & Psychology*, **70**, 482–492.

Larson, J., & Lynch, G. (1986). Induction of synaptic potentiation in hippocampus by patterned stimulation involves two events. *Science*, **232**, 985–988.

Larson, J., & Lynch, G. (1989). Theta pattern stimulation and the induction of LTP: The sequence in which synapses are stimulated determines the degree to which they potentiate. *Brain Research*, **489**, 49–58.

Larson, J., Wong, D., & Lynch, G. (1986). Patterned stimulation at the theta frequency is optimal for induction of long-term potentiation. *Brain Research*, **368**, 347–350.

Larson, J., Xiao, P., & Lynch, G. (1993). Reversal of LTP by theta frequency stimulation. *Brain Research*, **600**, 97–102.

Lynch, G., & Granger, R. (1991). Serial steps in memory processing: Possible clues from studies of plasticity in the olfactory-hippocampal circuit. In H. Eichenbaum, & J. L. Davis (Eds.), *Olfaction as a model system for computational neuroscience*. Cambridge, MA: MIT Press.

Lynch, G., & Granger, R. (1992). Variations in synaptic plasticity and types of memory in cortico-hippocampal networks. *Journal of Cognitive Neuroscience*, **4**, 189–199.

Lynch, G., Larson, J., Kelso, S., Barrionuevo, G., & Schottler, F. (1983). Intracellular injections of egta block induction of hippocampal long-term potentiation. *Nature*, **305**, 719–721.

Macrides, F. (1975). Temporal relationships between hippocampal slow waves and exploratory sniffing in hamsters. *Behavioral Biology*, **14**, 295–308.

Mori, K. (1987). Membrane and synaptic properties of identified neurons in the olfactory bulb. *Progress in Neurobiology*, **29**, 275–320.

Morris, R., Anderson, E., Lynch, G., & Baudry, M. (1986). Selective impairment of learning and blockade of long-term potentiation by an N-Methyl-d-Aspartate receptor antagonist, AP5. *Nature*, **319**, 774–776.

Mott, D., & Lewis, D. (1991). Facilitation of the induction of long-term potentiation by GABAb receptors. *Science*, **252**, 1718–1720.

Mulkey, R., & Malenka, R. (1992). Mechanisms underlying induction of homosynaptic long-term depression in area ca1 of the hippocampus. *Neuron*, **9**, 967–975.

Muller, D., & Lynch, G. (1988). Long-term potentiation differentially affects two components of synaptic responses in hippocampus. *Proceedings of the National Academy of Sciences (USA)*, **85**, 9346–9350.

Muller, D., Joly, M., & Lynch, G. (1988). Contributions of quisqualate and NMDA receptors to the induction and expression of LTP. *Science*, **242**, 1694–1697.

Nicoll, R. (1969). Inhibitory mechanisms in the rabbit olfactory bulb: Dendrodendritic mechanisms. *Brain Research*, **14**, 157–172.

Otto, T., Eichenbaum, H., Weiner, S., and Wible, C. (1991). Learning-related patterns of ca1 spike trains parallel stimulation parameters optimal for inducing hippocampal long-term potentiation. *Hippocampus*, **1**, 181–192.

Price, J. L. (1973). An autoradiographic study of complementary laminar patterns of termination of afferent fiber to the olfactory cortex. *Journal of Comparative Neurology*, **150**, 87–108.

Roman, F., Staubli, U., and Lynch, G. (1987). Evidence for synaptic potentiation in a cortical network during learning. *Brain Research*, **418**, 221–226.

Rumelhart, D. E., and Zipser, D. (1985). Feature discovery by competitive learning. *Cognitive Science*, **9**, 75–112.

Schneider, S. P., & Scott, J. W. (1983). Orthodromic response properties of rat olfactory bulb mitral and tufted cells correlate with their projection patterns. *Journal of Comparative Physiology*, **50**, 319–333.

Standing, L. (1973). Learning 10,000 pictures. *Quarterly Journal of Experimental Psychology*, **25**, 207–222.

Stanton, P., & Sejnowski, T. J. (1989). Associative long-term depression in the hippocampus induced by Hebbian covariance. *Nature*, **339**, 215–218.

Staubli, U., Larson, J., & Lynch, G. (1990). Mossy fiber potentiation and long-term potentiation involve different expression mechanisms. *Synapse*, **5**, 333–335.

Staubli, U., & Lynch, G. (1987). Stable hippocampal long-term potentiation elicited by "theta" pattern stimulation. *Brain Research*, **435**, 227–234.

Staubli, U., Rogers, G., & Lynch, G. (1994). Facilitation of glutamate receptors enhances memory. *Proceedings of the National Academy of Sciences*, **91**, 777–781.

Staubli, U., Thibault, O., DiLorenzo, M., and Lynch, G. (1989). Antagonism of NMDA receptors impairs acquisition but not retention of olfactory memory. *Behavioral Neuroscience*, **103**(1), 54–60.

Stevens, C. (1990). A depression long awaited. *Nature*, **347**, 16.

von der Malsburg, C. (1973). Self-Organization of orientation sensitive cells in the striate cortex. *Kybernetik*, **14**, 85–100.

Wyss, J. (1981). Autoradiographic study of the efferent connections of entorhinal cortex in the rat. *Journal of Comparative Neurology*, **199**, 495–512.

Zalutsky, R. A., and Nicoll, R. A. (1990). Comparison of two forms of long-term potentiation in single hippocampal neurons. *Science*, **248**, 1619–1624.

Zimmer, J. (1971). Ipsilateral afferents to the commissural zone of the fascia dentata demonstrated in decommissurated rats by silver impregnation. *Journal of Comparative Neurology*, **23**, 393–416.

3

Information Processing Strategies and Pathways in the Primate Visual System

David C. Van Essen and Charles H. Anderson

INTRODUCTION

The mammalian visual pathway is an amazingly complex and intricate system, capable of processing vast amounts of sensory information for use in the precise control of both immediate and long-term behavior. In many respects, the system is very well engineered for efficient extraction and encoding of information. To give just one example, it is well known that the spacing between photoreceptors in the region of highest acuity is closely matched to the limits of resolution imposed by the physical optics of the eye (Snyder & Miller, 1977). In this chapter, we consider how the notion of "good engineering" can help to understand information processing strategies used throughout the visual system. Although some aspects of human vision are mentioned, our analysis focuses on the macaque monkey, which has a superb visual system very similar to that of humans.

Our analysis starts in the retina, where there is a highly selective pruning of the information that enters the eye. In addition, a number of sophisticated data compression schemes are introduced to obtain efficient utilization of the information-carrying capacities of the optic nerve. It is important to understand the representations that emerge from these operations, because they impose major constraints on what can occur at subsequent stages of analysis in the visual cortex.

Several major pathways, including the *parvocellular* and *magnocellular* streams, originate within the retina and diversify within the cortex, producing a rich set of data pathways interconnecting a multitude of higher visual areas. The organization of visual cortex is discussed here at the coarse-grained level of large neural ensembles (areas, layers, and compartments) and also at the fine-grained level of individual neurons and local network architecture. Where possible, systems and computational issues that relate to neuronal architecture are considered. If evolutionary pressures have driven the entire system toward some near-optimal configuration, the properties of individual components should be related to system performance as a whole.

This chapter also emphasizes the need for dynamic control of the flow and form of information in the brain, an aspect of neural systems that is often overlooked. In general, our perceptions of the world are not derived from isolated, instantaneous glances. Rather, our eyes are in constant motion, moving between selected points of interest. Internal representations of the world are built up over the course of many of these saccades and are continuously updated as the external world changes. An analogous process takes place within the brain in the form of selective visual attention, in which detailed scrutiny is applied only to objects within a restricted, but dynamically changing, portion of the visual field (Julesz, 1984; Posner & Presti, 1987; Van Essen, Olshausen, Anderson, & Gallant, 1991). These and other observations (cf. Baron, 1987; Kosslyn, 1988) suggest that brain function involves a set of dynamically controlled and hierarchically organized processes. Knowledge of the structure of the visual areas of the cortex provides clues as to where and how this selective control might take place.

THE RETINO-GENICULATE PATHWAY

Overview

The basic layout of the first three stages of the visual pathway in the macaque monkey is illustrated in Figure 1. Each visual center (retina, lateral geniculate nucleus, and primary visual cortex) is a convoluted and/or curved sheet of tissue; they are shown here as flattened, two-dimensional representations that preserve the relative area of each structure. Light from the three-dimensional world is focused by the cornea and lens to form a two-dimensional image on each retina. This image is converted to electrical signals by a dense array of photoreceptors along the back surface of the retina. The number of cones, used for diurnal vision, is about 3×10^6 in the macaque and 6×10^6 in humans; the number of rods, used for nocturnal vision, is about 10^8 (Østerberg, 1935; Perry & Cowey, 1985).

After processing within several intermediate layers of the retina, information converges onto a population of one million retinal ganglion cells, which comprise several anatomically and physiologically distinct classes of output cell. About 90% of all ganglion cells project to the lateral geniculate nucleus (LGN), somewhat fewer than half going to the same side of the brain and somewhat more than half going to the opposite side (Perry, Oehler, & Cowey, 1984; Fukuda, Sawai, Watanabe, Wakakuwa, & Morigiwa, 1989). Individual LGN cells relay the outputs of retinal ganglion cells in approximately 1:1 fashion to the striate cortex (area V1). At the cortical stage there is a vast expansion in the neural machinery available to process these inputs, amounting to hundreds of cortical neurons for each LGN input (Schein & De Monasterio, 1987).

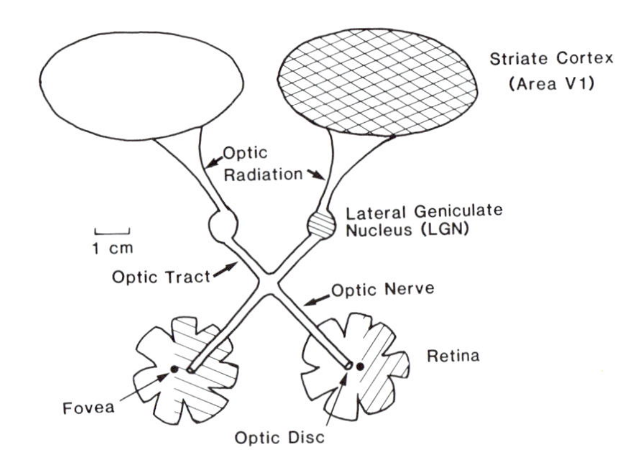

FIGURE 1 A schematic diagram of the retino-geniculo-striate pathway in the macaque monkey. Each visual center is shown as it appears after unfolding and flattening, with preservation of relative sizes. Each retina is about 600 mm^2 in surface area. Because they are hemispherical structures, cuts along the margin are made to flatten them. The LGN is a layered structure about 2 mm thick, and in surface area it is less than one-tenth the size of the retina (ca. 45 mm^2 for the outermost layer). The striate cortex (V1) is twice the size of the retina (1200 mm^2 in area in each hemisphere) and is about 2 mm thick. The right hemisphere receives inputs from both eyes, but only from those regions (hatched) subserving the left half of the visual field: approximately 60% of the left (contralateral) retina and 40% of the right (ipsilateral) retina. In the LGN, inputs from the two eyes are segregated into three pairs of layers, with the topmost subserving the contralateral eye. In the cortex, the left and right eye inputs initially remain segregated into ocular dominance stripes within layer 4 before intermixing occurs in the superficial and deep layers.

A Scale Invariant Sampling Strategy

In a standard CCD electronic camera, images are focused onto a uniform, rectangular array of nonoverlapping detectors (pixels). The resolution at which fine details can be analyzed is limited by the degree of blurring of the image and by the spacing of pixels in the array. Ideally, pixel spacing and image blur should be matched to one another. Sampling too coarsely can cause aliasing, that is, ambiguities in interpretation that can arise when an image is sampled more sparsely than prescribed by the Nyquist sampling theorem. Finer sampling tends to be wasteful, because it would exceed the optical resolution of the camera.

It is important to consider similar issues of image sampling in biological vision, even though the retina differs in many ways from an electronic camera. A particularly striking difference is that the primate retina is a variable resolution system, with extremely high acuity in the fovea and vastly poorer acuity in the periphery. In between these extremes, psychophysical data show an approximately linear change in resolution as a function of eccentricity in the visual field, that is, angular distance from the center of gaze (Westheimer, 1979). The linear nature of this relationship has interesting implications with regard to how much information the visual system transmits about objects in the visual field and how this depends on the distance from which the objects are viewed.

To illustrate this point, it is useful to consider the *sampling lattice* shown in Figure 2, which has a linear fall-off in resolution. After discussing the nature of this representation in the abstract, we will consider its relationship to the sampling strategy used by the retina, particularly at the level of retinal ganglion cells. The lattice contains an array of *sampling nodes*, indicated by dots in the figure. As noted by Koenderink and van Doorn (1978), this pattern has the appearance of a sunflower heart. The basic lattice can be described by two variables. One is the *sampling interval*, corresponding to the distance between adjacent nodes, which increases linearly with distance from the center. (There is some local scatter superimposed, which also occurs in the retina.) The second variable is the sampling area, corresponding to the domain over which information is integrated at a given node. We presume that the *sampling area* has a Gaussian sensitivity profile, with a width at half height indicated by the circles around some of the sampling nodes in the figure. By having these two variables scale together, all points on the retina are covered, and the degree of overlap between adjacent sampling areas is similar across the whole lattice.

Sampling with a linear decline in resolution (i.e., a linear increase in the sampling interval) leads to the property of scale invariance (van Doorn, Koenderink, & Bouman, 1972). In such a representation, the amount of information about any particular object in the visual field remains roughly constant as it is moved closer or further away from the observer, except for gain or loss of information near the center of the image. Consider as a specific example the outline of a hand that is projected onto the sampling lattice in Figure 2. (Imagine, for the moment, that this represents a sampling lattice for your retina and that you are fixating the palm of your outstretched hand.) When the hand is distant, it forms the smaller of two images on the lattice. When the hand is moved closer, cutting the distance in half, the image is enlarged by a factor of two; its margins are a factor of two farther into the periphery and it is sampled at half the resolution. For instance, in both images there are three or four sampling nodes spanning the tip of each digit. Hence, the amount of information transmitted about the shape of the hand is roughly the same at both scales. Obviously, there is more information about the texture of the center of the palm when the hand is closer. However, changing the distance alters the information content mainly in the region subserved by the center of gaze (or in the extreme periphery if the object becomes larger than the field of view). This analysis applies only when fixating a point near the center of the object, but that is indeed the case much of the time during normal visual behavior.

For a completely scale-invariant system, the sampling interval would decrease literally to zero at the

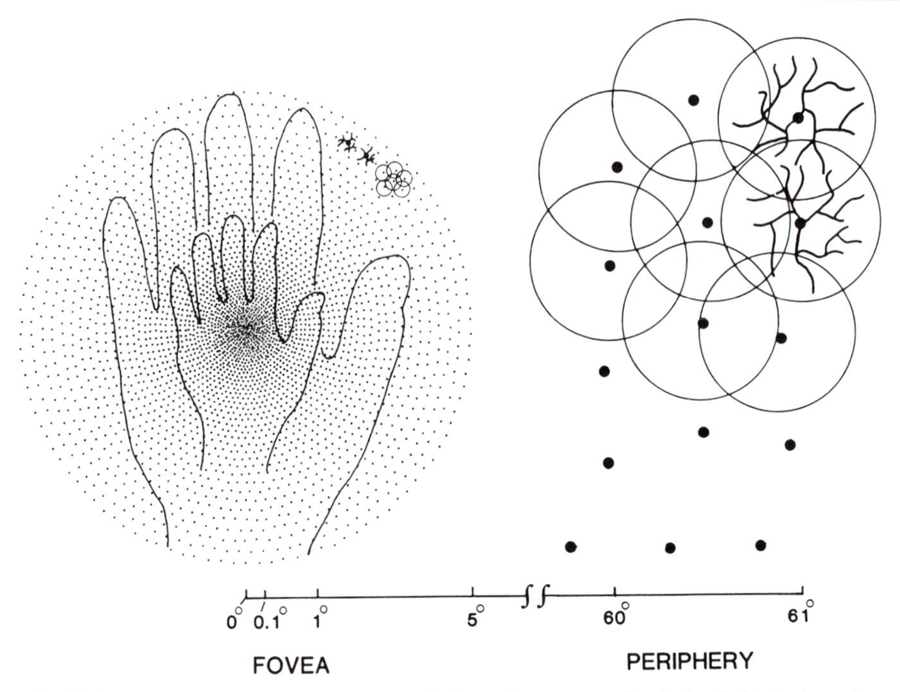

FIGURE 2 A scale-invariant sampling strategy similar to that represented at the level of retinal ganglion cells. On the left is a lattice in which the sampling interval between adjacent nodes increases linearly with distance from the center. The equation used to generate the lattice is $D = 0.06\ (E + 0.9\ E_{max})$, where E_{max} is the eccentricity at the perimeter of the lattice. To provide a visually distinct gradient, the slope for the equation was set much higher than that found in the primate retina (cf. Figures 3 and 4). Nonetheless, with the aid of the nonlinear scale shown at the bottom, the figure also conveys a quantitative indication of the retinal sampling density at different eccentricities. In the far periphery (60° eccentricity), the sampling interval (0.5°) is almost 50-fold lower than at the center of the fovea, as indicated on the right. The number of sampling nodes along a circumferential ring increases gradually with eccentricity, reaching an asymptotic value of $2\pi/\alpha$; this number is 105 for the example illustrated, but is 600–800 for the primate retina. Each sampling node conveys information about illumination in a small region, the sampling area. In a well designed system, each sampling area should have a smooth sensitivity profile, the width of which scales with the sampling interval, as indicated by the circles around a few of the nodes. A few of the nodes also have a neuron with a dendritic field drawn in to illustrate how neuronal geometry can be related to the sampling area. In the retina, more than one neuron is typically available to transmit information from each sampling node.

center of gaze. This is clearly impossible for biological vision, because there is a finite limit to resolution and a corresponding minimum size for photoreceptors. The compromise arrangement worked out by nature involves a sampling interval (D) that starts at a minimum value (δ) at the center of gaze and increases linearly with eccentricity with a slope (α):

$$D = \delta + \alpha E = \alpha(E_0 + E) \approx 0.01(1.3 + E) \text{ deg} \quad (1)$$

The estimated values for the primate visual system ($\delta \approx 0.01°$, $\alpha \approx 0.01°$) are based on human psychophysical data and on anatomical data from the macaque retina. The coarseness of sampling in the far periphery (60°) is illustrated on the right of Figure 2, where the spacing is about 50-fold greater than in the center of the fovea; this corresponds to a 2000-fold difference in the areal density of sampling nodes.

Using this strategy, the entire visual field can be represented with about 300,000 samples, which is comparable to the total number in a CCD camera having a 500 × 500 uniform sample array. To achieve a resolution of 0.01° uniformly over the whole field of view would require hundreds of times as many samples. This is by far the most important data reduction step taken by the visual system. More than 99% of the information available to the photoreceptors is discarded very early on. With inhomogeneous sampling, the entire field is available at varying degrees of resolution; detailed scrutiny of particular regions can be obtained by appropriate eye movements.

Retinal Sampling Strategies

The neural basis for scale-invariant sampling lies mainly within the retina. It reflects how image data are acquired by the cones and how they are further processed and encoded by the ganglion cells. It is particularly important to assess the issue at the ganglion cell level, because the optic nerve is a critical bottleneck through which all visual information must flow. Acuity and other aspects of vision are fundamentally constrained by the information represented at the level of retinal ganglion cells (cf. Rovamo & Virsu, 1979). However, it is also important to understand the nature of the initial neural representation provided by the cones.

Each node in a sampling lattice conveys information about what is happening in a small patch of the retina. In principle, this could be mediated by a single channel, as is suggested by the inclusion of only one neuron per sampling node in Figure 2. However, it is entirely reasonable to have more than one channel per node, especially because there are several types of information to be conveyed (e.g., distinctions of color and light vs dark features). In addition, real neurons operate using a limited range of firing rates, and more data can be conveyed by increasing the number of output neurons at each node. This multiple-neuron strategy does indeed appear to be used, and it leads to the notion of a sampling multiplicity (coverage factor) to signify the number of neural channels per sampling node. Another important factor is that there are two major ganglion classes that operate at different spatial scales. Hence, it is necessary to think in terms of two distinct sampling lattices within the retina that work together for the efficient encoding of spatial and temporal information.

Photoreceptor Distribution

Cone spacing and size increase relatively rapidly within the fovea (central 2°), but much more slowly at higher eccentricities (Figure 3, solid line). There is a progressive mismatch between cone spacing and the sampling interval for the main class of ganglion cells (dashed line). This signifies that the number of cones within each sampling area increases with eccentricity, approaching 20–30 cones per node in the periphery (Wässle & Boycott, 1991).

The center of the fovea is occupied exclusively by long-wavelength (L) and medium-wavelength (M) cones, whose broad absorption spectra are only slightly displaced from one another. Short-wavelength (S) cones, whose absorption spectrum is more substantially displaced toward short wavelengths, are more sparsely distributed. At their peak, they represent only about 7% of the cone population (De Monasterio, McCrane, Newlander, & Schein, 1985).

Rods, which have the high sensitivity needed for nocturnal vision, are absent in the center of the fovea, but rapidly increase in density outside the fovea, thereby providing a high collection efficiency for all but the central few degrees. In the periphery, rods are densely packed and cones are relatively sparsely spaced. Interestingly, the spacing between cones changes more rapidly than the decline in optical quality of the image in the periphery. Hence, peripheral images are spatially undersampled, and some aliasing can occur in principle. However, this aliasing is unlikely to introduce a significant contamination of the information represented at the ganglion cell level,

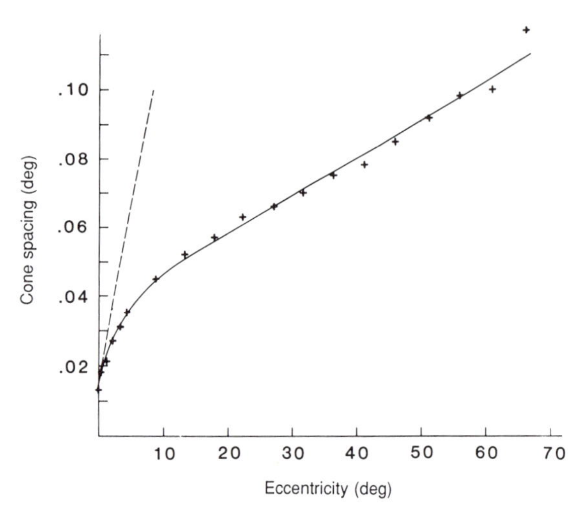

FIGURE 3 Cone spacing in the macaque retina. Values are calculated assuming hexagonal packing of cones and using published data on cone densities (Perry & Cowey, 1985). Numerical data was kindly provided by H. Perry. Only values for dorsal, ventral, and temporal retina are included because of the systematically higher cone density in the nasal quadrant. At the center of the fovea, cone spacing is about 0.013° (ca. 3 mm). The initial part of the curve is fit by the equation $D = 0.01(E + 1.3)°$, which provides one estimate for the canonical sampling lattice in the retina (dashed line). At higher eccentricities, cone spacing continues to increase, but much more slowly than the canonical sampling interval. Other estimates for how the sampling interval changes with eccentricity in macaques and humans generally show a slope (α) ranging from 0.007 to 0.01 and an x-intercept (E_0) between 0.8° and 2.5°; these are based on data for cone spacing (De Monasterio, McCrane, Newlander, & Schein, 1985; Williams, 1988), ganglion cell density (see Figure 4), and visual acuity (cf. Westheimer, 1979). The x-intercept ($E_0 = 1.3°$) indicates where the system makes the transition from scale invariance ($E >> E_0$) to translational invariance ($E << E_0$). Thus, vision is largely scale invariant outside the fovea (central 2°), largely translation invariant in the central 0.5°, and smoothly graded in the intermediate region.

because ganglion cells in the periphery integrate inputs from many cones and transmit only low resolution data (Williams & Collier, 1983).

Retinal Neural Layers

The highly stereotyped architecture of the retina includes only five major neuronal types. The most direct route for information flow is from photoreceptors to bipolar cells and from there to ganglion cells. Horizontal cells provide an alternate route from photoreceptors to bipolar cells, as well as feedback to photoreceptors. Amacrine cells mediate interactions in both directions between bipolar cells and ganglion cells. This complex network implements a variety of data compression schemes aimed at emphasizing information that is "useful" (behaviorally salient) at the expense of data that is relatively unimportant or which can be transmitted with lower fidelity (cf. Laughlin, 1987; Atick & Redlich, 1992).

The specific strategies described next reflect three important characteristics of information processing in the visual system. One is an emphasis on data compression techniques, such as differentiation in the domains of space, time, and spectral composition, which is done to reduce redundancy in the signals transmitted by neighboring neurons (cf. Adelson & Bergen, 1991). The second is division of labor by creation of multiple cell types. The third involves multiplexing of information so that each cell, although functionally specialized, is nonetheless able to carry more than one type of information. Our focus here is on the general nature of these strategies; useful reviews and references pertaining to the detailed neural circuitry are provided by Dowling (1987), Rodieck (1988), and Wässle & Boycott (1991).

Spatial Differentiation Both bipolar cells and ganglion cells have receptive fields with a characteristic center-surround organization, in which the direct input from one or more cones is counterbalanced by antagonistic inputs from a wider region that is generally presumed to be mediated by neighboring horizontal cells (but see Reid & Shapley, 1992). This spatial differentiation emphasizes local differences in light intensity at the expense of precise information about absolute intensities.

On and Off Channels All photoreceptors respond with a graded hyperpolarization (more negative membrane potential) to an increase in light intensity. In contrast, at the bipolar cell stage, only one half of the cells are hyperpolarized by light onset in the receptive field center ("off" bipolars); the other half are depolarized by light onset ("on" bipolars). This dichotomy between on and off channels is preserved at the ganglion cell stage and beyond. This is advantageous because ganglion cells convey information using a firing frequency that is modulated around a low to moderate level of spontaneous activity. Ganglion cells and, to an even greater degree, their targets in the LGN act as rectifiers. Because dark stimuli on a light background (e.g., the print in this book) are just as important as light stimuli on a dark background, the dual-channel arrangement of ganglion cells provides a conveniently balanced, symmetrical output representation.

Parvocellular and Magnocellular Channels Superimposed on the on/off dichotomy is another classification scheme that is based on cell size, projection patterns, and the type of information encoded by ganglion cells (Perry et al., 1984; Shapley & Perry, 1986). The vast majority of ganglion cells (approximately 80%) have relatively small cell bodies, axons, and dendritic arbors; they are commonly called parvocellular (P) cells. Cells in another important class, commonly called magnocellular (M) cells, have larger cell bodies, axons, and dendritic arbors. Both P and M cells project to the LGN; the remaining 10% of ganglion cells comprise a heterogeneous group with small axons, some of which probably project to the LGN (Conley & Fitzpatrick, 1989) and others to the superior colliculus (Perry & Cowey, 1985).

Temporal Differentiation Time-dependent changes in illumination at a single retinal location can be as meaningful as spatial differences in illumination, so it is not surprising that temporal differentiation is an important aspect of retinal processing. There are striking differences in temporal processing between P and M cells (Marrocco, McClurkin, & Young, 1982; Hicks, Lee, & Vidyasagar, 1983; Crook, Lange-Malecki, Lee, & Valberg, 1988). P cells tend to give sustained responses to a maintained stimulus; they typically respond best to patterns modulated at around 10 Hz; and they often cannot follow modulations more rapid than 20–30 Hz. M cells, in contrast, respond only transiently to a maintained stimulus; they usually respond best to modulation at 20 Hz or greater; and they continue to respond up to 60–80 Hz. Amacrine cells are suspected to be the major source of this temporal differentiation by the magnocellular pathway. Some M cells also respond to abrupt motion of patterns well outside the classical receptive field, indicating a more complex type of spatiotemporal processing (Kruger, 1977; Marrocco et al., 1982).

Spectral Differentiation P cells carry chromatic information by virtue of the spectral opponency of their receptive fields (De Monasterio, 1978; DeValois & DeValois, 1993; Reid & Shapley, 1992). As might be expected from the relative numbers of cone types, most P cells have red–green opponency, but a small minority have blue–yellow opponency. M cells are not overtly spectrally opponent, but they can respond in a nonlinear fashion to the presence of spectral contrast (Lee, Martin, & Valberg, 1989).

Sampling Multiplicity of P and M Cells

The sampling multiplicity, or retinal coverage factor, provides a measure of how many ganglion cell receptive fields of a given class overlap one another at each point on the retina (Peichl & Wässle, 1979). Anatomical data on dendritic field sizes of P and M cells (Perry et al., 1984; Watanabe & Rodieck, 1989) can be used to estimate how the sampling area varies with eccentricity. For both populations, dendritic field size increases linearly with eccentricity, but the slope is threefold greater for M than for P cells (Figure 4). The physiologically determined size of receptive field centers is also substantially greater for M cells than P cells (De Monasterio & Gouras, 1975; Derrington & Lennie, 1984). Some studies have emphasized that P and M cells have similar high-frequency cutoffs when acuity is measured using high-contrast gratings (Hicks et al., 1983; Blakemore & Vital-Durand, 1986; Crook et al., 1988). However, grating acuity is an unreliable measure of receptive field center size (see the following and legend to Figure 5).

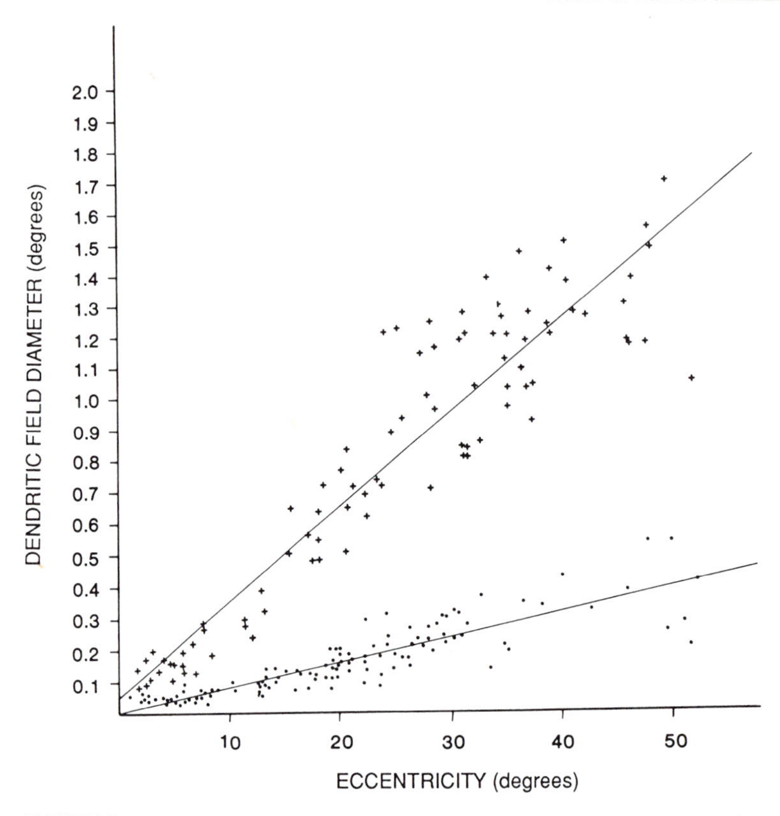

FIGURE 4 Dendritic field diameters of P and M ganglion cells in the macaque retina (based on Perry, Oehler, & Cowey, 1984). Outside the fovea, ganglion cells are more densely packed in the nasal retina than at the equivalent eccentricity in the temporal retina, and they have correspondingly smaller dendritic field arbors. To compensate for this systematic bias, we shifted the data points for nasal ganglion cells to eccentricities that were slightly closer to the fovea (34% closer for ganglion cells on the horizontal meridian and 23% closer for cells on the nasal side but away from the horizontal meridian). This adjustment was based on the magnitude of the reported ganglion cell density differences, and it noticeably improved the correspondence between dendritic diameter and eccentricity relative to their published results. The P cell data (dots) are fit by the regression line $D = 0.008 (E + 0.03)°$; the M cell data (crosses) by $D = 0.03 (E + 1.7)°$. Near the fovea, the data points are unreliable because there is substantial offset between ganglion cell bodies and the cones to which they are linked, and also because the P cell dendritic size reaches a minimum that reflects the size of cone cell synaptic terminals (Schein, 1988).

These findings, in conjunction with data on ganglion cell densities (Wässle, Grünert, Rohrenbeck, & Boycott, 1990), indicate that there are approximately three to five P cells per sampling area in the periphery, which allows for separate handling of information from the different chromatic channels as well as the "on" and "off" attributes. Earlier studies (Van Essen & Anderson, 1986; Schein 1988; Perry & Cowey, 1985, 1988) estimated the coverage factor in the center of the fovea to be two or less, where single L or M cones presumably drive the center receptive field of "on" and "off" P cells. Wässle et al. (1990), working with single retinas to eliminate uncertainties introduced by individual variation, found a notably higher value of three to four ganglion cells per cone in the center of the fovea. This suggests that the P cell sampling multiplicity may be roughly constant throughout the retina.

A similar analysis can be applied to the M cell sampling mosaic. The incidence of M cells (as a percentage of all ganglion cells) is 5–6% in the fovea and parafovea (Grünert, Greferath, Boycott, & Wässle, 1993) and increases to 10–12% over most of the retina (Perry et al., 1984), and approximately 20% in the far periphery (Silveira & Perry, 1991; see also Connolly & Van Essen, 1984). The ratio of M/P dendritic field sizes is about 3 over a wide range of eccentricities (see Figure 4 and text), but inside 10°, the ratio increases to about 5 in the macaque (Watanabe & Rodieck, 1989) and even higher in the human (Dacey & Petersen, 1992). The decreased incidence and increased relative size tend to cancel out, and, as a result, the estimated sampling multiplicity for M cells is approximately constant at approximately 3–4 over a wide range of eccentricities (Perry & Cowey, 1985; Grünert et al., 1993).

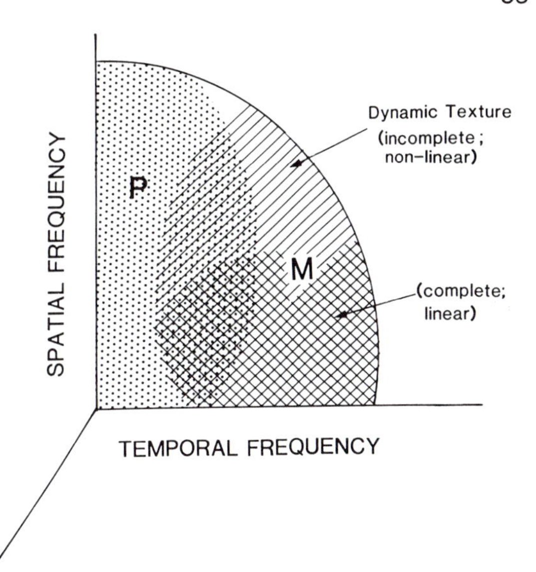

FIGURE 5 Subdivisions of information space mediated by the M and P subsystems. Each small patch of retina is subserved by a population of cones and bipolar cells, whose spatial and temporal bandwidth is represented by the area within the solid outline. Only one dimension of spatial frequency is shown, but the second dimension could be incorporated using the unlabeled axis included in the figures. Individual ganglion cells are only able to represent information across a portion of the entire domain. P cells subserve the domain of low temporal frequencies, including both low and high spatial frequencies (stippled area). M cells subserve mainly higher temporal frequencies, and they also span a wide range of spatial frequencies. At low spatial frequencies, responses are largely linear (cross-hatched region) and their high-frequency cutoff is on average only slightly less than that of P cells, despite their larger receptive field center size. This may reflect the higher contrast sensitivity of M cells compared with P cells (Purpura, Kaplan, & Shapley, 1988) and also the nonlinear properties of some M cells (Derrington & Lennie, 1984), which together should enhance the responses to gratings and textures that are much finer than the receptive field center. In addition, optical blurring (i.e., the point spread function) should affect the high-frequency cutoff of P cells more than M cells, because P-cell receptive fields are more closely matched to the optical limits of the retina. Responsiveness in the single-hatched region can signal the presence of dynamic texture (fine-grained, rapidly changing patterns in an image). However, complete fidelity in a representation of this region would necessitate as many M cells as P cells. Because M axons are thicker than P axons, this would more than double the diameter of the optic nerve, which in turn might impair ocular mobility.

Functional Division of Labor in the Retina

The differences between P and M cells represent a systematic division of labor in transmitting spatial, temporal, and chromatic information to the cortex. We believe that this reflects a strategy for packing the maximum amount of useful information into the optic nerve, given the finite transmission rates and finite diameters of real axons (Wolbarsht, Wagner, & Ringo, 1985; C. H. Anderson, 1986).

A convenient way of summarizing this strategy is illustrated in Figure 5. The region of information space associated with a small patch of retina can be conveniently represented in a two-dimensional plot showing temporal frequency along the x-axis and one dimension of spatial frequency along the y-axis. The solid line outlines the portion of this space from which information can, in principle, be transmitted, given the spatial constraints that are imposed by the retinal sampling lattices for cones and ganglion cells and the temporal constraints that are imposed by the firing rates, integrative properties, and intrinsic physiological noise of retinal neurons. Within this overall domain, P cells (stippling) and M cells (hatching) subserve partially overlapping, but substantially distinct regions as a result of their different temporal properties. P cells handle low temporal frequencies, M cells handle high temporal frequencies, and both types overlap considerably in the region of intermediate frequencies.

In the spatial domain, P cells act as relatively linear filters whose wide spatial bandwidth reflects a combination of small receptive field size and incomplete center-surround antagonism. The small size of receptive field centers ensures responsiveness to high spatial frequencies, whereas the incomplete balance between center and surround mechanisms results in some responsiveness even to very low spatial frequencies (Derrington & Lennie, 1984).

The situation for M cells is more complex. M cells provide a faithful representation of low spatial frequencies at high temporal frequencies, corresponding to coarse, rapidly changing patterns (Figure 5, double hatching, lower right). In addition, M cells respond to high spatial frequencies (Blakemore & Vital-Durand, 1986; Crook et al., 1988), but the representation of dynamic texture in this region is incomplete and of lower fidelity (single hatching, upper right). This is attributable to the nonlinear nature of responses in many M cells (De Monasterio, 1978; Kaplan & Shapley, 1982; Derrington & Lennie, 1984) and also to the fact that the overall number of M cells is insufficient for a complete, unambiguous representation at high resolution (Wässle & Boycott, 1991; Grünert et al., 1993). In short, M cells are able to signal the presence of moving or changing patterns over a wide range of spatial scales and temporal frequencies that can be used to signal the presence of interesting features, but leaves the details to be carried by the P cells.

In the spectral domain, the main points to emphasize are that P cells carry most of the color information about an image; yet, this is only a small percentage of the overall information relayed by P cells (cf. DeValois & DeValois, 1993). Psychophysical studies (Noorlander & Koenderink, 1983; Mullen, 1985) have shown that, at all eccentricities, acuity for isoluminant gratings (which differ only in color) is 5-fold to 10-fold worse than for luminance gratings (which differ only in brightness). By this measure, the chromatic information in a region of visual space may be as little as 1% ($1/10^2$) of the luminance information. This accounts for the ease with which color television signals have been made compatible with existing black and white standards, because relatively little additional information is required to match our capacity for perceiving color variations within an image (Buchsbaum, 1987).

Lateral Geniculate Nucleus

The LGN contains six main layers, which serve to segregate the inputs from P and M channels and to align the inputs from the two eyes while nonetheless keeping them physically segregated. The upper four layers (parvocellular or P) receive inputs from retinal P cells, whereas the lower two (magnocellular or M) receive inputs from retinal M cells. There is an approximately 1:1 coupling between ganglion cells and LGN relay cells (Lee, Virsu, & Creutzfeldt, 1983). There is also massive feedback to the LGN from a substantial fraction of the 10 million cells in layer 6 of striate cortex, as well as from various sources of nonvisual inputs. The functional significance of this feedback and of the various other inputs to the LGN has not been firmly established, although it is presumably somehow involved in the dynamic regulation of signal transmission (cf. Sherman & Koch, 1986; Casagrande & Norton, 1991). Interestingly, the rates of spontaneous activity and of visually evoked activity are generally much lower for LGN cells than for retinal ganglion cells (Kaplan, Purpura, & Shapley, 1987). An attractive hypothesis to account for this difference is that cortical feedback might suppress transmission in regions where the image is relatively smooth and featureless, while enhancing transmission for regions rich in features. A similar process, called *dynamic coring*, is used in some television receivers to reduce selectively noise levels in featureless regions of the image. The idea of differential suppression of LGN transmission has also been proposed by Crick (1984) and by Koch (1987) as a mechanism for directed visual attention. We suspect that visual attention, instead, involves dynamic mechanisms, operating mainly at the cortical level, and that the role of modulatory processes in the LGN is linked to basic signal-to-noise issues rather than attention per se.

Summary

The retina samples images in a spatially inhomogeneous fashion that involves a simple linear change in resolution with eccentricity. Luminance information at each sampling locus is compressed using spatial and temporal differencing. This information is distributed into the parvocellular and magnocellular systems, each with its own "on" and "off" subsystems. The P system handles low to moderate temporal frequencies and carries luminance information from the lowest resolution up to the highest allowed at a given eccentricity. The M system emphasizes luminance at lower spatial resolution, but higher temporal rates. Spectral information is multiplexed into the P system, but constitutes only a small percentage of the total amount of information that it normally carries. Multiplexed into the M system is a nonlinear measure of spatial and/or spectral energy, which may be especially useful for texture analysis, image segmentation, and preattentive visual processing.

VISUAL CORTEX AND NEURAL REPRESENTATIONS IN AREA V1

Overall Layout of Visual Areas

The cerebral cortex is the dominant structure of the mammalian brain and the principal site of higher levels of visual processing. The cortex is a thin, laminated sheet of tissue in which the various cortical layers differ in cell density, connectivity, and many other characteristics. As will become apparent, laminar organization is fundamental to understanding cortical function. In the macaque, the total cortical surface area (counting both hemispheres) is about 210 cm^2, equivalent to a large cookie (Felleman & Van Essen, 1991). Given a thickness of 1–2 mm and an average cell density of 10^5 cells/mm^2, this works out to a total population of approximately 2×10^9 neurons. More than one half of the cortex is related to vision; area V1 (striate cortex or primary visual cortex) alone is approximately 12% of the total. For comparison, cerebral cortex in the human is about 10 times larger in both surface area and total number of neurons (Blinkov & Glezer, 1968). Area V1 is only twice as large in the human as in the monkey, however, and it occupies only about 3% of the total cortical expanse.

In the macaque, visually responsive cortex occupies all of the occipital lobe, about one half of the temporal and parietal lobes, and a small portion of the frontal lobe. This large domain is divisible into many discrete visual areas that can be distinguished from one another on the basis of their overall pattern of inputs and outputs (connectivity). In addition, some areas can also be recognized by other criteria, including cortical architecture (structural appearance after appropriate staining), visual topography (a partial or complete representation of the visual field), and receptive field properties (e.g., selectivity for particular stimulus properties). Altogether, there are 32 identified areas that are largely or exclusively visual, including 7 areas that are associated with visuomotor control, polysensory processing, and/or attentional and cognitive processing (Felleman & Van Essen, 1991). Figure 6 shows the location and relative size of various areas, as displayed on a two-dimensional unfolded map of the cortex. All the boundaries between areas are drawn in, but for simplicity and clarity we have included only the names of the particular areas that are discussed in the following sections. These include areas V1, V2, V3, V4, the middle temporal area (MT), a cluster of six areas in the temporal lobe known collectively as the inferotemporal (IT) complex, and a cluster of areas in the parietal lobe known as the posterior parietal complex (PP).

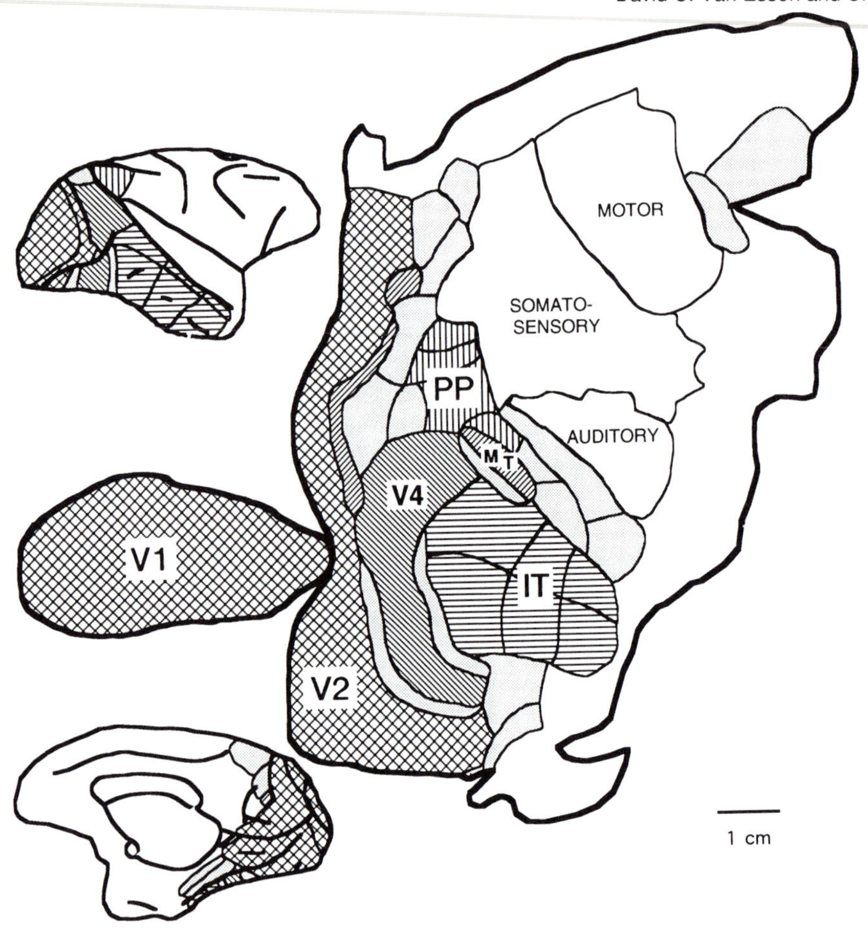

FIGURE 6 Visual areas in the cerebral cortex of the macaque, as seen in lateral and medial views of the right hemisphere (insets) and in an unfolded two-dimensional map. Visual cortex occupies more than one half of the entire cortex, mainly in the occipital, temporal, and parietal lobes (bold outline on the left on the map), but also in a portion of the frontal lobe. The major visual areas discussed in this chapter are shown by stippling. These include areas V1 (separated from the rest of the map by an artificial discontinuity, but physically contiguous in the intact hemisphere), V2, V4, the middle temporal area (MT), the posterior parietal complex (PP), and the inferotemporal complex (IT). IT and PP each contain many separate areas (Felleman & Van Essen, 1991). There are 14 additional visually related areas not labeled on the map. Modified, with permission, from Felleman & Van Essen (1991).

Broadly speaking, visual information that enters area V1 is processed in a highly localized fashion to generate a multitude of data representations for distribution to other areas subserving more specialized analyses (Hubel & Wiesel, 1977; Zeki, 1978). A useful distinction (but one to be made with caution) is that inferotemporal cortex is mainly concerned with object identification, pattern recognition, and aspects of "what" is in the visual world, whereas posterior parietal cortex is concerned with spatial relationships and the assessment of "where" things are (Ungerleider & Mishkin, 1982; Merigan & Maunsell, 1993).

Architecture, Connectivity, and the Sampling Lattice in V1

V1 receives a precisely organized array of projections from the LGN, which produces a representation of the visual field on the cortical surface shown on the two-dimensional map of Figure 7. There is a great emphasis on central vision, with nearly one half of the cortex devoted to the central 5°. To a first approximation, the amount of cortex devoted to the region between 5° and 10° eccentricity is similar to that from 10° to 20°, and from 20° to 40°, which is a manifestation of the scale-invariant sampling strategy that originates in the retina. This issue can be analyzed quantitatively by deriving expressions for the cortical magnification factor (mm^2 of cortex per degree^2 of visual field) as a function of eccentricity. Physiological mappings of several hemispheres by Van Essen, Newsome, and Maunsell (1984) yielded the relationship described in Equation 2 for a standard visual cortex:

$$M^2 = 140\,(0.78 + E)^{-2.2}\ \text{mm}^2/\text{deg}^2 \qquad (2)$$

This equation predicts a ratio of ~ 4000 in areal magnification factor for the center of the fovea versus 60° eccentricity. This is only slightly greater than the ratio of 2000 predicted from the retinal sampling lattice (Equation 1) or the ratio of 1000 predicted from the retinal ganglion cell density estimates of Wässle et al. (1990). Even closer agreement is obtained when the retinal data are compared with the results from a deoxyglucose mapping experiment of central fields in striate cortex (Tootell et al., 1988b). Given that there

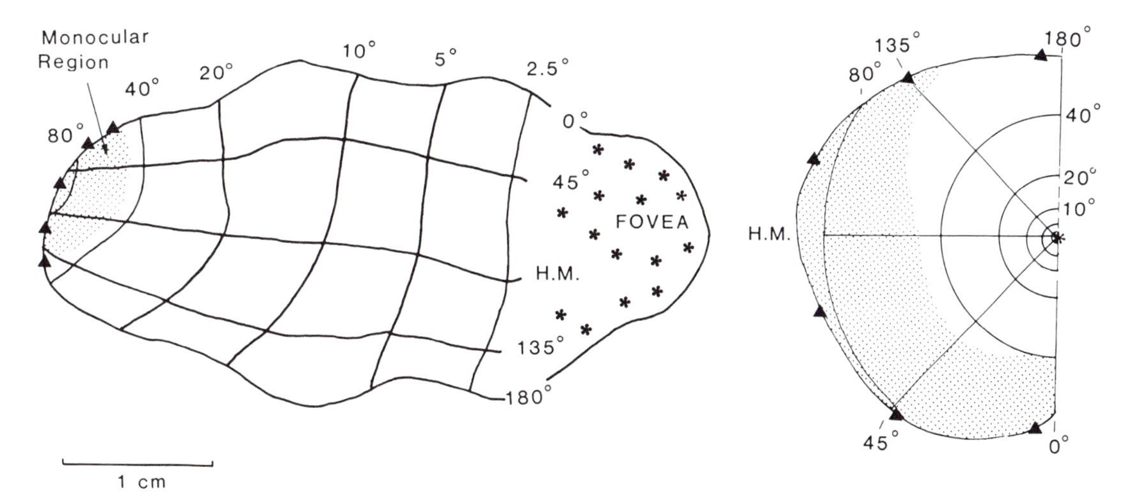

FIGURE 7 Topographic organization of area V1. Lines of constant eccentricity (semicircles in the visual field drawing on the right) map onto contours that run approximately vertically on the cortical map. Lines of constant polar angle (rays emanating from the center of gaze in the visual field) map onto contours that run approximately horizontally on the cortical map; there is also an inversion such that the lower and upper quadrants of the visual field are represented in dorsal and ventral halves of cortex, respectively. The foveal representation (asterisks), corresponding to the central 2° radius, occupies slightly more than 10% of V1. Fitting this data to a modified log polar mapping gives a linear cortical magnification factor of $M \propto (0.8 + E)^{-1.1}$. We believe this description fits the experimental data and the retinal sampling strategy better than the slightly different log conformal representation suggested by Schwartz (1980). However, there is considerable biological noise in actual cortical topography, both in terms of individual variability in topographic organization and in terms of significant local deviations from the mapping function to make these only approximate mathematical interpolation formulas. (Reprinted from *Vision Research*, **24**, Van Essen et al., The visual field representation in striate cortex of the macaque monkey: Asymmetries, anisotropies and individual variability, 429–448, copyright 1984, with kind permission from Elsevier Science Ltd., The Boulevard, Langford Kidlington OX5 1GB, UK.)

is substantial individual variability in all of these mappings, it is necessary to study both retinal and cortical representations in the same individual to obtain a substantially more precise understanding of these relationships.

In a standard Nissl stained section, the distribution of cells in V1 is relatively uniform within each layer, with no obvious periodicities running parallel to the cortical surface. Lurking beneath this apparent uniformity, however, is a modular organization revealed by several histochemical stains (most notably the enzyme cytochrome oxidase), and also reflected in anatomical connectivity patterns and in the physiological organization of receptive field properties (Hubel & Wiesel,

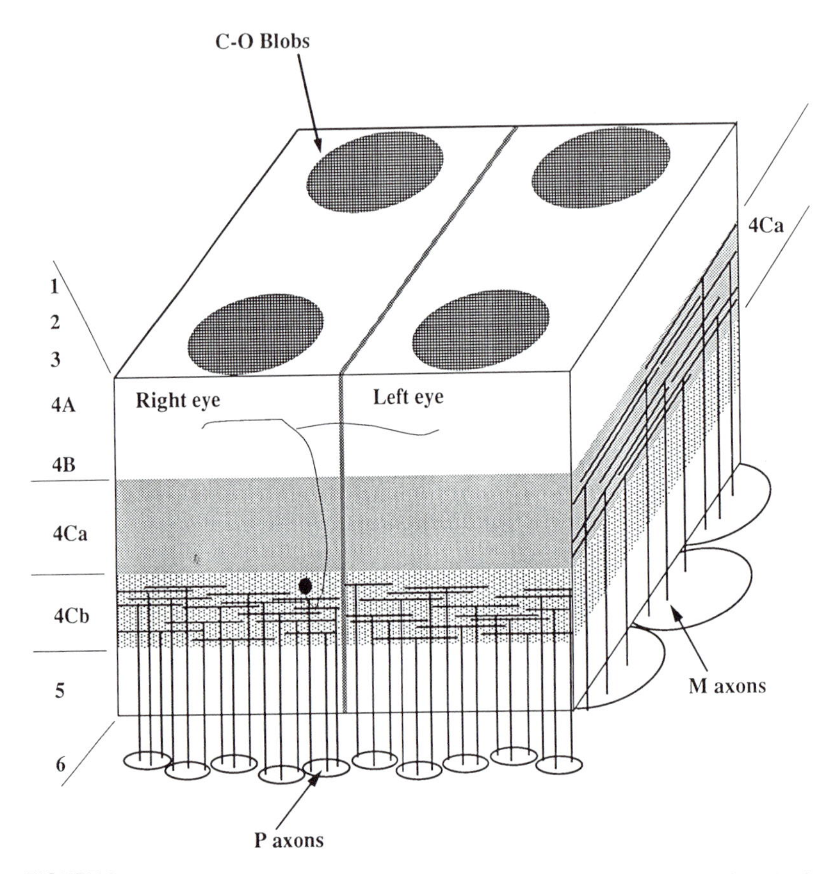

FIGURE 8 Architecture of P and M inputs to area V1. The diagram shows a 1-mm² patch of V1, with inputs from P cells terminating in layer 4Cb (*front face*) and inputs from M cells terminating in layer 4Ca (*right face*). Groups of axons associated with a single sampling node are bundled together by the small loops in the white matter. The actual density of P and M inputs and the extent and degree of overlap of axonal arbors within each layer are indicated semiquantitatively. This figure also shows two manifestations of modular organization within V1: ocular dominance stripes, in which there is sharp segregation of the inputs from left and right eyes in layer 4, and cytochrome oxidase "blobs," shown as circular patches on the surface of the cortex that are positioned along the center of each ocular dominance stripe. The spacing of lettering on the left provides a more accurate indication of the relative thickness of different layers than does the actual figure, in which layer 4C has been expanded for clarity.

1974; Livingstone & Hubel, 1984). Most of these patterns have a periodicity on the order of 1 mm. It is therefore of particular interest to ascertain the number and the distribution of cortical cells and of LGN inputs associated with each square millimeter of cortex. The key features of this arrangement are shown in Figure 8, which illustrates the pattern of LGN projections to a 1-mm^2 patch of V1. P cell inputs are displayed along the front face and the pattern of M cell inputs are displayed along the right side.

Each square millimeter of cortex receives on average approximately 800 P cell afferents (400 per eye) and about 80 M cell afferents (40 per eye). In relation to the P and M sampling lattices outlined previously, each square millimeter is supplied by about 120 sampling nodes from each eye for the P system (represented by loops around sets of P cell inputs) and about 12 sampling nodes from each eye for the M system.

V1 has many more distinct layers than are found in other cortical areas, which stems in large part from the segregation of P and M pathways (Hubel & Wiesel, 1977; Fitzpatrick, Lund, & Blasdel, 1985). P inputs terminate mainly within a thin sheet known as layer 4Cb, whereas M inputs terminate immediately above, in layer 4Ca. In each of these layers there is also a clear segregation of inputs for the two eyes. This is manifested by an alternating series of ocular dominance stripes, each approximately 0.5 mm wide and receiving inputs exclusively from one eye.

The axonal arbors of P cell afferents are relatively compact, with a typical arbor diameter of 150–250 μm, which is reflected in the horizontal extent of axons drawn in the figure (Blasdel & Lund, 1983). Dendritic arbors of cells in layer 4Cb are also approximately 150–200 μm in diameter (Lund, 1988). On the basis of these numbers and on the density of P cell afferents, it follows that each cell in layer 4Cb potentially can be in physical contact with about 100 P cells. We estimate that each neuron in layer 4Cb receives about 100 geniculo-cortical synapses, based on the data of O'Kusky and Colonnier (1982) and Blasdel and Lund (1983). Likewise, each cell in layer 4Ca receives a total of about 100 geniculo-cortical synapses from up to 100 M cells, whose axons are much larger but whose innervation density is much lower. However, it is not known how different geniculate axons contribute to the innervation of each cortical cell.

The segregation of M and P pathways is remarkably sharp at subcortical levels and in the geniculocortical afferent terminations. However, within the cortex, there is considerable intermixing that begins even within layer 4C by virtue of dendritic and local axonal arbors that cross between layers 4Ca and 4Cb. In the supragranular layers of cortex, there are three major compartments: layer 4B and the cytochrome oxidase (CO)-enriched "blobs" and CO-sparse "interblobs" of layers 2 and 3. Layer 4B is dominated by inputs from the M pathway, but the blobs and interblobs each receive major influences from both P and M inputs.

Physiological Properties in V1

Single unit recordings have revealed that neurons in V1 comprise a diverse and heterogeneous population, in contrast to the more stereotyped nature of and limited number of cell classes in the retina and LGN. This is commensurate with the large increase in the number of cells available for processing information from each patch of the visual field. As a population, cells in V1 have more than a dozen important receptive field characteristics representing the emergence of new properties or the refinement of existing properties of LGN cells. Most of these reflect highly local computations that take advantage of the precise topography in the organization of the ascending inputs to V1. However, some properties reflect the operation of feedback pathways and/or long distance lateral interactions.

Orientation and Spatial Frequency Two particularly important and intertwined characteristics are orientation selectivity and spatial frequency tuning. Except in layer 4Cb and the CO blobs, the majority of V1 cells show at least some selectivity for stimulus orientation (often very sharp tuning) and they are much more narrowly tuned for spatial frequency than are their LGN inputs (Hubel & Wiesel, 1968; De Valois, Albrecht, & Thorell, 1982; Blasdel & Fitzpatrick, 1984; Livingstone & Hubel, 1984). In simple cells, whose receptive fields can be subdivided into separate excitatory and inhibitory subregions, the overall profile of sensitivity can be approximated by a Gaussian weighted sine wave (Daugman, 1985; Jones & Palmer, 1987). A population of such cells tuned for an appropriate range of orientations and spatial scales can provide a complete representation of an image down to the limit of spatial resolution. Moreover, this is an efficient form of representation insofar as the outputs of different cells in response to natural images will tend to be statistically independent of one another (Daugman, 1985; Field, 1987, 1992; C. H. Anderson, unpublished results).

Another major cell class is the *complex cell*, in which orientation selectivity is also present, but the cell is relatively insensitive to the exact position of the stimulus (Hubel & Wiesel, 1962; Pollen, Goska, & Jacobsen, 1988). This strongly nonlinear property suggests that complex cells signal a spatially averaged energy associated with a particular orientation. This is analogous to the nonlinear behavior of many M cells in the retina and LGN, as discussed already. As with M cells, such an energy measure may be useful for image segmentation and texture analysis.

Velocity Another property involves selectivity for stimulus velocity—the direction and speed of motion. Velocity selectivity can be regarded as a form of spatiotemporal filtering, in which a cell responds optimally to a target moving along a particular orientation in space–time coordinates (Adelson & Bergen, 1985). If the spatiotemporal filter is strictly linear, then the preferred direction should reverse with a change in stimulus contrast (i.e., from dark to light), as has been reported for some cells in layer 4B (Livingstone & Hubel, 1984). Nonlinearities at a subsequent processing stage can yield directional preferences that are contrast independent.

Binocular Interactions Interactions between the two eyes are obviously important for binocular fusion (a single percept emerging from two separate views of the world) and for stereopsis (judgments of depth based on the small binocular disparities arising from the different viewing angles of the two eyes). Cells in layer 4C are monocularly driven, just like their LGN inputs, but in other layers, most cells can be activated through either eye. A substantial fraction of the cells in V1, particularly in layer 4B, show tuning for binocular disparity (Poggio, 1984).

Spectral Composition Most cells in the blobs of V1 are wavelength selective, and many have properties that are more complex than the basic center-surround chromatic opponency characteristic of P cells in the retina and LGN. Also, many cells in the interblobs show wavelength selectivity in addition to orientation selectivity (Livingstone & Hubel, 1984; Michael, 1985; T'so & Gilbert, 1988).

Length and Curvature Many V1 cells have the property of end-stopping, whereby the response to a long bar is much less than that to a short bar (Gilbert, 1977). The activity of these cells may be useful in signaling information about terminations, corners, and even smooth curvature of contours in the visual field (Dobbins, Zucker, & Cynader, 1987). End-stopping could also contribute to the responses to texture patterns described in the next section.

Texture and Motion Contrast Natural images contain many texture patterns that are more complex and irregular than simple lines and edges. Studies of neural responses to textures have revealed a number of interesting properties that are not readily predictable from studies based on simpler visual stimuli. For example, many neurons are strongly influenced by texture patterns lying outside the classical receptive field, in that the surround pattern modulates the response to a central target even though the surround is ineffective on its own (Allman, Miezin, & McGuinness, 1985). Such modulatory effects have been found with static as well as moving texture patterns (Van Essen et al., 1989; Knierim & Van Essen, 1992).

General Comments

Despite the vast amount of experimental data available on the neurophysiology of V1, there are several fundamental issues about its function that remain largely unresolved. One issue concerns the sharpness of anatomical and physiological classifications. In many cases, it is not known to what extent various compartments, cell types, and receptive field types represent biologically discrete groupings rather than points along a biological continuum. Another issue concerns the relationship between various receptive field properties (e.g., wavelength selectivity or direction selectivity) and their roles in perception (e.g., color or motion perception). For instance, recognition of the color of an object must surely rely on the activity of many wavelength selective neurons, but it is misleading to assume the converse. Many wavelength selective neurons may have alternative functions that are unrelated to color perception per se (cf. DeYoe & Van Essen, 1988; Dobkins & Albright, 1993; Van Essen & DeYoe, 1993). Finally, our description of cortex thus far has dealt only with static, hardwired aspects of processing. Dynamic aspects of processing that may occur in V1 and elsewhere will be discussed separately.

INFORMATION FLOW THROUGH VISUAL CORTEX

Current ideas about the functions of extrastriate visual cortex are guided by several broad principles, including the notions of hierarchical organization, concurrent processing streams, and highly distributed networks of information flow. These principles are derived from a combination of anatomical studies of the pathways for information flow, physiological studies of receptive field properties in different areas, and behavioral studies of the effects of selective lesions.

Each visual area is richly interconnected with other cortical areas and with subcortical nuclei, particularly the pulvinar complex (and also the LGN in the case of V1). In general, these linkages are reciprocal, or bidirectional, in nature. Typically, each area is strongly interconnected with five to ten other cortical areas, and there are usually several minor pathways as well (Felleman & Van Essen, 1991). At the low end of the spectrum, area V1 has major connections with only three cortical targets (areas V2, V3, and MT, in particular). At the high end, area V4 is connected with 21 other areas, and most of these pathways are reasonably robust.

The total number of identified pathways among 32 cortical areas related to vision is 305 (Felleman & Van Essen, 1991). If visual cortex were a fully interconnected matrix (i.e., if each area were linked to all others), there would be $32 \times 31 = 992$ pathways. Thus, the number of identified pathways is currently about one-third of the theoretical limit and may reach 45–50% once all have been identified.

Hierarchical Organization and Concurrent Processing Streams

Several lines of anatomical and physiological evidence suggest that information processing in the cortex is fundamentally hierarchical in nature (Rockland & Pandya, 1979; Van Essen & Maunsell, 1983). By this, we mean that each area, and each cell population within an area, represents a particular stage of analysis relative to all other components in the system. However, this is neither a strictly serial scheme nor a strictly feedforward model involving only unidirectional information flow. Rather, there is extensive feedback as well as parallel processing, both of which are fully compatible with hierarchical organization in the broad sense.

The original version of the cortical hierarchy included 13 visual areas interlinked by 36 different pathways (Van Essen & Maunsell, 1983). As more areas and pathways have been identified,the hierarchy has been extended to incorporate the new data. The most recent versions of the hierarchy (Felleman & Van Essen, 1991; Van Essen, Anderson, & Felleman, 1992) span 10 cortical levels of processing, as well as several subcortical levels, and extend all the way from the retina to the outputs from the visual system in limbic areas of the temporal lobe (entorhinal cortex, the hippocampus, and the amygdala). For simplicity, we will not illustrate the full scheme, but rather a version that involves only the few key areas previously identified on the cortical map. Thus, Figure 9 shows two subcortical stages (retina and LGN): area V1 is the first cortical stage; areas V2, V3, V4, and MT represent intermediate cortical stages; and the inferotemporal

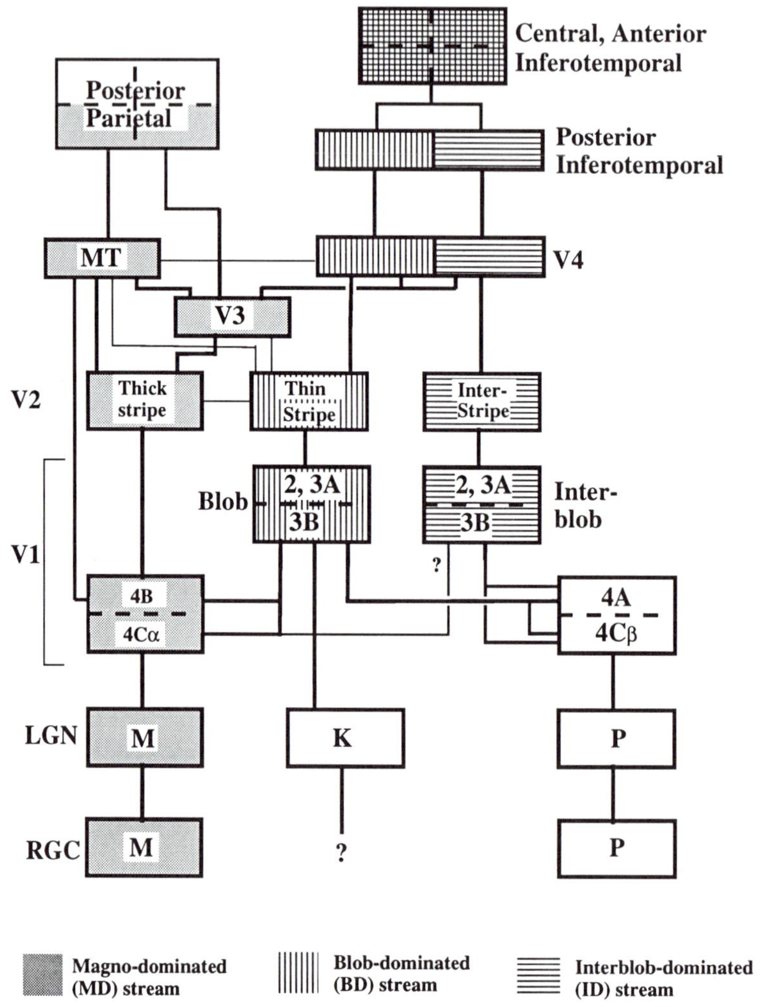

FIGURE 9 Hierarchical and compartmental organization of the visual pathway. This scheme illustrates information flow involving two subcortical levels and four cortical levels. Each of the lines between boxes represents a reciprocal pathway, and (except for the retino-geniculate projection) feedback connections are typically as robust as forward connections. Parallel processing is manifested at subcortical levels by the distinction between P, M, and K cells. In the cortex, there is convergence and divergence leading to three streams (MD, BD, and ID) that involve anatomically distinct compartments within V1, V4, and posterior inferotemporal cortex, as well as separate areas V3, MT, and areas of the posterior parietal complex streams.

and posterior parietal complexes represent the highest cortical stages.

This figure also illustrates the concurrent processing streams that are discernible at both subcortical and cortical levels of the hierarchy. At subcortical levels, there is a distinctive population of very small (koniocellular) neurons in the LGN that constitute a third processing stream, separate from the M and P streams that have already been extensively discussed. This koniocellular (K) stream receives major inputs from the superior colliculus (SC) as well as from small-caliber axons of presumed retinal origin (cf. Casagrande & Norton, 1991).

At the cortical level, there are again three major streams, but these are not simply one-to-one continuations of the subcortical M, P, and K streams. They have been described as the MD (magno-dominated), BD (blob-dominated), and ID (interblob-dominated) streams, which are established in V1 and can be followed through several additional stages of the cortical hierarchy (Van Essen & DeYoe, 1994; Van Essen & Gallant, 1994). The MD stream, which includes layers 4Cα and 4B in V1, appears on both anatomical and physiological grounds to be dominated by inputs from M cells in the LGN (Lachica, Beck, & Casagrande, 1992; Nealey & Maunsell, 1994). The BD stream receives convergent inputs directly or indirectly from M, P, and K cells of the LGN (Lachica et al., 1992; Casagrande & Lachica, 1992). Selective inactivation of LGN layers has provided physiological confirmation that the CO blobs (BD stream) and the CO interblobs (ID stream) in V1 are strongly influenced by both M and P inputs (Nealey & Maunsell, 1994).

In extrastriate cortex, the three processing streams are represented by a set of CO stripes in V2 (thick stripes, thin stripes, and interstripes) and at higher stages by a combination of distinct areas and compartments within areas. More specifically, the MD stream includes the thick stripes of V2, plus V3, MT, and areas in the posterior parietal complex. The BD stream includes the thin stripes of V2 and a set of compartments in V4 and posterior inferotemporal areas that cannot as yet be distinguished by their architecture. The ID stream includes the interstripes of V2 and the complementary set of compartments in V4 and posterior inferotemporal areas. Note that there is cross-talk between streams at several levels, even though they remain distinguishable by virtue of the dominant anatomical projections (DeYoe & Van Essen, 1988; Zeki & Shipp, 1988; Felleman & Van Essen, 1991; Merigan & Maunsell, 1993; Van Essen & DeYoe, 1994).

Physiologically, cells in the MD stream tend to be selective for stimulus direction, disparity, and orientation, but not for stimulus wavelength. These regions also tend to be heavily myelinated, which should contribute to fast axonal conduction velocities and enhance the capacity for carrying information at high temporal frequencies. The ID stream is characterized by a high incidence of orientation selectivity and end-stopping. Some cells are also selective for wavelength and binocular disparity, although there are discrepancies in the reported percentages of such cells (cf. DeYoe & Van Essen, 1988; Peterhans & von der Heydt, 1993). Cells in the BD stream tend to be selective for stimulus wavelength and also for low spatial frequencies (Tootell, Silverman, Hamilton, Switkes, & DeValois, 1988a). Hence, it is natural to suspect that this stream plays an important role in color perception (Livingstone & Hubel, 1988a,b; Zeki & Shipp, 1988). On the other hand, the various components of the BD stream are prominent in New World monkeys such as the owl monkey, which have little or no color vision. Moreover, it seems unlikely that color processing should exclusively occupy a large fraction of cortex, given that color information is such a small fraction of the information content we derive from natural images. Hence, it is likely that the BD stream is involved in other computational tasks, such as the analysis of brightness, surface textures, and/or shape based on shading cues.

Overall, these findings make sense, given what is known about the visual functions of the temporal and parietal lobes. As noted already, inferotemporal cortex has been implicated in pattern recognition and object identification (the "what" pathway) on the basis of the behavioral deficits occurring after lesions and on the presence of neurons selectively responsive to faces and other complex patterns in IT (Perrett, Mistlin, &

Chitty, 1987; Miyashita, 1993) and in V4 (Gallant, Braun, & Van Essen, 1993). It is logical that these functions should rely primarily on the high-resolution, spectrally coded information carried by the BD and ID streams. However, it is also appropriate that the MD stream should contribute to object identification to enhance our ability to recognize the shapes of moving objects (structure from motion) and objects having very low contrast.

Posterior parietal cortex has been implicated in the analysis of spatial relationships (the "where" pathway) and in the control of eye movements and of visual attention (Ungerleider & Mishkin, 1982; Posner & Presti, 1987; Andersen, 1989; Andersen, Snyder, Li, & Stricanne, 1993). It is not surprising that these processes should rely heavily on the information coming from the MD stream, with its high sensitivity to movement and change. However, it also makes sense for the BD and ID streams to contribute to these processes in order, for instance, to facilitate the analysis of spatial relationships between objects that are stationary or that are moving very slowly in the visual field.

In general, it seems unlikely that each processing stream simply handles the analysis of a single sensory cue (e.g., wavelength or velocity) for the purpose of contributing to a single aspect of perception. Rather, the cortex evidently acts as a concurrent processing system in which each stream carries out several interrelated tasks. For these tasks, each stage needs a diverse set of inputs and to distribute its outputs widely for subsequent stages of analysis (DeYoe & Van Essen, 1988; Van Essen & DeYoe, 1994).

Functionality of Different Cortical Layers

Each of the cellular layers within the cortex is distinctive in its connectivity and its neuronal architecture. How, then, do the layers differ from one another in their functions? One possibility is that all layers employ similar processing strategies, and that lamination is a simple way of stacking several closely related processing sheets on top of one another to minimize connection lengths. For example, it is often tacitly assumed that the different layers of visual cortex share a common function of processing ascending sensory signals to extract information about the presence of various features. We refer to this as *data analysis*; it is exemplified by orientation selectivity, direction selectivity, and other emergent receptive field properties of cells in V1 that reflect transformations of the properties of LGN cells.

In addition to data analysis, there are other computational problems that must also be faced to ensure proper cortical function. One that we suspect is of great importance involves the *dynamic routing* of information into and out of each cortical area. Dynamic control of information flow is critical for the operation of standard digital computers, and analogous principles may apply to the nervous system as well. Support for this notion arises from a variety of physiological, psychophysical, and computational considerations. These have been articulated elsewhere in the context of specific strategies that might be involved in stereoscopic depth perception, motion analysis, and directed visual attention (Anderson & Van Essen, 1987; Van Essen & Anderson, 1990). Here, we present the issue in a broader context that relates to the functionality of different cortical layers, irrespective of the particular cortical area in which they reside.

Our basic hypothesis is that the laminar organization of cortex reflects a tripartite division of labor, in which the data analysis stage is handled mainly within the superficial cortical layers; the middle layer (layer 4) is involved primarily in the gating or modulation of inputs for distribution to the superficial layers and the deeper layers mediate the control of this selection and routing process. We note from the outset that the hypothesis is speculative and is intended only as a first approximation to a fundamentally complex network. However, if even approximately correct, it has major implications for our understanding of information flow within the cortex.

Figure 10 schematically illustrates seven pathways for information flow within and between cortical areas. The black arrows are presumed to be data pathways that represent various transformations of the original image data arising from ascending information flow (left) or descending flow (right). Others (stippled arrows) reflect putative "control pathways," whose role is to regulate information flow along the data pathways (cf. Baron, 1987). The first stage (A) represents the entry of ascending inputs to the cortex, which terminate mainly within layer 4 but also have a minor spin-off into layer 6. For most areas, the ascending inputs

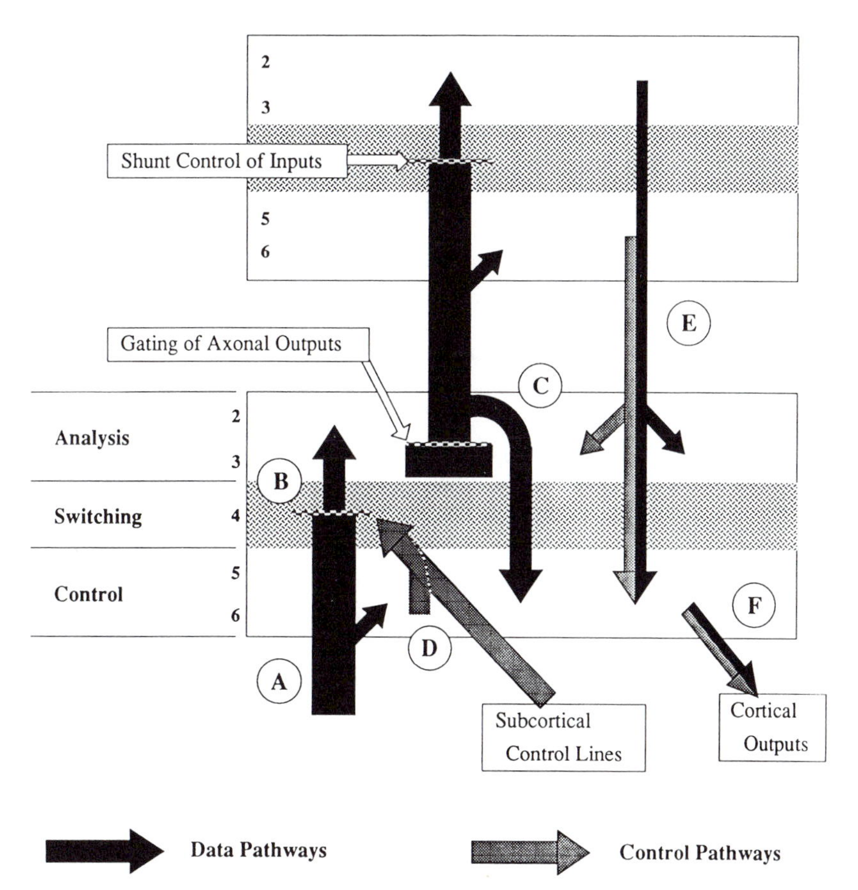

FIGURE 10 Proposed functionality of cortical layers. Arrows show the major routes of information flow within and between cortical areas as revealed by pathway tracing studies. Seven major stages can be distinguished, whose putative functions are suggested below. A. The cycle starts with ascending data from lower centers. B. Dynamic switching and filtering occurs within layer 4. (For area V1, this refers only to the geniculo-recipient layers 4A and 4C; layer 4B is considered one of the superficial layers in this context.) C. Data analysis occurs within the superficial layers, and the outputs are distributed to deeper layers within the same area and to layer 4 within higher areas. D. Appropriate control signals are generated within deep layers and in subcortical centers and are used to regulate dynamically the modulation process in layer 4. E. Descending signals (representing both data and control pathways) originate in both deep and superficial layers of a higher area and terminate in deep and superficial layers of a lower area, selectively avoiding layer 4. F. Cortical outputs are generated within the deep layers and are distributed to numerous subcortical centers.

originate from other cortical areas at lower hierarchical stages, but for V1 (or other primary sensory areas) the inputs are from the LGN (or other thalamic relay nuclei).

Layer 4 provides major ascending projections to the superficial layers. We suggest that layer 4 acts as a gating network, in which only a subset of the ascending inputs to the superficial layers are enabled at any given moment. This is schematized by the reduction in the size of the arrows between inflow and outflow from layer 4, reflecting a dynamic input selection process at locus B. A possible cellular mechanism for this putative gating process involves shunting inhibition of the dendrites of neurons within layer 4 (Koch, Poggio, & Torre, 1983; Anderson & Van Essen, 1987). It is controversial whether shunting inhibition actually occurs in visual cortex (Ferster & Jagadeesh, 1992). An alternative mechanism for achieving a similar outcome might involve nonlinear interactions among excitatory inputs, such as those mediated by NMDA receptors (Koch & Poggio, 1992).

Within the superficial layers, various types of data analyses are carried out, as appropriate for the particular area under consideration. The outputs of the superficial layers (C) include intrinsic projections to the deep layers within the same area and also extrinsic projections to higher cortical areas. These outputs may undergo another stage of selective gating which could be mediated by direct inhibition of axonal activity mediated by specialized cells known as *chandelier cells* (Somogyi, Freund, & Cowey, 1982; Schneider, 1985).

There are several possible sources for the putative control signals needed to determine how information is to be routed from one layer to the next. One major source (D) is layer 6, which provides feedback to layer 4 (Fitzpatrick et al., 1985). For extrastriate cortex, another source is the projection arising from the pulvinar nucleus (Benevento & Rezak, 1976; Ogren & Hendrickson, 1977). Finally, feedback from higher areas (E) could help regulate the axonal gating process hypothesized to occur in the outputs from superficial layers.

All cortical areas have major outputs to numerous subcortical structures in the thalamus, midbrain, and brain stem (including, e.g., the pulvinar, LGN, SC, basal ganglia, and pons). In general, these arise from cells in the deep cortical layers (F). Some of these pathways may represent feedback, whereas others may represent control pathways that feed directly into components of the motor system mediating eye, head, and limb movements.

The gating process in layer 4 presumably should operate in a coordinated fashion among neighboring cells. For example, in some situations it would be desirable to carry out a dynamic remapping process, involving translational shifts between the inputs and outputs of layer 4 mediated by a neural shifter circuit. We think that such a circuit could be constructed in a biologically plausible fashion and could aid in the solution of basic computational problems arising in the analysis of both motion and depth information (Anderson & Van Essen, 1987; Van Essen & Anderson, 1990). Analogous types of gating could be used for dynamic scaling and for remapping at higher stages in relation to directed visual attention (Olshausen, Anderson, & Van Essen, 1993; Van Essen et al., 1994). These can, in general, be regarded as reformatting operations that condition the input data before transmission to superficial layers.

The shifter hypothesis explicitly predicts that neuronal receptive fields should not be rigidly locked to absolute retinal coordinates, even at the earliest stages of visual cortex. Rather, receptive fields should shift in position over a significant range, in a manner dictated by motion, depth, or other types of information arising outside as well as inside the classical receptive field of the cells. The magnitude of these shifts should be very small at the initial stage, but would become larger at progressively higher stages of the cascade.

Evidence for dynamic shifts in sensory receptive fields has been reported in the primate SC (Jay & Sparks, 1984) and in area V4 and inferotemporal cortex (Moran & Desimone, 1985; Connor, Gallant, & Van Essen, 1993). For V1, the evidence is mixed as to whether dynamic receptive field shifts occur, with some evidence in support of the hypothesis (Motter & Poggio, 1982, 1990) and some evidence against (Gur & Snodderly, 1987).

The functional assignments we have proposed for different layers should not be regarded as rigid and all-or-none. For example, layer 4 of area V1 is likely to make some contribution to the data analysis stage in addition to whatever role it has in dynamic switching. This is suggested by the presence of significant numbers of orientation selective cells in layer 4Ca (Blasdel & Fitzpatrick, 1984). Also, a high incidence of orientation selectivity and even direction selectivity is found in layer 4 of V1 in the cat (Gilbert, 1977), suggesting that there are important species differences in laminar processing. For the superficial layers, we have already noted that they might carry out some dynamic switching functions in addition to their role in data analysis.

Despite this blurring of laminar distinctions, our basic hypothesis remains a simple one: A primary function of layer 4 is to regulate the incoming data in order to improve the efficiency and flexibility of computation in the superficial layers and to coordinate processing across cortical areas. Within this general framework, it is important to formulate specific experimental paradigms that can critically test the validity of the hypothesis and its generality across different species and different cortical areas.

Cortical Microcircuitry and Cascaded Network Architecture

The preceding two sections concentrated on information flow and neural representations in different areas and cortical layers. We now turn to a finer-grained analysis dealing with several aspects of the microcircuitry that underlies specific receptive field properties. Each cortical neuron typically receives between 10^3 and 10^4 synapses, and there is evidence that these usually reflect weak connections from hundreds or thousands of other neurons rather than massive inputs from just a few neurons (Braitenberg, 1978). Thus, there are literally millions of neurons that are only a few synaptic stages away from any given cortical cell. In this respect, the cortex is clearly a highly distributed neural network, and it may be profitable to apply some of the modeling approaches that have recently been applied to artificial neural networks. To make such models explicit and biologically realistic, it is important to know the spatial extent and relative strength of the major inputs and outputs of each cortical layer and each neuronal cell type, along with the degree of divergence and convergence as information flows in both directions through the cortical hierarchy. One useful way of representing some of these relationships is a connectivity matrix, analogous to those often used to illustrate connections in artificial neural networks (e.g., Hopfield, 1984). Two examples will be given using this approach, one concerning processing in the M stream and the other concerning visual attention and the pathway from V1 to IT.

A Connectivity Matrix for the M Stream

Figure 11 illustrates a connectivity matrix that covers several processing stages within the M stream, including the magnocellular layers of the LGN, layers 4Ca and 4B of V1, and area MT. The basic format is illustrated in the inset on the upper right. It involves a one-dimensional array of neurons whose cell bodies are aligned in a vertical row along the *y*-axis; for simplicity, only two neurons (i and j) are drawn. The inputs to each neuron impinge on a target dendrite that runs horizontally to the right of the cell body. Each neuron also has an axon that loops around and runs vertically, acting as a source that intersects with the dendrites of all of the neurons. The intersection between the axon of neuron i and the dendrite of neuron j is a synapse whose strength can be denoted as w_{ij}. More generally, the inputs to a given cell can be found by examining the synaptic weights along the appropriate row, and its outputs by examining the appropriate column.

In applying this analysis to the M stream, we create a one-dimensional array by imagining a line running along the representation of the horizontal meridian within each visual area. The entire visual representation, from fovea (F) to periphery (P), is included for each area, and successive hierarchical stages (LGN, V1-4Ca, V1-4B, and MT) are placed next to one another along each axis. Each rectangular region within

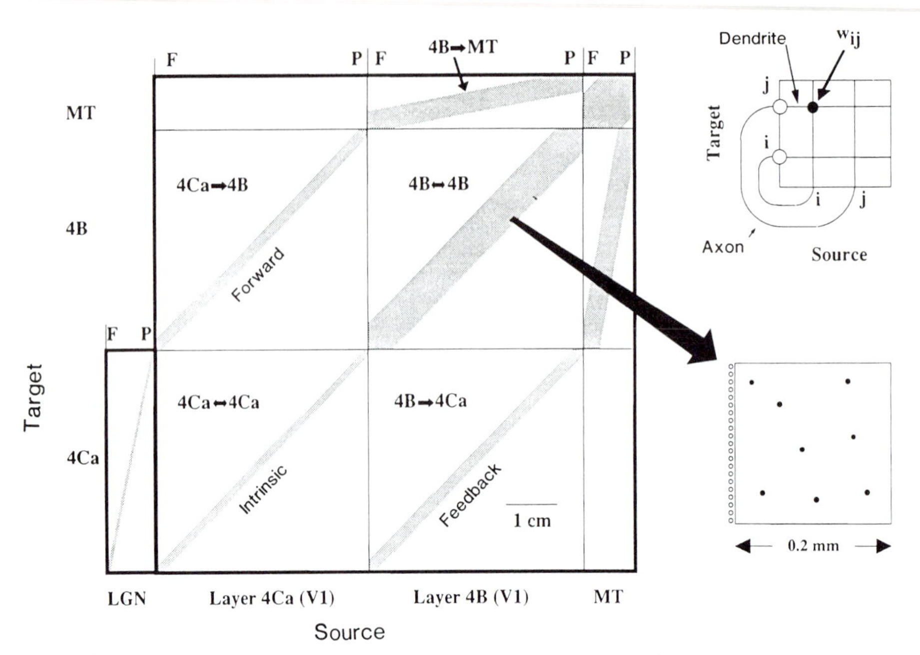

FIGURE 11 A connectivity matrix for representing circuitry within and between cortical areas. The inset on the upper right shows the format for representing connections within a one-dimensional array of neurons. The axes of the matrix are arranged to represent information flow from a source along the x-axis to a target along the y-axis. In neurobiological terms, cell bodies are arrayed along the y-axis, with dendrites running horizontally and axons curving around to run vertically. The main part of the figure shows a connectivity matrix for a one-dimensional slice through several major subdivisions of the M stream. The x-axis represents, in sequence, the axonal outputs of the LGN (magnocellular layers), layers 4Ca and 4B of V1, and area MT; the y-axis represents the inputs to layers 4Ca and 4B of V1 and to MT. Stippling indicates zones within which connections actually occur. Circuitry intrinsic to each area lies along the main diagonal through the matrix: feedforward pathways lie above and to the left, whereas feedback pathways lie below and to the right. This scheme does not show the ocular dominance stripes in layer 4C, but these and other features could readily be incorporated in a more refined version.

the overall matrix represents a single extrinsic or intrinsic pathway. Obviously, on such a coarse scale individual cells, axons, and connections cannot be resolved. For simplicity, we have used stippling to indicate regions where extensive synaptic contacts are made. These regions tend to run along diagonal bands within each rectangle, and the width of the band reflects the physical extent of divergence in the pathway. For example, the projection from the LGN to layer 4Ca of V1 is tightly organized topographically, insofar as individual magnocellular LGN axons arborize over a 1-mm zone within the 40-mm linear extent of V1 (narrow rectangle along far lower left). A comparable spread occurs in the intrinsic connections within layer 4Ca (large square in lower left) and in the projection from 4Ca to 4B, one square above (Blasdel, Lund, & Fitzpatrick, 1985; Fitzpatrick et al., 1985). Greater divergence occurs at higher stages, so that MT receives ascending inputs from a substantial portion of the visual field, and its intrinsic connections (small square, far upper right) actually cover most of the representation. These anatomical features reflect the gradual erosion of topography, as indicated by increases in classical receptive field size at progressively higher stages of the hierarchy. Feedback pathways, represented in rectangles along the lower right of the matrix, suggest an anatomical basis for the modulatory surround effects that have been demonstrated well outside the classical receptive field (Allman et al., 1985; Knierim & Van Essen, 1992).

The connectivity matrix can also be used to illustrate the actual density of synapses seen at a much finer grain. The inset on the lower right shows an expanded 1-mm segment from layer 4B. We estimate that only a small percentage of the total possible connections are actually present, making this a rather sparsely interconnected network.

Connectivity matrices can represent a wealth of anatomical detail in a way that can be related to information flow within and between areas. This provides a framework for generating models aimed at simulating specific aspects of network processing in the cortex (cf. Wehmeier, Dong, Koch, & Van Essen, 1989; Wilson & Bower, 1989). As more information about the extent and density of specific connections becomes available, the connectivity matrix can be refined to show much greater detail than the first-cut approximations shown here.

Inferotemporal Cortex and Directed Visual Attention

Discrimination and identification of complex patterns (e.g., faces) occurs readily over a wide range of positions, scales, and orientations. The underlying neural mechanisms are not well understood, but there is psychophysical and physiological evidence suggesting that pattern recognition is intimately related to the processes of figure–ground segregation and directed visual attention (Rolls & Baylis, 1986; Baron, 1987; Wise & Desimone, 1988; Van Essen et al., 1991). During attentive scrutiny, we can shift both the location and the spatial scale of the region of the visual field that is attended. Several observations suggest that the effective window of integration extends over approximately 15–30 resolvable steps (sampling nodes), irrespective of retinal position and the spatial scale of the analysis (Virsu & Rovamo, 1979; Jamar & Koenderink, 1983; Toet, van Eekhout, Simons, & Koenderink, 1987; Van Essen et al., 1991).

An attractive model for this process involves dynamic control of information flow into inferotemporal cortex (Anderson & Van Essen, 1987; Baron, 1987; Olshausen et al., 1993). In essence, the idea is that IT receives functionally effective inputs from only a limited portion of the retina, which varies from moment to moment and reflects the size and position of the window of attention. Figure 12 illustrates a modified version of our earlier model for routing of inputs related to visual attention. It involves a connectivity matrix that extends from V1 through V2, V4, and IT. The representation is highly simplified in several respects: it ignores the various compartmental and laminar subregions in V1, V2, and V4; it treats IT as a single area; and it ignores the intrinsic connections within each area.

The representation of the visual field is very sharp in V1, but it becomes eroded by the divergence of feedforward connections at each successive stage. As a consequence, each cell in IT has anatomical inputs that span nearly the entire retina. In keeping with this, receptive fields of IT neurons are known to be very large in anesthetized animals, often covering most of the visual field (Desimone & Gross, 1979). In alert animals, however, receptive fields tend to be restricted to the immediate region to which the animal is attending (Moran & Desimone, 1985) and they can shift with the position (but perhaps not the scale) at which attention is directed (Connor et al., 1993). This suggests that most of the physically available inputs to IT neurons are functionally ineffective in the alert animal, and that the effective inputs can be shifted dynamically in concert with changes in attention. This process is symbolized by the blackened regions within each of the diagonal bands in Figure 12. These represent subsets of enabled connections within the overall matrix which can give rise to a window of attention in a particular part of the visual field. Other connections are presumed to be rendered ineffective through the operation of control circuitry discussed earlier (cf. Figure 10).

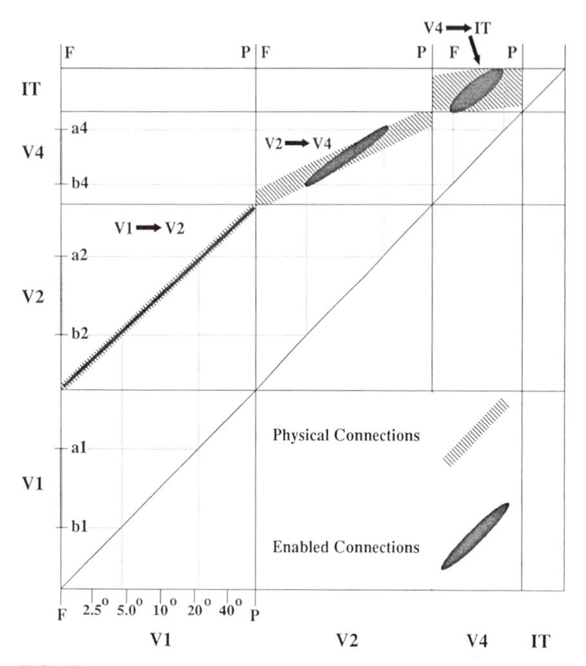

FIGURE 12 A circuit that could mediate directed visual attention, illustrated using a connectivity matrix leading from V1 to IT. For simplicity, only feedforward connections are shown; feedback and intrinsic circuitry are ignored. Also, only one of the numerous visual areas within the IT complex is illustrated, but the scheme could be readily extended to the entire complex. The extent of physical connections subserving each pathway is shown by hatching. Connections that are enabled at a given moment are indicated by dark shading; the rest are presumed to be disabled by shunting inhibition or some other nonlinear mechanism. The fine dotted lines allow one to trace the flow of information from a restricted portion of V1 (between points *a1* and *b1* on the left side of the matrix) to the entirety of IT. This divergence arises mainly from the gating postulated for the V2 to V4 and V4 to IT pathways, but it is useful to track the flow in a bottom-up sequence. Information flow from V1 to V2 involves relatively little divergence. The activated connections shown here occupy a narrow strip along the center of the band of physical connections. Consequently, V1 is mapped onto V2 in an orderly fashion that largely preserves local topographic order. In particular, points *a1* and *b1* map respectively onto *a2* and *b2*. Shifting the band of activated connections would produce a small amount of translation in the mapping, whereas broadening the band would blur the mapping by coupling each neuron in V2 to a wider area of V1. The mapping from V2 to V4 involves a zone of activated connections that is tilted relative to the diagonal formed by the physical connections. This has the consequence of expanding the representation of the region within the window of attention at the expense of the remainder of the representation: segment *a4-b4* occupies a larger fraction of V4 than the *a2-b2* segment does in V2. In going from V4 to IT, the zone of activated connections once again has an oblique orientation, which leads to all of IT being filled by the mapping from segment *a4-b4*. Working backward through the cascade demonstrates that IT as a whole receives an orderly mapping from segment *a1-b1* of V1, corresponding to a restricted window of attention. A similar analysis can be used to show that an individual cell in V4 or IT would have a restricted receptive field that shifts in accordance with changes in the window of attention. The size of the receptive field would reflect the aggregate amount of blurring that arises from the finite width of the activated zone at each level. Narrow zones would preserve positional information, whereas wider zones might achieve a better signal-to-noise by integrating across a larger domain. Neurons in V4 are known to be responsive when the animal is not attending or is attending a target outside the receptive field of the cell (Moran & Desimone, 1985). Presumably, this means that many of the connections serving regions outside the window of attention (i.e., outside the *a4-b4* segment) are activated, rather than suppressed, as shown in the figure for the sake of graphical clarity (see Olshausen et al., 1993). This diagram illustrates only the mapping within one hemisphere, corresponding to one visual hemifield. The scheme can readily be extended to cover both hemispheres and the full visual field. Extensive interhemispheric connections exist in extrastriate visual cortex, particularly at higher levels, which provides the anatomical basis for smooth integration across the entire visual field. Alternative schemes for obtaining coordinate transformations in inferotemporal cortex could be generated using this basic mapping scheme. For example, the representation in IT might be based on image segmentation and object-centered coordinates (cf. Baron, 1987). In that case, the mapping in IT would cover the region of the visual field corresponding to whatever is perceived as the "figure" in tasks of figure-ground segregation (a face, for example), even if the window of attention was involved in scrutiny of only portions of the object (e.g., the eyes and nose).

The way this scheme works can be visualized by tracing the cascade of activity arising from neurons in V1 within the small window from *a1* to *b1* shown on the lower left in Figure 12. This activates a set of neurons in V2 between points *a2* and *b2*, and neurons in V4 between *a4* and *b4*. Finally, this leads to a mapping across all of IT. The window of attention can be changed in position or scale by appropriate alterations in the pattern of effective connectivity at each stage, which would be mediated by top-down control signals rather than directly from ascending sensory inputs. Translations of the bloblike activated domains within the connectivity matrix would lead to shifts in the retinal position of the window; changes in orientation of the domains would modulate the effective size of the window. In the reverse direction, the receptive field of single neurons in IT and V4 can be mapped back to a small region within V1, which would change systematically when the window of attention is shifted.

In this simplified scheme, control of the routing process would require use of inhibitory circuitry to sculpt the position, orientation, and thickness of the activated domains that gate information flow through each pathway. More complex operations can be imagined, however, which could allow these domains to be curved, thereby warping the image representation. There might also be mechanisms for selection on the basis of cues other than position (such as depth or chromatic cues; Motter, 1992). The high-level control signals needed to mediate these processes might originate in the parietal lobe, frontal lobe, superior colliculus, and/or pulvinar, all of which have been implicated in the control of visual attention (Posner & Presti, 1987; Desimone, Wessinger, Thomas, & Schneider, 1990; Anderson & Van Essen, 1993).

This discussion is intended only to outline some of the possibilities for selective routing of information from the retina into the temporal lobe. There are numerous details to be worked out to establish this as a complete model of attentive processing (Olshausen et al., 1993), and many additional experiments are needed to establish whether the cortex indeed operates according to this strategy. There may be interesting generalizations to other sensory modalities, and to cognitive processing and motor control (Van Essen et al., 1994). There are also potential clinical implications, because some neurological and psychiatric disorders (e.g., dyslexia and schizophrenia) conceivably could reflect malfunctions in the circuitry involved in the dynamic control of information flow.

CONCLUDING REMARKS

In this chapter, we have reviewed many types of information about the anatomy and physiology of the primate visual system, and we have tried to place these data into the framework of systems-oriented computational models of visual processing. We believe that this type of interdisciplinary approach is crucial for any deep understanding of biological vision and for constructing successful artificial vision systems. Progress in this endeavor can be facilitated by bearing in mind several general points.

1. *Progressive complexity.* Model building is fundamentally an iterative process in which the models become more elaborate and complex to account more fully for the biological data. One cannot realistically expect full-scale models at the outset, but neither should one remain satisfied with using simple, qualitative schemes to account for the richness of biological systems.
2. *Good engineering.* Models should respect principles of good engineering. A model may be only as strong as its weakest engineering link. In this respect, it is important to specify what aspects of function a model is intended to simulate or represent. For sensory systems it is also important to ascertain how successful the model is in preserving information available in the raw sensory data. Once lost, information cannot be regained, and there should be a clear rationale when such losses occur in a model.
3. *Predictive nature.* Models that make strong, experimentally testable predictions have been relatively rare in neurobiology, which largely accounts for the lack of appreciation of neural modeling approaches in some quarters. There is a clear need for models that are cast in explicitly neurobiological terms so that they can be closely related to experimental data. One of the reasons we find the shifter hypothesis attractive is because of the specific predictions it makes about the dynamic nature of receptive fields. Models of this type can advance our understanding whether or not they eventually turn out to be correct in detail.
4. *Relevance to image processing.* Artificial vision systems must cope with problems that are fundamentally similar to those successfully met by biological vision systems after millions of years of evolution. In connection with this, models of visual processing provide an important two-way link between the fields of

biological vision and computer vision. Models developed in the study of biological vision can guide the fabrication of devices designed to extract useful information from complex natural images (cf. Mahowald & Mead, 1988). Principles derived from computer vision can, in turn, guide the formulation of models that represent specific biological processes. This interplay will in the long run serve as a powerful impetus to progress on both fronts.

Acknowledgments

Work from this laboratory was supported by ONR Contract N00014-89-J-1192 and by the McDonnell Center for Higher Brain Function at Washington University. We thank many of our colleagues for valuable discussions and S. Danker for help in preparation of the manuscript.

References

Adelson, E. H., & Bergen, J. (1985). Spatiotemporal energy models for the perception of motion. *Journal of the Optical Society of America*, **2**, 284–299.

Adelson, E. H., & Bergen, J. R. (1991). The plenoptic function and the elements of early vision. In M. Landy and J. Movshon (Ed.), *Computational models of visual processing*. Cambridge, MA: MIT Press.

Allman, J., Miezin, F., & McGuinness, E. (1985). Stimulus specific responses from beyond the classical receptive field: Neurophysiological mechanisms for local–global comparisons in visual neurons. *Annual Review of Neuroscience*, **8**, 407–430.

Andersen, R. A. (1989). Visual and eye movement functions of the posterior parietal cortex. *Annual Review of Neuroscience*, **12**, 277–403.

Andersen, R. A., Snyder, L. H., Li, C.-S., & Stricanne, B. (1993). Coordinate transformations in the representation of spatial information. *Current Opinion in Neurobiology*, **3**, 171–176.

Anderson, C. H. (1986). The transmission of information in X and Y ganglion cells. *Investigative Ophthalmology & Visual Science, Suppl.*, **27**, 242.

Anderson, C. H., & Van Essen, D. C. (1987). Shifter circuits: A computational strategy for dynamic aspects of visual processing. *Proceedings of the National Academy of Sciences of the United States of America*, **84**, 6297–6301.

Anderson, C. H., & Van Essen, D. C. (1993). Dynamic neural routing circuits. In D. Brogan, A. Gale, & K. Carr (Eds.), *Visual search 2*. London: Taylor & Francis.

Atick, J. J., & Redlich, A. N. (1992). What does the retina know about natural scenes? *Neural Computation*, **4**, 196–210.

Baron, R. J. (1987). *The cerebral computer*. Hillsdale, NJ: Erlbaum.

Benevento, L. A., & Rezak, M. (1976). The cortical projections of the inferior pulvinar and adjacent lateral pulvinar in the rhesus monkey (Macaca mulatta): An autoradiographic study. *Brain Research*, **108**, 1–24.

Blakemore, C., & Vital-Durand, F. (1986). Organization and postnatal development of the monkey's lateral geniculate nucleus. *Journal of Physiology*, **380**, 453–491.

Blasdel, G. G., & Fitzpatrick, D. (1984). Physiological organization of layer 4 in macaque striate cortex. *Journal of Neuroscience*, **4**, 880–895.

Blasdel, G. G., & Lund, L. S. (1983). Termination of afferent axons in macaque striate cortex. *Journal of Neuroscience*, **3**, 1389–1413.

Blasdel, G. G., Lund, L. S., & Fitzpatrick, D. (1985). Intrinsic connections of macaque striate cortex: Axonal projections of cells outside lamina 4C. *Journal of Neuroscience*, **5**, 3350–3369.

Blinkov, S. M., & Glezer, I. I. (1968). *The human brain in figures and tables*. New York: Basic Books.

Braitenberg, V. (1978). Cell assemblies in the cerebral cortex. *Lecture notes in biomathematics*, **21**, 171–188.

Buchsbaum, G. (1987). Color signal coding: Color vision and color television. *Color Research and Application*, **12**, 266–269.

Casagrande, V. A., and Lachica, E. A. (1992). What are the cytochrome (CO) blobs and interblobs really segregating? *Investigative Ophthalmology & Visual Science Supplement*, **33**, 900.

Casagrande, V. A., & Norton, T. T. (1991). Lateral geniculate nucleus: A review of its physiology and function. In A. G. Leventhal (Ed.), *Vision and visual dysfunction, Vol. 4, The neural basis of visual function*. New York: MacMillan Press.

Conley, M., & Fitzpatrick, D. (1989). Morphology of retinogeniculate axons in the macaque. *Visual Neuroscience*, **2**, 287–296.

Connolly, M., & Van Essen, D. (1984). The representation of the visual field in parvicellular and magnocellular layers of the lateral geniculate nucleus in the macaque monkey. *Journal of Comparative Neurology*, **226**, 544–564.

Connor, C. E., Gallant, J. L., & Van Essen, D. C. (1993). Effects of focal attention on receptive field profiles in area V4. *Society for Neuroscience Abstracts*. 19:974.

Crick, F. (1984). Function of the thalamic reticular complex: The searchlight hypothesis. *Proceedings of the National Academy of Sciences (USA)*, **81**, 4586–4590.

Crook, J. M., Lange-Malecki, B., Lee, B. B., & Valberg, A. (1988). Visual resolution of macaque retinal ganglion cells. *Journal of Physiology*, **396**, 205–224.

Dacey, D. M., & Petersen, M. R. (1992). Dendritic field size and morphology of midget and parasol ganglion cells of the human retina. *Proceedings of the National Academy of Sciences of the United States of America*, **89**, 9666–9670.

Daugman, J. G. (1985). Uncertainty relation for resolution in space, spatial frequency, and orientation optimized by two-dimensional visual cortical filters. *Journal of the Optical Society of America*, **2**, 1160–1169.

De Monasterio, F. M. (1978). Properties of concentrically organized X and Y ganglion cells of macaque retina. *Journal of Neurophysiology*, **41**, 1394–1417.

De Monasterio, F. M., & Gouras, P. (1975). Functional properties of ganglion cells of the rhesus monkey retina. *Journal of Physiology*, **251**, 167–195.

De Monasterio, F. M., McCrane, E. P., Newlander, J. K., & Schein, S. I. (1985). Density profile of blue-sensitive cones along the horizontal meridian of macaque retina. *Investigative Ophthalmology & Visual Science*, **26**, 289–302.

Derrington, A. M., & Lennie, P. (1984). Spatial and temporal contrast sensitivities of neurones in lateral geniculate nucleus of macaque. *Journal of Physiology*, **357**, 219–240.

Desimone, R., & Gross, C. G. (1979). Visual areas in the temporal cortex of the macaque. *Brain Research*, **178**, 363–380.

Desimone, R., Wessinger, M., Thomas, L., & Schneider, W., (1990). Attentional control of visual perception: Cortical and subcortical mechanisms. *Cold Spring Harbor Symposium on Quantitative Biology*, **40**, 963–971.

DeValois, R. L., Albrecht, D. G., & Thorell, L. G. (1982). Spatial frequency selectivity of cells in macaque visual cortex. *Vision Research*, **22**, 545–559.

DeValois, R. L., & DeValois, K. K. (1993). A multistage color model. *Vision Research*, **33**, 1053–1065.

DeYoe, E. A., & Van Essen, D. C. (1988). Concurrent processing streams in monkey visual cortex. *Trends in NeuroSciences*, **11**, 219–226.

Dobbins, A., Zucker, S. W., & Cynader, M. S. (1987). End-stopped neurons in the visual cortex as a substrate for calculating curvature. *Nature*, **329**, 438–441.

Dobkins, K. R., & Albright, T. D. (1993). What happens if it changes color when it moves?: Psychophysical experiments on the nature of chromatic input to motion detectors. *Vision Research*, **33**, 1019–1036.

Dowling, I. E. (1987). *The retina: An approachable part of the brain*. Cambridge, MA: Belknap.

Felleman, D. I., & Van Essen, D. C. (1991). Distributed hierarchical processing in the primate visual cortex. *Cerebral Cortex*, **1**, 1–47.

Ferster, D., & Jagadeesh, B. (1992). EPSP-IPSP interactions in cat visual cortex studied with *in vivo* whole-cell patch recording. *Journal of Neuroscience*, **12**, 1262–1274.

Field, D. J. (1987). Relations between the statistic of natural images and the response properties of cortical cells. *Journal of the Optical Society of America A*, **4**, 2379–2394.

Field, D. J. (1992). Scale-invariance and self-similar "wavelet" transforms: An analysis of natural scenes and mammalian visual systems. In M. Farge, J. Hunt, & J. C. Vassiclicos (Eds.), *Wavelets, fractals and Fourier transforms: New developments and new applications*. Oxford: Oxford University Press.

Fitzpatrick, D., Lund, I. S., & Blasdel, G. G. (1985). Intrinsic connections of macaque striate cortex: Afferent and efferent connections of lamina 4C. *Journal of Neuroscience*, **5**, 3329–3349.

Fukuda, Y., Sawai, H., Watanabe, M., Wakakuwa, K., & Morigiwa, K. (1989). Nasotemporal overlap of crossed and uncrossed retinal ganglion cell projections in the Japanese monkey (*Macaca fuscata*). *The Journal of Neuroscience*, **9**, 2353–2373.

Gallant, J. L., Braun, J., and Van Essen, D. C. (1993). Selectivity for polar, hyperbolic, and cartesian gratings in macaque visual cortex. *Science*, **259**, 100–103.

Gilbert, C. D. (1977). Laminar differences in receptive field properties of cells in cat primary visual cortex. *Journal of Physiology*, **268**, 391–421.

Grünert, U., Greferath, U., Boycott, B. B., & Wässle, H. (1993). Parasol (Pα) ganglion-cells of the primate fovea: Immunocytochemical staining with antibodies against $GABA_A$-receptors. *Vision Research*, **33**, 1–14.

Gur, M., & Snodderly, D. M. (1987). Studying striate cortex neurons in behaving monkeys: Benefits of image stabilization. *Vision Research*, **27**, 2081–2087.

Hicks, T. P., Lee, B. B., & Vidyasagar, T. R. (1983). The responses of cells in macaque lateral geniculate nucleus to sinusoidal gratings. *Journal of Physiology*, **337**, 183–200.

Hopfield, I. I. (1984). Neurons with graded response have collective computational properties like those of two-state neurons. *Proceedings of National Academy of Sciences of the United States of America*, **81**, 3088–3092.

Hubel, D. H., & Wiesel, T. N. (1962). Receptive fields, binocular interaction and functional architecture in the cat's visual cortex. *Journal of Physiology*, **160**, 106–154.

Hubel, D. H., & Wiesel, T. N. (1968). Receptive fields and functional architecture of monkey striate cortex. *Journal of Physiology*, **195**, 215–243.

Hubel, D. H., & Wiesel, T. N. (1974). Uniformity of monkey striate cortex: A parallel relationship between field size, scatter, and magnification factor. *Journal of Comparative Neurology*, **158**, 295–306.

Hubel, D. H., & Wiesel, T. N. (1977). Functional architect macaque monkey visual cortex. *Proceedings of the Royal Society of London*, **198**, 1–59.

Jamar, J. H. T., & Koenderink, J. J. (1983). Sine-wave gratings. Scale invariance and spatial integration at suprathreshold ct,q contrast. *Vision Research*, **23**, 805–810.

Jay, M. F., & Sparks, D. L. (1984). Auditory receptive fields in primate superior colliculus shift with changes in eye position. *Nature*, **309**, 345–347.

Jones, J. P., & Palmer, L. A. (1987). An evaluation of the two dimensional Gabor filter model of simple receptive fields in cs striate cortex. *Journal of Neurophysiology*, **58**, 1233–1258.

Julesz, B. (1984). Towards an axiomatic theory of preattentive vision. In G. M. Edelman, W. E. Gall, & W. M. Cowan (Eds.), *Dynamic aspects of neocortical function*. New York: Wiley.

Kaplan, E., Purpura, K., & Shapley, R. M. (1987). Contrast affects the transmission of visual information through the mammalian lateral geniculate nucleus. *Journal of Physiology*, **39**, 267–288.

Kaplan, E., & Shapley, R. M. (1982). X and Y cells in the lateral geniculate nucleus of macaque monkeys. *Journal of Physiology*, **330**, 125–143.

Knierim, J. J., & Van Essen, D. C. (1992). Visual cortex: Cartography, connections, and concurrent processing. *Current Opinion in Neurobiology*, **2**, 150–155.

Koch, C. (1987). The action of the corticofugal pathway on sensory thalamic nuclei: An hypothesis. *Neuroscience*, **23**, 399–406.

Koch, C., & Poggio, T. (1992). Multiplying with synapses and neurons. In *Single neuron computation neural nets: Foundations to applications* (pp. 315–345). New York: Academic Press.

Koch, C., Poggio, T., & Torre, V. (1983). Nonlinear interaction in a dendritic tree: Localization, timing and role in information processing. *Proceedings of the National Academy of Sciences of the United States of America*, **80**, 2799–2802.

Koenderink, J. J., & van Doorn, A. I. (1978). Visual detection of spatial contrast; influence of location in the visual field, target~ extent and illuminance level. *Biological Cybcrnetics*, **30**, 157–167.

Kosslyn, S. M. (1988). Aspects of a cognitive neuroscience of mental imagery. *Science*, **240**, 1621–1626.

Kruger, J. (1977). The shift-effect in the lateral geniculate body of the rhesus monkey. *Experimental Brain Research*, **29**, 387–391.

Lachica, E. A., Beck, P. D., & Casagrande, V. A. (1992). Parallel pathways in macaque monkey striate cortex: Anatomically defined columns in layer III. *Proceedings of the National Academy of Science of the United States of America*, **89**, 3566–3570.

Laughlin, S. M. (1987). Form and function in retinal processing. *Trends in NeuroSciences*, **10**, 478–483.

Lee, B. B., Martin, P. R., & Valberg, A. (1989). Sensitivity of macaque ganglion cells to luminance and chromatic flicker. *Journal of Physiology*, **414**, 223–243.

Lee, B. B., Virsu, V., & Creutzfeldt, O. D. (1983). Linear signal: Transmission from prepotentials to cells in the macaque lateral geniculate nucleus. *Experimental Brain Research*, **52**, 50–56.

Livingstone, M. S., & Hubel, D. H. (1984). Anatomy and physiology of a color system in the primate visual cortex. *Journal of Neuroscience*, **4**, 309–356.

Livingstone, M. S., & Hubel, D. H. (1988a). Do the relative mapping densities of the magno- and parvocellular systems vary with eccentricity? *Journal of Neuroscience*, **8**, 4334–4339.

Livingstone, M. S., & Hubel, D. H. (1988b). Segregation of form, color, movement, and depth: Anatomy, physiology, and perception. *Science*, **240**, 740–749.

Lund, J. S. (1988). Excitatory and inhibitory circuitry and laminar mapping strategies in primary visual cortex of the monkey. In G. M. Edelman, W. E. Gall, & W. M. Cowan (Eds.), *Signal and sense: Local and global order in perceptual maps*. New York: Wiley.

Mahowald, M., & Mead, C. A. (1988). A silicon model of early visual processing. *Neural Networks*, **1**, 91–97.

Marrocco, R. T., McClurkin, J. W., & Young, R. A. (1982). Spatial summation and conduction latency classification of cells of the lateral geniculate nucleus of macaques. *Journal of Neuroscience*, **2**, 1275–1291.

Merigan, W. H., & Maunsell, J. H. R. (1993). How parallel are the primate visual pathways? *Annual Review of Neuroscience*, **16**, 369–402.

Michael, C. R. (1985). Laminar segregation of color cells in the monkey's striate cortex. *Vision Research*, **25**, 415–423.

Miyashita, M. (1993). Inferior temporal cortex: Where visual perception meets memory. *Annual Review of Neuroscience*, **16**, 245–264.

Moran, J., & Desimone, R. (1985). Selective attention gates visual processing in the extrastriate cortex. *Science*, **229**, 782–784.

Motter, B. C. (1992). Selective activation of V4 neurons in a color and orientation discrimination task. *Investigative Ophthalmology & Visual Science, Supplement*, **33**, 1131.

Motter, B. C., & Poggio, G. F. (1982). Spatial invariance of receptive field location in the presence of eye movements of fixation for neurons in monkey striate cortex. *Society for Neuroscience Abstracts*, **8**, 707.

Motter, B. C., & Poggio, G. F. (1990). Dynamic stabilization of receptive fields of cortical neurons (VI) during fixation of gaze in the macaque. *Experimental Brain Research*, **83**, 37–43.

Mullen, K. T. (1985). The contrast sensitivity of human color vision to red–green and blue–yellow chromatic gratings. *Journal of Physiology (London)*, **359**, 381–409.

Nealey, T. A., & Maunsell, J. H. R. (1994). Magnocellular and parvocellular contributions to the responses of neurons in macaque striate cortex. *Journal of Neuroscience*, **14**, 2069–2079.

Noorlander, C., & Koenderink, J. J. (1983). Sensitivity to spatial-temporal color contrast in the periphery visual field. *Vision Research*, **23**, 1–11.

Ogren, M. P., & Hendrickson, A. E. (1977). The distribution of pulvinar terminals in visual areas 17 and 18 of the monkey. *Brain Research*, **137**, 343–350.

Olshausen, B. A., Anderson, C. H., & Van Essen, D. C. (1993). A neurobiological model of visual attention and invariant pattern recognition based on dynamic routing of information. *Journal of Neuroscience*, **13**, 4700–4719.

O'Kusky, J., & Colonnier, M. (1982). A laminar analysis of the number of neurons, glia, and synapses in the visual cortex (area 17) of adult macaque monkeys. *Journal of Comparative Neurology*, **210**, 178–290.

Østerberg, G. (1935). Topography of the layer of rods and cones in the human retina. *Acta Ophthalmologica*, **65**, 1–102.

Peichl, L., & Wässle, H. (1979). Size, scatter and coverage of ganglion cell receptive field centers in the cat retina. *Journal of Physiology*, **291**, 117–141.

Perrett, D. I., Mistlin, A. J., & Chitty, A. J. (1987). Visual neurones responsive to faces. *Trends in NeuroSciences*, **10**, 358–364.

Perry, V. H., & Cowey, A. (1985). The ganglion cell and cone distributions in the monkey's retina: Implications for central magnification factors. *Vision Research*, **12**, 1795–1810.

Perry, V. H., Oehler, R., & Cowey, A. (1984). Retinal ganglion cells that project to the dorsal lateral geniculate nucleus in the macaque monkey. *Neuroscience*, **12**, 1101–1123.

Peterhans, E., and von der Heydt, R. (1993). Functional organization of area V2 in the alert macaque. *European Journal of Neuroscience*, **5**, 509–524.

Poggio, G. F. (1984). Processing of stereoscopic information in primate visual cortex. In G. Edelman, W. E. Gall, & W. M. Cowan (Eds.), *Dynamic aspects of neocortical function*. New York: Wiley.

Pollen, D. A., Goska, J. P., & Jacobsen, L. D. (1988). Responses of simple and complex cells to compound sine-wave gratings. *Vision Research*, **28**, 25–39.

Posner, M. I., & Presti, D. E. (1987). Selective attention and cognitive control. *Trends in NeuroSciences*, **10**, 13–17.

Purpura, K., Kaplan, E., & Shapley, R. M. (1988). Background light and the contrast gain of primate P and retinal ganglion cells. *Proceedings of the National Academy of Sciences of the United States of America*, **85**, 4534–4537.

Reid, R. C., & Shapley, R. M. (1992). Spatial structure of cone inputs to receptive fields in primate lateral geniculate nucleus. *Nature*, **356**, 716–718.

Rockland, K. S., & Pandya, D. N. (1979). Laminar origins and termination of cortical connections of the occipital lobe in the rhesus monkey. *Brain Research*, **179**, 3–20.

Rodieck, R. W. (1988). The primate retina. In H. D. Steklis & J. Erwin (Eds.), *Comparative primate biology*, Vol. 4 (pp. 203–278). New York: Liss.

Rolls, E. T., & Baylis, G. C. (1986). Size and contrast have only small effects on the responses to faces of neurons in the cortex in the superior temporal sulcus. *Experimental Brain Research*, **65**, 38–48.

Rovamo, J., & Virsu, V. (1979). An estimation and application of the human cortical magnification factor. *Experimental Brain Research*, **37**, 495–510.

Schein, S. J. (1988). Anatomy of macaque fovea and spatial densities of neurons in foveal representation. *Journal of Comparative Neurology*, **269**, 479–505.

Schein, S. J., & De Monasterio, F. M. (1987). Mapping of retinal and geniculate neurons onto striate cortex of macaque. *Journal of Neuroscience*, **7**, 996–1009.

Schneider, W. (1985). Toward a model of attention and the development of automatic processing. In M. I. Posner & O. S. M. Marin (Eds.), *Attention and performance* (Vol. XI). Hillsdale, NJ: Erlbaum.

Schwartz, E. L. (1980). Computational anatomy and functional architecture of striate cortex: A spatial mapping approach to perceptual coding. *Vision Research*, **20**, 645–69.

Shapley, R., & Perry, V. H. (1986). Cat and monkey retinal ganglion cells and their visual functional roles. *Trends in NeuroSciences*, **9**, 229.

Sherman, S. M., & Koch, C. (1986). The control of retinogeniculate transmission in the mammalian lateral geniculate nucleus. *Experimental Brain Research*, **63**, 1–20.

Silveira, L. C. L., & Perry, V. H. (1991). The topography of magnocellular projecting ganglion cells (m-ganglion cells) in the primate retina. *Neuroscience*, **40**, 217–237.

Snyder, A. W., & Miller, W. H. (1977). Photoreceptor diameter and spacing for highest resolving power. *Journal of the Optical Society of America*, **67**, 696–698.

Somogyi, P., Freund, T. F., & Cowey, A. (1982). The axo-axonic interneuron in the cerebral cortex of the rat, cat and monkey. *Neuroscience*, **7**, 2577–2607.

Toet, A., van Eekhout, M. P., Simons, H. L., & Koenderink, J. J. (1987). Scale invariant features of differential spatial displacement discrimination. *Vision Research*, **27**, 441–451.

Tootell, R. B. H., Silverman, M. S., Hamilton, S. L., Switkes,E., & DeValois, R. L. (1988a). Functional anatomy of macaque striate cortex: V. Spatial frequency. *Journal of Neuroscience*, **8**, 1610–1624.

Tootell, R. B. H., Switkes, E., Silverman, M. S., & Hamilton, S. L. (1988b). Functional anatomy of macaque striate cortex: II. Retinotopic organization. *Journal of Neuroscience*, **8**, 1531–1568.

Ts'o, D. Y., & Gilbert, C. D. (1988). The organization of chromatic and spatial interactions in the primate striate cortex. *Journal of Neuroscience*, **8**, 1712–1727.

Ungerleider, L. G., & Mishkin, M. (1982). Two cortical visual systems. In D. J. Ingle, M. A. Goodale, & R. J. W. Mansfield (Eds.), *Analysis of visual behavior*. Cambridge, MA: MIT Press.

van Doorn, A. J., Koenderink, J. J., & Bouman, M. A. (1972). The influence of the retinal inhomogeneity on the perception of spatial patterns. *Kybernetic*, **10**, 223–230.

Van Essen, D. C., & Anderson, C. H. (1986). Sampling of the visual image in the retinal and LGN of the macaque: A quantitative model. *Investigative Ophthalmology & Visual Science Supplement*, **27**, 94.

Van Essen, D. C., & Anderson, C. H. (1990). Reference frames and dynamic remapping processes in vision. In E. Schwartz (Ed.), *Computational neuroscience*, (pp. 278–294). MIT Press, Cambridge, MA, pp. 278–294.

Van Essen, D. C., Anderson, C. H., & Felleman, D. J. (1992). Information processing in the primate visual system: An integrated systems perspective. *Science*, **255**, 419–423.

Van Essen, D. C., & DeYoe, E. A. (1993). Concurrent processing in the primate visual cortex. In M. S. Gazzaniga (Ed.), *The Cognitive Neurosciences*. Cambridge, MA: MIT Press.

Van Essen, D. C., DeYoe, E. A., Olavarria, J. F., Knierim, J. J., Sagi, D., Fox, J. M., & Julesz, B. (1989). Neural responses to static and moving texture patterns in visual cortex of the macaque monkey. In D. M. V. Lam & C. Gilbert (Eds.), *Neural mechanisms of visual perception* (pp. 135–153). Woodlands, TX: Portfolio Publishing.

Van Essen, D. C., & Gallant, J. L. (1994). Neural mechanisms of form and motion processing in the primate visual system. *Neuron*, **13**, 1–10.

Van Essen, D. C., & Maunsell, J. H. R. (1983). Hierarchical organization and functional streams in the visual cortex. *Trends in NeuroSciences*, **6**, 370–375.

Van Essen, D. C., Newsome, W. T., & Maunsell, J. H. R. (1984). The visual field representation in striate cortex of the macaque monkey: Asymmetries, anisotropies, and individual variability. *Vision Research*, **24**, 429–448.

Van Essen, D. C., Olshausen, B., Anderson, C. H., & Gallant, J. L. (1991). Pattern recognition, attention, and information bottlenecks in the primate visual system. In *Proceedings of SPIE Conference on Visual Information Processing: From Neurons to Chips*, **1473**, 17–28.

Van Essen, D. C., Olshausen, B., Gallant, J., Press, W., Anderson, C., Drury, H., Carman, G., & Felleman, D. (1994). Anatomical, physiological, and computational aspects of hierarchical processing in the macaque visual cortex. In C. Nothdurft (Ed.), *Structural and functional organization of the neocortex*.

Virsu, V., & Rovamo, J. (1979). Visual resolution, contrast sensitivity, and the cortical magnification factor. *Experimental Brain Research*, **37**, 475–494.

Wässle, H., & Boycott, B. B. (1991). Functional architecture of the mammalian retina. *Physiological Reviews*, **71**, 447–480.

Wässle, H., Grunert, U., Rohrenbeck, J., and Boycott, B. (1990). Retinal ganglion cell density and cortical magnification factor in the primate retina. *Vision Research*, **30**, 1897–1911.

Watanabe, M., & Rodieck, R. W. (1989). Parasol and midget ganglion cells of the primate retina. *Journal of Comparative Neurology*, **289**, 434–454.

Wehmeier, U., Dong, D., Koch, C., & Van Essen, D. C. (1989). Modeling the mammalian visual system. In C. Koch & I. Segev (Eds.), *Methods in neuronal modeling: From synapses to net works*. Cambridge, MA: MIT Press.

Westheimer, G. (1979). Scaling of visual acuity measurement. *Archives of Ophthalmology*, **97**, 327–330.

Williams, D. R. (1988). Topography of the foveal cone mosaic in the living human eye. *Vision Research*, **28**, 433–454.

Williams, D. R., & Collier, R. (1983). Consequences of spatial sampling by a human photoreceptor mosaic. *Science*, **221**, 385–387.

Wilson, M. A., & Bower, J. M. (1989). The simulation of large-scale neural networks. In C. Koch & I. Segev (Eds.), *Methods in neuronal modeling: From synapses to network*. Cambridge, MA: MIT Press.

Wise, S. P., & Desimone, R. (1988). Behavioral neurophysiology: Insights into seeing and grasping. *Science*, **242**, 736–741.

Wolbarsht, M. L., Wagner, H. G., & Ringo, J. L. (1985). Retinal mechanisms for improving visual acuity. In A. Fein & J. S. Levine (Eds.), *The visual system*. New York: Liss.

Zeki, S. M. (1978). Functional specialisation in the visual cortex of the rhesus monkey. *Nature*, **274**, 423.

Zeki, S., & Shipp, S. (1988). The functional logic of cortical connections. *Nature*, **335**, 311–317.

4

A Neurocomputational Theory of Hippocampal Function in Stimulus Representation and Learning

Mark A. Gluck and Catherine E. Myers

INTRODUCTION

Computational analyses of parallel-processing networks are an increasingly useful tool for understanding the neural bases of learning and memory. In particular, they provide an essential framework for exploring the behavioral implications of network-level characterizations of synaptic change in specific brain regions. Recently, such network models have begun to provide important insights into the functional role of the hippocampal region (see Gluck & Granger, 1993, for a review).

The hippocampus and adjacent cortical regions in the medial temporal lobe have long been implicated in human learning and memory via lesion data in humans (Scoville & Milner, 1957; Squire, 1982) and animals (Mishkin, 1982; Squire & Zola-Morgan, 1983). Nevertheless, there has been little consensus as to the precise specification of this brain region's functional role in learning and memory. Some hypotheses of hippocampal-region function have emphasized the critical role of this brain region in specific memorial tasks. Studies of lower (nonprimate) mammals have focused on place learning and spatial navigation as tasks that require an intact hippocampal region (McNaughton & Nadel, 1990; Morris, Garrud, Rawlins, & O'Keefe, 1982; O'Keefe & Nadel, 1978). Within cognitive neuroscience, hippocampal research has focused on explicit declarative memories in humans (Squire, 1987; Cohen & Squire, 1980). As distinct from procedural or implicit memories, these hippocampal-dependent memories are marked by their accessibility to conscious recollection.

Another approach to developing functional theories of hippocampal processing has been to postulate some underlying information-processing role(s) for the hippocampal region and then seek to derive a wider range of task-specific deficits. The goal here is to develop a comprehensive view of hippocampal-region function that illuminates common behavioral functions, and underlying neural mechanisms, in both animal and human memory behaviors. Such a broad characterization has proven elusive, yet some initial progress has been made in modeling the role of the hippocampus in certain associative learning paradigms. Here, we review some of our own efforts within this approach.

We have proposed a computational theory of the hippocampal region's function in mediating stimulus representation during learning (Gluck & Myers, 1993). New stimulus representations constructed in the hippocampal region are assumed to be biased by two constraints: first, to differentiate representations of stimuli that predict different future events, and second, to compress the representations of redundant cooccurring stimuli. Other brain regions, including cerebral and cerebellar cortices, are presumed to use these hippocampal representations to recode their own stimulus representations. In the absence of an intact hippocampal region, these other brain regions will attempt to learn associations using previously established fixed representations. This account has been instantiated as a connectionist network model that accurately predicts the behavior of intact and hippocampal-lesioned animals in a range of conditioning paradigms, including stimulus discrimination, reversal learning, stimulus generalization, latent inhibition, sensory preconditioning, and contextual sensitivity (Gluck & Myers, 1993; Myers & Gluck, in press).

We have also discussed how this proposed hippocampal-region role might be mapped onto anatomical structures within the region. The two representational biases described earlier can be dissociated, and may have distinct anatomical loci: We have hypothesized that stimulus compression may arise from the operation of circuitry in the superficial layers of entorhinal cortex, whereas stimulus differentiation may arise from the operation of constituent circuits of the hippocampal formation (Myers & Gluck, 1994; Myers, Gluck, & Granger, submitted). To the extent that relevant lesion data exist for simple associative learning tasks, our hypothesis is consistent with comparisons between selective lesions of the hippocampal formation and broader lesions that include damage to entorhinal cortex.

In this chapter, we describe in more detail our basic theory of hippocampal-region function in stimulus representation and then show how this representational function suggests a finer-grained functional division between the information processing roles of the hippocampus and the entorhinal cortex.

STIMULUS REPRESENTATION AND HIPPOCAMPAL FUNCTION

Two broad classes of hippocampal functions have been considered when theorists seek to identify an underlying information processing role for the hippocampal region. The first class includes temporal processing deficits that occur after hippocampal lesions; these include impairments with sequence learning (e.g., Buszaki, 1989), imprecise response timing (Akase, Alkon, & Disterhoft, 1989; Moyer, Deyo, & Disterhoft, 1990), and failure at intermediate term or working memory tasks (Rawlins, 1985; Olton, 1983), such as delayed nonmatching to sample (e.g., Zola-Morgan, Squire, & Amaral, 1989). A second class of hippocampal-dependent behaviors appear to concern aspects of stimulus representation; these focus on hippocampal-lesion deficits in learning complex tasks or in the flexible use of learned information in novel situations (Eichenbaum & Buckingham, 1990; Eichenbaum, Cohen, Otto, & Wible, 1992).

We have presented a computational theory of hippocampal function in associative learning in which the hippocampal region is presumed to play an information processing role in the representation of stimulus information (Gluck & Myers, 1993). Central to this account of hippocampal-region function is the idea of a *stimulus representation*, the pattern of activity evoked by a stimulus input. This representation is presumed to be distributed over many elements, which could be neurons in a brain or nodes in a connectionist network. Within this conceptual framework, learning about various stimuli is equivalent to associating the representations they activate with the appropriate behavioral outputs. In this way, learning about one particular stimulus will generalize to other stimuli as a function of how similar (or overlapping) are the different stimulus representations. If the representations of two stimuli are very similar, then associations that have accrued to one stimulus will tend to generalize to the other. If the representations of two stimuli are dissimilar, learning about one will generalize only minimally to the other.

From this perspective we can see how learning can be facilitated when internal representations are biased by several constraints. The first constraint, *predictive differentiation*, is a bias to maximally differentiate the representations of stimuli which should have minimal generalization. For example, if stimulus A reliably predicts reinforcement, whereas another stimulus, B, does not, the representations of A and B should be highly differentiated. This differentiation will ensure that what is learned about A will not generalize to—or interfere with—learning about B. Conversely, the representations of stimuli which predict similar outcomes should be compressed, or made more similar, to increase generalization between them. Another constraint, *stimulus–stimulus redundancy compression*, is a bias to compress, or make more similar, the representations of stimuli which should have maximal generalization. For example, if stimuli A and B reliably co-occur, they will logically tend to make the same predictions about future reinforcement. Through redundancy compression, their representations should become similar. The effect of this redundancy compression is that what is learned about A will tend to generalize strongly to B. Conversely, stimulus–stimulus differentiation should occur if stimuli never co-occur.

We have previously proposed that the hippocampal region has the ability to construct new stimulus representations biased by these constraints (Gluck & Myers, 1993). Other regions in the cerebral and cerebellar cortices, which are presumed to be the sites of long-term memory, may not be able to form these new representations themselves; these other regions can, however, adopt the new representations formed in the hippocampal region. The idea that the hippocampal region constructs new representations to facilitate learning is broadly consistent with a more qualitative characterization of the hippocampal-lesioned animal as being unable to flexibly apply representations to novel situations (Eichenbaum & Buckingham, 1990; Eichenbaum, Cohen, Otto, & Wible, 1992).

This theory can be instantiated, and tested, with a simple connectionist network model (Gluck & Myers, 1993). This hippocampal model, on the right in Figure 1A, learns to map from inputs representing the stimulus to outputs which reproduce that stimulus, plus a prediction of future reinforcement. This network also contains an internal layer of nodes, and these nodes develop an internal stimulus representation that is constrained by the biases to differentiate predictive information and compress redundant information. A second network, on the left in Figure 1A, represents a highly simplified model of some aspects of long-term memory in cerebral and cerebellar cortices (see Gluck, Myers & Thompson, this volume; Gluck, Goren, Myers, & Thompson, 1994). This network is not capable—on its own—of forming new stimulus representations at its internal layer. However, it is assumed to be able to acquire the representations formed in the hippocampal model. This cortical long-term memory network can, thereby, learn to map from these representations to an observable behavioral output response.

A hippocampal lesion is simulated in this model by disabling the hippocampal-region model (Figure 1B). The long-term memory network is left intact, but is no longer able to acquire new stimulus representations in its internal layer. Thus, it is left with the ability to do simple stimulus–response learning, based on the prior, and now permanent, static stimulus representations. These models of learning, with and without the hippocampal region, accurately predict a wide range of learning and generalization behaviors in which normal intact animals have been compared with hippocampal-lesioned animals (Gluck & Myers, 1993). A few of these results are presented here.

Successive Discrimination

Predictive differentiation is presumed to occur whenever two stimuli differentially predict reinforcement. The simplest task in which this occurs is *successive discrimination*. Two stimuli are presented successively; one (A+) reliably precedes some reinforcing event, whereas the other (B−) does not. Because this task does not require the construction of new stimulus representations, both the intact and lesioned models can learn the desired response of responding to A but

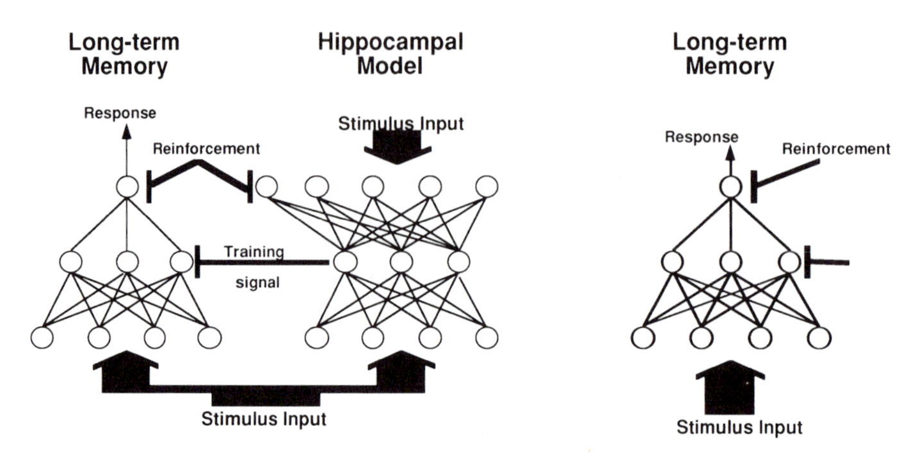

FIGURE 1 A. A model of the proposed hippocampal-region role in the formation of new stimulus representations (Gluck & Myers, 1993). Stimulus input is presented to a network representing a simplification of cerebellar substrates of conditioned learning; this network generates an output interpreted as the behavioral response. Simultaneously, the input is processed by a network modeling the ability of the hippocampal region to construct new stimulus representations. These representations are biased by predictive differentiation and redundancy compression, and may be acquired by the cerebellar network, which cannot otherwise develop new representations. B. Hippocampal lesion is simulated in this model by disabling the hippocampal network. The remaining cerebellar network is unable to form new stimulus representations, but can still learn simple associations based on its existing representations. Figure reprinted from Myers & Gluck (1994/in press).

not to B (as shown in Figure 2A). In fact, the lesioned model even shows a slight facilitation relative to the intact system. The intact model learns more slowly because it is learning more: It is not only mapping stimuli to responses, it is also constructing new, differentiated representations for those stimuli. Likewise, a facilitation in successive discrimination learning has often been observed in hippocampal-lesioned animals (Eichenbaum, Otto, Wible, & Piper, 1991; Eichenbaum, Fagan, Mathews, & Cohen, 1988; Eichenbaum, Fagan, & Cohen, 1986; Port, Mikhail, & Patterson, 1985; Schmaltz & Theios, 1972).

Reversal Facilitation

The intact model learns the successive discrimination slower than the lesioned model because the intact model is learning more: The intact model constructs new stimulus representations for A and B which minimize generalization interference between them. The utility of these new representations can be seen during a transfer task such as a reversal, in which, after training on A+, B−, the valences of the cues are suddenly reversed to A−, B+. This requires associating a new outcome with each stimulus. However, as the relevant stimuli are the same in both tasks, the initial differentiated representations will remain appropriate. This is expected to facilitate learning the reversal task, compared with learning the original discrimination; Figure 2B shows that in the intact model, the reversal is learned at least as quickly as the original task.

In the lesioned model, however, there is no new stimulus representation formed during the original task. The only learning that occurs is to associate the

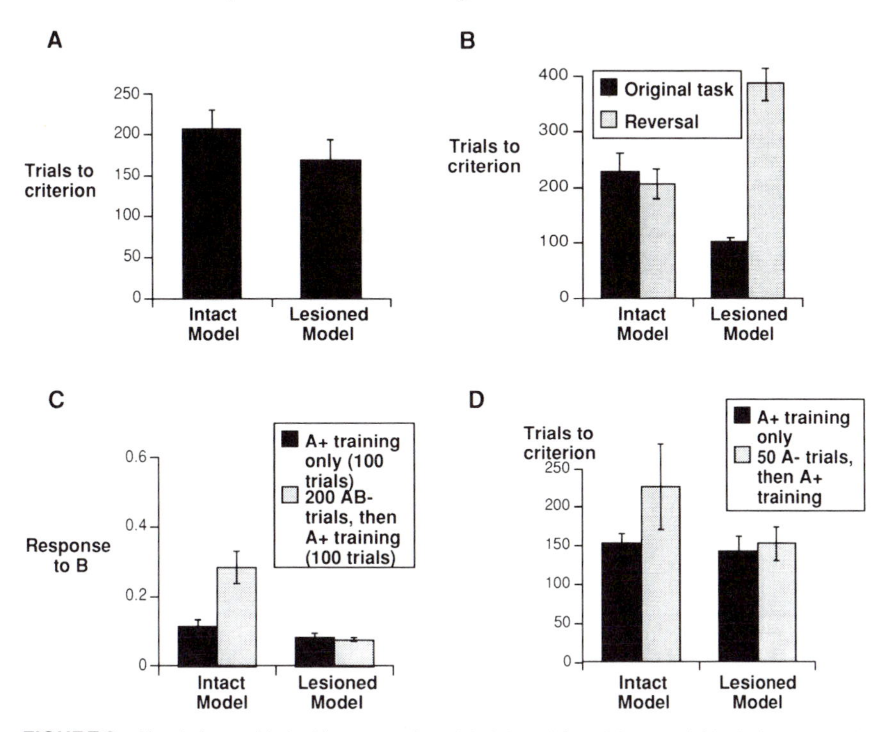

FIGURE 2 Simulations with the hippocampal model. Adapted from Myers and Gluck (in press) and Myers, Gluck, and Granger (submitted). All data shown are consistent with empirical data. A. *Successive discrimination*, alternating reinforced presentations of A and unreinforced presentations of B; system should learn to respond to A but not to B. No re-representations are needed, so both intact and lesioned models learn at similar rates; lesioned model shows slight but significant facilitation. B. *Reversal facilitation*. In the intact model, stimulus representations formed during the original (A+, B−) task are still appropriate for the reversed (A−, B+) task. Learning the reversal may therefore be facilitated. The lesioned model cannot construct new representations; all information is stored in the associations between stimulus and response. This information must be completely unlearned before the reversal can be learned, and so learning may be slowed relative to the original task. C. *Sensory preconditioning* similarly involves unreinforced preexposure, this time to a compound of two stimuli, AB. During this phase, the representations of A and B are compressed in the intact model. In a subsequent phase of learning to respond to A, some of this response will transfer to B, resulting in a greater response to B than if no preexposure had occurred. This effect is not seen in the lesioned model. D. *Latent inhibition*, whereby early unreinforced preexposure to a stimulus A retards later learning to respond to that stimulus. In the intact model, redundancy compression of the representations of A and the context occurs during the preexposure trials; this must be undone before a response to A can be learned. A system with no such preexposure will therefore learn faster. No latent inhibition is observed in the lesioned model.

existing representations with the correct responses. When the reversal takes effect, this learning must be completely undone to construct new, opposite associations. This may slow reversal learning dramatically, relative to the original discrimination. Figure 2B shows this retardation of reversal learning in the lesioned model. Likewise, hippocampal lesions have often been shown to impair reversal learning (e.g.,

Berger & Orr, 1983; Eichenbaum & Buckingham, 1990; Jones & Mishkin, 1972; Winocur & Olds, 1978).

Sensory Preconditioning

In the intact model, if two cues, A and B, consistently co-occur, their representations will tend to undergo redundancy compression. This increases the generalization between them. If this joint preexposure phase is followed by a subsequent phase of A+ training, some of the associations accrued to A will then transfer to B. As a result, the response to a test presentation of B will be stronger than in a control condition in which there was no joint AB preexposure (Figure 2C). This phenomenon is known as *sensory preconditioning*, and is observed in normal, intact animals (Thompson, 1972).

Because this effect is explained in terms of redundancy compression, it is not observed in the lesioned model (Figure 2C). Similarly, sensory preconditioning is eliminated in animals after hippocampal lesion (Port & Patterson, 1984).

Latent Inhibition

Redundancy compression can occur with contextual cues as well as with experimentally controlled conditioned stimuli. For example, if a phasic cue A is repeatedly presented in context X and not reinforced (XA− preexposure), its representation will become compressed with the tonic contextual cues with which it appears—because neither A nor the tonic cues are predictive of any reinforcement. This redundancy compression increases generalization between A and the tonic cues. In a subsequent phase in which the presence of A does predict reinforcement (XA+, X− training), the representation of A will have to be redifferentiated from the representation of the contextual cues alone (X− trials). This will tend to slow learning relative to a control condition in which there is no preexposure (Figure 2D). This effect is known as *latent inhibition*, and occurs in normal, intact animals (Lubow, 1973).

Again, because this effect is explained in terms of redundancy compression, it does not occur in the lesioned model (Figure 2D). Likewise, latent inhibition is eliminated by aspiration lesions of hippocampus (Solomon & Moore, 1975; Kaye & Pearce, 1987).

Contextual Sensitivity in Learning

In addition to the explicitly conditioned stimuli, there are typically many background or "contextual" cues present which can influence what is learned in a conditioning study. There is considerable evidence that the hippocampal region is involved in including contextual information in learned associations, leading to several theories that this contextual sensitivity is the primary functional role of the hippocampus. In contrast, our more general theory of hippocampal function in learning (Myers & Gluck, in press) can derive many of these hippocampal-dependent contextual effects as task-specific implications—including the latent inhibition results described previously.

Our hippocampal-region model does make two basic predictions regarding contextual processing. First, hippocampal lesions should not result in a contextual deficit per se; rather, contextual learning should be impaired in the same ways that learning about explicitly conditioned cues is impaired. This assumption is in many earlier conditioning models (e.g., Rescorla & Wagner, 1972), and predicts that a task involving contextual cues should be impaired after hippocampal lesion if and only if a parallel task involving explicitly conditioned cues is also impaired. For example, in our model, stimulus discrimination is not assumed to be impaired after hippocampal lesion (see Figure 1A). Therefore, our model predicts that hippocampal lesion should not impair learning simply that reinforcement is predicted in one context but not another. Hippocampal-lesioned animals, likewise, show no impairment at this contextual discrimination (Good & Honey, 1991). On the other hand, learning a configural task such as (A+, B+, AB−) is impaired in both hippocampal-lesioned animals (Rudy & Sutherland, 1989; Whishaw & Tomie, 1991) and our lesioned model (Gluck & Myers, 1993). Therefore, our model predicts a hippocampal-lesioned deficit in a configural task involving contexts (Myers & Gluck, in press); in fact, hippocampal-lesioned animals do show such a deficit (Good & Honey, 1991). These data support our prediction that hippocampal lesion should not specifically impair learning about contextual cues; instead, hippocampal lesion causes a representational deficit, which affects learning about both contextual cues and explicitly conditioned cues.

The second prediction of our connectionist model is that hippocampal representations include contextual information in every learned association. Because all stimuli present are included in the internal representation, the representation of any one stimulus is affected by the other stimuli present. This process underlies redundancy compression, but also ensures that the representation of a stimulus will include information about the context in which that stimulus occurred. Therefore, after the intact model learns to respond to one stimulus, A, in a given context, the response drops when A is presented in a new context (Myers & Gluck, in press). On the other hand, our lesioned model has no facility for forming new representations, and therefore no mechanism for including information about the context in which a response is learned. Therefore, the lesioned model shows no change in the strength of responding to A if A is presented in a new context (Myers & Gluck, in press). Under many circumstances, normal animals also show a decrement in the trained response to a stimulus after a salient context shift, whereas hippocampal-lesioned animals do not (e.g., Penick & Solomon, 1991; Antelman & Brown, 1972).

This account of contextual learning is also highly consistent with theories that context functions as an "occasion-setter" in normal animals (Hirsh, 1974; Bouton, 1991). It also relates to the idea that normal humans but not hippocampal-damaged amnesics can incorporate information about the context in which an event occurs to form declarative or episodic memories (Squire, 1992).

Other Behavioral Predictions

Our hippocampal-region model is able to predict accurately a wide range of conditioning phenomena, in addition to the examples described earlier. These include easy–hard transfer, compound preconditioning, and broadening of the generalization gradient after hippocampal lesion (Gluck & Myers, 1993). The model also makes several novel predictions which remain to be tested in intact and hippocampal-lesioned animals.

When Is the Hippocampus Involved?

Most of the earlier characterizations of hippocampal function reviewed at the start of this chapter focus primarily on differentiating between tasks that do—and do not—require an intact hippocampus. This approach has yielded a diverse and important empirical database which constrains all theories of hippocampal function. Some researchers have interpreted these data as implying that the hippocampus is not involved in behaviors that survive hippocampal lesions—including simple acquisition and stimulus discrimination.

In contrast, our theory suggests that the hippocampal region of an intact animal is involved in learning even the most elementary associations (see also Eichenbaum, Otto, & Cohen, 1992, and McNaughton, Leonard, & Chen, 1992, for related interpretations). Most of the empirical data analyzed in this chapter have been from learning paradigms that are variations and extensions of elementary acquisition and discrimination training. Although hippocampal damage may not always be clearly evident from analyses of error rates and trials-to-criterion during initial acquisition, we have interpreted behavioral measures of transfer performance and generalization as implying hippocampal involvement throughout all these types of learning.

This view of hippocampal involvement, in even the simplest forms of acquisition, suggests a possible interpretation of some seemingly paradoxical results. One example is the often-reported facilitation of simple discrimination learning in hippocampal-lesioned animals (see Figure 2A). A related result is the finding that disruption of hippocampal activity during conditioning—such as from inducing seizures with injections of penicillin, medial septal lesions, or electrical stimulation—impairs learning more than complete removal of the hippocampus (Solomon, Solomon, Van der Schaaf, & Perry, 1983). At first glance, these two findings seem contradictory. If removal of the hippocampus facilitates standard acquisition learning, why should disrupting the hippocampus impair learning?

Our account of hippocampal-region function provides one interpretation of these data. Hippocampal disruption may be approximated in our model by adding random noise to the activation levels of internal units in the hippocampal network. This will disrupt the representational "teaching signals" sent to the cortical network, and thereby cause the cortical network to develop an internal representation that is continually and randomly changing. In contrast, our model of a hippocampal-lesioned animal presumes that the cortical representations remain fixed throughout learning because of the absence of any representational teaching signals. We expect that learning with such a fixed stimulus representation should be faster (and more complete) than trying to learn with a randomly changing representation. As expected, the simulation results are broadly consistent with the empirical results reported by Solomon et al. (1983): Learning with several types of hippocampal disruption is severely impaired—worse, in fact, than in either the intact or lesioned animals.

Thus, these seemingly paradoxical results—that discrimination learning may be facilitated after hippocampal lesion, whereas a disrupted hippocampus is worse than learning with no hippocampus at all—can be reconciled within our account of hippocampal-region function. These results provide additional evidence that the hippocampal region is always involved in discrimination learning, even in the simplest tasks such as stimulus discrimination and acquisition.

This perspective suggests that the effects of hippocampal damage may be especially informative in studies of two-phase transfer tasks in which the effects of altered stimulus representations are most apparent, such as are observed in reversal learning, sensory preconditioning, and latent inhibition. This suggests that a profitable direction for future empirical studies may be to examine other relatively simple two-phase learning paradigms, in which both intact and hippocampal-lesioned animals behave similarly on an initial learning task, but respond differently on subsequent transfer and generalization tasks.

FROM ANATOMICAL FUNCTION TO PHYSIOLOGY

The hippocampal region encompasses several distinct structures, including hippocampus fields CA1–CA4, the dentate gyrus, the subicular complex, and overlying cortical areas including the entorhinal, perirhinal, and parahippocampal cortices (see Figure 3). Nevertheless, most theories of hippocampal function have viewed the hippocampal region as a "black box" in which these multiple structures are treated as a single functional unit (Eichenbaum, in press; Mishkin, 1982; Squire, 1987; Buzsaki, 1989; McNaughton & Nadel, 1990; Nadel, 1992; Gluck & Myers, 1993; Hirsh, 1974).

Recent advances in experimental methods, however, have made it possible to lesion selectively and study these individual regions (e.g., see Jarrard, in press; Zola-Morgan, Squire, Rempel, Clower, & Amaral, 1992). Concomitantly, researchers have begun to propose and test putative divisions of function among the distinct hippocampal-region structures (Lynch & Granger, 1992; Squire, 1992; Eichenbaum, Otto & Cohen, 1994). Here, we describe one such attempt, based on a preliminary mapping of our computational model onto distinct substructures of the hippocampal region. In particular, we seek to differentiate a functional role of the entorhinal cortex from a functional role of the hippocampus itself.

Dissociation of Hippocampal-Region Subfunctions

We start with the relatively abstract computational account of hippocampal-region function previously described (Gluck & Myers, 1993). This account presumes that the hippocampal region is essential for constructing new stimulus representations constrained by

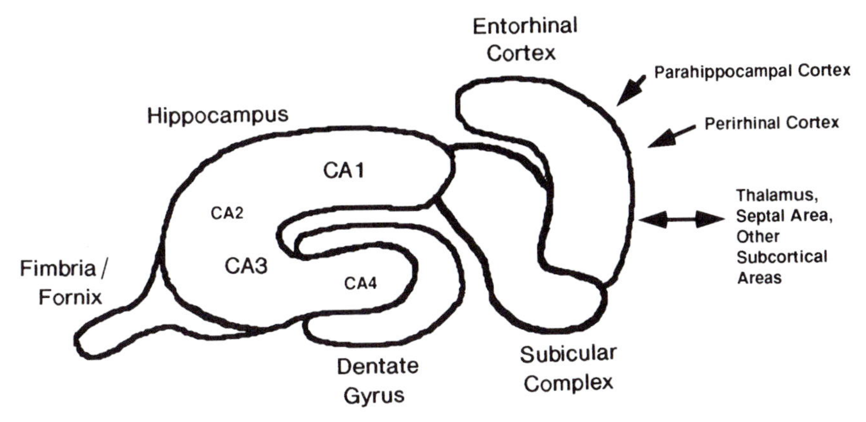

FIGURE 3 The hippocampal formation is usually defined as including fields CA1 through CA4 of the hippocampus, the dentate gyrus, and possibly the subicular complex. The hippocampal region includes these structures as well as the highly interconnected overlying cortical regions, including entorhinal cortex and possibly parahippocampal and perirhinal cortices. After Squire, Zola-Morgan, and Chen, 1988, and Witter and Amaral, 1991. Reproduced from Myers & Gluck (in press).

two biases: redundancy compression and predictive differentiation. Thus, the aggregate representational function of the hippocampal region could involve two subfunctions, each of which imposes one of these two constraints on the new stimulus representations that develop during learning (Gluck & Granger, 1993). Thus, the original hippocampal-region model described earlier can be recast as a pair of networks, one implementing redundancy compression and one implementing predictive differentiation. These two networks might then operate in series to recode stimulus representations (Figure 4).

Might these two representational constraints (recodings) likewise be localized in different anatomic substructures of the hippocampal region? One possible mapping is suggested by the similarity between redundancy compression, as proposed by Gluck and Myers (1993), and the stimulus-clustering capabilities of a recent computational model of olfactory (piriform) cortex proposed by Granger, Lynch, and co-workers (Ambros-Ingerson, Granger, & Lynch, 1990; Granger, Ambros-Ingerson, & Lynch, 1989; Granger & Lynch, 1991). Because superficial layers of piriform cortex are similar to superficial layers of entorhinal cortex, it is possible that the proposed stimulus–stimulus redundancy compression may be localized within the entorhinal cortex (see also Gluck & Granger, 1993).

We have hypothesized that the entorhinal cortex performs stimulus–stimulus redundancy compression, whereas other hippocampal-region structures—possibly the hippocampus proper—perform predictive differentiation (Myers et al., submitted). This hypothesized dissociation has several implications for interpreting recent empirical studies. In particular, it predicts that whereas lesions of the complete hippocampal region fully eliminate the ability to form new stimulus representations, more restricted lesions may result in more selective representational deficits—to the extent that these restricted lesions do not disrupt input and output pathways in other hippocampal-region structures. In particular, we hypothesize that lesions restricted to the hippocampal formation (H lesion) might not interfere with learning tasks that, in normal intact animals, depend only on stimulus–stimulus redundancy compression but not predictive differentiation. These behaviors should be disrupted by lesion either to the entorhinal cortex (EC lesion) or by broad lesions which include both hippocampus and entorhinal cortex, as well as surrounding structures (H++ lesion). In contrast, learning behaviors that

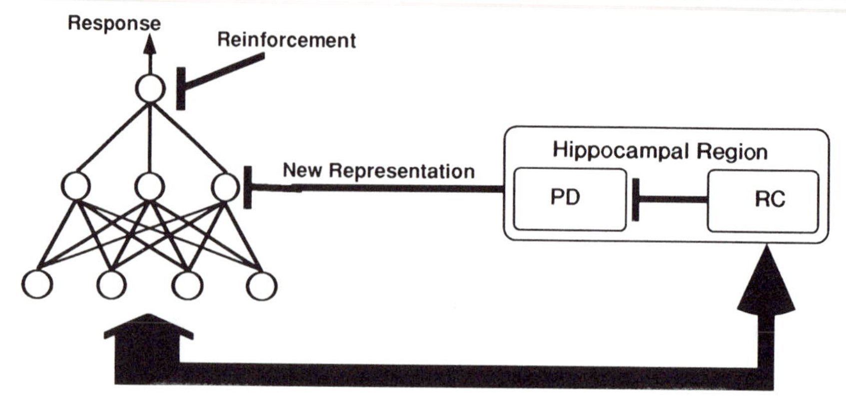

FIGURE 4 A serial version of the hippocampal-region model that dissociates the different stages of hippocampal-region processing. The biases of redundancy compression (RC) and predictive differentiation (PD) are separated into two serial-processing modules. The net effect of hippocampal-region processing is the same as in the original "black-box" model shown in Figure 1A. We have hypothesized that the redundancy compression function may be implemented in entorhinal cortex, whereas the predictive differentiation function may be implemented in the hippocampus proper.

result solely from predictive differentiation should be affected both by H lesion and by H++ lesion.

Comparisons with Empirical Data

We have compared the results of our current hypothesis—that the entorhinal cortex performs stimulus–stimulus redundancy compression, whereas the hippocampal formation performs predictive differentiation—with experimental data (Myers et al., submitted). We can simulate H lesions, EC lesions, or H++ lesions by appropriately disabling the components of the serial model. The model then accurately predicts that successive discrimination, which is expected to depend neither on predictive differentiation nor on redundancy compression, does not show a deficit after H lesions (Jarrard & Davidson, 1993), EC lesion (Otto, Schottler, Staubli, Eichenbaum, & Lynch, 1991), or H++ lesions (Port, Romano, & Patterson, 1986). Likewise, the serial model accurately predicts that latent inhibition, which is expected to depend on redundancy compression, is maintained after ibotenate lesions which are strictly limited to hippocampus (Honey & Good, 1993) but eliminated after less selective aspiration lesions (Solomon & Moore, 1975; Kaye & Pearce, 1987). Finally, the serial model predicts that simultaneous discrimination, dependent on both redundancy compression and predictive differentiation, should be sensitive to any form of hippocampal-region damage; again, this is consistent with empirical data (Eichenbaum et al., 1988). The model also accounts for data showing that the lesioned rats, even when they do solve some simultaneous discriminations, still fail to generalize appropriately (Bunsey & Eichenbaum, submitted). Finally, we have suggested that a novelty detection function, dependent on redundancy compression mechanisms in the parahippocampal region, could predict the pattern of impairment on delayed nonmatching tasks, after various hippocampal-region lesions (Myers et al., submitted).

This serial processing hypothesis draws support from two separate lines of reasoning. First, this hypothesis was suggested by an integration and comparison between two types of neural models: one which incorporates physiological and anatomical detail of a restricted brain region (Ambros-Ingerson et al., 1990) and another which inferred an information processing function for a more diffuse brain region from a wide range of behavioral studies comparing intact and lesioned animals (Gluck & Myers, 1993). A comparison of these two models supports the inference that the superficial layers of entorhinal cortex could, in principle, implement the stimulus–stimulus redundancy compression function suggested by the Gluck & Myers model. The second line of support for this serial model comes from existing selective lesion data that are consistent with this proposed dissociation of function between the entorhinal cortex and the hippocampus.

The most important conclusion to be drawn from this analysis is that there is a huge vacuum in the experimental literature with respect to the effects of selective hippocampal lesions. Only a few classical conditioning tasks have been thoroughly explored under conditions of hippocampal-formation lesion, parahippocampal lesion, or both. Even the several operant tasks described in this section are mostly incomplete in this regard. Only delayed nonmatching and successive discrimination have, to date, been carefully explored under the full range of lesion conditions in multiple species.

Application to Human Learning

One problem with comparing the results of hippocampal-region damage in humans and animals is that in humans the damage is often diffuse and imprecisely characterized. Structures within the hippocampal region may be only partially or unilaterally damaged, and damage almost always extends to other structures outside the region. However, there has been a good deal of success recently at equating learning deficits in humans with medial temporal lobe damage with monkeys having H++ lesions (Squire, 1992). Thus, human amnesics generally show the same failures as H++ monkeys: For example, impairment at delayed nonmatching and concurrent object discrimination, but not at motor skill learning (see Squire, 1992). There has also been some work on classical conditioning of eyeblink conditioning in medial temporal amnesia; the patients show effects very much like H++ animals: For example, normal acquisition but impaired reversal (Daum, Channon, & Canavan, 1989). Thus, it seems likely that a theory of hippocampal-region function in animals will be very similar when applied to humans. We note, in fact, that research on behavioral correspondences between classical conditioning and category learning (e.g., Gluck & Bower, 1988a,b; Gluck, Bower, & Hee, 1989; Gluck, 1991) may suggest possible avenues for integrating animal and human models of hippocampal-region function (see, e.g., Knowlton, Squire & Gluck, 1993).

Because of the diffuseness of human amnesic damage, it is extremely difficult to argue that particular hippocampal-region structures are responsible for different components of a hippocampal-region function. There have, however, been several cases in which damage has been limited to the hippocampal formation alone, including patient RB, who suffered an ischemic episode resulting in circumscribed damage to his hippocampal field CA1 (see Squire, 1992). Patients like RB show less impairment than patients who have sustained damage to the entire medial temporal lobe (Squire, 1992). These and other data imply that there are functions subserved by hippocampal-region structures such as entorhinal cortex and parahippocampal cortex, which the hippocampal formation alone cannot perform.

DISCUSSION

In this chapter, we have reviewed our original proposal that the hippocampal region is critically involved in the construction of internal representations of stimuli (Gluck & Myers, 1993). These representations are assumed to be constrained by biases toward **compression** or chunking of the representations of cooccurring or redundant stimuli while **differentiating** representations of stimuli according to how they reflect or predict outcomes or response behaviors. This account has been instantiated as a simple connectionist model, and suffices to accurately predict a wide range of conditioning data in intact and hippocampal-lesioned animals.

More recently, we have noted that it is theoretically possible to dissociate the two representational constraints of compression and differentiation (Myers et al., submitted). We have further hypothesized that they may be localized in distinct subregions of the hippocampal region: stimulus–stimulus redundancy compression in the entorhinal cortex and predictive differentiation in the hippocampus. Simulations in which the dissociated constraints operate serially were shown to maintain the entire range of hippocampal-dependent conditioned behaviors generated in the original aggregate model. More importantly, this dissociation is consistent with data showing that selective lesions of the hippocampal formation differentially disrupt tasks which we claim depend predominantly on outcome prediction, whereas selective entorhinal cortex damage differentially impairs performance if chunking or clustering of stimuli is crucial.

One salient aspect of this modeling is its integration of two different levels of modeling and theory development. Our original account of hippocampal-region function is a top-down model that begins with observed behaviors linked to particular anatomical structures and which attempts to identify neurobiological mechanisms that implement the behaviors. In contrast, the Ambros-Ingerson et al. (1990) model of olfactory cortex is an example of bottom-up modeling in that it attempts to identify emergent circuit function from biological details. Our serial hippocampal model described above integrates these two models—and, thus, these two methodological approaches: the network that implements redundancy compression, as associated specifically with superficial entorhinal cortex, is derived from physiologically based descriptions of circuitry within that brain region. The convergence of top-down and bottom-up methodologies in modeling is, we feel, a promising development and suggests that both approaches—along with further empirical studies—can contribute to the development of a unified understanding of hippocampal-region function in learning and memory.

Acknowledgments

This research was supported by the Office of Naval Research through the Young Investigator Program and Grant N00014-88-K-0112 (M.G.) and by NIMH National Research Service Award 1-F32-MH10351-01 (C.M.).

References

Akase, E., Alkon, D. L., & Disterhoft, J. F. (1989). Hippocampal lesions impair memory of short-delay conditioned eye blink in rabbits. *Behavioral Neuroscience*, **103**(5), 935–943.

Ambros-Ingerson, J., Granger, R., & Lynch, G. (1990). Simulation of paleocortex performs hierarchical clustering. *Science*, **247**, 1344–1348.

Antelman & Brown (1972).

Berger, T. W., & Orr, W. B. (1983). Hippocampectomy selectively disrupts discrimination reversal learning of the rabbit nictitating membrane response. *Behavioral Brain Research*, **8**, 49–68.

Bouton, M. (1991). Context and retrieval in extinction and in other examples of interference in simple associative learning. In L. Dachowski & C. F. Flaherty (Eds.), *Current topics in animal learning: Brain, emotion and cognition* (pp. 25–53). Hillsdale, NJ: Erlbaum.

Bunsey, M., & Eichenbaum, H. (submitted). Critical role of the parahippocampal region for paired association learning in rats.

Buszaki, G. (1989). Two-stage model of memory trace formation: A role for "noisy" brain states. *Neuroscience*, **31**(3), 551–570.

Cohen, N. J., & Squire, L. R. (1980). Preserved learning and retention of pattern analyzing skill in amnesia: Dissociation of knowing *how* and knowing *that*. *Science*, **210**, 207–209.

Daum, I., Channon, S., & Canavan, A. (1989). Classical conditioning in patients with severe memory problems. *Journal of Neurology, Neurosurgery and Psychiatry*, **52**, 47–51.

Eichenbaum, H. (in press). The Hippocampal system and declarative memory in humans and animals. Experimental analysis and historical origins. In D. L. Schacter & E. Tulving (Eds.), *Memory systems 1994*. Cambridge MA: MIT Press.

Eichenbaum, H., & Buckingham, J. (1990). Studies on hippocampal processing: Experiment, theory and model. In M. Gabriel & J. Moore (Eds.), *Learning and computational neuroscience: Foundations of adaptive networks* (pp. 171–231). Cambridge, MA: MIT Press.

Eichenbaum, H., Cohen, N. J., Otto, T., & Wible, C. (1992). Memory representation in the hippocampus: Functional domain and functional organization. In L. R. Squire, G. Lynch, N. M. Weinberger, & J. L. McGaugh (Eds.), *Memory Organization and Locus of Change* (pp. 163–204). Oxford: Oxford University Press.

Eichenbaum, H., Fagan, A., & Cohen, N. (1986). Normal olfactory discrimination learning set and facilitation of reversal learning after medial-temporal damage in rats: Implication for an account of preserved learning abilities in amnesia. *Journal of Neuroscience*, **6**(7), 1876–1884.

Eichenbaum, H., Fagan, A., Mathews, P., & Cohen, N. (1988). Hippocampal system dysfunction and odor discrimination learning in rats: Impairment or facilitation depending on representational demands. *Behavioral Neuroscience*, **102**(3), 331–339.

Eichenbaum, H., Otto, T., & Cohen, M. (1994). Two components of the hippocampal memory system. Behavioral and Brain Sciences, *17*.

Eichenbaum, H., Otto, T., & Cohen, N. (1992). The hippocampus: What does it do? *Behavioral and Neural Biology*, **57**, 2–36.

Eichenbaum, H., Otto, T., Wible, C., & Piper, J. (1991). Building a model of the hippocampus in olfaction and memory. In J. Davis & H. Eichenbaum (Eds.), *Olfaction as a model for computational neuroscience*. Cambridge, MA: MIT Press.

Gluck, M. A. (1991). Stimulus generalization and representation in adaptive network models of category learning. *Psychological Science*, **2**(1), 1–6.

Gluck, M. A., & Bower, G. (1988a). From conditioning to category learning: An adaptive network model. *Journal of Experimental Psychology: General*, **117**(3), 225–244.

Gluck, M. A., & Bower, G. (1988b). Evaluating an adaptive network model of human learning. *Journal of Memory and Language*, **27**, 166–195.

Gluck, M. A., Bower, G. H., & Hee, M. R. (1989). A configural-cue network model of animal and human associative learning. *Eleventh Annual Conference of the Cognitive Science Society*, Ann Arbor, MI.

Gluck, M. A., Goren, O., Myers, C., & Thompson, R. (1994). A higher-order recurrent network model of the cerebellar substrates of response timing in motor-reflex conditioning. Unpublished manuscript.

Gluck, M. A., & Granger, R. (1993). Computational models of the neural bases of learning and memory. *Annual Review of Neuroscience*, **16**, 667–706.

Gluck, M. A., & Myers, C. E. (1993). Hippocampal mediation of stimulus representation: A computational theory. *Hippocampus*, **3**, 491–516.

Gluck, M. A., Myers, C. E., & Thompson, R. F. (this volume). A computational model of the cerebellum and motor-reflex conditioning.

Good, M., & Honey, R. (1991). Conditioning and contextual retrieval in hippocampal rats. *Behavioral Neuroscience*, **105**, 499–509.

Granger, R., Ambros-Ingerson, J., & Lynch, G. (1989). Derivation of encoding characteristics of layer II cerebral cortex. *Journal of Cognitive Neuroscience*, **1**(1), 61–87.

Granger, R., & Lynch, G. (1991). Higher olfactory processes: Perceptual learning and memory. *Current Biology*, **1**, 209–214.

Hirsh, R. (1974). The hippocampus and contextual retrieval of information from memory: A theory. *Behavioral Biology*, **12**, 421–444.

Honey, R., & Good, M. (1993). Selective hippocampal lesions abolish the contextual specificity of latent inhibition and conditioning. *Behavioral Neuroscience*, **107**(1), 23–33.

Jarrard, L. (in press). On the role of the hippocampus in learning and memory in the rat. *Behavioral and Neural Biology*.

Jarrard, L., & Davidson, T. (1993). The hippocampus and complex, nonspatial discrimination: Is learning still "not possible"? In N. Spear, L. Spear, & M. Woodruff (Eds.), *Neurobiological plasticity: Learning, development and response to brain insults*.

Jones, B., & Mishkin, M. (1972). Limbic lesions and the problem of stimulus-reinforcement association. *Experimental Neurology*, **36**, 362–377.

Kaye, H., & Pearce, J. (1987). Hippocampal lesions attenuate latent inhibition and the decline of the orienting response in rats. *Quarterly Journal of Experimental Psychology*, **39B**, 107–125.

Knowlton, B., Squire, L. R., & Gluck, M. (1994). Probabilistic classification learning in amnesia. *Learning and Memory*, 1–15.

Lubow, R. (1973). Latent Inhibition. *Psychological Bulletin*, **79**, 398–407.

Lynch, G., & Granger, R. (1992). Variations in synaptic plasticity and types of memory in cortico-hippocampal networks. *Journal of Cognitive Neuroscience*, **4**, 189–199.

McNaughton, B. L., & Nadel, L. (1990). Hebb-Marr networks and the neurobiological representation of action in space. In M. A. Gluck & D. E. Rumelhart (Eds.), *Neuroscience and Connectionist Theory*. Hillsdale, NJ: Erlbaum.

McNaughton, B., Leonard, B., & Chen, L. (1992). Cortical-hippocampal interaction and cognitive mapping: An hypothesis based on reintegration of the parietal and inferotemporal pathways for visual processing. *Psychobiology*, 236–246.

Mishkin, M. (1982). A memory system in the monkey. *Philosophical Transactions of the Royal Society of London Series, B*, **298**, 85–92.

Morris, R. G. M., Garrud, P., Rawlins, J. N. P., & O'Keefe, J. (1982). Place navigation impaired in rats with hippocampal lesions. *Nature*, **297**, 681–683.

Moyer, J. R., Deyo, R. A., & Disterhoft, J. F. (1990). Hippocampectomy disrupts trace eye-blink conditioning in rabbits. *Behavioral Neuroscience*, **104**(2), 243–252.

Myers, C., & Gluck, M. (in press). Context, conditioning and hippocampal re-representation. *Behavioral Neuroscience*.

Myers, C. & Gluck, M. (1994). A neurocomputational theory of hippocampal function in associative learning. Proceedings World Congress on Neural Networks 1994, Vol. II, pp. 717–722.

Myers, C. Gluck, M., & Granger, R. (submitted). Dissociation of hippocampal and entorhinal function in associative learning: A computational approach.

Nadel, L. (1992). Multiple memory systems: What and why. *Journal of Cognitive Neuroscience*, **4**(3), 179–188.

O'Keefe, J., & Nadel, L. (1978). *The Hippocampus as a Cognitive Map*. Oxford, UK: Claredon University Press.

Olton, D. (1983). Memory functions and the hippocampus. In W. Seifert (Ed.), *Neurobiology of the Hippocampus*, (pp. 335–373). London: Academic Press.

Otto, T., Schottler, F., Staubli, U., Eichenbaum, H., & Lynch, G. (1991). Hippocampus and olfactory discrimination learning: Effects of entorhinal cortex lesions on olfactory learning and memory in a successive-cue, go-no-go task. *Behavioral Neuroscience*, **105**(1), 111–119.

Penick, S., & Solomon, R. (1991). Hippocampus, context and conditioning. *Behavioral Neuroscience*, **105**, 611–617.

Port, R., Mikhail, A., & Patterson, M. (1985). Differential effects of hippocampectomy on classically conditioned rabbit nictitating membrane response related to interstimulus interval. *Behavioral Neuroscience*, **99**, 200–208.

Port, R., & Patterson, M. (1984). Fimbrial lesions and sensory preconditioning. *Behavioral Neuroscience*, **98**, 584–589.

Port, R., Romano, A., & Patterson, M. (1986). Stimulus duration discrimination in the rabbit: Effects of hippocampectomy on discrimination and reversal learning. *Physiological Psychology*, **4**(3&4), 124–129.

Rawlins, J. (1985). Associations across time: The hippocampus as a temporary memory store. *Behavioral and Brain Sciences*, **8**, 479–496.

Rescorla, R., & Wagner, A. (1972). A theory of Pavlovian conditioning: Variations in the effectiveness of reinforcement and nonreinforcement. In, A. Black & W. Prokasy (Eds.) *Classical Conditioning II: Current Research and Theory*, New York, Appleton-Century-Crofts.

Rudy, J., & Sutherland, R. (1989). The hippocampal formation is necessary for rats to learn and remember configural associations. *Behavioral Brain Research*, **34**, 97–109.

Schmaltz, L. W., & Theios, J. (1972). Acquisition and extinction of a classically conditioned response in hippocampectomized rabbits (*Oryctolagus coniculus*). *Journal of Comparative and Physiological Psychology*, **79**, 328–333.

Scoville, W. B., & Milner, B. (1957). Loss of recent memory after bilateral hippocampal lesions. *Journal of Neurology, Neurosurgery, & Psychiatry*, **20**, 11–21.

Solomon, P., & Moore, J. (1975). Latent inhibition and stimulus generalization of the classically conditioned nictitating membrane response in rabbits (*Oryctolagus cuniculus*) following dorsal hippocampal ablation. *Journal of Comparative and Physiological Psychology*, **89**, 1192–1203.

Solomon, P., Solomon, S., Van der Schaaf, E., & Perry, H. (1983). Altered activity in the hippocampus is more detrimental to classical conditioning than removing the structure. *Science*, **220**, 329–331.

Squire, L. R. (1982). The neuropsychology of human memory. *Annual Review of Neuroscience*, **5**, 241–273.

Squire, L.R. (1987). *Memory and brain*. New York: Oxford University Press.

Squire, L. (1992). Memory and the Hippocampus: A synthesis from findings with rats, monkeys, and humans. *Psychological Review*, **99**(2), 195–231.

Squire, L. R., & Zola-Morgan, S. (1983). The neurology of memory: The case for correspondence between the findings for man and nonhuman primate. In J. A. Deutsch (Ed.), *The Physiological Basis of Memory*. New York: Academic Press. 199–268.

Squire, L., Zola-Morgan, S., & Chen, K. (1988). Human amnesia and animal models of amnesia: Performance of amnesic patients on tests designed for the monkey. *Behavioral Neuroscience*, **102**(2), 210–221.

Thompson, R. (1972). Sensory preconditioning. In R. Thompson & J. Voss (Eds.), *Topics in Learning and Performance* (pp. 105–129). New York: Academic Press.

Whishaw, I., & Tomie, J. (1991). Acquisition and retention by hippocampal rats of simple, conditional and configural tasks using tactile and olfactory cues: Implications for hippocampal function. *Behavioral Neuroscience*, **6**, 787–797.

Witter, M., & Amaral, D. (1991). Entorhinal cortex of the monkey: V. Projections to the dentate gyrus, hippocampus and subicular complex. *Journal of Comparative Neurology*, **307**, 437–459.

Winocur, G., & Olds, J. (1978). Effects of context manipulation on memory and reversal learning in rats with hippocampal lesions. *Journal of Comparative and Physiological Psychology*, **92**(2), 312–321.

Zola-Morgan, S., Squire, L., Rempel, N., Clower, R., & Amaral, D. (1992). Enduring memory impairment in monkeys after ischemic damage to the hippocampus. *Journal of Neuroscience*, **12**(7), 2582–2596.

Zola-Morgan, S., Squire, L., & Amaral, D. (1989). Lesions of the amygdala that spare adjacent cortical regions do not impair memory or exacerbate the impairment following lesions of the hippocampal formation. *Journal of Neuroscience*, **9**, 1922–1936.

5

A Computational Model of the Cerebellum and Motor-Reflex Conditioning

Mark A. Gluck, Catherine E. Myers, and Richard F. Thompson

INTRODUCTION

A major goal of research in neuroscience and cognitive science is to understand how the mammalian brain codes, stores, and retrieves memories. It has become increasingly clear in recent years that the essential memory traces for a given form of learning cannot be localized or analyzed until the entire essential memory trace network has been identified and the activity patterns of its key elements have been characterized. Having done so, it is equally necessary to show that the biological network and the processes of memory trace formation embedded within it can, in fact, generate the emergent phenomena of learning and memory. This mandates a theoretical, computational model of the functioning of the biological neural network. This computational model must be constrained by the neurobiological properties of the biological neural network on the one hand, and by the emergent behavioral phenomena on the other hand (Gluck & Thompson, 1987; Donegan, Gluck, & Thompson, 1989).

To date, great progress has been made in the study of simple forms of learning, particularly classical conditioning of the eye-blink reflex (see, e.g., Thompson, 1986, 1988; Thompson & Gluck, 1990). Following this progress, a computational model has been developed that is based on the cerebellar circuitry that generates the basic phenomena of eye-blink conditioning (Gluck, Goren, Myers, & Thompson, 1994).

In this chapter, we review this computational model of the cerebellar circuitry underlying simple motor-reflex conditioning. We begin by reviewing the anatomical data that constrain the model. Next, we present the computational model itself, and show how it successfully generates the basic behavioral phenomena. The model does not, however, generate hippocampal-dependent conditioned behaviors. This is to be expected because the model includes no characterization of hippocampal-dependent processes. We conclude by noting how future work will need to consider the interactions between the hippocampus and the cerebellum in motor-reflex learning.

THE CEREBELLUM AND MOTOR-REFLEX CONDITIONING

Significant progress in the study of learning systems has been made in the area of classical conditioning (see, e.g., Thompson, 1986; Thompson & Gluck, 1990). Classical conditioning is a particularly valuable paradigm for the analysis of brain substrates of sensory-motor learning because of the high degree of experimental control possible. In a typical experiment, a neutral stimulus is presented in close succession with a response-evoking stimulus called the unconditioned stimulus (US). After repeated pairings, the previously neutral stimulus, called the conditioned stimulus (CS), comes to evoke a conditioned response (CR), which is similar to the US-evoked response (UR). The time between the CS onset and the US onset is defined as the interstimulus interval (ISI). By varying the ISI and the intensity, duration, number, and type of stimuli, a large number of conditioning phenomena can be explored.

A popular paradigm for studying conditioning behaviors is the rabbit eye-blink task, which has been well studied over nearly three decades (see Gormezano, Kehoe, Marshal, 1983, for review). In this preparation, the rabbit is restrained during presentation of stimuli while the movement of the nictitating membrane (third eyelid) is measured. Most commonly, the CS is a light or tone, and the US is a corneal air-puff. Eye-blink conditioning can occur with ISIs from 100 msec to well over 1 sec. After approximately 200 trials, rabbits will give CRs on more than 90% of trials. The CR onset initially develops near the US onset, and gradually adjusts to begin earlier in the trial. In a well-trained animal, the CR generally peaks near the onset of the US.

Work by Thompson and colleagues argues strongly that the cerebellum and its associated brain stem circuitry are both necessary and sufficient for classical conditioning of the eye-blink response, as well as for other responses to aversive stimuli (see, e.g., Thompson, 1986, 1988). A very localized region of the cerebellar interpositus nucleus is especially critical; small electrolytic lesions of this region completely and permanently abolish the eye-blink conditioned response, but do not affect the unconditioned reflex eyelid closure to the air-puff US. These and similar studies have helped us to determine the cerebellar circuitry for conditioned learning of motor reflexes. The resulting model of cerebellar circuitry for motor-reflex conditioning is shown in Figure 1.

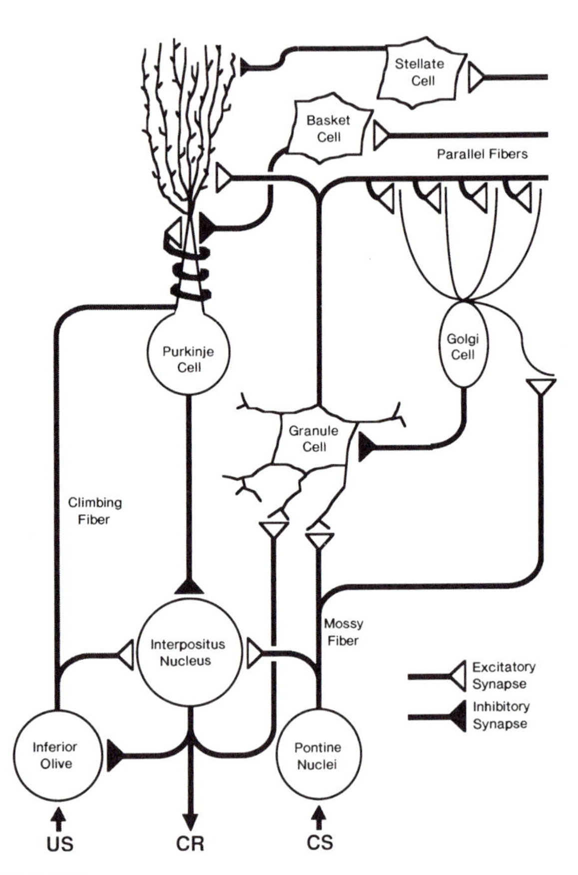

FIGURE 1 Simplified schematic of the neural circuits identified as subserving the conditioned eye-blink response. Information about CS occurrence is assumed to travel from pontine nuclei via mossy fiber projections to the cerebellum. The reinforcement pathway, which carries information about the US, is taken to reach the cerebellum via the climbing fibers from the inferior olive. The efferent pathway driving the CR consists of projections from the interpositus nucleus, which ultimately drive motor neurons. The CR pathway also projects to the inferior olive and to the cerebellum via mossy fiber inputs to granule cells.

Information about CS occurrence enters the cerebellum via mossy fiber projections from the pontine nuclei (Steinmetz, Rosen, Chapman, Lavond, & Thompson, 1986). A second input arrives via the climbing fibers, which are assumed to originate in the inferior olive (McCormick, Steinmetz, & Thompson, 1985; Yeo, Hardiman, & Glickstein, 1985). All these inputs contact Purkinje cells in the cerebellum, which develop patterned changes in discharge frequency that precede and predict the occurrence and form of the learned response within trials and which predict the development of learning over training trials (McCormick & Thompson, 1984b). Purkinje cell afferents pass through the interpositus nucleus and eventually reach descending pathways which act on motor neurons (Thompson, 1986). Neural unit recordings from the interpositus nucleus show the development of a pattern of increased frequency of unit discharges that models the amplitude time-course and development of the learned response (the CR), but not the reflex response to the US (McCormick & Thompson, 1984a,b). The interpositus nucleus also regulates the inferior olive via an inhibitory pathway, thereby exerting a modulatory influence on the reinforcement (US) pathway.

Further behavioral, anatomical, and neurophysiological evidence for this model circuit is discussed in Gluck et al. (1994).

A MODEL OF CEREBELLAR CIRCUITRY

Our computational model incorporates elements of the cerebellar circuit described above. For present purposes, we have adopted the simplification of representing the cerebellar cortex and interpositus nucleus as a single adaptive processing element or node (labeled CC/IN in Figure 2). Future work will, of course, be required to separate the individual contributions of these two structures. Input to the CC/IN is provided by nonadaptive processing elements representing the pontine nuclei, one node representing each CS input. The CC/IN output represents the firing of the interpositus nucleus, which closely approximates and precedes the behavioral response. The inferior olive is modeled by a single nonadaptive node which receives excitatory input representing the presence of the US.

This model is trained via a formal learning rule derived from behavioral analysis. This rule, the Rescorla-Wagner rule (Rescorla & Wagner, 1972), addresses numerous conditioning experiments that suggest that the effectiveness of a US for producing associative learning depends on the relationship between the US and the expected outcome (Rescorla, 1968; Wagner, 1969; Kamin, 1969). The Rescorla-Wagner rule assumes that the association that accrues between a stimulus and its outcome on a trial is proportional to the degree to which the outcome is unexpected (or unpredicted) given all the stimuli present on that trial. Formally, the change in the association

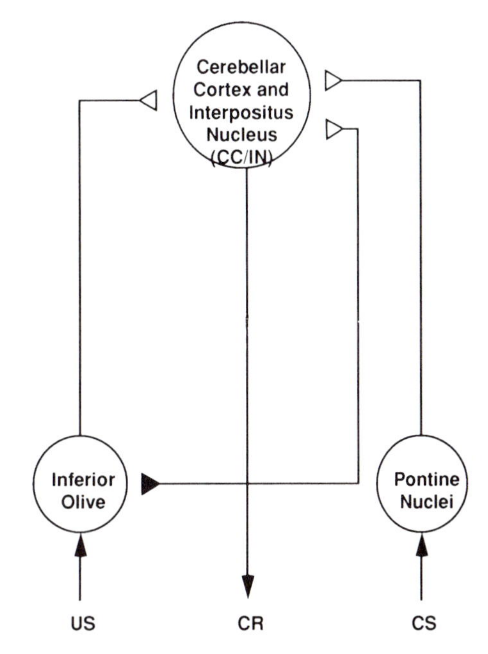

FIGURE 2 The cerebellar model, including neural circuits and pathways from Figure 1 (Gluck et al., 1994). A single higher-order adaptive element corresponds to the CC/IN node. The CR output drives the motor response, but also travels back to the inferior olive, where it inhibits US activation, thereby reducing the reinforcement signaled to the cerebellum. It also returns, via recurrent connections, to the CC/IN, allowing for complex timing of peak CR.

strength V_i between one CS, CS_i , and the US can be described by

$$\Delta V_i = \beta(\lambda - \sum_k V_k) \quad (1)$$

where β reflects the learning rate, λ is the total reinforcement supported by the US (or 0 if no US is present), and $\sum_k V_k$ is the sum of the associative strengths between the US and all stimuli k appearing on the trial. If CS_i is paired with the US, V_i will tend to increase; if CS_i is presented without the US, V_i will tend to decrease. If the US is perfectly predicted, and $\lambda = \sum_k V_k$, then there will be no further change to V_i.

Our circuit-level model of cerebellum can implement this learning rule. In the Rescorla-Wagner rule, during acquisition of a response to CS_i, the amount of associative strength added decreases (as V_i approaches λ). One way to accomplish this is to reduce the amount of reinforcement available as training progresses (Wagner, 1978, 1981; Wagner & Donegan, 1989). In our cerebellar model, this would mean that climbing fiber activations of Purkinje cells and interpositus neurons in the CC/IN would be reduced as a function of learning. This can be accomplished if, in addition to driving a behavioral response, the CC/IN output also exerts an inhibitory influence on the US pathway passing through the inferior olive. Then, as the CR is increasingly well learned, activation of the US pathway by the external US is increasingly attenuated (see also Donegan & Wagner, 1987; Donegan et al., 1989). This idea is supported by recent anatomical and physiological evidence showing the existence of a powerful inhibitory pathway from the interpositus directly to the inferior olive (Andersson, Gorwitz, & Hesslow, 1987; Nelson & Mugnaini, 1987), as well as by cell recordings which show an attenuation of US-onset-evoked activity in the dorsal accessory olive neurons (Sears & Steinmetz, 1991).

Our model also addresses the temporal specificity of the conditioned response, which allows the maximal response to be delayed after CS onset so as to coincide accurately with US onset. Some researchers have proposed delay lines for this purpose (Desmond & Moore, 1988; Moore & Blazis, 1989); however, we believe a more biologically plausible explanation is that this temporal specificity is mediated by modifiable recurrencies in the CS–CR pathway (see Figures 1, 2). The CR output from the cerebellar cortex and interpositus nucleus feeds back as mossy fiber input to the cerebellum and, via parallel fibers, activates Purkinje cells in cerebellar cortex which have extensive and elaborate dendritic trees; these dendritic trees may allow the cells to receive higher-order analyses of their inputs, that is, noting not only which inputs are present, but which configurations are present. The presence of configural, as well as simple, inputs dramatically increases the computing power of a node (Shepard, 1987, 1989; Gluck & Bower, 1988; Gluck, 1991; Rumelhart, Hinton, & Williams, 1986). Taken together, these two properties—recurrent connections and higher-order inputs—may be sufficient to account for the role of the cerebellum in controlling the precise timing of the conditioned response (Gluck et al., 1994). We next discuss some simulation results that support this claim.

SIMULATION RESULTS WITH THE CEREBELLAR MODEL

Our cerebellar model has been tested under a variety of conditioning paradigms. In this section, we review its performance with two key behavioral paradigms: delay conditioning and blocking.

In delay conditioning, a CS is repeatedly paired with the US with some specified delay between CS onset and US onset (the ISI), and CS and US terminate together. Figure 3A shows the response topography from our cerebellar model; after training, the peak CR is timed to coincide with US onset, for either a long or short ISI. Furthermore, the increased complexity of the long-ISI paradigm increases learning time and results in a broader CR shape; again, this is consistent with the behavioral data (Gormezano et al., 1983).

A blocking experiment (Kamin, 1969) involves three phases. First, a single CS, A, is trained to predict the US. Next, compound training pairs both A and a second CS, B, with the US. Finally, the response to B

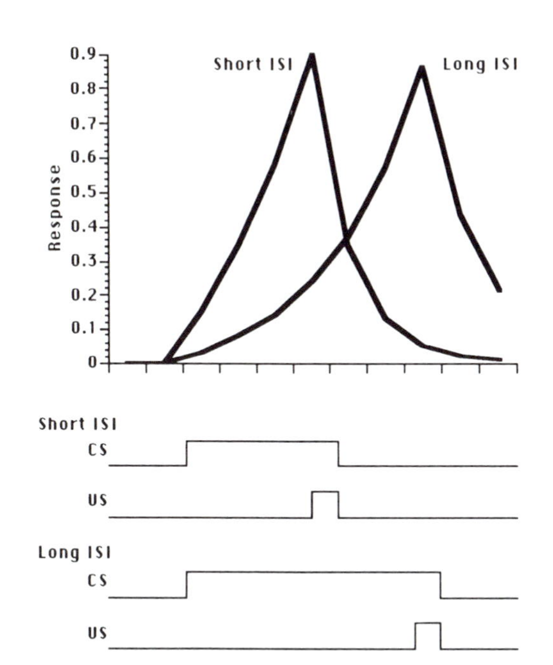

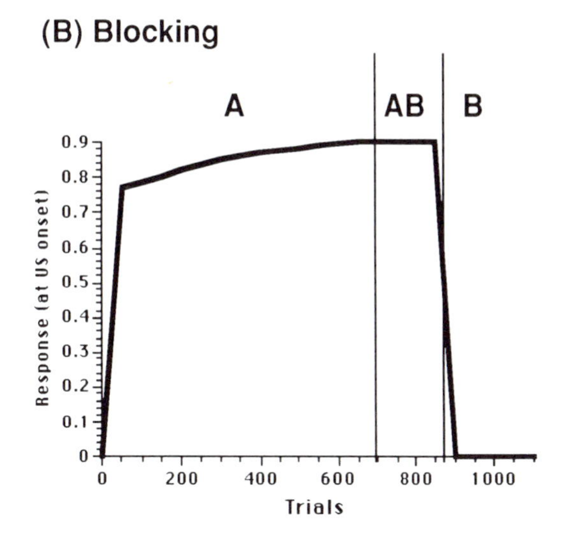

FIGURE 3 Empirical results with the cerebellar model (Gluck et al., 1993). Results are consistent with animal data. A. Delay conditioning response topography: The peak CR is timed to coincide with US arrival, after training with either a long or short ISI. B. Blocking: CS B, presented in compound with a previously trained CS A, is "blocked" from acquiring associative strength. This is shown by a lack of response to B-alone presentation.

alone is tested. In animals, B generates little or no response: as if the associability of B has been "blocked" by the presence of A, already sufficient to predict the US. This effect is accounted for by the Rescorla-Wagner rule (1972) as follows: associative strength changes are proportional to the difference between the actual reinforcement (US or no US) and expected reinforcement (the CR). If the difference is zero, there is no further learning. Our model also exhibits a blocking effect (Figure 3B) caused by the negative feedback loop from the CR pathway, via the inferior olive, to the US pathway (Figure 2).

These simulations (and others discussed in Gluck et al., 1994) illustrate how even a simplified computational characterization of cerebellar circuitry can generate a wide range of trial-level and real-time characteristics of the conditioned eye-blink behavior in animals. There are, however, many salient limitations to this model, in addition to the lack of physiological detail. These are reviewed next.

FUNCTIONAL ROLE OF OTHER BRAIN STRUCTURES

Although the cerebellum is believed to be the substrate of motor-reflex conditioning, other brain structures are engaged during conditioning. One critical group of structures is known collectively as the hippocampal region (including hippocampus, dentate gyrus, subiculum, and entorhinal cortex). Hippocampal-region lesions do not impair learning or retention of the CR for simple delay conditioning (see Port, Mikhail & Patterson, 1985). However, some more complex forms of conditioning are hippocampal-dependent. It is logically expected that a model of the cerebellum alone should exhibit deficits on hippocampal-dependent tasks, because the model presumably lacks the processing that would be contributed by the hippocampus.

In fact, our cerebellar model fails on just these tasks. For example, animals with large hippocampal-region lesions show impairment at learning tasks that require configural sensitivity—such as the negative patterning task in which either of two CSs predicts the US but their compound does not (Rudy & Sutherland, 1989). The cerebellar model is also unable to learn this task (Gluck et al., 1994). Animals with hippocampal-region damage also show marked impairment in training procedures in which unreinforced preexposure affects later learning; such paradigms include sensory preconditioning (Port & Patterson, 1984) and latent inhibition (Solomon & Moore, 1975). Consistent with this data, the cerebellar model also fails to exhibit this class of effect. Furthermore, animals with hippocampal-region damage are impaired in training tasks requiring a sensitivity to challenging temporal factors, such as very long ISI conditioning and trace conditioning (Moyer, Deyo, & Disterhoft, 1990). Appropriately, the cerebellar model cannot learn in trace conditioning or very long ISI conditioning (Gluck et al., 1994).

GENERAL DISCUSSION

We have presented a computational model of the cerebellar circuitry believed to mediate eye-blink conditioning and other discrete sensory-motor learning tasks. Our primary interest has been to understand the circuit's ability to generate a well-timed response that accurately anticipates the temporal interval between the CS and the US. The model emphasizes the importance of two known recurrent efferent collateral projections from the response (CR) pathway. The rich dendritic branching of Purkinje cells are interpreted as providing a high-dimensional ("higher-order") expansion of input signals sufficient to compute necessary timing behaviors.

This model correctly predicts a range of conditioning behaviors previously shown to be localized in the cerebellum. As expected, it fails to show conditioning behaviors that are known to depend on the hippocampal region (e.g., configural learning, preexposure effects, trace conditioning, and long ISI conditioning). We have therefore suggested an analogy between this model and conditioning in the hippocampal-damaged animal.

The cerebellum has long been known to be involved in motor control, movement, and coordination. The anatomical uniformity and seeming simplicity of the cerebellar cortical circuitry (see Figure 1) has repeatedly proven to be an attractive target for theorists seeking to understand, and model, neuronal function. The model presented here can be seen as an elaboration and extension of a long line of cerebellar theories dating back to Marr (1969), who proposed that the two main afferents converging on Purkinje cells in the cerebellar cortex—mossy fibers and climbing fibers—form a simple associative learning system. Mossy fibers, in Marr's theory and in our own, provide sensory inputs to the Purkinje cells via parallel fiber inputs, whose synaptic efficacy are modified by the presence (or absence) of co-occurring climbing fiber input. Thus, Marr first suggested that the climbing inputs can be characterized as teaching signals that reorganize sensory-motor associations conveyed along mossy and parallel fibers. Successively more refined and elaborated versions of Marr's theory were presented by Albus (1971), Eccles (1977), Calvert and Meno (1972), and Hassul and Daniels (1977). Our current model, described here and in Gluck et al. (1994), builds on these earlier cerebellar models, integrating them with the extensive neural and behavioral data constraining theories of the neural substrates of classical conditioning.

Acknowledgments

This research was supported by the Office of Naval Research through the Young Investigator Program and Grant N00014-88-K-00112 (M.G.), by NIMH National Research Service Award 1-F32-MH10351-01 (C.M.), and by the Office of Naval Research (R.F.T.).

References

Albus, J. S. (1971). A theory of cerebellar function. *Mathematical Biosciences*, **10**, 25–61.

Anderson, E., Gorwitz, M., & Hesslow, G. (1987). Inferior olive excitability after high frequency climbing fiber activation in the cat. *Behavioral Neuroscience*, **103**(5), 935–943.

Calvert, T. W., & Meno, F. (1972). Neural systems modeling applied to the cerebellum. *IEEE Transactions on Systems, Man & Cybernetics SMC-2*, 363–374.

Desmond, J. E., & Moore, J. W. (1988). Adaptive timing in neural networks: The conditioned response. *Biological Cybernetics*, **58**, 405–415.

Donegan, N. H., Gluck, M. A., & Thompson, R. F. (1989). Integrating behavioral and biological models of classical conditioning. In R. D. Hawkins & G. H. Bower (Eds.), *Computational models of learning in simple neural systems (Vol. 22 of the Psychology of Learning and Motivation)*. New York: Academic Press.

Donegan, N. H., & Wagner, A. R. (1987). Conditioned diminution and facilitation of the UCR: A sometimes-opponent-process interpretation. In I. Gormezano, W. Prokasy, & R. Thompson (Eds.), *Classical conditioning: II. Behavioral, neurophysiological, and neurochemical studies in the rabbit*. Hillsdale, NJ: Erlbaum.

Eccles, J. C. (1977). An instruction-selection theory of learning in the cerebellar cortex. *Brain Research*, **127**, 327–352.

Gluck, M. A. (1991). Stimulus generalization and representation in adaptive network models of category learning. *Psychological Science*, **2**(1), 1–6.

Gluck, M. A., & Bower, G. (1988). Evaluating an adaptive network model of human learning. *Journal of Memory and Language*, **27**, 166–195.

Gluck, M., Goren, O., Myers, C., & Thompson, R. (1994). A higher-order recurrent network model of the cerebellar substrates of response timing in motor-reflex conditioning. Unpublished manuscript.

Gluck, M. A., & Myers, C. E. (1992). Hippocampal function in representation and generalization: A computational theory. *Proceedings of the 14th Annual Meeting of the Cognitive Science Society*. Hillsdale, NJ: Erlbaum.

Gluck, M. A., & Thompson, R. F. (1987). Modeling the neural substrates of associative learning and memory: A computational approach. *Psychological Review*, **94**, 176–191.

Gormezano, I., Kehoe, E. K., & Marshal, B. S. (1983). Twenty years of classical conditioning research with the rabbit. *Progress in Psychobiology and Physiological Psychology*, **10**, 197–275.

Hassul, M., & Daniels, P. D. (1977). Cerebellar dynamics: The mossy fiber input. *IEEE Transactions on Biomedical Engineering BME-24*, 449–456.

Kamin, L. J. (1969). Predictability, surprise, attention and conditioning. In B. Campbell & R. Church (Eds.), *Punishment and aversive behavior*. New York: Appleton-Century-Crofts.

Marr, D. (1969). A theory of cerebellar cortex. *Journal of Physiology*, **202**, 437–470.

McCormick, D. A., & Thompson, R. F. (1984a). Cerebellum: Essential involvement in the classically conditioned eyelid response. *Science*, **223**, 296–299.

McCormick, D. A., & Thompson, R. F. (1984b). Neuronal responses of the rabbit cerebellum during acquisition and performance of a classically conditioned nictitating membrane-eyelid response. *Journal of Neuroscience*, **4**(11), 2811–2822.

McCormick, D. A., Steinmetz, J. E., & Thompson, R. F. (1985). Lesions of the inferior olivary complex cause extinction of the classically conditioned eyeblink response. *Brain Research*, **359**, 120–130.

Moore, J. W., & Blazis, D. E. J. (1989). Stimulation of a classically conditioned response: A cerebellar neural network implementation of the Sutton-Barto-Desmond model. In W. O. Berry (Ed.), *Neural models of plasticity: Experimental and theoretical approaches*. New York: Academic Press.

Moyer, J. R., Deyo, R. A., & Disterhoft, J. F. (1990). Hippocampectomy disrupts trace eye-blink conditioning in rabbits. *Behavioral Neuroscience*, **104**(2), 243–252.

Nelson, B., & Mugnaini, E. (1987). GABAergic innervation of the inferior olivary complex and experimental evidence for its origin. *In Symposium: The olivocerebellar system in motor control*. Turin, August 9–12, No. 9.

Port, R., Mikhail, A., & Patterson, M. (1985). Differential effects of hippocampectomy on classically conditioned rabbit nictitating membrane response related to interstimulus interval. *Behavioral Neuroscience*, **99**, 200–208.

Port, R., & Patterson, M. M. (1984). Fimbrial lesions and sensory preconditioning. *Behavioral Neuroscience*, **98**, 584–589.

Rescorla, R. A. (1968). Probability of shock in the presence and absence of CS in fear conditioning. *Journal of Comparative and Physiological Psychology*, **66**, 1–5.

Rescorla, R., & Wagner, A. (1972). A theory of Pavlovian conditioning: Variations in the effectiveness of reinforcement and nonreinforcement. In A. Black & W. Prokasy (Eds.), *Classical conditioning: II. Current research and theory*. New York: Appleton-Century-Crofts.

Rudy, J. W., & Sutherland, R. J. (1989). The hippocampal formation is necessary for rats to learn and remember configural discriminations. *Behavioral Brain Research*, **34**, 97–109.

Rumelhart, D. E., Hinton, G. E., & Williams, R. J. (1986). Learning internal representations by error propagation. In D. Rumelhart & J. McClelland (Eds.), *Parallel distributed processing: Explorations in the microstructure of cognition (Vol. 1. Foundations).* Cambridge, MA: MIT Press.

Sears, L. L., & Steinmetz, J. E. (1991). Dorsal accessory inferior olive activity diminishes during acquisition of the rabbit classically conditioned eyelid response. *Brain Research*, **545**, 114–122.

Shepard, R. N. (1987). Toward a universal law of generalization for psychological science. *Science*, **237**, 1317–1323.

Shepard, R. N. (1989). *A law of generalization and connectionist learning.* Talk presented at the Annual Meeting of the Cognitive Science Society, Ann Arbor, MI, August 17–18.

Solomon, P. R., & Moore, J. W. (1975). Latent inhibition and stimulus generalization of the classically conditioned nictitating membrane response in rabbits (*Oryctolagus cuniculus*) following dorsal hippocampal ablation. *Journal of Comparative and Physiological Psychology*, **89**, 1192–1203.

Steinmetz, J. E., Rosen, D. J., Chapman, P. F., Lavond, D. G., & Thompson, R. F. (1986). Classical conditioning of the rabbit eyelid response with a mossy fiber stimulation CS. I. Pontine nuclei and middle cerebellar penduncle stimulation. *Behavioral Neuroscience*, **100**, 871–880.

Thompson, R. F. (1986). The neurobiology of learning and memory. *Science*, **233**, 941–947.

Thompson, R. F. (1988). The neural basis of associative learning of discrete behavioral responses. *Trends in Neurosciences*, **11**(4), 152–155.

Thompson, R. F., & Gluck, M. A. (1990). Brain substrates of basic associative learning and memory. In H. J. Weingartner & R. F. Lister (Eds.), *Cognitive Neuroscience*. New York: Oxford University Press.

Wagner, A. R. (1969). Stimulus selection and a "modified continuity theory." In G. H. Bower & J. T. Spence (Eds.), *The psychology of learning and motivation (Vol. 3).* New York: Academic Press.

Wagner, A. R. (1978). Expectancies and the priming of STM. In S. H. Hulse, H. Fowler, and W. K. Honig, (Eds.), *Cognitive Aspects of Animal Behavior*. Hillsdale, NJ: Erlbaum.

Wagner, A. R. (1981). SOP: A model of automatic memory processing in animal behavior. In G. Miller (Ed.), *Information processing in animals: Memory mechanisms.* Hillsdale, NJ: Erlbaum.

Yeo, C. H., Hardiman, M. J., & Glickstein, M. (1985). Classical conditioning of the nictitating membrane response of the rabbit: II. Lesions of the cerebellar cortex. *Experimental Brain Research*, **60**, 99–113.

6

The Design of Intelligent Robots as a Federation of Geometric Machines

Rolf Eckmiller

INTRODUCTION

René Descartes (1632) was probably the first scientist who dared to postulate that both structure and function of living organisms can (in principle) be described by the laws of physics. The revolutionary potential of such a postulate was so large in his time, which was dominated by the traditional belief in the existence of special *vis vitales* (vital forces), that Descartes kept his book manuscript in a drawer and ordered it to be published only after his death, thus avoiding unpleasant discussions similar to those Galileo was experiencing at about the same time.

Descartes dealt in one of his book chapters with the question of how to build a machine that is similar to our body. Since then, physiologists, physicists, mathematicians, and engineers have been studying the neural control of biological systems (such as flies, frogs, and humans) and trying to develop technical systems with somewhat similar capabilities.

How does one build "intelligent" robots? Why is it impossible even today to order an artificial fly which can easily land on the ceiling, or an artificial frog which can track a fly and catch it in mid-flight? Several reasons come to mind.

1. Many computer scientists are still convinced that information systems should be serial, sequential, digital computers which require algebraic representations of any given mathematical problem.
2. Most engineers are trying to solve current problems of, for example, machine vision or robotics by applying those conventional computers and conventional concepts of control theory.
3. Neuroscientists are typically flooded with barely analyzed biological data and are not well trained in mathematics or computer science.
4. Mathematicians, physicists, and other neural net scientists are just beginning to identify the basic requirements for the design of intelligent robots including partitioning of sensorimotor functions of biological or technical systems into a federation of geometrical machines (Koenderink, 1987), geometric representations (rather than algebraic representations) of the various sensory or motor func-

tions in neural nets, and information processing by means of self-organizing neural networks with parallel, asynchronous, analog functions rather than by means of software-driven conventional computers.

In acknowledgment of these reasons and requirements we are presently witnessing the development of a new interdisciplinary research field: Neural Networks for Computing, also known as Neuroinformatics (Denker, 1986; Rumelhart & McClelland, 1986; Eckmiller & von der Malsburg, 1988).

The concepts of brain function, artificial intelligence, cellular automata, complex network topology, and self-organization are being merged. This approach of transferring concepts of biological neural control systems to artificial intelligent robots and the design of neural computers will ultimately yield a new breed of special-purpose computers with self-organizing neural net hardware.

BIOLOGICAL NEURAL CONTROL SYSTEMS AS A FEDERATION OF BIOLOGICAL NEURAL NETS

Neural Processing Elements

The human central nervous system (CNS) as well as commercially available computers are information processing systems. However, several fundamental differences are listed in Table 1.

The CNS of vertebrates or invertebrates can be reduced (in a highly simplified form) to large numbers of neurons, synapses, and connections. Neurons generate an output signal as a sequence of impulses (pulse duration of about 1 msec) or as slow continuous changes of the corresponding nerve endings (axons). The contacts we know best between neurons are called synapses. There exist, however, several other types of contacts (junctions), which have hardly been considered so far in neural net research (Kuffler & Nicholls, 1976; Lynch, 1986; Edelman, Gall, & Cowan, 1987).

Typical Features of Neural Information Processing

A neuron carries out a continuous (analog) evaluation of all positive (excitatory) and negative (inhibitory) input signals and represents the result as the time course of its membrane potential. As long as this fluctuating membrane potential (in the range of about -40 to -90 mV between the intracellular and extracellular) remains below (at more negative values than) the membrane threshold potential (e.g., at -60 mV), the neuron does not generate impulses as an output signal. If, however, the membrane potential crosses the membrane threshold (i.e., more positive than approximately -60 mV) impulses are generated (Hodgkin & Huxley, 1952). The inverse of the impulse intervals is approximately proportional to the suprathreshold membrane potential. In other words, the instantaneous impulse rate (impulses per second) encodes the suprathreshold time course of the membrane potential. There are, however, many neurons (e.g., in the retina)

TABLE 1 Neural Processing Elements

	Brain	Computer
Hardware	10^{11} neurons with 10^3–10^4 variable synapses at the input per neuron and 10^3–10^4 output synapses per neuron; 10^{15} connections	10^7 transistors with one input and one output each; 32 connections between CPU and memory
Programming memory	Self-organization during learning Represented in various functional parameters of neurons and their synapses; analog	Software as a sequence of instructions 10^{11} transistors
Information processing	Parallel analog (discrete or continuous), asynchronous (without a clock)	Sequential, digital, synchronous (with a clock)

which communicate with neighboring neurons directly via slow variations of the membrane potential without encoding output signals into impulse trains.

The impulse train typically leaves the neuron via its axon. Depending on the axonal properties, the sequence of individual impulses travels with propagation velocities ranging from less than 1 m/sec to more than 100 m/sec and reaches many other neurons due to the many synapses at the axonal arborization (Rall, 1977).

There are at least two types of synapse. Excitatory synapses transform an individual impulse into an excitatory transient membrane modulation or excitatory postsynaptic potential (EPSP) toward the membrane threshold; inhibitory synapses transform an impulse into an inhibitory transient membrane modulation or inhibitory postsynaptic potential (IPSP). EPSPs and IPSPs can have a roughly similar time course with a time constant of about 10 msec; however, typically IPSPs have a somewhat longer decay. A single neuron is capable of receiving an asynchronous sequence of many IPSPs and EPSPs from many neurons and of generating a resulting membrane potential time function. The synapses act not only as decoders for the incoming impulse trains but also as evaluators regarding their sign and amplitude (Kuffler & Nicholls, 1976).

For purposes of self-organization, the weighting factors of individual synapses (EPSP or IPSP amplitude up to 10 mV in response to a single impulse) must be modifiable. Functional changes of synapses have been associated with learning in numerous biological studies (Brazier, 1979; Selverston, 1985; Lynch, 1986; Edelman et al., 1987; *Trends in NeuroSciences*, 1988).

Typical Neural Net Topologies in the Primate Brain

The CNS in primates (monkeys, apes, humans) consists of many brain regions with (simplified) various characteristic neural net topologies. Please note the important distinction between architecture, topography, and topology. The term "neural net topology" refers not only to the neuroanatomically evident architecture (or wiring pattern) of a given set of neurons but also to the functional influence of each neuron on the other neurons. It is conceivable that the neural net topology of a given brain region is not static (or only gradually changing during learning) but rather rapidly modifiable.

Neural Net Topology of the Retina

The retina develops during early ontogeny as an outgrowth of a part of the brain called the diencephalon and represents one of the best studied brain regions, partly due to its direct accessibility (Rodieck, 1973; Dowling, 1987). The retina has a neural net topology with five separate layers for photoreceptors, horizontal cells, bipolar cells, amacrine cells, and retinal ganglion cells. The photoreceptor layer in humans consists of about 120×10^6 receptor cells (cones and rods). The retinal output which is represented by the spatio-temporal pattern of the impulse rates of all ganglion cells (about 1.2×10^6 neurons in humans) travels along the optic nerve toward the thalamus.

In spite of more than 30 years of intensive neurobiological research, the retinal microcircuitry and its information processing power have not been completely revealed. Buzz words such as receptive field properties, contrast enhancement, preprocessing, low-level vision, and lateral inhibition barely mask the fact that we still don't know why one needs more than 200×10^6 retinal neurons and what the appropriate mathematical description of the retinal function is. The retina has been the object of both hardware and software simulations (Rosenblatt, 1961; Eckmiller, 1975; Fukushima, 1980; Koenderink, 1987; Sivilotti, Mahowald, & Mead, 1987; Mead, 1989).

Neural Net Topology of the Visual Cortex

The cerebral cortex belongs to the phylogenetically youngest brain region, the telencephalon. The percentage of total brain matter devoted to cerebral cortex gradually increased during evolution from lower vertebrates such as fish (cerebral cortex not existent) to rep-tiles and birds (with a rather small cerebral cortex), to rats, cats, and finally to primates (Thompson, 1967).

Each hemisphere of the cerebral cortex in primates can be thought of as a large, multi-layered surface, which had to be folded in order to fit into the limited space within the skull. The cortex is functionally subdivided into many separate regions with different multilayered neural net topologies. One large portion is called visual cortex and consists of several subregions, each containing an entire retinotopic representation (map). The remaining part of the cerebral cortex consists of other sensory, motor, and association regions (Brodal, 1981; Hyvärinen, 1982; Edelman, Gall, & Cowan, 1984).

The primary visual cortex (area 17) in the cat has a neural net topology with 6–8 layers (depending on the definition). The afferent visual activity pattern from the retina reaches area 17 via the lateral geniculate nucleus (LGN). In fact, the retinal information is already split into at least two independent parallel preprocessed patterns traveling along separate nerve fibers projecting separately to two separate cortical layers (layer 4Ab and layer 4C; see Gilbert, 1983). The corresponding area V1 in macaque monkeys can be subdivided into at least nine layers (Hubel & Wiesel, 1977; Van Essen, 1979; Swadlow, 1983; Eckmiller, 1987b). There seems to be a regular pattern of interlaminar connections (e.g., from layer 4 to the superficial layers 1–3, from layers 2–3 to layer 5, from layer 5 to layer 6, and from layer 6 back to layer 4 (Blasdel, Lund, & Fitzpatrick, 1985; Fitzpatrick, Lund, & Blasdel, 1985). These finding have been taken as morphological evidence for intracortical reciprocal connections and even loops. Functionally different parallel visual pathways for X and Y cells (Eckmiller, 1987b) connect specific layers in the LGN with specific separate layers in area V1.

There is some recent evidence that the neural net topology might dynamically modulate to adjust its function to various modes of the visual system ranging from sleep to alertness with attention to nonvisual sensory inputs to attentive visual pattern recognition or pursuit eye movements. The neural pathways mediating such topology modulation signals could possibly reach area V1 from subcortical regions such as the central thalamus (Schlag & Schlag-Rey, 1984).

Neural Net Topology of the Cerebellar Cortex

The cerebellum consists of the cerebellar cortex with many multilayered subregions for specific sensorimotor functions, which project to one of the deep cerebellar nuclei or the vestibular nuclear complex. Due to its characteristic and surprisingly regular neural net architecture, the cerebellar cortex with its molecular cell layer, granular cell layer as input structures, and the Purkinje cell layer as output (Brodal, 1981) has invited various simulations and functional considerations (Eccles, Ito, & Szentágothai, 1967; Marr, 1969; Albus, 1979; Pellionisz & Llinas, 1979).

In recent years it has become clear that cerebellar inputs typically project in parallel to cerebellar cortical neurons and to the target neurons of the corresponding Purkinje cells in the vestibular or deep cerebellar nuclei. Several possible wiring diagrams for these corticonuclear microcomplexes (CNMC) have been discussed on the basis of a vast body of neurobiological data (Ito, 1984). This CNMC concept is supported by the finding that any lesions of the cerebellar cortex may modify (if only temporarily) the signal flow processing in the cerebellum, whereas lesions in the vestibular or deep cerebellar nuclei result in a total, permanent interruption of this signal flow (Eckmiller & Westheimer, 1983).

Communication between Various Brain Regions

The great variety of brain structures and cortical regions (Brodal, 1981; Baron, 1987) can be described as densely packed ensembles of neurons with various characteristic network topologies, which communicate with each other via numerous nerve fiber bundles (neural pathways or nerves). Signal transport along those pathways (consisting of thousands of axons) occurs by

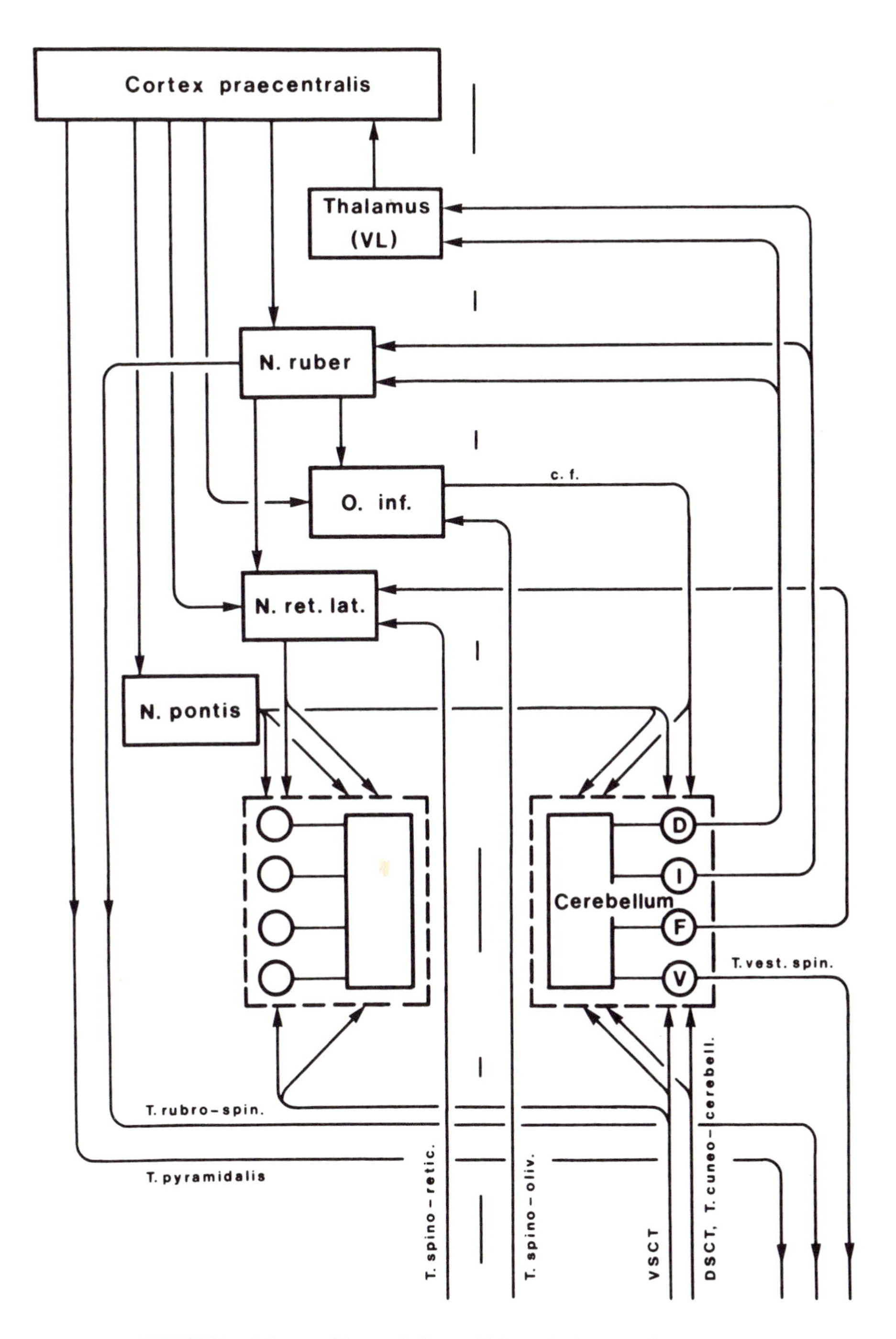

FIGURE 1 Scheme of the cerebellum with its major inputs and outputs.

means of many simultaneous impulse trains. In contrast to conventional digital computers with a clearcut separation between central processing unit, memory, and registers, the CNS can rather be described as a federation of many special-purpose parallel processing networks. Each of these special-purpose networks is capable of independent processing and storage of information for various sensory, associative, or motor functions. According to this hypothesis, the neural impulse rate patterns traveling along the various pathways represent only intermediate results, which are often very difficult to decipher, particularly at the single neuron level. For this reason, several labs have recently developed analysis methods for simultaneous recordings from an ensemble of neurons (Eckmiller, Blair, & Westheimer, 1974; Krüger, 1983; Gerstein & Aertsen, 1985). It is not surprising that in various cases only the final motor pattern (e.g., for speech or gesticulation movements) reveals the processing result, which could not have been predicted even by analyzing the neural activity of the corresponding motor nerve.

Although it is hopeless at the present time to design a circuit diagram of the primate CNS, two schemes are presented here to demonstrate certain kinds of communication networks (of course at a highly simplified level) linking various brain regions. Figure 1 gives a simplified connectivity scheme for the cerebellum with its major inputs and outputs. The cerebellum is subdivided into two halves (separated by the broken vertical lines) to indicate connections with the various ipsilateral (same side) and contralateral brain regions. Each cerebellar half (depicted as a dashed line of a square) consists of the cerebellar cortex (rectangular block) and the corresponding cerebellar nuclei (D, dentate; I, interpositus; F, fastigial; V, vestibular). Whereas the cerebellar output signals arise exclusively in one of these cerebellar nuclei, the input signals reach both cortex and corresponding nuclei in parallel. These input signals reach the cerebellum either as climbing fibers (c.f.) arising in the contralateral inferior olive (O. inf.) or as mossy fibers from various brain regions such as the ipsilateral or contralateral pontine nuclei (N. pontis), as well as the ipsilateral dorsal spinocerebellar tract (DSCT), and the ipsilateral and contralateral ventral spinocerebellar tract (VSCT) ascending from the spinal cord (see also Gluck et al., this volume, Chapter 5).

Figure 1 indicates that the cerebellar nuclei on one side communicate with the contralateral cerebral motor cortex (cortex praecentralis) via the thalamus and that this motor cortex communicates with both halves of the cerebellum via the inferior olive (input to contralateral cerebellum), via the pontine nuclei or even via the cascade of red nucleus (N. ruber) and lateral reticular nucleus (N. ret. lat.). Furthermore both motor cortex (by way of the pyramidal tract) and cerebellum (by way of the vestibular–spinal tract) have direct descending pathways to control or modify various movements at the spinal cord level (Brodal, 1981). Please note that this connectivity scheme is by no means complete and the fact of reciprocal connections between any two brain regions should not be taken as evidence for simple feedback loops.

Whereas Figure 1 shows the major pathways between cerebellum and its inputs and outputs, Figure 2 is a schematic circuit plan (based on about 400 neurobiological studies) of one particular sensorimotor system, namely, for the control of visually guided pursuit eye movements (PEM) in primates (Eckmiller, 1987b). This circuit plan not only depicts connections between the various participating brain regions but it also indicates the corresponding types of neural activity that have been found at these locations in the trained macaque monkey during PEM. The layout of this circuit plan is deliberately open ended to allow the inclusion of additional connections and brain regions without changing the existing basic circuit. The chosen sequence of brain regions (from top to bottom) from retina (RET) via thalamus (LGN, PGN, PUL, IML), pretectum and tectum (NOT, SC), cerebral cortex (V1, V2, MT, MST, 7a, 6, 8), pontine nuclei (PON, DLPN), cerebellum (FL, VER, CN, VN), reticular formation (MRF, PPRF), to the oculomotor nuclei (III, IV, VI) is arbitrary, although it reflects the general direction from sensory to motor regions. Connections are always arranged in a clockwise direction to distinguish downstream pathways on the right (e.g., from RET to SC) from upstream pathways on the left.

The different symbols at the various brain regions in Figure 2 indicate characteristic types of neural activity at these locations:

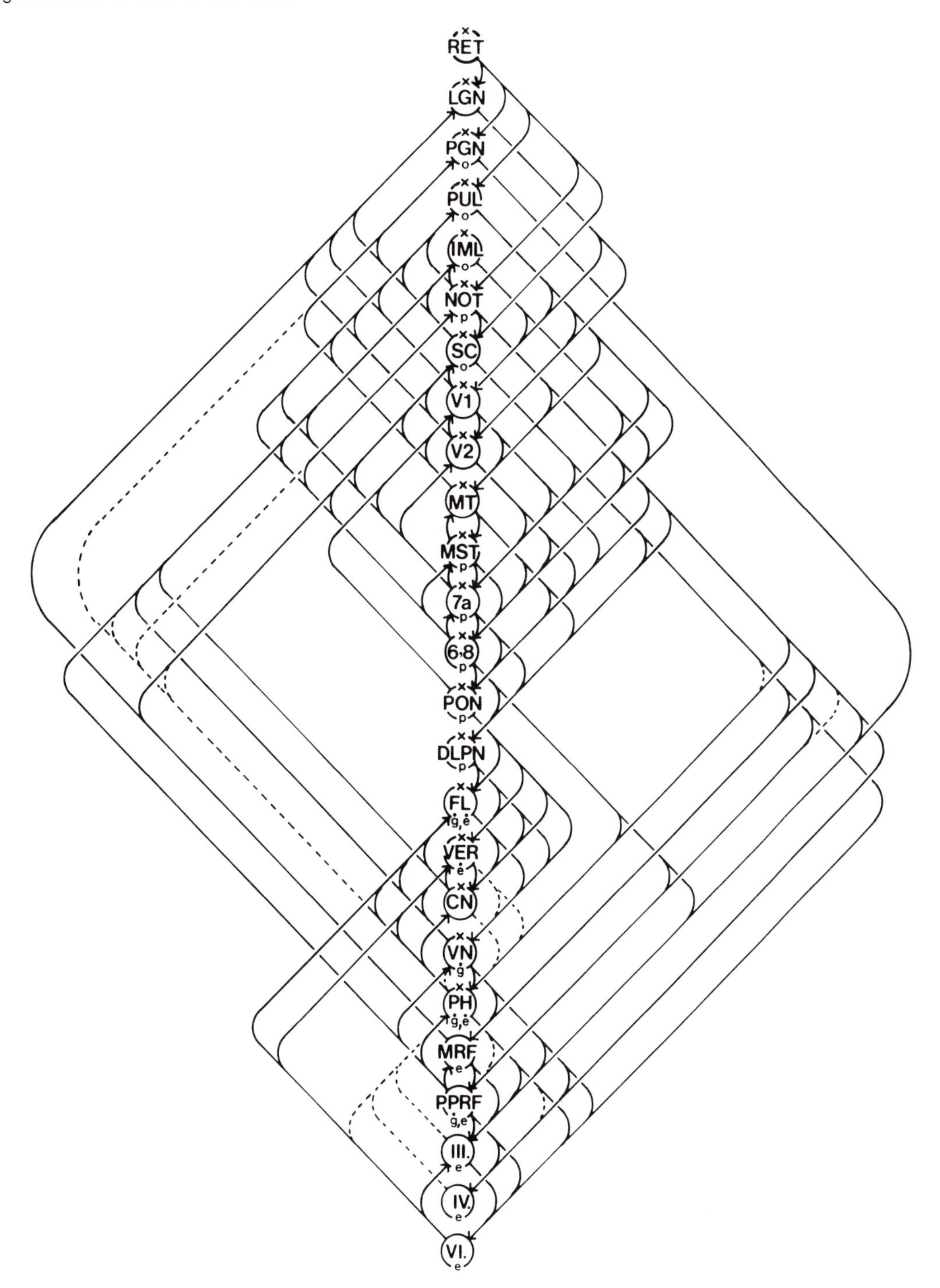

FIGURE 2 Schematic circuit plan of the oculomotor pursuit system in primates.

x: afferent visual activity
o: oculomotor-related (however unrelated with PEM) activity
p: pursuit-related (but not encoding PEM velocity) activity
$\dot{g}$: gaze velocity correlated activity during PEM or head movements
$\dot{e}$: pursuit eye velocity coded activity
e: eye position (superimposed on eye velocity) coded activity

Figure 2 ignores the fact that the various neural processing stages typically consist of subdivisions or layers that may be the exclusive source or target of individual pathways. Pursuit related activity (p) has been found in alert monkeys to occur in area MST, posterior parietal cortex area 7a, the precentral cortex area 6 and area 8, as well as in some pontine nuclei in the brain stem. These pursuit-related neurons are specifically activated during PEM in a given direction (e.g., right or left), even during temporary target disappearance. Since only the precentral (motor) cortex has direct projections (see Figure 2) to all other brain regions (including NOT) where p activity occurs, it has been proposed as the most likely source of its origin (Eckmiller, 1987b). This postulate (which needs further experimental verification) implies that the motor program for PEM arises from a first draft signal p in the oculomotor related regions of the motor cortex. Figure 2 indicates that this first draft signal p reaches the pontine nuclei (PON, DLPN) as input stages for the cerebellum (cerebellar cortex flocculus, FL; vermis, VER; cerebellar nuclei, CN; vestibular nuclei, VN) and in parallel both nucleus prepositus hypoglossi (PH) and pontine reticular formation (PPRF). Interestingly, the refined pursuit motor signal occurs as gaze velocity ($\dot{g}$) and than as eye velocity ($\dot{e}$) activity in the cerebellar cortex as well as in PH and PPRF. Afferent visual activity is available all the way down to PH, possibly to optimize the pursuit motor program by providing retinal error signals (position error and slip velocity). The final step requires neural integration of the eye velocity signal ($\dot{e}$) into an eye position signal (e) as control signal for the ocular motoneurons in one of the oculomotor nuclei (III, IV, VI) which project to one of the 12 eye muscles.

These two examples for neurobiological communication schemes (Figures 1 and 2) emphasize several concepts of information processing in the CNS.

1. Many brain regions are reciprocally interconnected and thereby do not allow a clear logical separation and localization of the various sensory and motor functions.
2. An adequate control theory for neural net federations such as the CNS does not yet exist, not even for small parts such as retina or cerebellar cortex.
3. The general finding of fault tolerance and recovery from significant brain lesions is probably based on neural plasticity, that is, the capability of a given brain region to adjust its topology to new or changing tasks and even to take over the function of another, damaged brain region.

INTELLIGENT ROBOTS AS A FEDERATION OF ARTIFICIAL NEURAL NETS

The vertebrate CNS with its many distinct brain regions can be considered to be a federation of special-purpose neural nets (e.g., visual cortex, cerebellum). The hierarchy and communication lines within this federation as well as the special purpose functions of its members evolve during early stages of ontogeny on the basis of genetically defined initial neural net topologies, learning rules for self-organization, and learning opportunities.

One could argue that it is impossible to decode (and copy) an information processing system such as the CNS of a primate, which required many million years to evolve. However, the rapidly growing research in the field of neural computers (Feldman & Ballard, 1982; Eckmiller & von der Malsburg, 1988; see also von der Malsburg, this volume, Chapter 22) is based on the optimistic assumptions that

1. it is possible to technically implement various artificial neural net topologies having the capability of self-organization.
2. it is possible to design and subsequently imprint learning rules that lead to robust and rapidly

converging self-organization of a neural net with a given initial topology. Such self-organization may even lead to changes (updates) in a previously established self-organization.

3. it is possible to generate, and possibly even predict, a desired information processing function by exposing an initial neural net topology with its imprinted learning rules to a set of learning opportunities including experience of imperfect performance in the learning phase.
4. following a certain amount of learning experience, a given neural net is capable of generalizing from a limited set of correctly learned functions to an entire class of special purpose functions (e.g., robot pointing to any marked point within the entire pointing space independent of initial robot position).

Intelligent robots are currently being conceived and designed that consist of various neural net modules with special purpose functions such as pattern recognition, associative memory, internal representation of patterns and trajectories, generation of motor programs, and sensory as well as motor coordinate transformation (Kuperstein, 1987; Barhen, Dress, & Jorgensen, 1988; Eckmiller, 1988).

The ear and eye modules in Figure 3 represent modules for visual and auditory trajectory detection (not necessarily including pattern recognition). Both sensory input modules receive signal time courses, which occur as spatio-temporal trajectories (Figure 4) at the level of the corresponding sensory receptor array (retinal or basilar membrane in vertebrates). For example, the vision module (eye) monitors the event of a letter "b" being written by a teacher on the x–y plane, whereas the hearing module (ear) is thought to monitor the acoustic event of a spoken "b".

It is assumed here that each sensory module is connected with the internal representation module via a specific sensory coordinate transformation module (depicted as three-layered structure). The function of these coordinate transformation modules is to generate a normalized spatio-temporal trajectory to be

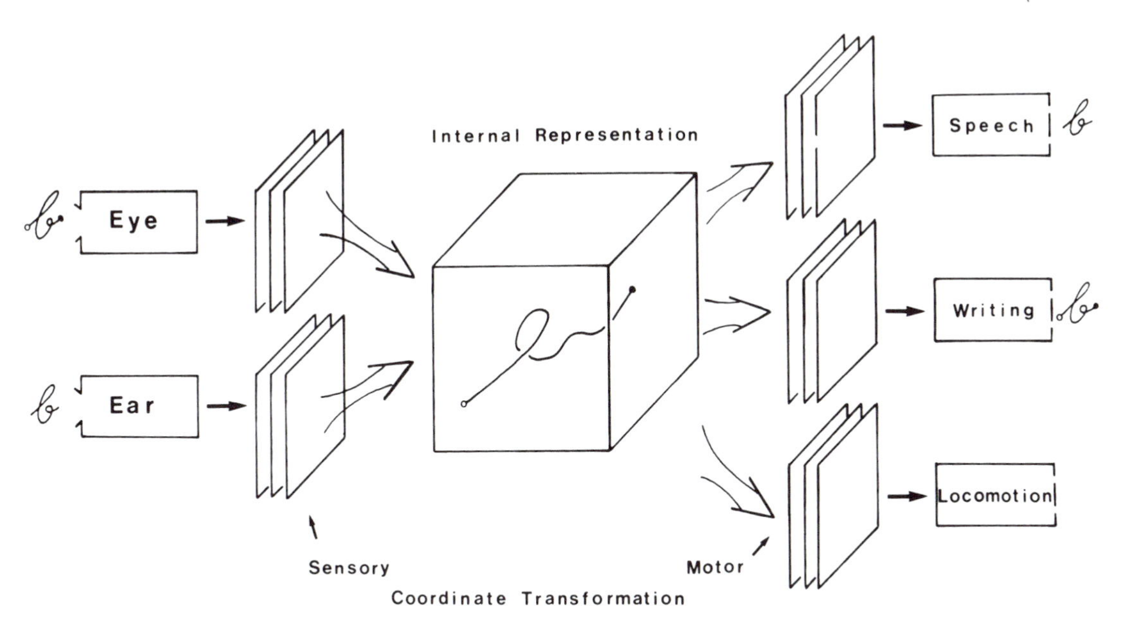

FIGURE 3 Scheme of an intelligent robot consisting of several neural net modules.

stored in the internal representation module. For the sake of simplicity, one might think of these internally stored trajectories as paths in the three-dimensional space of a neural net, along which a particle, neural activity, peak, or soliton (Lamb, 1980) can travel with constant or varying speed. The internal representation module can store very large amounts of different trajectories, which could share neural elements in its three-dimensional network. This model contains elements of the engram hypothesis (Lashley, 1950).

The subsequent generation of a corresponding movement trajectory for drawing a letter "b" or speaking a "b" is assumed here to require the activation of the appropriate stored trajectory. Please note that pattern recognition is not required for these two sensorimotor tasks. In fact, the pattern recognition process might operate on the basis of internally stored trajectories, which would be available for iterative recognition procedures after the end of the sensory event. For the purpose of motor program generation, an activity peak travels along the stored path on the neural net of the internal representation module with a predefined velocity, thus generating various time courses for the spatial component of the desired trajectory.

The desired trajectory can be thought of as the three-dimensional trajectory of the fingertips while drawing a "b" or as the virtual center of contraction of all speech muscles while vocalizing a "b".

In such typically redundant motor systems, specific motor coordinate transformation modules serve to map the internally generated desired trajectory onto a set of simultaneous control signals for the individual actuators (motors, muscles). The scheme of an intelligent robot in Figure 3 emphasizes the notion that large portions of information processing in both primates and robots with neural net control may primarily involve *handling of spatio-temporal trajectories* for the purpose of detection, recognition, storage, generation, and mapping. The time parameter in these trajectory-handling neural nets is not defined by a central clock but is implicitly represented by the propagation velocity of neural activity within the neural net.

Obviously the presently available general purpose digital computers are not particularly well suited for trajectory handling, whereas special purpose neural nets with a clear advantage for trajectory handling are particularly ill equipped for most algebraic analytical operations. This principle difference between self-organizing neural computers and software controlled digital computers should, however, not be interpreted as a problem but rather as indication for two applications areas with minimal overlap.

Neural Nets for the Management of Sensory Trajectories

Sensory stimuli, be they auditory, visual, tactile, or vestibular, arrive at specific sense organs or arrays of receptor cells and generate trajectories or space–time functions of neural activity in real time. The available information can be decoded as a pattern of stimulation intensity, I, at sensory location, s, as a function of time, t. Such an intensity-modulated trajectory $I\ (s,t)$ as depicted in Figure 4 may occur just once, but has to be stored for purposes of recognition, association, or later generation of corresponding motor trajectories. Think, for example, of a young child's ability to perform a vocalization on the basis of a previously heard new word without the need to make several false attempts and without even recognizing the meaning of this word. Location, s, in this scheme may correspond with retinal location or with sound frequency (location on the basilar membrane). Stimulus intensity, I, is represented in Figure 4 by the trajectory width. In order to store the time parameter along the path s, we require the representation of time as spatial location in a specialized neural map. According to my hypothesis, sensory trajectories are being stored such that their spatial parameters are stored as a path on one neural map, whereas their temporal parameters are stored as a separate path on another neural map. Both neural maps can be thought of as two-dimensional surfaces consisting of neurons which have connections only to their neighbors.

One important aspect of sensory trajectory analysis concerns velocity detection of the moving object (e.g., visual target). Various types of velocity-sensitive visual neurons in the vertebrate visual system have

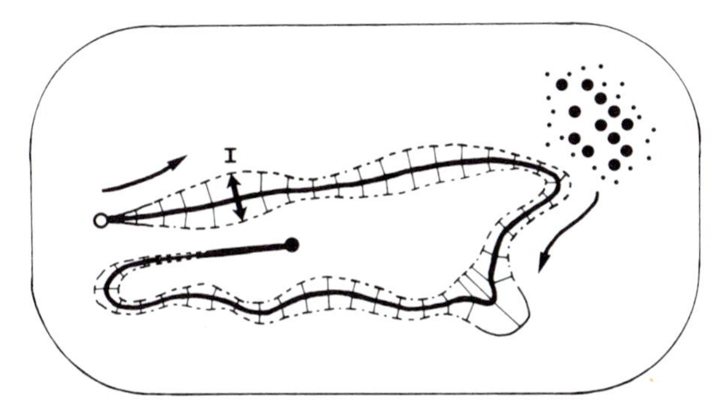

FIGURE 4 Intensity-modulated sensory trajectory $I(s,t)$ on a sensory neural net.

been simulated by means of a special-purpose analog computer (Eckmiller, 1975). This electronic retina consists of discrete electronic circuits for 115 photoreceptors (with phototransistors as light-sensitive input elements), 115 bipolar cells, 25 horizontal cells, 25 amacrine cells, and 25 retinal ganglion cells. The topology of this small portion of the five-layered retinal network could be selected by means of various patchboard connections. While signal processing occurs in the first four layers by evaluating the graded membrane potential time courses of various neurons as positive or negative, the ganglion cells in the fifth layer generate impulse rate time courses as retinal output. More recent hardware models of the retina have used analog VLSI technologies (Sivilotti et al., 1987; Mead, 1989; Faggin & Mead, this volume, Chapter 15).

An example of capability of such an artificial neural network for the simulation of biological data is given in Figure 5 (Eckmiller & Grüsser, 1972). Class 2 neurons in the frog retina are not only qualitatively move-

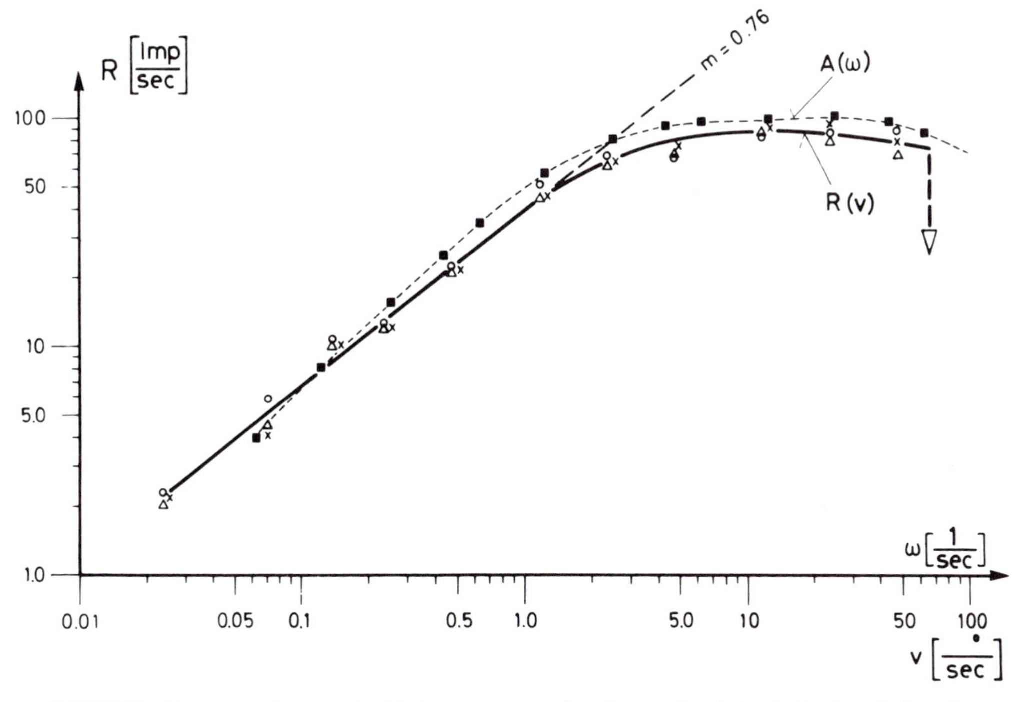

FIGURE 5 Simulation of the relationship between average impulse rate R and visual stimulus velocity v for a class 2 neuron in the frog retina. Slope m gives the exponent of the power function. At high stimulus velocities, the ganglion cell stops generating impulses (in agreement with neurophysiological data) thus leading to an abrupt drop of the characteristic (vertical arrow).

ment sensitive but encode the target velocity in the average impulse rate R (impulses per second) as a power function over a velocity range of more than two log units. The velocity characteristic in Figure 5, which fits that of biological class 2 neurons (with m = 0.7 as average exponent of the power function) very well was achieved by connecting 7 neighboring photoreceptors (as receptive field center) and the surrounding 30 photoreceptors (as the corresponding antagonistic periphery of the receptive field) directly with one ganglion cell. The spontaneous impulse rate was adjusted to $R_0 = 0.9$ (impulses per second).

Neural Nets for the Management of Motor Trajectories

Biological motor control systems for rhythmic functions such as those controlling heartbeat, locomotion, and respiration (Miller & Scott, 1977; Grillner & Wallén, 1985; Selverston, 1985) are assumed to be based on a small number of neurons. In these cases the time parameter is represented by oscillations and time constants of neural activity within a small neural net that is capable of internally generating the required time functions even without sensory feedback from muscles or joints.

For quasiballistic goal directed movements (e.g., reaching, pointing, saccadic eye movements) various mathematical and neural models have been proposed to explain the typical correlation between amplitude, velocity profile, and duration of the movement (Miles & Evarts, 1979; Arbib, 1981; Eckmiller, 1983; Hogan, 1984; Haken, Kelso, & Bunz, 1985; Berkinblit, Feldman, & Fukson, 1986).

A quite different kind of neural motor program generator is required for nonrhythmic and nonballistic smooth movements such as speech movements, voluntary limb movements (dancing, drawing, or gesticulating), or pursuit eye movements. Biological data on neural networks that act as function generators for such smooth movement trajectories are scarce (Morasso & Mussa Ivaldi, 1982; Soechting, Lacquaniti, & Terzuolo, 1986; Baron, 1987; Eckmiller, 1987b).

Various research groups have recently begun to consider self-organization in neural nets for robot control by using a combination of initial neural net topology, learning rules, and learning opportunities (Grossberg, 1969, 1970; Raibert, 1978; Albus, 1979; Pellionisz & Llinas, 1985; Kawato, Furukawa, & Suzuki, 1987; Kuperstein, 1987; Barhen et al., 1988; Eckmiller, 1988; Ritter & Schulten, 1988).

Generation of Velocity Time Functions with a Neural Triangular Lattice

The generation of nonrhythmic and nonballistic smooth velocity time courses requires a flexible neural net acting as a function generator. A neural net model (Eckmiller, 1987a) for internal representation and generation of two-dimensional movement trajectories including PEM, is described in the next section. This model accounts for sensory updating, prediction, and storage, and it is biologically (at least in principle) plausible, though not based on experimental data. The key features of this neural function generator can be summarized as follows.

1. The velocity time course in one movement direction is represented by the trajectory of a neural activity peak (AP) that travels with constant velocity from neuron to neuron.
2. The neural net is arranged as a neural triangular lattice (NTL) on a circular surface.
3. The NTL output signal is proportional to the eccentricity of AP relative to the NTL center.

NTL Topology Consider a large number of identical processing elements (neurons) with analog features. These neurons are arranged in a neural triangular lattice (NTL) in which they are connected only to their six immediate neighbors.

Such a NTL with a radius of 50 neurons from NTL center to its periphery would consist of about 8,000 neurons. The connectivity strength, C_T, of tangential synaptic connections between neurons located along concentric circles about the NTL center is slightly larger than C_R, for radial synapses. Each neuron has a subthreshold potential, P, similar to the membrane potential for biological neurons. P is always assumed to represent the average of the potential values of all immediately neighboring neurons due to continuous equilibration of possible potential gradients via neural connections (junctions). External potential changes can only be applied to the NTL center thus yielding a center-symmetrical potential field. Such an input-dependent potential field can be thought of as an elastic circular membrane whose center is being pushed up or down, while being held at the circular rim.

NTL Dynamics A suprathreshold activity peak (AP) can be initiated (or extinguished) only at the NTL center, and becomes superimposed on the potential field. For sake of simplicity AP can be thought of as a neural action potential (Rall, 1977) or a soliton (Lamb, 1980), which travels away from the NTL center (following the potential field gradient during a positive potential input) in the same angular direction (3 o'clock).

AP travels with constant propagation velocity from one neuron to one of its neighbors, and its activity does not otherwise spread. Every time AP arrives at a neuron, a decision has to be made concerning its next destination. Due to certain transient constraints imposed on the adjacent synapses by the traveling AP, only those three of the six neighboring neurons can be selected that are located in forward (straight, right, or left) continuation of the trajectory. The decision of which of the three possible target neurons is selected is based on a combined evaluation of the connectivity strength, c, and the potential gradient dP/dR between the neighboring neurons. If the potential field P is flat throughout the NTL, AP will select the tangentially adjacent neuron, since C_T is slightly larger than C_R for the radial connections. In case of a positive potential input at the NTL center, however, the potential field exhibits a radial gradient toward the periphery, which in effect pulls AP toward the NTL periphery (equivalent to higher velocities at the NTL output). Similarly, a negative input attracts AP toward the center portion of a neural triangular lattice (NTL) with a traveling activity peak.

A portion of the NTL together with AP (filled circle) and the last portion of the memory trace of its trajectory is shown in Figure 6. Horizontal connections

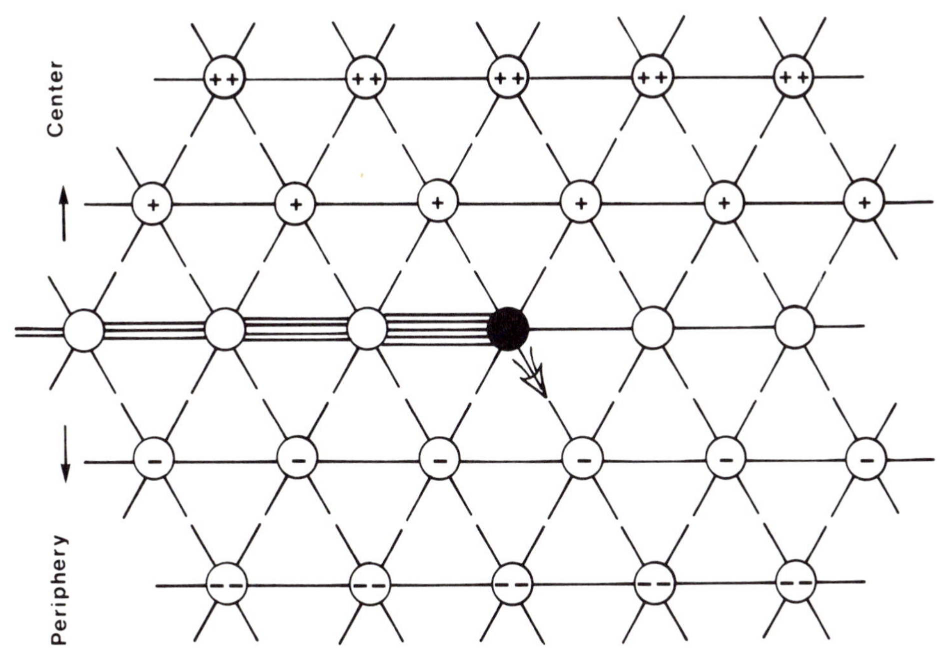

FIGURE 6 Portion of a neural triangular lattice (NTL) with a traveling activity peak (AP).

in Figure 6 are assumed to follow circles about the NTL center. AP had traveled from the left and is about to move toward the NTL periphery due to an assumed sudden positive step input at the NTL center.

A memory trace of the most recently traversed portion of the trajectory is created by means of a temporary increase of the connectivity strength as indicated by the number of connecting lines in Figure 6. It is assumed that the memory trace, which gradually fades with a time constant of about 1 sec, serves as a trajectory guide for subsequent movements, for example during postpursuit eye movements after sudden target disappearance, and for neural prediction during pursuit of a periodically moving target. Once a memory trace exists, it can be used to reduce the amount of time necessary for updating during periodical movements in one dimension (e.g., horizontal) which are generated by repeated creation of the same alternating velocity trajectories on the two velocity NTLs (Figure 7).

The potential fields on the two velocity NTLs, including the location of AP and the gradually increasing memory trace, are shown in Figure 7 at four successive times. AP trajectories always start and end in the NTL center, which corresponds to zero velocity. When the NTL input signal changes from zero to a positive or negative value, the topology of the two NTLs changes by equal amounts in opposite directions. It is assumed here that AP had been initiated in the center of the NTL for velocities to the right (left half of Figure 7) at time zero and was traveling toward the periphery (starting in the 3 o'clock direction) in response to a positive input. The memory trace is indicated as a dotted line. At 0.4 sec, AP is traveling on a circle due to the flat potential field (zero input). At 0.6 sec AP travels toward the periphery in response to a positive input and, at 0.8 sec, toward the center due to a negative input. Please note that the potential field of the other NTL is always identical except for the sign and the absence of an AP. At 1.0 sec, AP had already reached the center of the NTL and become extinguished there. Simultaneously AP became initiated at the opposite NTL for velocities to the left (right half of Figure 7) and is now traveling toward the periphery. The radial location of AP is continuously being monitored by neurons, which serve as eccentricity detectors (see Figure 4).

The correlation between two typical velocity time courses $\dot{\theta}(t)$ and the corresponding trajectories on an NTL are depicted in Figure 8. The following assumptions have been made for this example: the velocity function generator should be able to generate any smooth velocity time course for velocities up to 50°/sec and accelerations up to 200°/sec^2.

The constant time it takes for AP to travel from one neuron to the selected neighbor is $T = 5$ msec, which corresponds to a velocity $v = 200$ neurons/sec. The direct radial distance between the center and the circular NTL periphery border shall be covered by 50 neurons.

NTL Mechanisms for Storage and Retrieval of Trajectories

The detailed synaptic connectivity pattern of three adjacent NTL neurons is shown in Figure 9 to describe the architecture of an element of this triangular lattice. NTL neurons (large open circles with axon-like projections and synapses) are reciprocally connected via synapses. Two additional neurons with presynaptic synapses (synapse on another synapse), a pattern retrieval neuron (PR), and a velocity modulation neuron (VM) have contacts with the NTL. The PR neuron belongs to a large population of identical PR neurons for storage and retrieval of entire trajectories that always begin and end in the NTL center. It is assumed that each PR neuron has excitatory synapses (small circles) on each of the NTL synapses. In contrast, the VM neuron has inhibitory synapses. (depicted as open blocks) on each of the NTL synapses in order to control the propagation velocity, v, of AP on the NTL.

When a given trajectory has to be learned (stored) while being generated by means of input modulation of the potential field, one of the PR neurons (that has not been used before) is first selected for this task, as indicated by the large filled arrow at the bottom of Figure 9.

In the present example, it is assumed that AP travels along a given trajectory (from bottom left to right) including the two hatched NTL synapses. The connec-

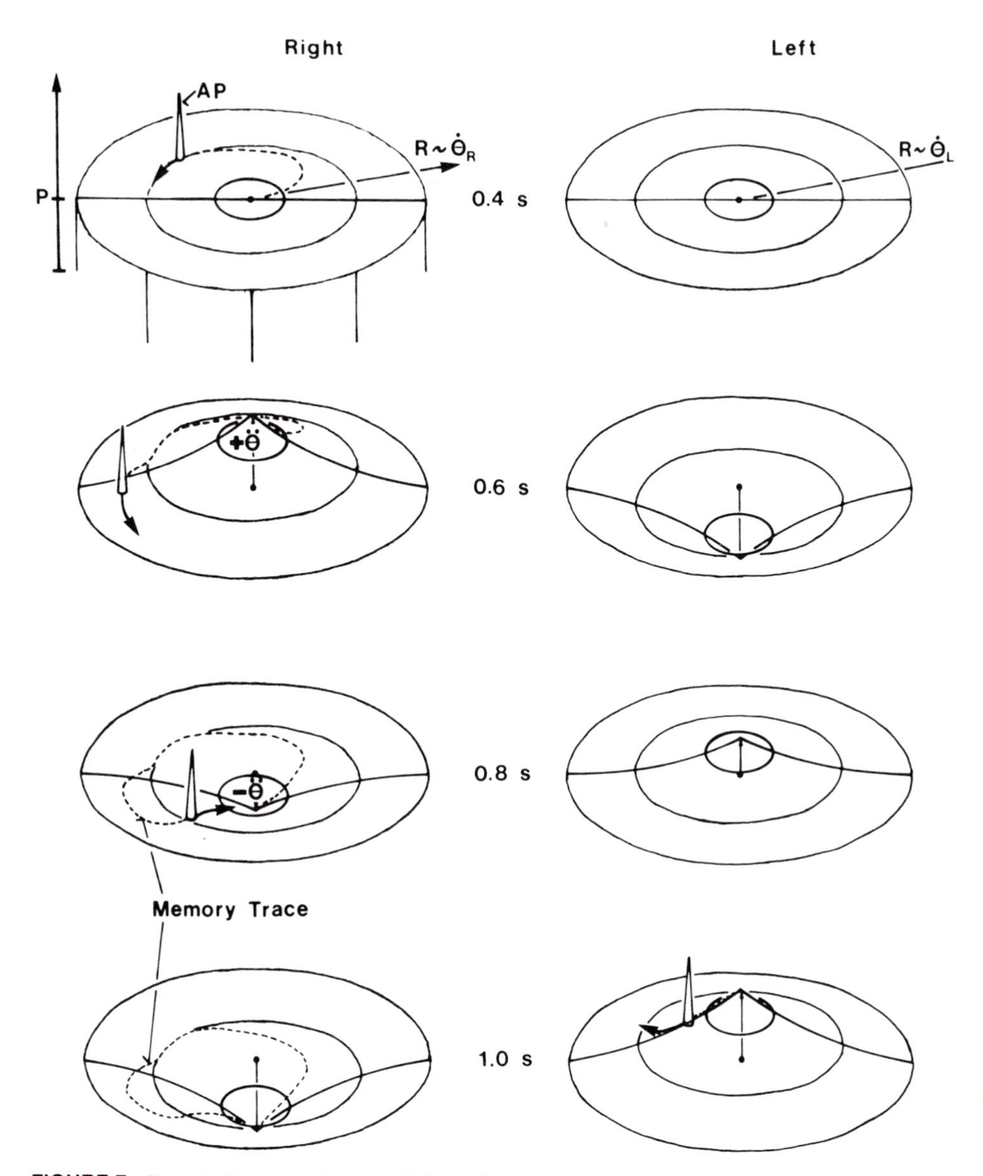

FIGURE 7 Potential fields (P) of a pair of NTLs at four consecutive times. Activity peak (AP) had been initiated in the center of the left NTL at time $t = 0$. Development of a trajectory of a traveling AP on a pair of NTLs encoding velocity to the right or to the left at four subsequent times. The membrane-like NTL surfaces indicate the potential fields P.

tivity strength of those few synapses, which belong to the selected PR neuron and have a presynaptic connection with the NTL synapses along the AP trajectory (here, the two hatched synapses), becomes permanently increased. This process is assumed here to represent learning (memory) and is indicated by an enlargement

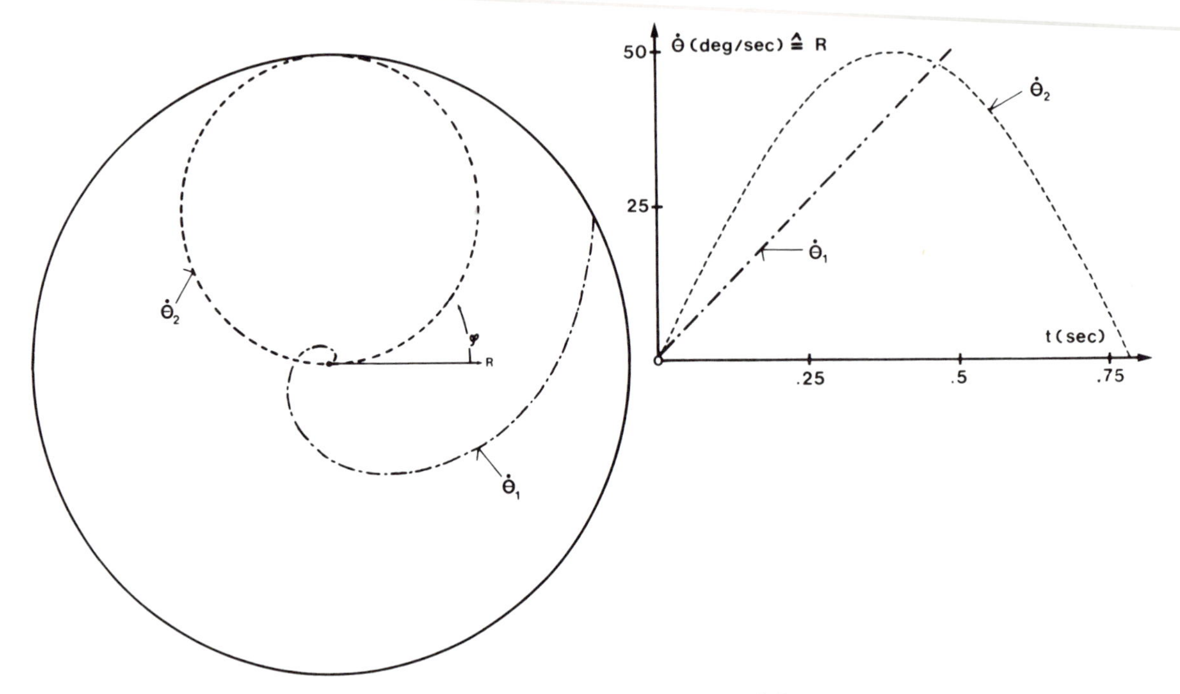

FIGURE 8 Examples for the representation of two velocity time courses ($\dot{\theta}_1, \dot{\theta}_2$) as trajectories on a NTL.

of the two corresponding PR synapses (fiiled circles). The synapse connectivity of the other synapses as well as of all synapses of the nonselected neurons remains low (except for those synapses that were involved in storing other trajectories). In this way a single PR neuron stores one trajectory during the process of its first generation on the NTL. The latter retrieval of this permanently stored trajectory is implemented by a brief activation pulse for the corresponding PR neuron. This activation pulse is assumed to yield a strong temporary enhancement of those NTL synapses that compose the desired trajectory, thus creating a memory trace. Immediately afterward AP becomes initiated in the NTL center. AP travels (although the potential input to the NTL center is zero) along this temporarily existing memory trace as in the groove of a record. Again the NTL output monitors the radial AP distance during constant propagation velocity of AP and thereby generates the previously stored velocity time function.

The concept of a motor program generator with NTL topology and a traveling activity peak incorporates elements of soliton theory (Lamb, 1980) and of cellular automata theory (Toffoli & Margolus, 1987).

Generation of Visually Acquired Two-Dimensional Trajectories

The NTL function generator (Figures 6–9) is equipped with mechanisms for storage and retrieval of numerous smooth velocity time courses. A pair of NTLs would be required to generate one-dimensional movements in both directions plus subsequent neural integrators to generate various movement trajectories. However, the NTL concept can be further optimized to allow for generation of two-dimensional trajectories. Consider an intelligent robot with neural net modules (Figure 3) which has to learn how to draw visually monitored patterns.

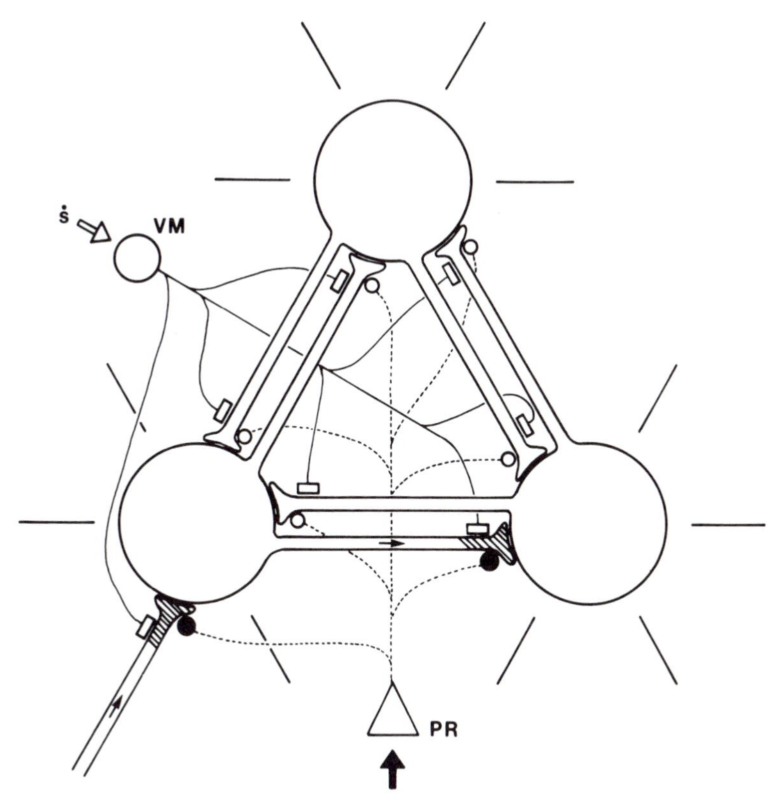

FIGURE 9 Neural net topology of three adjacent NTL neurons together with a pattern retrieval (PR) neuron and a velocity modulation (VM) neuron.

The proposed process is schematically demonstrated in Figure 10 by means of four diagrams (Eckmiller, 1987c). The upper left diagram gives a typical writing trajectory for letter "b" in the *x–y* plane starting at the open circle and stopping at the filled circle. It is assumed that the intelligent robot has the following modules available.

1. Vision module with retinotopic internal representation and with neural detectors of target eccentricity relative to the fovea, that is, position error as well as target slip velocity and target slip acceleration. The target is the tip of a writing element, which the teacher uses to write/draw on the *x–y* plane.
2. Velocity NTL with potential input at the NTL center as shown in the upper right of Figure 10. AP travels with constant velocity, $v = c$.
3. Position NTL (lower right in Figure 10) with a velocity modulation (VM) neuron at the input to modulate the travel velocity of AP. In this case, AP position on the NTL is not monitored as radial distance from the NTL center but as horizontal (x^*) and vertical (y^*) eccentricity. The VM neuron has a tonic activity, which normally inhibits AP movement on the position NTL. However, the output of the velocity NTL inhibits the VM neuron activity, thus yielding a propagation velocity of AP on the position NTL proportional to the output signal of the velocity NTL.

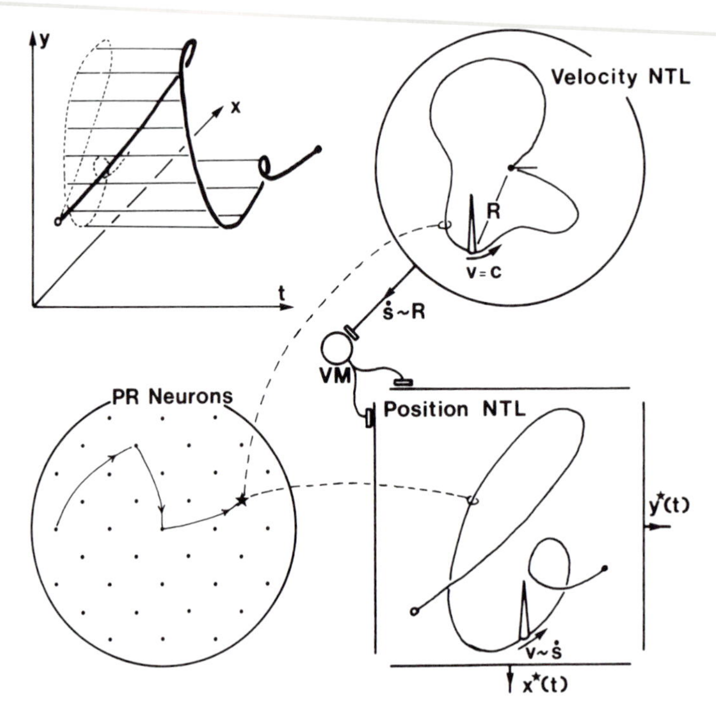

FIGURE 10 Scheme for generation of the movement trajectory for a letter "b" by means of a velocity NTL and a position NTL.

Learning and subsequent drawing or writing of a two-dimensional trajectory such as the letter "b" consist of the following functions.

1. Teacher draws letter "b" with writing element in x–y plane.
2. Vision module detects acceleration, $\ddot{s}$, of writing element in the direction of drawing trajectory, $s(t)$ and uses $\ddot{s}(t)$ to modulate the potential input of the velocity NTL. This operation yields a velocity $\dot{s}\,(t)$, whose trajectory is immediately stored by a selected PR neuron (depicted as star in group of PR neurons; lower left in Figure 10).
3. Upon completion of the têacher's drawing task, the vision module projects the entire pattern of letter "b" onto the position NTL, where the same selected PR neuron (having synapses on both NTLs) stores the pattern.
4. If the robot has to draw the newly stored letter "b" sometime later, the corresponding PR neuron (star) becomes briefly activated, thus generating the memory traces on both NTLs. The start location of AP on the position NTL is not the center but the start position (open circle) of the trajectory for letter "b". The simultaneously traveling APs on both NTLs generate the appropriate horizontal and vertical components of the desired trajectory to drive a two-dimensional drawing device (e.g., x–y plotter).

This scheme does not require a neural integrator. The spatial parameters of the trajectory are stored on the position NTL, whereas the temporal aspects are stored separately on the velocity NTL. This important concept can be compared with the situation of a car on a race track, in which the driver is only told how to change his speed with time, but not how the track is formed. Only when he actually drives along the race track (equivalent to a two-dimensional path) can the spatio-temporal trajectory unfold and be monitored by two orthogonally placed eccentricity detectors.

Acknowledgments

This work was supported in part by the Deutsche Forschungsgemeinschaft (SFB 200/B10) and the Bundesministerium für Forschung und Technologie (BMFT).

References

Albus, J. S. (1979). Mechanisms of planning and problem solving in the brain. *Mathematical Biosciences*, **45**, 247–293.

Arbib, M. A. (1981). Perceptual structures and distributed motor control. In J. M. Brookhart, V. B. Mountcastle, & V. B. Brooks (Eds.), *Handbook of physiology, section 1: the nervous system* (Vol. II. *Motor control*, Part 2). Baltimore: Williams & Wilkins.

Barhen, J., Dress, W. B., & Jorgensen, C. C. (1988). Applications of concurrent neuromorphic algorithms for autonomous robots. In R. Eckmiller & C. von der Malsburg (Eds.), *Neural computers*. Heidelberg: Springer-Verlag.

Baron, R. J. (1987). The high-level control of movements. In R. J. Baron (Ed.), *The cerebral computer—An introduction to the computational structure of the human brain*. Hillsdale, NJ: Erlbaum.

Berkinblit, M. B., Feldman, A. G., & Fukson, O. I. (1986). Adaptability of innate motor patterns and motor control mechanisms. *Behavioral and Brain Science*, **9**, 585–638.

Blasdel, G. G., Lund., J. S., & Fitzpatrick, D. (1985). Intrinsic connections of macaque striate cortex: Axonal projections of cells outside lamina 4C. *Journal of Neuroscience*, **5**, 3350–3369.

Brazier, M. A. B. (Ed.) (1979). *Brain mechanisms in memory and learning: From the single neuron to man*. New York: Raven.

Brodal, A. (1981). *Neurological anatomy*. New York: Oxford University Press.

Denker, J. S. (Ed.) (1986). Neural networks for computing. *AIP Conference Proceedings*, 151.

Descartes, R. (1632). *Traité de l'homme*, (translated and commented by K. E. Rothschuh, *Über den Menschen*, 1969). Heidelberg: Verlag Lambert Schneider.

Dowling, J. E. (1987). *The retina—An approachable part of the brain*. Cambridge, MA: Harvard Univ. Press.

Eccles, J. C., Ito, M., & Szentágothai, J. (1967). *The cerebellum as a neuronal machine*. New York: Springer-Verlag.

Eckmiller, R. (1975). Electronic simulation of the vertebrate retina. *IEEE Transactions on Biomedical Engineering*, **BME-22,** 305–311.

Eckmiller, R. (1983). Neural control of foveal pursuit versus saccadic eye movements in primates-single unit data and models. *IEEE Transactions on Systems, Man, and Cybernetics*, **SMC-13,** 980–989.

Eckmiller, R. (1987a). Computational model of the motor program generator for pursuit. *Journal of Neuroscience Methods*, **21**, 127–138.

Eckmiller, R. (1987b). Neural control of pursuit eye movements. *Physiological Reviews*, **67**, 797–857.

Eckmiller, R. (1987c). Neural network mechanisms for generation and learning of motor programs. *Proceedings of the IEEE First International Conference on Neural Networks, IV*. 545–550.

Eckmiller, R. (1988). Neural networks for motor program generation. In R. Eckmiller & C. von der Malsburg (Eds.), *Neural computers*. Heidelberg: Springer-Verlag.

Eckmiller, R., Blair, S. M., & Westheimer, G. (1974). Oculomotor neuronal correlations shown by simultaneous unit recordings in alert monkeys. *Brain Research*, **21**, 241–250.

Eckmiller, R., & Grüsser, O. J. (1972). Electronic simulation of the velocity function of movement-detecting neurons. *Bibliotheca Ophthalmologica*, **82**, 274–279.

Eckmiller, R., & Westheimer, G. (1983). Compensation of oculomotor deficits in monkeys with neonatal cerebellar ablations. *Experimental Brain Research*, **49**, 315–326.

Eckmiller, R., & von der Malsburg, C. (Eds.) (1988). *Neural computers*. Heidelberg: Springer-Verlag.

Edelman, G. M., Gall, W. E., & Cowan, W. M. (Eds.) (1984). *Dynamic aspects of neocortical function*. New York: Wiley.

Edelman, G. M., Gall, W. E., & Cowan, W. M. (Eds.). (1987). *Synaptic function*. New York: Wiley.

Feldman, J. A., & Ballard, D. H. (1982). Connectionist models and their properties. *Cognitive Science*, **6**, 205–254.

Fitzpatrick, D., Lund, J. S., & Blasdel, G. G. (1985). Intrinsic connections of macaque striate cortex: Afferent and efferent connections of lamina 4C. *Journal of Neuroscience*, **5**, 3329–3349.

Fukushima, K. (1980). Neocognitron: A self-organizing neural network model for a mechanism of pattern recognition unaffected by shift in position. *Biological Cybernetics*, **36**, 193–202.

Gerstein, G., & Aertsen, A. (1985). Representation of cooperative firing activity among simultaneously recorded neurons. *Journal of Neurophysiology*, **54**, 1513–1528.

Gilbert, C. D. (1983). Microcircuitry of the visual cortex. *Annual Review of Neuroscience*, **6**, 217–247.

Grillner, S., & Wallén, P. (1985). Central pattern generators for locomotion, with special reference to vertebrates. *Annual Review of Neuroscience*, **8**, 233–261.

Grossberg, S. (1969). Some networks that can learn, remember and reproduce any number of complicated space–time patterns, I. *Journal of Mathematics and Mechanics*, **19**, 53–91.

Grossberg, S. (1970). Some networks that can learn, remember, and reproduce any number of complicated space–time patterns, II. *Studies in Applied Mathematics*, **49,** 135–165.

Haken, H., Kelso, J. A. S., & Bunz, H. (1985). A theoretical model of phase transitions in human hand movements. *Biological Cybernetics*, **51**, 347–356.

Hodgkin, A. L., & Huxley, A. F. (1952). A quantitative description of membrane current and its application to conduction and excitation in nerve. *Journal of Physiology*, **117**, 500–544.

Hogan, N. (1984). An organizing principle for a class of voluntary movements. *Journal of Neuroscience*, **4**, 2745–2754.

Hubel, D. H., & Wiesel, T. N. (1977). Ferrier lecture. Functional architecture of macaque monkey visual cortex. *Proceedings of the Royal Society of London, Series B*, Biological Sciences, **198**, 1–59.

Hyvärinen, J. (1982). *The parietal cortex of monkey and man*. Berlin: Springer-Verlag.

Ito, M. (1984). *The cerebellum and neural control*. New York: Raven.

Kawato, M., Furukawa, K., & Suzuki, R. (1987). A hierarchical neural network model for control and learning of voluntary movement. *Biological Cybernetics*, **56**, 1–17.

Koenderink, J. J. (1987). Representation of local geometry in the visual system. *Biological Cybernetics*, **55**, 367–375.

Krüger, J. (1983). Simultaneous individual recordings from many cerebral neurons: Techniques and results. *Review of Physiology, Biochemistry and Pharmacology*, **98**, 177–233.

Kuffler, S. W., & Nicholls, J. G. (1976). *From neuron to brain*. Sunderland, MA: Sinauer.

Kuperstein, M. (1987). Adaptive visual–motor coordination in multijoint robots using parallel architecture. *Proceedings of the IEEE International Conference on Robotics and Automation*, **3**, 1595–1602.

Lamb, G. L., Jr. (1980). *Elements of soliton theory*. New York: Wiley.

Lashley, K. S. (1950). In search of the engram. *Symposium of the Society for Experimental Biology*, **4**, 454–482.

Lynch, G. (1986). *Synapses, circuits, and the beginnings of memory*. Cambridge, MA: MIT Press.

Marr, D. (1969). A theory of cerebellar cortex. *Journal of Physiology*, **202**, 437–470.

Mead, C. (1989). *Analog VLSI and neural systems*. Reading, MA: Addison-Wesley.

Miles, F. A., & Evarts, E. V. (1979). Concepts of motor organization. *Annual Review of Psychology*, **30**, 327–362.

Miller, S., & Scott, P. D. (1977). The spinal locomotor generator. *Experimental Brain Research*, **30**, 387–403.

Morasso, P., & Mussa Ivaldi, F. A. (1982). Trajectory formation and handwriting: A computational model. *Biological Cybernetics*, **45**, 131–142.

Pellionisz, A., & Llinas, R. (1979). Brain modeling by tensor network theory and computer simulation. The cerebellum: Distributed processor for predictive coordination. *Neuroscience*, **4**, 323–348.

Pellionisz, A., & Llinas, R. (1985). Tensor network theory of the metaorganization of functional geometries in the central nervous system. *Neuroscience*, **16**, 245–273.

Raibert, M. H. (1978). A model for sensorimotor control and learning. *Biological Cybernetics*, **29**, 29–36.

Rall, W. (1977). Core conductor theory and cable properties of neurons. In E. Kandel (Ed.), *Handbook of physiology, section 1: the nervous system* (Vol. 1, Part 1).

Ritter, H., & Schulten, K. (1988). Extending Kohonen's self-organizing mapping algorithm to learn ballistic movements. In R. Eckmiller & C. von der Malsburg (Eds.), *Neural computers*. Heidelberg: Springer-Verlag.

Rodieck, R. W. (1973). *The vertebrate retina—Principles of structure and function*. San Francisco: Freeman.

Rosenblatt, F. (1961). *Principles of neurodynamics: Perceptrons and the theory of brain mechanisms*. Washington, DC: Spartan.

Rumelhart, D. E., & McClelland, J. L. (Eds.) (1986). *Parallel distributed processing* (Vol. 1. *Foundations*). Cambridge, MA: MIT Press.

Schlag, J., & Schlag-Rey, M. (1984). Visuomotor functions of central thalamus in monkey. II. Unit activity related to visual events, targeting, and fixation. *Journal of Neurophysiology*, **51**, 1175–1195.

Selverston, A. I. (Ed.) (1985). *Model neural networks and behaviour*. New York: Plenum.

Sivilotti, M. A., Mahowald, M. A., & Mead, C. A. (1987). Real-time visual computations using analog CMOS processing arrays. In P. Losleben (Ed.), *Advanced research in VLSI*. Cambridge, MA: MIT Press.

Soechting, J. F., Lacquaniti, F., & Terzuolo, C. A. (1986). Coordination of arm movements in three-dimensional space. Sensorimotor mapping during drawing movement. *Neuroscience*, **17**, 295–311.

Swadlow, H. A. (1983). Efferent systems of primary visual cortex: A review of structure and function. *Brain Research Review*, **6**, 1–24.

Thompson, R. F. (1967). *Foundations of physiological psychology*. New York: Harper & Row.

Toffoli, T., & Margolus, N. (1987). *Cellular automata machines—A new environment for modeling*. Cambridge, MA: MIT Press.

Trends in NeuroSciences (1988). Special issue on learning and memory. *Trends in NeuroSciences*, **11**, 125–181.

Van Essen, D. C. (1979). Visual areas of the mammalian cerebral cortex. *Annual Review of Neuroscience*, **2**, 227–263.

7

New Approaches to Nonlinear Concepts in Neural Information Processing: Parameter Optimization in a Large-Scale, Biologically Plausible Cortical Network

Walter J. Freeman and Koji Shimoide

INTRODUCTION

Models of biological neural networks necessarily contain large spatially distributed arrays of interconnected nodes. Each node performs a set of operations on its multiple inputs and has an array of targets for its output. Specification of these connections and operations by the modeler depends on the fitness criteria by which the performance of the model is to be evaluated with respect to the biological function. Owing to the very large number of parameters, systematic search through parameter space is not feasible. Hence, it is necessary to group in advance the parameters by type and function, and to make decisions about which of these are to be fixed at common values throughout the model, and which are to be varied and over what ranges. The aim of these decisions is to simplify the task of engineering a model that captures the essence of the neural dynamics. An example developed here is taken from the vertebrate olfactory system, in which the number of neurons ranges upward of 100 million, but the degrees of freedom needed for modeling are far less because of the cooperativeness among the neurons. Through the adoption of appropriately defined state variables and operators at nodes, the complexity of modeling chaotic cortical dynamics is reduced, and the grouping of parameters is facilitated, along with the ordering of them according to their sensitivities in determining the function of the models.

A look at the histological structure of a representative sample of any part of the cerebral cortex will give the impression of unimaginable complexity. The cell bodies of the neurons in Nissl stains in each cubic millimeter are uncountable and can only be estimated by statistical sampling. In Golgi stains, it becomes clear that each neuron is connected to thousands of other neurons by axonal and dendritic branches over distances ranging from fractions of a millimeter to several centimeters. The spatial patterns of branching are endlessly varying, and the possibilities for feedback connections through loops between neurons in chains appear to be unlimited. Biologists report that each dendritic and axonal tree can be modeled with hundreds, even thousands of compartments, and the membrane of each compartment can be equipped with many varieties of chemically and electrically dependent, ion-specific channels. Studies of the properties of single neurons have revealed unlimited complexity in the functional characteristics. Many of these properties have been successfully modeled using high order nonlinear differential equations that reveal multiple solutions sets, including fixed point, limit cycle, and chaotic attractors (Thompson & Stewart, 1988). There is no end in sight to the increasing complexity of the neuron and of the cortical population, as studies of their cellular and molecular properties continue in laboratories around the world.

Yet, what is needed is the formulation of some generalizations that will support reduction in complexity, so that the way in which cortex functions can be grasped in its ultimate simplicity. One of the key insights toward this goal is the recognition that the element of cortical function, which is modeled by a node in a neural network simulation, is not a neuron but a local population of neurons (Freeman, 1975, 1979a–c, 1987a, 1991a; Freeman & Skarda, 1985; Skarda & Freeman, 1987). The evidence is both structural and functional. Each neighborhood contains on the order of 10^4 to 10^5 neurons that maintain synaptic interconnections, each of them transmitting to and receiving from 10^3 to 10^4 others. There are no privileged pairs of neurons, and the likelihood of action onto the self is vanishingly small. In normal function, each neuron fires less than 10 pulses/sec, and these pulses occur at intervals distributed according to a Poisson process, with a refractory period. Intracellular recording reveals that neurons spend 99% of their life span with randomly fluctuating membrane potential just below their threshold for firing, that is, near rest and not in domains that are far from equilibrium, having exotic states manifested in endogenously repetitive firing. Each neuron depends in its firing on the return of its own discharge through its action on others in its neighborhood, which comes back through innumerable channels on chains of varying length, so that its feedback channel can be described by a one-dimensional diffusion process (Freeman, 1964). Hence, the degree of interaction with neighbors determines the mean rate of firing for each neuron, but the local timing is randomized, both for the single pulse trains and the cross-correlation between pairs of pulse trains. The result of this interaction in the absence of input is a sheet of randomly firing neurons. Any covariance that is induced by input can be expected to stand out clearly under spatiotemporal integration of an activity measure taken over the whole cortical array.

What emerges from this widespread, weak synaptic interaction is the local population, which can be construed to correspond to the cortical column or hypercolumn, as defined by means of unit recording (Mountcastle, 1966). The derivation of a local mean field quantity for estimating the activity of the population is done by recording the local electrical field potential (Freeman, 1975, 1992), commonly known as the electroencephalogram (EEG), which reflects the sum of potentials established by dendritic current over the tissue resistance of the column in those cortical architectures in which the neurons are aligned in palisades, such that the currents of individual neurons form a scalar sum (Mitzdorf, 1987). The dendritic current for each neuron has the form of a closed loop, in accordance with Kirchoff's current law, with an electromotive force across the membrane in one direction at a synapse, and an impedance-matched site of action in the flow in the opposite direction of the current at the trigger zone. The intracellular limb of the current determines the probability of the firing of a neuron. The extracellular limb establishes an ohmic potential difference over the tissue resistance. The contributions from the local neurons sum to approximate the local mean field, which in turn is proportional to the summed firing rates of neurons in the column, and can be measured as its pulse density.

These extracellular dendritic currents do not play a causal role in the formation of a cooperative population. That unification emerges from the synaptic interactions of its component neurons. The EEG does provide an observable, indirect epiphenomenon for estimating the local wave and pulse densities. The local population provides a form of multiplexing in which the contribution from each neuron is elicited at random and is equivalent to that of any other. The readout by neurons to which cortical output is then directed is by spatiotemporal integration, in which the identities of the transmitting neurons are lost by smoothing. These inferences are based on the statistical analyses of pulse probability, conditional on the amplitude of the EEGs of local populations (Freeman, 1975; Eeckman & Freeman, 1990, 1991) and on the anatomical properties of the transmission pathways. The overriding conclusion of these studies is that the element of cortical function is the macroscopic local population. The input and output have the form of trains of action potentials, which are treated as microscopic point processes (Brillinger, 1975, 1992). The main cognitive operation of cortex is to construct a macroscopic activity pattern from a spatial pattern of input action potentials during a brief time window, and to express the macroscopic construct in a selection of neural pulses that are then transmitted to cortical targets. Therefore, in modeling cortex, the appropriate macroscopic state variable is a continuous time function that expresses local mean field amplitude. Input is on parallel lines in the form of binary pulses or steps, and output is likewise expressed in spatial patterns of binary amplitudes derived over the time window. Both input and output are in the form of binary vectors, but the intervening operations must have the form of continuous time dynamics that is best expressed in nonlinear ordinary differential equations (ODE).

LOCAL OPERATIONS OF CORTICAL POPULATIONS

Each local population receives axonal pulses at synapses, converts the pulses to dendritic current, and sums the current at the cell body. At the initial segment, the sum of the current is reexpressed as a probability of firing for the single neuron and as a pulse density for the population. There are three main types of neurons, which are defined by their actions: excitatory, inhibitory, and modulatory. The conversion from pulse density to current density at excitatory synapses is expressed by a positive gain coefficient k_e, and that at inhibitory synapses is expressed by a negative gain coefficient k_i. The action of modulatory neurons is expressed by a state-dependent change in either k_e or k_i. The conversion of the sum of current density is described by a static nonlinear operator in the form of a sigmoid curve (Figure 1) that has a single parameter Q_m specifying the slope and maximal asymptote of the curve (Freeman, 1979a). It is important to note that the sigmoid is a property of the population and not of the single neuron. Its shape and the value of Q_m are determined experimentally from the pulse probability of cortical neurons and are conditional on wave

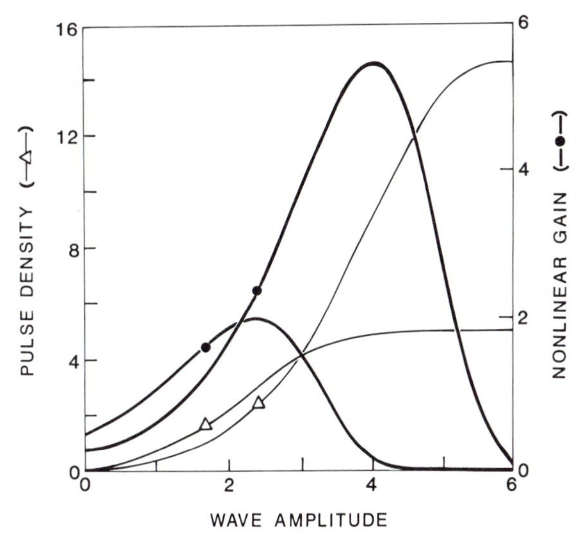

FIGURE 1 The biologically derived sigmoid curve relating pulse density, p, to wave density, v (light trace), is asymmetric about its rest point (triangle). The steepest point of the slope, dp/dv (dark trace, lies to the excitatory side of the rest point. This feature is the basis for the input-dependent gain by which state transitions are induced by input. The steepness is determined by a parameter Q_m denoting the maximal asymptote. Curves are shown for $Q_m = 2$ in a quiescent subject and $Q_m = 6$ in an aroused, motivated subject. Average and maximal values in the bulb and pyriform cortex are 5 and 12, respectively (Eeckman & Freeman, 1991). The equations (Freeman, 1979a) are

$$Q = Q_m \{1 - \exp[-(e^v - 1)/Q_m]\},\ v > -u_o$$

$$Q = -1,\ v \leq -u_o$$

$$u_o = -\ln[1 - Q_m \ln(1 + 1/Q_m)]$$

$$p = u_o (Q + 1)$$

$$dp/dv = u_o \exp[v - (e^v - 1)/Q_m]$$

amplitude of the local EEG (Freeman, 1975; Eeckman & Freeman, 1991).

The operator for the integration of dendritic current is a second-order ODE that is derived by observing and measuring the compound extracellular postsynaptic potential of a population of neurons that is not interactive and that is denoted a "K0 set" (Freeman, 1975). Examples of K0 sets are provided by olfactory receptors and by cortical neurons under deep anesthesia. This impulse response (Figure 2) under deep anesthesia has the form of a transient, with a rapid rise in potential to a peak followed by an exponential decay (Freeman, 1975). The rise time is determined by the serial delays imposed by terminal axonal dispersion, synaptic diffusion, and the cable properties of the dendrites. Each of these three operations can be described as a one-dimensional diffusion process for which the impulse response is a gamma distribution of order 1/2. This function is invariant in form under convolution with itself, so the serial delays are combined in a single transcendental operator. This is replaced by a rational approximation of poles and zeroes. Because the rise time is so brief in comparison with later events, a single pole and zero in a first-order ODE suffices. The small-signal passive membrane property is approximated by another first-order ODE in series with the first term (Freeman, 1964). The two rate parameters are experimentally evaluated by measuring the open loop impulse response of cortex under very deep anesthesia, when the synaptic interactions within the

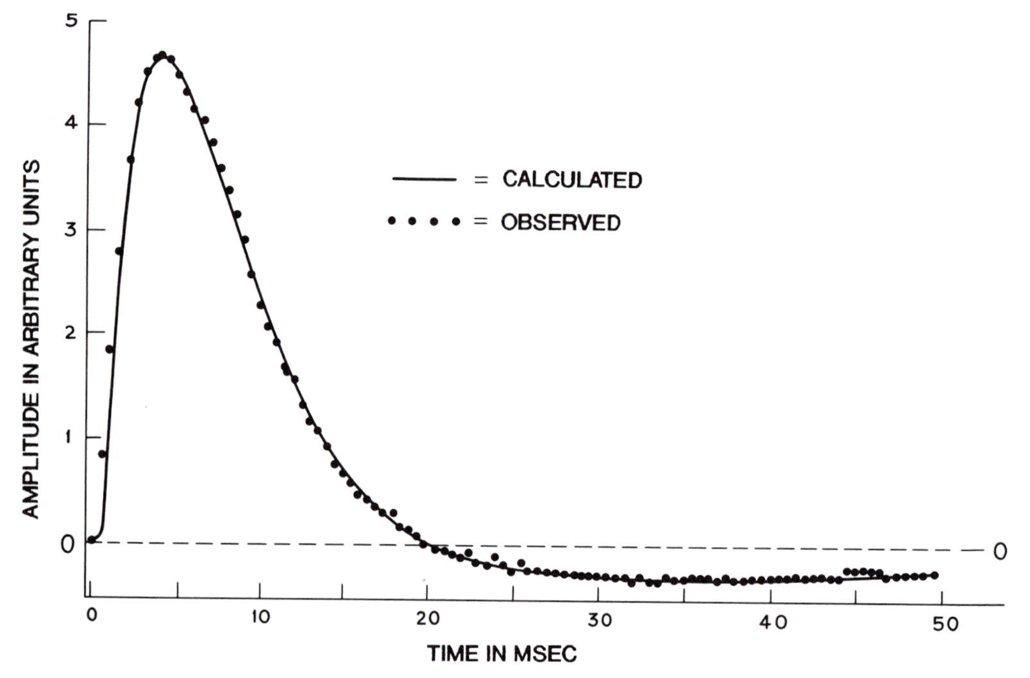

FIGURE 2 The dots show the open loop impulse response of the pyriform cortex under deep anesthesia. The fitted curve is generated by a sum of exponential terms consisting of the average passive membrane decay rate, $a = 220$/sec, and a set of poles and zeroes constituting a rational approximation for the cascaded synaptic and dendritic cable delays at the input. In K0 models, the finite rise time is approximated by a single pole, $b = 720$/sec, as the basis for a second-order ODE that approximates this impulse response in the small-signal linear range of dendritic function. The limitation on cortical activity that is imposed by the sigmoid at the trigger zones keeps cortical neurons within this linear range for synaptic transmission. (From Freeman, 1964.)

cortex are suppressed (Figure 2). Although a second-order ODE is commonly avoided in network modeling, it is essential for modeling cortical dynamics because the serial delays enhance the destabilization of cortical networks into oscillatory domains that are essential for its performance (Freeman, 1975, 1991b).

In summary, the three serial operations by a local population are the pulse-to-wave conversion expressed by a gain coefficient (with a sign inversion for inhibition), linear integration with two time parameters, and conversion of wave-to-pulse density by a sigmoid function governed by a single parameter, Q_m. The state variable, v, in the ODE (v_e for excitatory and v_i for inhibitory) is used to specify the amplitude of activity of the local population in either the pulse or wave modes of activity. Simulated time series are compared with recordings of EEGs and of multiunit activity from microelectrodes in the cortex in order to evaluate the adequacy of simulation by the models.

KI AND KII NETWORKS THAT MODEL CORTICAL POPULATIONS

Each local population is homogeneous with respect to its accessibility by a parallel input pathway and the sign of action of its neurons: excitatory, inhibitory, or modulatory. Cortex contains all three types of neurons, and they are arranged in layers, so that networks must be constructed in accordance with the architecture and connectivity of cortex. There are some simplifying rules. First, in the noninteractive state imposed by deep anesthesia, the neurons lie in parallel, and their response to impulse input is a compound postsynaptic potential (PSP) that constitutes a time ensemble average of the time-locked PSPs of the individual neurons. This population is represented in a model by a K0 set (Freeman, 1975), which is the second-order ODE, and for which $Q_m = 0$ (Figure 3). Second, in the normal state of cortex, the excitatory neurons interact by mutual excitation. There is no self-excitation. This interactive population is represented by a KI_e set, which is formed by two $K0_e$ sets. The feedback gain is given by the square of the forward gain k_{ee} after operation of the sigmoid function on the wave amplitude of the transmitting population. Third, the inhibitory neurons interact by mutual inhibition, without self-inhibition, and the cooperative population is represented by a KI_i set from two $K0_i$ sets, in which the feedback gain is the square of k_{ii} applied to the output of the sigmoid operator. These KI sets are both modeled by positive feedback loops, for which the impulse response decays monotonically in time from an initial peak value, but in which there is cooperative spatial coherence for the KI_e set and competitive spatial contrast for the KI_i set (Freeman, 1975). Finally, interaction of excitatory and inhibitory neurons in cortex results in oscillations. The model that is formed by the interaction of a KI_e set with a KI_i set is denoted a KII set. Its impulse responses typically have the form of a damped sine wave (Freeman, 1975, 1991b). The characteristic frequencies of oscillation are determined by the time constants of the ODEs of the component neurons, which are evaluated for the K0 sets in the open loop state. The frequency of each KII set is modulated by the values of its synaptic gain parameters over a range called the gamma range of the EEG (35–90 Hz). The gamma activity was first seen in the olfactory system, and is now found in the visual cortex (Freeman & van Dijk, 1987; Eckhorn et al., 1988; Gray, Koenig, Engel, & Singer, 1989), auditory cortex (Ribary et al., 1992), and somatosensory cortex (Freeman & Maurer, 1989; Murthy & Fetz, 1992).

Cortical neuron populations exist in the form of sheets that can be represented in a model by the two dimensions of a surface. To minimize edge effects, a toroidal boundary condition is commonly used to model a collection of KII sets, each with an input channel to one of its component $K0_e$ sets and an output line from that same set. Two questions arise. How many KII sets are needed to represent the distributed dynamics of an area of cortex, and what are the length constants over which each acts onto its neighbors for the three types of interaction: excitatory, inhibitory, and negative feedback? The number N of nodes in a KII model is selected by balancing two constraints. The first is the number that is required to simulate the

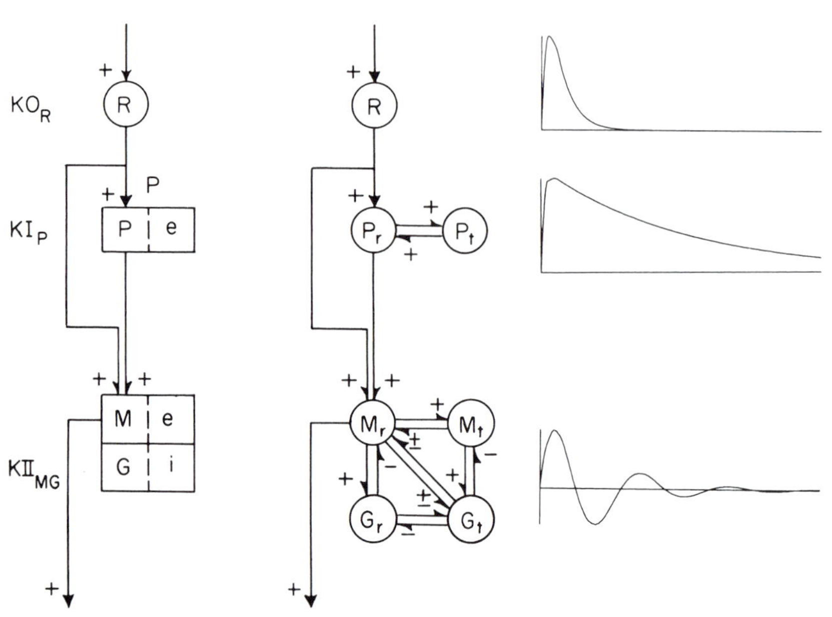

FIGURE 3 The first stages of the olfactory system consist of the receptor layer, R, which is represented by a $K0_e$ set; the periglomerular interneurons in the glomerular layer of the olfactory bulb, P, that form a KI_e set; the mitral cells, M, in the bulbar depth that also form a KI_e set; and the granule cells, G, that form a KI_i set. The reciprocal synaptic action between the mitral and granule cells forms a KII set. The impulse response of the K0 set is calculated from the second-order ODE: $d^2/dt^2\ v(t) + (a + b)\ d/dt\ v(t) + v(t) = ab\ \delta(t)$, where $a = 220$/sec, $b = 720$/sec, and $\delta(t)$ is the input delta function. The impulse responses of the KI sets show prolonged monotonic decay. The impulse response of the KII set conforms to a damped sine wave. (From Freeman, 1975.)

observed spatial dynamics of local areas of the sensory cortex. The second constraint is the amount of computer time that is required to solve the ODEs to simulate the space–time patterns of observed activity. Most of the extant experimental data have come from 8×8 arrays of 64 electrodes and amplifiers. The available computational resources have made it possible to simulate cortical dynamics with this number of nodes for some purposes, but for others, such as parameter optimization, an N of 4, 8, or 16 has often sufficed. In regard to connection distances, it is known that the density of neuronal connections falls exponentially with distance from each neuron over typical distances of a millimeter (Sholl, 1956), whereas recording arrays have revealed widespread spatial coherence of cortical oscillations over distances greater than a centimeter (Freeman & van Dijk, 1987). This experimental finding of substantial spatial coherence provides the basis for yet another powerful simplification in the modeling. Each node can be connected to every other node globally with respect to mutual excitation and mutual inhibition, thereby preserving an important degree of spatial symmetry; the negative feedback connections are restricted to those within nodes as the basis for preserving the degrees of freedom that are required for the expression of differing cortical states by distinctive spatial patterns of amplitudes of oscillation, with the spatially coherent waveform that constitutes a common carrier of cortical information (Freeman, 1991a).

The parameters have now been grouped as follows: time, nonlinearity, distance, and synaptic gain. For benchmark purposes, there are 64 nodes, each composed of a $K0_e$ subset and a $K0_i$ subset requiring two second-order ODEs (equivalent to four first-order ODEs for a total of 256 equations). The two time parameters, a = 220/sec and b = 720/sec, for the two types of populations are the same everywhere and for all times. The value for Q_m is fixed at $Q_m = 8$, for all nodes, both excitatory and inhibitory, and for all times. Distance parameters disappear with global connectivity and the toroidal boundary condition. The gain parameters, k_{ij}, cannot be measured directly in experimental preparations. When the open loop rates, a, b, have been fixed, the values for the four types of forward gain, k_{ee}, k_{ei}, k_{ie}, and k_{ii}, are estimated from measurements of the closed loop rate constants of cortical impulse responses in normal waking subjects (Freeman, 1975, 1991b). At the onset of modeling, all of the k_{ij} are set equal to unity. Thereafter, they are adjusted globally to new values to achieve performance of the model that is consistent with the cortical impulse responses to be simulated. Parameter adjustment is done by incremental changes and repeated testing, on the premise that the synaptic event in the normal range of cortical function (e.g., not epileptic or chemically altered) is linear (independent of amplitude), time-invariant, and spatially uniform with respect to node (cortical column). The questions to be answered later are which synapses change with learning, how much, and in what spatial patterns?

It is important to note that the aim here is reduction in complexity without loss of the essential features of cortical operation. The time parameters are essential. The sigmoid nonlinearity is essential. The types of neuron are essential. The connectivity patterns are essential. What is gained by moving to the level of the population for modeling is the systematic submergence of the complexities of the single neurons and of their detailed spatial patterns of connections resulting from the immense numbers of neurons, the continual divergence and convergence in parallel transmission, and the spatiotemporal integration that is taking place concomitantly in the many thousand neurons comprising a local neighborhood. The parameters for time, distance, and nonlinear conversion are fixed over space, time, and neuron type. Only the strengths of the input and of the four synaptic gain coefficients need to be adaptively modified, and, on first pass, the values can be fixed spatially over the surface of the model and for the time durations needed to solve the equations to derive recognizable events. The KII set thus constructed represents a model of a small area of cortex that contains a few hundred local populations or columns with a radius of a few millimeters. It also constitutes a module for incorporation into a KIII network of KI and KII sets to represent the dynamics of cortical systems.

KII NETWORKS THAT MODEL LEARNING IN CORTEX

Experience has shown that there is no one set of parameters for the various embodiments of KII or KIII dynamic models that is absolutely optimal. A realistic approach with a new embodiment or a new program, machine, or computer language is to fix the time, gain, and length constants at values taken from the biological experiments, and to adjust the four k_{ij} parameters globally until the form of the impulse response of each KII array conforms to an observed averaged evoked potential of the part being modeled. The conditions of recording and measurement include use of low-stimulus intensity to keep the area of cortex in its small signal near linear range, with verification that superposition (additivity and proportionality) holds by use of paired shock testing (Biedenbach & Freeman, 1965; Freeman, 1975). Examples shown in Figure 4 are taken from modeling three parts of the olfactory system: the olfactory bulb, anterior nucleus, and pyriform cortex (Freeman, 1987b). The important features of these impulse responses are (1) the damped sine wave generated by the negative feedback in each component with the synaptic gain $k_{ei}*k_{ie}$; (2) the differing characteristic frequencies of the three parts, which are incommensurate; (3) the decay rates of the envelopes, which are similar for the three parts; and (4) the deviations from zero of the baseline of the oscillations, which are modeled by differing asymmetries in the values of k_{ee} and k_{ii} in the three parts.

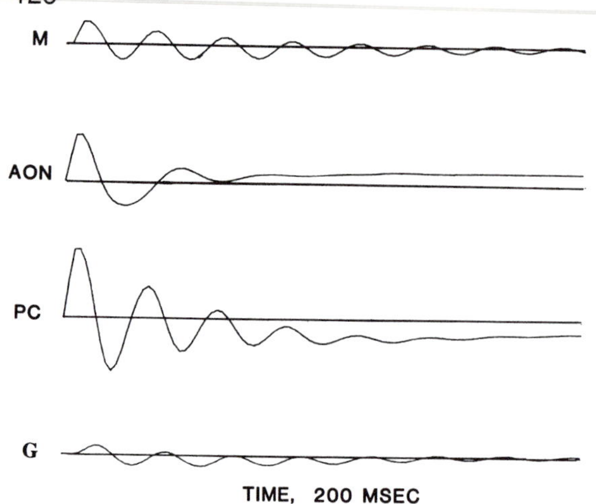

FIGURE 4 Closed-loop impulse responses are shown for three components of the olfactory system in the small-signal linear range as simulated by KII sets with waveforms set by the four gain parameters (k_{ee}, k_{ei}, k_{ie}, and k_{ii}) for the mitral KI_e set (M) of the bulb (0.25, 1.50, 1.50, and 1.80), the KI_e sets of the anterior olfactory nucleus (1.50, 1.50, 1.50, and 1.80) and pyriform cortex (0.25, 1.40, 1.40, and 1.80), and the KI_i set (granule cells of the bulb, OB). In each KII set, the output of the KI_i set lags the output of the KI_e set by one-quarter cycle of the oscillation, as shown by comparing M and OB. These oscillations occur in the gamma range of the EEG (35–90 Hz). The base frequency of 40 Hz is determined by the four steps for each cycle (excitation of KI_e, excitation of KI_i, inhibition of KI_e, and inhibition of KI_i), each lasting 6.25 msec to give a cycle duration of 25 msec. (From Freeman, 1987b. Reprinted with permission from Springer-Verlag.)

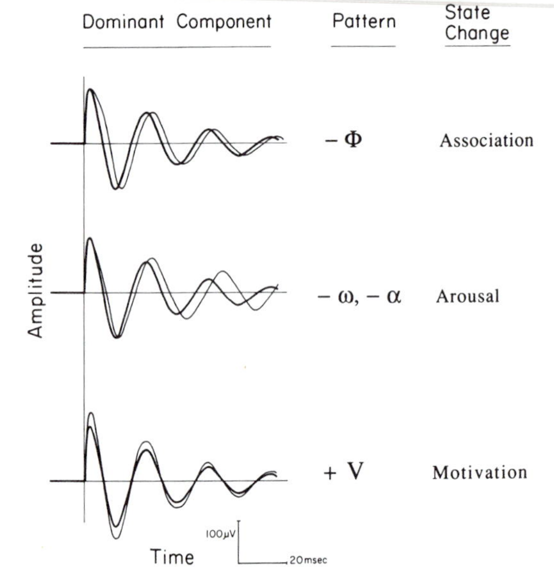

FIGURE 5 Three components of change in cortical impulse responses with associative learning are simulated with a KII set. The overall change is elicited by presenting the electrical stimulus used to evoke the responses paired with an unconditioned reinforcement stimulus, in this case water for a thirsty subject. The component of change that is specific for the conditioned stimulus is a decrease in phase, $-\phi$. There is an accompanying decrease in frequency $-\omega$ and decay rate $-\alpha$ of the oscillation, which also can accompany nonspecific arousal in pseudoconditioning. An increase in amplitude without change in frequency can be partialed out by changes in the degree of motivation, for example, increasing water deprivation. The overall change is simulated with the KII set by increasing k_{ee}. (Modified from Emery & Freeman, 1969. Reprinted with kind permission from Pergamon Press, Ltd., Headington Hill Hall, Oxford OX3 0BW, UK.)

These results are obtained experimentally by using electrical stimuli of afferent pathways to the three parts of the olfactory system in subjects with electrodes for stimulation and recording fixed during surgery, and also with recording during the normal waking state. The simplest approach for identifying the changes in synaptic strengths with learning is to train the subjects by pairing the delivery of the electrical stimuli as a conditioned stimulus with a reinforcing unconditioned stimulus. As the subject acquires a conditioned reflex that demonstrates learning to identify the electrical stimulus, the waveform of the impulse response changes (Emery & Freeman, 1969). The initial peak amplitude decreases slightly, and the duration of the initial peak increases significantly, which is expressed by a decrease in the phase lead of the sine wave (Figure 5). There is also a slight decrease in the characteristic frequency, and the decay rate of the envelope increases significantly, indicating that the change with learning has decreased the stability of cortex with respect to the conditioned stimulus. This response configuration is simulated with the KII model by increasing the gain k_{ee} between the $K0_e$ subsets, which represents the strength of associative action between cortical excitatory cells (Emery & Freeman, 1969; Freeman, 1975; Bressler, 1988; Haberly & Bower, 1984; Freeman & Grajski, 1987; Grajski & Freeman, 1989), and it is not achieved by varying other k_{ij} or by the strength of action of the afferent axons onto the excitatory cells. The amount of increase in k_{ee} is postulated to be less than 40% (Freeman, 1979b).

When a subject has been induced to orient to the electrical stimulus but the unconditioned stimulus is omitted, the subject learns to ignore the irrelevant conditioned stimulus, and habituation takes place. The form of the impulse response again changes, in this case with a slight increase in the characteristic frequency and a substantial increase in the decay rate of the envelope of the sine wave, showing that cortical stability in respect to the afferent stimulus has been increased. Simulation of this pattern in the performance of the model is achieved by concomitant reduction in k_{ee} and k_{ei} using a multiplicative, neuromodulatory parameter, denoted δ, which represents a reduction in the synaptic effectiveness of the cortical excitatory neurons receiving the input, δk_{ee}, δk_{ei}, irrespective of whether their action is onto other excitatory neurons or onto inhibitory neurons (Freeman, 1979b).

An important subsidiary change in the waveform of the impulse response is seen in a spectral shift of the activity away from the characteristic frequency near 40 Hz in the gamma range to a low frequency band called the theta range (Figure 6). The significance of this spectral shift is that the cortical area is stabilized with respect to an oscillation in the gamma range for the unwanted background stimulus, but it is now *destabilized* with respect to oscillation in the gamma range for a desired concomitant foreground input that has been reinforced during past presentations. Stimuli from the environment always consist of a mixture of the component that is significant, which is identified by the reinforcement, and the bulk of the input components that are not so identified. Learning consists, in part, in the increased sensitivity to the one and the decreased sensitivity to all of the others. Yet, the actions of those others are not filtered and removed but are redirected to enhance the sensitivity to the desired input in a nonspecific excitation of the excitatory neurons in the cortical area. This phenomenon appears to be especially useful when the input is a weak stimulus that is embedded in substantial background clutter, by which the background serves to enhance the response to the foreground.

Further comment is needed at this point concerning the nature of the cortical instability, as it is modeled by a KII set. In conformance with biological observations, the isolated cortical area with no input typically has a point attractor manifested by a state of nonzero steady state activity. With an excitatory step input, it can be driven through a state change to a limit cycle attractor, in which periodic oscillation is observed. This input-dependent state change, which is reversed by termination of the input, is caused by the asymmetry of the sigmoid curve (Freeman, 1979a). The steepest part of the curve is located to the excitatory side of the rest point, so that excitatory input increases the forward gain of the excited neurons. This increase also increases the feedback gains, which depend on the product of the synaptic gains and the sigmoid gains of all pairs of K0 subsets. Owing to the delays introduced by the synaptic and dendritic operators incorporated into the second-order ODE, any increase in feedback

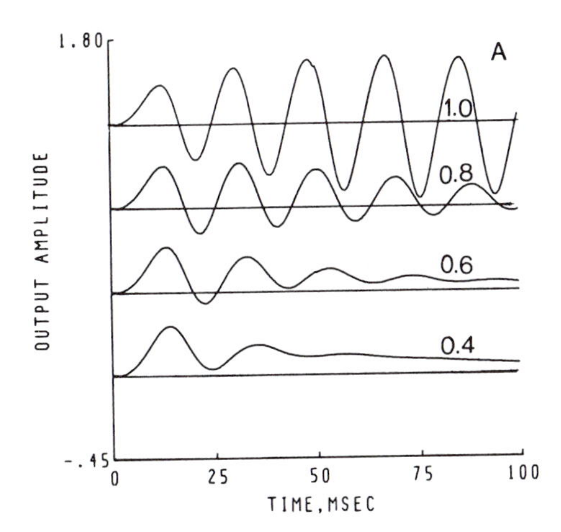

FIGURE 6 The pattern of change in cortical impulse responses with habituation is simulated with induced bursts in a KII set using a step input that simulates receptor input during an inhalation. Progressive decrease in δ (modulating the output k_{ee} and k_{ei} gain parameters from a node receiving unwanted input to its excitatory and inhibitory targets) attenuates the oscillation in the gamma range and accentuates an excitatory output in the theta range (the upward baseline shift). (From Freeman, 1979b. Reprinted with permission from Springer-Verlag.)

gain will decrease the decay rates of the damped sine impulse responses. Three factors operate to enhance the destabilization. One is the value of Q_m, which determines the steepness of the sigmoid and which is increased by neuromodulation during increased arousal, as with hunger, thirst, fear, or other conditions of reinforcement. The second factor is the effect of habituation through δ, by which background input serves to deliver nonspecific excitation that enhances the foreground step input. The third and most important factor is the synaptic change in k_{ee} with associative learning (Freeman, 1979b), which enhances mutual excitation. The result of input to excitatory nodes having this property is regenerative positive feedback that, in combination with the asymmetric sigmoid, can lead to an explosive destabilization. In behavioral terms, the presentation of a weak but significant stimulus to a motivated and attentive subject can trigger a strong conditioned response. The "amplification" occurs by virtue of the state change, as the weak stimulus causes a state change in which the cortical area receiving the input switches from a quasi-steady state to a strongly oscillatory state (Figure 7).

Owing again to the asymmetry of the sigmoid curve, the state change is fully and quickly reversed on cessation of the triggering stimulus, such that the oscillation occurs in a brief time window on the order of 100 msec. This event is called a *burst*. There is a significant parallel here between the single neuron and the local population. In the normal subject, both are bistable, the neuron in the sense that it is either subthreshold or generating an impulse, and the cortex in the sense that it is either in a burst or an interburst state. In both cases, it is the voltage-dependent sodium conductance of the neuronal membranes that is primarily responsible for the bistability (Freeman, 1979b), but its forms of manifestation differ between the single neuron and the population.

Input to the bulb from receptors that are selectively activated by an odorant is transmitted simultaneously on multiple axons in parallel. The strength of action of individual receptor axons onto bulbar neurons is automatically scaled down in proportion to the local density (number of active axons in each glomerulus) by an electrochemical attenuation that, in effect, transforms the input to a logarithmic scale (Freeman, 1975, 1991c) at the synapses between the receptors and the bulbar neurons. The synaptic changes with learning that take place between bulbar neurons are governed by two rules. (1) In associative learning, the synapses between neurons that are coactivated in response to the stimulus are strengthened (Hebb, 1949), provided that a neuromodulator signal is given under the control of the reinforcement (Gray, Freeman, & Skinner, 1986). The synaptic change appears to take place within each reinforced trial and appears to be permanent. (2) In habituation, there is no reinforcement, and the efficacy of the neurons receiving the input is gradually reduced over several trials irrespective of the activity of the

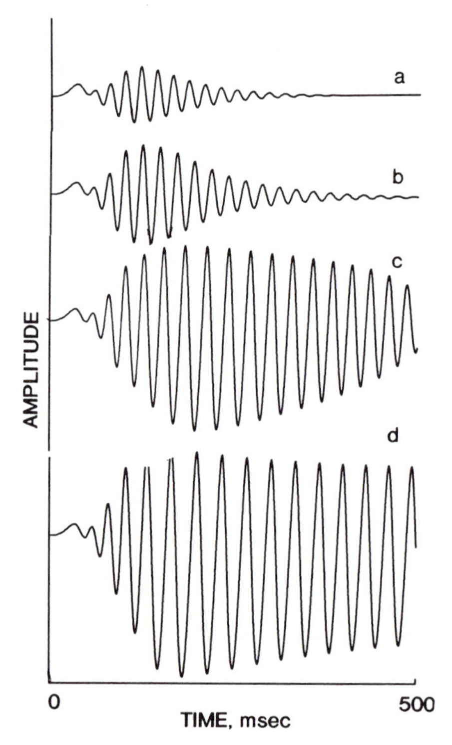

FIGURE 7 The effect on a step response of a KII set of increasing k_{ee} is shown: (a) 1.125; (b) 1.15; (c) 1.175; and (d) 1.20. The impulse response is a damped sine wave with decreasing phase, frequency, and decay rate (Figure 5). A step input is required for a state change to take place from a point attractor to a limit cycle attractor. The step is initiated at $t = 0$ msec and terminated at $t = 120$ msec, after which, the KII set returns to its point attractor. (From Freeman, 1979b. Reprinted with permission from Springer-Verlag.)

targets. Unlike that for associative learning, the process is fully reversible upon dishabituation (Gray & Skinner, 1988; Grajski & Freeman, 1989).

These two learning rules are implemented in the KII and KIII models by cross-correlation of the outputs of the array of excitatory populations, and by use of global commands to "associate" or to "habituate," which means to modify the matrix of connection strengths selectively in accordance with measurements of the outputs (Freeman, 1979c). Each specified class of input for discrimination on simultaneous or repeated presentation leads to the emergence of a subset of nodes with strengthened interconnections that is called a "template" or "nerve cell assembly" (Hebb, 1949). The overlap is suppressed by habituation. In preliminary work, the values for k_{ee} (reciprocally for pairs of nodes) and δ (modulating both k_{ee} and k_{ei} for single nodes) were changed on a graded scale in proportion to correlation coefficients. During testing it became apparent that in large models the use of graded values is unnecessary, and that only three parameters need specification. The value for k_{ee} before learning and for those connections not changed is typically $0.5 < k_{lo} < 1.0$, on the criterion that all nodes are required to have the same waveform, though at differing phase and amplitude. The value for k_{ee} after learning is typically $2.0 < k_{hi} < 3.0$, on the criterion that burst formation should be facilitated, but the system should not go to a steady state at maximal excitation. The value for the habituation parameter $0.3 < \delta < 0.6$ is based on the criteria that cross-activation between coexisting nerve cell assemblies should be suppressed, and that the contribution of a low frequency baseline shift on activation of the habituated nodes by the undesired input should facilitate the destabilization of the network from its rest state to an oscillatory state.

KIII NETWORKS THAT MODEL CORTICAL SYSTEMS

Records of EEGs from the olfactory system in the absence of sensory input reveal background activity that appears to be random. Records lasting several seconds or more are characterized by broad power spectra, in which the logarithm of power decreases monotonically with increasing log frequency; amplitude histograms closely conform to the normal density function; and autocovariance functions approach zero amplitude usually within the 100 msec duration, which conforms to the typical averaged evoked potential (the cortical impulse response). This configuration of findings holds for many other parts of the cortex, for example, the visual cortex (Freeman & van Dijk, 1987). If the subjects are adequately motivated, such that the sigmoid curve is elevated by a value of $Q_m > 4$, olfactory EEGs change dramatically. With olfactory input having the form of repetitive driving by respiration (typically in the theta range of 3–7/sec in rabbits and rodents, but lower in the delta range, < 3/sec in large subjects), each volley of receptor discharge on inhalation causes a wave of excitation in the bulb. This excitation is typically accompanied by a burst of oscillation in the gamma range (Figure 8), so that two peaks emerge in the spectrum of the bulbar EEG (Freeman, 1975; Bressler, 1988), one in the theta range of driving by input and the other in the range of the characteristic olfactory oscillations. The burst is not periodic, but typically shows both amplitude and frequency modulation throughout its duration of approximately 100 msec. The center frequency varies unpredictably from each burst to the next, and in the spectra of most bursts there are two or more peaks in the gamma range. The spectra of time series including multiple burst and interburst segments (Figure 9) show an elevation of power in the gamma range above a linear relationship between log power and log frequency ("1/f") that typifies the background activity in the absence of bursts (Kay, Shimoide, & Freeman, in press).

It is clear that Fourier analysis constitutes an arbitrarily linear decomposition of an intrinsically nonlinear and chaotic event. The multiple Fourier components of each burst all have the same spatial pattern of amplitude and of time-dependent amplitude modulation of the burst (Freeman & Viana Di Prisco, 1986), indicating that spectral components of bursts do not indicate that bursts arise from periodic oscillators, but serve merely as convenient basis functions for measuring phase and amplitude. Each component has a

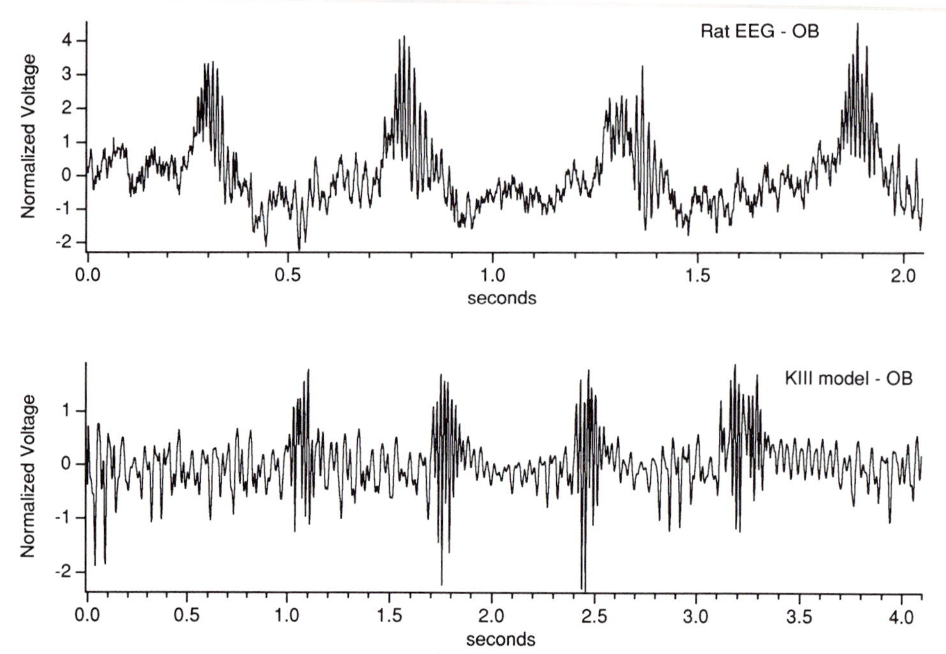

FIGURE 8 The EEG from a rat OB (2 sec) is simulated with the KIII model (4 sec). Burst frequencies for the KIII model were adapted by parameter adjustment to match the frequencies found in the rabbit, which typically are lower than those in the rat (Bressler & Freeman, 1980). Unpublished data from Kay, Shimoide, & Freeman, in press.

phase gradient in the form of a cone in spherical coordinates (Freeman & Baird, 1987), and the spatial location and sign (maximal phase lead or lag) are the same for the largest two Fourier components, the others being too small to measure. Despite the complexity of the burst waveforms, the same time series appears on all recording channels although at differing phase and amplitude values of the main components. These phase patterns show that the aperiodic waveforms are generated endogenously through the interactions of the bulbar excitatory and inhibitory neurons. Input receptor neurons do not have periodic discharges in the gamma range, and the temporal smoothing that is imposed on receptor input, through the olfactory pathway operating as a low pass filter, could not transmit driving input at gamma frequencies.

The variability in location and sign of the conic phase pattern shows that the aperiodic oscillation does not result from a "pacemaker" neuron or focus in the bulb. The widespread spatial coherence of the common aperiodic carrier wave and the tendency for recurring spatial patterns of its amplitude modulation support the conclusion that the EEGs manifest a chaotic attractor that is supported by the olfactory system. Moreover, it is the entire olfactory system that supports the attractor, because surgical interruption of the pathways that connect the bulb to the anterior nucleus and pyriform cortex cause the chaotic background activity to disappear and be replaced by steady states or by limit cycle activity of the bulb in response to receptor input (Gray & Skinner, 1988).

Although the KII set modeling the bulb or pyriform cortex is capable of supporting a chaotic attractor and a limited basin of attraction in its parameter space (Freeman & Jakubith, 1993), the chaotic activity of the KII model yields a central peak and not the "1/f" form

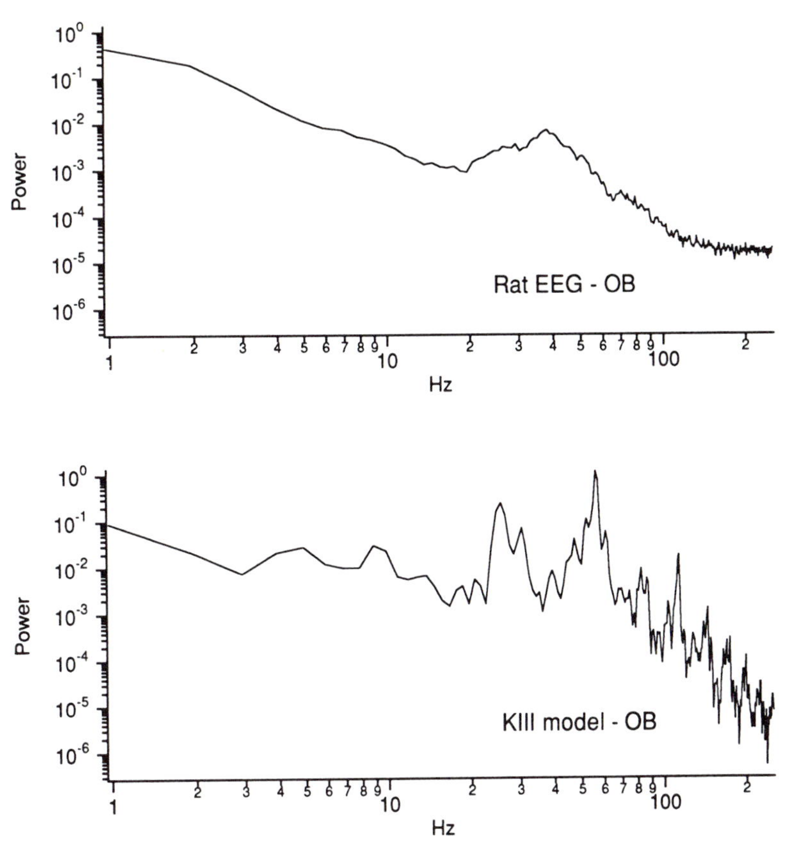

FIGURE 9 The power spectrum of a rat OB EEG is simulated with the KIII model. Unpublished data from Kay, Shimoide, & Freeman, in press.

observed in the EEGs of the normal, intact system. To simulate the time series of the EEGs, it is necessary to construct a KIII model of the bulb, nucleus, and pyriform cortex (Figure 10), each with its own characteristic, incommensurate frequency, and then to interconnect them with forward and feedback pathways (Freeman, 1987b; Shimoide, Greenspon, & Freeman, 1993). The axonal delays in the forward limb are negligible, because in the olfactory system these axons have high conduction velocities and narrow temporal dispersions. The feedback limbs are formed with large numbers of small axons that have distributed conduction velocities (Freeman, 1975), and therefore they contribute not only delay but temporal dispersion that operates as a low pass filter to smooth the gamma activity in the feedback limbs. Four long feedback delay paths have been identified in the olfactory system (Freeman, 1987b), each with a minimum and a maximum of delay and a continuous distribution between the extrema. The values of the delays are adjusted by trial and error to approximate the "1/f" spectra (Figure 9) that are observed in the EEGs (Kay et al., in press).

Three of the connections are from excitatory to inhibitory nodes, and their roles are to dampen the global activity of the system. The fourth and most important is an excito-excitatory connection from the anterior nucleus to the outer layer of the bulb, where it

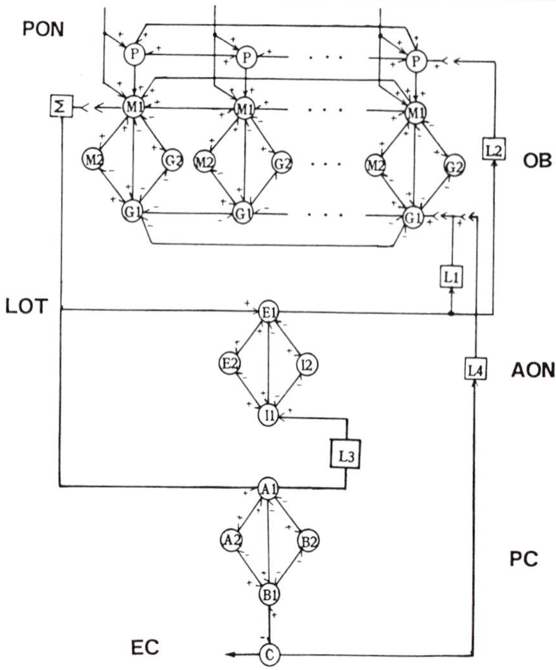

FIGURE 10 A flow diagram of the olfactory system for *N* nodes (bulb, OB) shows parallel input (the primary olfactory nerve, PON) to excitatory interneurons (periglomerular cells, P) and to mitral cells (M) with their inhibitory granule cells (G). The output of the OB is delivered by the axons of the lateral olfactory tract (LOT) to the anterior olfactory nucleus (AON) and pyriform cortex (PC), which are not distributed in the present embodiment of the KIII model. Each of the long feedback pathways has a distributed delay operator (L1, . . . , L4). (From Yao & Freeman, 1990. Reprinted with kind permission from Pergamon Press, Ltd.)

drives the external granule cells known as the "periglomerular" neurons. It turns out that these neurons play a crucial role in the genesis of chaotic activity of the olfactory system, because they provide an excitatory bias to the negative feedback loops comprising the bulb, nucleus, and pyriform cortex (Freeman, 1975, 1987b). The positive feedback, through dispersive delay from the nucleus, provides the basis for continual expansion of the activity of the olfactory system, whereas the other long dispersive delay paths provide for continual compression of the activity. The incommensurate characteristic frequencies of the three main components make it impossible for the system to settle into a limit cycle and thereby ensure the aperiodicity of the output as observed at the macroscopic level. The ongoing pulse activity at the microscopic level is also maintained by the global feedback over the long delay lines, because when these pathways are surgically cut, the activity of single neurons in the nucleus and pyriform cortex disappears, and the neurons become silent. This silence is characteristic of the isolated cortical slab (Burns, 1958) and slice preparations, unless they are subjected to an excitatory bias by sustained chemical or electrical input, in which case a limit cycle pattern may appear, but not the broad spectrum "1/f" activity found in the normal state.

However, the long feedback delay does not provide the full story. A discrepancy emerges from the fact that the largest distributed feedback delay found in the olfactory system (L4) is on the order of 25 msec, but both the system and its KIII model show a tendency for peak activity near 3/sec (Freeman, 1987a), which implies a loop delay near 300 msec. A source of such long time constants might be sought in slow EPSPs and slow IPSPs found for single neurons in a wide variety of nervous systems. These cannot account for the performance of the KIII model, because they are not included. Hence, an alternative explanation can be entertained. The most likely candidates are the KI_e and KI_i sets having feedback gains near unity, which give rise to closed loop rate constants near zero, that is, long time constants. A main candidate for a key role in the genesis of the basal spectrum is the set of periglomerular cells (P in Figure 10). Its internal feedback gain, k_{pp}, may be set to unity, but its input gain, k_{ap}, and its output gain, k_{pm}, must be reduced well below unity. The selection of useful operating values depends on meeting the criteria of having a system that maintains a stable basal chaotic state after initiation of the computation by an impulse input, which responds to simulated receptor input with a burst, and which returns to the same basal state when the input has been terminated (Freeman, 1987b; Yao & Freeman, 1990).

In conjunction with the use of software embodiments to generate simulations of chaotic activity patterns, it is essential to take note of a computational parameter not already mentioned, which is the time step, Δt, that is used for solving ODEs by numerical integration. Numerical instabilities can arise for time steps that are either too long or too short. The difficulties in dealing with chaotic solutions are that (1) their spectra are broad, leading to the possibilities of round-off errors at low rates of change and inadequate extrapolations at high rates of change, which may cause computational instability, and (2) the form of the erroneous output may be indistinguishable from the aperiodic and broad spectrum solutions that are desired. The approach used here to solve this problem was to adopt a conservatively small time step and to determine a range across which chaotic solutions were unaffected by changes in the time step. The values of Δt were pushed in either direction until failure of the simulation was observed (Figure 11). The optimal step size was then chosen within the range with respect to computational speed (Figure 12).

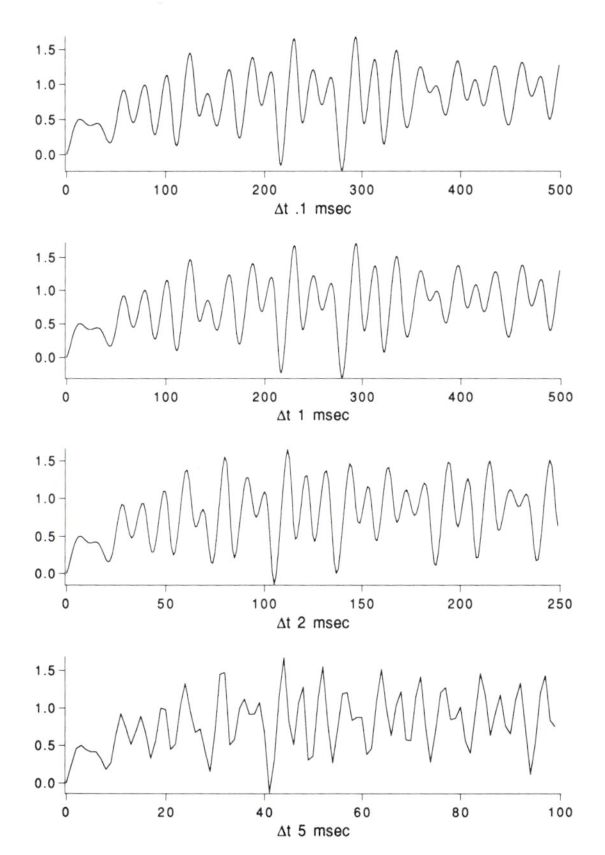

FIGURE 11 An example is shown of the simulation of the basal chaotic output of a mitral node in a KIII model. The range of the spectrum depends on the values of the two rate constants, $a = 220/$sec (4.5 msec) and $b = 720/$msec (1.3 msec), and the range determines the choice of the time step, Δt, for numerical integration of the ODEs. This form of output was used for evaluation of Δt over a range of values starting with a conservatively small step of 0.1 msec and ranging up to 10 msec. The tests were conducted with a small KIII set consisting of four nodes operating in a chaotic domain. The optimal value of $\Delta t = 1$ msec was chosen by the criteria (1) that up to that value the computed waveforms were insensitive to the step size, and (2) that the speed of computation was substantially improved up to that level but much less so for larger step sizes (see Figure 12).

DISCUSSION

Emphasis has been placed on the use of biological data as the basis for optimizing parameters in models used to simulate brain function. In particular, the properties of the EEG reflect the characteristics of cortical

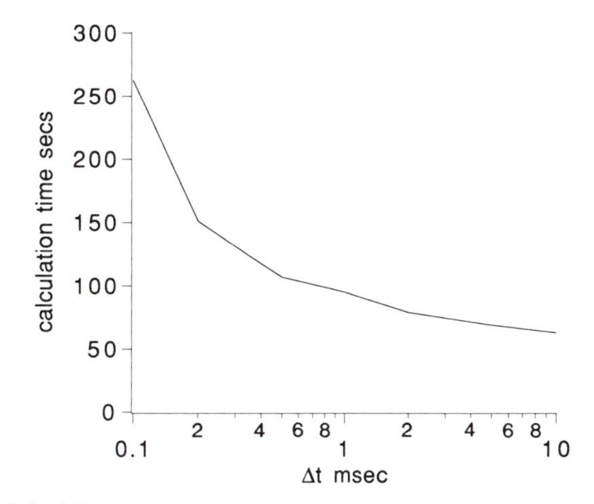

FIGURE 12 An evaluation is shown of the dependence of computation time on the duration of the time step, Δt. The test was done on a Macintosh IIci with cache card for a computed segment of 500 msec duration. The numerical integrator was the Livermore solver for ordinary differential equations (LSODE) based on the Gear and Gearb packages (Hindmarsh, 1983), which proved to be superior to the fifth-order Runge-Kutta algorithm.

dynamics that are among the main determinants and constraints of the information processing done by cerebral cortex. Key features to emerge are the aperiodic and largely, although not entirely, chaotic time series from both the cortex and the models; the widespread spatial coherence of the activity that allows it to serve as a carrier of information; and the broad spectrum, with predominance of spectral energy in the low end. Surgical isolation shows that cortical slabs and slices lose this characteristic activity. Modeling reveals that the low frequency components as well as the robustness of chaotic activity stem from the interaction of multiple parts of the forebrain over long pathways with distributed feedback delays, with intermediate KI_e and KI_i sets having rate constants near zero (very long time constants not based on slow EPSPs and IPSPs).

The use of biological data as a source of criteria for parameter optimization can lead to conflict and confusion in the field of neural networks and neurocomputation. The conflict arises between biologists working from the "bottom up" and engineers and computer scientists working from the "top down." Most biologists' models consist of simulated neurons that operate with variable synaptic conductances, spike trains governed by the Hodgkin-Huxley equations, and chains of compartments that represent the distributed linear dynamics of dendritic trees. Their aim is to achieve verisimilitude to the properties of neurons. Most engineers' models employ simplified elements such as integrators for dendrites, variable resistors for synaptic weights, and continuous variables to represent pulse densities rather than pulse trains. Their aim is to simulate the operations of biological neural systems in pattern classification and learning.

The confusion arises because many biologists attempt to model and explain central neurons in the same way as peripheral neurons. This is a mistake, because the sensory, motor, and autonomic neurons in the peripheral nervous system operate largely independently of other neurons in the periphery, in contrast to central neurons, which receive input from many thousands of others and transmit output to many thousands of others. Under these circumstances, ensembles of interactive central neurons emerge that are characterized by macroscopic properties, including time and distance scales that are much larger than the homologous microscopic properties of single neurons. The macroscopic activities, in turn, control the interactions and activity patterns of single neurons at the microscopic level. Such macroscopic ensembles are not found in the peripheral nervous system.

The macroscopic patterns determine the output of cortex, even though the activities are conveyed by pulses on individual axons. This is because most central pathways have substantial axonal divergence, which implies that neurons to which the pathways project perform on-line, real-time spatial integration of their input. The trigger zones of the integrating neurons have no information about which subsets of dendritic inputs are controlling their firing rates. In this circumstance, the macroscopic pattern is enhanced by the receiver, and the microscopic activity of the transmitter is smoothed and discarded. In a word, the cortical signal is "laundered" by axonal divergence and spatial integration (Freeman, 1990, 1991a).

The conflict and confusion are resolvable, because the kinds of equations that are used by engineers to design and construct neural networks conform to the macroscopic properties of ensembles, and these equations are fully applicable to the dynamics of central neural ensembles. The key problems for biologists are (1) to measure the neural activities of populations in both their microscopic and macroscopic forms, and (2) to describe the neural operations by which each is transformed into the other. It is apparent that sensory information is brought to the central nervous system in microscopic form, that it is transformed to macroscopic information by sensory cortex, and that it is retransformed back to microscopic form when it is sent by sensory cortex to association cortices and into the motor systems. The key problems for engineers are how to (1) build into their models the capacities to accept large quantities of data from arrays of sensors in parallel simultaneously, (2) maintain a substantial number of chaotic attractors in the state space of a model, and (3) give rapid and unbiased access to the basin of each attractor in conformance to the characteristics of the input in each time frame of observation. The reduction to practice inevitably will require searching through parameter spaces of very high dimension. The distinction between microscopic and macroscopic levels of neural function and the resultant choices of the state variables of the models should play major roles in defining and regulating that search.

SUMMARY

The simulation of a cortical network of neural populations requires use of a large number of second-order ODEs, each comprising a node in the KIII model that is equivalent to a cortical column containing 10^4 to 10^5 cells, and requires the appropriate networks to contain a large number of parameters. Owing to the sensitivity of chaotic systems to seemingly minor differences in their structures and parameters, and to differences in computer systems by which they are embodied, there are no set parameters for all embodiments, and parameterization is an ongoing process. The parameters must be evaluated with respect to criteria that are derived from observations on the spatial and temporal properties of the macroscopic activity of the cortex in normal, waking subjects that are engaged in sensory information processing. The salient properties of this macroscopic activity, as revealed in the EEGs, are the spatial coherence of its waveform over areas containing hundreds and even thousands of columns, the broad "1/f" spectrum of the activity, and the spatial patterning of the amplitude modulation of the common carrier, which varies during training to perform sensory discriminations. The requisite parameters are grouped into classes. The two time constants of the second-order ODEs are fixed for all nodes and all times at values derived from measuring open loop rate constants of cortical impulse responses. The steepness of the sigmoid curve matching the operation of the trigger zones in converting dendritic current density to neuronal pulse density is governed by a single fixed parameter Q_m, which is evaluated in waking subjects from the statistical dependence of pulse probability conditional on the EEG amplitude. The four connectivities among the excitatory and inhibitory neurons, k_{ee}, k_{ei}, k_{ie}, and k_{ii}, are initially set to unity. They are modified in such a way as to simulate the small signal, near linear impulse responses (the averaged evoked potentials that resemble damped cosines) of the KII sets that make up the cortical system, and the changes in the waveforms of the impulse responses with associative learning and habituation. The long feedback paths that dominate the interactions between components in cortical systems provide dispersive delays, which are required to approximate the "1/f" spectrum of the basal olfactory system. Of even greater importance, are the excito-excitatory connections through the outer layer of the olfactory bulb, which must be adjusted with special care to replicate the pseudorandom properties of the EEGs in both burst and interburst periods.

Acknowledgments

This work was supported by Grant N66365 from the Office of Naval Research.

References

Biedenbach, M. A., & Freeman, W. J. (1965). Linear domain of potentials from the prepyriform cortex with respect to stimulus parameters. Experimental. *Neurology*, **11**, 400–417.

Bressler, S. L. (1988). Changes in electrical activity of rabbit olfactory bulb and cortex to conditioned odor stimulation. *Journal of Neurophysiology*, **102**, 740–747.

Bressler, S. L., & Freeman, W. J. (1988). Frequency analysis of olfactory system EEG in cat, rabbit and rat. *Electroencephalography and Clinical Neurophysiology*, **50**, 19–24.

Brillinger, D. R. (1975). The identification of point process systems. *Annals of Probability*, **3**, 909–929.

Brillinger, D. R. (1992). Nerve cell spike train data analysis: A progression of technique. *Journal of the American Statistical Association*, **87**, 260–271.

Burns, B. D. (1958). *The mammalian cerebral cortex*. Baltimore, MD: Williams & Wilkins.

Eckhorn, R., Bauer, R., Jordan, W., Brosch, M., Kruse, W., Munk, M., & Reitboeck, H. J. (1988). Coherent oscillations: A mechanism of feature linking in visual cortex? *Biological Cybernetics*, **60**, 121–130.

Eeckman, F. H., & Freeman, W. J. (1990). Correlations between unit firing and EEG in the rat olfactory system. *Brain Research*, **528**, 238–244.

Eeckman, F. H., & Freeman, W. J. (1991). Asymmetric sigmoid nonlinearity in the rat olfactory system. *Brain Research*, **557**, 13–21.

Emery, J. D., & Freeman, W. J. (1969). Pattern analysis of cortical evoked potential parameters during attention changes. *Physiology and Behavior*, **4**, 67–77.

Freeman, W. J. (1964). A linear distributed feedback model for prepyriform cortex. *Experimental Neurology*, **10**, 525–547.

Freeman, W. J. (1975). *Mass action in the nervous system.* New York: Academic Press.

Freeman, W. J. (1979a). Nonlinear gain mediating cortical stimulus-response relations. *Biological Cybernetics*, **35**, 237–247.

Freeman, W. J. (1979b). Nonlinear dynamics of paleocortex manifested in the olfactory EEG. *Biological Cybernetics*, **35**, 21–37.

Freeman, W. J. (1979c). EEG analysis gives model of neuronal template-matching mechanism for sensory search with olfactory bulb. *Biological Cybernetics*, **35**, 221–234.

Freeman, W. J. (1987a). Techniques used in the search for the physiological basis of the EEG. In A. Gevins & A. Remond (Eds.), *Handbook of EEG and clinical neurophysiology* (Vol. 3A, Part 2, Ch. 18, pp. 583–664). Amsterdam: Elsevier.

Freeman, W. J. (1987b). Simulation of chaotic EEG patterns with a dynamic model of the olfactory system. *Biological Cybernetics*, **56**, 139–150.

Freeman, W. J. (1990). On the problem of anomalous dispersion in chaoto-chaotic phase transitions of neural masses, and its significance for the management of perceptual information in brains. In H. Haken & M. Stadler (Eds.), *Synergetics of cognition* (Series in Cognition, Vol. 45, pp. 126–143). Berlin: Springer-Verlag.

Freeman, W. J. (1991a). The physiology of perception. *Scientific American*, **264**, 78–85.

Freeman, W. J. (1991b). Colligation of coupled cortical oscillators by the collapse of the distributions of amplitude-dependent characteristic frequencies. In C. W. Gear (Ed.), *First NEC research symposium* (pp. 69–103). Philadelphia: SIAM.

Freeman, W. J. (1991c). Nonlinear dynamics in olfactory information processing. In J. L. Davis & H. B. Eichenbaum (Eds.), *Olfaction: A model system for computational neuroscience* (pp. 225–250). Cambridge, MA: MIT Press.

Freeman, W. J. (1992). Tutorial in Neurobiology: From single neurons to brain chaos. *International Journal of Bifarcation and Chaos*, **2**, 451–482.

Freeman, W. J., & Baird, B. (1987). Relation of olfactory EEG to behavior: Spatial analysis. *Behavioral Neuroscience*, **101**, 393–408.

Freeman, W. J., & Grajski, K. A. (1987). Relation of olfactory EEG to behavior: Factor analysis. *Behavioral Neuroscience*, **101**, 766–777.

Freeman, W. J., & Jakubith, S. (1993). Bifurcation analysis of continuous time dynamics of oscillatory neural networks. In A. Aertsen & W. von Seelen (Eds.), *Brain theory—Spatiotemporal aspects of brain function* (pp. 183–207). Amsterdam: Elsevier.

Freeman, W. J., & Maurer, K. (1989). Advances in brain theory give new directions to the use of the technologies of brain mapping in behavioral studies. In K. Maurer (Ed.), *Conference on topographic brain mapping* (pp. 118–126). Berlin: Springer-Verlag.

Freeman, W. J., & Skarda, C. (1985). Nonlinear dynamics, perception, and the EEG; the neo-Sherringtonian view. *Brain Research Reviews*, **10**, 147–175.

Freeman, W. J., & Van Dijk, B. (1987). Spatial patterns of visual cortical fast EEG during conditioned reflex in a rhesus monkey. *Brain Research*, **422**, 267–276.

Freeman, W. J., & Viana Di Prisco, G. (1986). Relation of olfactory EEG to behavior: Time series analysis. *Behavioral Neuroscience*, **100**, 753–763.

Grajski, K. A., & Freeman, W. J. (1989). Spatial EEG correlates of nonassociative and associative learning in rabbits. *Behavioral Neuroscience*, **103**, 790–804.

Gray, C. M., Freeman, W. J., & Skinner, J. E. (1986). Chemical dependencies of learning in the rabbit olfactory bulb: Acquisition of the transient spatial-pattern change depends on norepinephrine. *Behavioral Neuroscience*, **100**, 585–596.

Gray, C. M., Koenig, P., Engel, K. A., & Singer, W. (1989). Oscillatory responses in cat visual cortex exhibit intercolumnar synchronization which reflects global stimulus properties. *Nature*, **338**, 334–337.

Gray, C. M., & Skinner, J. E. (1988). Field potential response changes in the rabbit olfactory bulb accompany behavioral habituation during repeated presentation of unreinforced odors. *Experimental Brain Research*, **73**, 189–197.

Haberly, L. B., & Bower, J. M. (1984). Analysis of association fiber system in piriform cortex with intracellular recording and staining techniques. *Journal of Neurophysiology*, **51**, 90–112.

Hebb, D. O. (1949). *The organization of behavior: A neuropsychological theory.* New York: Wiley.

Hindmarsh, A. C. (1983). A systematized collection of ODE solvers. In R. S. Stepleman et al. (Eds.), *Scientific computing* (pp. 55–64). Amsterdam: North Holland. LSODE is available via the anonymous ftp-server: research.att.com.

Kay, L., Shimoide, K., & Freeman, W. J. (in press). Comparison of EEG time series from rat olfactory system with model composed of nonlinear coupled oscillators. *International Journal of Bifurcation and Chaos.*

Mitzdorf, U. (1987). Properties of the evoked potential generators: Current source-density analysis of evoked potentials in cat cortex. *International Journal of Neuroscience*, **33**, 33–59.

Mountcastle, V. B. (1966). The neural replication of sensory events in the somatic afferent system. In J. C. Eccles (Ed.), *Brain and conscious experience* (pp. 85–115). Berlin: Springer-Verlag.

Murthy, V. N., & Fetz, E. E. (1992). Coherent 25–35 Hz oscillations in the sensorimotor cortex of awake behaving monkeys. *Proceedings of the National Academy of Sciences of the United States of America*, **89**, 5670–5674.

Ribary, U., Llinas, R., Lado, F., Mogilner, A., Jagow, R., Nomura, M., & Lopez, L. (1992). The spatial and temporal organization of the 40 Hz response in human brain: An MEG study. In M. Hoke et al. (Eds.), *Biomagnetism: Clinical aspects* (pp. 158–163). Amsterdam: Elsevier.

Shimoide, K., Greenspon, M. C., & Freeman, W. J. (1993). Modeling of chaotic dynamics in the olfactory system and application to pattern recognition. In F. H. Eeckman (Ed.), *Neural systems analysis and modeling* (pp. 365–372). Boston: Kluwer.

Sholl, D. A. (1956). *The organization of the cerebral cortex*. New York: Wiley.

Skarda, C. A., & Freeman, W. J. (1987). How brains make chaos in order to make sense of the world. *Brain and Behavioral Science*, **10**, 161–195.

Thompson, J. M. T., & Stewart, H. B. (1988). *Nonlinear dynamics and chaos*. New York: Wiley.

Yao, Y., & Freeman, W. J. (1990). Model of biological pattern recognition with spatially chaotic dynamics. *Neural Networks*, **3**, 153–170.

Yao, Y., Freeman, W. J., Burke, B., & Yang, Q. (1991). Pattern recognition by a distributed neural network: An industrial application. *Neural Networks*, **4**, 103–121.

II

Emulated and Simulated Systems

8

Neural Computation of Visual Images

Paul Mueller, David Blackman, and Roy Furman

INTRODUCTION

How do we see? The vertebrate vision system is perhaps the most complex neural assembly known. Although more details are known about vision than about any other neural system, it is not yet fully understood how we analyze the images in our environment nor what computational strategies are used for this task. However, to build machines that can visually analyze their surroundings and move in a natural environment, we have to employ, in all likelihood, the principles and algorithms used by biological vision systems.

As a step toward this goal, we shall discuss the design of a simple neural system based on our current understanding of the initial computations performed by the retina and primary visual cortex on a stationary image. Our system is concerned only with the analysis of form and not with motion, color or binocular disparity.

Some of the operations performed by retina and primary visual cortex have been fairly well understood for some time (Kuffler, 1953; Hubel & Wiesel, 1962, 1965; Enroth-Cugell & Robson, 1966) and are, in fact, what one would expect from first principles, that is, the decomposition of the image into a set of pattern primitives and their individual representation by local neural activity. It is our premise that the retina and primary visual cortex decompose the image into spatial primitives such as edges and lines or bars of different spatial scales, orientations, and contrast parameters. We argue that each neuron performs a specific computational task and that a complete and systematic representation of all spatial primitives at the neural level forms the basis for further analysis at higher stages. This argument is based on the fact that the vertebrate vision system can recognize these primitives within an image, suggesting that they have individual representations. Furthermore, single cell recordings by Hubel and Wiesel (1962, 1965) have identified neurons in the primary visual cortex tuned to such primitives, although the extent to which the different primitives are represented is not known. Beyond this we can only guess at how these primary representations are utilized at subsequent stages to yield a more global "understanding" and recognition of a scene.

With eventual hardware implementation in mind, it is our first task to derive the simplest connection architecture that would generate sets of neurons tuned to different primitives and to compare the performance of these units with cortical neurons.

The second problem to be addressed is the integration of such units into a functional system. Representation of different spatial scales and orientations, receptor field overlap, representational density as a function of eccentricity, and tiling geometry shall be discussed. Unfortunately, biological data provide only minimal background for these problems, and we have to rely on arguments of parsimony and computational utility.

SYSTEM DESIGN

Overview

Our system, shown in Figure 1, consists of arrays of simple neuron analogs arranged in layers. The first layer represents an array of receptors. The second layer receives input from the receptors and forms two separate arrays of "on-center" and "off-center" units modeled after the on-center and off-center retinal X ganglion cells. The output from this layer converges on neurons in the third layer which is analogous to parts of area V_1 to V_3 in the visual cortex. This layer is

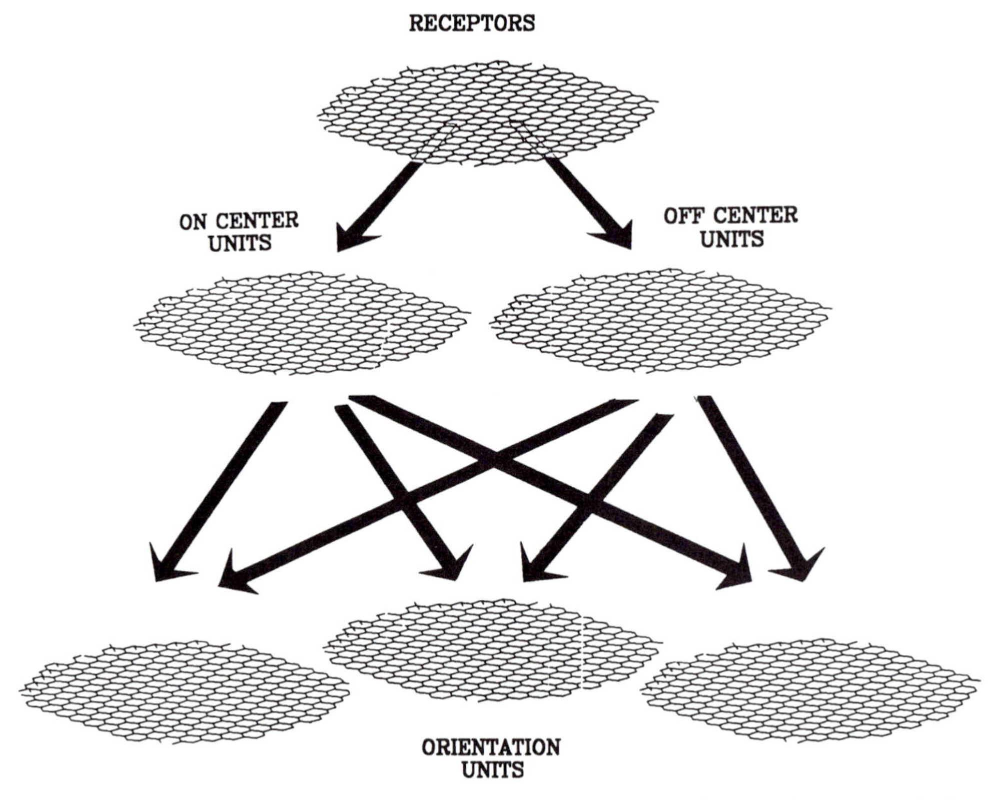

FIGURE 1 The system is organized in several layers. The first layer consists of an array of receptors (typically 160 x 160 units) which are connected to two arrays (100 x 100) of on-center and off-center units forming the second layer. Units in these arrays are connected to a third layer consisting of a large number of separate arrays (60 x 60). Each array is dedicated to the extraction of a specific primitive (oriented edges, lines, etc.). In order to avoid boundary artifacts, the number of units in each array is smaller than the number of units in arrays in the preceding layer. The units are arranged in hexagonal geometry. Each unit in the orientation arrays receives excitatory and inhibitory inputs from both on- and off-center units in a push–pull fashion.

a composite of arrays of units tuned to different primitives. The following primitives are represented as separate arrays: oriented edges, oriented lines (or bars), and end-stopped oriented edges or bars. In addition, lines or bars are represented at different spatial scales (length and width). The number of orientation angles and of spatial scales can be selected on the basis of computational needs. For reasons of simplicity we do not include a separate representation of the lateral geniculate nucleus.

Neuron Properties

The input–output relations of our neurons are idealized versions of a typical biological neuron (see Figure 2). They have been discussed in more detail elsewhere (Mueller & Lazarro, 1986; Mueller, Spiro, Blackman, & Furman, 1987). Each unit has a threshold, V_t, which is adjustable from negative to positive values, an adjustable minimum output at threshold, E_x, and a maximum output, E_{max}. Over the dynamic range of the neuron, the transfer function, that is, the relation between the sum of inputs and the output, can be either linear with a slope controlled by a constant, s,

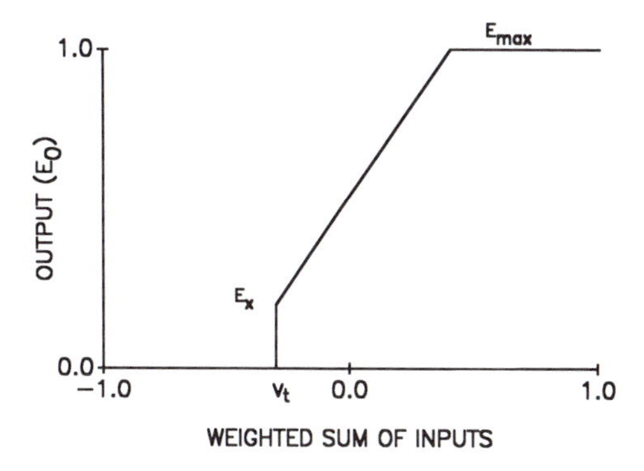

FIGURE 2 Steady state input–output function of an individual model neuron used in the simulations. Each unit has an adjustable threshold, V_t, an input-output relation above threshold, adjustable by a slope constant s, and extra output constant, E_x, and a maximum output value E_{max}. The output can be positive or negative and each unit can therefore excite or inhibit other units. Within the dynamic range, the input-output function was usually assumed to be linear as shown here and defined by equations (1-2), but logarithmic or exponential functions have also been explored.

(which was the case for most computations shown below) or nonlinear, for example, exponential or logarithmic. Logarithmic transfer characteristics broaden and exponential characteristics compress the dynamic range of the units. In the linear case, the input–output function of the units is determined by the following relationships:

Define the input activity of the jth neuron to be

$$E_{in,j} = \sum_i E_{out,i} \sigma_{ij} \tag{1}$$

where $E_{out,i}$ = output from i^{th} neuron and σ_{ij} = gain of the synapse between neuron neuron i and j. Then

$$E_{in,j} = \begin{cases} 0 \text{ if } E_{in,j} < V_T \\ s(E_{in,j} - V_T) + E_x \text{ if } V_T < E_{in,j} < E_{max} \\ E_{max} \text{ if } E_{in,j} > E_{max} \end{cases} \tag{2}$$

Permitting negative thresholds allows units to output in the absence of external input, as is typical for the retinal ganglion cells. The minimum output at threshold, E_x is equivalent to the initial firing rate of a biological neuron at threshold. Large values of E_x can be used to implement logic operations. Our units do not generate action potentials, but their continuous output is comparable to the time averaged firing frequency of biological neurons.

Connection Architectures

Receptor Array

Receptor units are arranged in a densely packaged array. Their input–output relation is a bounded linear function, and there are no lateral connections between units. Input images are set either by special programs that generate simple geometric shapes or from an image scanner with an 8-bit gray scale resolution.

On- and Off-Center Arrays

The units of these arrays receive direct input from the receptors. Their receptor fields have a hexagonal center–surround organization. The connections and synaptic gains of the center have one sign while those of the surround have the opposite sign. In the on-center plane, the center is excitatory and the surround is inhibitory. The reverse is true for the off-center plane. The diameter of the center and of the surround can be adjusted. In the field's center, the gains are set so that the sum of the gains in each ring is equal to 1. In the center and surround, the gains decrease exponentially with distance from the center. The surround gain values are calculated so that the sum of values over the entire center is proportional to the negative sum over the entire surround. The constant of proportionality is usually 1. The constant can be set to be greater than 1 so that increasing the background level of the receptors causes an increase in the activity of the on/off-center units.

The connection architecture implemented here differs from its biological counterpart where the surround field is generated by horizontal cells (Werblin & Dowling, 1969; Attwell, 1986) and was chosen to avoid expensive iterative, digital computations. However, the resulting input–output relations are equivalent. We should also mention that the analog VLSI retina built by Carver Mead (Mead & Mahowald, 1988; see also Mead, 1989) conforms to the biological model. Analog methods have no problem solving simultaneous nonlinear equations in real time. Note that problems of receptor field size, overlap (tiling), and density of receptors relative to on- or off-center units already enter at this stage.

Orientation Arrays

A number of parallel arrays receive input from the on- and off-center arrays. These are orientation arrays, which respond to a primitive oriented at specific angles. Each orientation unit receives symmetrical inputs from on-center and off-center units with opposite signs for the synaptic gains. This push–pull arrangement, which may also exist in the biological system (Heggelund, 1981; Schiller, 1982; Ferster, 1988), improves the performance within the dynamic range of the orientation units and contributes to generating the Gabor-like output function of the orientation units seen in the biological system (Gabor, 1946; Marcelja, 1980; Daugman, 1985; Jones & Palmer, 1987). There are six types of orientation arrays in all: the edge, the line, and four end-stop detection arrays. The edge detection units respond best to an edge situated in the center of their receptive fields. The line detectors respond best to a line or bar of a certain width. End-stop units respond when a line or edge terminates at a certain location within their receptive fields. There are four types of end-stop arrays: one each for line and edges, which detect an end-stop at either end of the image, and two array types which detect end-stops at both ends of the edge or line. Units in the latter arrays detect edges and lines of specific lengths and widths.

There are a number of arrays for each type. All detect the same primitive, but each responds to a different angle of orientation. The number of angles is programmable. In addition to arrays which detect the primitive at these angles, there is a corresponding set of arrays which respond to the primitive of opposite contrast but at the same angle (light areas are dark and dark areas are light). Finally, all primitives and orientations are represented in different arrays at varying spatial scales of length and width. As a result of these multiple representations, each receptor projects to many orientation units in multiple arrays. The exact number depends on the number of different spatial scales and orientations that are chosen.

The receptive field of an orientation unit is rectangular and typically receives input from 8 x 14 on- and off-center units. The rectangle is aligned with the axis of orientation which passes through the center of the field. The strength of the synaptic gains varies with distance from the orientation axis. Depending on field size and the ratio of receptor density to on- and off-center density, each unit receives 150 to 280 inputs. This is similar to the biological system where the input field of a simple cell in the fovea is approximately 20 x 12 receptors wide (Dow, Snyder, Vautin, & Bauer, 1981).

Setting of Synaptic Gains

The connection strength (synaptic gain or weight) from receptors to on- or off-center units is set manually to compute a center–surround function as outlined above. Modifications of center and surround width as well as the center–surround gain ratio are under program control which automatically adjusts the gain values. Field size and gain values for the connections from on- and off-center units to the orientation units can also be set by hand using values that approximate the activity of the on- and off-center units in response to a particular primitive. However, it is more efficient to use the following one-step forward learning algorithm for this process.

To compute the synaptic gain from a unit in the on- or off-center arrays to a unit in an orientation array, the primitive that is to be recognized by the orientation unit is projected onto the receptor array. The products of the receptor array output and the on- and off-center synaptic gains are then used to form a weighted sum to compute the desired orientation unit synaptic gain, σ_{ij}.

$$\sigma_{ij} = \sum_{h} E_{oh}\sigma_{hi}\, pm \qquad (3)$$

where E_{oh} is the output of the receptor at location h and σ_{hi} is the synaptic gain between the receptor at h and the on- or off-center cell at i. The orientation synaptic gain is also scaled by a factor, p, to allow the maximum output of the entire orientation receptive field to fall within the neuron's dynamic range, and a factor, m, that is a function of distance along the orientation axis. This latter factor is used to improve the performance of end-stopped units.

This algorithm results in the gain fields shown in Figure 3. Since these gain fields are independent of the image position in the receptor field, the same gain field is applied to all units in an orientation layer. The algorithm also could have been used to set the synaptic gains of the on- and off-center units by projecting light and dark dots on the receptor array if there had been a distance-weighted lateral inhibition between receptors.

In a sense the procedure resembles the matched filter techniques used in image processing except that it accounts for the special input–output characteristics of neural units. The algorithm adjusts both inhibitory and excitatory gains even though the input units (on- or off-center) may be inhibited below threshold and their output cannot serve as an indicator of the inhibitory level.

The size and shape of the input receptor fields obtained by this procedure differ for edge and line units. Inputs to line units are symmetric whereas inputs to edge units are asymmetric, differing in their inhibitory and excitatory regions by a 90° phase angle. The synaptic gain fields shown in Figure 3 were obtained by training with "ideal" edges and bars that had vertical contrast borders. Training with bars or edges that have sloping intensity profiles or bars with unequal contrast amplitudes at their borders results in synaptic gain fields with intermediate phase angles and unequal width and amplitudes of inhibitory and excitatory regions. There is evidence that biological units can have intermediate phase angles, although the extent of representation of different phase angles is unknown. For simplicity, we have restricted our system to four phases. This affords optimal tuning to "ideal" bars of opposite contrast signs, that is, light or dark, and to edges of opposite contrast directions normal to the edge.

Units tuned to different spatial scales are obtained by presenting lines of different width and edges of different intensity gradients during the training procedure. The inhibitory gains in the end-stopped regions are adjusted for optimal performance and suppression of unwanted output from units lying outside the feature. Single end-stopped line units are not very sensitive to the length of the field since they only detect single endpoints of the feature. The double end-stopped units, on the other hand, respond to edges or lines of a particular length which is determined by the length of the receptor field between the end-stopped regions. Extraction of features of different length and width thus requires units with different axial ratios of their receptor fields. We have not yet explored the optimal extent to which different axial ratios are to be represented.

Representation and Resolution

Assuming that we have chosen a set of primitives that is adequate for primary image decomposition, and that we can construct units that are tuned to these primitives, we are still faced with the unsolved issues of their optimal representational density. Little is known from biology about the representation of different spatial scales, receptor field overlap, axial ratios, and representational density as a function of eccentricity. For

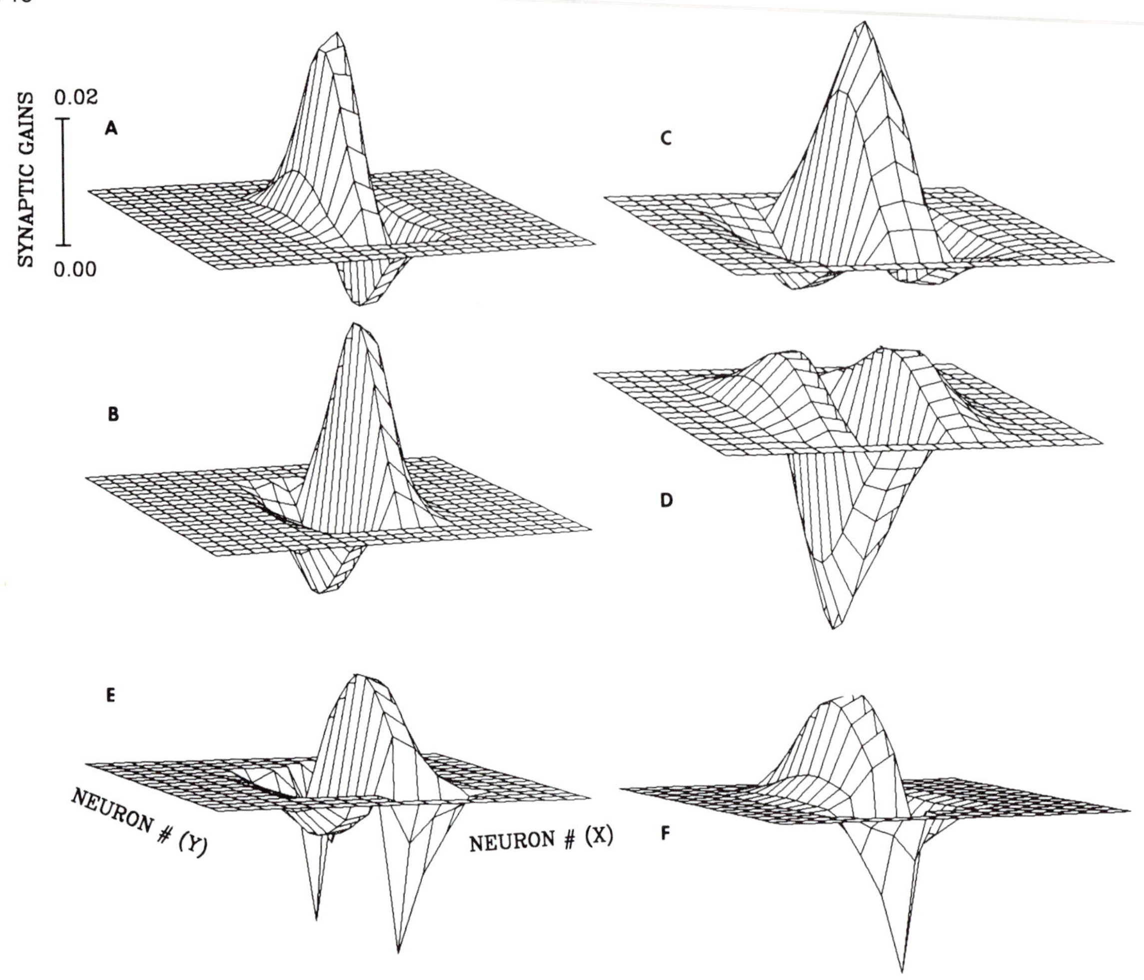

FIGURE 3 Three-dimensional plots of the synaptic gains (weights) for connections from on-center (A,C) and off-center (B,D) units to an edge unit (A,B) and a line unit (C,D). For line units tuned to opposite contrast sign, the sign of the synaptic gains from on- and off-center units are interchanged. E and F show examples of gains from on-center units to a double and single end-stopped edge unit. Positive values indicate excitation; negative values, inhibition. The gains shown here were obtained from "ideal" edges and lines by the procedure described in the text. Edges and lines with more complex contrast profiles generate gain fields of different width, as well as intermediate phase and amplitude relations between inhibition and excitation. G. Gray scale representation of synaptic gains from on-center units to edge units tuned to edges of different orientation.

a machine, orientation increments of 10° seem reasonable, but a coarser representation may also suffice. A fine grained spatial scale representation and dense receptor field tiling would lead to ratios of receptor to orientation units of more than 1000 to 1. Higher order processing of all this information over a large visual field would be practically impossible. Biology has solved aspects of this problem by restricting the finest spatial scales to a very small region of densely packed receptors in the center of the retina which is only 25 receptors wide in the macaque monkey (Perry & Cowey, 1985). It is most likely that the dense repre-

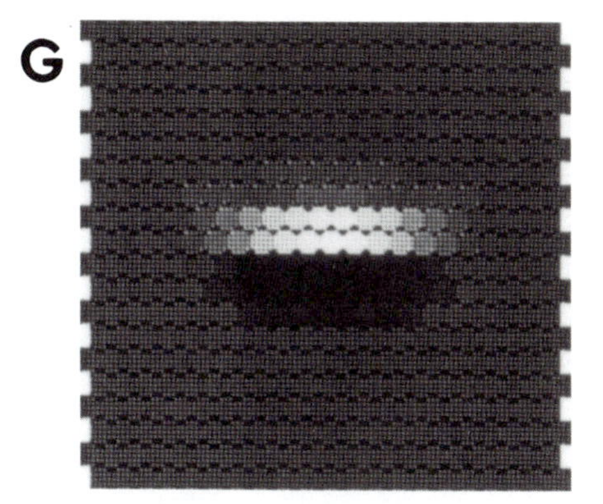

FIGURE 3 *(CONTINUED)*

sentation is restricted to this area and that the representational density rapidly decreases toward the periphery. The dramatic increase in the ratio of cortical cells to receptors in the center of the visual field is known as cortical magnification.

There are two representational issues that must be kept separate. One issue is the distribution of spatial scales as a function of eccentricity. The other is the density at which a given spatial scale is represented, that is, what is the receptor field overlap (tiling) and how does it vary with eccentricity. In the biological system, spatial scale representation, as well as receptor and x-cell density, decrease toward the periphery. In our system we can explore this issue by adjusting the ratios of the receptor to on- and off-center unit density as well as the on- and off-center to orientation unit density. The latter can be adjusted either by increasing the receptor field width but keeping the number of inputs constant (leading to an increase of spatial scale), or by keeping the density of inputs within the receptor field constant but reducing the overlap between the receptor fields, that is, moving the center of the receptor field by more than one receptor unit (see Figure 13 under Computation of Natural Images). This reduces the tiling density by a factor which itself can be made to increase as a function eccentricity so that the smaller spatial scales are represented by fewer units further out in the visual field.

SYSTEM PERFORMANCE

Receptors

The intensity value of images projected on the receptor array are scaled from 0 to 1 with a typical resolution of 8 bits. Geometric images of arbitrarily higher gray scale resolution can be obtained by computation. The input–output resolution function of the receptors is linear and there is no lateral interaction between receptors.

On- and Off-Center Units

Figure 4 shows the performance of the on- and off-center units in response to a single dot of light and a square projected on the receptor field. A single spot of light activating only one receptor is represented by a cluster of active units in the on- and off-center arrays. This cluster representation, which appears here as an initial property at the first stage of computation, is preserved at all subsequent stages, which also seems to be the

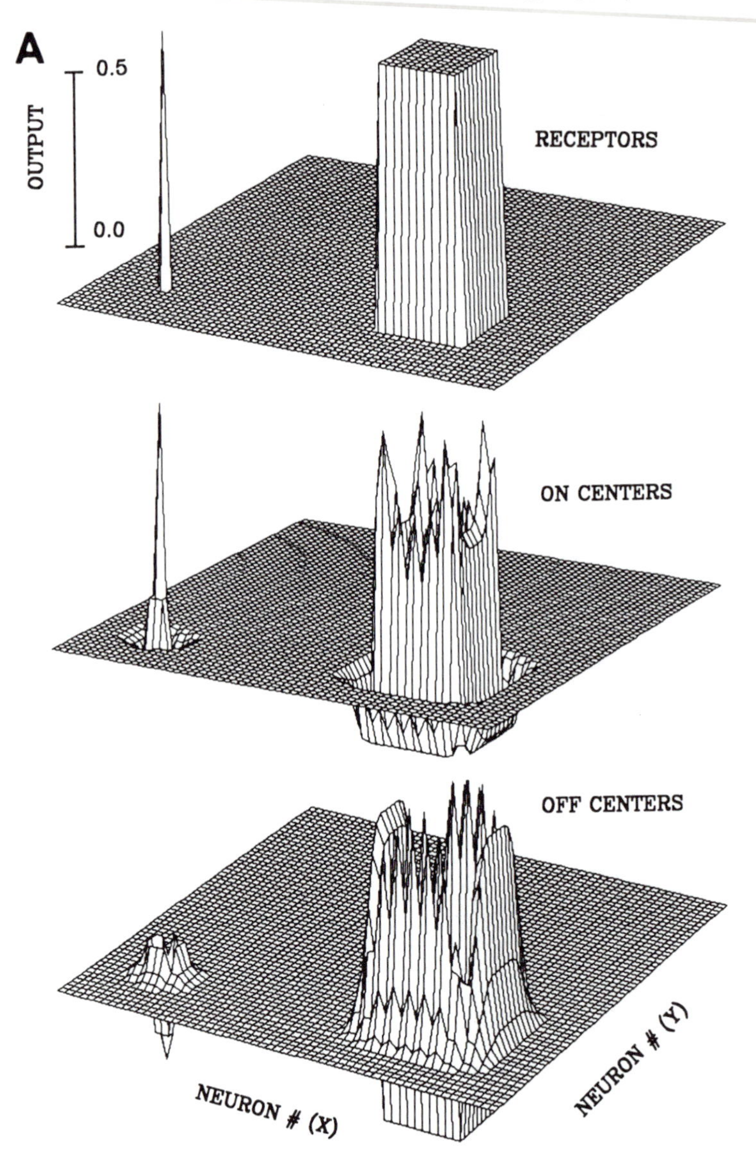

FIGURE 4 Three dimensional (A) and gray scale (B) representations of the outputs from receptors and on- or off-center units in response to a spot and a rectangle of light. The units have antagonistic center–surround connections with a center diameter of 3 receptor units and a total receptor field diameter of 9 units. The ratio of the sum of center and surround gains was 1. The neuron threshold V_t was set at -0.2 and E_x was set to 0. In the 3-D plots shown here and in subsequent figures, units are projected on a square array whereas in the actual calculations all units are arranged in hexagonal geometry. There is therefore a small mismatch between the two representations.

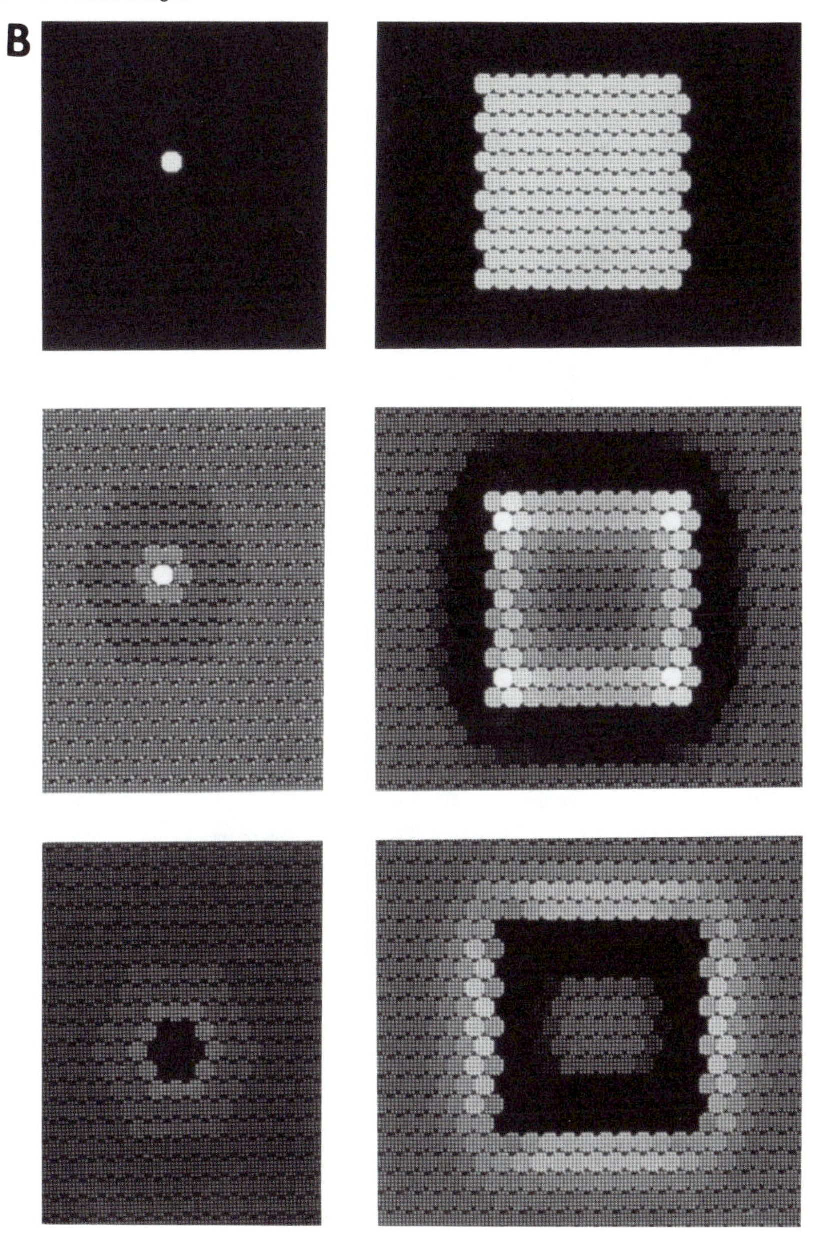

FIGURE 4 *(CONTINUED)*

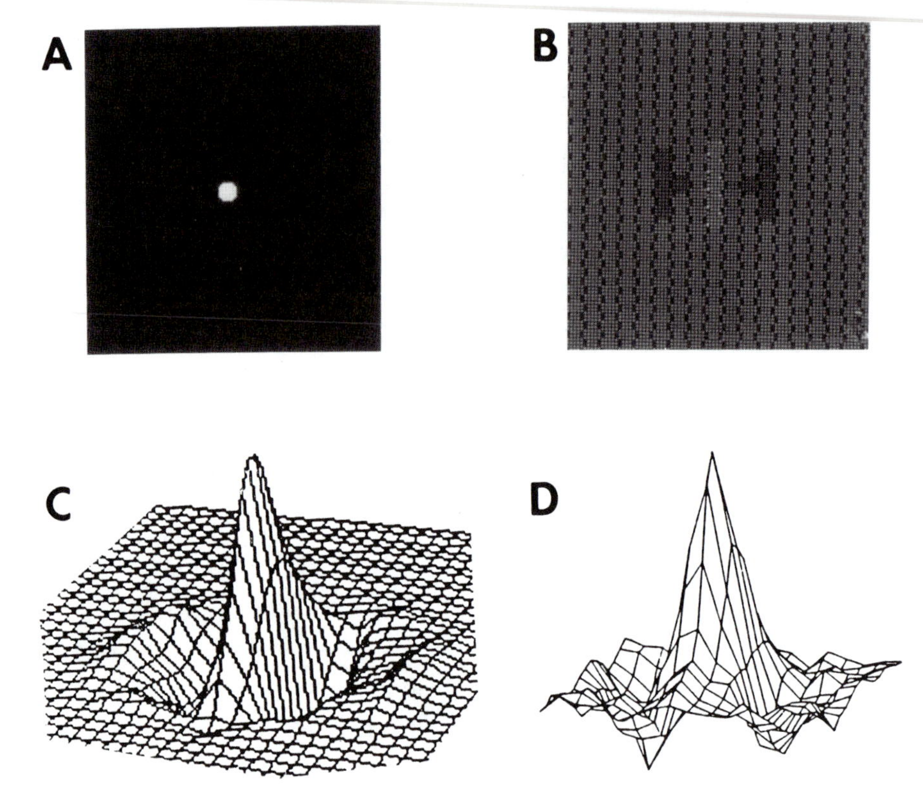

FIGURE 5 Outputs from oriented line units to a single spot of light on the receptor plane. A and B show the outputs of the receptor plane (A) and line unit plane (B) in gray scale. The activity of the line units is barely visible in B but a 3-D plot at higher gain shows a distinct field of active units, some of which are excited whereas others are inhibited (C). This output pattern is similar to a 2-D Gabor function and is almost identical to the output from a simple cell in cat striate cortex in response to a spot of light applied at different positions within its receptor field as recorded by Jones and Palmer (1987), and shown in D. Notice the inhibitory and excitatory side bands which enhance the spatial frequency tuning of such cells. E. Examples of outputs generated by line and edge units in response to two single spots of light or dark in the receptor plane in the presence of 0.5 background illumination. Each of these responses resembles a 2-D Gabor function with differing phase relations. The response fields were obtained with synaptic gains as shown in Figure 3.

case for biological systems. For the square image, the outputs from on-center or off-center units show the well known enhancement of boundary contrast. One interesting point is the increase of the on-center output and suppression of off-center output at the corners of the figure. The receptor field width of the on- and off-center units determines the cluster diameter and affects to some extent the resolution. However, since the position of maximal activity at the cluster center can always be extracted by subsequent processing, for example, by subtraction of neighboring clusters, the maximal resolution can be higher than the cluster diameter. In fact, the resolution is mainly determined by the tiling of the receptor fields, that is, by the ratio of the receptor density to the on- and off-center unit density (see Figure 9 under Orientation Units).

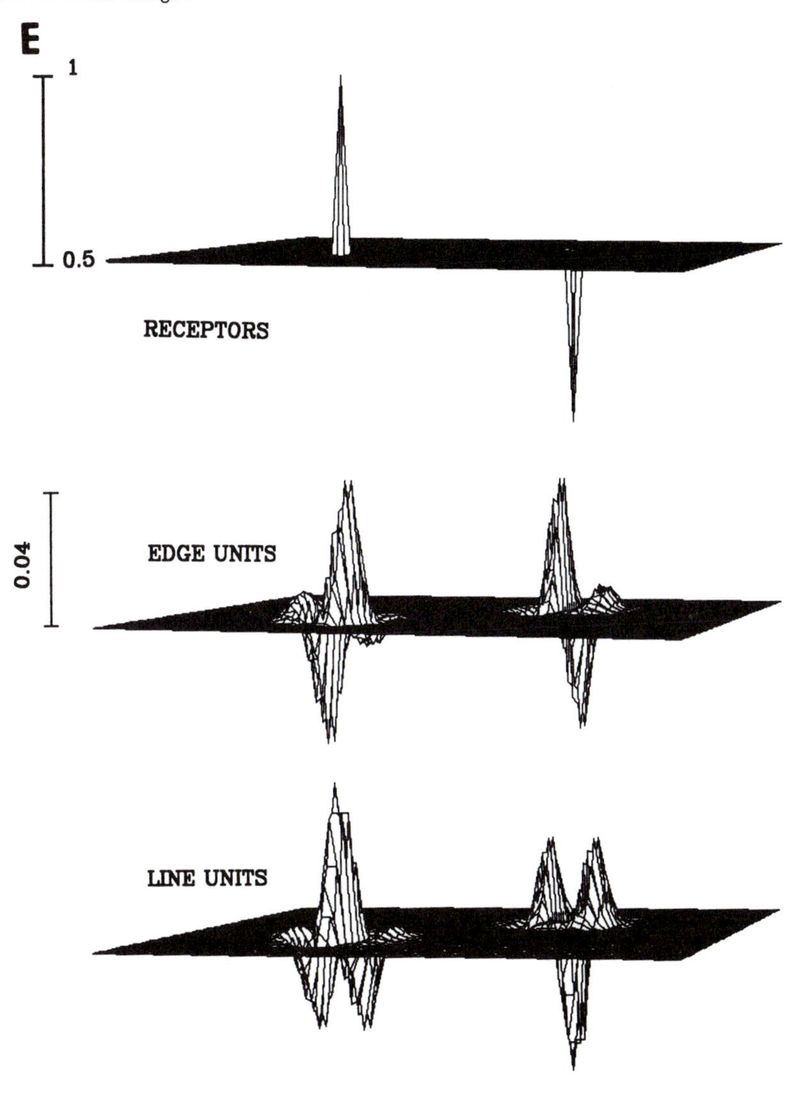

FIGURE 5 *(CONTINUED)*

The contrast sensitivity of on- and off-center units is determined by the synaptic gain and the input–output transfer function chosen for the units. Our system is sufficiently flexible to match contrast sensitivity relations observed in the biological retina.

The thresholds of the units were adjusted such that they had a 20% dark activity. This dark activity is a useful feature which not only assures that the units operate in the linear range but is also of computational significance because inhibition by units with resting activity can be utilized to implement logical "AND" functions. For all computations presented in the figures the initial output at threshold (E_x) was set to zero to maximize the linear dynamic range of the units.

Orientation Units

There is a large parameter space determining the performance of the orientation units. Neuron characteristics, that is, threshold, minimum output at threshold, and input–output transfer function, form only one set of these parameters. More important are the synaptic gains and receptor fields of the on- and off-center units as well as the gains and receptor fields of the orientation units. We can discuss here only briefly the effects of these parameters on the performance of the orientation units and shall mainly present the results using a set of standardized units. Within limits, modifications of the above parameters change only the quantitative aspects of the system.

Figures 5 through 11 illustrate the performance of the different orientation units. Figure 5 shows the response of a line unit to a single spot of light. As was the case at the previous stage, a single input by a spot of light is represented by a cluster of active units with maximal activity occurring at the position of the spot. In this case the response is no longer radially symmetric but has a long axis corresponding to the orientation axis of these particular units. Moreover, the field of active units now has the shape of a 2-D Gabor function, which is typically observed for a simple cell in the primary visual cortex. Other examples of the response fields for edge and line units are shown in Figure 5E.

The shape of the response function depends on the synaptic gains connecting the on- and off-center units to the orientation unit. As described above, these gains are determined during training from the response of the on- and off-center units to an image. In this case the stimulus was a white line with vertical edges. Through the training algorithm, different images, such as sloping edges (Marr & Hildreth, 1980) or unequal intensity levels on two sides of a

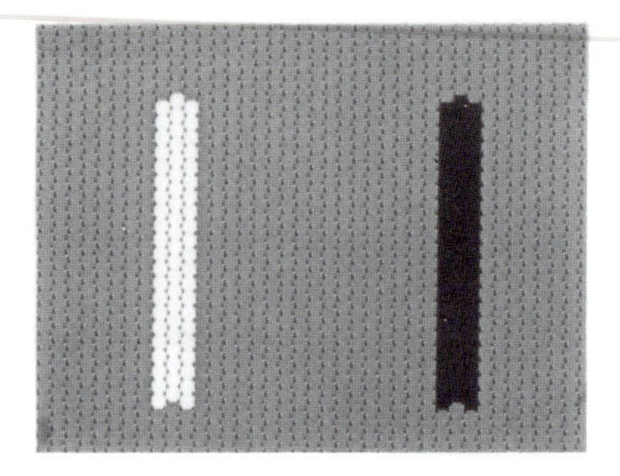

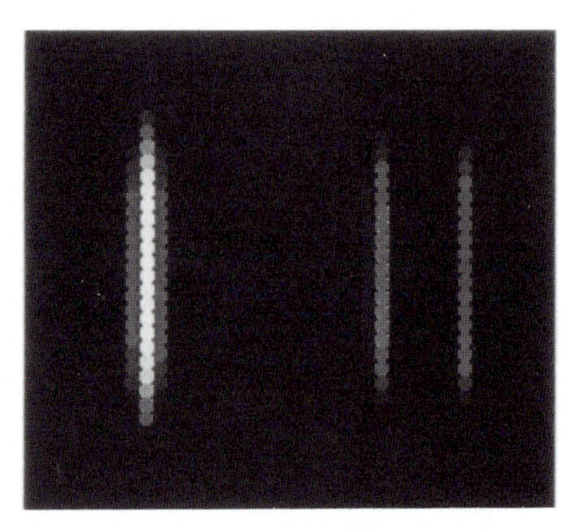

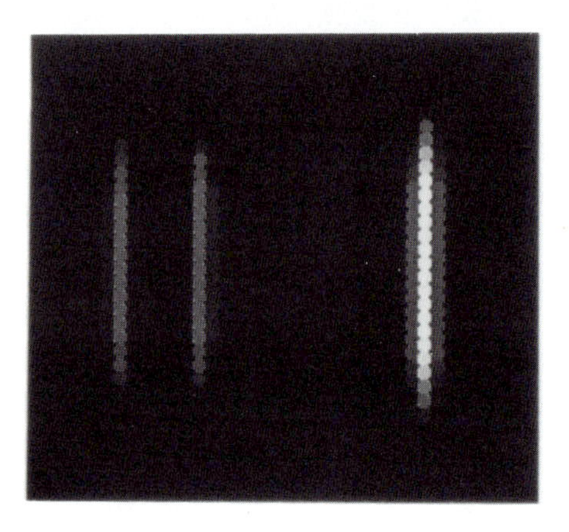

FIGURE 6 Outputs from line units tuned to opposite contrast sign in response to light and dark bars projected on the receptor plane. The line unit tuned to the light bar had synaptic gain fields shown in Figure 3C and D, whereas, for line units tuned to inverse contrast, the signs of the synaptic gains were interchanged. Notice that the units tuned to a light bar give weak outputs at the borders of a dark bar and vice versa. These side bands generate computational ambiguities. However, they can be suppressed by thresholding, contrast sensitivity adjustment, or extension of inhibitory regions of the receptor fields (see Figure 7). *Top*, receptor; *center*, line units; *bottom*, inverse line units.

line, generate different gain fields resulting in response fields in which inhibitory and excitatory regions can have different peak amplitudes, widths, and phase angles. Moreover, the number of excitatory side lobes in the response field are a function of the ratio of the width of the on- and off-center receptor fields to the width of the image used to compute the orientation unit gain fields. Variations in the shape of the response functions of this kind have also been observed by Jones and Palmer (1987) and by Mullikin, Jones, and Palmer (1984) in simple cells in the cat cortex. We do not know the computational significance of these response field variations and if they are completely represented at every point in the visual field.

The response to dark and light lines by line unit arrays tuned to opposite contrast is shown in Figure 6. The responses exhibit excitatory side bands for the nontuned image which can be suppressed as discussed in the section on Computational Ambiguities.

Figure 7 shows the orientation tuning of an edge unit and Figure 8A that of line units. The orientation tuning curve is similar to that observed in biological systems but is somewhat less steep, perhaps due to the

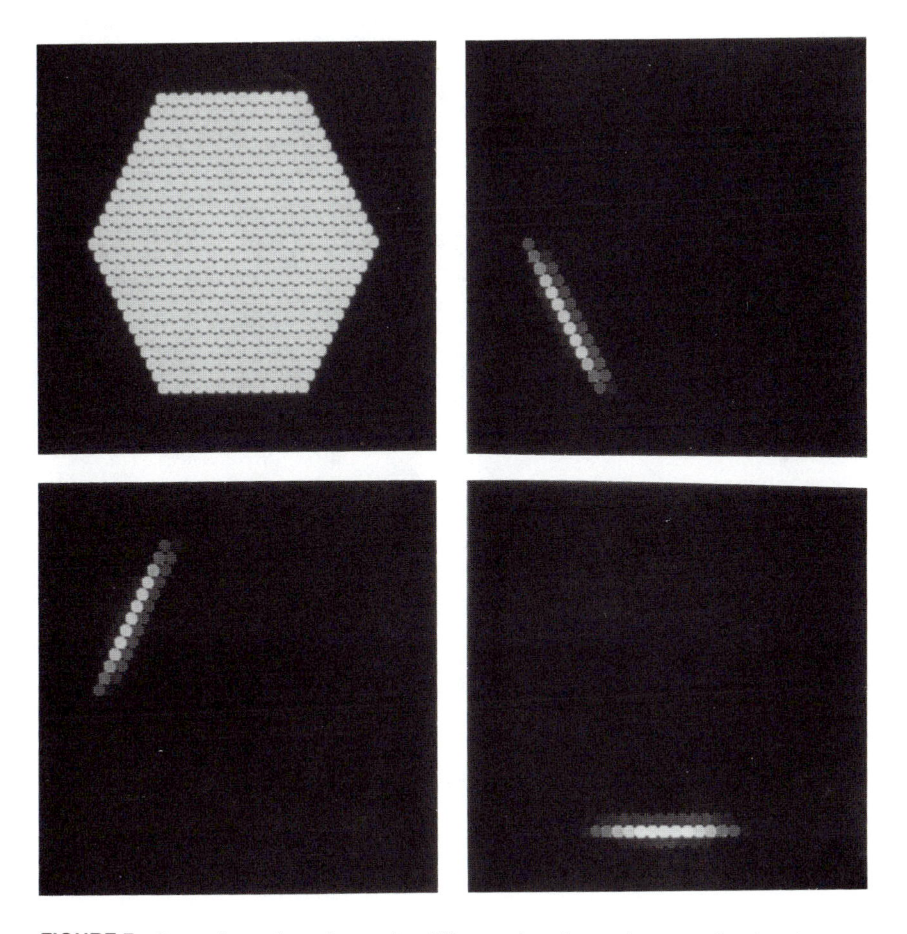

FIGURE 7 Output from edge units tuned to different orientations and contrast directions in response to a hexagon in the receptor plane.

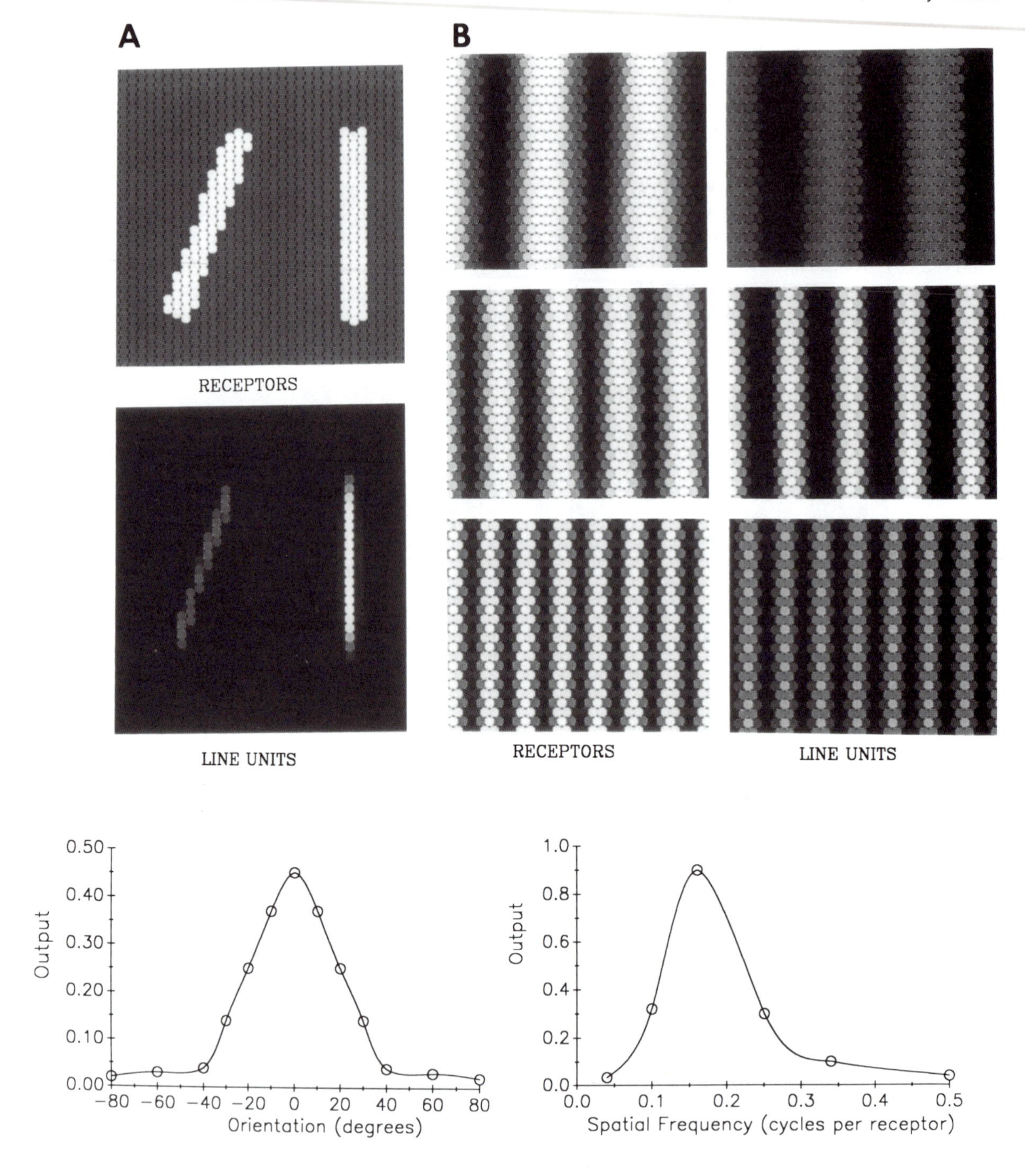
A
B
RECEPTORS
LINE UNITS
RECEPTORS
LINE UNITS
0.50
0.40
0.30
0.20
0.10
0.00
Output
-80 -60 -40 -20 0 20 40 60 80
Orientation (degrees)
1.0
0.8
0.6
0.4
0.2
0.0
Output
0.0 0.1 0.2 0.3 0.4 0.5
Spatial Frequency (cycles per receptor)

absence of mutual inhibition of units tuned to different orientations.

The shape of the gain fields determines the best width of a line or edge to which a unit is tuned. It also determines the frequency of a sinusoidal intensity grating which will produce the largest unit response (see Figure 8B). In fact the spatial frequency response curves are similar to those of simple cells in the visual cortex.

The orientation units shown here received inputs from a field of on- and off-center units, each of which measured 12 x 16 units. In spite of this wide receptor field, our units are able to resolve lines that are only one receptor unit wide (see Figure 9).

End-stopped Orientation Units

End-stopped units are inhibited if the length of the feature extends beyond one or both ends of the unit's receptive field (for review see Orban, 1984). Figure 10 shows the response of an array of single end-stopped edge units to a square image in the receptor field. The units respond only to the lower left hand corner of the image because they were tuned to an edge of 90° orientation and a contrast direction of increasing luminosity in the direction of the x axis. Responses of single end-stopped line units are also shown in this figure. Units of this type respond not only to end points and corners of images but also to local curvatures with an output amplitude that depends on the radius of curvature. Double end-stopped units are inhibited when either or both sides of the feature exceed the length of the receptive field and thus detect edges or lines of a particular length (see Figure 11A and B). Generally the tuning characteristics of these end-stopped units duplicate those reported by Hubel and Wiesel (1962, 1965) for certain hypercomplex cells in the visual cortex.

Computational Ambiguities

The response amplitude of the orientation units is a function of contrast, contrast sign, orientation, and position of the unit relative to the tuned feature. These n parameters are represented in n dimensional space by the multiple arrays of tuned units. Ambiguities occurring in the outputs of individual units can be resolved either by processing within this space or at subsequent stages. One prominent ambiguity is the relative lack of specificity of orientation units for the sign and direction of contrast which produces excitation side bands for images of opposite contrast (e.g., Figure 6). This ambiguity could be resolved by mutual inhibition in contrast sign space (Walters, 1987) but is computationally expensive. As shown in Figure 7, we have achieved considerable side band suppression by thresholding, increase of contrast sensitivity through adjustment of the slope constant, s, in Equation 2, and extension of inhibitory regions of the receptor fields without resorting to interarray inhibition.

Computation of Natural Images

The presentation of a natural image such as a face to the receptor plane generates some activity in most or all orientation arrays. For a typical natural image, only a few units in each array are activated and the total activity, as a fraction of all orientation units in the different arrays, is relatively sparse. Some examples are

FIGURE 8 A. Output from line units tuned to 90° orientation to two light bars on the receptor plane oriented at 75° and 90°. Notice the much smaller output for the 75° bar. The plot shows the orientation tuning of the line units. B. Response of receptors and line units to sinusoidal line patterns of different spatial frequencies. The optimal spatial frequency is a function of the center field width of the on-center and off-center units, and the width of the excitatory region of the input field from on- and off-center units. This parameter is derived by the procedures described earlier. In this case the widths of these fields were set to 3. As a result the units respond best to a line that is 3 units wide. Notice that the line units still respond somewhat to the highest spatial frequency (1 unit wide) even though their center field is much wider. The plot shows the maximal response amplitude as a function of spatial frequency.

shown in Figure 12 and 13. As seen in those figures, the activity again occurs in clusters and these clusters have typically a long axis coinciding with the orientation axis of the units. Because of this, the overlap of the receptor fields can be reduced in the direction of the orientation axis without significant loss of information. This leads to ratios of orientation units to receptor units that are less than unity and a corre-

FIGURE 9 Response of on-center (*top right*), off-center (*bottom left*), and line units (*bottom right*) (90°) to a square of lines on the receptor plane (*top right*). The on- and off-center units resolve the single lines even though their center field was 3 units wide. The line units were tuned to a line width of 3 units. Their receptor field was 12 x 14 units; thus each unit received projections from 168 receptors, yet the units were able to resolve the grid pattern. In addition, they also respond strongly to the vertical boundaries of the square, providing some degree of image segmentation.

sponding data reduction. An example of such a reduced orientation array is seen in Figure 13. The reduction in the number of orientation units simplifies the computation at subsequent stages.

DISCUSSION

Our system has been designed with minimal connectivity to avoid feedback and lateral connections wher-

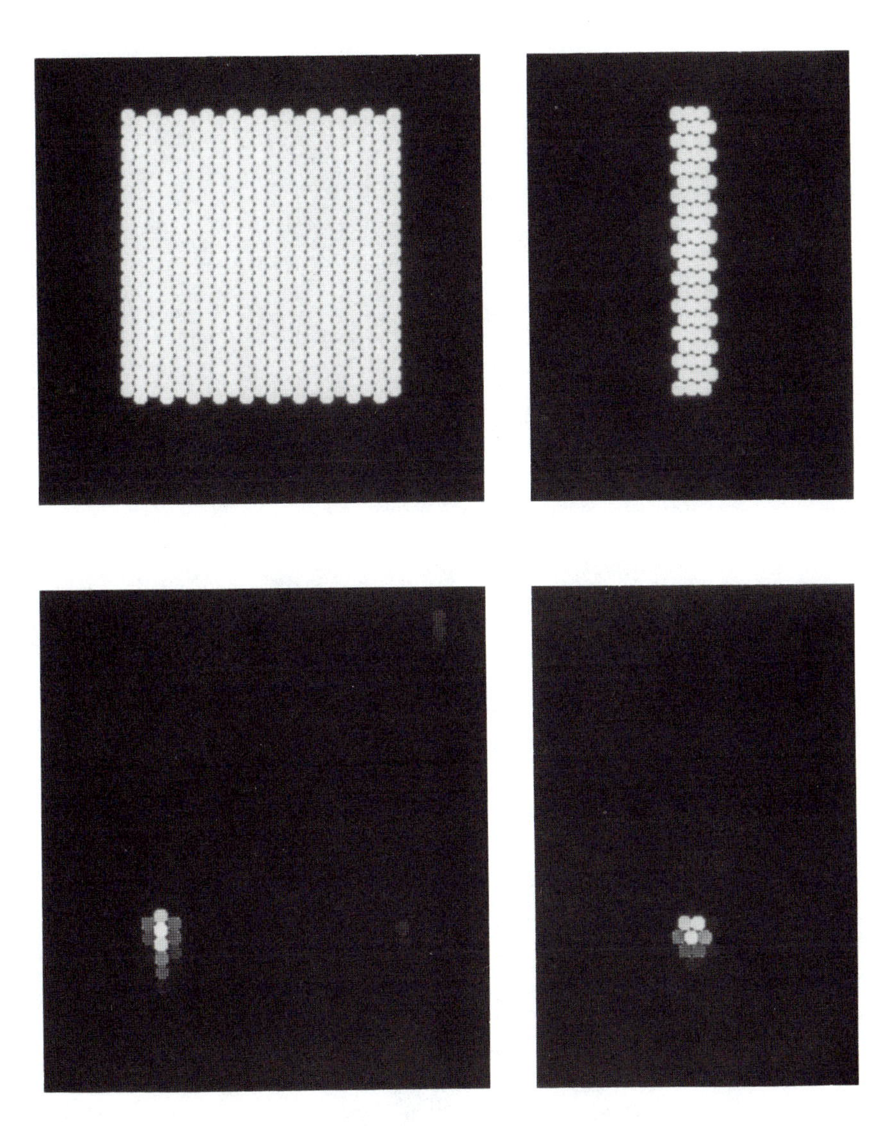

FIGURE 10 Outputs from single end-stopped edge (*bottom left*) and single end-stopped line units (*bottom right*) in response to the illustrated activity of the receptor planes (*top left and right*). The units were tuned to 90° orientation and require termination of the edge or line at the bottom of the receptor field.

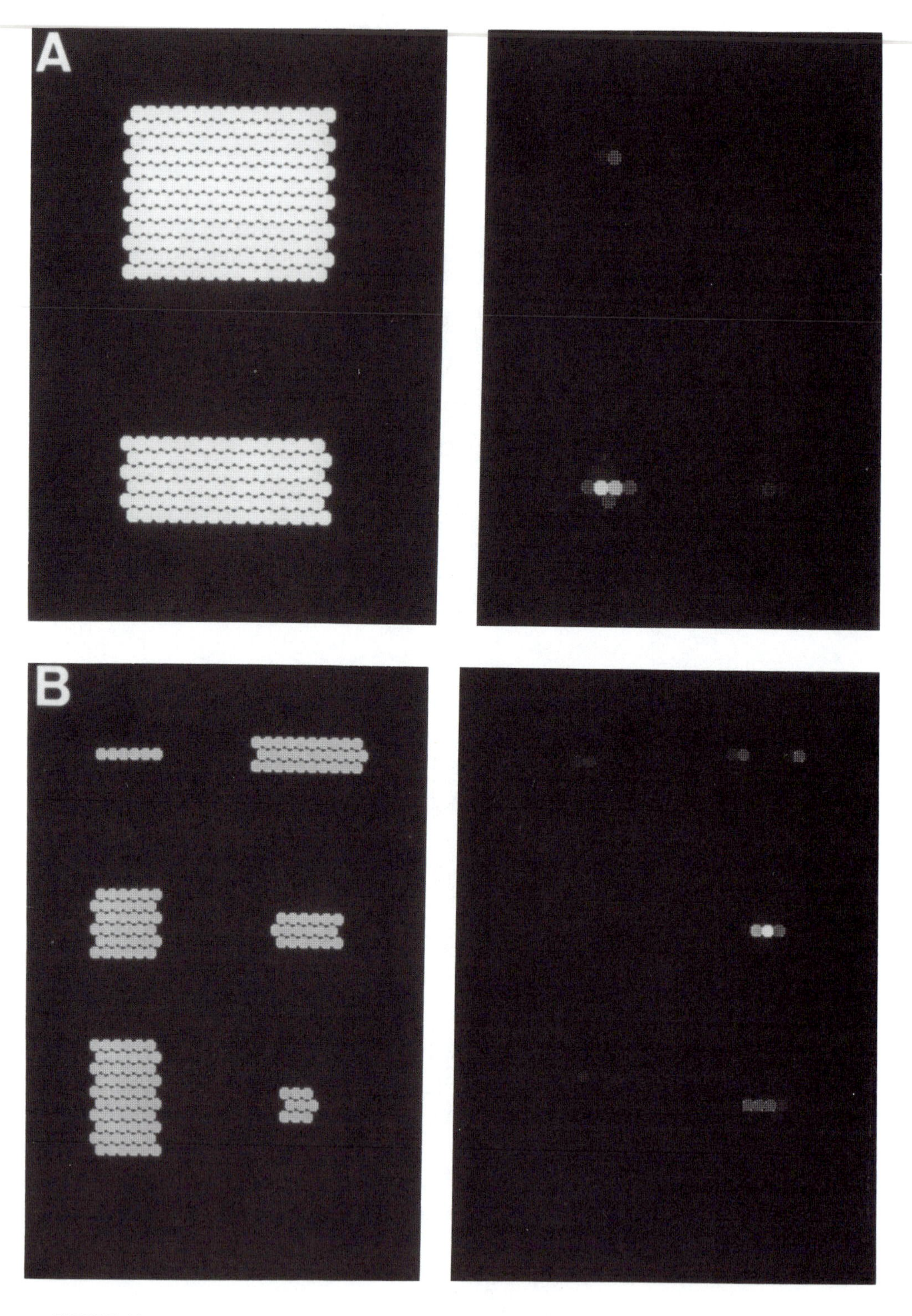

FIGURE 11 A. Responses of double end-stopped edge units (*right*) to edges of different lengths (*left*). These units were tuned to vertical edges 6 units long with a contrast direction of increasing light from left to right. They

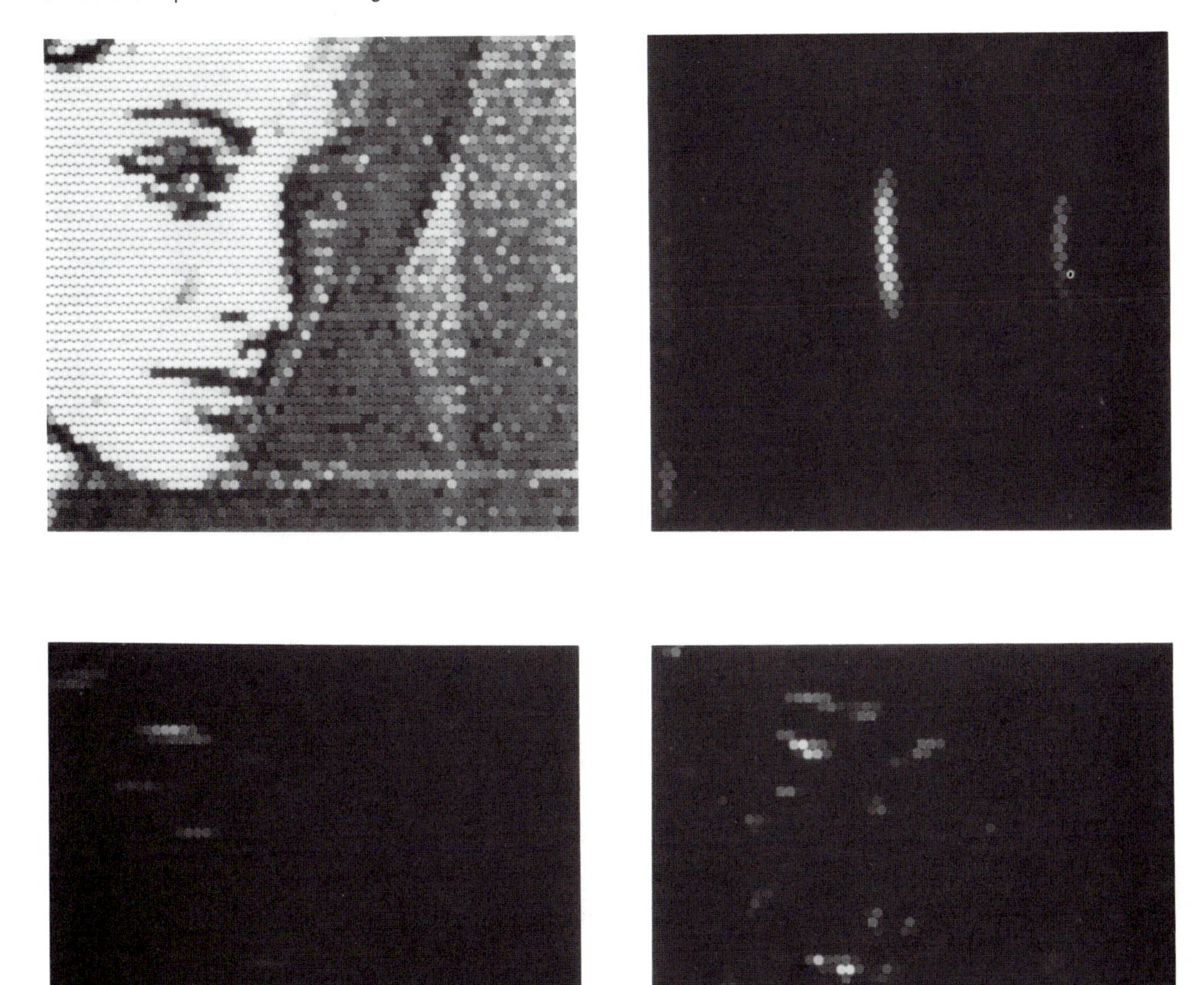

FIGURE 12 Output from receptors and from examples of orientation units for a natural image. Notice the relatively sparse output in the orientation plane, especially for the end-stopped units. Notice also that the active units form elongated clusters with a long axis in the direction of their tuning orientation. This elongated clustering is less evident for the end-stopped units. *Top left*, receptors scaled 6:1. *Top right*, edge units (90°). *Bottom left*, edge units (0°). *Bottom right*, single end-stopped edge units (0°)

FIGURE 11 *(CONTINUED)* do not respond to the longer edge of the larger rectangle on top. B. Responses of double end-stopped line units (*right*) to lines of varying length and width (*left*). These units were tuned to a line 3 units wide and 6 units long (middle right of receptor plane). Rectangles of other dimensions evoke much smaller responses.

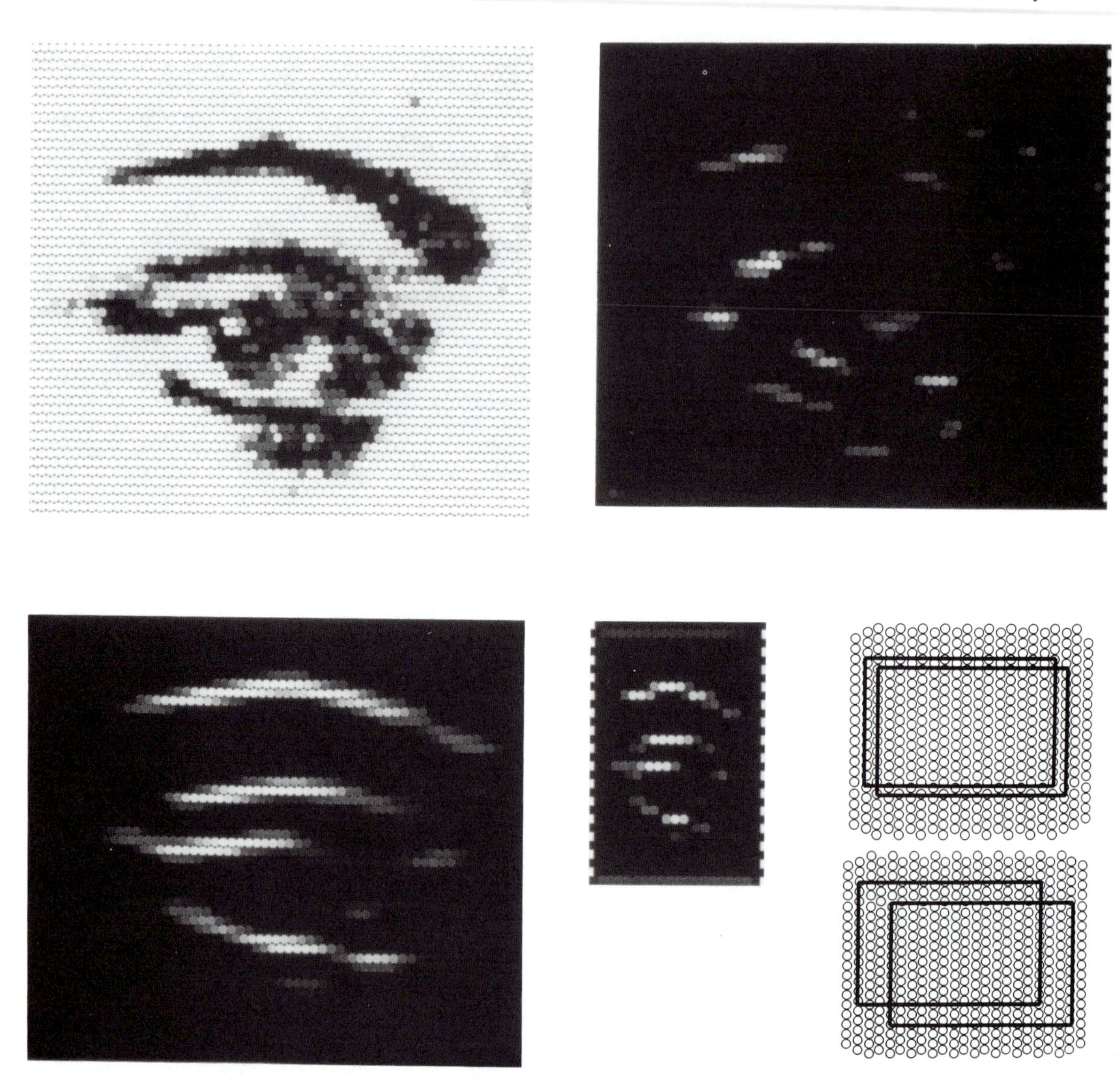

FIGURE 13 Outputs from receptor (*top left*, scaled 2:1) and orientation planes (*top right*, single end-stopped edge units (0°); *bottom left*, edge units (0°)) from a section of the same image shown in Figure 12 computed at a smaller spatial scale. Adjustment of the spatial scale was achieved by changing image magnification rather than changing the receptor field size of the neurons. The axial ratio of the spatial scales was kept unchanged. Notice that the output pattern of the orientation planes bears little resemblance to the corresponding outputs in the previous figure. When the receptor fields of the edge units are stepped by a distance of 3 units in the orientation direction and by 2 units in the direction normal to the axis of orientation (*bottom center*), the number of units in the array is only 1/6 of the receptors. The general pattern of activity and sparse representation is, however, preserved. The lower right hand insert shows the size and tiling of the receptor fields of two neighboring edge units as used for the computation at step sizes 1,1 and 3,2. The size of the receptor field as projected here on the receptor array closely matches the receptor field of a simple cell receiving projection from the center of the fovea in the monkey (Dow, Snyder, Vautin, & Bauer, 1981).

ever possible. Therefore, the system differs from its biological counterpart in several ways. For example, the center–surround organization is accomplished by direct connections from receptors to on- and off-center units. In the biological retina, however, lateral inhibition from the horizontal cells, followed by push–pull coupling through the on and off bipolar cells to the ganglion cells, serves this function. Geniculate neurons have essentially the same receptive fields as the retinal ganglion cells and may serve as a relay station. Hence we have omitted a separate representation of geniculate cells. Furthermore, we have omitted cross inhibition between units of different orientation or spatial scale at identical position, and the inhibition between units of identical orientation but different position which occurs in the cortex. There are several models of the connection architecture for orientation selectivity that incorporate such inhibition (Ferster & Koch, 1987; Koch, 1987). The functions of hypercomplex cells (end-stopped edges and lines) have been obtained directly from on- and off-center projections instead of connections via nonstopped cells. This difference may have computational consequences which need further investigation.

The push–pull combination of on- and off-center units produces output fields from simple line and edge orientation units resembling 2-D Gabor functions. The Gabor type output fields can also be implemented directly without intermediate processing by on- and off-center units (Turner, 1986; Daugman, 1988). However, the enhancement of spatial contrast provided by the on- and off-center preprocessing stage yields enhanced responses from the orientation units. This in turn permits better feature discrimination through thresholding, and ultimately leads to a sparser representation of the image.

The synaptic gains were derived in a sequential fashion from one layer to the next using the inputs to the previous stage as templates for the gains. This procedure creates a series of matched filters which are applied sequentially when computing the layers. It requires no iterative computations as used by Linsker (1986a–c) or by other learning procedures such as back propagation or other gradient descent methods (Rumelhart & McClelland, 1986).

We do not know if this method is used in the biological system. It is however tempting to speculate how it might be employed at subsequent stages of visual processing to arrive at units that are tuned to higher order features that occur in natural images. Such units have been found, for example, in the inferotemporal cortex of monkeys, where neurons respond to specific faces or certain views of faces (Gross, Desimone, Albright, & Schwartz, 1984; Perrett et al., 1985; Rolls, Baylis, & Leonard, 1985).

The initial image decomposition into primitives is a crucial step for further processing. First of all the machine itself must be able to detect the presence or absence of local pattern primitives, and the extraction of oriented lines and edges already provides some limited image segmentation (Grossberg & Mingolla, 1985; Turner, 1986; see also Figure 9). Perhaps equally important is the fact that, through this decomposition, the system arrives at a sparse representation of the input pattern. For a typical visual scene only a small subset of all units involved in decomposition is active, and subsequent decoding of the set of active primitives is much more manageable than would be the direct decoding of the raw input pattern. But what are the subsequent decoding strategies? It is unlikely that the entire collection of primitives representing a specific pattern is summed directly into highly specific neurons. Most likely there are intermediate stages that represent groups of primitives. For example the set of oriented edge units of a particular contrast sign and spatial scale activated by an image within an orientation array might be summed into specific neurons. This would become quite feasible if the array size is reduced by appropriate receptor field tiling as shown in Figure 13. Alternatively, units might be tuned to patterns in orientation space. For example, line crossings give rise to activity in units at the position of crossing in those arrays tuned to the different orientations of the crossing lines, and both position and angle of crossing are uniquely represented by the combination of these active units. Similar groupings occur in spatial scale space for overlapping objects, and units tuned to such groupings might be used to represent relations between areas and objects of different sizes. In any case, sequential processing at several stages as used in multilayer perceptrons or the neocognitron seems indispensable (Fukushima, 1980; Menon & Heinemann, 1988; Miyake & Fukushima, 1984).

Whereas we are fairly well satisfied with the performance of the individual units, the overall organization of the system requires further consideration. How many units would be needed for the decomposition of a visual image? For 15° increments of orientation tuning, that is, 12 different orientations, there are 24 arrays each of edge units (two contrast directions) and line units (two contrast signs), 48 arrays each for single end-stop line and edge units (these units are tuned to lines and edges of specific length) and 24 arrays each for double end-stopped line and edge units. Thus in order to decompose an image into its primitives at a single spatial scale, 192 separate arrays are needed. If the spatial scales are changed by powers of 2 from .03° to 60°, 11 times this number or 2112, would be needed to cover an entire visual scene. However, if the receptor field overlap is reduced as suggested in Figure 13, the number of units in each array may be only a fraction of the corresponding retinal receptor array. In addition, the highest spatial scales are only represented at the center of the visual field and the spatial resolution decreases with increasing eccentricity. Such a "pyramidal" representation of spatial scales is apparently implemented in the biological system both at the receptor (Perry & Cowey, 1985) and cortical level (Sakitt & Barlow, 1982) and has also been used in machine vision (Carlson, Klopfenstein, & Anderson, 1981). As a result, there is a further reduction of the number of units required for image decomposition. This reduction factor depends of course on the range of increments of spatial scales that are represented. In our current system the total number of units for primary image decomposition would be between 100 and 200 times the number of retinal receptors, that is, for a retinal array of modest size, say 40,000 units, one would need between 10^6 and 10^7 feature units each with 150 to 250 inputs. For robotics, a system of limited acuity in which the representation of spatial scales is restricted to a range from 2–60° would probably suffice, reducing the number of feature units into the range of 10^5 to 10^6.

VLSI implementation of such a system would obviously be a difficult task. However, as Carver Mead and his group (Mead & Mahowald, 1988; Silvilotti, Mahowald, & Mead, 1987) have demonstrated, it is possible to build large scale arrays of analog processing units that perform the function of the retinal receptors and on-center units described above. Furthermore, it may not be necessary to represent all orientations and spatial scales in parallel. Instead one might consider scanning the image at different magnifications and rotations with a few orientation arrays, storing the output for further processing. Even such a relatively simple system could already greatly improve the computational speed and point toward a real time vision system.

References

Attwell, D. (1986). Ion channels and signal processing in the outer retina. *Quarterly Journal of Experimental Physiology*, **71**, 497–536.

Carlson, C. R., Klopfenstein, R. W., & Anderson, C. H. (1981). Spatially inhomogeneous scaled transforms for vision and pattern recognition. *Optics Letters*, **6**, 386–388.

Daugman, J. G. (1985). Uncertainty relation for resolution in space, spatial frequency and orientation optimized by two-dimensional visual cortical filters. *Journal of the Optical Society of America*, 2, 1160–1178.

Daugman, J. G. (1988). Complete discrete 2D Gabor transforms by neural networks for image analysis and compression. *IEEE Transactions on Acoustics, Speech, and Signal Processing*, **36**, 1169–1178.

Dow, B. M., Snyder, A. Z., Vautin, R. G., & Bauer, R. (1981). Magnification factor and receptive field size in foveal striate cortex of the monkey. *Experimental Brain Research*, **44**, 213–228.

Enroth-Cugell, C., & Robson, J. G. (1966). The contrast sensitivity of retinal ganglion cells of the cat. *Journal of Physiology*, **187**, 517–555.

Ferster, D. (1988). Spatially opponent excitation and inhibition in simple cells of cat visual cortex. *Journal of Neuroscience*, **8**, 1172–1180.

Ferster, D., & Koch, C. (1987). Neural connections underlying orientation selectivity in cat visual cortex. *Trends in NeuroSciences*, **10**, 487.

Fukushima, K. (1980). Neocognitron: A self-organizing neural network model for a mechanism of pattern recognition unaffected by shift in position. *Biological Cybernetics*, **36**, 193–202.

Gabor, D. (1946). Theory of communication. *Journal of the Institution of Electrical Engineering, Part 3,* **93,** 429–457.

Gross, C. G., Desimone, R., Albright, T. D., & Schwartz, E. L. (1984). Inferior temporal cortex as a visual integration area. In F. Reinoso-Suarez & C. Ajmone-Marsan (Eds.), *Cortical integration.* New York: Raven.

Grossberg, S., & Mingolla, E. (1985). Neural dynamics of perceptual grouping: Texture boundaries, and emergent segmentations. *Perception & Psychophysics,* **38**(2), 141–171.

Heggelund, P. (1981). Receptive field organization of simple cells in cat striate cortex. *Experimental Brain Research,* **42,** 89–98.

Hubel, D. H., & Wiesel, T. N. (1962). Receptive fields, binocular interaction and functional architecture in the cat's visual cortex. *Journal of Physiology,* **160,** 106–154.

Hubel, D. H., & Wiesel, T. N. (1965). Receptive fields and functional architecture in two nonstriate visual areas (18 and 19) of the cat. *Journal of Neurophysiology,* **18,** 229–289.

Jones, J. P., & Palmer, L. A. (1987). An evaluation of the two-dimensional Gabor filter model of simple receptive fields in cat striate cortex. *Journal of Neurophysiology,* **58,** 1233–1258.

Koch, C. (1987). A network model for cortical orientation selectivity in cat striate cortex. *Investigative Ophthalmology & Visual Science, Suppl. 3* 28, 126.

Kuffler, S. W. (1953). Discharge patterns and functional organization of mammalian retina. *Journal of Neurophysiology,* **16,** 37–68.

Linsker, R. (1986a). From basic network principles to neural architecture: Emergence of orientation-selective cells. *Proceedings of the National Academy of Sciences of the United States of America,* **83,** 8390–8394.

Linsker, R. (1986b). From basic network principles to neural architecture: Emergence of spatial-opponent cells. *Proceedings of the National Academy of Sciences of the United States of America,* **83,** 7508–7512.

Linsker, R. (1986c). From basic network principles to neural architecture: Emergence of orientation columns. *Proceedings of the National Academy of Sciences of the United States of America,* **83,** 8779–8783.

Marcelja, S. (1980). Mathematical description of the responses of simple cortical cells. *Journal of the Optical Society of America,* **70,** 1297–1300.

Marr, D., & Hildreth, E. (1980). Theory of edge detection. *Proceedings of the Royal Society of London, Series B,* **207,** 187–217.

Mead, C. A., & Mahowald, M. (1988). A silicon model of early visual processing. *Neural Networks,* **1,** 91–97.

Mead, C. A. (1989). *Analog VLSI and neural systems.* New York: Addison-Wesley.

Menon, M. M., & Heinemann, K. G. (1988). Classification of patterns using a self-organizing neural network. *Neural Networks,* **1,** 201–215.

Miyake, S., & Fukushima, K. (1984). A neural network model for the mechanism of feature-extraction. *Biological Cybernetics,* **50,** 377–384.

Mueller, P., & Lazzaro, J. (1986). A machine for neural computation of acoustical patterns with application to real time speech recognition. *American Institute of Physics, Conference Proceedings,* **151,** 321–326.

Mueller, P., Spiro, P., Blackman, D., & Furman, R. (1987). Neural computation of visual images. *Proceedings of the IEEE 1st Annual International Conference on Neural Networks, IV,* 75–88.

Mullikin, W. H., Jones, J. P., & Palmer, L. A. (1984). Periodic simple cells in cat area 17. *Journal of Neurophysiology,* **52,** 372–387.

Orban, G. A. (1984). *Neuronal operations in the visual cortex.* New York: Springer-Verlag.

Perrett, D. I., Smith, P. A. J., Potter, D. D., Mistlin, A. J., Head, A. S., Milner, A. D., & Jeeves, M. A. (1985). Visual cells in the temporal cortex sensitive to face, view, and gaze direction. *Proceedings of the Royal Society of London, Series B,* **223,** 293–317.

Perry, V. H., & Cowey, A. (1985). The ganglion cell and cone distribution in monkey's retina: Implications for central magnification factors. *Vision Research,* **25,** 1795–1810.

Rolls, E. T., Baylis, G. C., & Leonard, L. M. (1985). Role of low and high spatial frequencies in the face-selective responses of neurons in the cortex in the superior temporal sulcus in the monkey. *Vision Research,* **25,** 1021–1035.

Rumelhart, D. E., & McClelland, J. L. (Eds.) (1986). *Parallel distributed processing* (Vol. 1. *Foundations*). Cambridge, MA: MIT Press.

Sakitt, B., & Barlow, H. B. (1982). A model for the economical encoding of the visual image in cerebral cortex. *Biological Cybernetics,* **43,** 97–108.

Schiller, P. H. (1982). Central connections of the retinal ON and OFF pathways. *Nature,* **297,** 580–583.

Silvilotti, M., Mahowald, M. A., & Mead, C. (1987). Real time visual computations using analog MOS processing arrays. In P. Losleben (Ed.), *Advanced research in VLSI Proceedings of the 1987 Stanford Conference* (pp. 295–312). Cambridge, MA: MIT Press.

Turner, M. R. (1986). Texture discrimination by Gabor functions. *Biological Cybernetics,* **55,** 71–82.

Walters, D. (1987). Rho-Space: A neural network for the detection and representation of oriented edges. *Proceedings of the Cognitive Science Society,* 445.

Werblin, F. S., & Dowling, J. E. (1969). Organization of the retina of the mudpuppy. *Journal of Neurophysiology,* **32,** 339–355.

9

Models of the Neural Basis of Insect Behavior

Randall D. Beer, Roy E. Ritzmann, and Hillel J. Chiel

INTRODUCTION

An animal's behavior fundamentally arises from the interaction between its nervous system, its body, and its environment. Although simulations of neural circuitry have received a great deal of attention in recent years, the relationship of these models to actual behavior has often gone unexamined. Simulating the neural basis of animal behavior has been termed *computational neuroethology* (Beer, 1990; Achacoso & Yamamoto, 1990; Cliff, 1991). To bridge the gap between neural circuitry and behavior, computational neuroethology simulates the generation of behavior through the interaction between explicit models of an animal's nervous system, its body, and its environment. Such simulations are carried out for two related, but distinct, reasons: (1) they can deepen our understanding of the neural control of animal behavior; (2) they can serve as the basis for abstracting biological control principles for technological applications such as autonomous robots.

Elucidating the neural basis of animal behavior is a technically and conceptually difficult endeavor. A computer simulation of a given neuroethological system can aid this endeavor in a number of important ways:

(1) It synthesizes diverse experimental data into an interactive model in which the relationship between neural and behavioral events can be directly observed and manipulated.
(2) It can be used to test our functional understanding of the system.
(3) It encourages synergistic interactions between experiment and modeling, as manipulations of the simulation suggest additional experiments which in turn serve to further refine the model.
(4) It can support observations or manipulations that are technically difficult *in vivo*.

There is currently a great deal of interest in the design of flexible autonomous robots. Most current robotic systems lack the versatility and robustness of even simpler animals such as insects. Because many of the behavioral capabilities required by an autonomous robot are very similar to those already exhibited by animals, there is the possibility of incorporating biological control principles into such robots. Abstracting the essential principles from the details of their biological implementation is a crucial step in this process. Computer models of behavior and its underlying neural control can play a crucial role in this abstraction.

In this chapter, we describe two examples of work in computational neuroethology. The first one involves a neural network architecture for hexapod locomotion that was abstracted from work on the neural basis of the walking of a cockroach. A successful robotic implementation of this locomotion controller is also described. Finally, we describe a biologically realistic computer model of the cockroach escape response.

HEXAPOD LOCOMOTION

Currently, there is a particular interest in the construction of legged robots. Legs possess obvious advantages for locomotion over complex and varied terrain. However, controlling the locomotion of legged robots has turned out to be a difficult and computationally demanding task. In contrast, insects are capable of robustly solving the complex coordination problems raised by legged locomotion in real-time for a wide variety of terrains (Graham, 1985). What can we learn from biology?

Background

During normal walking, insects exhibit a variety of statically stable gaits. Slow-walking insects often utilize gaits consisting of distinct metachronal waves, in which each leg begins its swing immediately following the swing of the leg behind it. Fast-walking insects, on the other hand, typically utilize the tripod gait, in which the front and back legs on each side of the body swing in phase with the middle leg on the opposite side. Wilson (1966) suggested that the entire range of observed insect gaits could be explained by assuming that fixed, antiphasic metachronal waves on each side of the body increasingly overlap as walking speed increases.

The overall neural organization of the walking system of the American cockroach (*Periplaneta americana*) has been studied by Pearson and his colleagues (Pearson, Fourtner, & Wong, 1973; Pearson, 1976). This work led to the development of the *flexor burst-generator* model for cockroach locomotion. In this model, the basic swing and stance movements of each leg are generated by a central pattern generator whose operation is modulated by feedback from sensory structures in the insect's legs. The overall speed of walking is set by descending influences from higher command centers. Interleg coordination is achieved by inhibitory interactions between adjacent pattern generators.

Model

Inspired by Pearson's flexor burst-generator model, we designed a distributed neural network architecture for hexapod locomotion (Beer, Chiel, & Sterling, 1989; Beer, 1990). This circuit was used to control the walking of a simulated insect whose body model is illustrated at the top of Figure 2. Each of the model's six legs can swing forward or backward, and each leg possesses a foot which can be either up or down. A leg whose foot is up can swing freely. When its foot is down, any forces generated by a leg are applied to the body. The body can only move when it is statically stable (i.e., the center of mass falls within the polygon of support formed by the stancing legs).

Our locomotion circuit is shown in Figure 1. Each of the model neurons is capable of capacitatively integrating the weighted sum of its inputs and passing this sum through a saturating linear threshold function. At the center of each leg controller is a pacemaker neuron whose output rhythmically oscillates due to the presence of a simplified voltage-dependent intrinsic current (for details, see Beer, 1990). The interval between bursts depends linearly on the tonic level of excitation that the pacemaker receives, with excitation decreasing this interburst interval, and inhibition increasing it. In addition, a strong excitatory pulse between bursts or a strong inhibitory pulse within a burst can reset a pacemaker's burst rhythm. These pacemaker neurons implement the flexor burst-generator in Pearson's model.

A pacemaker burst initiates a swing by inhibiting the foot and backward swing motor neurons, and exciting the forward swing motor neurons, causing the foot to lift and the leg to swing forward. Between pacemaker bursts, the foot is down and tonic excitation from the command neuron moves the leg backward. The output of the central pattern generator is fine-tuned by feedback from two sensors that signal when a leg is nearing its extreme forward or backward position. A forward angle sensor encourages a pacemaker to terminate a burst by inhibiting it, whereas a backward angle sensor encourages a pacemaker to initiate a burst by exciting it. The forward angle sensor also

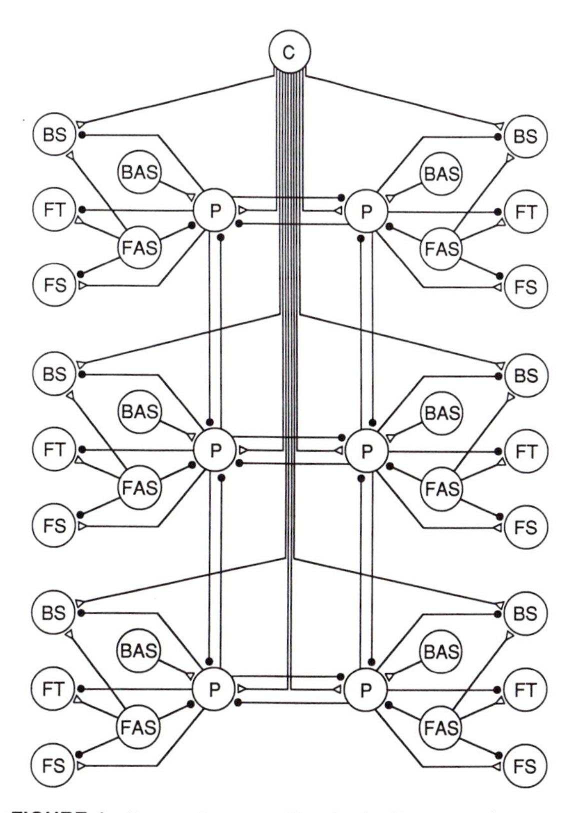

FIGURE 1 Locomotion controller circuit. C, command neuron; FT, foot motor neurons; FS and BS, forward swing and backward swing motor neurons; P, pacemaker neurons; and FAS and BAS, forward and backward angle sensors. Excitatory connections are shown as open triangles and inhibitory connections are shown as filled circles.

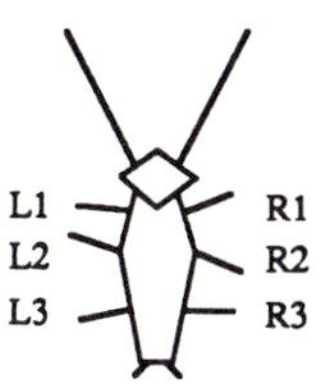

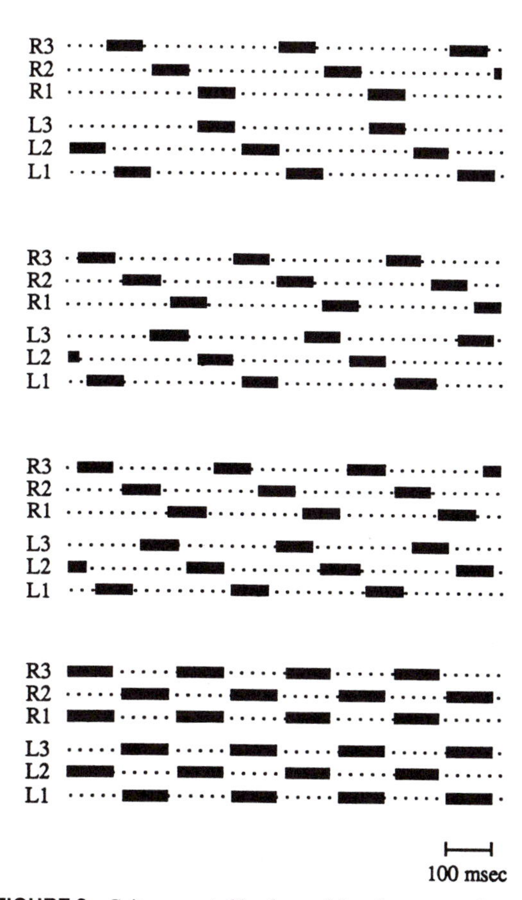

FIGURE 2 Gaits generated by the model as the command neuron activity is varied from lowest (*top*) to highest (*bottom*). The gaits shown correspond to command neuron outputs of 0.25, 0.3, 0.35, and 1, respectively. Black bars denote the swing phase of a leg. The body model and leg labeling conventions are shown at top.

makes direct connections to the motor neurons, modeling leg reflex pathways described in Pearson's model.

To generate statically stable gaits, the swings of the individual legs must be coordinated in some way. Following Pearson's model, we inserted mutually inhibitory connections between the pacemaker neurons of adjacent legs. At high frequencies of stepping, this architecture produces the commonly observed tripod gait. At slower stepping frequencies, however, the network is underconstrained. Statically stable gaits are not reliably generated. In particular, no metachronal waves are produced. Thus, we augmented Pearson's model with a mechanism for generating metachronal waves. The basic idea was to phase-lock the pattern generators on each side of the body into a metachronal relationship by lowering the burst frequency of the rear pattern generators (Graham, 1977). We implemented this entrainment by slightly increasing the angle ranges of the rear legs, thereby lowering their stepping frequency relative to the middle and front legs. We found that this phase locking could also be accomplished by lowering the strengths of the connections from the command neuron to the rear pacemakers, thus lowering their natural frequency relative to the middle and front pacemakers for any given command neuron setting (Chiel & Beer, 1989; Beer, 1990).

Simulation Results

When embedded within the body model just described, this locomotion circuit is capable of producing a continuous range of statically stable insect-like gaits (Figure 2). These gaits range from the tripod gait at high stepping frequencies to slower metachronal gaits at lower stepping frequencies. All of these gaits are produced by the same circuit simply by varying the tonic level of activity of the command neuron. In addition, smooth transitions between these gaits can be generated by continuously varying the command neuron activity.

To explore the robustness of this locomotion circuit, we performed a series of lesion studies on it (Chiel & Beer, 1989; Beer, 1990). Specifically, we examined the controller's response to such perturbations as the removal of selected sensors or connections. We found that the ability of this circuit to generate statically stable hexapod gaits was quite robust to such perturbations. For example, lesioning any single sensor or interpacemaker connection did not generally disrupt locomotion. Indeed, the controller could often tolerate the random removal of up to half of the interpacemaker connections.

Our locomotion circuit was part of a larger project to construct a simulated insect whose behavior was controlled by an artificial nervous system (Beer, 1990). The goal of this project was to demonstrate that neurobiological control principles could be effectively applied to artificial agents. The complete simulated insect could walk, wander, follow the edges of obstacles and seek out and consume food. These behaviors were organized into a simple behavioral hierarchy. Like the locomotion controller, the neural circuit responsible for the artificial insect's consummatory behavior was directly inspired by a biological system, in this case the feeding circuitry of the marine mollusc *Aplysia*. Circuits controlling the remaining behaviors were designed to be as consistent as possible with known neurobiological principles, but were not directly based on any existing circuit.

Implementation in a Robot

To demonstrate the practical utility of our locomotion circuit, we embedded it in an actual hexapod robot (Figure 3; Beer, Chiel, Quinn, & Espenchied, 1992; Quinn & Espenschied, 1992). This physical implementation allowed us to explore whether the gaits that the circuit generates in simulation persist in the presence of such physical effects as noise, friction, inertia, and delays. Not only did the robot exhibit the same range of gaits observed in simulation, but it was similarly robust to perturbations (Chiel, Beer, Quinn, & Espenschied, 1992). These results demonstrate that the observed gaits are not simply artifacts of the many physical simplifications of the simulated body, but are in fact robust properties of the locomotion circuit. Of course, it is important to point out that this controller

FIGURE 3 A photograph of the hexapod robot. Its dimensions are 50 cm long by 30 cm wide, and it weighs approximately 1 kg. Each leg has two degrees of freedom: an angular motion responsible for swing and stance movements and a radial motion involved in raising and lowering the leg. The swing motion has a range of over 45° from vertical in either direction. The radial motion is accomplished by means of a rack-and-pinion transmission. Both degrees of freedom are driven by two watt dc motors with associated integral transmissions. Position sensing for each degree of freedom is provided by potentiometers mounted in parallel with the motors. This robot was constructed by Ken Espenschied and Roger Quinn at CWRU.

only addresses the problem of gait control. Equally important problems in legged locomotion include postural control, directional control, negotiation of complex terrain, and compensation for leg damage or loss.

It is interesting to compare our locomotion controller to one that has been proposed by Brooks (1989). The two controllers share a common emphasis on distributed approaches to legged locomotion with low computational demands. In Brooks' controller, each leg is driven by a chain of peripheral reflexes. For example, whenever a leg is lifted, it is swung forward and whenever it reaches an extreme forward position, it is put down. The movements of the individual legs are coordinated by a centralized mechanism that tells each leg when to lift. Different gaits are produced by explicitly modifying this mechanism. Such a controller is inherently less robust than ours because damage to a single sensor in a reflex chain or to the centralized gait sequencer may compromise the robot's locomotion.

Given the results summarized in this section, it appears that we have succeeded in abstracting from the American cockroach a locomotion controller that offers a number of potential advantages over conventional approaches. First, conventional controllers are often centralized (i.e., optimal control decisions are made by a single processor after simultaneously considering all sensory inputs and all performance requirements). On the other hand, our controller is fully distributed. No single component is uniquely responsible for the observed gaits. Rather, these gaits are actively constructed from the dynamics of interaction between the coupled pattern generators and the sensory feedback that they receive. Second, whereas centralized control approaches are often brittle, the lesion studies have demonstrated that our controller is robust to a variety of realistic perturbations, including sensor loss and partial communication failure between individual leg controllers. Third, whereas centralized controllers are usually computationally expensive, our controller is relatively efficient, consisting of only 37 neurons that could easily be implemented in discrete analog components. Because of its simplicity, this circuit could be made to operate at high stepping frequencies. Indeed, our robotic implementation of this controller operates in real time even though it is simulated on a personal computer.

COCKROACH ESCAPE RESPONSE

We have illustrated how one can abstract biological control principles and apply them to the design of autonomous agents. We now illustrate the use of computer simulation to deepen our understanding of the neural control of a particular animal behavior: the cockroach escape response. One of the crucial problems that both animals and robots must solve is orientation toward and away from objects. In addition to basic walking movements, *P. americana* also escapes from predators. The escape system was once thought to be a simple reflex that maximized speed by utilizing relatively few neurons. However, as we have studied the behavior and underlying neural circuitry, we have come to appreciate the complexity of the system.

The cockroach must solve several problems in order to execute a proper escape movement. It must distinguish between sensory activity that represents a real threat and that which represents benign activities in its environment. Then, it must locate the threat, and assay other factors that are also important. Is the animal near a wall? Is it in a particularly vulnerable situation, such as a well-lighted environment? Is the stimulus coming from a conspecific that is trying to mate or from a threatening predator? The escape decision must include these and probably many more pieces of information. Moreover, it must act in a period of approximately 60 msec. Any delay could be fatal.

Behavior of the Escape Turn

The most basic requirement of the escape system is to detect faithfully the presence and location of threats. The cockroach must then use this information to direct escape movements that carry it out of harm's way. These can be divided into an initial turn directed away from the threat followed by a more random run. We focus only on the initial turn. A lunging predator creates a wind front, which the cockroach detects with sensory apparatus located on its cerci, two antenna-like appendages on the rear of the animal's abdomen (Camhi & Tom, 1978; Camhi, Tom, & Volman, 1978). The ventral surface of each cercus is covered with very sensitive mechanoreceptive hairs. Winds as small as 12 mm/sec evoke an escape response (Camhi & Nolen, 1981). The directional nature of the turn ensures that the animal will at least start its escape by moving away from the lunging predator. The latency between wind stimulation and the onset of turning movements is approximately 58 msec for a standing cockroach (Roeder, 1967). This drops to 14 msec for an animal that is walking slowly at the time of stimulation (Camhi & Nolen, 1981). The actual turning movement takes between 20 and 30 msec (Nye & Ritzmann, 1992).

Leg Movements

Because the turning movements are completed in 20–30 msec, high-speed movie cameras or video systems are required to visualize the leg movements in any detail. With data from these systems, it is now clear that the escape turn is a unique behavior distinct from the conventional tripod gait described earlier for running insects (Camhi & Levy, 1988). Moreover, there are actually three types of turns that are generated, depending on the front–back orientation of the wind source (Nye & Ritzmann, 1992).

The escape turn can be observed either in a free-ranging animal or in one that has been tethered to a rod and is walking on a slippery glass plate. Observations of free-ranging animals demonstrate how leg movements actually turn the animal away from a wind stimulus. However, a tethered preparation allows for more precise quantitative analysis of each leg movement. A comparison of the data in these two conditions indicates that the active leg movements are essentially the same (Camhi & Levy, 1988).

The principal joints that are used in all turns are the coxa-femur (CF) joint and femur-tibia (FT) joint of each leg (Figure 4, A and B). In general, the CF joint develops forces in the anterior-posterior direction, driving the animal forward or backward. The FT joints, especially in the prothoracic and mesothoracic legs, cause lateral movements which turn the animal's body to the right or to the left. In the simplest escape turns (type I), the prothoracic (T1) and mesothoracic (T2) legs extend posteriorly and laterally toward the wind source (Camhi & Levy, 1988; Nye & Ritzmann, 1992). The backward force of each leg occurs as the CF joint extends (Figure 4C1). If the tarsi remain stationary, as is the case in freely moving animals, the CF movements push the animal's body forward (Figure 4C2). In the T2 legs (and probably also in the T1 legs), the FT joint that is ipsilateral to the wind source invariably extends while the FT joint of the contralateral leg flexes. As a result, both legs push laterally toward the wind source. The resulting forces pivot the animal away from the stimulus. At the same time, both of the T3 legs push backward as a result of coordinated extensions of CF and FT joints. This movement simply provides forward thrust. The combined movement of all of the legs drives the animal forward and away from the wind source.

The anterior-posterior orientation of the wind source influences the direction of movement of some joints (Nye & Ritzmann, 1992). In the T2 legs, the FT joint movements remain essentially the same. However, with more anterior positions, the CF joints of the contralateral T2 and T3 legs tend to flex rather than extend

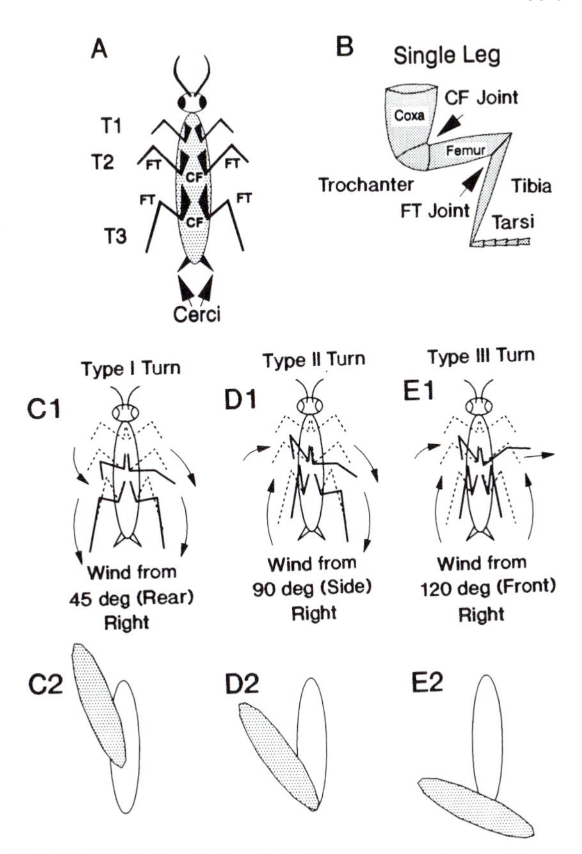

FIGURE 4 A description of the leg movements involved in escape turns. A. The principal joints used in the escape turn are shown on a diagram of the cockroach. Only T2 and T3 legs have been analyzed to date. CF, coxa femur joints, FT, femur tibia joints. B. A more detailed representation of the cockroach leg to show the location of CF and FT joints. C1-E1 summarize leg movements recorded in a tethered preparation in which the legs slip on a lightly oiled glass surface. C2-E2 show the resulting body movements that would occur in a freely moving animal. C. With wind from the right rear, the cockroach usually responds with a type I turn. All CF joints extend, while the T2FT joints provide turning direction. As a result, the animal moves forward and to the left. D. With wind from the side, the cockroach makes type II turns. In this movement the contralateral CF joints flex moving the leg forward. This will pull the animal forward on the contralateral side while pushing the animal back on the ipsilateral side. As a result the cockroach pivots in place through a greater angle. E. With wind from the front, the animal flexes both CF joints, pulling the animal backward while it turns to the left. A subsequent type I turn completes a very large angle turn.

(Figure 4D1). Nevertheless, the tarsi remain in contact with the substrate. As a result, the leg that is contralateral to the wind source pulls the animal backward, while the ipsilateral leg pushes that side forward. The effect is similar to that seen in a rowboat when an oarsman pulls on one oar while pushing on the other. In this way, a larger turn angle is accomplished than that which occurs with the type I turn (Figure 4D2). This is referred to as a type II turn.

With the wind placed at even more anterior positions, yet another turning strategy is revealed, which is referred to as a type III turn (Figure 4E1). In a type III turn, *both* T3 legs move forward as a result of CF flexion. Under free ranging conditions, this actually pulls the animal backward (Figure 4E2). However, the FT movements continue to turn the cockroach away from the wind source. With the animal's body drawn back and over its legs, it executes a second movement similar to a type I turn. The effect resembles the movements made as one attempts to back up a car and turn it around in a driveway. In free-ranging animals, type III turns can rotate the animal in excess of 90 deg.

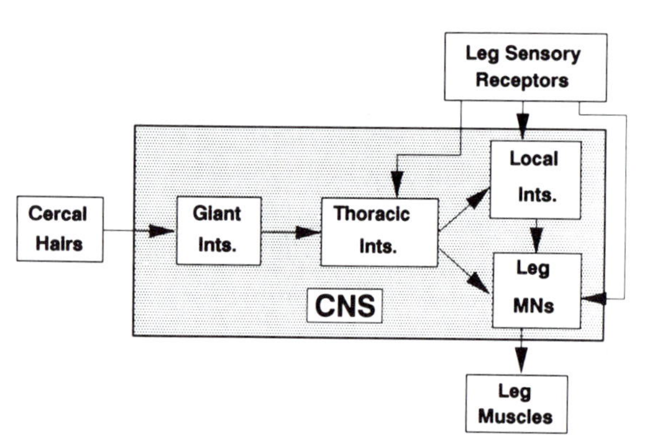

FIGURE 5 The basic escape circuit. This diagram describes the pathway between populations of neurons that make up the escape circuit. Cercal afferents project to giant interneurons within the central nervous system (CNS). The ventral giant interneurons project to type A thoracic interneurons. These produce motor effects either by direct connections with motor neurons or via local interneurons. Leg sensory receptors can influence the circuit via connections with type A thoracic interneurons, local interneurons, or in some cases direct connections to motor neurons.

Basic Circuit for the Escape Turn

The basic neural circuit is responsible for sensing wind stimulation, conducting that information to the motor control centers in the thoracic ganglia, and ultimately generating the appropriate motor responses. The neural components of this circuit and their connections have been documented by extensive observations, primarily employing dual intracellular recording and dye filling techniques (Figure 5). There are two main sites of processing that take place in this circuit. One is in the terminal abdominal ganglion where sensory information is integrated by a population of giant interneurons. The other is in the thoracic ganglia where directional wind information and other sensory data are integrated by thoracic interneurons and then used to direct motor activity. We describe the neural circuitry and our models of these two integration centers separately in the following sections.

Processing in the Terminal Ganglion

Mechanosensitive hairs located on the ventral surface of the cerci detect the wind puffs. The receptor hairs sit in asymmetrical sockets that confer a preferred plane of deflection on each hair (Nicklaus, 1965). The sockets are arranged in nine major columns such that every hair in a particular column has the same preferred direction of stimulation. Bipolar sensory neurons innervate each hair, and their axons project to the terminal ganglion through the cercal nerves. Intracellular recordings from the axons of these sensory neurons reveal receptive fields that are biased to one broadly tuned direction (Westin, 1979). Considering the entire population of sensory neurons in the cercal nerve, wind stimuli can be detected from any direction around the animal.

Within the terminal ganglion, cercal afferents make contact with the giant interneurons via cholinergic synapses (Sattelle, 1985). The pattern of synaptic connections with the ventral giant interneurons (vGIs) has been documented (Daley & Camhi, 1988). Again this pattern reflects the columns of hairs on the cerci. Afferents within each column make essentially the same type of connection with each vGI (Table 1). The strength of connections varies from column to column and the pattern of connectivity varies with each of the vGIs.

An individual GI can be identified either by its dendritic morphology within the terminal ganglion (Daley, Vardi, Appignani, & Camhi, 1981) or by the position of its axon relative to other GIs as they pass through the abdominal ganglia (Westin, Langberg, & Camhi, 1977). The GIs can be divided into morphologically and physiologically meaningful subgroups. The axons of the four pairs of vGIs reside in the ventral intermediate tract (VIT) of the abdominal nerve cord. Those of the three pairs of dorsal giant interneurons (dGIs) are found in the dorsal intermediate tract (DIT).

The number of action potentials recorded from a GI as wind is presented from various different directions forms a reproducible wind field for each GI (Westin et al., 1977). Much of the information on GI wind fields has been reviewed previously (Ritzmann, 1984). For our discussion, it is sufficient to point out the various reproducible wind biases of each of the vGIs (Figure 6A). GIs 2 and 4 are essentially omnidirectional. However, Camhi and Levy (1989) have presented evidence that, under certain stimulus conditions, a left–right bias can be detected in GI2 responses. GI1 wind fields are consistently biased toward winds originating from the side on which its axon is located. GI3 wind fields are biased toward wind from the front of the animal.

A Model of the Ventral Giant Interneurons

As we have just described, a great deal is known about the overall response properties of the vGIs, as well as the pattern of connectivity between the cercal hair

TABLE 1 Cercal to vGI Connectivity Pattern That Was Used to Construct the Computer Model[a]

		Left				Right			
		vGI1	vGI2	vGI3	vGI4	vGI1	vGI2	vGI3	vGI4[1]
Left Cercal Columns	a	+	+	n/c	+	+	+	n/c	+
	d	+	+	+	+	n/c	+	+	+
	g	+	+	+	+	n/c	+	+	+
	f	+	+	+	+	n/c	n/c	+	n/c
	h	+	+	−	+	+	+	−	+
	i	+	+	+	+	+	+	+	+
	k	+	+	+	+	n/c	+	+	+
	l	+	+	n/c	+	n/c	+	n/c	+
	m	+	+	−	+	+	+	−	+
Right Cercal Columns	a	+	+	n/c	+	+	+	n/c	+
	d	n/c	+	+	+	+	+	+	+
	g	n/c	+	+	+	+	+	+	+
	f	n/c	n/c	+	n/c	+	+	+	+
	h	+	+	−	+	+	+	−	+
	i	+	+	+	+	+	+	+	+
	k	n/c	+	+	+	+	+	+	+
	l	n/c	+	n/c	+	+	+	n/c	+
	m	+	+	−	+	+	+	−	+

[a]This table was based on data in Daley and Camhi (1988). + indicates an excitatory connection; − indicates an inhibitory connection; and n/c indicates pairs where no connection was detected in neurobiological recordings. The pattern of connections for vGI4 was not reported by Daley and Camhi (1988). However, because the wind field of vGI4 is qualitatively similar to that of vGI2, we utilized the vGI2 pattern for both of these neurons.

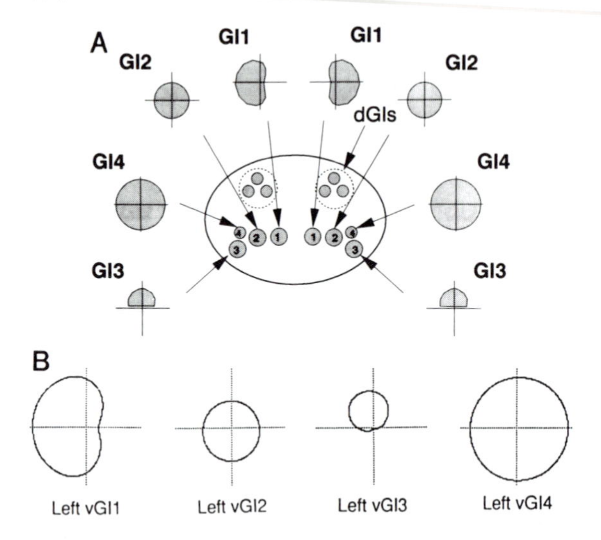

FIGURE 6 A summary of the wind fields of vGIs. A diagram of a cross-section of an abdominal ganglion shows the typical location of vGI (1–4) and dGI axons. The wind fields for each vGI summarize data on the number of action potentials evoked by winds from various different directions as plotted on polar coordinates (based on Westin et al., 1977). See text for details. B. Wind fields of the four left vGIs in the model.

columns and the vGIs. However, the data necessary for a detailed biophysical model of these cells is not currently available. For this reason, our initial work has employed simplified neural network models and learning techniques. This approach has proven to be effective for analyzing a variety of neuronal circuits (e.g., Lockery, Wittenberg, & Kristan, 1989; Anastasio & Robinson, 1989). Specifically, using backpropagation, we trained sigmoidal model neurons to reproduce the observed wind response properties of the vGIs. It is important to emphasize that we employed backpropagation solely as a means for finding the appropriate connection weights given the known structure of the circuit, and no claim is being made about the biological validity of backpropagation.

In the model, cercal hairs were connected to vGIs in the following manner. We used a single directional input for each of the nine major columns on each cercus, for a total of 18 inputs. This assumed homogeneous responses within each column. Although the amplitude of the vGI response to individual sensory neurons actually has a proximal-distal gradient (Hamon, Guillet, & Callec, 1990), the directional properties within each column are consistent (Daley & Camhi, 1988), justifying our simplification. The wind field for each column of cercal hairs was modeled as a cardioid oriented in that column's preferred direction (Figure 7).

Within our model, the pattern of synaptic connectivity between cercal inputs and vGIs incorporates Daley and Camhi's intracellular data on connectivity (see Ta-

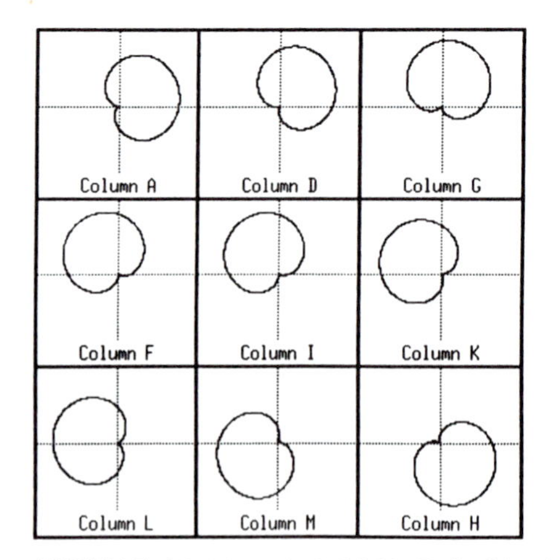

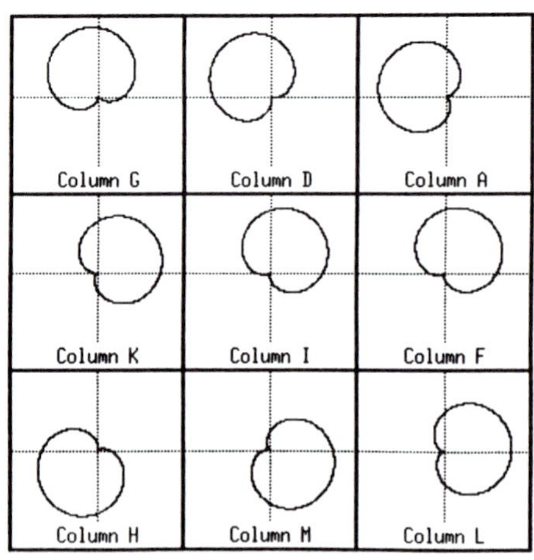

FIGURE 7 Model cercal wind fields. Each of the nine major hair columns on the left and right cerci are represented by a single model input.

ble 1). These data provide two types of information. First, each vGI receives input from a restricted set of hair columns. Second, the sign of connection can be either positive (excitatory) or negative (inhibitory). This information was used to constrain the connection weights generated by backpropagation to be biologically relevant.

Using this approach, we have reconstructed the observed wind fields of the eight vGIs. Each model vGI was trained to reproduce the corresponding wind field by constrained backpropagation. The resulting responses of the left four model vGIs are shown in Figure 6B. The responses of the right model vGIs are mirror images. These model wind fields closely approximate those observed in the cockroach (Westin, et al., 1977).

Processing within the Thoracic Ganglia

To direct the turn, there must be an element within the circuit that integrates the information contained within the eight vGIs and which uses this to direct activity in the appropriate leg motor neurons. The most logical site for such integration would be in the thoracic ganglia. The organization within each of the thoracic ganglia includes at least two levels of interneurons. Although stimulation of vGIs can evoke motor activity (Ritzmann & Camhi, 1978), this response is very weak, occurs at long latencies, and requires high frequency trains of activity in the vGIs. Several laboratories have spent considerable time searching for direct connections between vGIs and leg motor neurons, with no success. However, vGIs do make constant, short latency connections with thoracic interneurons, which in turn activate leg motor neurons either directly or via additional thoracic interneurons (Ritzmann & Pollack, 1986).

Input to Type A Thoracic Interneurons

The thoracic interneurons that receive input from vGIs have been classified as type A thoracic interneurons (TI_As; Westin, Ritzmann, & Goddard, 1988). So far, 13 TI_As have been identified. Because they have been found on either side of both the mesothoracic (T2) and the metathoracic (T3) ganglia, these individuals represent a population of at least 52 cells. Given that they probably exist in the prothoracic ganglion (T1) as well, a conservative estimate would put the total population at between 76 and 100 interneurons. All but one of the interneuron types are intersegmental. The axon of each TI_A is located contralateral to its soma. The somata are located in one of two areas of the ganglion. In one subpopulation, which is referred to as the dorsal posterior group (DPG), the somata are found bilaterally in the dorsal, posterior region of the ganglion. In the other subpopulation, the somata are located on the ventral surface on either side of the midline. These interneurons are referred to as ventral median cells (VMCs).

Perhaps the most interesting morphological feature of the TI_As is the presence of prominent ventral branches on one or both sides of the midline (Figures 8 and 9). We have referred to these as ventral median (VM) branches. Their location near the ventral intermediate tract, which contains the vGI axons, suggests a possible point of contact. This is supported by morphological studies as well as by developmental rearrangements (Casagrand & Ritzmann, 1991).

The VM branch morphology predicts the pattern of vGI connections to each TI_A. We have performed numerous experiments in which individual vGIs were stimulated intracellularly while recording from various identified TI_As. Out of these observations, we have arrived at a consistent pattern of vGI-to-TI_A connections (Ritzmann & Pollack, 1988). The pattern matches perfectly with VM branch morphology. TI_As with bilateral VM branches always receive inputs from vGIs on both sides of the ganglion (Figure 8). TI_As with only one VM branch only receive vGI inputs on the side that contains the VM branch, whether that be ipsilateral to the soma or to the axon (Figure 9). The wind response of each TI_A is also correlated to the VM branch morphology (Westin et al., 1988). TI_As with bilateral VM branches have symmetrical responses to right and left winds. However, TI_As with only one VM branch have a stronger response to wind on the side that contains that branch. This is again regardless of

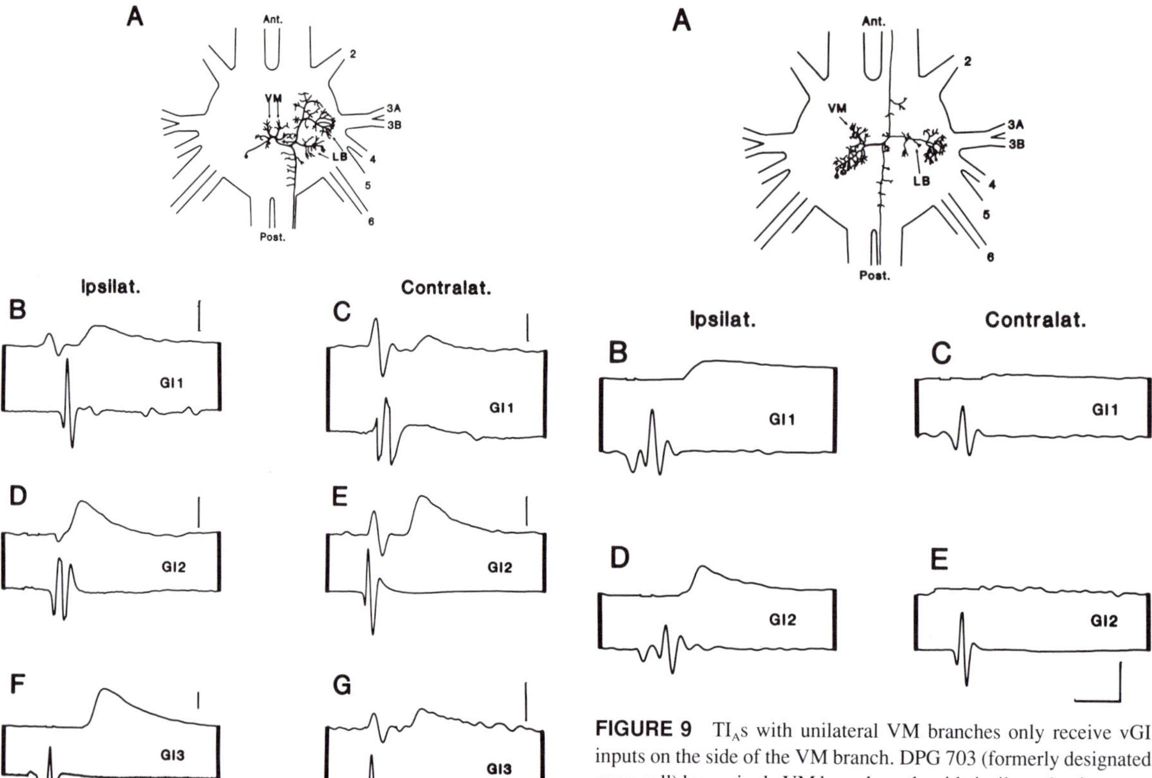

FIGURE 8 TI_As with bilateral VM branches receive vGI inputs on both sides. All vGIs excite DPG 301 (formerly labeled reverse J cell). A. Whole mount of a DPG 301 cell showing bilateral VM branches. B–G. Responses to intracellular stimulation of individual ipsilateral and contralateral vGI axons. In all records, the top trace is an intracellular recording from a DPG 301. The bottom trace is an extracellular recording of the abdominal cord used to verify intracellular activation of the vGI axon. The GI that was stimulated is indicated above each abdominal cord record. The amplitude calibration at each record represents 4 mV for each trace except G, where it represents 2 mV. The time calibration represents 2 msec for all records. This figure is reproduced from Ritzmann and Pollack (1988) with the permission of John Wiley & Sons, Inc.

FIGURE 9 TI_As with unilateral VM branches only receive vGI inputs on the side of the VM branch. DPG 703 (formerly designated cross cell) has a single VM branch on the side ipsilateral to its soma. It only receives vGI inputs on that side. A. Morphology of DPG 703 in whole mount, showing location of VM branch and lateral motor branch (LB). B–E. Intracellular records taken from DPGs 703 in response to intracellular stimulation of various vGI axons. Setup is as in previous figure. Note that only those GIs on the ipsilateral side (the same as the VM branch) evoked a response in this cell. Calibrations: 10 mV, 2 msec for all records. This figure is reproduced from Ritzmann and Pollack (1988) with the permission of John Wiley & Sons, Inc.

whether the VM branch is on the soma side or on the axon side. Both of these observations are consistent with the assumption that the VM branches are the sites of synaptic contact with vGI axons.

It is important to note that all documented connections between vGIs and TI_As have been excitatory. Moreover, we have not detected any inhibitory connections between individual TI_As. We have found weak excitatory connections between left and right homologous pairs of bilateral TI_As. Because the wind fields of bilateral TI_As are not biased in a left–right orientation, weak excitatory interactions would only serve to potentiate or synchronize the signal, but would not affect directionality. The total lack of mutual inhibition between left and right TI_As suggests that each TI_A acts as an independent entity, reaching a level of activity that is dependent on the pattern of vGI connections and the direction of wind activity. The output of the entire system, then, would be a tally of activity in all of the TI_As. Thus, the TI_A network may act much like the congress of cells that has been proposed for primate motor control (Georgopoulos, Kettner, & Schwartz, 1988).

Output from Type A Thoracic Interneurons

The primary connections from the TI_As are made on the side opposite the soma (Ritzmann & Pollack, 1986). We have documented connections between TI_As and several motor neurons and local interneurons (Ritzmann & Pollack, 1990). In these data, we found yet another interesting correlation with morphology. In all of our records, TI_As in the DPG subgroup (dorsal somata) excite postsynaptic cells (Figure 10), whereas TI_As in the VMC subgroup (ventral cell bodies) inhibit postsynaptic neurons. There is precedence for this arrangement in insects. In the locust flight system, intersegmental interneurons with dorsal cell bodies are consistently excitatory, whereas those with ventral cell bodies located near the midline are inhibitory (Pearson & Robertson, 1987).

For both subgroups, the influence on motor activity follows several parallel pathways (Ritzmann & Pollack, 1990). Because most of the TI_As are intersegmental interneurons, they reach cells both in their ganglion of origin and in adjacent ganglia (Figure 10). In addition, they will evoke motor neuron activity either directly or via local interneurons. The local interneurons that are involved are morphologically similar to interneurons in locust that serve to group activity in several motor neurons into coordinated movements (Burrows, 1980; Siegler & Burrows, 1979). The strength of connections for each of these parallel pathways in the cockroach is similar; that is, intersegmental and intrasegmental pathways appear to be of equal strength, as are direct motor connections and polysynaptic pathways, including local interneurons.

Hypotheses on Orientation in the TI_A Circuit

Given the patterns of connectivity that have been elucidated for both input and output pathways of the TI_As, we can begin to develop hypotheses on how the TI_As process wind information and generate the appropriate turning movements. The organization of motor connections could fall into one of two different patterns. The most straightforward pattern is a dedicated system. Each TI_A could be dedicated to one or more roles in generating the turn. This could be as major as commanding a complete turn to the right, or it could be a small subset of turning movements such as extending the FT joints. In the opposite extreme, the pattern of output could be distributed among the entire population of TI_As. Under this form of organization, several TI_As might make motor connections that are not totally consistent with the turning movement. However, taken as a total population, the responses of all TI_As would average out to evoke an appropriate turn.

Given the pattern of connections from vGIs to TI_As, and the information we have acquired on joint angle changes associated with the turning movements, we can predict the connections that would be made from various TI_As in a dedicated system. The TI_As can be segregated into smaller groups based on their pattern of vGI inputs and the side of the animal on which their motor effects occur. For example, if we consider a response to wind from the right rear, we will have to account for excitatory connections from four subgroups of TI_As (Figure 11). Two of the subgroups include TI_As with VM branches on both sides of the midline. These include interneurons with axons on the right side and those with axons on the left side. Because cells with bilateral VM branches are excited equally well by wind from the right and wind from the left, we would expect excitation of these cells to cause joint movements that will occur regardless of left or right wind direction. Such movements include

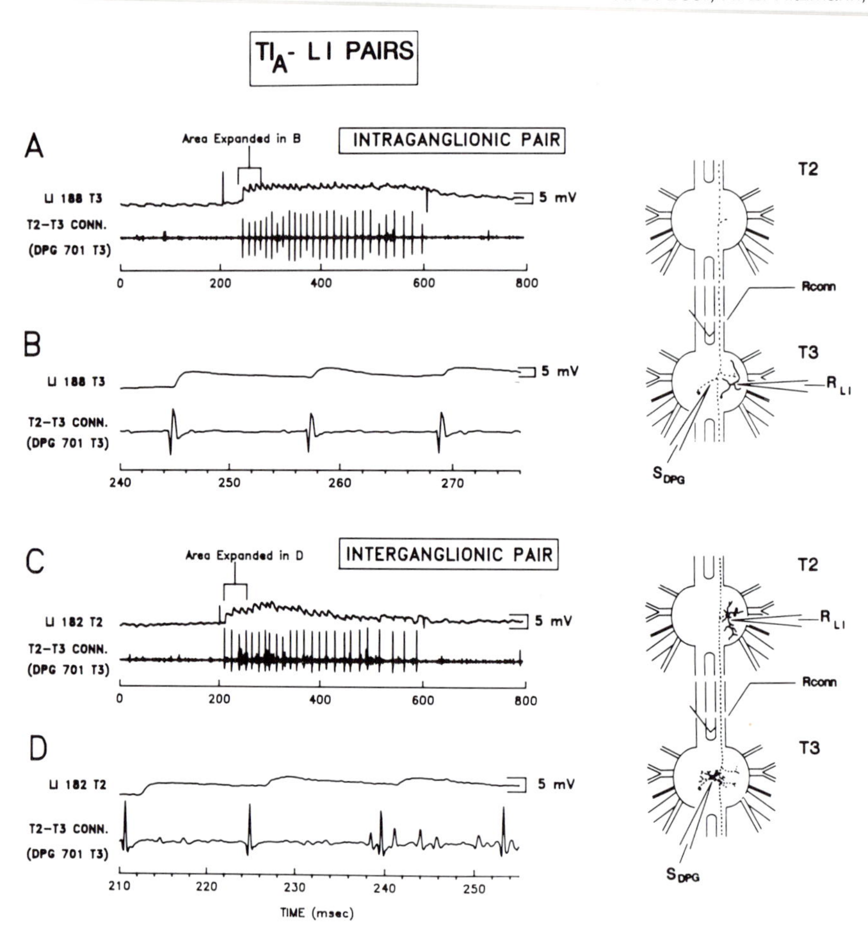

FIGURE 10 PSPs evoked in interganglionic and intraganglionic pairs are similar. A and B. Recordings from local interneuron (LI) 188 (top traces) in T3 in response to intracellular stimulation of DPG 701 in T3 (see insets for recording setup). The area bracketed in A is expanded in B to show the consistent relationship between DPG 701 action potentials in the T2-T3 connective and PSPs in LI 188. C and D. Results from a similar experiment in which an interganglionic pair was monitored. In this case, DPG 701 in T3 was stimulated in conjunction with a recording from LI 182 in T2. Note that the responses in the two pairs are similar. The drawing of DPG 701 in the inset for the intraganglionic pair (A and B) shows only the primary branches to avoid obscuring the drawing of the motor neuron. This figure is reproduced from Ritzmann and Pollack (1990) with the permission of John Wiley & Sons, Inc.

extension of CF joints in both T2 and T3. These movements provide power to the turn, but do not enter into directionality unless the wind comes from the front of the animal.

Left–right directionality would arise from TI$_A$s with unilateral VM branches. These interneurons receive inputs only from vGIs on the side on which the VM branch is located. The resulting directional wind bias

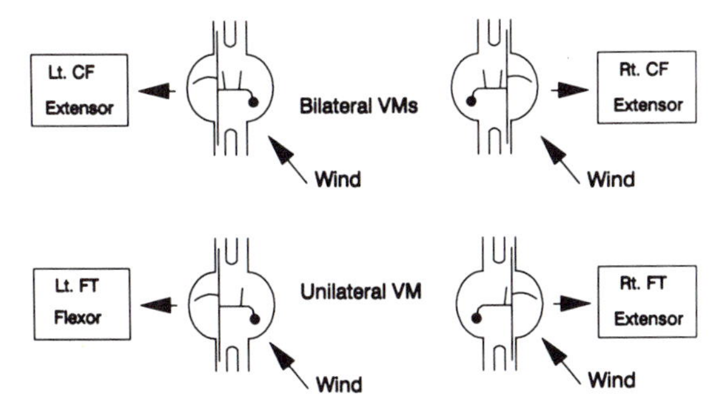

FIGURE 11 In a dedicated system, these excitatory cells must be accounted for: Cells with BILATERAL VM branches respond to wind from either direction and excite motor neurons that are active regardless of turn direction. Cells with UNILATERAL VM branches act on motor neurons that provide directional movements.

suggests that these cells would be associated with movements that influence the direction of turning, for example, the FT joints of the legs in T2. Two subgroups of unilateral TI_As must be considered. These are TI_As that have a VM branch ipsilateral to their axon and those that have a VM branch contralateral to their axon. Because direct motor outputs are restricted to the side on which the axon is located, we can restrict our discussion of motor output to that side. Given the current example (wind from the right rear), we need only consider TI_As with VM branches on the right side. The two relevant subgroups are those that respond to wind from the right and evoke motor output on the right, and those that respond to wind from the right and evoke motor output on the left (Figure 11). We know from our video analysis that wind from the right causes the FT joint of the right T2 leg to extend. If the system has dedicated connections, we would, therefore, predict that TI_As with VM branches on the right side and ipsilateral to their axon would excite extensor motor neurons of the FT joint. Our video analysis would also predict that TI_As with VM branches on the right side, but contralateral to the axon, would excite flexor motor neurons of the left FT joints.

In addition to excitatory connections, we must account for inhibitory connections from TI_As. This is facilitated by the correlation between soma location and sign of connection. As a result, we would predict that TI_As with ventral somata would inhibit motor neurons that are antagonistic to those excited by similar TI_As with dorsal somata.

Neural Network Model of the Escape System

The population of thoracic neurons that integrates vGI information is quite large. Therefore, a complete understanding of the nature of its control function would be greatly facilitated by the use of computer modeling techniques. With this in mind, we extended our earlier vGI model to include circuitry in the thoracic ganglia. This was then incorporated into a model of the insect's body to form a complete simulation of the escape system.

Our three-dimensional, kinematic model of the insect's body accurately represents the essential degrees of freedom of the legs during escape turns. These include CF and FT joints of each leg which are primarily responsible for turning the animal. The leg segment lengths and orientations, as well as the joint angles and axes of rotation, were derived from actual measurements (Nye & Ritzmann, unpublished data). Our motion analysis data were taken from tethered insects which were suspended by a rod above a greased plate. Because an insect thus tethered is neither supporting its own weight nor generating appreciable forces with its legs, a kinematic body model can be defended as an adequate first approximation.

The leg movements of the simulated body were controlled by a neural network model of the entire escape circuit (Figure 12). Where sufficient data was available, the structure of this network was constrained appropriately. The first layer of this circuit was described previously and was not subject to further modification. The remaining layers were structured as follows. There are six groups of six representative TI_As, one group for each leg. Within a group, representative members of each identified class of TI_A were modeled. The individual classes of TI_As were based on VM branch morphology and underlying vGI connectivity (Figures 8 and 9). Where known, the connectivity

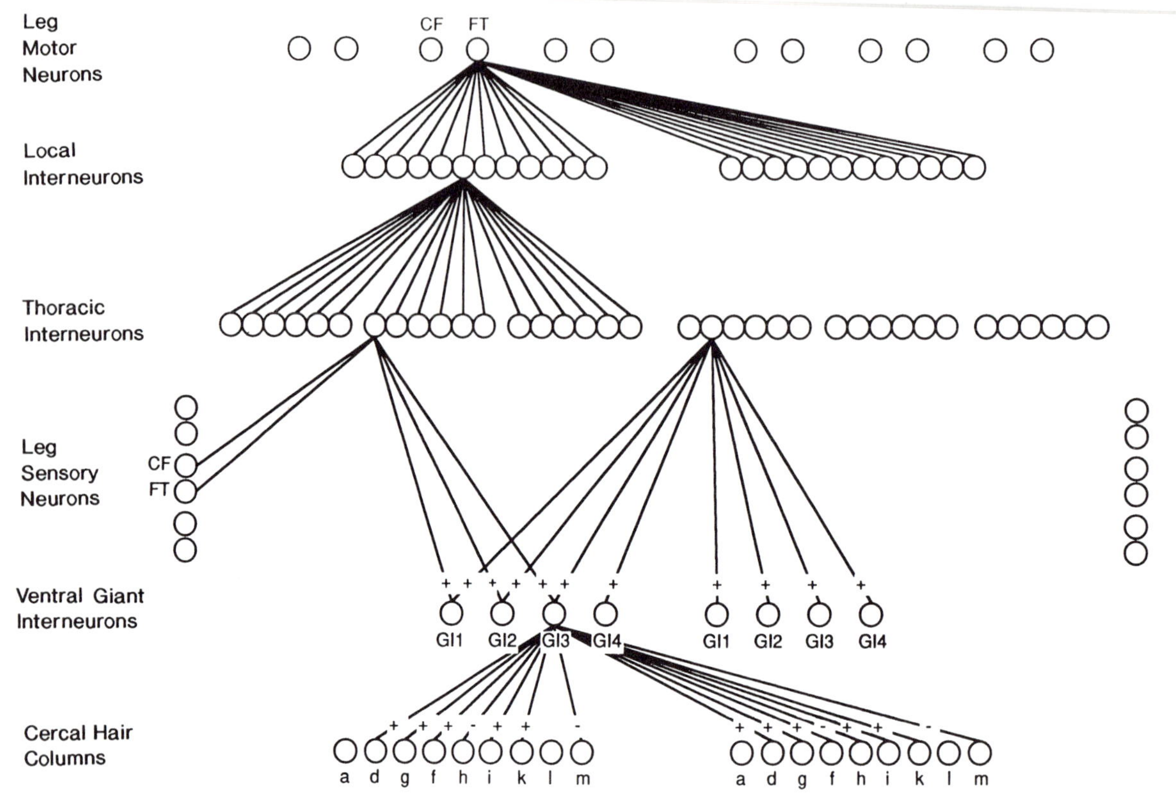

FIGURE 12 Neural network model of the cockroach escape circuitry. Only representative connection patterns between the layers of the model are shown. The signs of certain connections were constrained, as shown.

from the vGIs to each class of TI_As was enforced and all connections from vGIs to TI_As were constrained to be excitatory (Ritzmann & Pollack, 1988), as described earlier. Model TI_As also received inputs from leg proprioceptors encoding joint angle (Murrain & Ritzmann, 1988). The TI_A layer for each side of the body was fully connected to 12 local interneurons, which were in turn fully connected to motor neurons. Again, these constraints followed the general pattern of connectivity revealed by intracellular recordings (Figure 10). The output of each motor neuron was interpreted as a change in angle of the corresponding joint in the body model. The data on leg movement which were described earlier (Nye & Ritzmann, 1992) were used as training data for the model escape circuit. Only movements of the middle and hind legs were

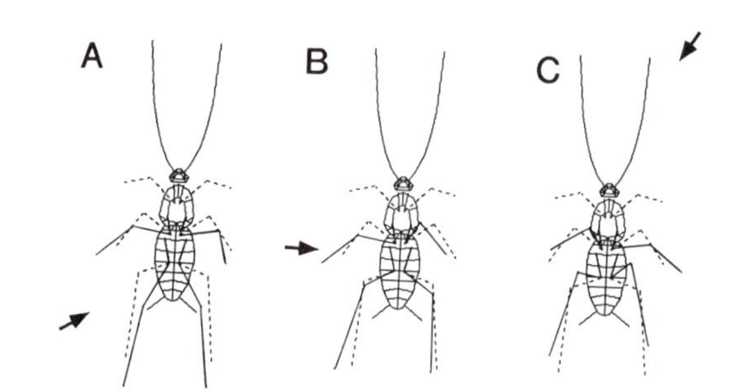

FIGURE 13 Ventral view of the leg movements produced by the model for wind puffs from different directions. Dotted lines indicated the leg positions before the simulated wind puff and solid lines indicated their position after the wind puff has been delivered. The arrows indicate the direction of the wind puff. See text for details.

considered because data were not available for the front legs.

After training with constrained backpropagation, the model successfully reproduced the essential features of these data (Figure 13). Wind from the rear (Figure 13A) always caused the rear and middle legs of the model to thrust back, whereas the middle FT joints directed leg movements toward the "wind source." This is characteristic of a type I turn (Figure 4C1), which would propel the body forward and away from the wind source in a freely moving insect. A simulated wind from the side (Figure 13B) caused similar FT joint movements except that the contralateral T2 leg was drawn forward instead of extending backward. This is typical of a type II turn (Figure 4D1). However, in the actual insect, the T3 legs underwent a similar switch which was not observed in the model. A simulated wind from the front caused both T3 legs to be drawn forward, in addition to the contralateral T2 leg (Figure 13C). This is consistent with a typical type III turn in the actual insect (Figure 4E1).

Manipulation of the Model

The results just described demonstrate that several neuronal and behavioral properties of this system can be reproduced using simplified, but biologically constrained, neural network models. However, to serve as a useful tool for understanding the neuronal basis of the cockroach escape response, it is not enough for the model to reproduce simply what is already known about the normal operation of the system. To test and refine the model, we must also examine its responses to various lesions and compare them to the responses of the insect to analogous lesions. Here, we report the results of two such experiments.

Immediately after removal of the left cercus, cockroaches make a much higher proportion of incorrect turns (i.e., turns toward rather than away from the wind source) in response to wind from the left, whereas turns in response to wind from the right are largely unaffected. (Vardi & Camhi, 1982a). These results suggest that, despite the redundant representation of wind direction by each cercus, the insect integrates information from *both* cerci in order to compute the appropriate direction of movement. As shown in Figure 14, the response of the model to a left cercal ablation is consistent with these results. In response to wind from the unlesioned side, the model generates leg movements that would turn the body away from the source of wind. However, in response to wind from the lesioned side, the rear legs thrust back, the middle leg ipsilateral to the ablation pulls in, and the contralateral leg pushes out. These leg movements would turn the body toward the wind in a free-ranging insect.

It is interesting to note that after a 30-day recovery period the directionality of a lesioned insect's escape response is largely restored (Vardi & Camhi, 1982a). Although the mechanisms underlying this adaptation are not yet fully understood, they appear to involve a reorganization of the vGI connections from the intact cercus (Vardi & Camhi, 1982b; Volman, 1989). After a cercal ablation, the wind fields of the vGIs on the ablated side are significantly reduced. After the 30-day recovery period, however, these wind fields are largely restored. We have also examined these effects in the model. After cercal ablation, the model vGI wind fields show some resemblance to those of similarly lesioned insects. In addition, using vGI retraining to simulate the adaptation process, we have found that

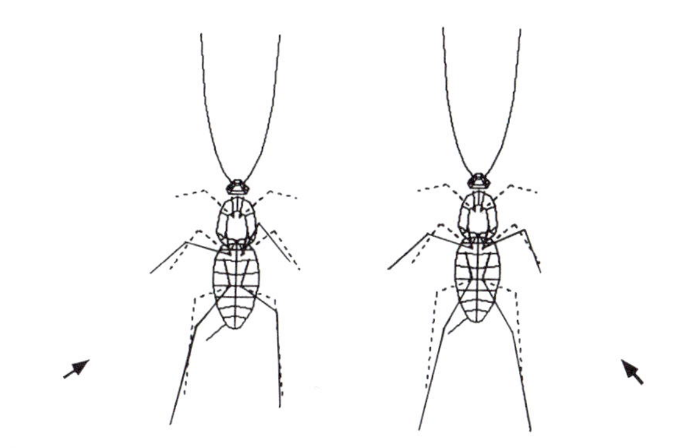

FIGURE 14 Ventral view of the model leg movements following left cercal ablation. (Left) Wind from the unlesioned side evokes leg movements consistent with a turn away from the wind direction. (Right) Wind from the side of the cercal ablation evokes leg movements consistent with a turn toward the direction of the wind.

the model can effect a similar recovery of vGI wind fields by adjusting the connections from the intact cercus.

A second experimental manipulation that has been performed on this system is the selective lesion of individual ventral giant interneurons (Comer, 1985). The only result that we will describe here is the lesion of vGI1. In the animal, this results in a behavioral deficit similar to that observed with cercal ablation. Correct turns result for wind from the unlesioned side, but a much higher proportion of incorrect turns are observed in response to wind from the lesioned side. The response of the model to this lesion is also similar to its response to a cercal ablation (Figure 14) and is thus consistent with these experimental results.

Future Directions

As encouraging as the initial results reported in this section are, the current model is limited by the extreme simplicity of both its neural and peripheral components. Indeed, rather than being the final word, we view this model as only an initial step in a continuing cycle of refinement in which experimental and simulation work must continuously interact. We are currently including additional constraints in our model. For example, vGI models incorporating experimentally determined firing properties are being used to explore the mechanisms responsible for the vGI wind fields and their adaptation to cercal ablation (Beer, Kacmarcik, Chai, Ritzmann, & Chiel, 1991). In addition, we have begun to incorporate physical dynamics into our kinematic body model as a step toward making the periphery of our simulated cockroach more realistic.

Moreover, as more complex properties are uncovered in the system, models such as the one presented here can be useful in understanding how these properties are incorporated into the basic circuit. For example, recent findings have established that the escape system incorporates much more sensory information into its escape decisions than that found in the wind stimulus alone. The animal can adjust its escape to compensate for initial leg position (Camhi & Levy, 1988; Nye & Ritzmann, 1992), and it can alter its turn when it is positioned near a wall (Ritzmann, Pollack, & Hudson, 1991). The population of TI_As can account for much of this interaction. Most if not all TI_As are multimodal. They respond to auditory, tactile, visual, and proprioceptive information, in addition to descending wind stimuli (Ritzmann et al., 1991). Thus, we have proposed that the network of TI_As could interpret wind information in the context of all of the other data that they integrate. The resulting response could then be altered to reflect a changing environmental context.

The problem in testing this hypothesis stems from the large number of parameters that are involved in a relatively large population of approximately 100 TI_As. To monitor all of these properties across the TI_A population would be a heroic task. However, as the various sensory parameters are determined for representatives of each class of TI_A, it will be possible to incorporate this into our present model and eventually test how these factors influence the escape turn. By comparing the results to actual turning movements, we can come to appreciate how the individual personalities of each TI_A within the entire population affects the behavior.

CONCLUSION

Computational neuroethology has much to offer to both robotics and neuroethology. Building and manipulating simulations of neuroethological systems can provide new insights into the neural mechanisms of animal behavior, as well as suggest new approaches to the control of autonomous mobile robots.

In this chapter, we have presented two examples of such simulations. First, we described a novel neural network architecture for hexapod locomotion that was abstracted from work on the neural basis of insect walking. This circuit is fully distributed, efficient and robust, and has been used to control the locomotion of an actual hexapod robot.

Second, we described a model of the cockroach escape response that was capable of reproducing not only the normal leg movements observed during the initial turn, but also those observed after two different lesions. Our experience with this simulation is currently serving as the basis for a second cycle of experimental work and model refinement in what we hope will ultimately become an ideal model system for studying the neural basis of distributed sensorimotor control and context-dependent responses. Although these two models are currently limited in a number of ways, they both illustrate the exciting potential that lies at the interface between neuroethology and robotics, and the role that computational neuroethology can play in fostering this interaction.

Acknowledgments

This work was supported in part by ONR Grant N00014-90-J-1545 to R.D.B. and NIH Grant NS-17411 to R.E.R. Additional support was provided by the Howard Hughes Medical Institute and the Cleveland Advanced Manufacturing Program. H.J.C. thanks the NSF for its support with grant BNS-8810757.

References

Achacoso, T. B., & Yamamoto, W. S. (1990). Artificial ethology and computational neuroethology: A scientific discipline and its subset by sharpening and extending the definition of artificial intelligence. *Perspectives in Biology and Medicine*, **33**, 379–389.

Anastasio, T. J., & Robinson, D. A. (1989). Distributed parallel processing in the vestibulo-oculomotor system. *Neural Computation*, **1**, 230–241.

Beer, R. D. (1990). *Intelligence as adaptive behavior: An experiment in computational neuroethology*. New York: Academic Press.

Beer, R. D., Chiel, H. J., Quinn, R. D., & Espenschied, K. S. (1992). *Neural Computation*, **4**, 356–365.

Beer, R. D., Chiel, H. J., & Sterling, L. S. (1989). Heterogeneous neural networks for adaptive behavior in dynamic environments. In D. S. Touretzky (Ed.), *Advances in neural information processing systems 1* (pp. 577–585). Palo Alto, CA: Morgan Kaufmann.

Beer, R. D., Kacmarcik, G. J., Chai, S., Ritzmann, R. E., & Chiel, H. J. (1991). Ventral giant interneuron wind fields in the cockroach modeled with constrained backpropagation. *Society for Neuroscience Abstracts*, **16**, 759.

Brooks, R. A. (1989). A robot that walks: Emergent behaviors from a carefully evolved network. *Neural Computation*, **1**, 253–262.

Burrows, M. (1980). The control of sets of motoneurons by local interneurones in the locust. *Journal of Physiology*, **298**, 213–233.

Camhi, J. M., & Levy, A. (1988). Organization of a complex motor act: Fixed and variable components of the cockroach escape behavior. *Journal of Comparative Physiology*, **163**, 317–328.

Camhi, J. M., & Levy, A. (1989). The code for stimulus direction in a cell assembly in the cockroach. *Journal of Comparative Physiology A*, **165**, 83–97.

Camhi, J. M., & Nolen, T. G. (1981). Properties of the escape system of cockroaches during walking. *Journal of Comparative Physiology*, **142**, 339–346.

Camhi, J. M., & Tom, W. (1978). The escape behavior of the cockroach Periplaneta americana. I. Turning response to wind purffs. *Journal of Comparative Physiology*, **128**, 193–201.

Camhi, J. M., Tom, W., & Volman, S. (1978). The escape behavior of the cockroach Periplaneta americana. II. Detection of natural predators by air displacement. *Journal of Comparative Physiology*, **128**, 203–212.

Casagrand, J. L., & Ritzmann, R. E. (1991). Localization of ventral giant interneuron connections to the ventral median branch of thoracic interneurons in the cockroach. *Journal of Neurobiology*, **22**, 643–658.

Chiel, H. J., & Beer, R. D. (1989). A lesion study of a heterogeneous artificial neural network for hexapod locomotion. *Proceedings of the First International Journal Conference on Neural Networks*, pp. 407–414, IEEE.

Chiel, H. J., Beer, R. D., Quinn, R. D., & Espenschied, K. S. (1992). Robustness of a distributed neural network controller for a hexapod robot. *IEEE Transactions on Robotics and Automation*, **8**, 293–303.

Cliff, D. (1991). Computational neuroethology: A provisional manifesto. In J. A. Meyer & S. W. Wilson (Eds.), *From animals to animats* (pp. 29–39). Cambridge, MA: MIT Press.

Comer, C. M. (1985). Analyzing cockroach escape behavior with lesions of individual giant interneurons. *Brain Research*, **335**, 342–346.

Daley, D. L., & Camhi, J. M. (1988). Connectivity pattern of the cercal-to-giant interneuron system of the American cockroach. *Journal of Neurophysiology*, **60**, 1350–1368.

Daley, D. L., Vardi, N., Appignani, B., & Camhi, J. M. (1981). Morphology of the giant interneurons and cercal nerve projections of the American cockroach. *Journal of Comparative Neurology*, **196**, 41–52.

Georgopoulos, A. P., Kettner, R. E., & Schwartz, A. B. (1988). Primate motor cortex and free arm movements to visual targets in three-dimensional space. II. Coding of the direction of movement by a neuronal population. *Journal of Neuroscience*, **8**, 2928–2937.

Graham, D. (1977). Simulation of a model for the coordination of leg movement in free walking insects. *Biological Cybernetics*, **26**, 187–198.

Graham, D. (1985). Pattern and control of walking in insects. *Advances in Insect Physiology*, **18**, 131–140.

Hamon, A., Guillet, J. C., & Callec, J. J. (1990). A gradient of synaptic efficacy and its presynaptic basis in the cercal system of the cockroach. *Journal of Comparative Physiology A*, **167**, 363–376.

Lockery, S. R., Wittenberg, G., & Kristan, W. B., & Cottrell, G. W. (1989). Function of identified interneurons in the leech elucidated using neural networks trained by back-propagation. *Nature*, **340**, 468–471.

Murrain, M. P., & Ritzmann, R. E. (1988). Analysis of proprioceptive inputs to DPG interneurons in the cockroach. *Journal of Neurobiology*, **19**, 552–570.

Nicklaus, R. (1965). Die Erregung einzelner Fadenhaare von Periplaneta americana in Abhangidkeit von der Grosse und Richtung Auslenkung. *Z. Vergl. Physiology*, **50**, 331–362.

Nye, S. W., & Ritzmann, R. E. (1992). Motion analysis of leg joints associated with escape turns of the cockroach. *Journal of Physiology A*, **171**, 183–194.

Pearson, K. G. (1976). The control of walking. *Scientific American*, **235**, 72–86.

Pearson, K. G., & Robertson, R. M. (1987). Structure predicts synaptic function of two classes of interneurons in the thoracic ganglia of Locusta migratoria. *Cell Tissue Research*, **250**, 105–114.

Pearson, K. G., Fourtner, C. R., & Wong, R. K. (1973). Nervous control of walking in the cockroach. In R. B. Stein, K. G. Pearson, R. S. Smith, and J. B. Redford (Eds.), *Control of posture and locomotion* (pp. 491–514). New York: Plenum Press.

Quinn, R. D., & Espenschied, K. S. (1992). Control of a hexapod robot using a biologically inspired neural network. In R. D. Beer, R. E. Ritzmann, & T. McKenna (Eds.), *Biological neural networks in invertebrate neuroethology and robotics* (pp. 365–382). New York: Academic Press.

Ritzmann, R. E. (1984). The cockroach escape response. In R. C. Eaton (Ed.), *Neural mechanisms of startle behavior* (pp. 93–131). New York: Plenum Press.

Ritzmann, R. E., & Camhi, J. M. (1978). Excitation of leg motor neurons by giant interneurons in the cockroach Periplaneta americana. *Journal of Comparative Physiology*, **125**, 305–316.

Ritzmann, R. E., & Pollack, A. J. (1986). Identification of thoracic interneurons that mediate giant interneuron-to-motor pathways in the cockroach. *Journal of Comparative Physiology A*, **159**, 639–654.

Ritzmann, R. E., & Pollack, A. J. (1988). Wind activated thoracic interneurons of the cockroach. II. Patterns of connection from ventral giant interneurons. *Journal of Neurobiology*, **19**, 589–611.

Ritzmann, R. E., & Pollack, A. J. (1990). Parallel motor pathways from thoracic interneurons of the ventral giant interneuron system of the cockroach, Periplaneta americana. *Journal of Neurobiology*, **21**, 1219–1235.

Ritzmann, R. E., Pollack, A. J., Hudson, S. E., & Hyvonen, A. (1991). Convergence of multi-modal sensory signals at thoracic interneurons of the escape system of the cockroach, Periplaneta americana. *Brain Research*, **563**, 175–183.

Roeder, K. D. (1967). *Nerve cells and insect behavior*. Cambridge, MA: Harvard University Press.

Sattelle, D. B. (1985). Acetylcholine receptors, In *Comprehensive Insect Physiology, Biochemistry and Pharmacology* (Vol. 11, pp. 395–434). G. A. Kerkut and G. I. Gilbert (eds.) Pergamon Press, New York.

Siegler, M. V. S., & Burrows, M. (1979). The morphology of local non-spiking interneurons in the metathoracic ganglion of the locust. *Journal of Comparative Neurology*, **183**, 121–148.

Vardi, N., & Camhi, J. M. (1982a). Functional recovery from lesions in the escape system of the cockroach I. Behavioral recovery. *Journal of Comparative Physiology*, **146**, 291–298.

Vardi, N., & Camhi, J. M. (1982b). Functional recovery from lesions in the escape system of the cockroach II. Physiological recovery of the giant interneurons. *Journal of Comparative Physiology*, **146**, 299–309.

Volman, S. F. (1989). Localization of the enhanced input to cockroach giant interneurons after partial deafferentation. *Journal of Neurobiology*, **20**, 762–783.

Westin, J. (1979). Responses to wind recorded from the cercal nerve of the cockroach Periplaneta americana I. Response properties of single sensory neurons. *J. Comp. Physiol.* **133**, 97–102.

Westin, J., Langberg, J. J., & Camhi, J. M. (1977). Response properties of giant interneurons of the cockroach Periplaneta americana to wind puffs of different directions and velocities. *J. Comp. Physiol.* **121**, 307–324.

Westin, J., Ritzmann, R. E., & Goddard, D. J. (1988). Wind activated thoracic interneurons of the cockroach: I. Responses to controlled wind stimulation. *Journal of Neurobiology*, **19**, 573–588.

Wilson, D. M. (1966). Insect walking. *Annual Review of Entomology*, **11**, 103–122.

10

A Silicon Model of Auditory Localization

John Lazzaro and Carver Mead

The principles of organization of neural systems arose from the combination of the performance requirements for survival and the physics of neural elements. From this perspective, the extraction of time-domain information from auditory data is a challenging computation; the system must detect changes in the data which occur in tens of microseconds, using neurons which can fire only once per several milliseconds. Neural approaches to this problem succeed by closely coupling algorithms and implementation, unlike standard engineering practice, which aims to define algorithms that are easily abstracted from hardware implementation.

The echolocation system of the bat and the passive localization system of the barn owl are two neural systems which use extensive time-domain processing to create an accurate representation of auditory space. While the bat has developed specialized algorithms to implement an active sonar system, the barn owl and mammals use similar techniques to perform auditory localization. We are building silicon models of the passive localization system of the barn owl to explore the general computational principles of time-domain processing in neural systems.

The barn owl (*Tyto alba*) uses hearing to locate and catch small rodents in total darkness. The owl localizes the rustles of the prey to within 1–2° in azimuth and elevation (Knudsen, Blasdel, & Konishi, 1979). The owl cannot localize sounds when one ear is completely occluded; the auditory cues are differences in sound received by the two ears. The owl uses different binaural cues to determine azimuth and elevation. The elevational cue for the owl is binaural intensity difference. This cue is a result of a vertical asymmetry in the placement of the barn owl's ear openings, as well as a slight asymmetry in the left and right halves of the barn owl's facial ruff, which acts as a sound-collecting device (Knudsen & Konishi, 1979). The azimuthal cue is binaural time difference. The binaural time differences are in the microsecond range, as expected from the short interaural distance of the owl, and vary as a function of azimuthal angle of the sound source (Moiseff & Konishi, 1981).

Unlike those in the visual system, the primary sensory neurons of the auditory system do not directly record the spatial location of a stimulus. Neuronal circuits must compute the position of sounds in space from acoustic cues in the stimulus. In the barn owl, the external nucleus of the inferior colliculus (ICx) contains the neural substrate of this computation, a spatial map of auditory space (Knudsen & Konishi, 1979). Neurons in the ICx respond maximally to stimuli located in a small area in space, corresponding to a specific combination of binaural intensity difference and binaural time difference (Figure 1). Recordings of the ICx neurons made while binaural intensity difference (Knudsen & Konishi, 1980) and

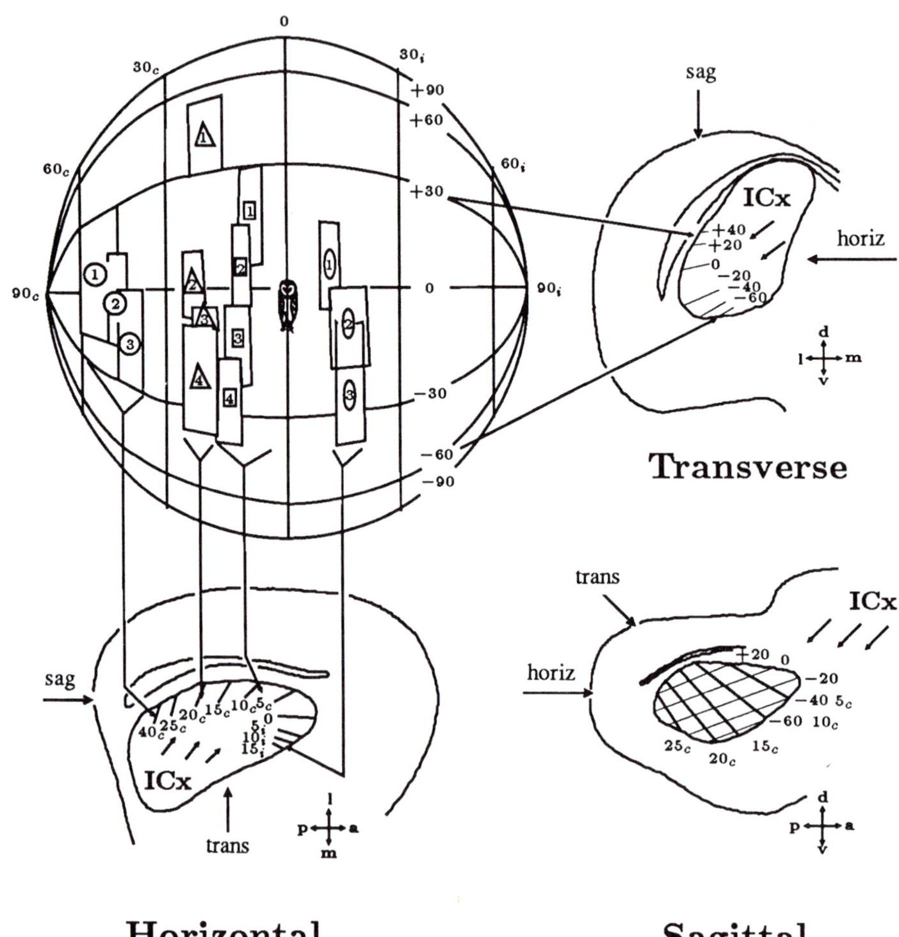

FIGURE 1 Neural map of auditory space in the barn owl. Auditory space is depicted as an imaginary globe surrounding the owl. Projected onto the globe are the receptive field best areas of 14 neurons. The best area, a zone of maximal response within a receptive field, remains unaffected by variations in sound intensity and quality. The numbers surrounding the same symbols (circles, rectangles, triangles, ellipses) represent neurons from the same electrode penetration; the numbers themselves denote the order in which the neurons were encountered. Below and to the right of the globe are illustrations of three histological sections through the inferior colliculus; arrows point to the external nucleus, the ICx. Iso-azimuth contours are shown as thick lines in the horizontal and sagittal sections; iso-elevation contours are represented by thin lines in the transverse and sagittal sections. a, anterior; c, contralateral; d, dorsal; i, ipsilateral; l, lateral; m, medial; p, posterior; v, ventral. From Knudsen and Konishi (1978) and Konishi (1986).

binaural time difference (Moiseff and Konishi, 1981) are varied separately confirm the role of these acoustic cues.

There are several stages of neural processing between the primary sensory neurons at each cochlea and the computed map of space in the ICx. Each primary auditory fiber initially divides into two distinct pathways (Figure 2). One pathway processes intensity information, and thus encodes elevation cues, whereas the other pathway processes timing information, and thus encodes azimuthal cues. The two pathways recombine in the ICx to produce a complete map of space (Takahashi & Konishi, 1988b).

Recent anatomical and physiological studies provide a model of the structure and function of the time-coding pathway of the barn owl (Carr & Konishi, 1988; Takahashi & Konishi, 1988a; Fujita & Konishi, 1988, unpublished). This pathway has clear anatomical and physiological homologues in other avian species and in mammals, although the most completely studied example of time-coding processing is that done in the barn owl.

We have built an integrated circuit that models the time-coding pathway of the barn owl, using analog, continuous-time processing. The chip receives two inputs, corresponding to the sound pressure at each ear of the owl. The chip computes as its output a map of the interaural time difference between the input signals, encoding the azimuthal localization cue. The architecture of the chip directly reflects the anatomy and physiology of the neural pathway; intermediate outputs of the chip correspond to different neurons in this pathway.

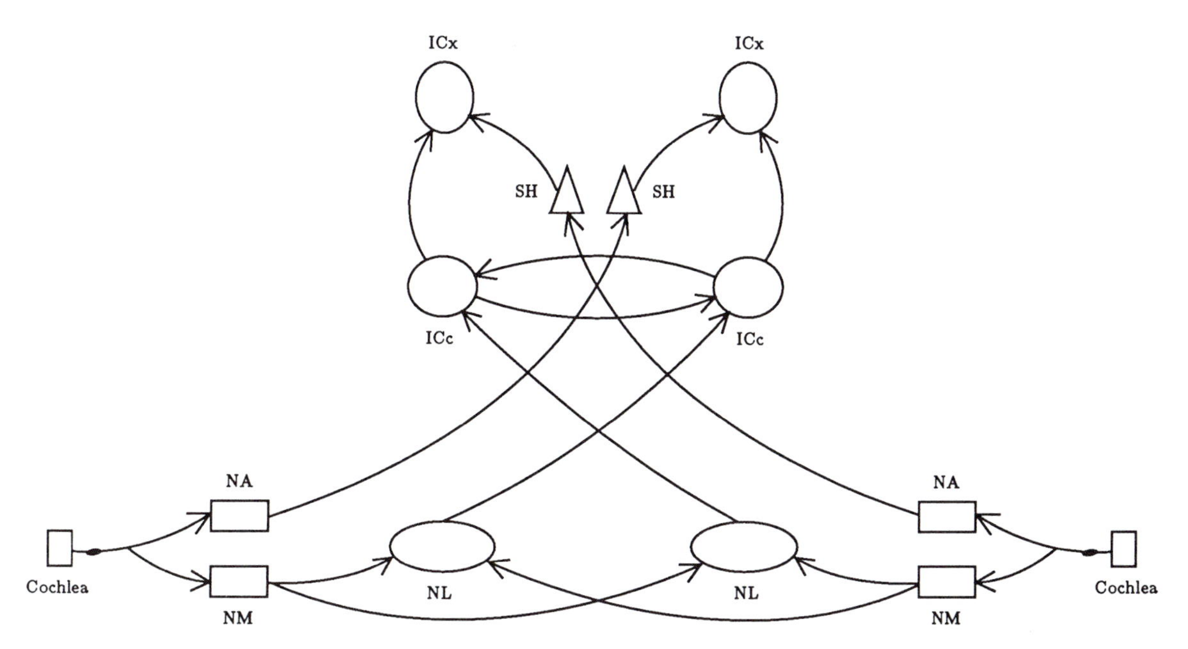

FIGURE 2 Schematic drawing of the auditory pathway of the barn owl. The pathway begins at the cochlea, the sound transducer. The cochlea projects to the nucleus angularis (NA), to start the amplitude-coding pathway, and to the nucleus magnocellularis (NM), to start the time-coding pathway. The NA projects contralaterally to the shell (SH) of the central nucleus of the inferior colliculus, which projects to the external nucleus (ICx) of the inferior colliculus, the location of the complete map of auditory space. The NM projects bilaterally to the nucleus laminaris (NL); the NL projects contralaterally to the central nucleus (ICc) of the inferior colliculus. Fibers connect the ipsilateral and contralateral sides of the ICc, returning information to the original side. The ICc then projects to the ICx, the location of the complete map of auditory space. Adapted from Fujita and Konishi (1988, unpublished) and from Takahashi and Konishi (1988b).

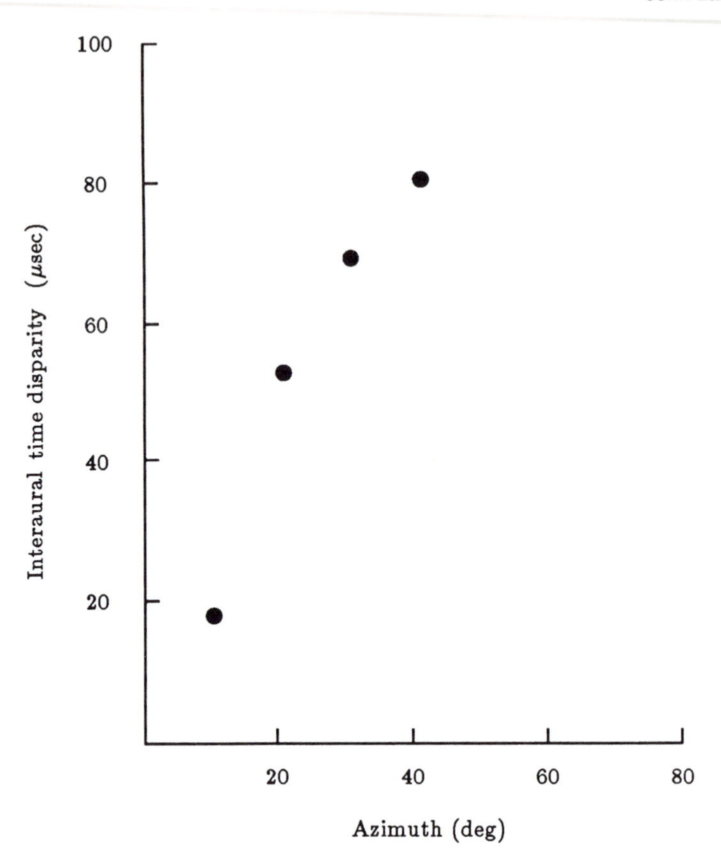

FIGURE 3 Binaural time difference in the owl, as a function of azimuthal angle. Small microphones installed near the ear drums were used to measure the interaural time differences in the microsecond range. From Moiseff and Konishi (1981).

THE TIME-CODING PATHWAY OF THE OWL

By installing small microphones near the eardrums of the owl, Moiseff and Konishi (1988) have demonstrated the dependence of binaural time difference on the azimuthal angle of a sound source (Figure 3). The time differences are in the microsecond range, as expected from the short interaural distance of the owl. The time-coding pathway of the owl computes azimuth, using this binaural time difference as a cue.

The time-coding pathway receives as input a collateral of the primary auditory fibers. The primary auditory fibers originate at the cochlea, the organ that converts the sound energy present at the eardrum into the first neural representation of the auditory system. Both mechanical and electrical processing occur in the cochlea. The sound energy at the eardrum is coupled into a traveling wave structure, the basilar membrane, which converts time-domain information into spatially encoded information by spreading out signals in space according to their time scale (or frequency). The velocity of propagation along the structure decreases exponentially with distance. The structure contains active elements; outer hair cells act as motile elements to reduce the damping of the passive basilar membrane, allowing weaker signals to be heard (Kim, 1984).

Inner hair cells contact the basilar membrane at discrete intervals. Each inner hair cell acts as a transducer, converting basilar membrane energy into a graded electrical signal. Spiral ganglion neurons connect to each inner hair cell, and produce action potentials in response to inner-hair-cell electrical activity. The primary auditory fibers are the axons of the spiral ganglion neurons, which travel to the brain stem auditory nuclei (Pickles, 1982).

The primary auditory fibers encode characteristics of a sound present at the eardrum. When pure tones are presented as stimuli, a nerve fiber is most sensitive to tones of a specific frequency. This characteristic frequency corresponds to the basilar membrane location of the inner hair cell associated with the nerve fiber. The spiral trunk of the auditory nerve preserves this ordering; the primary nerve fibers are cochleotopically and tonotopically mapped (Evans, 1984).

The mean firing rate of a primary auditory fiber encodes sound intensity. The temporal pattern of nerve firings reflects the shape of the filtered and rectified sound waveform; this phase locking does not diminish at high intensity levels (Evans, 1984). This temporal patterning preserves the timing information needed to encode interaural time differences.

In the owl, the primary auditory fibers project to the nucleus magnocellularis (NM), the first nucleus of the time-coding pathway (Figure 2). Neurons in the NM preserve the temporal patterning of nerve firings conveyed by the primary auditory fibers (Sullivan & Konishi, 1984). The NM projects bilaterally to the nucleus laminaris (NL), the first nucleus in the time-coding pathway that receives inputs from both ears. The mammalian homologue of the NL is the medial superior olive (MSO), which also receives inputs from both ears.

Neurons in both the NL (Moiseff & Konishi, 1983) and the MSO (Goldberg & Brown, 1969) are most sensitive to binaural sounds that have a specific interaural time delay (Figure 4). In 1948, Jeffress proposed

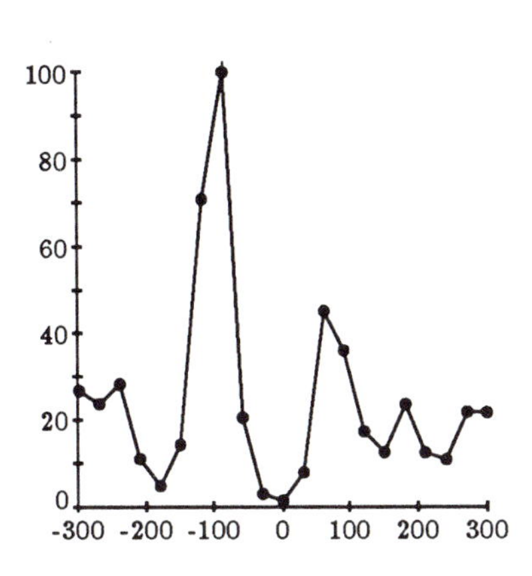

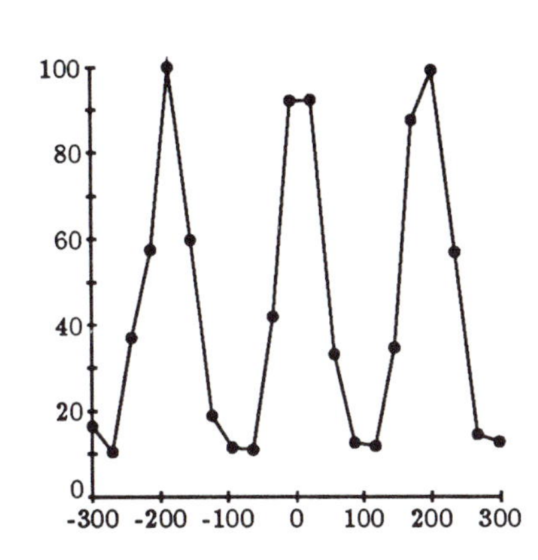

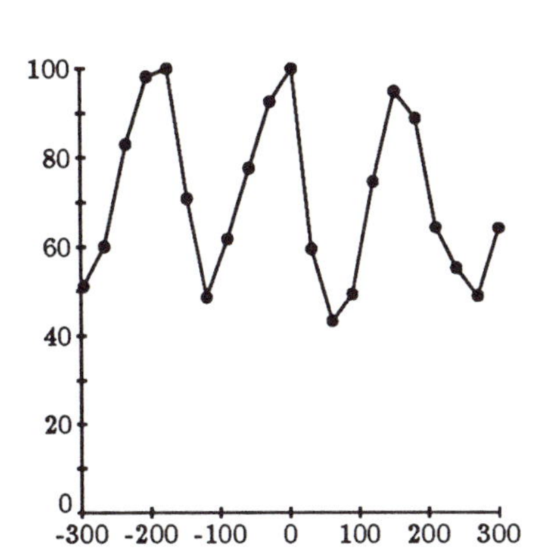

FIGURE 4 The delay curves obtained under physiological conditions from single units in the NL, the ICc, and the ICx. Broadband noise was used as a stimulus. Neurons with similar best frequencies (5.4–5.7 kHz) are shown. Neuronal selectivity markedly improves between the NL and the ICc. Responses are phase ambiguous in the NL and the ICc, but not in the ICx, when stimulated with noise. From Fujita and Konishi (1988, unpublished).

a model to explain the encoding of interaural time difference in neural circuits (Jeffress, 1948). In the model, as applied to the anatomy of the owl, axons from the ipsilateral (same side) NM and the contralateral (opposite side) NM, with similar characteristic frequencies, enter the NL from opposite surfaces (Figure 5). The axons travel antiparallel, and action potentials counterpropagate across the NL; the axons act as neural delay lines. NL neurons are adjacent to both axons. Each NL neuron receives synaptic connections from both axons, and fires maximally when action potentials present in both axons reach that particular neuron at the same time. In this way, interaural time differences map into a neural place coding; the interaural time difference that maximally excites an NL neuron depends on the position of the neuron in the NL (Konishi, 1986).

There is considerable evidence to support the application of this theory to the barn owl. Anatomical surveys of the NL of the owl show antiparallel axons, which enter from opposite sides, and which originate from the ipsilateral and contralateral NMs (Figure 6). Physiological data from NL neurons show a linear relationship between axonal time delay and axonal position in the NL (Carr & Konishi, 1988). Note in Figure 4, NL neurons respond maximally to several different interaural time delays. As predicted by the Jeffress model, NL neurons respond to all interaural time differences that result in the same interaural phase difference of the characteristic frequency of the

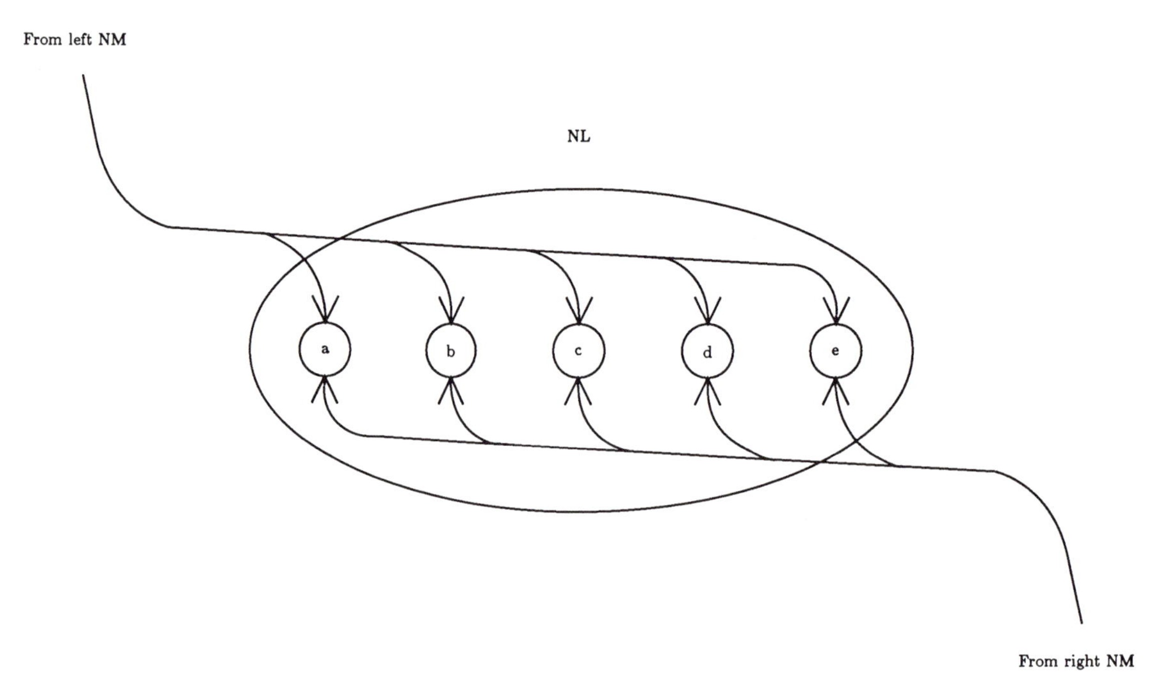

FIGURE 5 The Jeffress model for measuring and encoding interaural time differences, as applied to the barn owl. Fibers from the left and right NM converge on the NL; the NM fibers encode timing information in the temporal patterning of their action potentials. Each fiber has a uniform time delay. The NL neurons (circles labeled a–f) act as coincidence detectors, firing maximally when signals from two sources arrive simultaneously. The pattern of innervation of the NM fibers in the NL create left–right asymmetries in transmission delays; when the binaural disparities in acoustic signals exactly compensate for the asymmetry of a particular neuron, this neuron fires maximally. The position in the array thus encodes interaural time difference. Adapted from Konishi (1986).

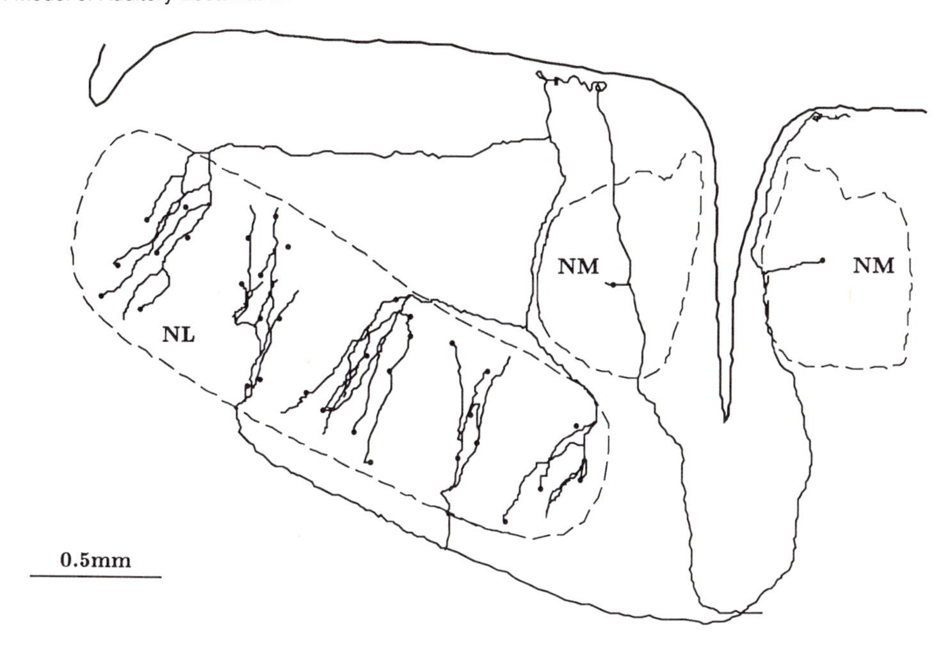

FIGURE 6 Innervation of the nucleus laminaris. Axons from the ipsi- and contralateral NM enter, respectively, the dorsal and ventral surfaces of the NL. Each axon runs along the respective surface until it comes to the isofrequency band which it is destined to innervate. Within this band, the axon sends two or three collaterals into the NL, each then dividing into several branches. These fibers, which may be as long as 1 mm, run along a relatively straight course toward the opposite surface. The fibers appear to establish contacts with the NL cell bodies either by *en passant* endings or by small branchlets. From Carr and Konishi (1988) and Konishi (1986).

input fibers. In mammals, neuronal recordings in the MSO are also consistent with the Jeffress model (Goldberg & Brown, 1969).

NL outputs project contralaterally to a subdivision of the central nucleus of the inferior colliculus (ICc). ICc neurons are more selective for interaural time differences than are NL neurons (Figure 4); application of bicuculline methiodide (BMI, a selective GABA antagonist) to the ICc reduces this improved selectivity. BMI acts as a blocker of inhibition; the action of BMI suggests that the increased selectivity of ICc neurons is a result of inhibitory circuits between neurons that are tuned to different interaural time differences (Fujita & Konishi, 1988, unpublished). Inhibitory circuits between neurons tuned to the same interaural time differences may also be present. In addition, evidence suggests that ICc neurons temporally integrate information to improve selectivity for interaural time differences. The sensitivity of ICc neurons to interaural time differences increases as a function of the duration of the sound stimulus for sounds lasting less than 5 msec (H. Wagner & M. Konishi, 1988, unpublished).

The ICc projects to the external nucleus of the inferior colliculus (ICx) after crossing sides in the ICc. The ICx integrates information from the time-coding pathway and from the amplitude-coding pathway to produce a complete map of auditory space. Neurons in the ICx, unlike those in earlier stations of the time-coding pathway, respond to input sounds over a wide range of frequencies. ICx neurons are more selective for interaural time differences than are ICc neurons for low frequency sounds. In addition, neurons in earlier stations of the time-coding pathway respond to all interaural time differences that result in the same interaural phase difference of the neuron's characteristic frequency; neurons in the ICx respond maximally to one interaural time difference (Figure 4). This behavior suggests that ICx neurons combine information over many frequency channels in the ICc, to disambiguate interaural time differences

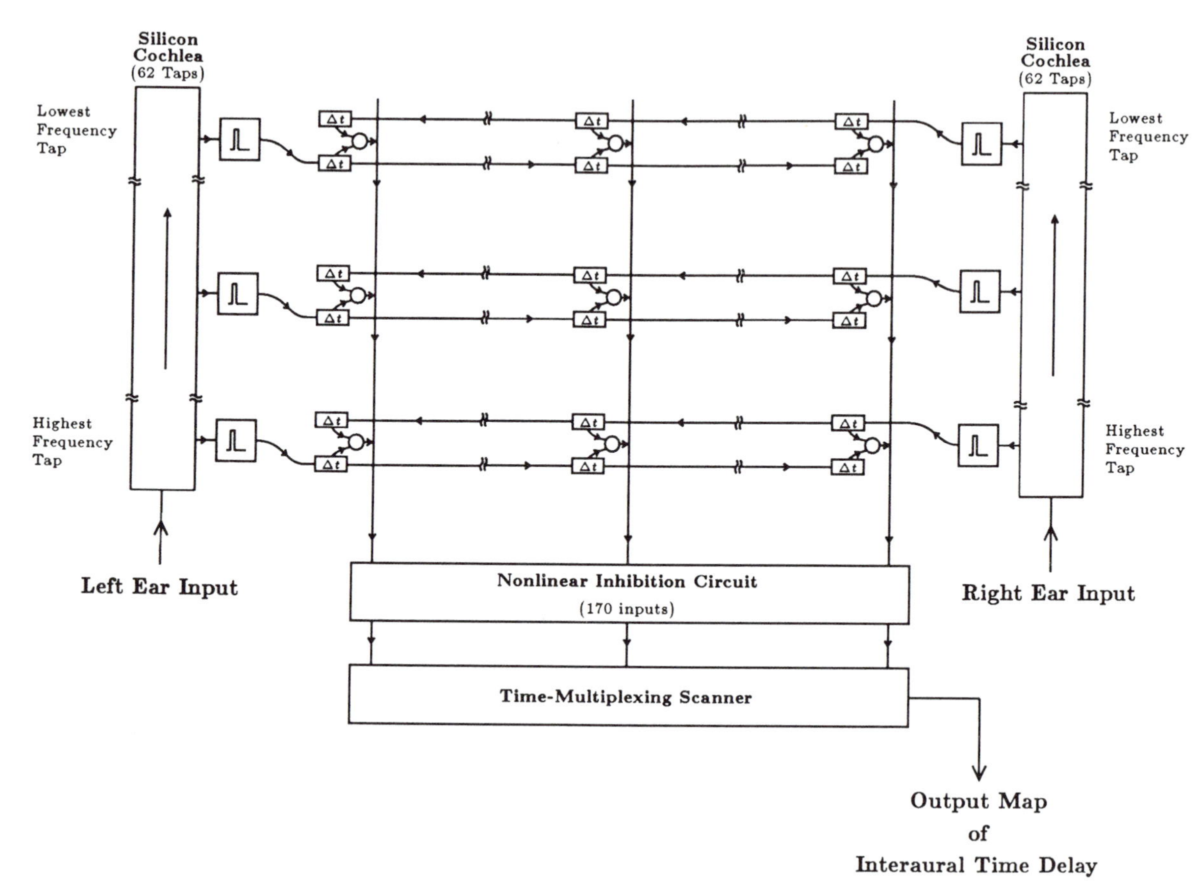

FIGURE 7 Floorplan of the silicon model of the time-coding pathway of the owl. Sounds for the left ear and right ear enter the respective silicon cochlea at the lower left and lower right of the figure. Silicon inner hair cells tap each silicon cochlea at 62 equally spaced locations; each silicon inner hair cell connects directly to a silicon spiral ganglion cell. The square box marked with a pulse represents both the silicon inner hair cell and silicon spiral ganglion. Figure 8 describes this box in more detail. Each silicon spiral ganglion cell generates action potentials; these signals travel down silicon axons, which propagate from left to right for silicon spiral ganglion cells from the left cochlea, and from right to left for silicon spiral ganglion cells from the right cochlea. The rows of small rectangular boxes, marked with the symbol Δt, represent the silicon axons. 170 silicon NL cells, represented by small circles, lie between each pair of antiparallel silicon axons. Each silicon NL cell connects directly to both axons, and responds maximally when action potentials present in both axons reach that particular neuron at the same time. In this way, interaural time differences map into a neural place code. Each vertical wire that spans the array combines the response of all silicon NL neurons which correspond to a specific interaural time difference. These 170 vertical wires form a temporally smoothed map of interaural time difference, which responds to a wide range of input sound frequencies. The nonlinear inhibition circuit near the bottom of the figure increases the selectivity of this map. The time-multiplexing scanner transforms this map into a signal suitable for display on an oscilloscope.

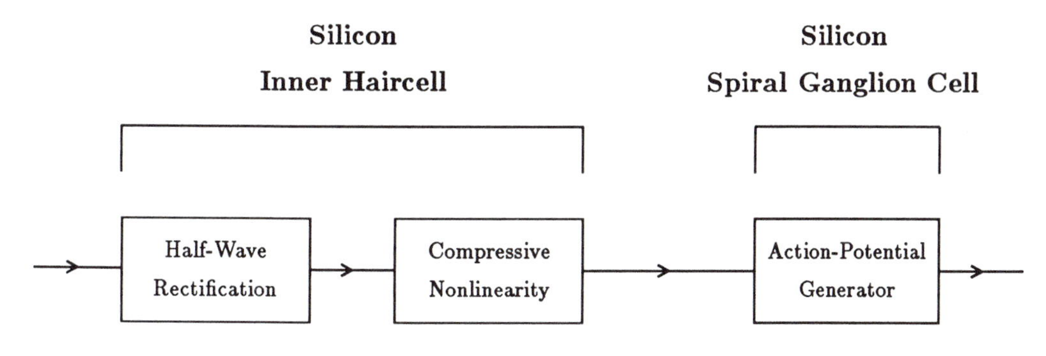

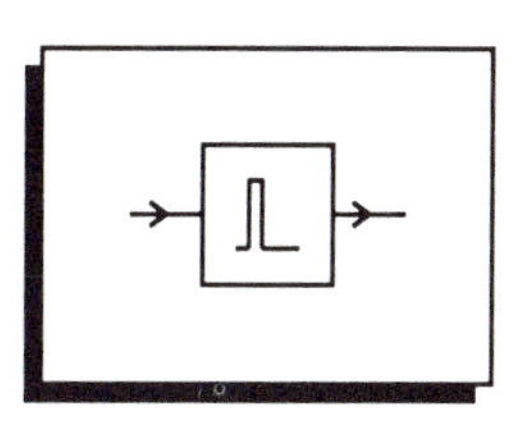

Figure 8 Functional block diagram of a silicon inner hair cell, connecting to a silicon spiral ganglion cell. Silicon inner hair cells half-wave rectify and nonlinearly compress the input signal. Silicon spiral ganglion cells convert the analog output of the silicon inner hair cell into fixed-width, fixed-height pulses. The timing of these pulses preserves the temporal characteristics of the input waveform, and the mean frequency of these pulses encodes the intensity of the input signal.

from interaural phase differences. Application of BMI to the ICx results in phase ambiguity and decreased selectivity, suggesting an inhibitory mechanism for this computation (Fujita & Konishi, 1988, unpublished).

A SILICON MODEL OF THE TIME-CODING PATHWAY

We have built a silicon model of the time-coding pathway of the owl. The integrated circuit models the structure as well as the function of the pathway; most subcircuits in the chip have an anatomical correlate. The chip computes all outputs in real time, using analog, continuous-time processing.

Figure 7 shows the architecture of the chip. The chip receives two inputs, corresponding to the sound pressure at each ear of the owl. Each input connects to a silicon model of the mechanical processing of the cochlea. The cochlea circuit is a one-dimensional physical model of the traveling wave structure formed by the basilar membrane. The damping of the circuit is controllable, modeling the effect of the outer hair cells. The velocity of propagation along the structure decreases exponentially with distance; exponential gradients are easily implemented in a silicon medium. In engineering terms, the model is a cascade of second-order sections, with exponentially scaled time constants (Lyon & Mead, 1988, 1989).

Silicon models of inner hair cells connect to the cochlea circuit at constant intervals. 62 inner-hair-cell circuits connect to each silicon cochlea. Inner hair cells act as transducers, converting mechanical energy into electrical signals. During this transduction, signal-processing operations occur (Figure 8). Inner hair cells half-wave rectify the mechanical signal, responding to motion in only one direction. Inner hair cells also compress the mechanical signal nonlinearly, reducing a large range of input sound intensities to a manageable excursion of signal level (Pickles, 1982). The silicon inner hair cell model computes these operations.

The output of each inner hair cell model connects directly to a spiral ganglion neuron model (Figure 8). The spiral ganglion neuron produces action potentials in response to inner-hair-cell electrical activity. The temporal pattern of action potentials reflects the shape of the analog input; the mean firing rate encodes the intensity of the input signal. The silicon spiral ganglion neuron performs this function, converting the analog output of the inner hair cell model into fixed-width, fixed-height pulses.

In the owl, the spiral ganglion neurons project to the NM. The NM acts as a specialized relay station; neurons in the NM preserve the temporal characteristics of the primary auditory fibers, and project bilaterally to the NL. For simplicity, our integrated circuit does not model the NM; each spiral ganglion neuron directly connects to the NL.

In both owls and mammals, two instances of each nucleus in the auditory pathway appear in the brain stem, symmetric about the midline. Obviously, for binaural processing, two cochleas and two NMs are required. The NL, however, receives bilateral projection from the NM; each NL primarily encodes one half of the azimuthal plane. This partitioning continues at higher stations; for simplicity, our integrated circuit has only one copy of the NL and of later stations, which encodes all azimuthal angles.

The chip models the anatomy of the NL directly. The two silicon cochleas lie at opposite ends of the chip; the sounds in each silicon cochlea travel in parallel up the chip. Spiral ganglion neuron circuits from each ear, which receive input from inner-hair-cell circuits at the same cochlear position, project to separate axon circuits, which travel antiparallel across the chip. When a sound is presented to both chip inputs, action potentials counterpropagate across the chip.

The axon circuit is a discrete neural delay line; for each action potential at the axon's input, a fixed-width, fixed-height pulse travels through the axon, section by section, at a controllable velocity (Mead, 1989). After excitation of the circuit with a single action potential, only one section of the axon is firing at any point in time. NL neuron circuits lie between each pair of antiparallel axons at every discrete section, and connect directly to both axons. Simultaneous action potentials at both inputs excite the NL neuron circuit; if only one input is active, the neuron generates no output. This simplified model differs from the owl in several ways. The silicon NL neurons are perfect coincidence detectors; in the owl, NL neurons also respond, with reduced intensity, to monaural input. In addition, only two silicon axons converge on each silicon NL neuron; in the owl, many axons from each side converge on an NL neuron.

For each pair of antiparallel axons, there is a row of 170 NL neuron circuits across the chip. These neurons form a place encoding interaural time difference. If a sound signal, offset in one ear by an interaural time difference, excites the silicon cochleas, both inputs to each pair of antiparallel axons receive a similar temporal pattern of action potentials, although one input's signal is offset by the interaural time delay. The maximally excited NL neuron in the owl is at the position at which the axonal time delays exactly cancel the interaural time delay. In engineering terms, each NL neuron computes the cross-correlation function of a filtered version of the sound inputs; the value of this function is the relative probability of the sounds being offset by a particular interaural time difference (Yin & Kuwada, 1984).

In the owl, the NL projects to the ICc, which in turn projects to the ICx. The ICx integrates information from the time-coding pathway and from the amplitude-coding pathway to produce a complete map of auditory space. The final output of our integrated circuit models the responses of ICx neurons to interaural time differences.

In response to interaural time differences, ICx neurons act differently from NL neurons. Experiments suggest mechanisms for these differences; our integrated circuit implements several of these mechanisms to produce a neural map of interaural time difference.

Neurons in the NL and ICc respond to all interaural time differences that result in the same interaural phase difference of the neuron's characteristic frequency;

neurons in the ICx respond to only the one true interaural time difference. This behavior suggests that ICx neurons combine information from many frequency channels in the ICx to distinguish interaural time differences from interaural phase differences; indeed, neurons in the NL and ICc reflect the frequency characteristics of spiral ganglion neurons, whereas ICx neurons respond equally to a wide range of frequencies.

In our chip, all NL neuron outputs corresponding to a particular interaural time delay are summed to produce a single output value. NL neuron outputs are current pulses; a single wire acts as a dendritic tree to perform the summation. In this way, a two-dimensional matrix of NL neurons reduces to a single vector; this vector is a map of interaural time difference, for all frequencies. In the owl, inhibitory circuits between neurons tuned to the same interaural time differences may also be present before summation across frequency channels. Our model does not include these circuits.

Neurons in the ICc are more selective to interaural time differences than are neurons in the NL; in turn, ICx neurons are more selective to interaural time differences for low frequency sounds than are ICc neurons. At least two separate mechanisms join to increase selectivity. The selectivity of ICc and ICx neurons increases with the duration of the sound stimulus, for sounds lasting less than 5 msec implying that the ICc and perhaps the ICx may use temporal integration to increase selectivity (H. Wagner & M. Konishi, 1988, unpublished). Our chip temporally integrates the vector that represents interaural time difference; the time constant of integration is adjustable.

Nonlinear inhibitory connections between neurons tuned to different interaural time differences in the ICc and ICx also increase sensitivity to interaural time differences; application of an inhibitory blocker to either the ICc or ICx decreases sensitivity to interaural time difference (Fujita & Konishi, 1988, unpublished). In our chip, a shunting inhibition circuit (Lazzaro, Ryckebusch, Mahowald, & Mead, 1988) processes the temporally integrated vector that represents interaural time difference. The circuit performs a winner-take-all function, producing a new map of interaural time difference in which only one neuron has significant energy. The circuit uses global inhibition across all time-disparity channels, rather than neighborhood inhibition, to produce a more selective map of interaural time difference. This map is the final output of the chip. The chip time-multiplexes this final map of interaural time difference on a single wire for display on an oscilloscope; the time-multiplexing circuit is not part of the owl model.

COMPARISON OF RESPONSES

We presented periodic click stimuli to the chip; one input received the sound directly while the other input received a time-delayed replica of the sound (Figure 9). We recorded chip response at several intermediate outputs and at the final output map of interaural time difference, and compared chip response with similar responses recorded from the time-coding pathway of

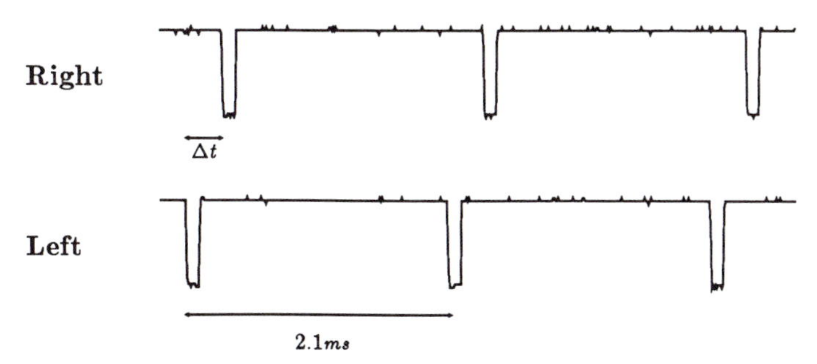

FIGURE 9 Input stimulus for the chip. Both left and right ears receive a periodic click waveform, at a frequency of 475 Hz. The time delay between the two signals, notated as Δt, is variable.

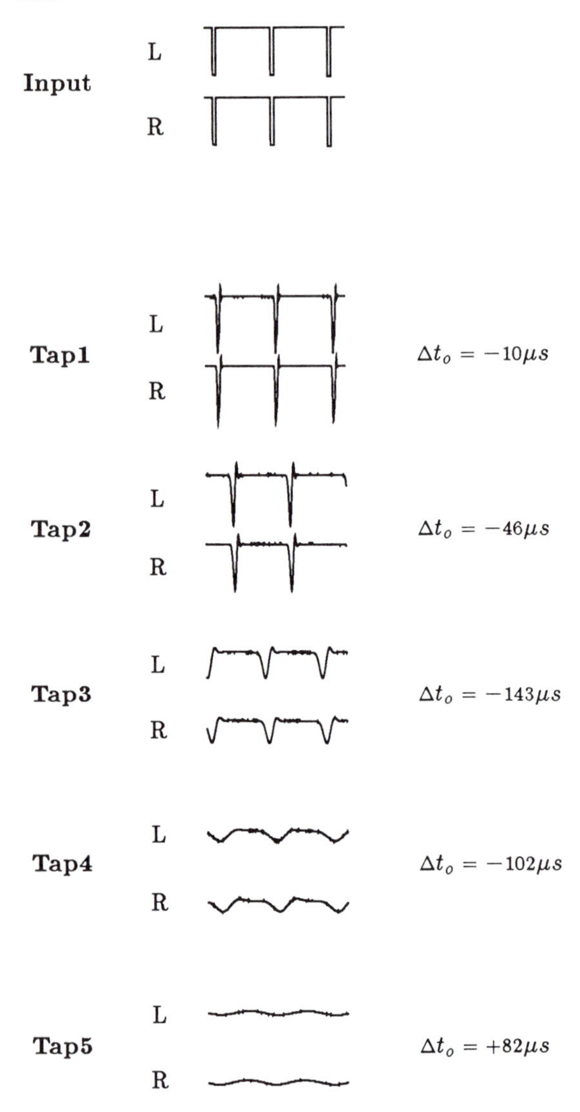

FIGURE 10 Response of both cochleas to the input stimulus of Figure 9; this stimulus is reproduced at the top of this figure. The horizontal axis is time; each vertical axis is cochlear response. The figure shows the output response at five cochlear positions, for both left and right cochleas. Tap 1 is closest to the beginning of the cochlea, and retains most of the high frequency content of the signal. Later taps show less and less high frequency content; tap 5 contains only the fundamental frequency of 475 Hz. Each tap accentuates a particular frequency, the characteristic frequency of the cochlear position; this frequency decreases with increasing tap number. The cochleas also act as exponentially tapered delay lines; notice that the response of each tap occurs progressively later from the input waveform. For tap 4 and tap 5, the delay of the cochlea exceeds a single period of the input waveform. The cochleas are not exactly matched; a difference in time delay between the two cochleas is noted as Δt_O, on the right of each pair of outputs. For lower-numbered taps, the delay is small compared with the time delay of the silicon axons; for higher-numbered taps, it is significant.

the barn owl. We also separately fabricated and tested several parts of the complete model.

Figure 10 shows the basilar membrane response of the chip to click stimuli, with no interaural time difference. The figure shows output responses at several equally spaced basilar membrane locations for both left and right cochleas. The frequency content of the response progressively decreases at locations farther from the beginning of the cochlea; in addition, the response at each location has a slight resonant overshoot. The response also shows a time delay between each pair of adjacent locations; for more distant locations, this time delay exceeds the period of the input waveform. These observations match the behavior of the traveling wave structure of physiological cochleas; a comparison of the responses of the physiological and silicon cochleas was made by Lyon and Mead (1988).

The left and right cochlea responses are nearly identical; there are small differences in the velocity of propagation as a result of circuit element imperfections. Because of these differences, the time delays in the left and right cochleas at a particular location are not equal; these differences, shown in Figure 10, are a potential source of localization error.

The temporal firing patterns of spiral ganglion neurons encode timing information, which is vital for the correct function of the time-coding pathway. To examine this property, the same sound is presented to the cochlea many times, and the response of the spiral ganglion neuron is recorded. This data is reduced to a poststimulus-time histogram, in which the height of each bin of the histogram indicates the number of spikes occurring within a particular time interval after the presentation of the sound.

Figure 11a shows a poststimulus-time histogram of a mammalian spiral ganglion neuron, in response to a

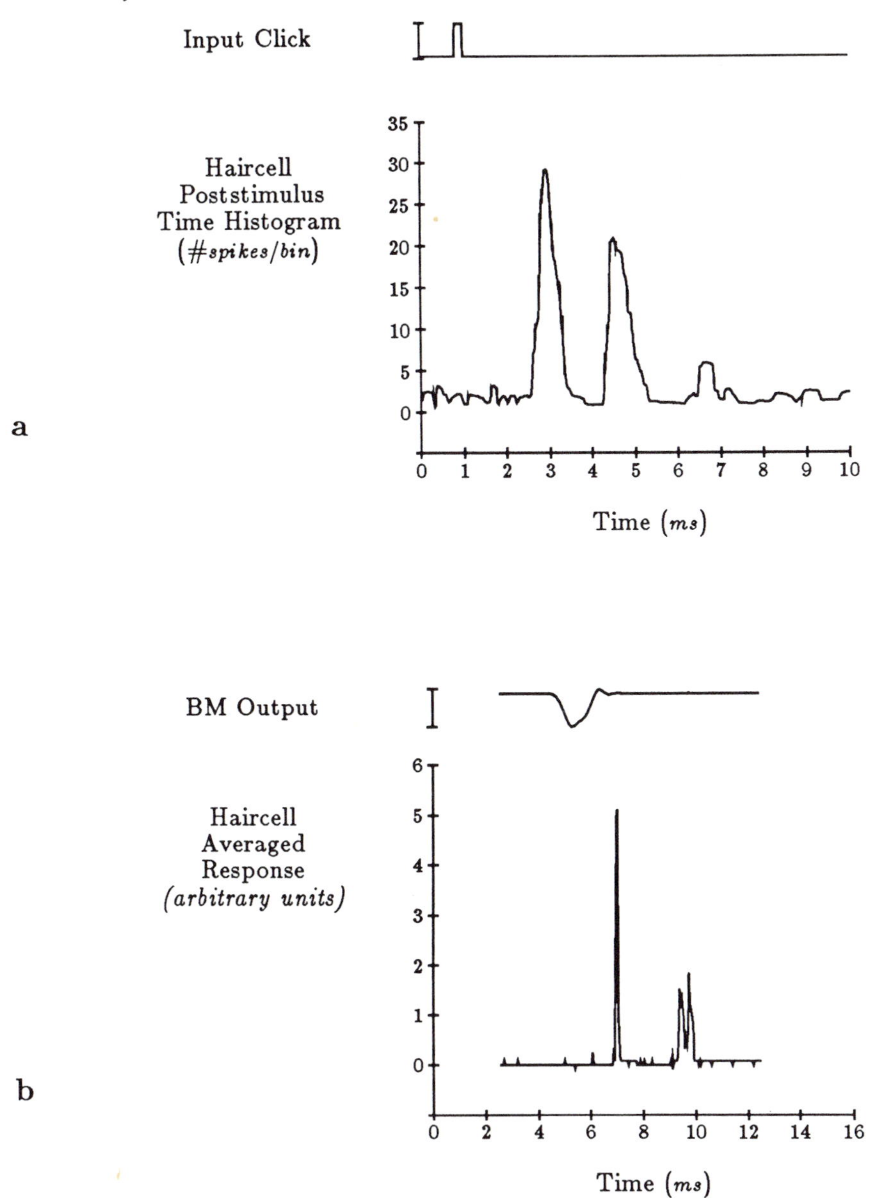

FIGURE 11 Comparison of spiral ganglion response of mammalian primary nerve fiber and silicon spiral ganglion cell. a. Poststimulus time histogram of many responses to a click stimulus, in a mammalian primary nerve fiber. Note that the nerve firings maintain the temporal qualities of the stimulus. The histogram shows the frequency shaping of the cochlea, and the half-wave rectification of the inner hair cell. From Evans (1984). b. Average of many responses to a click stimulus, in a silicon spiral ganglion cell. The top trace shows a cochlea tap which is slightly before the spiral ganglion position, showing the contribution of frequency shaping by the cochlea. The graph shows this frequency shaping, and the half-wave rectification of the silicon inner hair cell.

click. The histogram shows the resonant-overshoot response of the cochlea, and the half-wave-rectified response of the inner hair cell; the phase-locking property of the spiral ganglion neuron preserves these temporal details. Figure 11b shows the averaged response of a silicon spiral ganglion neuron to a click stimulus; the circuit similarly preserves temporal details.

Figure 12 shows the output of several sections of the axon circuit in response to a single input pulse. Only one section is active at any point in time; the pulse width does not grow or shrink systematically, but stays roughly constant. Figure 13 shows the variation in axon pulse width over about 100 sections of axon, due to circuit element imperfections. In this circuit, a variation in axon pulse width indicates a variation in the velocity of axonal propagation; this variation is a potential source of localization error.

The final output of the chip is a map of interaural time difference. Three signal-processing operations, computed in the ICx and ICc, improve the original encoding of interaural time differences in the NL: temporal integration, integration of information over many frequency channels, and inhibition among neurons tuned to different interaural time delays. In our

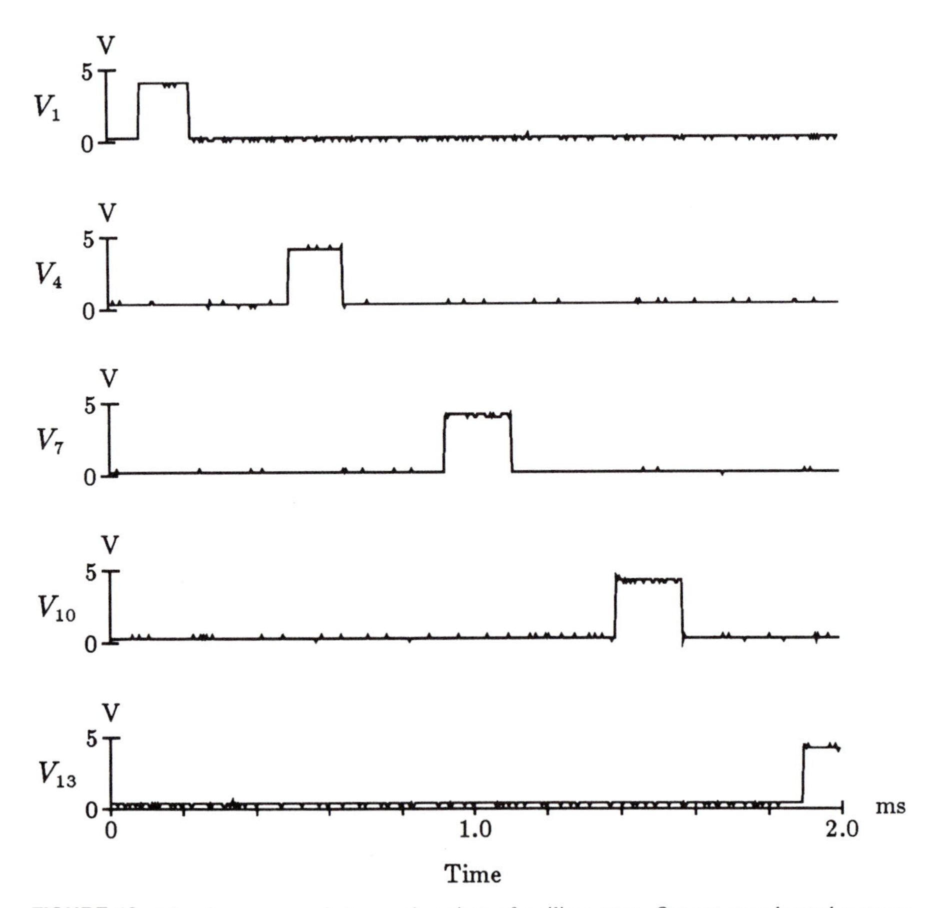

FIGURE 12 Waveform measured at several sections of a silicon axon. Outputs were located at every third section; thus, two outputs are not shown for every one that is. Note that the pulse height and pulse width stay fairly constant, and that only one axon section is active at any point in time. From Mead (1989).

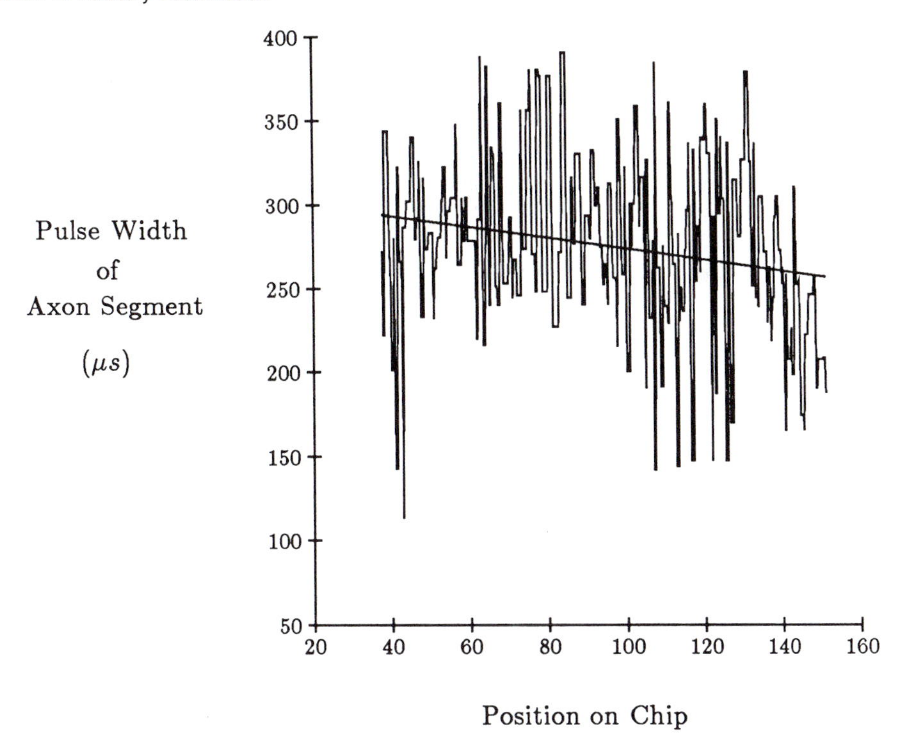

FIGURE 13 Variation in the pulse width of a silicon axon, over about 100 axonal sections. Axons were set to fire at a slower velocity than in the owl model, for more accurate measurement. The plot shows the actual section-by-section pulse width.

chip, we can disable the inhibition and temporal integration operations and observe the unprocessed map of interaural time difference.

Figure 14 (top) shows these unprocessed maps of interaural time difference, created in response to click stimuli. Each map corresponds to an interaural time difference that is 100 μsec greater than that for the previous map; the first map corresponds to simultaneous excitation of both ears. The vertical axis of each map corresponds to neural activity level, whereas the horizontal axis of each map corresponds to linear position within the map. Each map is an average of several maps recorded at 100 msec intervals; averaging is necessary to capture a representation of the quickly changing, temporally unsmoothed response. The encoding of interaural time difference is present in the maps, but false correlations add unwanted noise to the desired signal. By combining the outputs of 62 rows of NL neurons, each tuned to a separate frequency region, the maps in Figure 14 (top) correctly encode interaural time difference, despite variation in axonal velocity, mismatches in cochlear delay, and noise.

Figure 14 (bottom) shows maps of interaural time difference taken with inhibition and temporal integration operations enabled, using stimuli identical to those for Figure 14 (top). For most interaural time differences, only one peak exists. Off-chip averaging of maps over time is not necessary, as the temporal integration operation on the chip smooths the time response of the chip. The maps do not reflect the periodicity of the individual frequency components of the sound stimulus; experiments with a noise stimulus confirm the phase-clarification property of the chip.

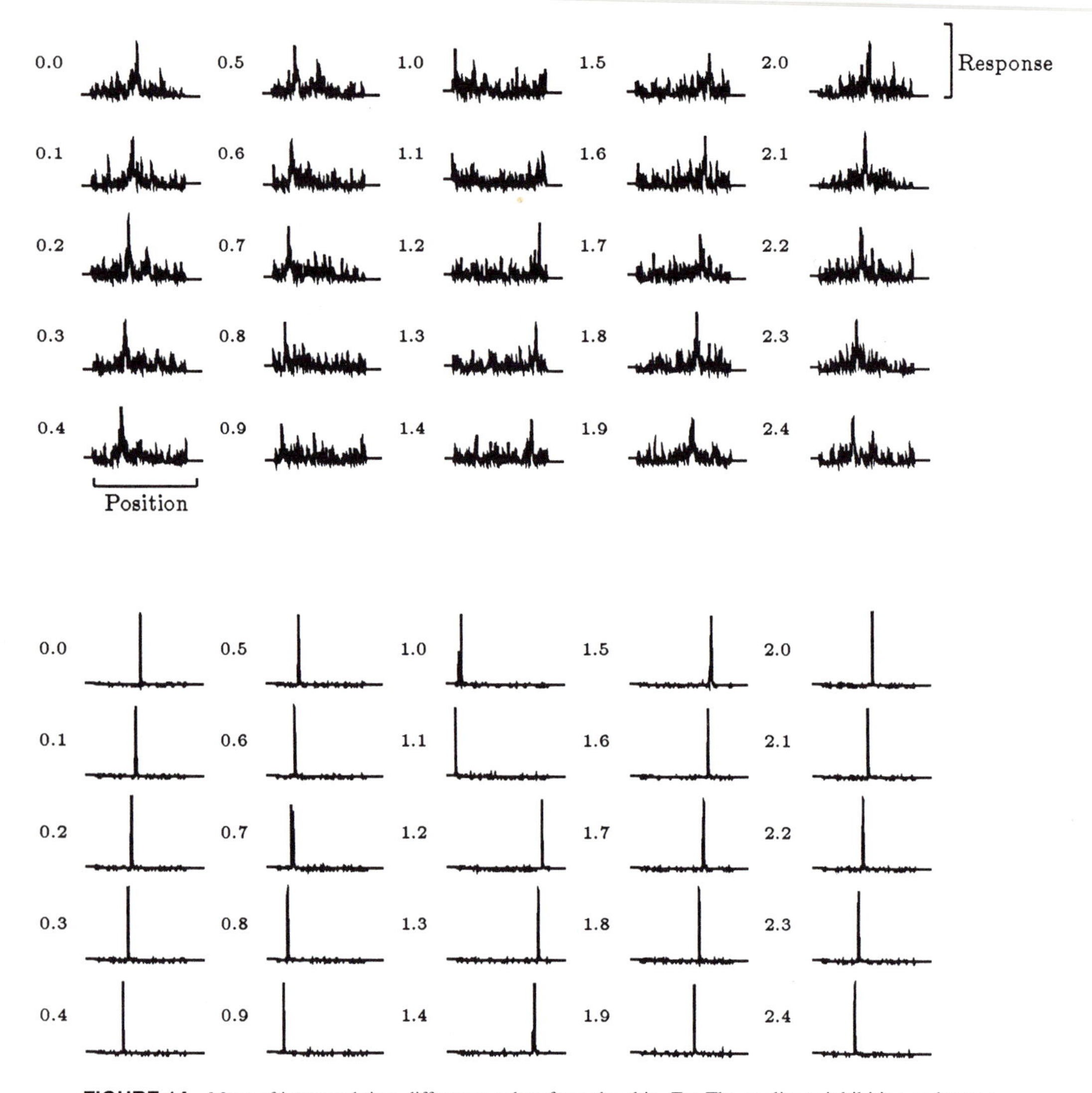

FIGURE 14 Maps of interaural time difference, taken from the chip. *Top* The nonlinear inhibition and temporal smoothing operations were turned off, showing the unprocessed map of interaural time delay. Each individual plot shows the neural response at every position across the map. The stimulus for each plot is the periodic click waveform of Figure 9, offset by an interaural time difference, shown in the upper left corner of each plot, measured in milliseconds. These responses are the averaged result of four maps, at 100 msec intervals, using a digital oscilloscope, since the on-chip temporal smoothing is disabled. The encoding of interaural time difference is present in the maps, but false correlations add unwanted noise to the desired signal. Since we are using a periodic stimulus, large time delays are interpreted as negative delays, and the map response wraps from one side to the other at an interaural time difference of 1.2 msec. *Bottom* The nonlinear inhibition and temporal smoothing operations were turned on, showing the final output map of interaural time delay. Format is identical to top. Off-chip averaging was not used, since the chip temporally smooths the data. Each map shows a single peak, with little activity at other positions, due to nonlinear inhibition.

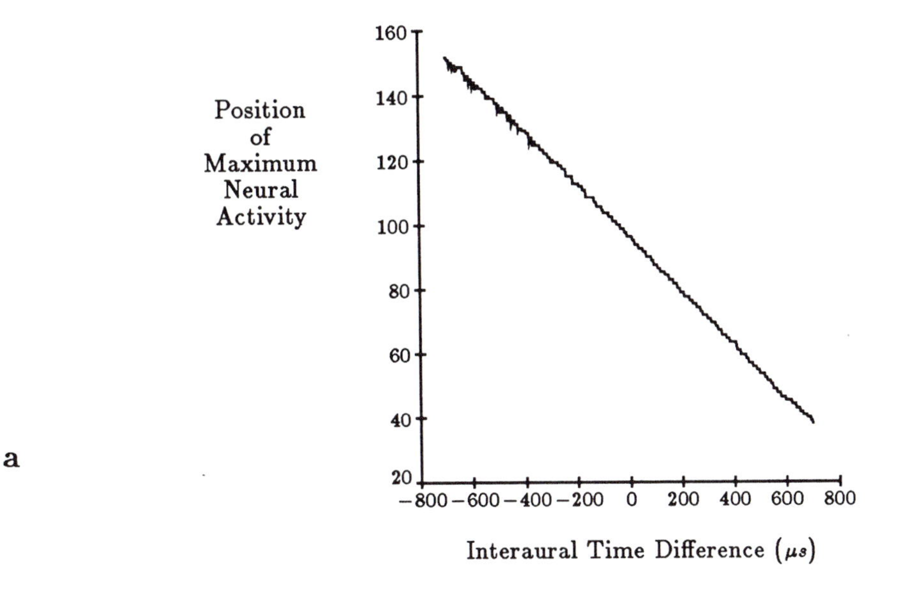

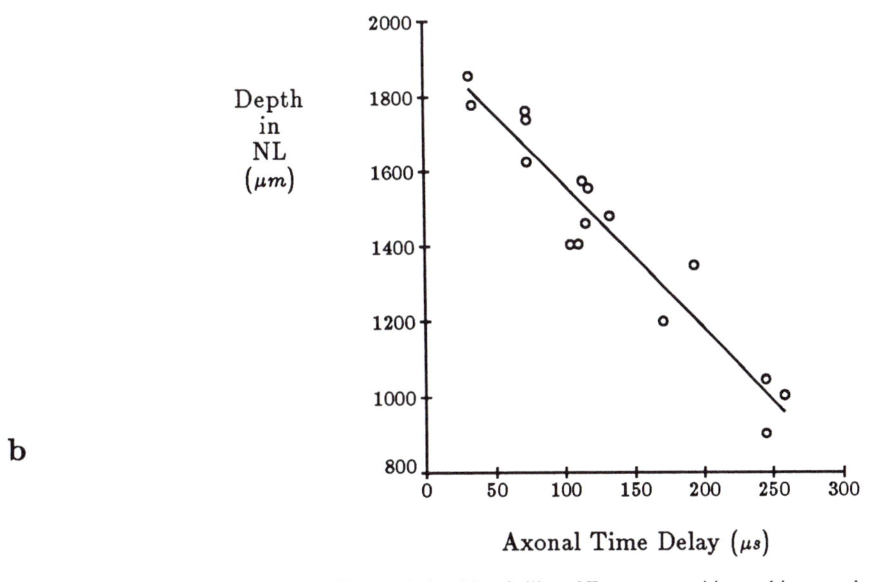

FIGURE 15 a. Chip data showing the linear relationship of silicon NL neuron position and interaural time difference. For each interaural time difference presented to the chip, the output map position with the maximal response is plotted. The linearity shows that silicon axons have a uniform mean time delay per section. b. Recordings of the NM axons innervating the NL in the barn owl. The figure shows the mean time delays of contralateral fibers recorded at different depths during one penetration through the 7 kHz region. From Carr and Konishi (1988).

Figure 15a is an alternative representation of the map of interaural time difference computed by the chip. We recorded the map position of the neuron with maximum signal energy for different interaural time differences using periodic click stimuli. Carr and Konishi (1988) performed a similar experiment in the NL of the barn owl (Figure 15b), mapping the time delay of an axon innervating the NL as a function of position in the NL. The linear properties of our chip map are the same as those of the owl map.

CONCLUSIONS

Traditionally, scientists have considered analog integrated circuits and neural systems to be two disjoint disciplines. The two media are different in detail, but the physics of computation in silicon technology and in neural technology are remarkably similar. Both media offer a rich palette of primitives in which to build a structure; both pack a large number of imperfect computational elements into a small space; both are ultimately limited not by the density of devices, but by the density of interconnects.

We have built a 220,000 transistor chip that models, to a first approximation, a small but significant part of a spectacular neural system. The architecture of the chip reflects the anatomy of the owl; the algorithms of the chip model the algorithms of the owl. These algorithms are relatively insensitive to imperfections in computational elements; the velocity of our silicon axons varies by a factor of two, without seriously degrading system output. This model is a first step in auditory modeling. Each individual circuit in the chip is only a first approximation of its physiological counterpart. In addition, there are other auditory pathways to explore: the intensity-coding localization pathway, the elevation localization pathway in mammals, and most formidably, the sound-understanding structures that receive input from these pathways.

We have built this chip on a foundation of 50 years of neuroscience; in return, we offer our chip and silicon technology as a tractable testbed for new theories in auditory neuroscience. We have also built this chip on a foundation of 50 years of electronics and computer science; in return, we offer interesting applications for the 50-million-transistor chips of the next decade. The territory is vast, and is largely unexplored. The rewards are great for those who simply press forward.

Acknowledgments

This chapter was originally published as "A Silicon Model of Auditory Localization" in *Neural Computation*, **1,1**. We wish to thank M. Konishi and his entire research group, in particular S. Volman, I. Fujita, and L. Proctor, for critically reading and correcting the manuscript, and for consultation throughout the project. In addition, we wish to thank D. Lyon, M. Mahowald, T. Delbruck, L. Dupré, J. Tanaka, and D. Gillespie for critically reading and correcting the manuscript, and for consultation throughout the project. D. Gillespie and D. Speck aided in writing software to generate stimuli. We wish to thank Hewlett-Packard for computing support, and DARPA and MOSIS for chip fabrication. This work was sponsored by the Office of Naval Research.

References

Carr, C. E., & Konishi, M. (1988). Axonal delay lines for time measurement in the owl's brainstem. *Proceedings of the National Academy of Sciences of the United States of America*, **85**, 8311–8315.

Evans, E. F. (1984). Functional anatomy of the auditory system. In H. B. Barlow & J. D. Mollon (Eds.), *The senses*. Cambridge, England: Cambridge Univ. Press.

Fujita, I., & Konishi, M. (1988). *The role of GABAergic inhibition in coding of interaural time difference in the owl's auditory system*. Unpublished manuscript.

Goldberg, J. M., & Brown, P. B. (1969). Response of binaural neurons to dog superior olivary complex to dichotic tonal stimuli: Some physiological mechanisms of sound localization. *Journal of Neurophysiology*, **32**, 613–636.

Jeffress, L. A. (1948). A place theory of sound localization. *Journal of Comparative and Physiological Pyschology*, **41**, 35–39.

Kim, D. O. (1984). Functional roles of the inner and outer-haircell subsystems in the cochlea and brainstem. In C. I. Berlin (Ed.), *Hearing Science*. San Diego: College-Hill Press.

Knudsen, E. I., Blasdel, G., G., & Konishi, M. (1979). Sound localization by the barn owl measured with the search coil technique. *Journal of Comparative Physiology*, **133**, 1–11.

Knudsen, E. I., Konishi, M., & Pettigrew, J. D. (1977). Receptive fields of auditory neurons in the owl. *Science*, **198**, 1278–1280.

Knudsen, E. I., & Konishi, M. (1978). A neural map of auditory space in the owl. *Science*, **200**, 795–797.

Knudsen, E. I., & Konishi, M. (1979). Mechanisms of sound localization in the barn owl (*Tyto alba*). *Journal of Comparative Physiology*, **133**, 13–21.

Knudsen, E. I., & Konishi, M. (1980). Monaural occlusion shifts receptive field locations of auditory midbrain units in the owl. *Journal of Neurophysiology*, **44**, 687–695.

Konishi M., (1986). Centrally synthesized maps of sensory space. *Trends in NeuroSciences*, **4**, 163–168.

Lazzaro, J. P., Ryckebusch, S., Mahowald, M. A., & Mead, C. A. (1988). Winner-take-all networks of O(n) complexity. *Proceedings of the IEEE Conference on Neural Information Processing Systems*, 703–711.

Lyon, R. F., & Mead, C. (1988). An analog electronic cochlea. *IEEE Transactions on Acoustics, Speech, and Signal Processing*, **36**, 1119–1134.

Lyon, R. F., & Mead, C. (1989). Electronic cochlea. In C. Mead (Ed.), *Analog VLSI and neural systems*. Reading, MA: Addison-Wesley.

Mead, C. (Ed.) (1989). *Analog VLSI and neural systems*. Reading, MA: Addison-Wesley.

Moiseff, A., & Konishi, M. (1981). Neuronal and behavioral sensitivity to binaural time differences in the owl. *Journal of Neuroscience*, **1**, 40–48.

Moiseff, A., & Konishi, M. (1983). Binaural characteristics of units in the owl's brainstem auditory pathways: Precursors of restricted spatial fields. *Journal of Neuroscience*, **3**, 2553–2562.

Pickles, J. O. (1982). *An introduction to the physiology of hearing*. London: Academic Press.

Sullivan, W. E., & Konishi, M. (1984). Segregation of stimulus phase and intensity coding in the cochlear nucleus of the barn owl. *Journal of Neuroscience*, **4**, 1787–1799.

Takahashi, T. T., & Konishi, M. (1988a). Projections of the cochlear nuclei and nucleus laminaris to the inferior colliculus of the barn owl. *Journal of Comparative Neurology*, **274**, 221–238.

Takahashi, T. T., & Konishi, M. (1988b). Projections of the nucleus angularis and nucleus laminaris to the lateral lemniscal nuclear complex of the barn owl. *Journal of Comparative Neurology*, **274**, 221–238.

Yin, T. C. T., & Kuwada, S. (1984). Neuronal mechanisms of binaural interaction. In G. M. Edelman, W. E. Gall, & W. M. Cowan (Eds.), *Dynamic aspects of neocortical function*. New York: Wiley.

11

Selective Recognition Automata

George N. Reeke, Jr., Leif H. Finkel, and Gerald M. Edelman

INTRODUCTION

Since the first tentative hypotheses by the pre-Socratic philosophers nearly 2500 years ago linking the brain with the origins of our perceptions, thoughts, and actions, the neural basis of behavior has remained one of the central questions of both science and philosophy. The last several decades have witnessed a remarkable flowering of study in both basic neuroscience and in the allied fields of psychology related to learning, memory, and cognitive activities. Yet despite important gains in all these areas, the link between the structure and function of nervous tissue and even the most rudimentary perceptual or cognitive capabilities remains elusive. Our failure to understand the neural basis of behavior is, in part, attributable to the complex, nonlinear, and parallel organization of the brain; however, the situation may also be due in part to the prevailing "bottom up" paradigm in the neurosciences, which focuses on the properties of single neurons.

We have recently developed a new approach to the linking of neuroscience and psychology, called *synthetic neural modeling* (SNM). SNM involves large-scale computer simulations of neural networks that are rigorously based upon known anatomical, physiological, and pharmacological properties of the nervous system. The goal of SNM is to develop a number of component models, each focusing on a particular aspect of the neural basis of behavior, and to use these models to generate testable experimental predictions. We describe here two of these component models—one dealing with the organizational properties of cortical maps and the other with the development of a selective recognition automaton, a synthetic "creature" that has a nervous system, a body-form, and a set of rather interesting behaviors.

Biologically Based Properties

Despite the advent of large-scale computers, neural network simulations must still make judicious simplifications if they are to involve large numbers of elements over thousands of iterations. Choices must be made regarding the degree of detail or realism to be maintained at the various levels of organization, from molecular to behavioral. It is currently not feasible, in general, to incorporate a high level of detail at more than two such levels (Sejnowski, Koch, & Churchland, 1988). For example, large numbers of neural connections in multiple simulated brain areas make any complexity at the synaptic level computationally prohibitive. In addition, effects at one organizational level may interact with those at another. For instance, we have found that the properties and utility of various synaptic modification rules depend upon the anatomical characteristics of the network in which they are embedded (Finkel & Edelman, 1985). It is for this reason that it is desirable to construct and study a number of separate component models.

In all such models, there are several important aspects of neural organization which, even if not emulated in detail, must be considered. We now enumerate these basic properties.

Anatomy We typically employ large numbers of connections between neural units (thousands to millions), since anatomy, and particularly variability in anatomy, provides the most basic constraint on neural function. We usually use orderly long-range connections with a degree of randomness in the positions of the fine terminations. The structure of axonal and dendritic trees, that is, the position of individual connections on a cell, plays a crucial role in many models (Rall, 1964; Koch, Poggio, & Torre, 1983; Fleshman, Segev, & Burke, 1988).

Physiology Several recent models have begun to incorporate a number of distinct cell types with characteristic ion currents (Traub, Knowles, Miles, & Wong, 1984; Finkel & Edelman, 1987; Wilson & Bower, 1988; Granger, Ambros-Ingerson, Henry, & Lynch, 1988). However, the diversity and richness of *in vivo* neuronal properties is beyond current simulation capabilities. A rudimentary set of realistic neuronal properties is essential and we have emulated voltage decays, firing thresholds, and various synaptic elements such as transmitters, receptors, ion channels, and second messengers.

Synaptic Plasticity One of the most critical elements of any neural model is the rule or rules chosen for modification of synaptic connection strengths. Since Hebb's (1949) influential book, most models have opted for a rule in which correlated activity between presynaptic and postsynaptic cells gives rise to an increase in synaptic strength (the original Hebb rule) and uncorrelated activity decreases connection strengths (one version of a modified Hebb rule). The Hebb rule is simple and displays a number of useful properties. However, it also has a number of weaknesses that have been less widely recognized. First, there is little experimental evidence at the synaptic level that correlated cell "firing" leads to synaptic modifications. In fact, several experiments directly demonstrate that correlated firing in itself does not lead to synaptic strengthening (Wigstrom & Gustafsson, 1983; Carew, Hawkins, Abrams, & Kandel, 1984). From the point of view of function, the Hebb rule cannot account for context-dependent changes, that is, changes dependent upon a particular pattern of synaptic inputs, because synaptic change is based on activity at a single synapse only.

For these reasons, we have developed several new synaptic rules that are based on recent biochemical and biophysical principles of synaptic function (Finkel & Edelman, 1985, 1987). We distinguish changes in presynaptic efficacy (the release of transmitter) and postsynaptic efficacy (the transduction of transmitter binding into a voltage change). The postsynaptic rule, in particular, accounts for so-called heterosynaptic interactions, whereby one input to a neuron can modulate the efficacy of other inputs to that cell. This heterosynaptic property, which depends upon both electrical and biochemical properties of the cell, endows the postsynaptic rule with context-sensitive properties.

Figure 1 shows the basic principles involved in the proposed postsynaptic rule. Four synapses are shown on a branching dendritic tree; each synapse contains both transmitter receptors (rectangles) and voltage-dependent ion channels (triangles). Initially, a transmitter, T1, binds to receptors at the lower right synapse, evoking a voltage change, but also initiating

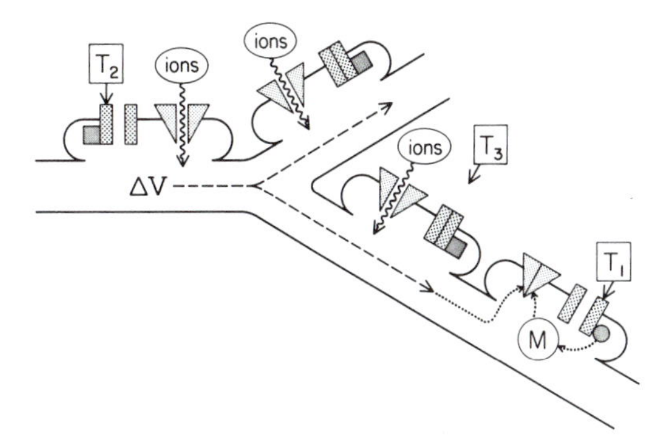

FIGURE 1 The postsynaptic modification rule. Schematic diagram of four synapses on a branching dendritic tree. Each synapse contains both receptors (rectangles) and voltage-dependent ion channels (triangles). Transmitter (T_1) binds to the lower right synapse, and produces a substance, M, which modifies voltage dependent channels (providing that the channels are in a modifiable or "depolarized" state). Inputs to other synapses (e.g., T_2) create a depolarization that electrotonically propagates through the cell. If the wave of depolarization reaches the bottom right synapse and increases the fraction of channels in the modifiable state when the concentration of M is still high, a larger change in postsynaptic efficacy will occur. This is the basis of heterosynaptic interactions in the synaptic rule. Reproduced with permission from Finkel & Edelman (1985).

a local biochemical cascade that gives rise to the production of M, a modifying substance meant to represent an intracellular second messenger such as Ca^{2+} or cAMP. M acts on voltage-dependent channels at the synapse and increases their conductance. The main assumption of the postsynaptic rule is that the efficacy with which M modifies these channels depends upon their functional state. In particular, we assume that transitions between channel states are voltage dependent, so that significant numbers of channels are modifiable only when depolarized. Therefore, if at a later time, another transmitter, T2, binds to receptors at the upper left synapse, the following sequence of events ensues. A depolarization due to the new input propagates electrotonically through the cell, changing the state of voltage-dependent channels as it reaches them. If the depolarization reaches the bottom right synapse and increases the fraction of channels in the modifiable state at a time when the local concentration of M is still high (from the recent arrival of a local input), then the number of channels, and thus the local postsynaptic efficacy, is increased to a greater extent. Thus, distant inputs can modulate the efficacy of local synaptic changes, a heterosynaptic effect.

We have used the postsynaptic rule with success in accounting for a number of features of map organization (Pearson, Finkel, & Edelman, 1987). However, this particular scheme is only one example of a large family of such rules that couple electrical and biochemical events within the neuron. In fact, the *N*-methyl-D-aspartate (NMDA) receptor, which has recently been implicated in the induction of long-term potentiation (LTP) in the hippocampus (Collingridge & Bliss, 1987), appears to combine voltage-dependent and transmitter-dependent mechanisms in a single molecule (MacDermott & Dale, 1987).

Neural Maps and Higher Order Structures One of the triumphs of modern neuroscience has been the discovery of supraneuronal units of functional organization. Foremost among these is the cortical column (Mountcastle, 1957; Hubel & Wiesel, 1962) and the menagerie of blobs, stripes, puffs, barrels, bands, and zones which have subsequently been found throughout many cortical and subcortical areas. These structures highlight the central role of mapping in the nervous system. While many models have used random or complete connectivity, the actual connections between cortical areas appear to be neither; they are orderly, largely topographic, and divisible into recognizable pathways (Van Essen, 1985). Neural maps can be viewed as the main operational device of the nervous system, allowing organized interactions to occur among functionally specialized networks. Numerous examples from the auditory (Knudsen & Konishi, 1978; Suga, 1984), visual (Zeki, 1983; Lee, Rohrer, & Sparks, 1988), somatosensory (Mountcastle, 1984), and associative areas (Schwartz, Zheng, & Goldman-Rakic, 1988) all emphasize the power and centrality of these structures.

Reentry The coordination of the operation of various maps, distributed throughout the nervous system, requires an integrative process which we have termed *reentry* (Edelman, 1978, 1981, 1987). Reentry is the dynamic, ongoing process of signaling between maps along ordered anatomical connections in a parallel and recursive fashion. A number of anatomical architectures can subserve reentry; reciprocal, asymmetric, or indirect mappings are possible, and the particular functional properties of reentry depend upon the properties of the maps being connected. However, in general, reentry can mediate associative or correlative functions as well as resolving conflicts in the responses of different areas.

Output and Behavior The final consideration in any model must be its output. Hopes for the emergence of some useful property are too often submerged in the hidden complexities of large parameter spaces, and ample attention must be paid at the outset to "what the model does." Many models are subject to the criticism of hidden "homunculi"—unspecified observers who must interpret the activity patterns of the network to yield a meaningful response. The selective recognition automaton described in the following sections counters this problem by having an easily observed motor output which acts upon a simulated environment. The behavioral feedback loop—by which a creature's actions influence its subsequent sensory inputs—adds a new dimension to the stability and operational criteria of the model and provides the model with the basis for adaptive responses.

With these preliminaries in hand, we turn now to a discussion of the central principles guiding our own choice of models.

PRINCIPLES OF NEURONAL GROUP SELECTION

Selection as a Paradigm

The models we have constructed are all based upon the theory of neuronal group selection (TNGS), which proposes that the nervous system operates by a form of selection, akin to natural selection in evolution but taking place within the brain and during the lifetime of each animal. Selection is a powerful biological principle, with central applications in both evolution and development. It is at the heart of the best understood biological recognition system, the immune system (Edelman, 1973). Selective systems are natural categorization machines—for example, the elaborate taxonomy of plant and animal species was created by natural selection—and categorization is, in our view, the central operation of the nervous system.

Categorization in Adaptive Behavior

Categorization, the ability to create classes and to segregate objects into them, is a prior function to perception, learning, and memory. Every animal must develop a set of categorization schemes which allow it to deal in an adaptive fashion with its environment. Traditional approaches to categorization, many of which subtly underlie current neural models, assume that a set of singly necessary and jointly sufficient conditions can define a category. This view is derived from Platonic essentialism, wherein the attributes of an object reflect those of a perfect "exemplar" object. However, the real world is a good deal more complicated. Trying to write down the descriptive attributes of any common class of objects, such as chairs, animals, or games (Wittgenstein, 1953), is a futile and frustrating experience. For as Bongard (1970) has pointed out, the descriptive borders between classes are not akin to hyperplanes in some large-dimensional attribute space, but are more like the boundaries between water and solid in a sponge—totally intermixed and inseparable.

There have been a few recent attempts at developing richer definitions of categorization (Rosch, 1978; Smith and Medin, 1981). One particular construct which appears promising is the notion of polymorphous sets (Ryle, 1949; Wittgenstein, 1953). A polymorphous set is defined as those objects which possess any m out of n given attributes, when $m < n$. Thus for example, for $m = 2$ and $n = 4$, the set of attributes could be "round," "edible," "inorganic," and "green." Then a round, edible object (e.g., a cantaloupe) would be in the class, as would an inorganic, green object (e.g., a green car), even though the two objects might share no properties in common.

The approach to neural function starting with categorization is an attempt to circumvent problems associated with the more conventional information processing approach. The problem with information processing in a biological context is to explain how the animal "knows" what the information is unless it is defined beforehand. This *a priori* property of information (epitomized by the agreed codes of Shannon theory) is replaced by *a posteriori* selection of adaptive responses in a selective system. Furthermore, although information processing systems can show combinatorial richness, they are fundamentally incapable of creative behavior. For animals, *de novo* synthesis of new responses is critical to survival.

Development and Evolution

One of the chief advantages of a selective approach is the natural link it provides to the development and evolution of the nervous system. One of the major characteristics of the fine structure of the brain is its tremendous degree of variability, both between individuals and between similar units within the same animal (Allman, 1988). For example, cortical area V1, the primary visual cortex in monkeys, varies in size by 50% from animal to animal (Van Essen, 1985). Such variability is even more apparent at the neuronal and synaptic level. Even identified neurons in invertebrates show dramatic differences between individuals, including individuals that are genetically identical (reproduced by parthenogenesis; Macagno, Lopresti, & Levinthal, 1973; Pearson & Goodman, 1979).

These considerations suggest that the nervous system is not built as a rigidly designed, point-to-point wired system. Rather, the selective point of view maintains that anatomical variability is unavoidable from a developmental point of view (Edelman, 1987) and, moreover, is essential to the functioning of the system—for variability provides the substrate upon which selection acts. As we will discuss in the following sections, the nervous system probably continues to generate functional variability throughout the lifetime of the organism.

Neuronal Groups—The Units of Selection

According to the TNGS, the basic operational units of the nervous system are not single neurons but local groups of strongly interconnected cells. These groups represent an intermediate stage between neurons and the supraneuronal anatomical units mentioned above. They are functional units, not anatomical units, and the membership of neurons in a group can be changed by alterations in synaptic connection strengths. Groups are introduced into the theory because it is difficult to see how individual neurons could show the necessary cooperative properties. Local competition and cooperation among cells are clearly necessary to generate local order in the cortex (e.g., maps in which neighboring cells have similar receptive fields), and the notion of groups only requires separating these interactions into localized aggregates. As we will show in the following sections, such a process occurs readily in simulated neural networks.

Although groups have not yet been experimentally documented, there are two sources of experimental evidence that support their existence. The experiments on somatosensory cortex map organization discussed here (Kaas, Merzenich, & Killackey, 1983) are difficult to explain without invoking neuronal groups. These same investigators have described an experiment in which the median nerve in a monkey was sectioned and allowed to regenerate, and tangential electrode penetrations were made in the somatosensory cortex after recovery. It was found that cells recorded in local small regions of cortex tended to have very similar receptive fields, but that adjacent small regions varied widely in their receptive fields. This result supports the notion of local groups containing cells with similar receptive field properties.

Another source of evidence for groups comes from the work of Gray and Singer (1989), Gray et al. (1989), and Eckhorn et al. (1988), who have found that local groups of cells in the striate cortex tend to oscillate in phase at a common frequency, usually between 20 and 50 Hz. Adjacent groups of cells (usually corresponding to orientation columns) can be distinguished by their temporal oscillation patterns. Simulated neuronal groups operating according to principles of the TNGS indeed produce similar oscillations under appropriate conditions of external stimulation (Sporns et al., 1989).

The Role of Degeneracy

Neuronal groups exemplify another main principle of neuronal group selection, the concept of degeneracy, according to which nonisomorphic structures can perform similar functions. Thus, each neuronal group has a manifold of possible functions (exactly which of these are expressed depends upon the current state of its intrinsic and extrinsic connections); conversely, any particular function can be carried out, more or less well, by a large collection of different groups. Degeneracy endows the nervous system with a great deal of combinatorial richness and allows selection to guide the evolution of the network through the adaptive landscape.

Degeneracy also exists at the level of the cortical area. For example, the many (up to 20 in some species) areas of the visual cortex are each specialized for particular selectivities to motion, color, shape, or depth. However, there is a great deal of overlap in these functional specializations and a number of areas have relatively good orientation selectivity, directional selectivity, etc. (Felleman & Van Essen, 1987). Functional degeneracy may allow the various cortical areas to communicate with each other more efficiently, as well as to serve similar functions under a wide range of conditions.

Higher Order Functions

The TNGS has recently been extended to account for a number of higher order functions of the nervous system including motor control, memory, and language (Edelman, 1987). It is beyond our purview here to discuss these areas of the theory. However, the point should be made that one of the strengths of synthetic neural modeling is that it is based upon a larger theoretical construct. We believe that it is only by the study of a number of detailed, computer based simulations of component models of the nervous system combined with the elaboration of a general theory of brain function that we can ever hope to understand the basic operations of the nervous system.

We now consider as a first example one of these component models having to do with the organization of maps in the somatosensory cortex of the monkey.

ORGANIZATION OF TOPOGRAPHIC MAPS IN SOMATOSENSORY CORTEX

Physiological Data

The existence of cortical maps has been known since the early part of this century and, although early mapping experiments demonstrated the lability of cortical representations (Leyton & Sherrington, 1917), the general concept nevertheless arose that the maps were static structures, totally dependent upon the underlying anatomy. Recent evidence from a number of species has shown, however, that many cortical maps continually reorganize in an ongoing dynamic process that continues throughout adult life. Perhaps the most elegant demonstration of this phenomenon comes from the somatosensory cortex of the owl and squirrel monkey (Kaas et al., 1983; Mountcastle, 1984). Cortical areas 3b and 1 of the monkey both contain a complete somatotopic representation of the body surface; neurons in these areas respond to light touch to the skin. Multiple, closely spaced microelectrode recordings reveal that maps vary greatly among individuals and, furthermore, that they vary over time within the same individual (Merzenich et al., 1983). More dramatic changes are seen after perturbation of the sensory stimulation to the skin. For example, if a cutaneous nerve is transected, if the fingers are amputated, or if a local region is continuously stimulated, the locations of map borders (i.e., the borders between representations of different body parts) can move hundreds of microns, and entirely new representations can emerge (Jenkins, Merzenich, & Ochs, 1984).

We have proposed that the dynamic, competitive process which gives rise to map reorganization is the process of neuronal group selection (Edelman & Finkel, 1984). More recently, we have developed a detailed, biologically realistic computer simulation of a small patch of somatosensory cortex, which successfully reproduces many of the experimental findings through the use of selective mechanisms acting on neuronal groups (Pearson et al., 1987).

Construction of a Selective Model

Figure 2 shows a schematic of the network used in the simulations. The network comprises 1536 cells of two cell types, excitatory cells (triangles) and inhibitory cells (circles), each with characteristic connection patterns. Each cell is connected to neighboring cells in the network (excitatory–excitatory, excitatory–inhibitory, inhibitory–excitatory, but no inhibitory–inhibitory connections), with the probability of connection between two cells decreasing with distance according to a Gaussian distribution. Excitatory cells have a smaller mean distance of connection than inhibitory cells. Altogether, there are approximately 70,000 intrinsic connections in the network.

Excitatory cells also receive inputs from an external receptor array, meant to correspond to the hand of the monkey. These inputs are topographically mapped onto the network (i.e., adjacent regions of the receptor sheet map onto adjacent regions of the network). The receptor sheet is divided into four fingers and a palmar region. Connections from the back of the hand (the right half of the receptor sheet) are also mapped topographically onto the network but, as shown in Figure 2, this mapping is mirror symmetrical to the first, so that the front and back of a digit both map to approximately the same region of the network. Each cell receives equal numbers of connections from the glabrous and dorsal aspects of the receptor sheet; altogether, there are approximately 100,000 extrinsic connections.

The anatomical connections are generated at the start of the simulation and, once generated, are never changed. However, the strengths of the connections, both intrinsic and extrinsic, are modifiable according to the postsynaptic rule discussed earlier. The simulated cells are described by several parameters (see Pearson et al., 1987, for details) including a voltage decay rate, a sigmoidal output function, and a type of shunting inhibition which was found to confer a remarkable degree of dynamical stability upon the network over a wide range of stimulation intensities. Separate subroutines allowed the receptor sheet to be stimulated with a wide range of stimuli, and the activity of every cell to be monitored throughout the simulation.

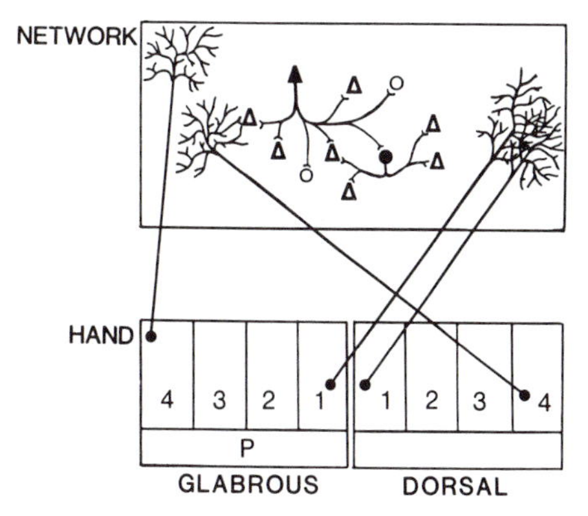

FIGURE 2 Schematic diagram of network used for somatosensory map simulations. The network contains two types of cells: excitatory (triangles) and inhibitory (circles). Each cell contacts neighboring cells with a probability depending on the distance between them (a few connections of the two darkened cells are shown). Also shown is a receptor sheet representing the hand, which contains glabrous and dorsal surfaces of four fingers and a palmar region. 1024 pressure sensitive receptors project to the network in a topographic fashion, each arborizing over roughly 10% of the network. Receptors from the corresponding glabrous and dorsal positions project to the same network location, so that each cell receives equal numbers of connections from the front and back of the hand. Reproduced with permission from Pearson, Finkel, & Edelman (1987).

The Formation of Neuronal Groups

Figure 3 (top) shows the connection strengths of the intrinsic connections in the network at the start of the simulation. A straight line has been drawn between all excitatory cells that are anatomically connected; the shade of gray indicates the strength of each connec-

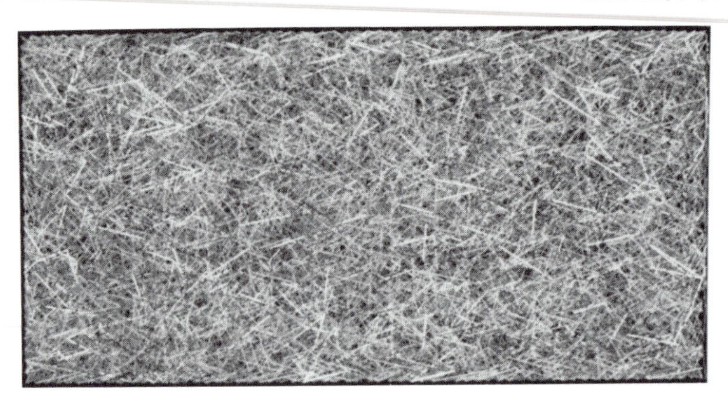

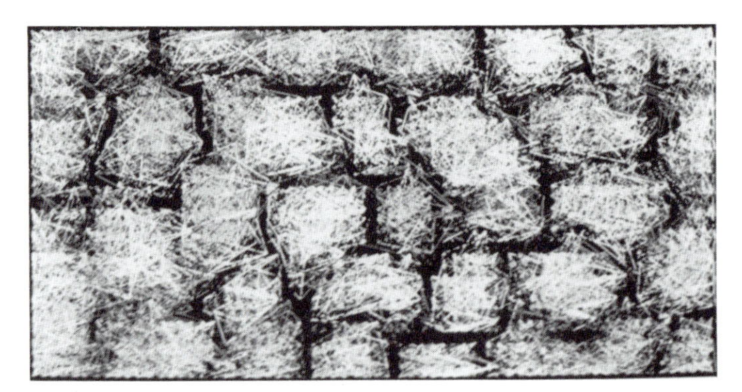

FIGURE 3 Neuronal group formation. (*Top*) Initial network before group formation. Straight lines have been drawn between anatomically connected cells in a portion of the network containing 144 cells, and the lines have been shaded according to the strengths of the connections (darker, stronger; lighter, weaker). Approximate positions of cells are indicated by small circles. Note the absence of any global organization in the connection strengths. (*Bottom*) After group formation. Intrinsic connection strengths in the network after four passes of stimulation (see text). Neuronal groups have formed which are local collections of cells with strengthened mutual connections and weakened connections to outside cells. Reproduced with permission from Pearson, Finkel, & Edelman (1987).

tion. As can be discerned, there is no global organization to the distribution of connection strengths. Instead, one sees the effect of the local Gaussian distribution used to assign initial connection strengths. Starting with this network configuration, the receptor sheet was stimulated at random positions with a small square stimulus until all locations had been stimulated. Four passes of this random stimulation were made. Each pass consisted of six cycles of activation followed by four cycles of rest and this sequence was repeated three times at each location before moving the stimulus to the next location. Figure 3 (bottom) shows the strengths of the intrinsic connections after this stimulation. As can be seen, the neurons organized into small neuronal groups, typically including 10–15 cells (*in vivo* we would predict groups to contain hundreds to thousands of cells). Cells within a group have strengthened their mutual connections and weakened their connections to cells outside the group.

This phenomenon of group formation is robust, and occurs over a wide range of stimuli, anatomical architectures, and neuronal parameters. The critical features appear to be (1) local inhibition must affect a larger area than local excitation, (2) stimulation must be relatively even over the receptor surface, without favoring local areas, and (3) the synaptic modification rule must be voltage-dependent, and it is particularly efficient if higher voltages are required to weaken connections than are required to strengthen them.

Once neuronal groups have formed, they demonstrate a number of interesting functional properties, preeminent among which is a large degree of cooperativity among the cells of a group and a tendency toward competition between cells in different groups. Groups compete for cells belonging to other groups (some evidence of this can be seen in the intermediate strength connections between some of the groups in Figure 3), and more importantly, they compete for the location and size of their receptive fields. Each group tries to expand its receptive field at the expense of neighboring groups, and the ensuing competition is responsible for many basic aspects of map organization.

Map Organization and Plasticity

Group competition is best observed by constructing receptive field maps of the network. Figure 4A shows the receptive field map corresponding to the network in Figure 3 (top) at the start of the simulations. The fingers and palm of the hand are distinguished by shading as shown in the small diagram on the right. The topographic nature of the underlying anatomy manifests itself in this initial map, in which the four fingers and palm are topographically represented. Note that, at this stage, each cell responds to stimulation of regions on both the front and the back of the hand.

Figure 4B shows the receptive field map after the receptor sheet has been stimulated to produce the groups shown in Figure 3 (bottom). The map retains its global topographic organization; however, the locations of the borders between the fingers and within the palm have moved due to competition between groups. Also, segregated regions of representation of the front of the hand (black dots) and the back of the hand (white dots) have arisen. The map in Figure 4B bears a good qualitative resemblance to the receptive field maps found in monkey somatosensory cortex, with segregated regions of dorsal and glabrous representation. The borders between these regions correspond to borders between neuronal groups, and the sharpness of these borders is, in turn, due to the sharpness of those group borders. Group competition allows the borders to remain sharp while their locations move over significant distances. Also note that some regions of the map in Figure 4B contain neither black nor white dots—cells in these regions retain mixed receptive fields and respond to stimulation of both sides of the hand. Most of these cells are located at the edges of the network and reflect a boundary condition artifact having to do with the density of connections at the network border.

Figure 4C shows the results of a perturbation experiment in which continuous, focal stimulation was applied to the front of digit 3 (indicated by the X in the diagram). As can be seen, the representation of the stimulated surface has greatly increased in the map (by nearly 14-fold), occupying areas formerly devoted to adjacent fingers, palm, and the back of the hand. If a more even distribution of stimulation is once again applied for some time to the receptor sheet, the map again reorganizes to a configuration resembling that in Figure 4B.

The final simulation, shown in Figure 4D, emulates median nerve transection. The connections from the right half of the front of the receptor sheet have been interrupted (black area in the diagram of the hand) and normal stimulation has been continued to the entire receptor sheet. As can be seen, there is no longer any representation of the denervated surface (i.e., there are no black dots in the right half of the map). In place of the former representation, however, is a new topographic representation of the back of the hand (white dots). In addition, several large silent areas (black regions) have appeared; cells in these regions are unresponsive to all stimulation. However, we have found that, with continued stimulation, these silent areas gradually disappear, leaving an intact organized map.

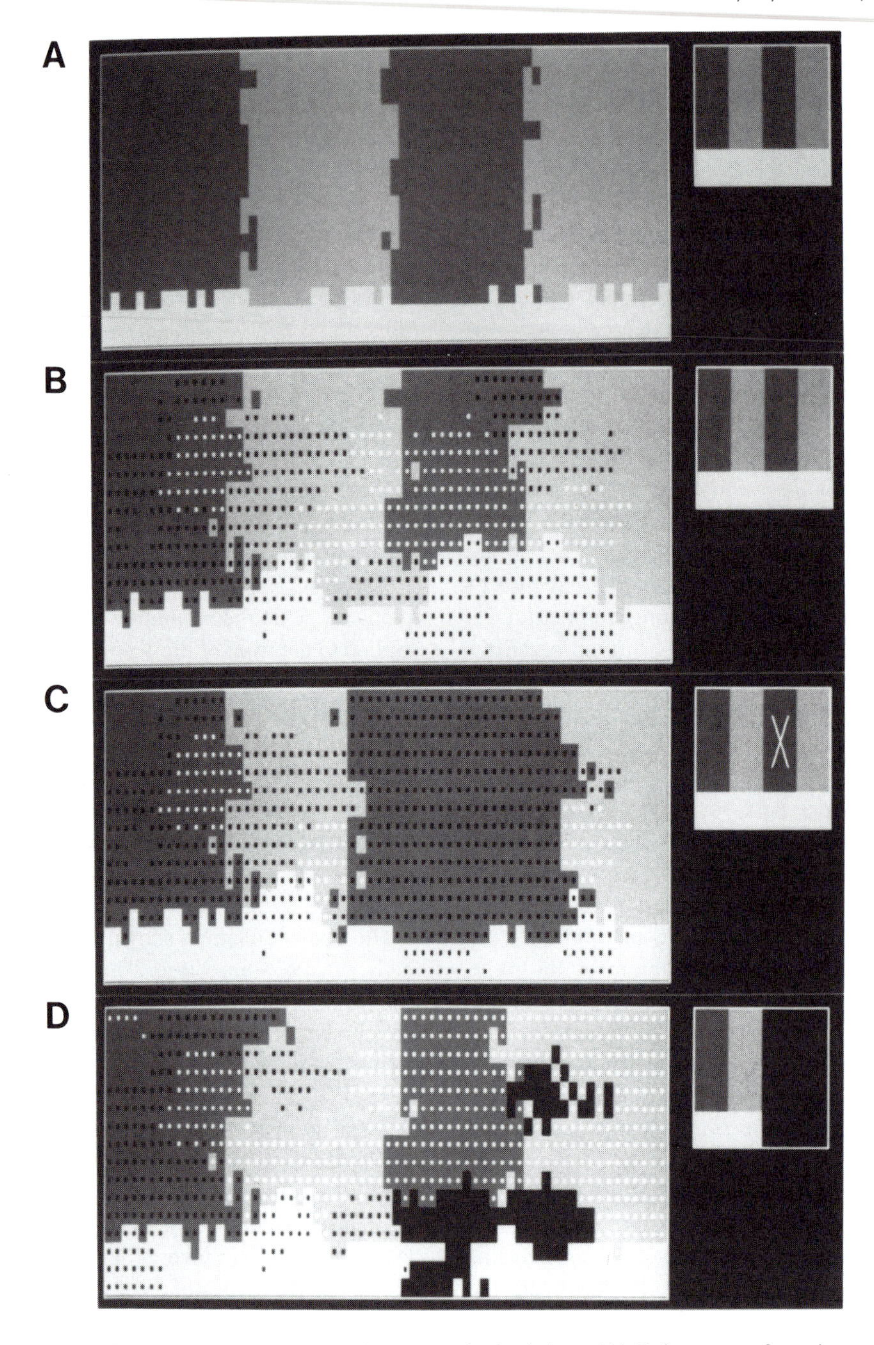

FIGURE 4 Receptive field maps from network simulations. (A) Before group formation. Receptive fields were determined for each excitatory cell in the network. The area corresponding to the position of the cell in the network is shaded according to the key given in the diagram of the

Thus, these computer simulations, with their extreme simplifications of the anatomical and physiological properties of somatosensory cortex, nevertheless capture many of the basic elements of map organization and reorganization. The underlying anatomy is capable of supporting a manifold of physiological maps. Which particular map is present depends upon the individual variabilities in the anatomy and connection strengths of the particular network, as well as upon the history of stimulation received by that individual. Selective and competitive processes constantly act to reorganize the map into one which maximizes the amount of cortical area devoted to more highly stimulated areas. This allows maximum use of cortex for those areas which, from an adaptive point of view, are most heavily used.

We have used this computer model to generate a number of experimental predictions (Pearson et al., 1987), several of which have been tentatively confirmed (Merzenich, Recanzone, Jenkins, Allard, & Nudo, 1988). We have also incorporated the principles learned from these simulations in our more abstract models of recognition automata, which we now address.

HOW A SELECTIVE NERVOUS SYSTEM CAN LEARN

Types of Learning in Selective Systems

In the example just considered, relatively long term changes in neuronal responses occurred as a result of changes in stimulus patterns reaching the somatosensory cortex, suggesting that the system has somehow "learned." However, the application of the term "learning" to this situation seems to us inappropriate. It is instructive to examine the reasons for this reluctance. There is, in fact, no general agreement among psychologists as to exactly what phenomena are included in the term "learning." Many authors in the artificial intelligence field follow Newell, Shaw, and Simon (1958) in taking learning to be "any more or less lasting change in the response of the system to successive presentations of the same stimulus," a view which omits essential intentional aspects. Most authors who deal with "neural net" models make a distinction between "supervised" and "unsupervised" learning—the former, but not the latter, generally involving an omniscient teacher capable either of "clamping" the outputs of the nervous system to a set of "correct" values for a given input, or of supplying exact corrections to be applied to the outputs when an error occurs. Neither of these models is possible in biological organisms, which have neither the means for behavior to be imposed from outside nor the exact reciprocal nervous connectivities needed to propagate error corrections backward from output units to the interior.

The "supervised"/"unsupervised" distinction is thus not of much use in biology, where all examples are necessarily "unsupervised." In working with selective models, we find it more useful to distinguish three cases according to the source of the signals which may contribute to the modification of particular synaptic weights: (1) simple tuning (local signals), (2) tuning biased by value (heterosynaptic signals originating from a global evaluation of behavior), and (3) true learning (signals resulting from reentrant compar-

hand (*right*). All cells responded to both glabrous and dorsal stimulation. The initial map shows good topography. (B) Receptive field map after group formation. Black dots indicate glabrous receptive fields, white dots indicate dorsal receptive fields. (Undotted cells retain mixed receptive fields.) Map is now largely segregated into glabrous and dorsal regions. Locations of map borders have changed, but topography is maintained. (C) Map obtained after repeated focal stimulation. Glabrous surface of digit 2 was repeatedly stimulated. (D) Map obtained after "median nerve" section. All inputs from glabrous surface of digits 1 and 2 were transected, and the entire receptor sheet stimulated in a normal fashion (see text). Map shows no representation of denervated surfaces; instead, a new representation of the dorsal surfaces of digits 1 and 2. Large black regions are silent areas, unresponsive to stimulation. With continued stimulation, these silent areas become responsive, yielding an intact topographic representation. Reproduced with permission from Pearson, Finkel, & Edelman (1987).

ison of present and past states). We reserve the term "learning" to the third case, which alone allows the possibility of actions to be arbitrarily dissociated from their immediate ecological consequences, and thus to acquire symbolic meaning. We use the word "training" to describe the process of encounter with the environment which is required for any of the three modes to occur.

Simple Tuning The first type of learning is *simple tuning*. Simple tuning is defined as a form of selection in which synaptic change depends only on levels of pre- and postsynaptic activity at the synapse in question. Its main role is to enhance the selectivity of responses that occur spontaneously in a crude form. Simple tuning includes Hebbian learning, but encompasses other kinds of synaptic changes that depend only on local activity, for example, homeostatic mechanisms in which connection strength *decreases* rather than *increases* with increasing activity. Simple tuning is only capable of enhancing (or repressing) responses that have a local basis; it cannot be coupled to overall goals of the organism. Nonetheless, its power as a mechanism for adaptation should not be underestimated; we have shown with a model called Darwin II that a system with purely local synaptic change rules is capable, after training, of recognizing familiar objects, categorizing them, generalizing on known categories, and even associating different objects that it has placed in a common category (Edelman & Reeke, 1982; Reeke & Edelman, 1984).

Tuning Biased by Value The second form of learning in selective systems is *tuning biased by value*. To the purely local synaptic change rules involved in simple tuning is added a *heterosynaptic* term, by which the magnitude of the changes in strength of any given synapse may be modulated by the presence or absence of correlated activity in other nearby synapses (either through voltage-dependent mechanisms or through the activity of second messengers). Heterosynaptic effects provide a mechanism for the modulation of tuning by nonlocal factors. In order to achieve tuning biased by value, the heterosynaptic inputs are connected to repertoires, the activity of which reflects a simple, internal estimate of the value to the organism of any behaviors it may generate. Tuning is no longer confined to the blind amplification of all responses, but now reflects the biased selection of those responses that are judged, by the organism itself, as being of adaptive value to it. (The synaptic changes in the somatosensory model are considered to be examples of simple tuning. Although heterosynaptic effects are present, the key element of bias in synaptic change according to the overall value of the generated behavior, needed for tuning biased by value, is not present.)

The mechanism by which adaptive value is estimated is called a "value scheme." Value schemes are closely related to "drives" in psychology in that they reflect innate, evolutionarily determined behavioral biases. Value schemes play somewhat the role of the teacher in the "supervised" learning schemes of other "neural net" models, but differ in two critical respects. First, value is internal to the organism, not imposed from outside, and second, value provides only a general bias to select behavior that is found *ex post facto* to be adaptive to the organism—it does not "know" correct answers as does the "teacher." The concept of value will be expanded with examples in the next section when we discuss particular value schemes used in connection with the Darwin III selective automaton.

True Learning For all its ability to amplify behaviors that have selective advantage for an organism, tuning biased by value is inadequate to generate the richness of behavior seen in higher animals because it does not permit actions to be divorced from their immediate consequences and analyzed in terms of their place in a sequence of behaviors. For this, a third form of learning, which we shall call *true learning*, is required. True learning comprises a whole series of conscious and unconscious abilities but at its simplest is distinguished by the ability to evaluate behavioral sequences differently depending on their temporal and spatial context.

At the mechanistic level, this ability appears to require that signals mediating recognition and behavior must at some level be filtered through dynamic couplings of networks, couplings which themselves are dependent on responses obtained during previous couplings. The most obvious way for this to be accomplished is for the efficacy of anatomically reentrant connections between repertoires to be modulated on a rapid time scale by heterosynaptic signals resulting from earlier responses occurring in the same or different loci. Provided the establishment of these dynamically "gated" pathways is itself subject to selection under a more primitive mechanism, such as tuning biased by value, the gated pathways would provide a medium for the exchange of the kinds of signals needed for the context-sensitive control of behavior.

Regardless of the detailed mechanism, which remains uncertain, sequence-dependent learning is clearly a necessary condition for language and other higher cognitive functions involving the manipulation of symbol strings. At a more basic level, this form of learning permits behavior to be modified by the external manipulation of value, as in classical and operant modes of conditioning. Experiments aimed at demonstrating these modes of behavior in selective automata are underway, but will not be described here.

Implementation of Learning in Selective Automata

In this section, we present some of the factors which must be considered when implementing tuning or learning protocols in selective automata. The principle ground rules which flow from the TNGS are the following: (1) Prior specification of procedures for generating correct responses by computation is not allowed; all evaluation of behavior in selective systems is *ex post facto*. (2) The modes must be consistent with the known facts of neurophysiology. In particular, bidirectional connections may not be used to propagate error information backward through the network. (3) Modification of individual synapses must be based only on signals available at the sites of those synapses. (Heterosynaptic contributions from nearby synapses are allowed.) (4) External means may not be used to impose correct responses on the output units of the system. (5) Behavior must be evaluated internally by parts of the neural network system that are sufficiently simple to have been evolutionarily specified.

Some methods of proceeding consistent with these rules are described in the following section. Detailed mathematical descriptions of the synaptic rules are contained in the papers by Reeke et al. (1989) and by Finkel and Edelman (1985).

The Role of Noise The basic rule that behavior in selective systems can only be selected *ex post facto* implies that there must initially be some behavior to select *from*. In an animal with some experience, this behavior might well result from activity in previously selected neuronal groups, responding more or less well to a new stimulus with features partially overlapping those of some earlier stimulus. However, in a naive animal, this mechanism is unavailable. It is also not possible to rely on activity resulting from connections between sensory and motor areas that initially happen to be strong; if only a small fraction of such connections were initially strong, the likelihood of getting useful responses by chance would be very small; if, on the other hand, that fraction were large, the animal would be subject to uncontrollable seizures.

The solution to this apparent dilemma is to make use of the spontaneous activity, or "noise," that is present in all neural circuits, and which is problematic to most nonselective theories of nervous system function. Supposing the incoming connections to each motor area to be generally weak at the outset, noise provides the activity needed to generate initial (aimless) behavior which serves as a substrate for selection. Once a set of behaviors has been selected, and connections into the groups which initiate them have been strengthened, the effects of the noise will diminish in a natural and automatic fashion as groups excited through their specific inputs compete with those excited less strongly by noise. In situations in which no selected response is available, noise will still provide spontaneous activity upon which further selection can act.

The Role of Inhibition Selection does not always have to operate upon excitatory circuits. In complex motor control systems such as the arm-reaching system described in a later section, there can be very large numbers of possible motions that are counterproductive in a given situation, but which would take a very long time to select against individually because of their low *a priori* probability of occurrence when excited by noise. In such situations, it may be more efficient for selection to act upon inhibitory circuits such as those between the "cerebellum" and motor pathways in the model. Supposing these circuits to be set up between individual "Purkinje" cells and large regions of motor pathways, a "Purkinje" cell firing in response to a particular combination of sensory inputs (conveyed to it by its "parallel fiber" inputs) will tend to inhibit most motor activity in that region. However, given a simple tuning synaptic rule that weakens inhibition in response to correlated pre- and postsynaptic activity, the cerebellar inhibition will be most weakened where it impinges upon motor circuits that are most active, thus permitting selected behaviors to get through while the larger numbers of rare motor patterns remain inhibited in those situations in which they are unselected and thus inappropriate; these same motor patterns may still be selected under other circumstances in which the particular "Purkinje" cells that inhibit them are not active.

The Role of Temporal Integration In all learning situations (other than simple tuning) there is an inevitable time delay between the production of a behavioral act and the evaluation of that act by a teacher (in "neural net" models) or a value scheme (in selective models). During this time delay, the neuronal activity responsible for initiating the act may well decay, leaving the problem of assigning changes to the particular synapses that were involved in generating the act. In most "neural net" models using "supervised learning," this problem is avoided because the teacher, in addition to being omniscient, is infinitely fast, providing its error correction information between two cycles of the neural net, before activity has decayed. This approach to dealing with the problem is acceptable as long as the "correct" behaviors are all known to the teacher in advance, but cannot be adequate in the general case in which behaviors must be evaluated by their consequences. In selective systems, this latter case always applies, and the time delay problem must be faced from the outset.

There are two interesting scales of time delay, depending on whether additional behavior is generated between the completion of an act and the response of a value repertoire. The first case represents a difficult and as yet unsolved problem for behavioral modeling because, as soon as it is possible for more than one action to occur before a value response is obtained, the problem arises of determining which of the synapses active in generating the various actions should be subject to biasing by the value signal. The only general solution to this problem that seems workable is for entire *sequences of actions* to be subject to selection as a whole, the sequences being built up hierarchically from shorter component sequences that are themselves subject to selection in simpler tasks prior to being included in any longer sequences. This process does not require any special neural mechanisms; it is only necessary to adjust thresholds such that only the relatively stronger responses of already selected lower-order groups are sufficient to excite cells in target repertoires that will become involved in the higher-order sequences.

In the simpler case in which no behavior intervenes between an action and the corresponding value response, it suffices to assume the existence of a "modifying substance," a chemical, the concentration of which effectively represents a temporally integrated record of recent activity at each synapse. The concentration of "modifying substance" may then be used in place of current activity in the application of whatever synaptic change rule is being used. In this way, selective amplification can occur at synapses that are not currently active, but which were active in the recent past, particularly at the time the most recent behavior was being shaped in the nervous system. Finkel and Edelman (1985) have discussed possible physiological candidates for modifying substances and have presented a set of equations describing synaptic changes dependent on them.

DARWIN III

In a previous section we presented a model of somatosensory cortex that enables us to test whether a network of neural groups operating with rules for cellular response and synaptic change based on the TNGS could reproduce the experimental behavior of an isolated brain region. We now proceed to consider a broader and more abstract model, Darwin III, in which we apply similar principles to a set of perceptual and motor control problems that are not so readily accessible to experimental study. In the course of describing Darwin III, we indicate the reasons for our assumptions and show how the general theory of selective learning given earlier works in a number of particular examples. We then conclude this chapter by comparing the assumption and results of our models with those of other "neural net" models in an effort to identify key features that we believe are important to using any of them to obtain a better understanding of brain function.

The World, the Phenotype, and the Nervous System

In accord with the principles of synthetic neural modeling, Darwin III incorporates the three domains of the environment, the phenotype of a particular creature, and the nervous system of that creature, all in a single computer simulation. The environment is a two-dimensional world of objects which move about in accord with protocols devised by the experimenter. The forces that create, destroy, and move these objects are generally inaccessible to the automaton, but the objects obey simple physical laws of impulse and momentum when the automaton acts upon them. The phenotype of Darwin III is that of a simple, sessile creature with an eye and a single multijointed arm (more arms can be added at will) with which it can explore and act upon the environment. It has senses of vision, touch, and kinesthesia (the ability to sense the motions of its own joints). It is able to act upon the environment in only two ways, by moving its eye or by moving its arm. Its nervous system can be constructed in any number of ways; current versions include some 50,000 cells of about 50 different kinds connected by about 620,000 synaptic junctions. Like an experimental animal, the behavior of Darwin III can be evaluated by an observer; unlike any animal, its complete internal state can also be monitored at will.

The immediate goal of our studies with Darwin III is to demonstrate that automata can be constructed according to the principles of the TNGS that have, in a crude form, some of the behavioral capabilities of animals, without the use of *a priori* definitions of categories, codes, or information-processing algorithms. Accordingly, the nervous system of Darwin III is simulated at the level of single cells. These cells may be allowed to form groups as a result of their competitive interactions (Finkel & Edelman, 1985) or, for convenience, the group structure may be specified *a priori*. The connections, initial connection strengths, and rules for cell responses and synaptic modifications are specified by the experimenter for each kind of cell, but no prior information about particular stimuli and no algorithms for determining behavior are specified. Responses are determined solely by what components actually do, and not by any built-in rules of performance at the system level.

Software for Simulation of Darwin III

We simulate Darwin III with a computer code that permits the construction of very general neural network systems (Reeke & Edelman, 1987). An experimenter can establish any number of neural repertoires with cells of one or more kinds arranged in layers within each repertoire. Each cell may have connections selected from three generic types: specific (constructed according to some specific rule, e.g., to form a topographic mapping), geometric (constructed to a given density versus distance relationship, often to provide lateral inhibition), or modulatory (receiving input proportional to the overall level of activity in some region). There are a variety of possible rules for synaptic amplification, but only specific connections can be plastic.

The activity level of every cell in Darwin III is determined by a "response function," which takes into account excitatory and inhibitory (hyperpolarizing or shunting) synaptic inputs, noise, decay of previous activity, depression and refractory periods, and long-term potentiation (LTP). The relative magnitudes of these terms can be varied parametrically. The mathematical details will be presented elsewhere (Reeke et al., 1989).

Rules for synaptic modification contain features reflecting the complex properties of real neurons and permitting any of the three major modes of selection-based learning already described. These synaptic rules constitute a simplified version of the rules employed in the somatosensory cortex model in a previous section (Finkel & Edelman, 1985; Pearson et al., 1987). The simplifications which are made in Darwin III for reasons of computational economy preserve the functionally critical heterosynaptic effects included in the more detailed rules.

Subsystems of Darwin III

In its present version, Darwin III includes four neuronal subsystems, each aimed at studying a different task: (1) a saccade and fine-tracking oculomotor system, (2) a reaching system using the arm, (3) a touch-exploration system using a different set of "muscles" in the same arm, and (4) a reentrant categorizing system. These systems have been combined to form an automaton able to use its sensorimotor apparatus in an adaptive way for autonomous behavior and categorization (Figure 5).

A word is in order about the biological basis of the neuronal repertoires which compose the subsystems of Darwin III. In some cases, the names of these repertoires correspond to the names of cytoarchitectonic areas in real brains; the coincidence of names is intended to be suggestive of the general function of these repertories as well as their relative position in an overall scheme and is not meant to imply that they are accurate models of the corresponding real brain functions.

Oculomotor System The simplest part of Darwin III is its oculomotor system. This set of repertoires is designed to let us study selective mechanisms in a simple motor control system. In it, visual signals impinge on a retina-like visual repertoire which is mapped in turn to a second repertoire of similar construction that represents the superior colliculus. "Colliculus" cells are densely connected to ocular motor neurons, which in turn innervate the simulated muscles that move the eye in its orbit. Initial connection strengths are assigned at random, so that eye motions are uncorrelated to visual stimulation.

An innate value scheme is used to bias the selection of eye movements: only those motions which bring visual stimuli into the foveal part of the retina, thus stimulating the value repertoire, are selected. Connections into eye motor neurons which are active in a short time interval before foveation occurs are selected and strengthened. This selection is strictly *a posteriori*. Motion initially occurs only because of spontaneous activity in the motor neurons; what is selected are populations of connections that favor the reoccurrence of motions that happen to give foveation. There is no *a priori* analysis indicating which connections may have adaptive value in this regard, and no error signal is computed and "back propagated" into the network.

In a standard training protocol, an object is presented at changing locations in the environment and eye motions occur. The improvement of the system is measured by the ability to foveate a stimulus quickly after its onset. Initially, there is no systematic eye motion, but after some 150 presentations of an object at different locations (during each presentation the object remains stationary for 16 cycles), the eye becomes able to center quickly on any stimulus that is presented peripherally.

Reaching System The reaching model contained in Darwin III provides a more complicated example of a basic motor control system. It enables us to investigate how multiple sensory pathways may contribute to a single motor control system and how degeneracy in neuronal group responses may act effectively to manage degeneracy in a mechanical effector system (the arm). It uses both vision and kinesthesia to guide

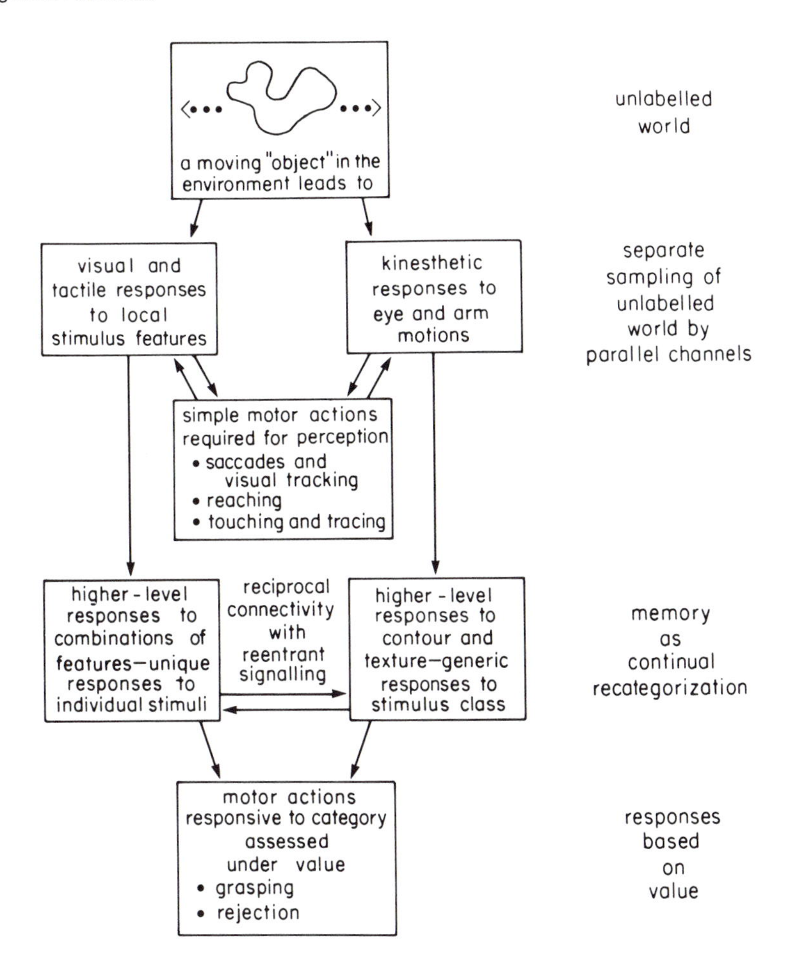

FIGURE 5 Overall schematic showing functional relationships of the components of Darwin III. The box at the top represents the environment, in which objects are presented and moved according to an experimental protocol. The other boxes represent functions carried out by the automaton. Each function may involve the activity of several interconnected neuronal repertoires (details not shown). The arrows indicate causal relationships between functions, and generally reflect the existence of anatomical connections among the various regions. The visually and kinesthetically based sampling systems that contribute to the categorization of stimuli are at the left and right, respectively. In the model, simple motor actions such as visual saccades and arm reaching are initiated in response to activity in primary sensory repertoires (second row of boxes). Additional connections would have to be added to reflect the influence of top down attentional mechanisms. Categorization is based on the reentrant correlation of responses in both visual and kinesthetic pathways; in general, other pairs (or higher-order combinations) of sensory modalities are involved in categorization, and the choices used here are intended to be merely exemplary of the possibilities. Reprinted by permission of *Daedalus*, Journal of the American Academy of Arts and Sciences, "Artificial Intelligence," **117,** 162, Cambridge, Massachusetts.

selection of smooth grasping movements from a prior repertoire of spontaneous gestural motions of the arm.

The anatomy of one version of the reaching system is shown schematically in Figure 6. A neuronal repertoire representing "motor cortex" initiates gestures. Its activity is fed via intermediate repertoires, representing brain stem nuclei, to the motor neurons that move arm muscles. A cerebellum-like structure filters gestures originating in this direct pathway by selective inhibition. The "cerebellum" is here simplified to just "granule cells" and "Purkinje cells." Granule cell inputs come from both vision and kinesthesia, such that granule cell firing corresponds to particular combinations of target positions and joint positions (i.e., conformations of the arm). The system has to associate these combinations with appropriate and inappropriate gestures through modification of synaptic connections, primarily those between granule cells and Purkinje cells. Purkinje cells then act to inhibit inappropriate gestures at the motor cortical level (deep cerebellar nuclei are omitted in this model).

Nothing in these circuits prejudices the arm to move in a coordinated way, or even in a particular direction, before training occurs. Instead, motions are selected from spontaneous movements when these are successful in getting the arm closer to the object. Success is reflected in the activity of a simple value repertoire in which signals from a visual area responsive to the creature's own hand and signals from another visual area responsive to stimulus objects are combined in a common map. Activity in this value repertoire increases when the two visual responses

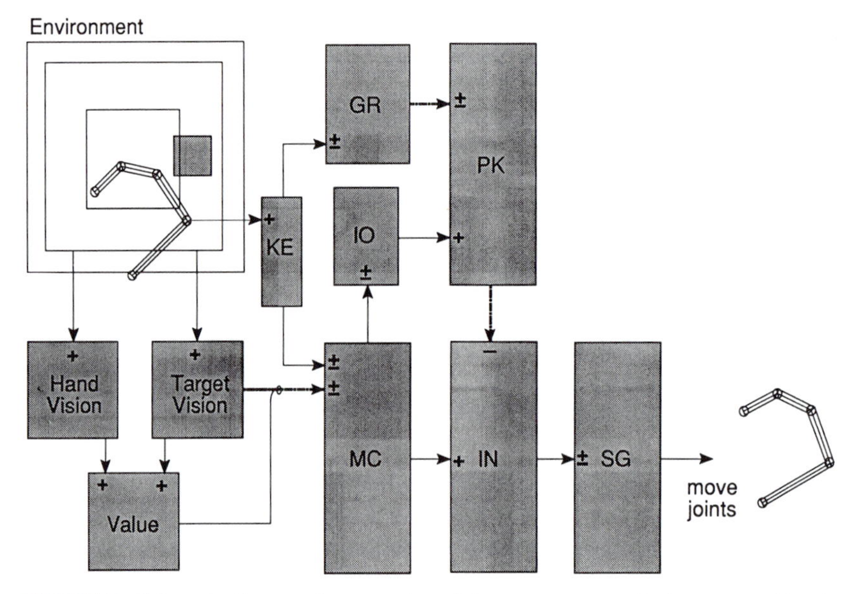

FIGURE 6 Schematic diagram of one version of the arrangement mediating arm reaching in Darwin III. The two large squares in the environment represent peripheral and foveal visual fields; the smaller shaded square is a stimulus object. The bent structure with circular links is the arm. The main motor pathways across the bottom of the diagram are responsive to both visual signals (indicating the location of the target) and kinesthetic signals (indicating the conformation of the arm). The symbols +, -, and ± indicate excitatory, inhibitory, and mixed connections, respectively. Heavy dashed arrows indicate modifiable connections. The biasing of such modification by value is indicated by lines from value repertoires terminating in circles around the connections receiving the bias. The model "cerebellum" (*top*) acts to inhibit inappropriate gestures and thus improves the directional specificity and smoothness of reaching.

map near a common location. This activity is carried to the "motor cortex," where it provides heterosynaptic bias for the amplification of synaptic strengths there, as well as to the "inferior olive," where it gates signals that enter the "cerebellum." Responses leading to motion of the arm closer to the object are thus selectively favored. As in the case with saccades, no calculation is required to identify the most effective connections; unlike the case with saccades, there are degenerate solutions to the problem corresponding to different positions of the joints that give the same final hand position; separate responses of degenerate neuronal groups provides separate pathways to achieve these different but equivalent solutions.

Training of Darwin III's reaching system leads to significant increases in performance, as seen in Figure 7. The large variance in individual motions that exists initially is progressively reduced, leading to a narrow envelope of motions, most of which move the end of the arm toward the object (Figure 7). When trained without the "cerebellum," or without the guidance of the value scheme, the system produces only broadly defined gestures that lack the precision needed to accomplish the reaching task. When trained properly, the system acts as a gestural module; several such modules could be combined in a map to accomplish reaching movements over a wide range of space.

Touch-Exploration System We have also used the arm of Darwin III to study possible neural mechanisms of exploratory behavior. For the present, we have examined only a limited mode in which the arm is guided by the sense of touch in the single distal digit. The arm assumes a canonical straightened posture during such exploration to reduce the computational burden of the model. Exploratory motions are generated initially in random directions by spontaneous neural activity in the relevant motor repertoires. A scheme that assigns positive value to the presence of touch signals of an intermediate pressure level on the digit is used to modify the random activity when it begins to lose touch with the object being palpated. The system at present has no place memory that would enable it to detect completion of a trace

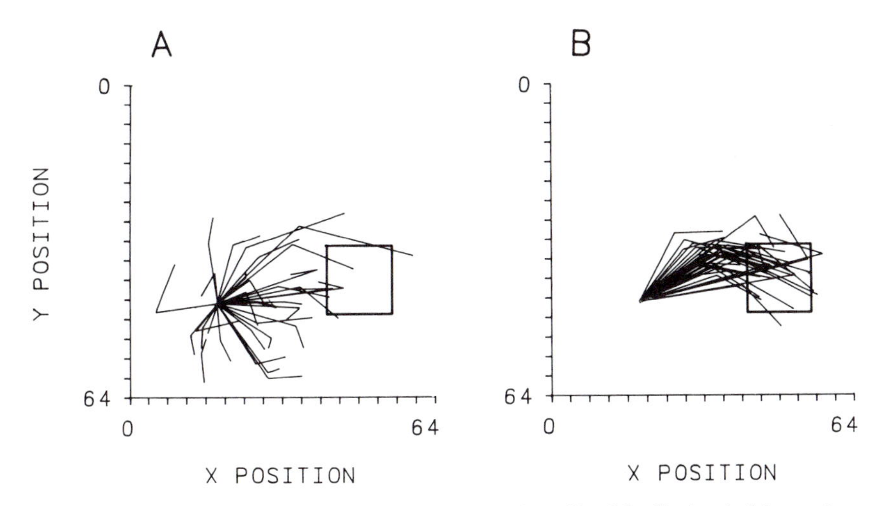

FIGURE 7 Traces of multiple reaching trials showing trajectories of the distal end of the arm from a fixed starting position (nexus of lines at left). The target area is indicated by a circle at the right. (A) Trajectories before training. Trajectories proceed in all directions and endpoints bear no relationship to target position. (B) Trajectories after training. Endpoints now center on the target position. There are multiple pathways from the starting position to the target, reflecting the degeneracy of the arm–motor system at both the mechanical and neuronal levels. Reproduced with permission from Reeke & Edelman (1988b).

around a closed object, but it is able to detect repetition in the patterns of motion sensed via kinesthesia. The continuing presence of novel kinesthetic patterns indicates that new regions of the object are still being explored; these patterns inhibit categorical responses to the current stimulus (see section on Darwin III's Categorization System), permitting exploration to be completed before attention passes on to another object.

This simple touch-exploration system is perhaps enough for a sessile creature. In order to model the more elaborate exploratory behavior of free-ranging animals, further effort will be needed to incorporate full-fledged mechanisms for motivation, random search, place memory, and recognition of novelty as well as completion.

Categorization System The various parts of Darwin III come together in the networks for categorization and response to stimuli. Categorization is needed to provide discrimination as a basis for purely consummatory functions as well as for all forms of learning and is thus a critical ability for all animals. To deal with nontrivial categorizations, a recognition system must be responsive to stimulus features of more than one kind. Coupled subsystems responsive to different sensory modalities (or submodalities) constitute particularly good examples; such systems are referred to as "classification couples" (Edelman, 1987). In Darwin III, we have chosen to combine vision and touch in a reentrant classification system.

For an animal in a natural environment, apparently arbitrary combinations of features may be indicative of certain categories having ethological value, for example, "good to eat." Tuning biased by value may be used to associate these categories with the appropriate responses. Because the relationship is essentially impossible to predict in advance, an association network capable of arbitrary couplings must be interposed between the perceptual and motor systems. Forms of behavior beyond mere association, including classical conditioning, may be obtained from such a system by experimental manipulation of the identifying features and values of the stimuli presented.

To demonstrate some of these effects in a simple way, we began with sensory networks capable of distinguishing just two categories. The categories are "rough striped" (bad for this species) and all others (good). The result of a "rough striped" categorization is one of the simplest possible behaviors, a reflex, in this case a "swat" at the "noxious" object with the arm. The usual result of the "swat" is to remove the stimulus from the creature's vicinity.

The visual component of the classification couple consists of three repertoires arranged hierarchically. The first of these is an LGN-like repertoire composed of on- and off-center units. It supplies excitation to a repertoire analogous to V1 which is responsive to oriented line segments. This in turn is connected to a higher-level repertoire containing groups that are responsive to combinations of V1 groups responsive to the same or different orientations. This last repertoire thus signals the presence of stimuli of particular visual classes, for example, those with stripes or other visual textures.

The touch component consists of a repertoire which contains cells responsive to two broad classes of stimulus boundaries—smooth and rough. The inputs to these cells come from kinesthetic receptors in the touch-exploration motor system and not from touch itself, because the intent is to categorize on the basis of gross shape as revealed by tracing rather than on the basis of tactile textures. The latter could equally well have been used.

The basis for association between the visual and touch-sensitive systems is a set of reentrant connections in both directions between higher-order cells in the two major divisions of the categorization system. Firing thresholds in these cells are affected by a form of long term potentiation, leading to lowered thresholds in cells receiving continued correlated stimulation from an object with an appropriate combination of visual and tactile characteristics. A "triggering" repertoire gives a "completion" response when visual inspection and tracing of a stimulus object reveal no more novel features.

Firing in the triggering repertoire, which indicates completion of the examination of a stimulus but is indifferent as to category, is coupled back to the two sensory repertoires. Triggering reexcites only those cells that were previously sensitized by long-term potentiation and that have reentrant connections from their opposite numbers in the other modality that were also selectively strengthened. The category-specific combined output enters an ethological value association area that in turn couples it to a motor center that generates the swat reflex.

As a result of all these connections, stimuli are subject to both visual and tactile examination, which are terminated as soon as novel features cease to be found. When this happens, a pattern is evoked in the reentrant categorization area that is characteristic of the particular class to which the object belongs. This pattern, if it is one to which the ethological value system responds, evokes the swat reflex to remove the stimulus. Stimuli not recognized as noxious are left undisturbed. The value scheme that selects rough-striped objects for rejection is here built in as an evolutionary imperative, but present work is aimed at selective formation of context-sensitive category response associations based on experience with past stimuli. In order for responses to go beyond mere instinct in such cases, there obviously must be an ability to evaluate consequences of behavioral acts. The results obtained so far give us confidence that we can model the formation of categories from collections of stimuli having arbitrary categories not reflected in the genetic makeup of the automaton, as well as the classification of individual stimuli according to these categories.

CONCLUSIONS

Comparison with "Neural Nets"

The goal of the research described in this chapter differs from that of the models commonly known as "neural nets" in that it aims to understand how perceptual, cognitive, and motor control functions are actually carried out in real nervous systems. This goal leads us to consider first those abilities of animals that appear to be most fundamental for their survival, namely, the ability to categorize objects and events in the world and respond in an adaptive manner to them. Matters of language, symbolic processing, and logic are left to later study, as these abilities seem to us to require the simpler ones as a necessary substrate for formation of the concepts with which they deal.

Given this outlook, we choose to avoid approaches which make use of unphysiological assumptions purely for convenience of mathematical analysis or because of analogies with interesting physical systems. We believe it is more profitable to analyze the consequences of realistic assumptions, by exhaustive computer modeling when necessary, rather than to make unrealistic assumptions purely for ease of analysis. For this reason, we omit from our models a number of common features of "neural net" models, including complete connectivity between cells in different layers, symmetric bidirectional connections, and an omniscient "teacher" that immediately evaluates the responses of the system and provides correction vectors for accurate correction of synaptic weights. We also avoid learning paradigms which involve the clamping of output units to correct response values or in which cyclic changes in operating parameters (e.g., "simulated annealing"; see Kirkpatrick, et al., 1983) are required during training (Kienker, Sejnowski, Hinton, & Schumacher, 1986). We add lateral connectivity within layers, more complex cellular dynamics (depression, refractory period, long-term potentiation, etc.), and more complex rules for synaptic modification (heterosynaptic effects, temporally integrated effects, etc.). These features are integrated into model automata that are required to accomplish tasks in a world with which they must interact through imperfect sensors and effectors which present some of the same problems that are faced by real animals. By so doing, we hope to create models that are consistent not just with the computational requirements of the behaviors they generate, but with evolutionary, developmental, and physiological constraints as well.

The Development of Non-von Neumann Computers

"Neural net" models are often put forward as paradigms for the development and programming of a new generation of non-von Neumann computers. They bring to this task the promise of doing away with much of the detailed "if–then" analysis that is such a central part of conventional programming, and of replacing it with a form of "learning by example" that can be more easily automated. This is to be done, however, within the conventional information processing framework, which assumes the existence of prior criteria for the discrimination of categories in the world, of codes capable of representing those categories in neural nets, and of algorithms for the computation of behavior from sensory input using the codes. These assumptions in many ways tie the new "neural net" models to their earlier counterparts and impose upon them the same shortcomings.

To our way of thinking, the fundamental reason for the inability of conventional computers to deal with novelty and complexity in the real world is not just the programming methodology employed, but rather the very assumption that novelty and complexity *can* be dealt with adequately by some prespecified procedure (algorithm; Reeke & Edelman, 1988a). If this view is correct, the "neural net" approach can do little to alleviate the problems of existing AI approaches and, in fact, may prove to be less effective because of the relative poverty of the logical operations available in "neural net" models relative to LISP programs.

Selective systems, on the other hand, do not assume the existence of prior descriptions of categories, codes for the abstract representation of information, or algorithms for the transformation of information into behavior. A response can thus be produced in the absence of appropriate prior instructions for that response. Everything is dependent on the history of the individual organism (or automaton) and its ability to generalize its experience to new situations by invoking appropriate combinations of overlapping and degenerate units previously selected on the basis of their ability to contribute to responses of adaptive value for that organism. Of course, this process can fail, but its modes of failure are likely to be much less catastrophic than those of a programmed system for which some particular contingency happened to be omitted. The construction of non-von Neumann machines based on selection should thus give us some unique and more animal-like devices useful in a great many practical applications.

Role of Synthetic Neural Modeling in the Study of the Nervous System

In the Introduction, we discussed some of the reasons we believe an overall theory of brain function is necessary if we are to begin to bridge the current gaps in understanding between the physiological and psychological levels of brain function. We have presented one such theory, the theory of neuronal group selection, and have demonstrated with computer simulations that the theory provides a self-consistent and productive picture of neural network function that can be used to explain simple patterns of behavior in animals. We believe that further elaboration of these models, carried out under the general rubric of "synthetic neural modeling," will lead to new insights into the way the fundamental architectural principles of the brain give rise to a hierarchy of perceptual and behavioral capabilities of increasing complexity and abstraction. When properly coupled to experimental brain science via the verification of experimental predictions, synthetic neural modeling should give us a powerful method of relating physiological properties, which can generally be observed only in small local regions of the nervous system and only over short periods of time, to those complex and relatively inaccessible higher functions which are our real interest.

Acknowledgments

This research has been supported in part by the Office of Naval Research, the John D. and Catherine T. MacArthur Foundation, the Lucille P. Markey Charitable Trust, the Pew Charitable Trusts, the van Ameringen Foundation, and the IBM Corporation. Some of this research was conducted using the Cornell National Supercomputer Facility, a resource of the Center for Theory and Simulations in Science and Engineering at Cornell University, which receives funding from the National Science Foundation and IBM Corporation, with additional support from New York State and members of the Corporate Research Institute.

References

Allman, J. (1988). Variations in visual cortex organization in primates. In P. Rakic & W. Singer (Eds.), *Neurobiology of neocortex*. New York: Wiley.

Bongard, M. (1970). *Pattern recognition*. New York: Spartan.

Carew, T. J., Hawkins, R. D., Abrams, T. W., & Kandel, E. R. (1984). A test of the Hebb postulate at identified synapses which mediate classical conditioning in *Aplysia*. *Journal of Neuroscience*, **4**, 1217–1224.

Collingridge, G. L., & Bliss, T. V. P. (1987). NMDA receptors—Their role in long-term potentiation. *Trends in Neuro-Sciences*, **10**, 288–293.

Eckhorn, R., Bauer, R., Jordan, W., Brosch, M., Kruse, W., Munk, M., & Reitboeck, H. J. (1988). Coherent oscillations: A mechanism of feature linking in the visual cortex: Multiple electrode and correlation analyses in the cat. *Biological Cybernetics*, **60**, 121–130.

Edelman, G. M. (1973). Antibody structure and molecular immunology. *Science*, **180**, 830–840.

Edelman, G. M. (1978). Group selection and phasic reentrant signalling: A theory of higher brain function. In G. M. Edelman & V. B. Mountcastle (Eds.), *The mindful brain*. Cambridge, MA: MIT Press.

Edelman, G. M. (1981). Group selection as the basis for higher brain function. In F. O. Schmidt, F. G. Worden, G. Adelman, & S. G. Dennis (Eds.), *Organization of the cerebral cortex*. Cambridge, MA: MIT Press.

Edelman, G. M. (1987). *Neural Darwinism*. New York: Basic Books.

Edelman, G. M., & Finkel, L. H. (1984). Neuronal group selection in the cerebral cortex. In G. M. Edelman, W. E. Gall, & W. M. Cowan (Eds.), *Dynamic aspects of neocortical function*. New York: Wiley.

Edelman, G. M., & Reeke, G. N., Jr. (1982). Selective networks capable of representative transformations, limited generalizations, and associative memory. *Proceedings of the National Academy of Sciences of the United States of America*, **79**, 2091–2095.

Felleman, D. J., & Van Essen, D. C. (1987). Receptive field properties of neurons in area V3 of macaque monkey extrastriate cortex. *Journal of Neurophysiology*, **57**, 889–920.

Finkel, L. H., & Edelman, G. M. (1985). Interaction of synaptic modification rules within populations of neurons. *Proceedings of the National Academy of Sciences of the United States of America*, **82**, 1291–1295.

Finkel, L. H., & Edelman, G. M. (1987). Population rules for synapses in networks. In G. M. Edelman, W. E. Gall, & W. M. Cowan (Eds.), *Synaptic function*. New York: Wiley.

Fleshman, J. W., Segev, I., & Burke, R. E. (1988). Electrotonic architecture of type-identified alpha-motoneurons in the cat spinal-cord. *Journal of Neurophysiology*, **60**, 60–85.

Granger, R., Ambros-Ingerson, J., Henry, H., & Lynch, G. (1988). Partitioning of sensory data by a cortical network. In D. Z. Anderson (Ed.), *Neural information processing systems*. New York: American Institute of Physics.

Gray, C. M., König, P., Engel, A. K., & Singer, W. (1989). Oscillatory responses in cat visual cortex exhibit intercolumnar synchronization which reflects global stimulus properties. *Nature (London)*, **338,** 334–337.

Gray, C. M., & Singer, W. (1989). Stimulus-specific neuronal oscillations in orientation columns of cat visual cortex. *Proceedings of the National Academy of Sciences of the United States of America*, **86,** 1698–1702.

Hebb, D. O. (1949). *The organization of behavior*. New York: Wiley.

Hubel, D. H. & Wiesel, T. N. (1962). Receptive fields, binocular interaction and functional architecture in the cat's visual cortex. *Journal of Physiology*, **160**, 106–154.

Jenkins, W. M., Merzenich, M. M., & Ochs, M. T. (1984). Behaviorally controlled differential use of restricted hand surfaces induce [*sic*] changes in the cortical representation of the hand in area 3b of adult owl monkeys. *Society for Neuroscience Abstracts*, **10**, 665.

Kaas, J. H., Merzenich, M. M., & Killackey, H. P. (1983). The reorganization of somatosensory cortex following peripheral-nerve damage in adult and developing mammals. *Annual Review of Neuroscience*, **6**, 325–356.

Kienker, P. K., Sejnowski, T. J., Hinton, G. E., & Schumacher, L. E. (1986). Separating figure from ground with a parallel network. *Perception*, **15**, 197–216.

Kirkpatrick, S., Gelatt, C. D., & Vecchi, M. P. (1983). Optimization by simulated annealing. *Science*, **220,** 671–680.

Knudsen, E. I., & Konishi, M. (1978). A neural map of auditory space in the owl. *Science*, **200**, 795–797.

Koch, C., Poggio, T., & Torre, V. (1982). Nonlinear interactions in a dendritic tree: localization, timing, and role in information processing. *Proceedings of the National Academy of Sciences of the United States of America*, **80**, 2799–2802.

Lee, C. K., Rohrer, W. H., & Sparks, D. L. (1988). Population coding of saccadic eye-movements by neurons in the superior colliculus. *Nature*, **332**, 357–360.

Leyton, A. S. F., & Sherrington, C. S. (1917). Observations on the excitable cortex of the chimpanzee, orangutan, and gorilla. *Quarterly Journal of Experimental Physiology*, **11**, 137–222.

Macagno, E. R., Lopresti, V., & Levinthal, C. (1973). Structure and development of neuronal connections in isogenic organisms: Variations and similarities in the optic system of *Daphnia magna*. *Proceedings of the National Academy of Sciences of the United States of America*, **70**, 57–61.

MacDermott, A. B., & Dale, N. (1987). Receptors, ion channels and synaptic potentials underlying the integrative actions of excitatory amino acids. *Trends in NeuroSciences*, **10**, 280–284.

Merzenich, M. M., Kaas, J. H., Wall, J. T., Nelson, R. J., Sur, M., & Felleman, D. J. (1983). Topographic reorganization of somatosensory cortical areas 3b and 1 in adult monkeys following restricted deafferentation. *Neuroscience (London)*, **8**, 33–55.

Merzenich, M. M., Recanzone, G., Jenkins, W. M., Allard, T. T., & Nudo, R. J. (1988). Cortical representational plasticity. In P. Rakic & W. Singer (Eds.), *Neurobiology of neocortex*. New York: Wiley.

Mountcastle, V. B. (1957). Modality and topographic properties of single neurons of cat's somatic sensory cortex. *Journal of Neurophysiology*, **20**, 408–434.

Mountcastle, V. B. (1984). Central nervous mechanisms in mechanoreceptive sensibility. In *Handbook of physiology—The nervous system* (Vol. 3,) pp. 789–878. Bethesda: American Physiological Society.

Newell, A., Shaw, J. C., & Simon, H. A. (1958). Elements of a theory of human problem solving. *Psychological Review*, **65,** 151–166.

Pearson, J. C., Finkel, L. H., & Edelman, G. M. (1987). Plasticity in the organization of adult cerebral cortical maps: A computer simulation based on neuronal group selection. *Journal of Neuroscience*, **7**, 4209–4223.

Pearson, K. G., & Goodman, C. S. (1979). Correlation of variability in structure with variability in synaptic connections of an identified interneuron in locusts. *Journal of Comparative Neurology*, **184**, 141–165.

Rall, W. (1964). Theoretical significance of dendritic trees for neuronal input-output relations. In R. Reiss (Ed.), *Neural theory and modelling*. Stanford, CA: Stanford University Press.

Reeke, G. N., Jr., & Edelman, G. M. (1984). Selective networks and recognition automata. *Annals of the New York Academy of Sciences*, **426**, 181–201.

Reeke, G. N., Jr., & Edelman, G. M. (1987). Selective neural networks and their implications for recognition automata. *International Journal of Supercomputer Applications*, **1**, 44–69.

Reeke, G. N., Jr., & Edelman, G. M. (1988a). Real brains and artificial intelligence. *Daedalus*, **117**, 143–173.

Reeke, G. N., Jr., & Edelman, G. M. (1988b). Recognition automata based on neural Darwinism. *Forefronts* (Ithaca, New York: Cornell University Center for Theory and Simulation in Science and Engineering) **3 (12),** 3–6.

Reeke, G. N., Jr., Finkel, L. H., Sporns, O., & Edelman, G. M. (1989). Synthetic neural modeling: A new approach to the analysis of brain complexity. In G. M. Edelman, W. E. Gall, & W. M. Cowan (Eds.), *Signal and sense: Local and global order in perceptual maps*. New York: Wiley.

Rosch, E. (1978). Principles of categorization. In E. Rosch & B. B. Lloyd (Eds.), *Cognition and categorization*. New York: Academic Press.

Ryle, G. (1949). *The concept of mind*. London: Hutcheson.

Schwartz, M. L., Zheng, D. S., & Goldman-Rakic, P. S. (1988). Periodicity of GABA-containing cells in primate prefrontal cortex. *Journal of Neuroscience*, **8**, 1962–1970.

Sejnowski, T. J., Koch, C., & Churchland, P. S. (1988). Computational neuroscience. *Science*, **241**, 1299–1306.

Smith, E. E., & Medin, D. L. (1981). *Categories and concepts*. Cambridge, MA: Harvard University Press.

Sporns, O., Gally, J. A., Reeke, G. N., Jr., & Edelman, G. M. (1989). Reentrant signaling among simulated neuronal groups leads to coherency in their oscillatory activity. *Proceedings of the National Academy of Sciences of the United States of America*, **86,** 7265–7269.

Suga, N. (1984). The extent to which biosonar information is represented in the bat auditory cortex. In G. M. Edelman, W. E. Gall, & W. M. Cowan (Eds.), *Dynamic aspects of neocortical function*. New York: Wiley.

Traub, R. D., Knowles, W. D., Miles, R., & Wong, R. K. S. (1984). Synchronized after discharges in the hippocampus—simulation studies of the cellular mechanism. *Neuroscience (London)*, **12**, 1191–1200.

Van Essen, D. C. (1985). Functional organization of primate visual cortex. In A. Peters & E. Jones (Eds.), *Cerebral cortex:* (Vol. 3. *Visual cortex*). New York: Plenum.

Wigstrom, H., & Gustafsson, B. (1983). Heterosynaptic modulation of homosynaptic long-lasting potentiation in the hippocampal slice. *Acta Physiologica Scandinavica*, **119**, 455–458.

Wilson, M. A., & Bower, J. M. (1988). A computer simulation of olfactory cortex with functional implications for storage and retrieval of olfactory information. In D. Z. Anderson, (Ed.), *Neural information processing systems*. New York: American Institute of Physics.

Wittgenstein, L. (1953). *Philosophical investigations* (English text of the 3rd ed.). New York: Macmillan.

Zeki, S. M. (1983). The distribution of wavelength and orientation selectivity in different areas of monkey visual cortex. *Proceedings of the Royal Society of London, Series B* **207**, 239–248.

12

An Overview of Neural Networks: Early Models to Real World Systems

Douglas L. Reilly and Leon N Cooper

INTRODUCTION

Scientific fashion, like the length of women's skirts and the width of men's ties, changes with the seasons. From the old belief that neural networks could do nothing, we have now, among current opinions, the suggestion that they can do everything. Truth, as "the Master of those that know" proposed, might be closer to the golden mean.

We look to biology for inspiration. Neocortex by itself (disconnected from sensory inputs) functions at a distinctly reduced level. (Classifying circuitry from retina through visual cortex as a neural network begs the question since these are largely genetically programmed so that their learning takes place on evolutionary time scales.) Such considerations lead us to conclude that if neural networks can function at all to do useful things it seems very likely that they will do so by being incorporated into systems containing many more or less conventional elements so that they can solve real world problems economically. It is likely that in the future such networks will be components of complex systems involving classification, computation, and reasoning.

Neural networks are inspired by biological systems in which large numbers of nerve cells, that individually function rather slowly and imperfectly, collectively perform tasks that even the largest computers have not been able to match. They are made of many relatively simple processors connected to one another by variable memory elements whose weights are adjusted by experience. They differ from the now standard von Neumann computer in that they characteristically process information in a manner that is highly parallel rather than serial, and they learn (memory weights and thresholds are adjusted by experience). Since neural networks learn, they differ from the usual artificial intelligence systems in that the solution of real-world problems requires much less of the expensive and elaborate programming and knowledge engineering required for such products as rule-based expert systems. Thus, neural network systems seem to some of us to represent the next generation of computer architecture: systems that combine the enormous processing power of von Neumann computers with the ability to make sensible decisions and to learn by ordinary experience, as we do ourselves.

The various neural networks that are currently in fashion differ in their ability to make accurate distinctions, their ability to learn quickly, efficiently, and without extensive retraining, their complexity—the level of interconnectivity (an important consideration for realization in hardware), and in the size of the conventional computer required to simulate the network. An important criterion in differentiating neural networks from one another is whether they learn quickly and accurately enough to be of use in real time situations. A related question is whether they require retraining on an entire data set to learn one new item. (It is necessary for a system to be able to learn new information rapidly and accurately in order to be able to adapt to changes that occur.)

At this point it is perhaps appropriate to make what should be an obvious (but surprisingly not universally appreciated) remark. Neural networks that learn accurately and rapidly are *not* the result of random connections of many elements that learn according to just any rule. Although random networks that adjust themselves in a random manner can be made to learn, this learning, that possibly might be called evolutionary learning, takes place in what might be called evolutionary time. It is therefore important to distinguish between the time required to design an efficient and rapidly learning network and *the time it takes such a network to internalize the rules necessary to deal with a particular environment.* Further, it is necessary to distinguish the generic network architecture and learning rules from the learned internalized rules specific to a particular environment.

Further, one should not be misled into judging the value of neural networks based on their ability to solve certain hard problems such as picking out a mouse in a complicated background fifty yards away. While it is true that some animals can do this, they very likely do this by a complex combination of information processing, pattern recognition, and feedback and -forth between cognitive acts and pure pattern recognition. While this may be an eventual goal for neural network systems, it should not be used as a criterion for their value. Neural networks can be of great value in helping to solve real world problems without duplicating everything that animals do. There are many situations in which the biological system does better, at present, than anything we can build even though we understand how the biological system functions.

In this chapter we present an overview of neural networks from early models to present systems that can solve real world problems.

Definitions and Notation

A neural network can be defined as a distributed computational system composed of a number of individual processing elements operating largely in parallel, interconnected according to some specific topology (architecture) and having the capability to self-modify connection strengths and processing element parameters (learning). In general a neural network performs some information processing function (pattern recognition, data compression, etc.). Neural networks can learn complex "rules." They differ in their efficiency and rapidity of learning, their ability to make distinctions, their capacity to generalize, and the type of machine and/or hardware on which they can run.

Among the problems that are difficult to solve using conventional rule-based techniques are those that might be characterized as having a high degree of entropy or variability as in Figure 1. A neural network processing element has inputs and synaptic weights that produce an integrated potential as an output as shown in Figure 2. The output is a sigmoidal function of the integrated potential as shown in Figure 3. A neural network is composed of such individual processing elements connected to one another according to some architecture (Figure 4). Input vectors are denoted by d, e, or f. Output vectors by c, g, or t. Transfer or memory storage matrices are A, M, W, or R. $f^1 \ldots f^\alpha$ are the N components of the αth vector.

Neural networks are characterized by their *architecture*: what is connected to what. They can be fully connected, sparsely connected, or feedforward. They can also be characterized by their *dynamics* (dynamical or equilibrium systems) and by their *learning rules*, that is, which network parameters (weights, thresholds, number of connections, etc.) change over time and how.

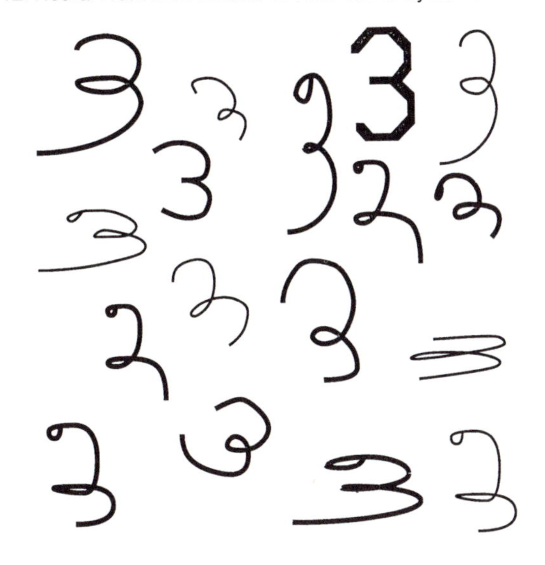

FIGURE 1 Which of the above characters is a three?

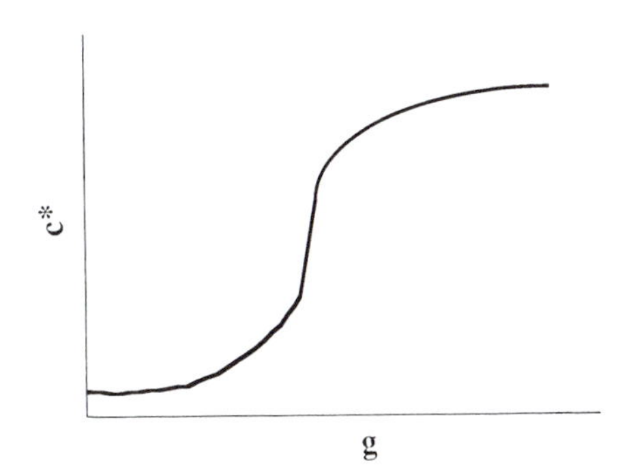

FIGURE 3 Sigmoidal function of integrated potential. $g_i = \sum_{j=1}^{N} A_{ij} f_j$; $c = c^*(\sum_{j=1}^{N} A_{ij} f_j) = c^*(g)$.

Among the architectural choices, we have to decide between a richly interconnected neural network as shown in Figure 5 or a simple feedforward neural network. Analysis of biological systems (Cooper & Scofield, 1988) suggests that complex completely interconnected neural networks can be approximated by such simple feedforward networks.

EARLY NETWORK LINEAR MODELS

We begin by discussing some of the properties of early linear neural network models. In the linear

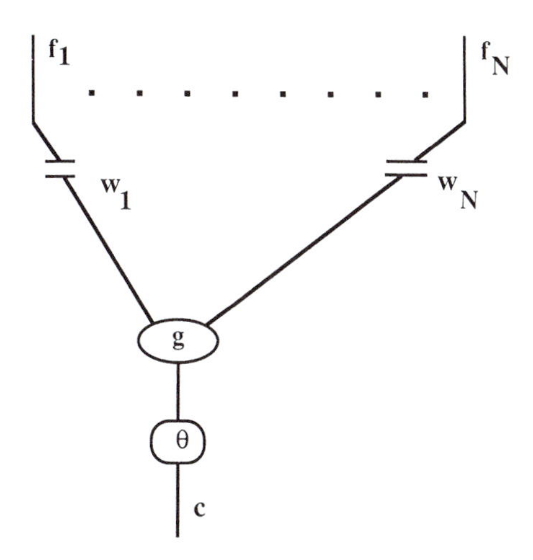

FIGURE 2 A typical neural network element. f_i, inputs; w_n, weights; $g = \sum_{j=1}^{N} w_j f_j$, integrated potential; θ, threshold function; c, output.

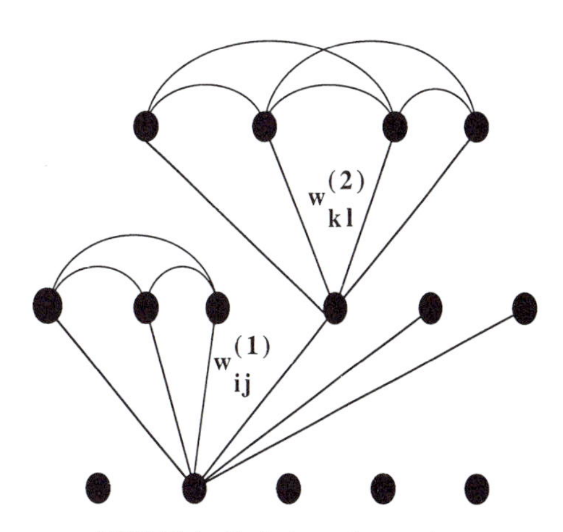

FIGURE 4 Typical neural network.

region let g_i be the cell output

$$g_i^{\alpha} = \sum_{j=i}^{N} A_{ij} f_j^{\alpha} \tag{1}$$

where $f_1^{\alpha} \ldots f_N^{\alpha}$ are the components of the αth vector.

We may then regard $[A_{ij}]$ (the synaptic strengths of the N^2 ideal junctions) as a matrix or a mapping which takes us from a vector in the F space to one in the G space. This maps the neural activities $f = (f_1, f_2 .. f_N)$ in the F space into the neural activities $g = (g_1, ... g_N)$ in the G space and can be written in the compact form

$$g = Af \tag{2}$$

In the earliest neural network models, it was proposed that it was in modifiable mapping of the type A that the experience and memory of the system are stored. In contrast with machine memory, which is, at present, local (an event stored in a specific place) and addressable by locality (requiring some equivalent of indices and files), animal memory is likely to be distributed and addressable by content or by association. In addition, for animals there need be no clear separation between memory and "logic."

It is convenient to write the mapping A in the basis of vectors the system has experienced

$$A = \sum_{\mu\nu} c_{\mu\nu} \mathbf{g}^{\mu} \times \mathbf{f}^{\nu} \tag{3}$$

Here g^{μ} and f^{ν} are output and input patterns of neural activity while the $c_{\mu\nu}$ are coefficients reflecting the degree of connection between various inputs and outputs. The symbol x represents the "outer" product between the input and output vectors. The ij^{th} element of A gives the strength of the ideal junction between the incoming neuron j in the F bank and the outgoing neuron i in the G bank (Figure 6). Thus, if only f_j is nonzero, g_i, the firing rate of the ith output neuron, is $g_i = A_{ij} f_j$. Since

$$A = \sum_{\mu\nu} c_{\mu\nu} \mathbf{g}^{\mu} \times \mathbf{f}^{\nu} \tag{4}$$

the ij^{th} junction strength is composed of a sum of the

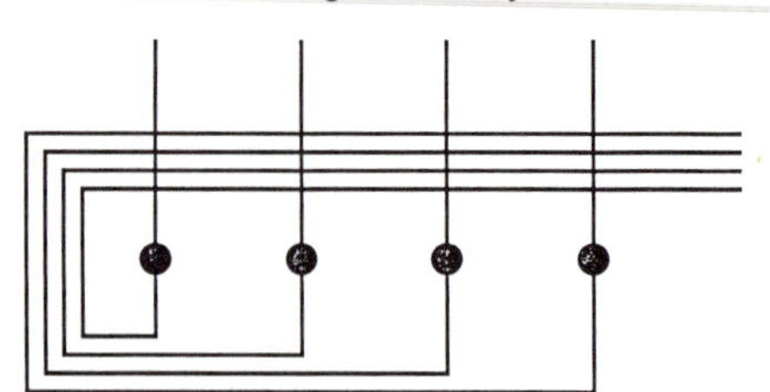

FIGURE 5 A fully interconnected network.

entire experience of the system as reflected in firing rates of the neurons connected to this junction. Each experience or association ($\mu\nu$), however, is stored over the entire array of N x N junctions. This is the essential meaning of a distributed memory: each event is stored over a large portion of the system, while at any particular local point many events are superimposed.

We have shown elsewhere that the nonlocal mapping, A, can serve in a highly precise fashion as a memory that is content addressable and in which "logic" is a result of association and an outcome of the nature of the memory itself (Cooper, 1973).

The matrix A gives perfect recall if the f^{ν} are orthogonal. If f^{ν} are not orthogonal, they can be orthogonalized by various techniques. This leads to what Kohonen has called an optimal mapping

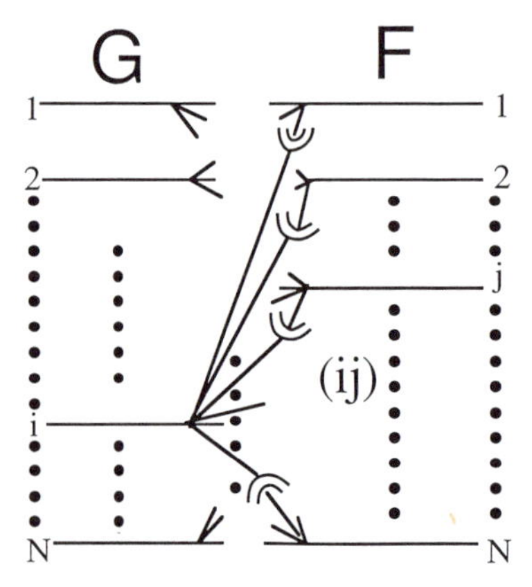

FIGURE 6 A typical neural network containing a distributed memory.

(Kohonen, 1984). The argument in a very simple form is as follows.

For all inputs, f^{μ}, and desired outputs, g^{ν}, we want

$$g^{\nu} = A * f^{\nu} \quad \nu=1,2,...K \tag{5}$$

In general, we want

$$G = A^{*}F \tag{6}$$

where G and F are the matrices

$$G = \begin{bmatrix} g_1^1 & \cdots & g_1^K \\ . & \cdots & . \\ . & \cdots & . \\ . & \cdots & . \\ g_N^1 & \cdots & g_N^K \end{bmatrix} \tag{7}$$

$$F = \begin{bmatrix} f_1^1 & \cdots & f_1^K \\ . & \cdots & . \\ . & \cdots & . \\ . & \cdots & . \\ f_N^1 & \cdots & f_N^K \end{bmatrix}$$

If $K = N$ (all matrices are square and inverses exist), then

$$A^{*} = GF^{-1} \tag{8}$$

For nonsquare matrices and for those with no inverses one can obtain A^{*} using the method of pseudoinverses (Kohonen, 1984).

LEARNING

Among the more interesting properties of neural networks is their ability to learn. These are not the only systems that learn, but it is their learning ability coupled with the distributed processing inherent in neural network systems that distinguishes these systems.

Among the various classical learning rules, the oldest and the most famous is that proposed by Hebb. "When an axon of cell A is near enough to excite a cell B and repeatedly or persistently takes part in firing it, some growth process or metabolic change takes place in one or both cells such that A's efficiency as one of the cells firing B is increased" (Hebb, 1949). In our notation, this could be written:

$$\delta A_{ij} \sim g_i f_j \tag{9}$$

where $g = Af$ is the actual output and f the actual input. We then have

$$A(t+1) = \gamma A(t) + \delta A(t) \tag{10}$$

where

$$\delta A(t) = \eta(g \times f) = \eta Af \times f \tag{11}$$

This yields

$$A(t+1) = \gamma A(t)\,[1+ \tfrac{\eta}{\gamma}\, f(t) \times f(t)\,] \tag{12}$$

and leads to *exponential growth of recognition of* f *if* f *is a repeated input.* Obviously we will need some limit on synaptic strengths. One approach to limiting exponential growth of synapses is "anti-Hebbian" unsupervised learning:

$$\delta A \sim -\eta\, g \times f \tag{13}$$

$$A(t+1) = \gamma A(t)\,[1- \tfrac{\eta}{\gamma}\, f(t) \times f(t)\,] \tag{14}$$

This learning rule projects a repeated incoming vector, f, to zero. The BCM model (Bienenstock, Cooper, & Munro, 1982) combines Hebbian and anti-Hebbian unsupervised learning; it has been applied to many situations in developing visual cortex and appears to explain normal rearing as well as the many deprivation and pharmacological experiments and has been extensively discussed (Cooper, Bear, & Ebner, 1987).

Supervised Learning

In supervised learning, one tries to make adjustments to the set of synaptic weights so that the actual output is guided to a "desired" or "target" output. Such techniques were explored as far back as the late 1950s and early 1960s by Rosenblatt (1958) and Widrow and Hoff (1960). Typically the modification algorithm takes the form

$$A(t+1) = \gamma A(t) + \delta A(t) \qquad (10)$$

where

$$\delta A(t) = \eta(t^{a} - g^{\alpha}) \text{ x } f^{\alpha} \qquad (15)$$

and again $g^{\alpha}(t) = A(t) f^{\alpha}(t)$ is the actual output for the input $f^{\alpha}(t)$ while t^{α} is the target output for pattern α. This yields

$$\delta A(t) = \eta(t^{\alpha} - A(t)f^{\alpha}) f^{\alpha} \qquad (16)$$

Note $\delta A(t) = 0$ if $A(t)f^{\alpha} = t^{\alpha}$, so that the correct A is a fixed point (denoted A^*). This A^* corresponds to the *pseudoinverse* discussed above.

We can see the connection with gradient descent if we define an "energy," an "error," or "cost" function:

$$E = \frac{1}{2}\sum_{\alpha}(t^{\alpha} - Af^{\alpha})^2 \qquad (17)$$

The variation of E with respect to A is:

$$\delta E = -(\delta A)\sum_{\alpha}(t^{\alpha} - Af^{\alpha}) \text{ x } f^{\alpha} \qquad (18)$$

If

$$\delta A \sim \sum_{\alpha}(t^{\alpha} - g^{\alpha}) \text{ x } f^{\alpha} \qquad (19)$$

as above, $\delta E \le 0$ under this variation.

We can picture what is happening in two dimensions as a shift of f^1 and f^2 to f^{1*} and f^{2*} as in Figure 7. With this, mapping A goes to A^* as

$$g^1 \text{ x } f^1 + g^2 \text{ x } f^2 \longrightarrow g^1 \text{ x } f^{1*} + g^2 \text{ x } f^{2*} \qquad (20)$$

Note that

$$f^{1*} \cdot f^1 = 1 \qquad f^{2*} \cdot f^1 = 0 \qquad (21)$$
$$f^{1*} \cdot f^2 = 0 \qquad f^{2*} \cdot f^2 = 1$$

Therefore

$$g^1 = A^* f^1 \qquad (22)$$
$$g^2 = A^* f^2$$

Thus the *actual output* becomes the *target output*. With this we can provide separation, recognition *of up to N vectors* in an N dimensional space. This leads to interesting results but has obvious problems. We get separation only if the number of inputs, K, is equal to or smaller than the dimensionality of the space, N. In general with such linear systems we obtain only linear separability. We must introduce some nonlinearity. Otherwise we get increasing confusion with more and more inputs.

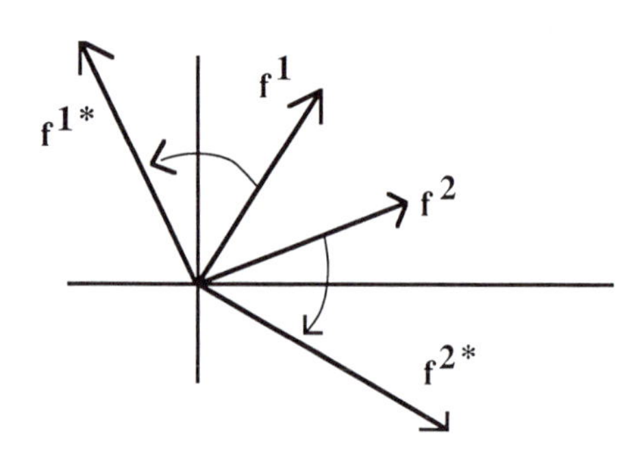

FIGURE 7 Process by which A → A*.

Nonlinear Systems

One way to characterize a neural network is as a nonlinear mapping from input f to output g:

$$g = M[f]$$

M is a nonlinear mapping of the input, f, into the output, g. A theorem of Kolmogorov (1957) indicates that nonlinear threshold devices as arranged in a neural network can yield essentially any nonlinear mapping. Therefore, there is reason to believe that if a problem can be solved by some nonlinear mapping of inputs into outputs, a neural network can provide the solution. Many problems can be solved this way.

Perceptrons

One of the earliest nonlinear neuron-like learning devices was Frank Rosenblatt's (1958) Perceptron

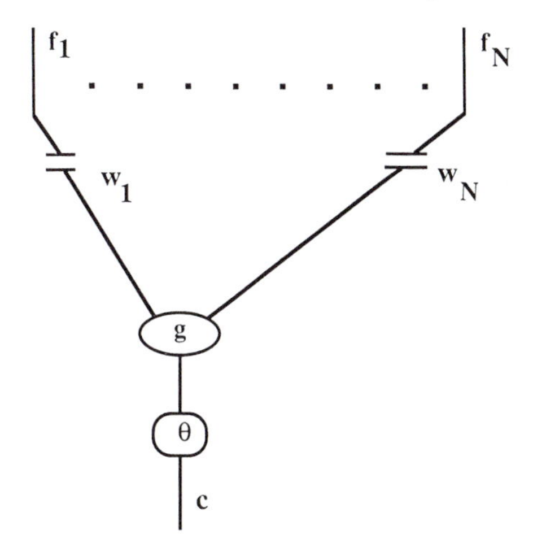

FIGURE 8 A perceptron.

(Figure 8). Perceptron learning is based on the difference between the actual output and the desired output. Other than such irrelevant features as limiting weights and adjustments to ± 1, it is very similar to gradient descent; it uses a simple threshold so that it incorporates a nonlinearity.

The update rule is

$$c \Rightarrow 1 \text{ if } \sum_j w_j f_j \geq \theta \qquad (23)$$
$$\Rightarrow 0 \text{ if } \sum_j w_j f_j < \theta$$

with c being the output, f_j the inputs, and w_j the synaptic weights. The learning rule is:

$$w_j \rightarrow w_j + \delta w_j \qquad (24)$$

with

$$\delta w_j = \eta(t - c) f_j, \quad (0 < \eta \leq 1)$$

where t is the target response, c the actual response.

An important result is the perceptron learning theorem which states that by this learning mechanism a proper decision hyperplane (Figure 9) will be found if it exists. It was always clear that the Perceptron cannot solve all problems since it is only capable of linear separations (Figure 10). However, multiple layer per-

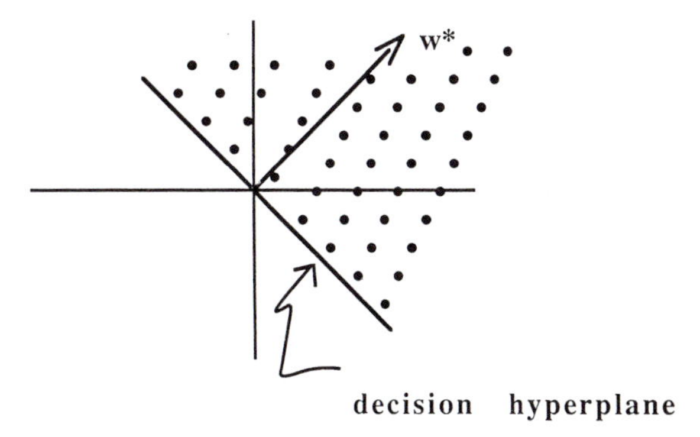

FIGURE 9 Decision hyperplane defined by a weight vector **W***.

ceptrons are not limited in this way. If fact, it is now known that with three layers (input, hidden, and output as shown in Figure 11) arbitrary nonlinear separability is possible (see RCE, the network to be discussed later; Lippman, 1987). But multilayer perceptrons suffer from the *credit assignment problem*. How should the weights be adjusted so that the proper decisions are made? In spite of the fact that Rosenblatt proposed a solution like that presently called back propagation of error, the supposed lack of a solution to this problem was a major argument used by Minsky and Pappert (1969) to discredit learning systems.

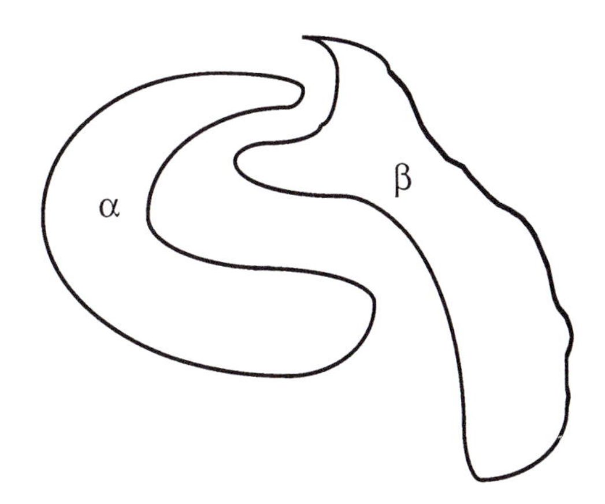

FIGURE 10 The decision regions α and β cannot be separated by a single perceptron.

At present, the credit assignment problem has been solved in various ways. Some of the advantages and disadvantages of the various methods will be discussed next.

A very popular method of dealing with the credit assignment problem in multilayer neural networks is to construct a cost function. This often takes the form:

$$E = \tfrac{1}{2}\sum_{\alpha}(t^{\alpha} - c^{\alpha})^2 \qquad (25)$$

$$= \tfrac{1}{2}\sum_{\alpha}(t^{\alpha} - M[f^{\alpha}])^2$$

with t^{α} the target output and f^{α} the input for pattern α. The object is to find some procedure for adjusting M (a function of all of the weights and thresholds ω^{β}_{ij} and θ^{β}_{i} $M[(\omega_{ij}),\theta_i])$ so that E is minimized. If E=0 exists then a solution exists (Figure 12).

Various Learning Rules

Although learning is one of the primary characteristics of neural networks, the ability to learn is not particularly mysterious. This may be illustrated by what we might call evolutionary learning, a procedure that learns, but rather slowly. Make random variations in the weights ω^{β}_{ij} and the thresholds θ^{β}_{i} in the nonlinear mapping presented earlier, and retain only those variations that reduce E. Thus we adjust weights and thresholds at random. When such an adjustment results in a lower cost, retain it. When it raises the cost, discard and try it again. Repeat this procedure on all inputs (the entire environment) or a statistically equivalent sample over and over again. If a solution exists (energy equals zero), eventually one will arrive at it unless one gets stuck in a local minimum (to get out of a local minimum, one can introduce noise). In any case of reasonable complexity, such a procedure takes a very long time.

Backward Propagation of Error, Generalized Widrow–Hoff, or Gradient Descent

To improve the very slow speed of random learning, various methods of directed learning have been proposed. These clearly increase speed of descent, but increase the computation in each adjustment. The most famous is called backward propagation of error (Parker, 1985; Rumelhart, Hinton, & Williams, 1986; Werbos, 1974). In this much discussed procedure, the synaptic weights are modified according to

$$\Delta w_{ij} = \eta d_i c_j \qquad (26)$$

where δ_i is the error signal, η is a small constant, and c_j is the output of cell j, and $\delta_i = (t_i - c_i) \cdot c_i'$ if i is an output unit, and where c_i' is the derivative of transfer function.

The error propagates backward through the network from the last layer (Figure 13). An iterative formula (backward propagation) can be defined relating the

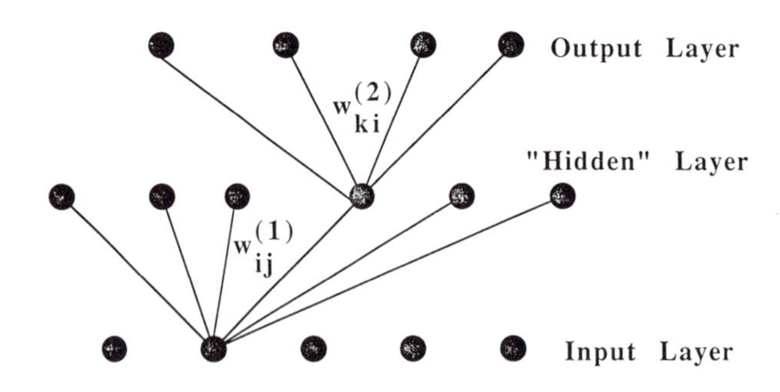

FIGURE 11 Multilayer network.

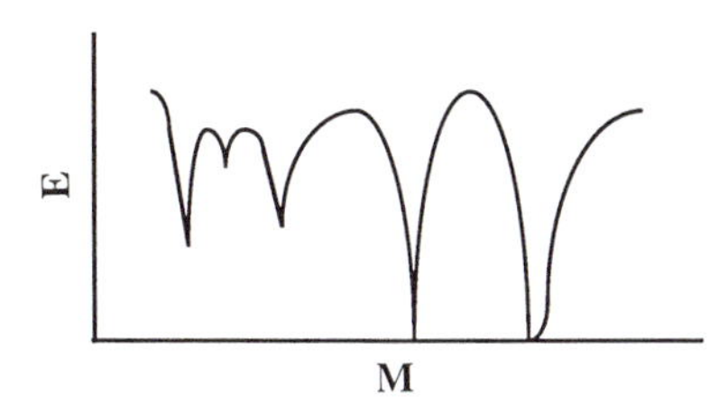

FIGURE 12 The cost as a function of M.

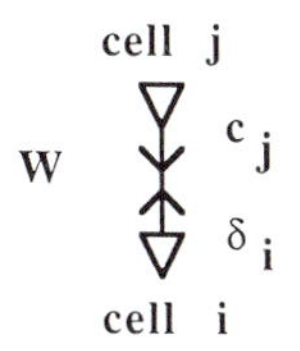

FIGURE 13 Backward propagation of error.

error signal at a given connection to the error signal of the layer after.

$$\delta_k = c'_k \sum_i \delta_i w_{ik} \tag{27}$$

Although this procedure speeds the learning process, it suffers from a problem common to all methods based on cost reduction. In many applications all data must be represented to learn the solution (Figure 14). In essence, the problem of which minimum is found depends on the starting point (initial conditions). Once in an incorrect minimum it is very difficult, if not impossible, to find a way out.

RELAXATION MODELS AND NEURAL NETWORKS

A network of neurons is a very nonlinear dynamic system. Because of recurrent connections, time delays, etc., a given input will, when regarded in detail, produce a very complicated response that in time may or may not settle down to some stable state. The evolution of the neural network state in such short time intervals (≈ 1 sec) can be and has been described by sets of coupled nonlinear equations (Wilson & Cowan, 1973; Edelman & Reeke, 1982; Grossberg, 1982) or can be approximated by some discrete updating mechanism (Anderson, Silberstein, Ritz, & Jones, 1977) perhaps the best known of which is Hopfield's (1982) model. In what follows, we describe the Hopfield model which illustrates in a fairly transparent manner some of the properties of such systems (Figure 14).

Hopfield's Model

Figure 15 is a schematic representation of the Hopfield model. Note that the architecture is fully interconnected. The updating procedure is random and asychronous. Learning was not emphasized in the original model, although some modification procedures do exist (Hopfield, Feinstein, & Palmer, 1983; Potter, 1987). The updating procedure is

$$c_i \rightarrow \theta(\sum_{j=i} w_{ij} c_j) \tag{28}$$

The corresponding Hopfield "energy" is:

$$E = -\tfrac{1}{2} \sum_j w_{ij} c_i c_j \tag{29}$$

Because the relaxation procedure is the discrete analog of gradient descent, the network will always descend to a local minimum.

$$\Delta E = -(\sum_{j=1}^{N} w_{ij} c_j) \Delta c_i \tag{30}$$

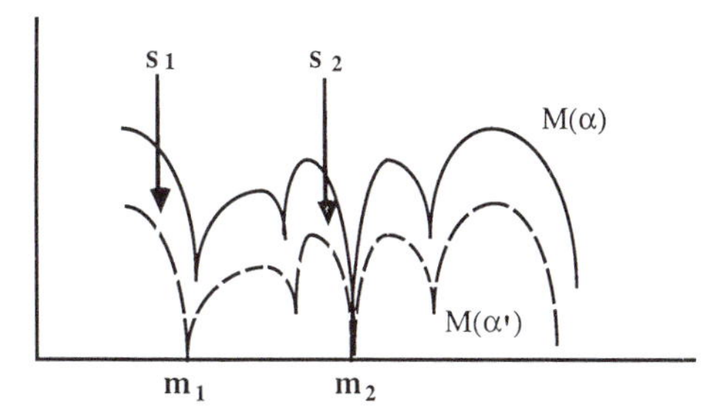

FIGURE 14 If the starting point is S_1, the minimum m_1 may be found for the restricted data set, α'. For the full data set, α, the only true minimum is m_2. This would be found with the restricted data set, α', only if the starting point were S_2.

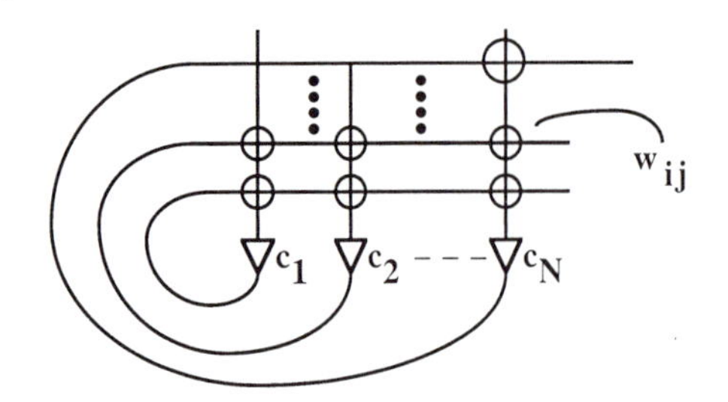

FIGURE 15 Schematic representation of the Hopfield model. w_{ij}, weights; $c_i = \pm 1$.

Since $(\sum_{j=1}^{N} w_{ij}c_j)$ and Δc_i always have the same sign, when $\Delta c_i \neq 0$ due to the updating procedure, we have $\Delta E \leq 0$. A frequent problem with relaxation procedures is that local mimima may be true memories or *spurious minima* (Figure 16).

Also, the number of spurious minima increases dramatically with the number of stored memories. Some problems of spurious minima in relaxation models may be overcome, however slowly, through the use of Boltzmann machines (Ackley, Hinton, & Sejnowski, 1984). The architecture of the Boltzmann machine is fully interconnected. Hidden units are used. The updating is not only random and asychronous, but also probabilistic. The system may escape from spurious minima due to noise introduced via the concept of temperature. In the update rule the probability to have c_i go to 1 is:

$$P(c_i \to 1) = \frac{1}{1 + e^{-\Delta E_i / T}} \tag{31}$$

Relaxation Model with High Storage Density

Whether it is more appropriate to describe the succession of stages in cortex by equilibrium feedforward or dynamic relaxation mechanisms is not yet known. It is quite possible that both have their regions of applicability. However, from the point of view of modeling the succession of states that correspond to recall and association of stored memory states, little seems to be gained by the use of elaborate relaxation mechanisms. All of the results can be obtained in a simpler and more transparent manner using network equilibrium methods that will be described in the following sections.

The problem of spurious memories in the Hopfield model is due to unwanted local or absolute minima in the energy equation (29). One might therefore ask whether it is possible to construct an energy function with minima only at the desired memory sites. This problem was solved in a model proposed by Bachmann, Cooper, Dembo, and Zeitouni (1987).

In this model, memory sites are represented by r^i, i=1...K. A network state is given by r. An "electrostatic" energy (Figure 17) is defined as a function of the memories and the network state. A simple case in three dimensions is

$$E = -\tfrac{1}{2}\sum_{i=1}^{K} Q_i \left| r - r^i \right|^{-1} \tag{32}$$

Bachmann et al, (1987) show that this energy has minima only at the designated sites. Such memory

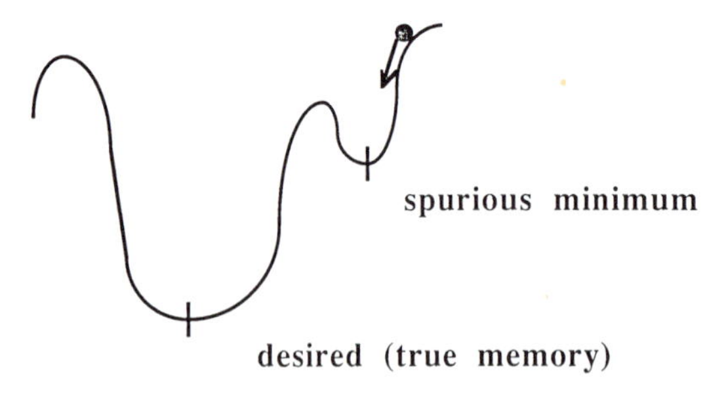

FIGURE 16 Spurious minima in the Hopfield model.

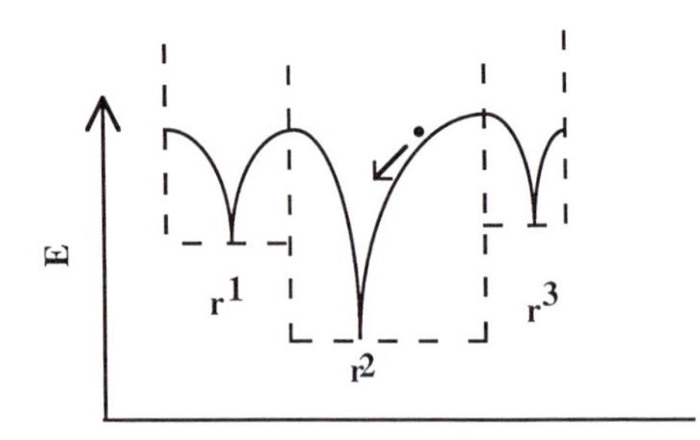

FIGURE 17 Electrostatic energy.

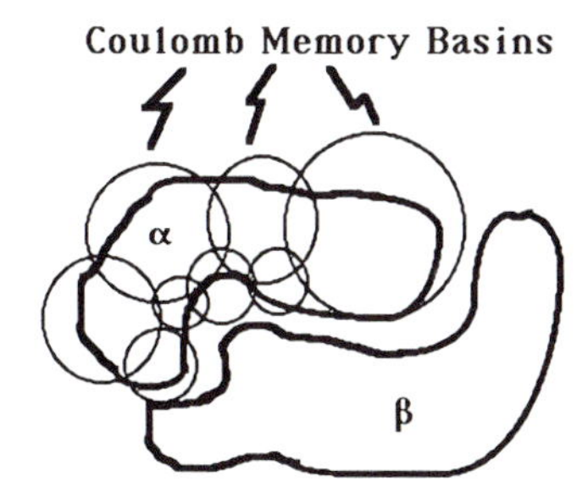

FIGURE 18 Coulomb memory basins.

sites with variable-width attractive basins can be used to outline a class territory as in Figure 18.

Now, we present an *equilibrium system* mapping, R, based on these ideas with *low connectivity*, *high density memory storage*, and *excellent ability to make distinctions* (separations in decision space and rapid learning) in which an arbitrary input e is mapped into output c by

$$c = R^{-1}e \text{ or } Rc = e \tag{33}$$

RCE Neural Network

We can describe the RCE neural network (Cooper, Elbaum, & Reilly, 1982; Reilly, Cooper, & Elbaum, 1982) in terms of its architecture, the transfer functions of its elements, and the learning laws that govern how the values of the weights and processing element parameters change over time.

The architecture of the RCE network specifies three processing layers: an input layer, an internal layer, and an output layer (Figure 19). Each node in the input layer registers the value of a feature describing an input event. If the application is character recognition, these features might be counts of the number of line segments present at various angles of orientation. If the application is signal classification, the features might be the power in a signal at various frequency bandwidths. If the application is emulating the judgment of a mortgage underwriter, the features would represent information derived from the mortgage application.

Each cell in the output layer corresponds to a pattern category. The network assigns an input to a category if the output cell for that category "fires" in response to the input. If an input causes only one output cell to become active, the decision of the network is said to be "unambiguous" for that category. If multiple output cells are active, the network response is "ambiguous." Though confused about the identity of the pattern, the network nonetheless offers a set of likely categorizations. Cells in the middle or internal layer of the network construct the mapping that ensures that the output cell for the correct category fires in response to a given input pattern.

The internal layer is fully connected to the input layer; each cell on the internal layer is connected to every cell on the input layer. The output layer is sparsely connected to the internal layer; each internal layer cell projects its output to only one output layer

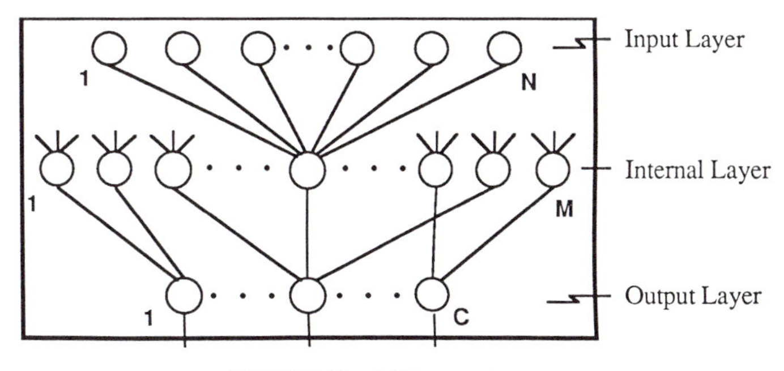

FIGURE 19 RCE network.

cell. Note that there are no recurrent connections in the RCE network, nor are there any lateral connections between cells in any given layer. The RCE is a reduced connectivity, feedforward network. An input pattern on the input layer creates a pattern of activity on the internal layer which in turn produces some set of firing cells on the output layer. All this occurs in a single pass through the network.

The connections between the ith internal layer cell and the input layer cells define a weight vector, w_i. If f represents the activity pattern of the input layer, then the transfer function of the ith internal layer cell in the RCE is given by Equation (34)

$$g_i = \Lambda_\lambda (D(w_i, f)) \tag{34}$$

$D(w_i, f)$ is some defined distance metric between vectors w_i and f (e.g., a Cartesian distance between real-valued vectors; a Hamming distance between binary-valued vectors; an inner product between normalized pattern vectors, etc.). Λ_λ is some threshold function, chosen appropriately for D. (If D is a Hamming distance, then $\Lambda_\lambda(x) = 1\ x \leq \lambda$; 0, otherwise).

The transfer function of a cell in the output layer of the RCE is such that an output layer cell performs a logical "OR" function on its inputs. Its inputs are unit strength signals from the firing internal cells to which it is connected. The connection strengths between an output cell and its suite of internal cells are all unit valued. Thus, if at least one of the internal layer cells connected to an output layer cell is firing, the output cell will fire.

The transfer function (Equation 34) associates each internal layer cell with a region of the input feature space (the cell's "influence field" or "attractive basin"). The location of the region is specified by the vector of weights coupling the cell to the input layer nodes. The size of the region is determined by the firing threshold of the cell. The geometry of the region is determined by the choice of distance metric. (For example, a Cartesian distance metric in a continuous-valued pattern space, R^N, results in the geometry of an N-dimensional sphere). Any input pattern falling within the influence region of the cell will cause the cell to fire.

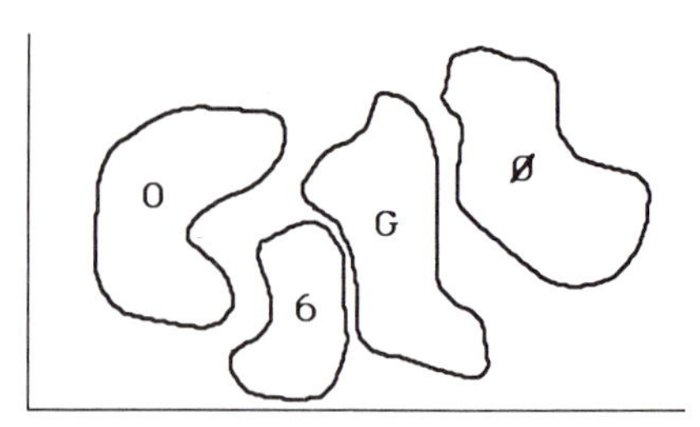

FIGURE 20 Hypothetical category territories for 0, 6, G, ϕ in a two-dimensional feature space.

The essence of the function of the network can be seen by regarding a pattern category as a collection of points in the N-dimensional feature space defined by the N cells of the input layer. A pattern of activity among the cells of the input layer corresponds to the location of a point in this feature space. All examples of a pattern category define a set of points in this feature space that can be characterized as a region (or a set of regions) having some arbitrary shape (Figure 20).

Just as a category of patterns defines a region (or regions) in the feature space of the system, a cell in the internal layer of the RCE network is associated with a set of points in the feature space. The geography of this set of points is defined by the transfer function of the internal cell.

For purposes of illustration, consider an RCE network with only two cells on the input layer. (In actual applications, a user defines as many input cells as he needs to represent his input feature vector to the system.) The transfer function can be thought of as defining a disk-shaped region centered at the feature space point w_i (the vector of weights coupling the ith internal cell to the input layer), with a radius λ_i around the w_i (Figure 21). Any point (feature vector) falling within this region will cause this internal cell to become active.

Thus in an RCE network, the internal layer cells define a collection of disks in the space of input patterns (Figure 21). These disks represent the memories built up in the RCE. Some of the disks may overlap. If a pattern falls within the attractive basins of several internal cells, they will all fire and fire their associated output cells. If all these internal cells are projecting

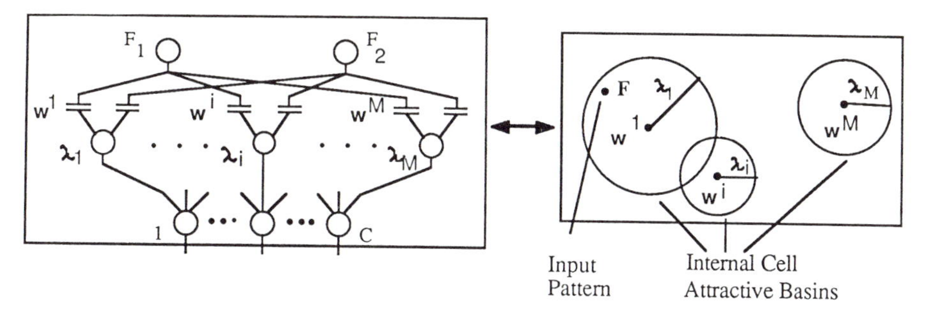

Figure 21 Feature space for a two-dimensional input pattern space showing input pattern (**F** = F_1, F_2) and RCE memories as attractive basins.

their outputs to the same output cell, then only that output cell will fire and the response of the network will be an unambiguous classification of the input. On the other hand, if the firing internal layer cells project their outputs to more than one output cell, then all of those output cells will fire, resulting in an ambiguous assignment of several possible classifications to the input pattern.

Learning involves two distinct mechanisms in the network. The first is cell commitment. Cells are "committed" to the internal layer as well as, though less frequently, to the output layer. When cells are committed, they are "wired up" according to the RCE network paradigm. Each cell in the internal layer is connected to the outputs of each of the cells in the input layer. Each cell in the internal layer projects its output to only one cell in the output layer. The second learning mechanism in the RCE network is the modification of the thresholds associated with cells in the internal layer. Thus, each internal processing element has its own weight vector w_i and threshold λ_i and their values are changed under separate modification procedures.

Both the commitment of cells and the adjustment of internal cell thresholds are controlled by training signals that move from the output layer back into the system. If an output cell (representing a given category of patterns) is off (0) and should be on (1), an error signal of +1 is generated for that output cell. If an output cell is on (1) that should be off (0) an error signal of -1 travels from that cell back into the internal layer.

An error signal of +1 traveling from the kth output layer cell into the system causes a new internal processing element to be committed. Its output is connected to the kth cell in the output layer, and its vector

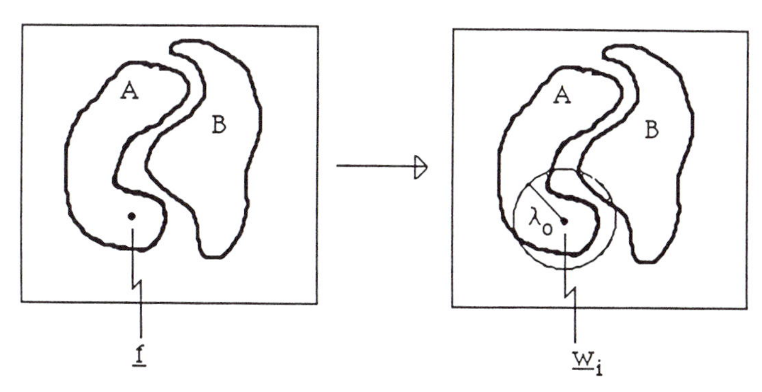

FIGURE 22 Feature space representation of cell commitment in internal layer.

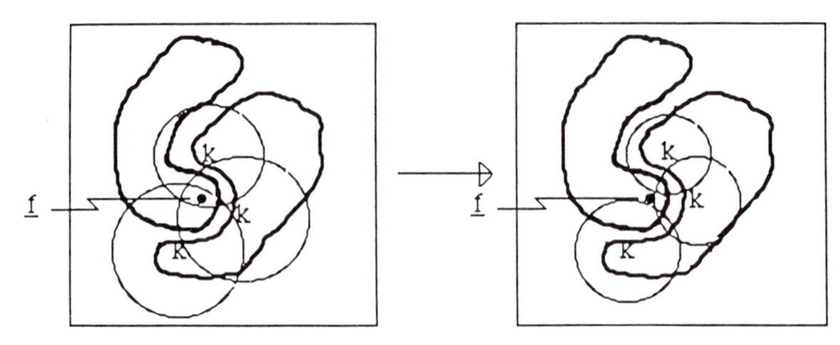

FIGURE 23 Threshold reduction for active internal layer cells associated with K^{th} output cell. After modification, f is no longer covered by any internal cell disk for k^{th} output cell.

of connections w_i to the input layer assumes the value of the current feature vector of the input layer. (w_{ij} <- f_j, $j = 1...N$) The threshold of the cell is assigned a value λ, where λ is defined as $\lambda = \max(\lambda_{max}, \lambda_{opp})$ where λ_{opp} is the distance to the nearest center of an internal cell influence field (the cells' weight vector w) for any class different from that of the current pattern, and λ_{max} is the maximum size of an influence field ever assigned an internal cell (a user-defined parameter of learning).

In the feature space of the system, this adds a new disk-shaped region that covers some portion of the class territory for this input's category (Figure 22).

If an error signal of -1 is sent from the kth output unit back into the system, then the system responds by reducing the threshold values (λ_i) of all the active internal units that are connected to the kth output cell. This has the effect of reducing the sizes of their disk-shaped regions so that they no longer cover the input pattern (Figure 23).

Table 1 summarizes the effect on the network of the various error signals traveling back from the output layer.

The RCE network can learn to distinguish pattern categories even when they are defined by complex relationships among the features characterizing the patterns. Figure 24 is taken from a simulation in which two hypothetical geometries were defined to represent two arbitrary pattern classes, α and β.

Training the network consisted of selecting a random set of points from the two class territories and presenting these points as input patterns to the system for training. From seeing the training samples, the system builds up a set of disks that covers the territories for the two hypothetical pattern classes α and β. Each disk corresponds to the "win region" of an internal unit and each disk is "owned" by a pattern category by virtue of its unique connection to one and only one output cell. The size of the disk is related to the threshold of the corresponding internal unit. Disk sizes that are too large cause the wrong internal cell to fire in response to a pattern input. The resultant "-1" training signal reduces the disk size to prevent it from firing for that pattern (or patterns like it) again. This process of committing disks and reducing their sizes allows the system to develop separating mappings even for

TABLE 1

Error signal from i^{th} output cell	Modification
+1	Commit an internal cell for this event and connect it to i^{th} output cell
-1	Reduce thresholds of all active internal cells connected to this output cell in order to turn the output cell off
0	No change to any of the internal cells connected to this output cell

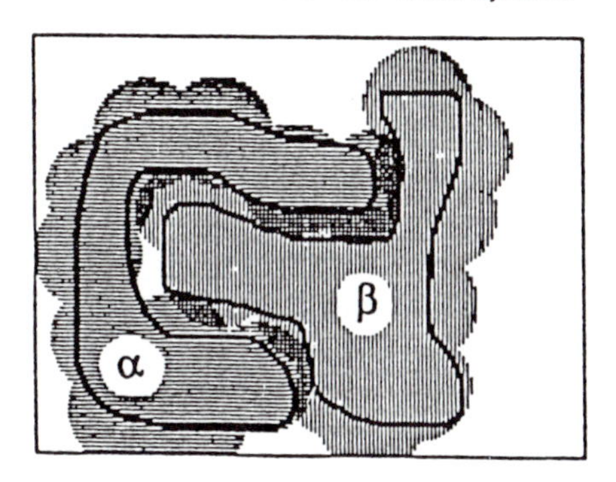

FIGURE 24 RCE network separating classes α and β.

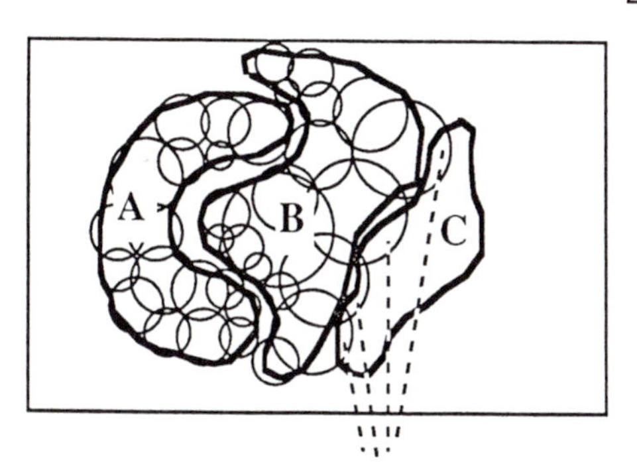

FIGURE 26 Dynamic category learning. New category C can be introduced after categories A and B are learned. Some examples of C will be classified incorrectly as B. Thus, the network needs to be shown examples of C and B, not of A.

nonlinearly separable pattern categories. In principle, an arbitrary degree of nonlinearity can be achieved.

Furthermore, this learning strategy is able to map out category regions even if they are disjoint in the feature space. This allows the system to learn concept formation as required for a problem. For example, the shapes of the images of "A" and "a" have little in common. In a typical feature space, their pattern territories are most likely not connected, or "disjoint" (Figure 25). Nonetheless, the system can map out both of these regions separately and, if so instructed during training, associate them with the same output cell representing the concept "A".

Dynamic category learning is possible with the RCE network. New classes of patterns can be introduced at arbitrary points in training without always involving the need to retrain the network on all of its previously learned data. Assume that the network has been fully trained on classes A and B. Imagine a new class of patterns C is to be introduced that lies near B in the feature space (Figure 26). To some examples of C the network will respond with B; some of the internal cells for B are generalizing more than they should. To train the network, examples of the new category C must be shown, as well as examples of the categories that are near C in the pattern space. The nearby categories are generally those that the network offers as incorrect answers to examples of the new class. Thus, in the above example, the network must be shown examples of class C and the previously learned class B. However, examples of A will not need to be shown to train the network on the new class C.

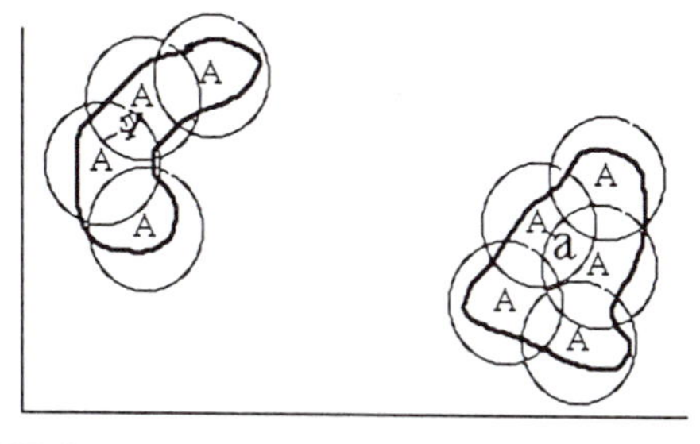

FIGURE 25 RCE network training to map subclasses "A" and "a" onto single category "A".

The RCE is a partially distributed network in the sense that the information about class A is not distributed over all the weights in the network, but rather over some subset of weights, namely, those associated with the internal cells that have their outputs targeted to the class A output cell. This still provides a reasonable degree of fault tolerance. If one of these internal cells fails to function, only a partial amount of the information about the given class is lost or degraded. This is because a class is generally represented by the overlap of disks associated with a number of internal cells.

The RCE can be trained to sharpen its understanding of a given class by showing it examples of pattern "noise", examples that have no class affiliation. Training on noise input allows the system to better

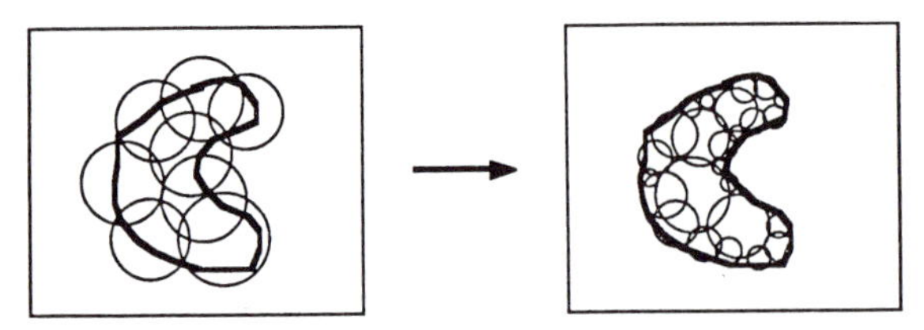

FIGURE 27 "Noise" inputs are any patterns taken from territory outside the class region outline. Training on noisy inputs improves initial approximate class mapping (*left*) by reducing the size of internal cell influence fields (*right*). No internal cells need be committed for noise.

define the extent of the boundary between the class and background noise (Figure 27). Furthermore, the net does not build up an explicit model of the noise, because no internal cells need to be committed for noise inputs.

In some pattern classification problems, there might be overlap between the class territories in the feature space (Figure 28). Within this confusion zone the pattern classes are not, strictly speaking, separable. To patterns falling within that territory, the network should respond with some measure of the likelihood of the most probable class affiliation.

In training, internal cells cannot have their influence field sizes reduced below some critical size, λ_c. Cells with influence field size reduced to this value are termed "probabilistic" cells; all other internal cells are referred to as "deterministic." When a probabilistic cell fires, it fires the output cell to which it is connected, but, weakly, in the sense that the output of the network is now officially "ambiguous." Even if only one output cell is firing, if it is being stimulated only by probabilistic internal cells, the answer of the network remains ambiguous. The output cell's classification is offered as a possible, but not definite classification of the input.

During learning, probabilistic cells block the commitment of deterministic cells in any space that they occupy. If an input pattern for class B falls only within the influence fields of one or more probabilistic cells for class A, the network will not commit a deterministic cell for this pattern. It will, however, commit a probabilistic cell for class B, centered at the input pattern site, with influence field equal to λ_c. The network allows the influence fields of probabilistic cells for one class to cover the center point of probabilistic cells for another class. This is another way in which they will differ importantly from deterministic cells. The commitment procedure will result in a layering of the confusion zone with probabilistic cells for A and a layering with probabilistic cells for B. In this way, any input pattern that falls within the confusion zone will fire at least one probabilistic cell for A and one for B. This will cause the network to respond with the ambiguous response, A or B. In a later evolution of the RCE, these probabilistic cells keep pattern and classification counts that allow them to estimate the probability that a pattern falling within their influence field is an example of the class they represent (Scofield, Reilly, Elbaum, & Cooper, 1987).

The memories that the system learns during training are stored in the weight vectors between the input layer and the internal layer. The number of cells in the internal layer grows automatically as a function of the complexity of the problem. This is in contrast to other network models in which the size of the network is

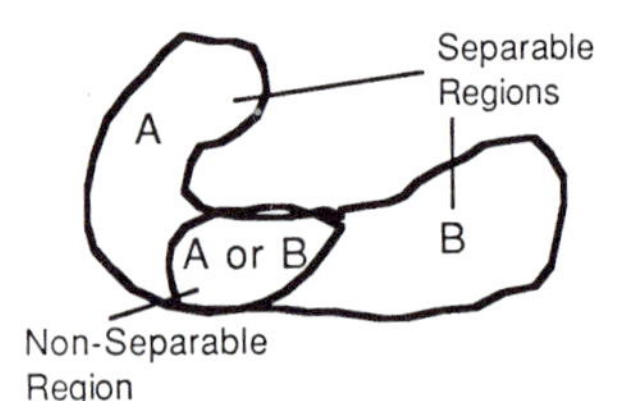

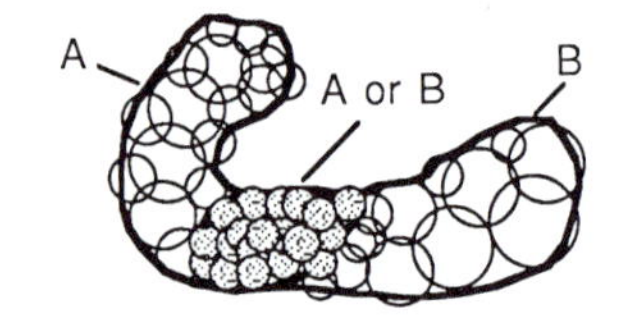

FIGURE 28 On the left, two class territories that overlap in a confusion zone. On the right, separable territories are mapped out with deterministic cells (open circles); the confusion zone, with probabilistic cells (filled circles).

fixed and can only be predetermined through trial and error. The number of memories that the RCE network can develop is limited, in principle, only by the number of available sites in the pattern space. In networks such as the fully interconnected Hopfield network, the number of memories that can be stored is typically limited to 10% of the number of nodes in the network. Various procedures have been studied to relax this memory storage limitation, although none approaches the 1 memory/1 cell ratio of the RCE net. Additionally, as more nodes are added to the Hopfield net, the number of connections grows as the square of the number of nodes. Due to the essentially feedforward connectivity of the RCE net, the number of connections grows linearly with the number of nodes.

We can summarize the features discussed for the RCE network. Its reduced internal connectivity and feedforward architecture reduces the number of computations that need to be performed within the net to compute a response. This has made it possible to develop software applications of the RCE net that run in real time. Furthermore, these architectural properties simplify the problem of designing special purpose silicon for chip level implementation of the net. The network can learn rapidly, generalizing quickly to the notion of a pattern category after only a few examples of the class are presented. This is due to the fact that changes to the weight vectors of internal layer cells are on the same order of magnitude as the input signal, and not some fraction of the signal magnitude. The network is able to resolve pattern classes that are not linearly separable, and even pattern classes represented by disjoint territories in the feature space. The network can store large numbers of memories if required for learning many complex categories. It automatically commits new cells in the internal and output layers as required to accommodate this need. Its partially distributed character allows it to make local changes to memory, thus providing it a capability for true dynamic category learning.

Multiple Network System

In many real-world pattern recognition problems a single representation is not appropriate for all the classification decisions the system must make. (The representation of the problem is defined by the set of features whose measurements characterize the pattern as an input to the system.) Often, decision making in a problem occurs on the basis of information contained in subsets of all the possible features that characterize the data. The definition of these feature subsets in effect partitions the entire feature space into subspaces. These partitions may arise naturally in a problem as a consequence of different sensor sources for the data. (A feature subset can be associated with each sensor.) Other partitions are suggested by common themes in the characteristics being measured for the pattern (description of a shape in Cartesian versus polar coordinates, measurements in length versus spatial frequency, etc). In other cases, feature subsets are a consequence of general knowledge about the structure of decision making in the given problem domain. Natural partitioning of the feature set can also result from the introduction of new features to the system late in the process of learning.

A multiple neural network system has been designed in which a number of RCE networks are combined together with a Controller module (Cooper, Elbaum, Reilly, & Scofield, 1988; Reilly, Scofield, Elbaum, & Cooper, 1987). The Controller integrates the responses of the RCE networks into a system response and, on the basis of corrective feedback from an instructor, directs the training of the networks in the system (Figure 29). Each RCE network processes a user-defined feature subset. By virtue of the feature subset it is processing, a particular RCE network may be able to be trained to make unambiguous, correct decisions about patterns belonging to certain categories. For other categories, the partitioning may be such that no single RCE network has enough information to reliably classify the event. In such cases, an RCE network can develop category mappings that at least allow it to indicate likely pattern categories for the input. The Controller then correlates the answers for several such networks in an attempt to identify the pattern.

In this multiple neural network system, the RCE nets are arranged in user specified groups or levels. A

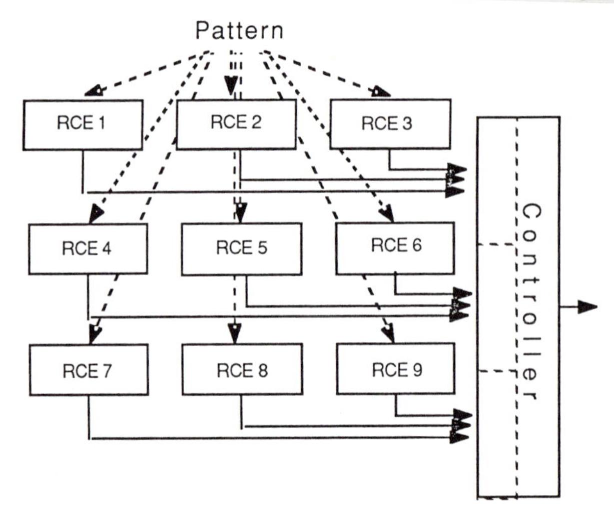

FIGURE 29 Architecture of a 3 x 3 multiple neural network system, showing Controller receiving classification outputs from each of the component 9 RCE neural nets.

level may contain as few as one or as many as all the nets in the system. The Controller polls the RCE networks, level by level, until it has enough information to decide about the pattern's identity. Networks at a certain level in the system are only accessed if the Controller has been unable to decide on an overall system response, given the responses of previous levels.

In a multiple neural network system, the ability of the RCE network to build up knowledge not only of the shapes of separable pattern class territories but also of the approximate shape of overlapping class territories allows the Controller to correlate ambiguous answers of several networks to produce an unambiguous response. For example, network 1 may have enough information to identify class α, but cannot distinguish classes β and γ. On the other hand, network 2 may be able to identify γ, but cannot distinguish between α and β. (Class territories for this problem may look like those shown in Figure 30.) Each RCE network maps out the shapes of the confusion zones in its feature spaces. Consequently, the system can identify examples of β on the basis of responses of both networks. Examples of β will produce a confused response of (β,γ) in network 1 and (α,β) in network 2. The Controller can integrate these ambiguous answers to decide upon β as the pattern's identity.

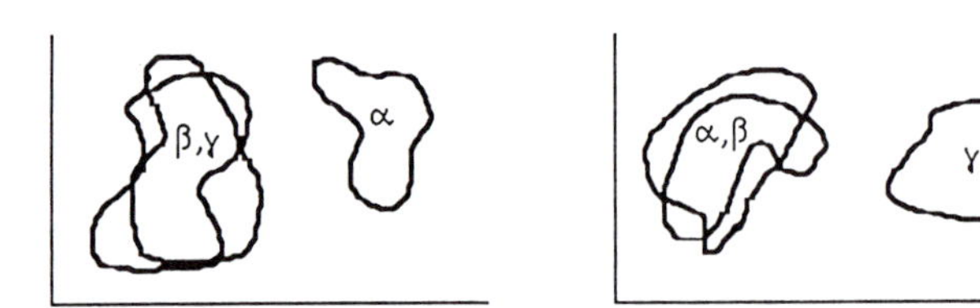

FIGURE 30 Hypothetical class territories for classes α, β, and γ on two different feature spaces. In the feature space on the left, β is largely indistinguishable from γ; in the feature space on the right, β is degenerate with α.

Component networks in the system that have, by virtue of their assigned feature subsets, a certain expertise at deciding for some set of categories must be able to develop that expertise during the course of training. This is ensured by several factors. First, the algorithm that the Controller uses to assemble the system response directs it to weight more heavily a network whose response is unambiguous over one whose response is a set of classes. After training, networks that cannot develop a separable mapping for a class of patterns tend to produce confused responses or no response to examples of that pattern class. Thus, this aspect of the Controller's function ensures that at the least, the confused responses of an inexpert network can be overlooked in favor of the unambiguous responses of an expert network.

Training ensures that the incorrect responses of an inexpert network are "trained out." This results from the fact that the Controller broadcasts its training signals in stages to the various networks. The "-1" signals are sent first to those active RCE networks whose outputs include incorrect classifications. If this modification to system memory does not produce a correct classification of the input pattern, then "+1" signals are sent to networks on a selective basis. If at a given level there is both an expert and an inexpert network (relative to a particular pattern class), this method of training will ensure that the mapping in the inexpert network will not survive.

This is illustrated in Figure 31. There, two hypothetical pattern classes are distinguishable by one network but are totally indistinguishable by another. (In the feature space of the latter network, their class territories are identical. Consequently, only one such territory is pictured.) After initial mappings have developed, training can correct any erroneous or confused system responses simply by cell threshold reduction in the networks. This occurs more often in network 2 than in network 1. Eventually, the internal cells of network 2 have their disk-shaped regions so reduced in size as to be completely ineffective for producing a response from the net. No additional internal cells are committed in network 2 because network 1 has developed a mapping that produces unambiguous and accurate responses. One can regard this as a type of "evolutionary" learning in which a given internal cell mapping of a class territory "survives" if it is able to produce accurate classifications on the part of the network for that class. The fracturing of those mappings which produce incorrect responses renders them ineffective for generating a response.

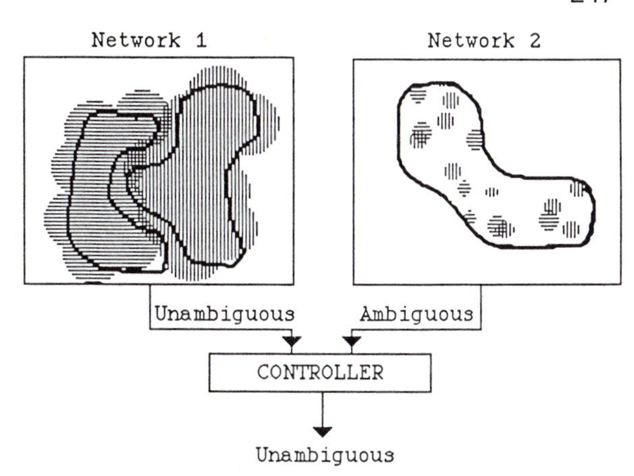

FIGURE 31 Memory states of RCE networks late in training. In network 1, the two pattern classes are represented by two different feature space regions. In network 2, the class regions are degenerate (both classes share a single region). Network 1 trains to dominate system response.

The ability of a multiple neural network learning system to learn pattern categories using feature space partitions has a number of important advantages. First, it allows the system to be configured to what is known about the problem. As an example, consider a problem in which information from two different sensors must be used to identify classes of signals. Some classes of signals may be identified solely on the basis of information from one of the sensors. Other inputs may be classified by appealing to data coming from the other sensor. For still other signals, it may only be possible to identify them by appealing simultaneously to data from both sources. A useful architecture for this problem would use three RCE networks. At the first level would be a network coupled to one sensor; at the next level, a net coupled to inputs from the other sensor. At the third level of the system would be a network coupled to inputs from both sensors.

Hierarchical filtering can be implemented in this multi-network system using a multi-level architecture. Feature subsets may carry the same kind of information, but at different degrees of resolution. For example, in a signal processing application, a representation can be generated by taking the Fourier transform of the input to yield the power spectrum of the signal. The power spectrum samples the energy in the signal at particular frequencies. That sampling can be coarse (gross averages of the power contained in a few frequency bands) or fine (more samples of the energy taken from a larger number of frequency ranges). Networks can be arranged in a hierarchy starting with the coarsest representation in the uppermost level, with successive levels carrying more detailed representations. Those signals which can be identified by appealing only to coarse information will be categorized by the upper levels. Those signal classes which can only be distinguished on the basis of very detailed information will exist as confusion zones in the upper levels. Upper level networks will pass these signals down to the deeper level nets which carry the fine scale information needed to resolve their identities.

With such hierarchical filtering, categorization knowledge can develop with significant reductions in storage space required for implementing the system. Associated with each internal cell is an amount of local storage memory determined, for the most part, by the number of weights per cell. The number of internal cell weights is determined by the number of input features and the precision to which each feature is represented. If upper levels carry coarse feature information, then the mappings that develop in these levels require less memory for storage.

In general, networks are accessed by the Controller only as it needs additional input from them to decide on the identity of a pattern. The order of accessing is determined by their place in the architecture of the system. This makes it possible to incorporate new RCE networks into a system that has already developed memories from some previous training. If the new networks are positioned at the deeper levels of the system, they are accessed only for those pattern

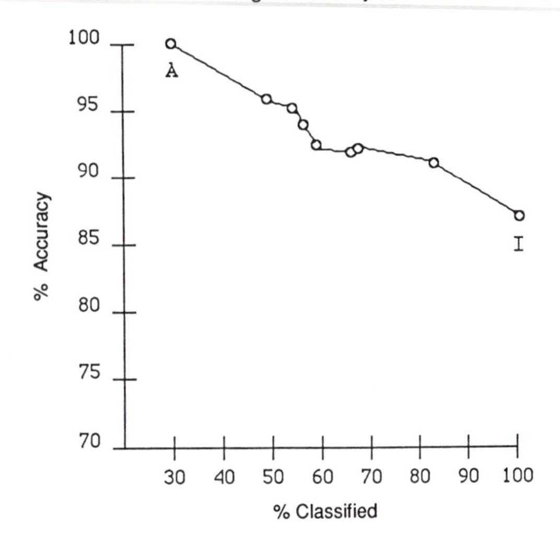

FIGURE 32 Performance of multiple neural network system trained to predict underwriter's Accept/Deny decisions on applications for home mortgages. A, minimum error mode; I, maximum response mode.

classes that the originally configured system could not identify. Importantly, the existing memory of the system is protected and the entire system does not have to be reconstituted by retraining on all previously learned categories. Thus, the system can be upgraded through the introduction of new input features without jeopardizing the current knowledge of the system.

The degree of decision-making or the "tendency for decisions" is adjustable in the system through the definition of Controller parameters. At one extreme, the system can be biased to avoid making an unambiguous decision on a pattern unless there is substantial agreement among its component networks as to the pattern's identity. This "minimum error" mode is desirable in those applications in which there is a high premium associated with each error made. At the other extreme, the system can operate in a "maximum response" mode. This mode is often useful in applications where the system is maximizing throughput, and occasional errors can be tolerated or detected by some filtering system downstream. Thus it is possible to set operating points along a continuum between strict minimum error and strict maximum response (Figure 32).

SUMMARY

In this chapter we have attempted to present an overview of neural networks from the earliest models to a current system.

Multiple RCE neural network systems can efficiently and rapidly learn to separate enormously complex decision spaces and coordinate many neural networks, each a specialist in dividing some portion of the decision space.

To demonstrate that neural networks can solve difficult real-world (as opposed to toy) problems, to show the advantages they offer, and to properly test existing neural network systems as well as to design future systems, these systems must be put to work on real world data. Often the effects we are looking for are small or obscured by noise and variability, so that we cannot be convinced that the system is functioning satisfactorily unless it is in a real world situation.

A number of applications have been developed with multiple RCE neural network systems in problem domains ranging from character recognition to vision (Rimey, Gouin, Scofield, & Reilly, 1986; Reilly et al., 1987) to decision making in financial services (Collins, Ghosh, & Scofield, 1988). Other applications for problems in signature recognition, process monitoring, multisensor fusion, and risk assessment are currently in various stages of exploration or development.

References

Ackley, D. H., Hinton, G. E., & Sejnowski, T. J. (1984). *Boltzmann machine: Constraint satisfaction networks that learn, Carnegie Mellon University, (*Technical Report no. CMU-CS 84-119).

Anderson, J. A., Silberstein, J. W., Ritz, S. A., & Jones, R. S. (1977). Brain state in a box. *Psychological Review*, **84**, 413–451.

Bachmann, C. M., Cooper, L. N., Dembo, A., & Zeitouni, O. (1987). A relaxation model for memory with high storage density. *Proceedings of the National Academy of Sciences of the United States of America*, **84**, 7529–7531.

Bienenstock, E. L., Cooper, L. N., & Munro, P. W. (1982). Theory for the development of neuron selectivity: Orientation and binocular interaction in visual cortex. *Journal of Neuroscience*, **2**, 32–48.

Collins, E., Ghosh, S., & Scofield, C. L. (1988). An application of a multiple neural network learning system to emulation of mortgage underwriting judgements. *IEEE International Conference on Neural Networks*, **2**, 459–466.

Cooper, L. N. (1973). A possible organization of animal memory and learning. In B. Lindquist & S. Lindquist (Eds.), *Proceedings of the nobel symposium on collective properties of physical systems*, (pp. 252–264). New York: Academic Press.

Cooper, L. N., Bear, M., & Ebner, F. (1987). Physiological basis for a theory of synapse modification. *Science*, **237**, 4248.

Cooper, L. N., Elbaum, C., & Reilly, D. L. (Awarded 1982). *Self organizing general pattern class separator and identifier.* U.S. Patent No. 4,326,259.

Cooper, L. N., Elbaum, C., Reilly, D. L., & Scofield, C. L. (Awarded 1988). *Parallel, multi-unit, adaptive nonlinear pattern class separator and identifier.* U.S. Patent No. 4,760,604.

Cooper, L. N., & Scofield, C. L. (1988). Mean field theory of a neural network. *Proceedings of the National Academy of Sciences of the United States of America*, **85**, 1973.

Edelman, G. M., & Reeke, G. N., Jr. (1982). Selective networks capable of representative transformations, limited generalizations and associative memory. *Proceedings of the National Academy of Sciences of the United States of America*, **79**, 2091–2095.

Grossberg, S. (1982). *Studies of mind and brain.* Dordrecht, The Netherlands: Reidel.

Hebb, D. O. (1949). *The organization of behavior.* New York: Wiley.

Hopfield, J. J. (1982). Neural networks and physical systems with emergent collective computational abilities. *Proceedings of the National Academy of Sciences of the United States of America*, **79**, 2554–2558.

Hopfield, J. J., Feinstein, D. I., & Palmer, K.G. (1983). Unlearning has a stabilizing effect in collective memories. *Nature*, **304**, 158–159.

Kohonen, T. (1984). *Self-organization and associative memory.* Berlin: Springer-Verlag.

Kolmogorov, A. K. (1957). On the representation of continuous functions of several variables by superposition of continuous functions of one variable and addition. Doklady Akademii Nauk. SSSR, **114**, 369–373.

Lippman, R. P. (1987). An introduction to computing with neural nets. *IEEE Transactions on Acoustics, Speech, and Signal Processing*, **4(2)**, 4–22.

Minsky, M. L., & Pappert, S. (1969). Perceptrons; an introduction to computational geometry. Cambridge, MA: MIT Press.

Parker, D. B. (1985). *Learning-logic* (REPORT TR-47). Cambridge MA: MIT Center for Research in Computation Economics Management Science.

Potter, T. W. (1987). *Storing and retrieving data in a parallel distributed memory system.* Ph.D. thesis. Binghamton: State University of New York.

Reilly, D. L., Cooper, L. N., & Elbaum, C. (1982). A neural model for category learning. *Biological Cybernetics*, **45**, 35–41.

Reilly, D. L., Scofield, C., Elbaum, C., & Cooper, L. N. (1987). Learning system architectures composed of multiple learning modules. *IEEE International Conference on Neural Networks*, **2**, 495–503.

Rimey, R., Gouin, P., Scofield, C., & Reilly, D. L. (1986). Real-time 3-D object classification using a learning system. *Proceedings of SPIE—the International Society of Optical Engineers*, **726,** 552–558.

Rosenblatt, F. (1958). *The perceptron, a theory of statistical separab in cognitive systems* (Report no.VG-1196-G-1). Cornell Aeronautical Laboratory, Ithaca.

Rumelhart, D. E., Hinton, G. E., & Williams, R. J. (1986). Learning internal representations by error propagation. In D. E. Rumelhart & J. L. McClelland (Eds.), *Parallel distributed processing* (Vol. 1. *Foundations*). Cambridge, MA: MIT Press.

Scofield, C. L., Reilly, D. L., Elbaum, C., & Cooper, L. N. (1987). Pattern class degeneracy in an unrestricted storage density' memory. In D. Z. Anderson (Ed.), *Neural information processing systems*. New York: American Institute of Physics.

Widrow, B., & Hoff, M. E. (1960). Adaptive switching circuits. In *Institute of Radio Engineers, Western Electronic Show and Convention, Wescon Convention Record,* **4.** 96–104.

Wilson, H. R., & Cowan, J. D. (1973). A mathematical theory for the functional dynamics of cortical and thalamic nervous tissue. *Kybernetik*, **13**, 55–80.

13

Neural Nets for Adaptive Filtering and Adaptive Pattern Recognition

Bernard Widrow and Rodney Winter

BACKGROUND

The fields of adaptive signal processing and adaptive neural networks have been developing independently, yet they have common ground in the adaptive linear combiner (ALC). When its inputs are connected to a tapped delay line, the ALC becomes a key component of an adaptive filter. When its output is connected to a quantizer, the ALC becomes an adaptive threshold element or adaptive neuron.

Adaptive filters have enjoyed great commercial success in the signal processing field. All high-speed modems in use today are equipped with adaptive equalization filters. Long distance telephone and satellite communications links are being equipped with adaptive echo cancelers for filtering out echo, allowing simultaneous two-way communications. Other applications include noise canceling and signal prediction. Adaptive threshold elements, on the other hand, are the building blocks of neural networks. Neural nets are currently the focus of widespread research interest, and applications in pattern recognition and trainable logic are being investigated. Uses of neural network systems have not yet had the commercial impact of adaptive filtering.

The commonality of the ALC to the fields of adaptive signal processing and adaptive neural networks suggests that the two fields have much to share with each other. It is the purpose of this chapter to describe some of the uses of the ALC in practical applications of signal processing and pattern recognition.

The Adaptive Linear Combiner

The adaptive linear combiner (ALC) shown in Figure 1 is the basic building block for most adaptive systems. The output is a linear combination of the many input signals. The weighting coefficients comprise a "weight vector." The input signals comprise an "input signal vector." The output signal is the inner product or dot product of the input signal vector with the weight vector. The output signal is compared with a special input signal called the "desired response," the difference being the error signal. The weighting coefficients or "weights" of the ALC are adjusted to optimize performance. Generally, the weights are adjusted

Reprinted, with permission, from *Computer*, Vol. 21, No. 3, pp. 24–39, March 1988. Copyright by IEEE.

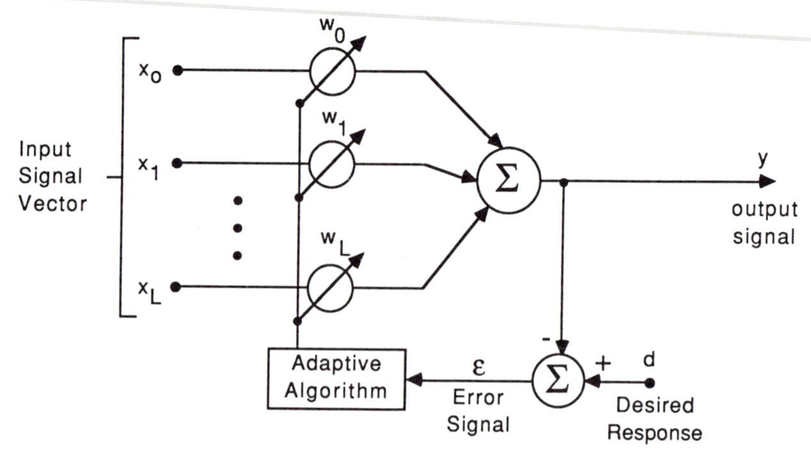

FIGURE 1 Adaptive linear combiner (ALC).

so that the mean square of the error signal is minimized. Many adaptive algorithms have been devised to automatically adjust the weights. The most popular has been the Widrow–Hoff LMS (least mean square) algorithm devised in 1959 (Widrow & Stearns, 1985).

Adaptive Filters

Digital signals generally originate from sampling continuous input signals by analog-to-digital conversion. Filtering digital signals is often done by means of a tapped-delay-line or transversal filter, as shown in Figure 2 (top). The sampled input signal is applied to a string of delay boxes, each delaying the signal by one sampling period. An ALC is seen connected to the taps between the delay boxes. The filtered output is a linear combination of the current and past input signal samples. By varying the weights, the impulse response from input to output is directly controllable. Since the frequency response is the Fourier transform of the impulse response, controlling the impulse response controls the frequency response. The weights are usually adjusted to cause the output signal to be a best least squares match over time to the desired-response input signal.

There are many other forms of adaptive filters reported in the literature (Widrow & Stearns, 1985). Some use feedback to obtain both poles and zeros. The filter of Figure 2 (top) realized only zeros. Some adaptive filters are based on lattice structures to achieve more rapid convergence under certain conditions. However, the simplest, most robust, and widely used filter is that of Figure 2 (top) adapted by the LMS algorithm.

Adaptive Threshold Element

An adaptive threshold element, a key component in adaptive pattern recognition systems, is shown in Figure 3. It is composed of an adaptive linear combiner cascaded with a quantizer. The output of the ALC is quantized to produce a binary "decision." Most often, the inputs are binary and the desired response is binary. As such, the adaptive threshold element is trainable and capable of implementing binary logic functions. The LMS algorithm was originally developed to train the adaptive threshold element of Figure 3. This element was called an "adaptive linear neuron" or "ADALINE" (Widrow, 1962).

The adaptive threshold element was an early neuronal model. The adaptive weights were analogous to synapses. The input vector components related to the dendritic inputs. The quantized output was analogous to the axonal output. The output decision was determined by a weighted sum of the inputs, much the same way real neurons were believed to behave.

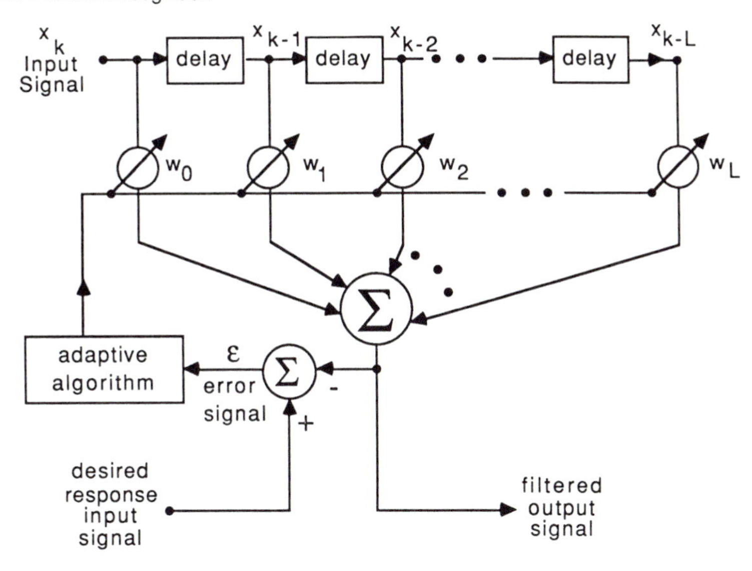

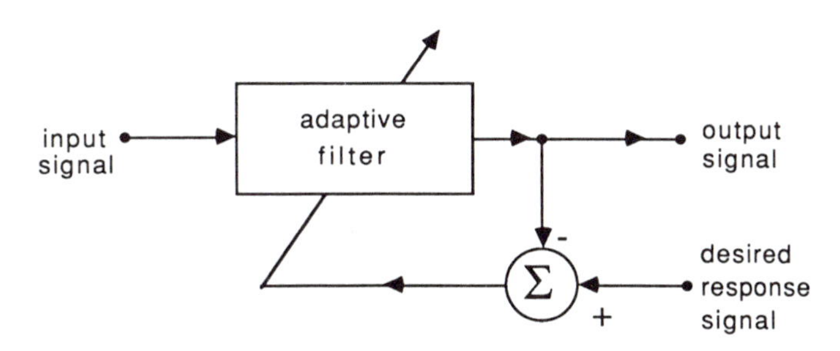

FIGURE 2 Adaptive digital filter. *Top,* details of a tapped-delay-line digital filter. *Bottom,* symbolic representation of an adaptive filter.

ADAPTIVE SIGNAL PROCESSING

A symbolic representation of an adaptive filter is shown in Figure 2 (bottom). The filter has an input signal and produces an output signal. The desired response is supplied during training. A question naturally arises. If the desired response were known and available, why would one need the adaptive filter? Put another way, how should one obtain the desired response in a practical application? There is no general way to answer these questions. Insight can be obtained, however, by study of successful application examples.

Example 1—System Modeling

In many engineering and scientific applications, a system of unknown structure is present whose input and output signals are observable. Knowledge of the dynamic response of the unknown system is desired.

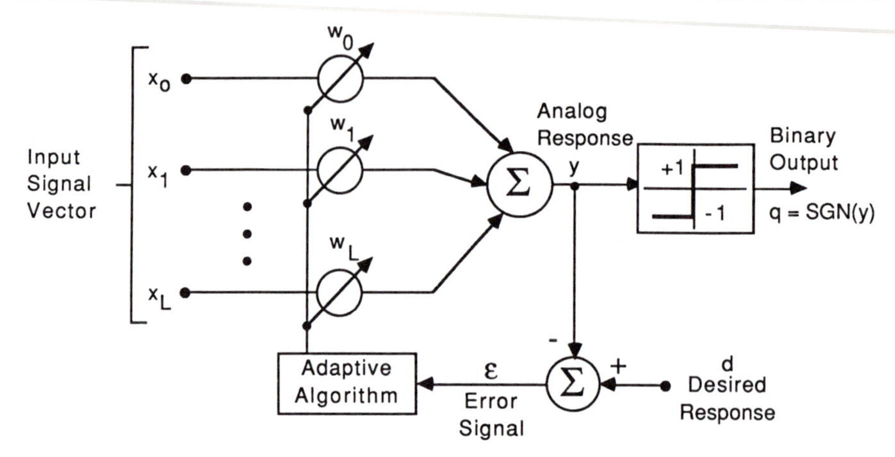

FIGURE 3 Adaptive threshold element (ADALINE).

One way to find such knowledge is shown in Figure 4. The unknown system's input is applied to an adaptive filter. The unknown system's output is used as the desired response of the adaptive filter. The adaptive filter develops an impulse response to match that of the unknown system since the filter and the system develop similar outputs when driven by the same input.

Example 2—Statistical Prediction

Future values of time-correlated digital signals can be estimated from present and past input samples. Optimal linear least squares filtering techniques for signal prediction were developed by Norbert Wiener (Kailath, 1981). Knowing the autocorrelation function of the signal to be predicted, Wiener's theory yields the impulse response of the optimal filter. More often than not, however, the autocorrelation function is unknown and is possibly time variable. One could use a correlator to measure the correlation function and plug this into Wiener theory to get the optimal impulse response, or one could get the optimal prediction filter directly by adaptive filtering. The latter approach is illustrated in Figure 5.

Referring to this figure, the input signal is delayed by Δ units of time and then fed to an adaptive filter. The undelayed input serves as the desired response for this adaptive filter. The filter weights adapt and con-

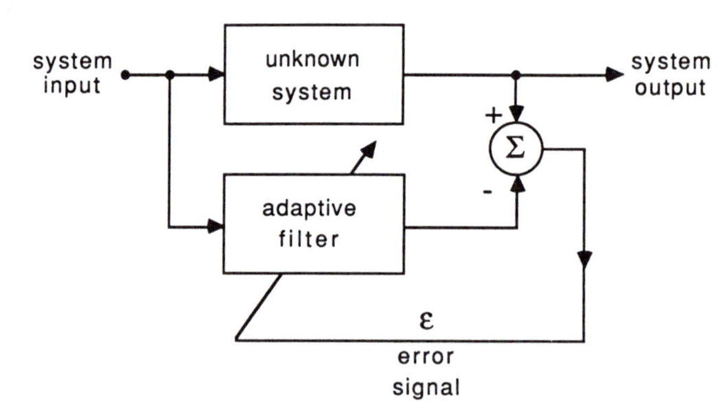

FIGURE 4 System modeling.

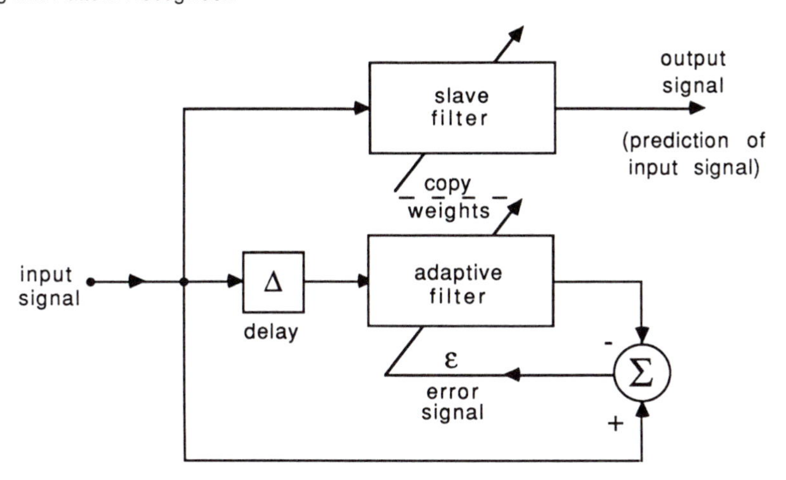

FIGURE 5 An adaptive statistical predictor.

verge to produce a best least squares estimate of the present input signal, given an input which is this very signal delayed by Δ. The optimal weights are copied into a "slave filter" whose input is delayed and whose output therefore is a best least squares prediction of the input Δ units of time into the future.

Example 3—Noise Canceling

A common problem in signal processing is that of separating a signal from additive noise. Figure 6 (top) shows a classical approach to this problem using optimal Wiener or Kalman (Kailath, 1981) filtering. The purpose of the optimal filter is to pass the signal s without distortion while stopping the noise n_0. In general, this cannot be done perfectly. Even when using the best filter, the signal will be distorted and some noise will go through to the output. Figure 6 (bottom) shows another approach to the problem using adaptive filtering. This approach is viable only when an additional "reference input" is available containing noise n_1 which is correlated with the original corrupting noise n_0. In Figure 6 (bottom), the adaptive filter receives the reference noise, filters it, and subtracts the result from the noisy "primary input," $s + n_0$. For this adaptive filter, the noisy input $s + n_0$ acts as the desired response. The "system output" acts as the error for the adaptive filter. Adaptive noise canceling generally performs much better than the classical approach since the noise is subtracted out rather than filtered out.

One might think that some prior knowledge of the signal s or of the noises n_0 and n_1 would be necessary before the filter could adapt to produce the noise canceling signal y. A simple argument will show, however, that little or no prior knowledge of s, n_0, or n_1, or of their interrelationships is required.

Assume that s, n_0, n_1, and y are statistically stationary and have zero means. Assume that s is uncorrelated with n_0 and n_1, and suppose that n_1 is correlated with n_0. The output is

$$\varepsilon = s + n_0 - y \tag{1}$$

Squaring, one obtains

$$\varepsilon^2 = s^2 + (n_0 - y)^2 + 2s(n_0 - y) \tag{2}$$

Taking expectations of both sides of Equation 2 and realizing that s is uncorrelated with n_0 and with y yields

$$\begin{aligned} E[\varepsilon^2] &= E[s^2] + E[(n_0 - y)^2] + \\ 2E[s(n_0 - y)] &= E[s^2] + E[(n_0 - y)^2] \end{aligned} \tag{3}$$

Adapting the filter to minimize $E[\varepsilon^2]$ will not affect the signal power, $E[s^2]$. Accordingly, the minimum

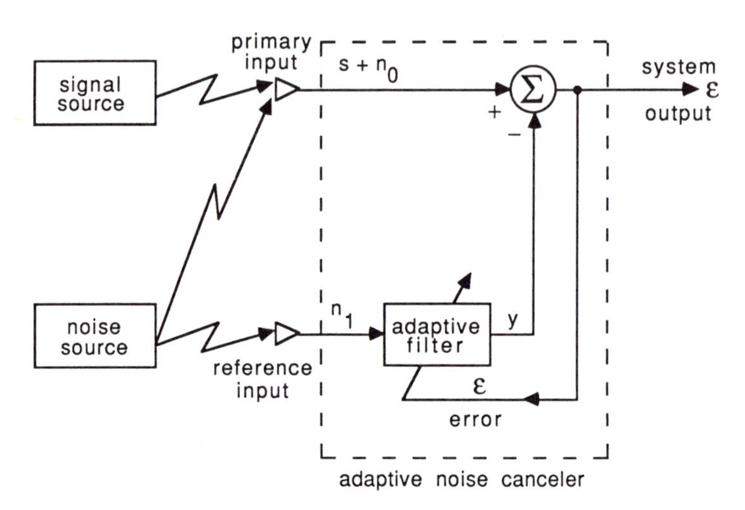

FIGURE 6 Separation of signal and noise. *Top*, classical approach. *Bottom*, adaptive noise canceling approach.

output power is

$$E_{min}[\varepsilon^2] = E[s^2] + E_{min}[(n_0 - y)^2] \tag{4}$$

When the filter is adjusted so that $E[\varepsilon^2]$ is minimized, $E[(n_0 - y)^2]$ is therefore minimized. The filter output y is then a best least squares estimate of the primary noise n_0. Moreover, when $E[(n_0 - y)^2]$ is minimized, $E[(\varepsilon - s)^2]$ is also minimized, since, from Equation 1,

$$(\varepsilon - s) = (n_0 - y) \tag{5}$$

Adjusting or adapting the filter to minimize the total output power is tantamount to causing the output ε to be a best least squares estimate of the signal s for the given structure and adjustability of the adaptive filter, and for the given reference input.

Many practical applications for adaptive noise canceling techniques have been found. One of these involves canceling interference from the mother's heart when attempting to record clear fetal electrocardiograms (ECG). Figure 7 shows the location of the fetal and maternal hearts, and the placement of the input leads. The abdominal leads provide the primary input (containing fetal ECG and interfering maternal ECG signals) and the chest leads provide the reference input (containing pure interference, the maternal ECG). Results are shown in Figure 8. The maternal ECG from the chest leads was adaptively filtered and subtracted from the abdominal signal leaving the fetal ECG. This was an interesting problem since the fetal and maternal ECG signals had spectral overlap. The two hearts were electrically isolated and worked independently, but the second harmonic frequency of the maternal ECG was close to the fundamental of the fetal ECG. Ordinary filtering techniques would have great difficulty with this problem.

Example 4—Adaptive Echo Canceling

Echo is a natural phenomenon in long distance telephone circuits since there is amplification in both

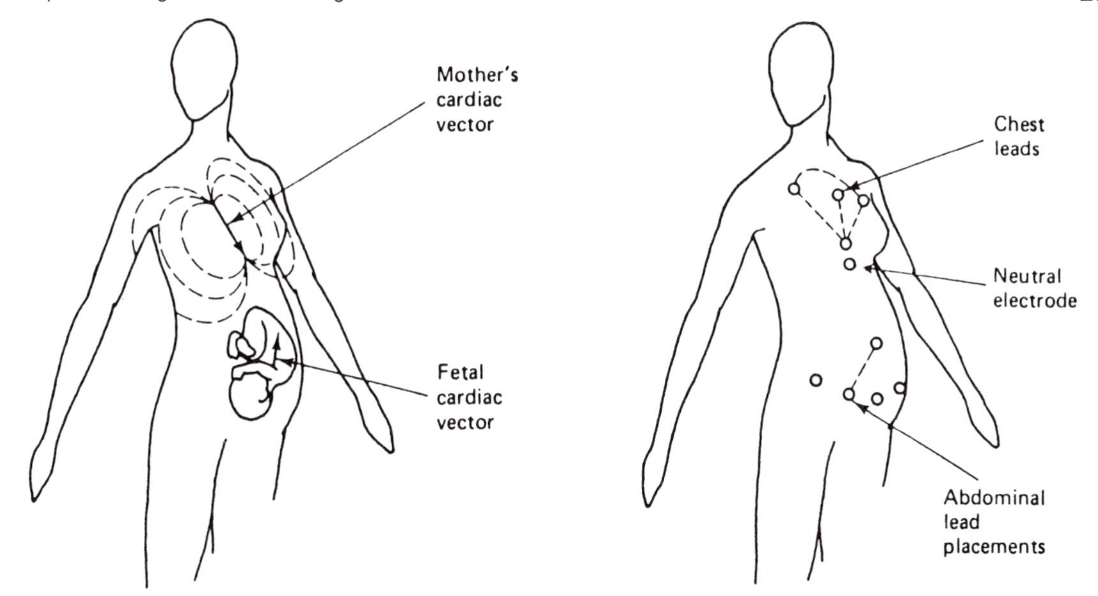

FIGURE 7 Canceling maternal heartbeat in fetal electrocardiography. *Left*, cardiac electric field vectors of mother and fetus. *Right*, placement of leads.

directions and since, at each end of the circuit, the telephone transmitters and receivers are coupled by series circuits. Echo suppressors have been used to break the feedback, giving one-way communication to the party speaking first. To avoid switching effects and to permit simultaneous two-way transmission of voice and data, echo suppressors are being replaced worldwide by adaptive echo cancelers as shown in Figure 9.

Referring to Figure 9, the delay boxes represent the transmission delays in the long line. Note that separate circuits are normally used in each direction because the repeater amplifiers used to overcome transmission loss are one-way devices. Hybrid transformers are used to prevent incoming signals from coupling through the telephone set and passing as outgoing signals. Hybrids are balanced to do this by designing them for the average local telephone circuit. Since each local circuit has its own length and electrical characteristics, the hybrid cannot do its job perfectly. Using an adaptive filter to cancel any incoming signal that might leak through the hybrid eliminates the possibility of echo. The circuit of Figure 9 works well, allowing simultaneous two-way communication without echo.

Example 5—Inverse Modeling

Using an adaptive filter for direct modeling of an unknown system to obtain a close approximation to its impulse and frequency responses was shown in Figure 4. By changing the configuration, it is possible to use the adaptive filter for inverse modeling to obtain the reciprocal of the unknown system's transfer function. The idea is illustrated in Figure 10. The output of the unknown system is the input to the adaptive filter. The input of the unknown system delayed by Δ units of time is the desired response of the adaptive filter. For simplicity, assume that Δ is set to zero delay. To make the error small, the cascade of the unknown system and the adaptive filter would need a unity transfer function. Therefore, using an adaptive algorithm to make the error small causes the adaptive filter to develop a transfer function which is the inverse of that of the unknown system.

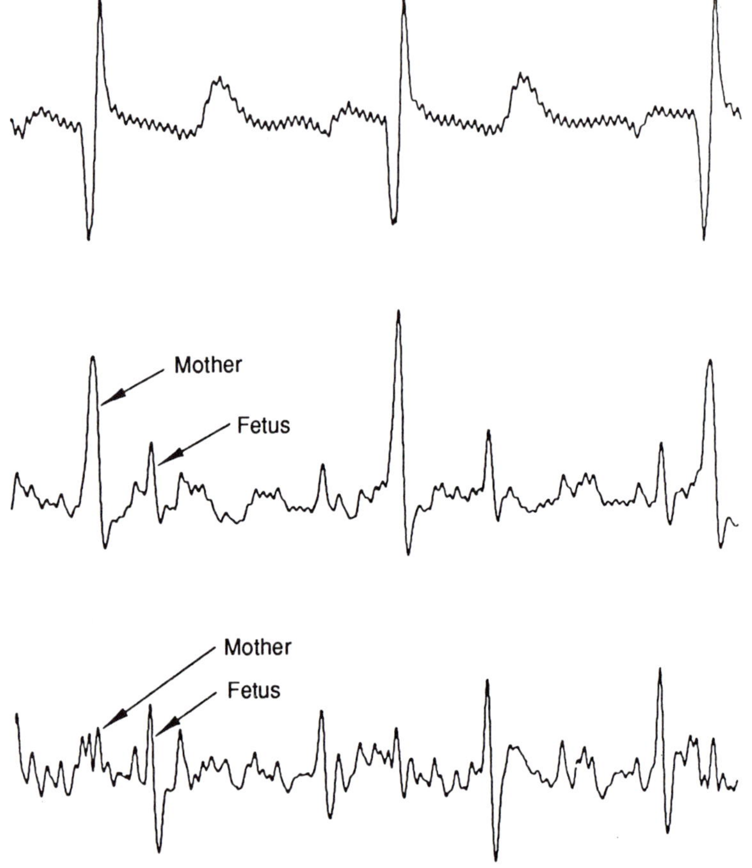

FIGURE 8 Result of fetal ECG experiment (bandwidth, 3–35 Hz; sampling rate, 256 Hz). *Top*, reference input (chest lead). *Center*, primary input (abdominal lead). *Bottom*, noise canceler output.

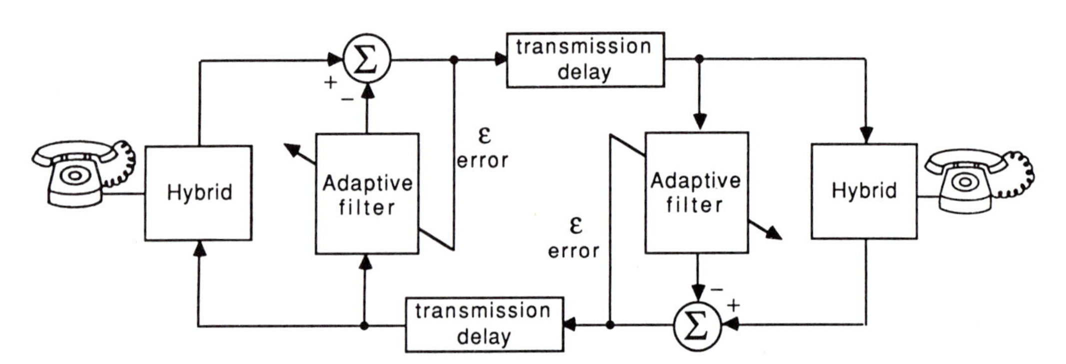

FIGURE 9 Long-distance system with adaptive echo cancellation.

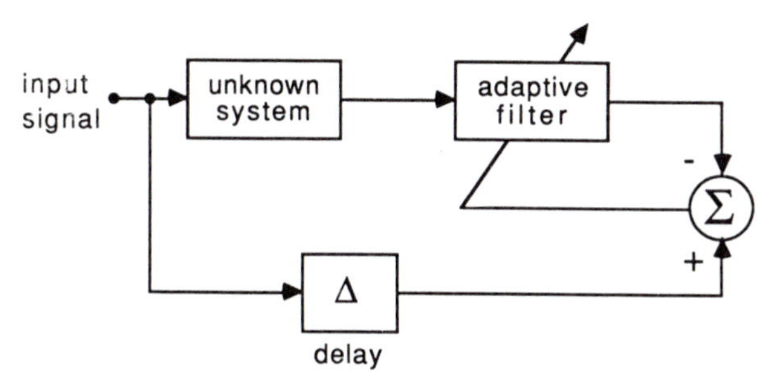

FIGURE 10 Inverse modeling.

If the response of the unknown system contains a delay or is nonminimum phase, allowing the delay Δ to be nonzero would be highly advantageous. The inclusion of Δ delays the inverse impulse response, but yields a much lower mean square error. In some applications, one would like to make this delay as small as possible. In other applications, this delay is of no concern except that it should be chosen to minimize the mean square error. Applications for inverse modeling exist in the field of adaptive control, in geophysical signal processing where it is called "deconvolution," and in telecommunications for channel equalization.

Example 6—Channel Equalization

Telephone channels, radio channels, and even fiber optic channels can have nonflat frequency responses and nonlinear phase responses in the signal passband. Sending digital data at high speed through these channels often results in a phenomenon called "intersymbol interference," caused by signal pulse smearing in the dispersive medium. Equalization in data modems combats this phenomenon by filtering incoming signals. A modem's adaptive filter can compensate for the irregularities in magnitude and phase response of the channel by adapting itself to become a channel inverse.

An adaptive equalizer is pictured in Figure 11. It consists of a tapped delay line with an adaptive linear combiner connected to the taps. Deconvolved signal pulses appear at the weighted sum which is quantized to provide a binary output corresponding to the original binary data transmitted through the channel. The ALC and its quantizer comprise a single ADALINE. Any least squares algorithm can be used to adapt the weights, but the LMS algorithm is used almost exclusively in the telecommunications industry.

In operation, the weight at a central tap is generally fixed at unit value. Initially, all other weights are set to zero so that the equalizer has a flat frequency response and a linear phase response. Without equalization, telephone channels can provide quantized binary outputs that reproduce the transmitted data stream with error rates of 10^{-1} or less. As such, the quantized binary output can be used as the desired response to train the neuron. It is a noisy desired response initially. Sporadic errors cause adaption in the wrong direction but, on average, adaptation proceeds correctly. As the neuron learns, noise in the desired response diminishes. The method is called "decision directed" learning and was invented by Dr. Robert W. Lucky of AT&T Bell Labs (Lucky, 1985). Once the adaptive equalizer converges, the error rate will typically be 10^{-6} or less.

Figure 12 (top) shows the analog response of a telephone channel carrying high-speed binary pulse data. Figure 12 (bottom) shows the same signal after going through a converged adaptive equalizer. This is called an "eye" pattern. Equalization opens the eye and allows clear separation of +1 and -1 pulses. When using a modem with an adaptive equalizer, it is possible to transmit about four times as much data through the same channel with the same reliability as would be without equalization.

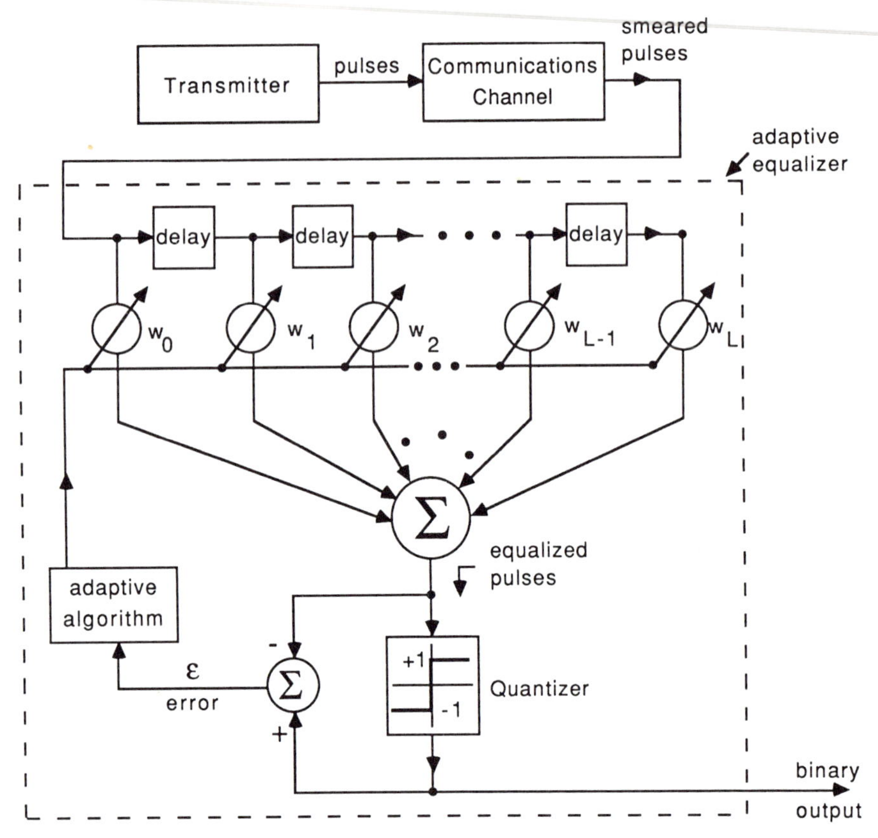

FIGURE 11 Adaptive channel equalizer with decision-directed learning.

A new concept called ISDN (integrated services digital network) is in development and deployment, making possible high-speed digital communication through ordinary local copper telephone circuits. ISDN requires both adaptive equalization and adaptive echo canceling at each line termination. The number of adaptive filters to be used in the world's telecommunications plant will be massive.

ADAPTIVE PATTERN RECOGNITION

The adaptive threshold element of Figure 3 can be used for pattern recognition as a trainable logic device. It can be trained to classify input patterns into two categories. For these applications, the zeroth weight, w_0, has a constant input $x_0 = +1$ which does not change from input pattern to pattern. Varying the zeroth weight varies the threshold level of the quantizer.

Linear Separability

With n binary inputs and one binary output, a single neuron of the type shown in Figure 3 is capable of implementing certain logic functions. There are 2^n possible input patterns. A general logic implementation would be capable of classifying each pattern as either +1 or -1, in accord with the desired response. Thus, there are 2^{2^n} possible logic functions connecting n inputs to a single output. A single neuron is capable of realizing only a small subset of these functions, known as the linearly

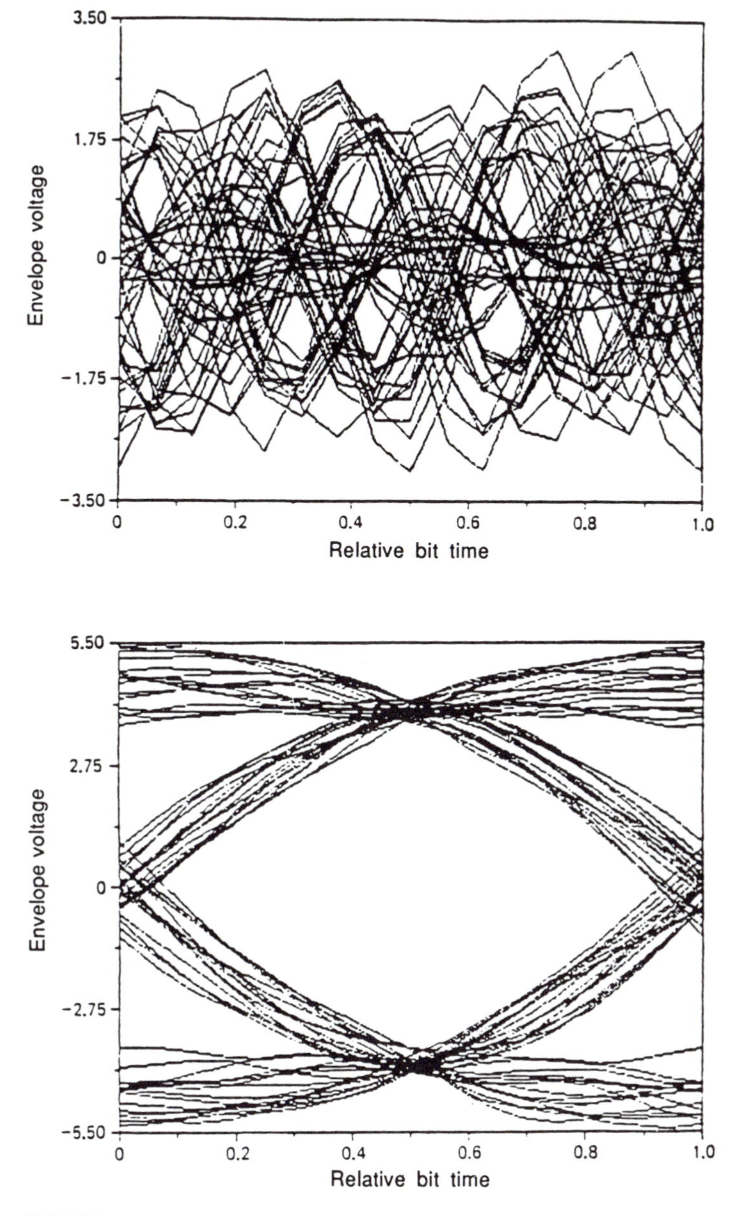

FIGURE 12 Eye patterns formed by overlaying cycles of the received waveform. *Top*, before equalization. *Bottom*, after equalization.

separable logic functions (Lewis & Coates, 1967). These are the set of logic functions that can be obtained with all possible settings of the weight values.

In Figure 13, a two-input neuron is shown. In Figure 14, all possible binary inputs for a two-input neuron are shown in pattern vector space. In this

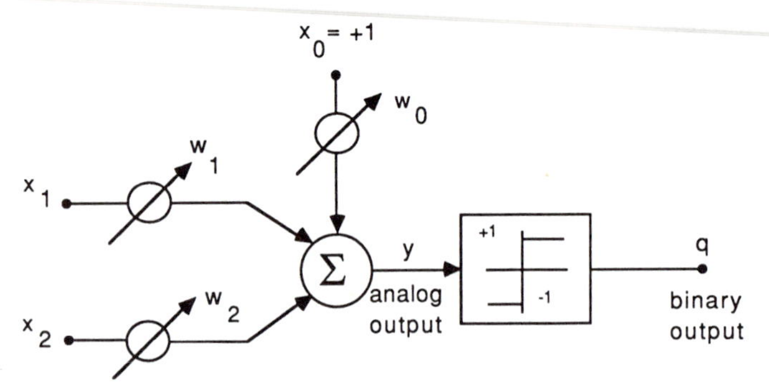

FIGURE 13 A two-input neuron.

space, the coordinate axes are the components of the input pattern vector. The neuron separates the input patterns into two categories, depending on the values of the input signal weights and the bias weight. A critical thresholding condition occurs when the analog response y equals zero.

$$y = x_1 w_1 + x_2 w_2 + w_0 = 0 \tag{6}$$

$$\therefore x_2 = -\frac{w_0}{w_2} - \frac{w_1}{w_2} x_1 \tag{7}$$

This linear relation is graphed in Figure 14. It comprises a separating line which has slope and intercept of

$$\text{slope} = -\frac{w_1}{w_2} \; ; \; \text{intercept} = -\frac{w_0}{w_2} \tag{8}$$

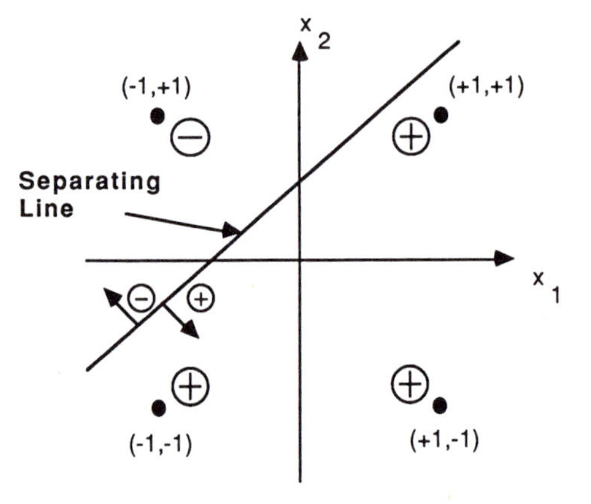

FIGURE 14 Separating line in pattern space.

The three weights determine slope, intercept, and the side of the separating line that corresponds to a positive output. The opposite side of the separating line corresponds to a negative output.

The input/output mapping obtained in Figure 14 illustrates an example of a linearly separable function. An example of a nonlinearly separable function with two inputs is the following:

$$\begin{aligned} (+1,+1) &\rightarrow +1 \\ (+1,-1) &\rightarrow -1 \\ (-1,-1) &\rightarrow +1 \\ (-1,+1) &\rightarrow -1 \end{aligned} \tag{9}$$

No single unit exists that can achieve this separation of the input patterns.

With two inputs, almost all possible logic functions can be realized by a single neuron. With many inputs, however, only a small fraction of all possible logic functions are linearly separable. The single neuron can only realize linearly separable functions and generally cannot realize most functions. However, combinations of neurons or networks of neurons can be used to realize nonlinearly separable functions.

Nonlinear Separability—MADALINE Networks

An approach to the implementation of nonlinearly separable logic functions was initiated at Stanford by W. C. Ridgway III (Ridgway, 1962) in the early 1960s.

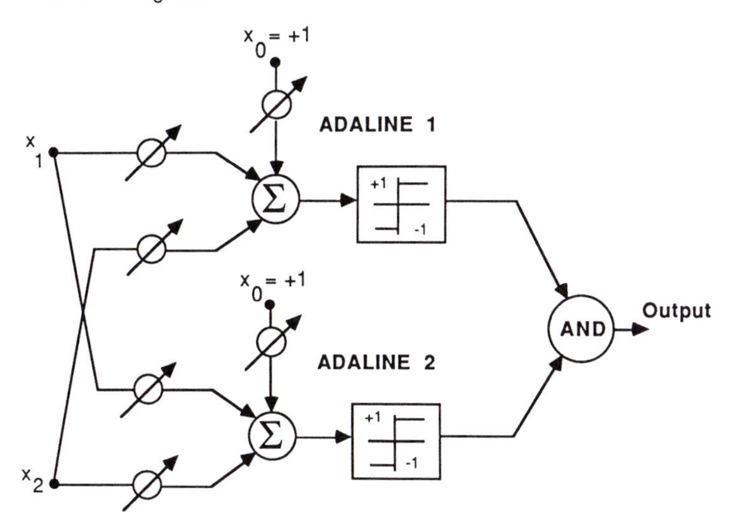

FIGURE 15 A two-neuron form of MADALINE.

Retinal inputs were connected to adaptive neurons in a single layer. Their outputs in turn were connected to a fixed logic device providing the system output. Methods for adapting such nets were developed at that time. An example of such a network is shown in Figure 15. Two ADALINES are connected to an AND logic device to provide an output. Systems of this type were called MADALINES (many ADALINES). Today such systems would be called small neural nets.

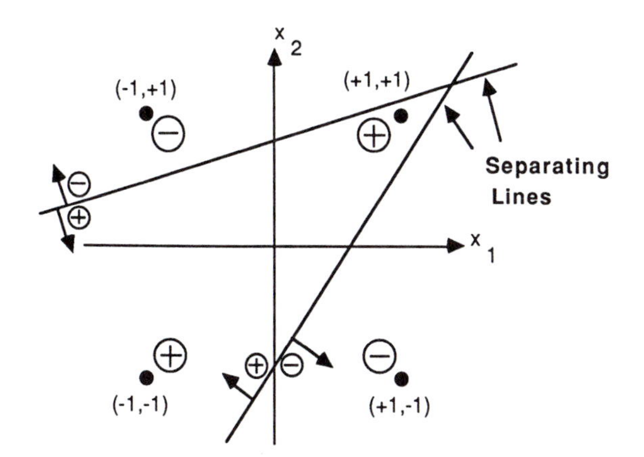

FIGURE 16 Separating boundaries for MADALINE of Figure 15.

With weights suitably chosen, the separating boundary in pattern space for the system of Figure 15 would be as shown in Figure 16. This separating boundary implements the nonlinearly separable logic function (Werbos, 1974).

MADALINES were constructed with many more inputs, with many more neurons in the first layer, and with various fixed logic devices in the second layer such as AND, OR, and MAJority vote-taker. These three functions are in themselves threshold logic functions, as illustrated in Figure 17. The weights given will implement these functions, but the weight choices are not unique.

Layered Neural Nets

The MADALINES of the 1960s had adaptive first layers and fixed threshold functions for the second (the output) layers (Ridgway, 1962; Nilsson, 1965). The feedforward neural nets of the 1980s have many layers, and all layers are adaptive. The back propagation method of Rumelhart & McClelland (1986) is perhaps the best known example of a multilayer network. A three-layer feedforward adaptive network is illustrated in Figure 18.

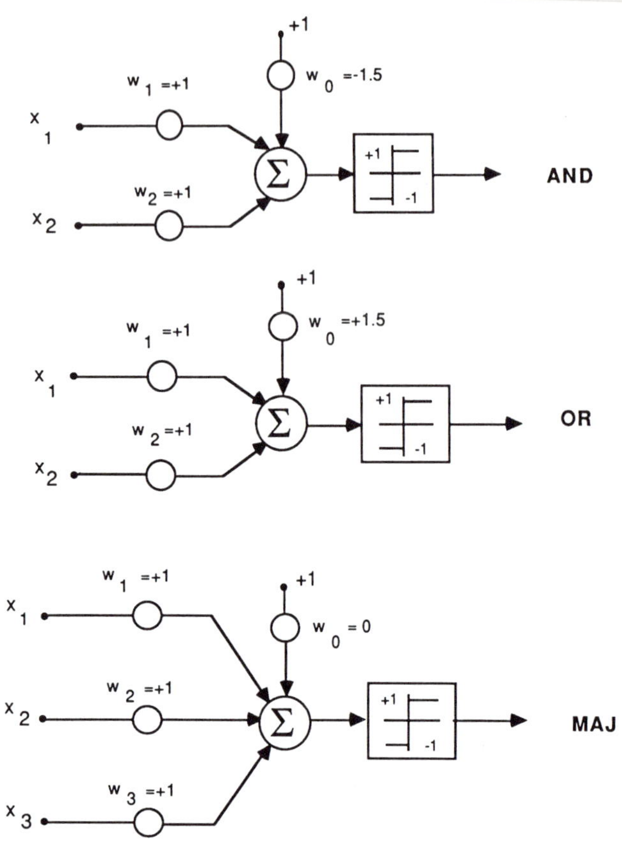

FIGURE 17 A neuronal implementation of AND, OR, and MAJ logic functions.

It is a simple matter to adapt the neurons in the output layer, since the desired responses for the entire network (which are given with each input training pattern) are the desired responses for the corresponding output neurons. Given the desired responses, adaption of the output layer can be a straightforward exercise of the LMS algorithm. The fundamental difficulty associated with adapting a layered network lies in obtaining desired responses for the neurons in the layers other than the output layer. The back propagation algorithm reported first by Paul Werbos (1974), then discovered by David Parker (1985), and again discovered by Rumelhart and McClelland (1986) is one method for establishing desired responses for the neurons in the "hidden layers," those layers with neuronal outputs that do not appear directly at the system output (refer to Figure 18). There is nothing unique about the choice of desired outputs for the hidden layers.

Generalization in layered networks is a key issue. The question is, how well do multilayered networks perform with inputs for which they were not specifically trained? The question of generalization will be important and some good examples are being developed where useful generalizations take place. Many different algorithms may be needed for the adaptation of multilayered networks to produce required generalizations. Without generalization, neural nets will be of little engineering significance. Merely learning the

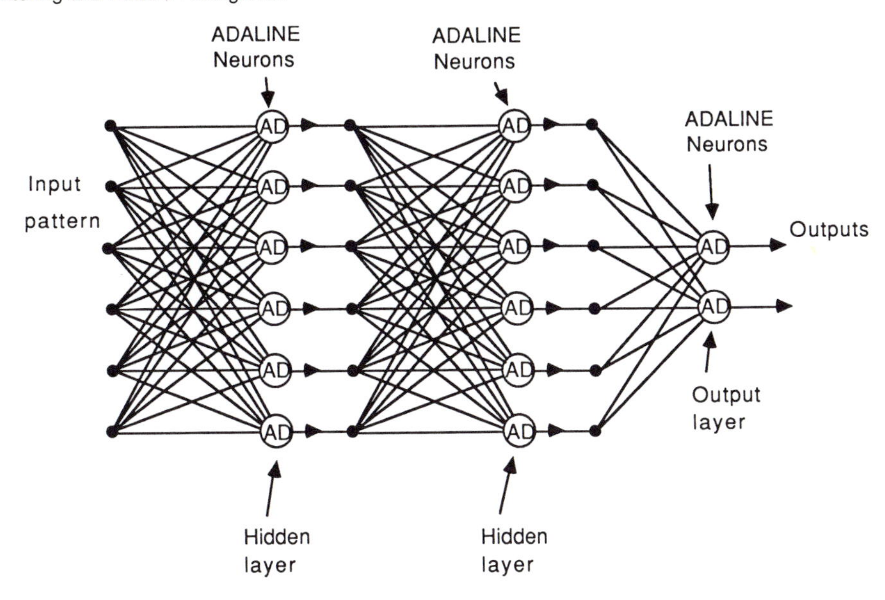

FIGURE 18 A three-layer adaptive neural network.

training patterns can be accomplished by storing these patterns and their associated desired responses in a look up table.

The layered networks of Rumelhart and McClelland (1986) utilize neuronal elements like the ADALINE of Figure 3, except that the quantizer of threshold device is a soft limiting "sigmoid" function rather than the hard limiting "signum" function of the ADALINE. The various back propagation algorithms for adapting layered networks of neurons require differentiability along the signal paths of the network, and cannot work with the hard limiter of the ADALINE element. The sigmoid function has the necessary differentiability. However, it presents implementational difficulties if the neural net is to be ultimately constructed digitally. For this reason, a new algorithm was developed for adapting layered networks of ADALINE neurons with hard limiting quantizers. The new algorithm is an extension of the original MADALINE adaptation rule (Ridgway, 1962; Nilsson, 1965) and is called MADALINE Rule II or MRII. The idea is to adapt the network to properly respond to the newest input pattern while minimally disturbing the responses already trained in for the previous input patterns. Unless this principle is practiced, it is difficult for the network to simultaneously store all of the required pattern responses.

LMS or Widrow–Hoff Delta Rule for the Single Neuron

The LMS algorithm applied to the adaption of the weights of a single neuron embodies a *minimal disturbance principle*. This algorithm can be written as

$$W_{k+1} = W_k + \frac{\alpha}{|X_k|^2} \varepsilon_k X_k \tag{10}$$

The time index or adaption cycle number is k. W_{k+1} is the next value of the weight vector, W_k is the present value of the weight vector, X_k is the present input pattern vector, and ε_k is the present error (i.e., the difference between the desired response d_k and the analog output before adaptation). The above recursion formula is applied to each adaptation cycle and the error is reduced as a result by the fraction α. This can be demonstrated as follows. At the kth cycle, the error is

$$\varepsilon_k = d_k - X_k^T W_k \qquad (11)$$

The error is changed (reduced) by changing the weights

$$\Delta \epsilon_k = \Delta (d_k - X_k^T W_k) = -X_k^T \Delta W_k \qquad (12)$$

In accord with the LMS rule (Equation 10), the weight change is

$$\Delta W_k = W_{k+1} - W_k = \frac{\alpha}{|X_k|^2} \varepsilon_k X_k \qquad (13)$$

Note that ΔW is proportional to X. For this reason, LMS performs better with the symmetric inputs +1 and -1 rather than the usual +1 and 0, when binary input patterns are used. Combining Equations 12 and 13 we obtain

$$\begin{aligned} \Delta \epsilon_k &= -X_k^T \frac{\alpha}{|X_k|^2} \epsilon_k X_k \\ &= -X_k^T X_k \frac{\alpha}{|X_k|^2} \epsilon_k \qquad (14) \\ &= -\alpha \epsilon_k \end{aligned}$$

Therefore, the error is reduced by a factor of α as the weights are changed while holding the input pattern fixed. Putting in a new input pattern starts the next adaptation cycle. The next error is then reduced by a factor of α, and the process continues. The choice of α controls stability and speed of convergence (Widrow & Stearns, 1985). Stability requires that

$$2 > \alpha > 0 \qquad (15)$$

Making α greater than 1 generally does not make sense, since the error would be overcorrected. Total error correction comes with $\alpha = 1$. A practical range for α is

$$1.0 > \alpha > 0.1 \qquad (16)$$

To verify that the LMS rule embodies the minimal disturbance principle, refer to Equation 13. The weight change vector ΔW_k is chosen to be parallel to the input pattern vector X_k. From Equation 12, the change in the error is equal to the negative dot product of X_k with ΔW_k. Thus, the needed error correction is achieved with the smallest magnitude of weight vector change. When adapting to respond properly to a new input pattern, the responses to previous training patterns are therefore minimally disturbed, on the average. The algorithm also minimizes mean square error (Widrow & Stearns, 1985), for which it is best known.

Adaptation of Layered Neural Nets by the MRII Rule

The minimal disturbance principle can be applied to the adaptation of the layered neural network of Figure 18 in the following way. Present an input pattern and its associated desired responses to the network. The training objective is to reduce the number of errors to as low a level as possible. Accordingly, when the first training pattern is presented, the first layer will be adapted as required to reduce the number of response errors at the final output layer. In accord with the minimal disturbance principle, the first layer neuron with an analog response closest to zero is given a trial adaptation in the direction to reverse its binary output. When the reversal takes place, the second layer inputs change, the second layer outputs change, and consequently the network outputs change. A check is made to see if this reduces the number of output errors for the current input pattern. If so, the trial change is accepted. If not, the weights are restored to their previous values and the first layer neuron with the analog response next closest to zero is trial adapted, reversing its response. If this reduces the number of output errors, the change is accepted. In not, the weights are restored and one goes on to adaptively switch the neuron with an analog response next closest to zero, and so on, disturbing the neurons as little as possible. After adapting all neurons whose output reversals reduced the number of output errors, neurons are then chosen in pair combinations and trial adaptations are made which can be accepted if output errors are reduced. After adapting the first layer neurons in singles, pairs, triples, etc., up to a predetermined limit in combination size, the second layer is adapted to further reduce the number of network output errors. Simulation results indicate that pairwise trials are sufficient in layers having up to 25 ADALINES. The method of choosing the neurons to

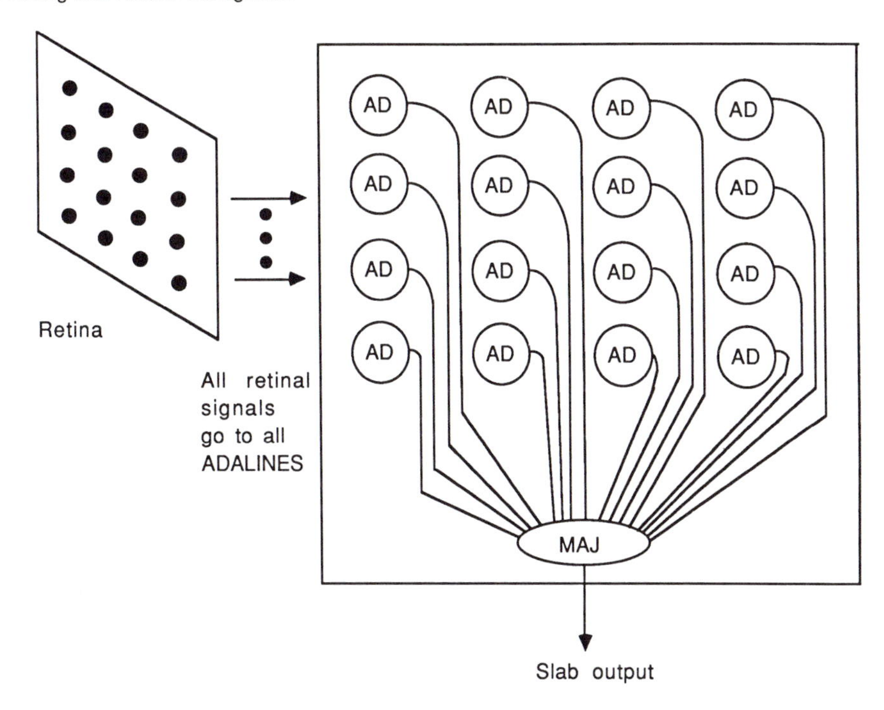

FIGURE 19 One slab of a left–right, up–down translation-invariant network.

be adapted in the second layer is the same as that for the first layer. If further error reduction is needed, the output layer is then adapted. This is straightforward, since the output layer desired responses are the given desired responses for the entire network. After adapting the output layer, the responses will be correct. The next input pattern vector and its associated desired responses are then applied to the neural network and the adaptive process resumes.

When training the network to respond correctly to the various input patterns, the "golden rule" is *give the responsibility to the neuron or neurons that can most easily assume it*. In other words, *don't rock the boat* any more than necessary to achieve the desired training objective. Simulation results using this minimal-disturbance MRII algorithm are presented later.

Application of Layered Networks to Pattern Recognition

It would be useful to devise a neural net configuration that could be trained to classify an important set of training patterns as required, but have these network responses be invariant to translation, rotation, and scale change of the input pattern within the field of view. It should not be necessary to train the system with the specific training patterns of interest in all combinations of translation, rotation, and scale.

The first step is to show that a neural network exists having these properties. The invariance methods that follow are extensions of results reported earlier by Widrow (1962). The next step is to obtain algorithms to achieve the desired objectives.

Invariance to Up–Down, Left–Right Pattern Translation

Figure 19 shows a planar network configuration (a "slab" of neurons) that could easily be used to map a retinal image into a single-bit output so that, with proper weights in the neurons of the network, the response will be insensitive to left–right and/or up–down translation. The same slab structure can be replicated, with different weights, to allow the retinal

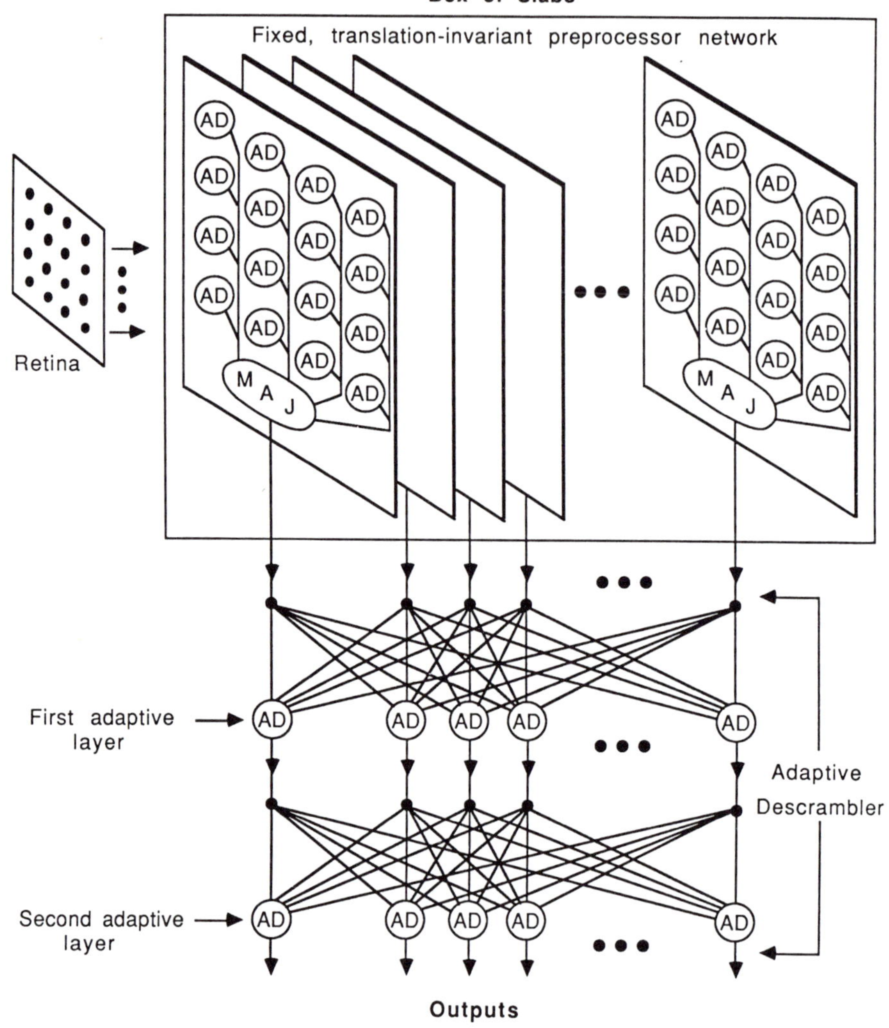

FIGURE 20 A translation-invariant preprocessor network and an adaptive two-layer descrambler network.

pattern to be independently mapped into additional single-bit outputs, all insensitive to left–right, up–down translation.

The general idea is illustrated in Figure 20. The retinal image having a given number of pixels can be mapped through an array of slabs into a different image that could have the same, more, or fewer pixels, depending on the number of slabs used. In any event, the mapped image would be insensitive to up–down, left–right translation of the original image. The mapped image in Figure 20 is fed to a set of ADALINE neurons that can be easily trained to provide output responses to the original image as required. This amounts to a "descrambling" of the preprocessor's outputs. The descrambler's output responses would classify the original input images and would at the same time be insensitive to the left–right, up–down translations.

In the systems of Figure 19 and 20, the elements labeled "AD" are ADALINES. Those labeled "MAJ" are majority vote-takers. (If the number of input lines to MAJ is even and there is a tie vote, these elements are biased to give a positive response.) The AD elements are adaptive neurons and the MAJ elements are fixed neurons, as in Figure 17.

How the weights are structured in the system of Figure 19 to cause the output to be insensitive to left–right and up–down translation needs some further explanation. Let the weights of each ADALINE in Figure 19 be arranged in a square array. Let the corresponding retinal pixels also be arrayed in a square pattern. Let the array of weights of the upper-left ADALINE be designated by the square matrix (W_1). Let the array of weights of the next lower ADALINE be $T_{D1}(W_1)$. The operator T_{D1} represents "translate down one." This set of weights is the same as that of the topmost ADALINE, except that they are en masse translated down by one pixel. The bottom row is wrapped around to comprise the top row. The patterns on the retina itself are wrapped around on a cylinder when they undergo translation. The weights of the next lower ADALINE are $T_{D2}(W_1)$, and those of the next lower ADALINE are $T_{D3}(W_1)$. Returning to the upper–left ADALINE, let its neighbor to the right be designated by $T_{R1}(W)_1$, T_{R1} being a "translate right one" operator. The pattern of weights for the entire array of ADALINES of Figure 19 is given by

$$\begin{array}{cccc} (W_1) & T_{R1}(W_1) & T_{R2}(W_1) & T_{R3}(W_1) \\ T_{D1}(W_1) & T_{R1}T_{D1}(W_1) & T_{R2}T_{D1}(W_1) & T_{R3}T_{D1}(W_1) \\ T_{D2}(W_1) & T_{R1}T_{D2}(W_1) & T_{R2}T_{D2}(W_1) & T_{R3}T_{D2}(W_1) \\ T_{D3}(W_1) & T_{R1}T_{D3}(W_1) & T_{R2}T_{D3}(W_1) & T_{R3}T_{D3}(W_1) \end{array} \tag{17}$$

As the input pattern is moved up, down, left or right on the retina, the roles of the various ADALINES interchange. Since the outputs of the ADALINES are all equally weighted by the MAJ element, translating the input pattern up–down and/or left–right on the retina will have no effect on the MAJ element output.

The set of "key" weights (W_1) can be randomly chosen. Once chosen, they can be translated according to Equation 17 to fill out the array of weights for the system of Figure 19. This array of weights can be incorporated as the weights for the first slab of ADALINES shown in Figure 20. The weights for the second slab would require the same translational symmetries but be based on a different randomly chosen set of key weights (W_2). The mapping function of the second slab would therefore be distinct from that of the first slab.

The translational symmetries in the weights called for in the system of Figure 19 could be manufactured in and fixed, or the ADALINE elements could arrive at such symmetries as a result of a training process. If, when designing a pattern recognition system for a specific application, one knew that translational variance would be a required property, it would make sense to manufacture the appropriate symmetry into a fixed weight system, leaving only the final output layers of ADALINES of Figure 20 to be plastic and trainable. Such a preprocessor would definitely work, would provide a very high speed response without requiring scanning and searching for the pattern location and alignment, and would be an excellent application of neural nets.

Invariance to Rotation

The system represented in Figure 20 is designed to preprocess retinal patterns with a translation-invariant fixed neural net followed by a two-layer adaptive descrambler net. This system can be expanded to incorporate rotational invariance in addition to translational invariance.

Suppose that all input patterns can be presented in "normal" vertical orientation, approximately centered within the field of view of the retina. Suppose that all input patterns can be presented when rotated by 90° from normal, and 180°, and 270° from normal. Thus each pattern can be presented in all four rotations and, in addition, in all possible left–right, up–down translations. The number of combinations would typically be large. The problem is to design a neural net preprocessor that is invariant to translation and to rotation by 90°.

Refer to Figure 19, which shows a single slab of ADALINE elements. This slab produces a majority output that is insensitive to translation of the input pattern on the retina. Next, replicate this slab four-fold and let the majority outputs feed into a single majority output element. In the first slab, the matrix of weights of the ADALINE in the upper left hand corner is designated by (W_1). The matrices of weights of all ADALINES in the first slab are shown by Equation 17. In the second slab, the weight matrix of the upper left hand corner ADALINE corresponds to the weight matrix in the first slab, except rotated clockwise 90°. This can be designated by $R_{C1}(W_1)$. The corresponding upper left hand corner ADALINE weight matrix of the third slab can be designated by $R_{C2}(W_1)$, and of the fourth slab $R_{C3}(W_1)$. Thus, the weight matrices of the upper left hand corner ADALINES begin with (W_1) in the first slab, and are rotated clockwise by 90° in the second slab, by 180° in the third slab, and by 270° in the fourth slab. The weight matrices of all of these slabs are translated right and down, in the fashion of Equation 17, starting with the upper left hand corner ADALINES. For example, the array of weight matrices for the second slab is represented by Equation 18.

$$\begin{array}{cccc} R_{C1}(W_1) & T_{R1}R_{C1}(W_1) & T_{R2}R_{C1}(W_1) & T_{R3}R_{C1}(W_1) \\ T_{D1}R_{C1}(W_1) & T_{R1}T_{D1}R_{C1}(W_1) & T_{R2}T_{D1}R_{C1}(W_1) & T_{R3}T_{D1}R_{C1}(W_1) \\ T_{D2}R_{C1}(W_1) & T_{R1}T_{D2}R_{C1}(W_1) & T_{R2}T_{D2}R_{C1}(W_1) & T_{R3}T_{D2}R_{C1}(W_1) \\ T_{D3}R_{C1}(W_1) & T_{R1}T_{D3}R_{C1}(W_1) & T_{R2}T_{D3}R_{C1}(W_1) & T_{R3}T_{D3}R_{C1}(W_1) \end{array} \tag{18}$$

It is clear that the majority output response will be unchanged by translation of the pattern on the retina. Rotation of the pattern by 90° causes an interchange of the roles of the slabs in making their responses but, since they are all weighted equally by the output majority element, the output response is unchanged by 90° rotation and translation. Insensitivity to 45° rotation can be accomplished by using more slabs. A complete neural network providing invariance to rotation and translation could be constructed. Each translation invariant slab of Figure 20 would need to be replaced by the rotation invariant multiple slab and majority element system described earlier.

Invariance to Scale

The same principles can be used to design invariance nets to be insensitive to scale or pattern size. Establishing a "point of expansion" on the retina so that input can be expanded or contracted with respect to this point, two ADALINES can be trained to give similar responses to patterns of two different sizes if the weight matrix of one were expanded (or contracted) about the point of expansion like the patterns themselves. The amplitude of the weights must be scaled in inverse proportion to the square of the linear dimension of the retinal pattern. Adding many more slabs, the invariance net can be built around this idea to be insensitive to pattern size as well as translation and rotation. Implementation of the systems described herein would require VLSI circuits, taking advantage of the abundance and low cost of VLSI electronics.

Simulation Results

The system of Figure 20 was computer simulated. The training set consisted of 36 patterns each arranged on a 5 x 5 pixel retina in "standard" position. Twenty-five slabs, each with 25 ADALINES, having weights fixed in accord with the symmetry patterns of 1 and 0 were used in the translation-invariant preprocessor. The preprocessor output represented a scrambled version of the input pattern. The nature of this scrambling was determined by the choice of the key weight matrices $(W_1),...,(W_{25})$. These key weights were chosen randomly, the only requirement was that the input pattern to preprocessor output may be one-to-one. This choice of weights produced a very noise tolerant mapping. Methods of training in the key weights using MRII to customize them to the training set are being investigated.

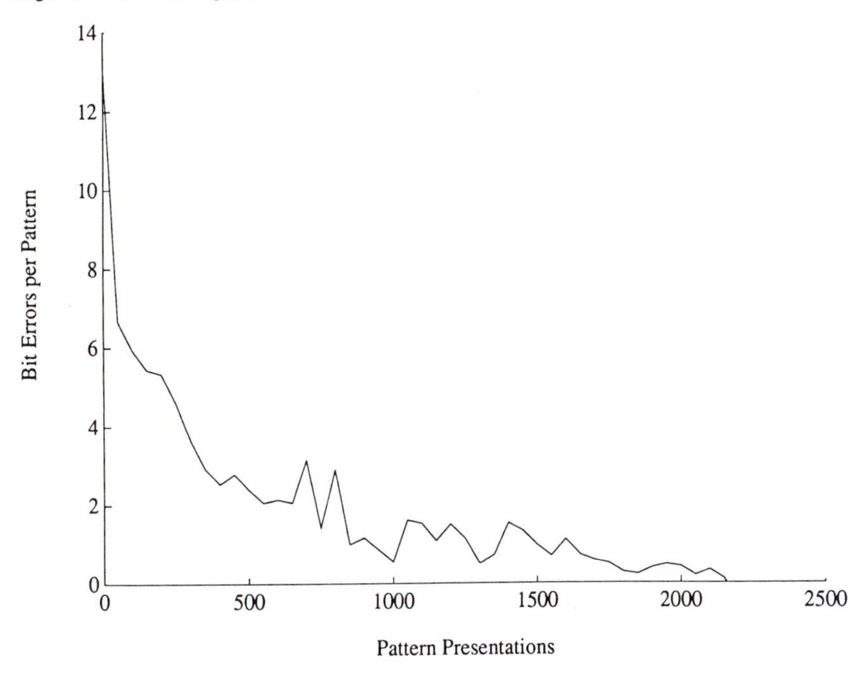

FIGURE 21 Learning curve for two-layer 25 x 25 adaptive descrambler.

MRII was used to train the descrambler. The descrambler was a two-layer system with 25 ADALINES in each layer. The initial weights of the descrambler were chosen randomly. All random weights in the system were chosen independently, identically distributed uniformly on the interval (-1, +1). Patterns were presented in random order, and each pattern was equally likely to be the next presented. The desired response used was the training pattern in standard position. The system as a whole would then recognize any trained in pattern in any translated position on the input retina and reproduce it in standard position at the output. A typical learning curve for the descrambler is shown in Figure 21. The graph shows the number of incorrect pixels at the output, averaged over the training set, after every 50 pattern presentations.

Much work on MRII remains to be done, including detailed studies of its convergence properties and its ability to produce generalization. Preliminary results are very encouraging. Applying the algorithm to problems will lead to insights that will hopefully allow a mathematical analysis of the algorithm.

The concept of using an invariance preprocessor followed by a descrambler is a potentially powerful one. The concept will soon be applied to the speech recognition problem. When a word is spoken by different people or even by the same person at different times, the sounds produced differ greatly, but remain recognizable as the same word, at least to a human. Therefore there must be properties about those sounds that are invariant from utterance to utterance. It is believed that a system similar to the one in Figure 20 would be useful in developing a multi user speech recognition system.

SUMMARY

Learning systems based on the adaptive linear combiner have many applications in signal processing and pattern recognition. Commercial signal processing

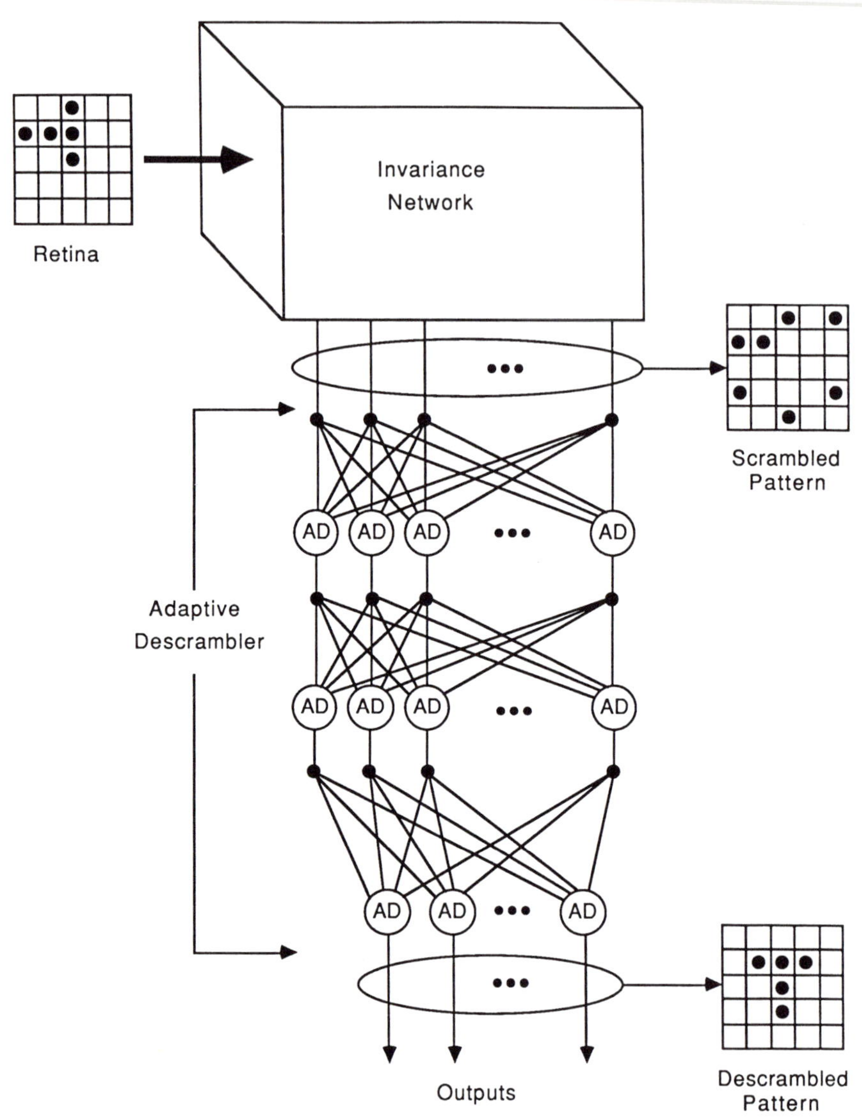

FIGURE 22 A MADALINE system for pattern recognition.

applications include adaptive noise canceling, adaptive echo canceling, and adaptive equalization. Adaptive pattern recognition is not yet commercial, but research is progressing.

A general concept for pattern recognition is described herein involving the use of an invariance net followed by a trainable classifier. The key ideas are illustrated in Figure 22. The invariance net can be trained or designed to produce a set of outputs that are insensitive to translation, rotation, scale change, etc., of the retinal pattern. These outputs are scrambled however. The adaptive layers can be trained to descramble the invariance net outputs and to reproduce the original patterns in "standard" position, orientation, scale, etc. Multilayer adaptation algorithms are essential to making such a scheme work. A new MADALINE adaptation rule (MRII) has been devised for such a purpose, and preliminary experimental results indicate that it works and is effective.

Acknowledgments

This chapter is a slightly expanded version of an article which appeared in *Computer,* **Vol. 21, No. 3,** pp. 24–39, March 1988.

References

Kailath, T. (1981). *Lectures on Wiener and Kalman Filtering*. New York: Springer-Verlag.

Lewis, P. M., II, & Coates, C. L. (1967). *Threshold logic*. New York: Wiley.

Lucky, R. W. (1965). Automatic equalization for digital communication. *Bell Systems Tecnology Journal,* **44,** 547–588.

Nilsson, N. (1965). *Learning machines*. New York: McGraw-Hill.

Parker, D. B. (1985). *Learning logic* (Technical Report no. TR-47). Cambridge, MA: Center for Computational Research in Economics and Management Science. MIT.

Ridgway, W. C., III (1962). *An adaptive logic system with generalizing properties*. Ph.D. thesis, and Report no. 1556–1 Stanford Electronics Labs., Stanford University, Stanford, CA.

Rumelhart, D. E., & McClelland, J. L. (Eds.) (1986). *Parallel distributed processing* (Vols. 1 and 2). Cambridge, MA: MIT Press.

Werbos, P. (1974). *Beyond regression: New tools for prediction and analysis in the behavioral sciences*. Ph.D. thesis, Harvard University, Cambridge, MA.

Widrow, B. (1962). Generalization and information storage in networks of adaline "neurons." In M. C. Yovitz, G. T. Jacobi, & G. D. Goldstein (Eds.), *Self organizing systems 1962* pp. 435–461). Washington, DC: Spartan.

Widrow, B., & Stearns, S. D. (1985). *Adaptive signal processing*. Englewood Cliffs, NJ: Prentice-Hall.

III

Electronic Networks

14

A Construction Set for Silicon Neurons

Rodney Douglas and Misha Mahowald

INTRODUCTION

Because neuronal circuits are complex, it is difficult to find a level for their analysis that is both tractable and biologically correct. Bold abstractions may make short-term gains by providing simplified mathematical or engineering systems, but if these abstractions lose biological accuracy, they also lose their predictive power in the experimental domain and become less useful in furthering neuroscience. On the other hand, retaining biological accuracy often leads to huge, unwieldy, and time-consuming simulations on digital computers. Finding a route through the parameter space of such simulations is an art rather than a science. Where, then, can we find a useful compromise?

Although there are many other aspects of neurobiology that may contribute to neuronal computations, the electrophysiology of neurons and the connectivity between them are most clearly understood, and they are certainly key features in understanding the computations. An artificial system that could easily express these features in very large networks that operate in real time would provide a powerful tool not only for neuroscientists seeking to preserve biological accuracy, but also for theoreticians theoricians seeking useful biological principles.

In this chapter, we describe our first step toward such realistic neuronal networks: a complete set of circuit modules for constructing analogs of individual neurons in complementary metal-oxide semiconductor (CMOS) very large scale integrated (VLSI) technology. Our approach to creating artificial neurons (Mahowald & Douglas, 1991) is not a black-box engineering approach, it is based on a physically motivated description of the neuron, rather than on a functional one. Instead of explicitly implementing a predefined input/output relationship, the construction set allows an analog neuron to be assembled on the basis of measured biophysical properties of a real neuron, whose input/output function need never be known. The assembly process consists in combining circuit elements in one-to-one correspondence with the physical components of the neuron that the circuits represent. The input/output relationship of the analog neuron results from the concurrent dynamical activity of its coupled circuit modules, just as the input/output relationship of the cell arises from the biophysical interaction of its primitive elements.

EMULATION OF NEURONAL ELECTROPHYSIOLOGY

Our approach to the emulation of neuron electrophysiology is similar to the compartmental strategies used by biophysicists to quantize the neuron for numerical simulation. These conventional simulations of neuronal biophysics represent the neuron as a finite number of discrete elements whose individual properties are known and expressed in a set of equations. The model neuron is divided into convenient cylindrical compartments that correspond roughly to physical slices of a real neuron (Rall, 1977; Koch & Segev, 1989). The cylinders are coupled to each other with the same organization as the slices of the cell to which they correspond. Each cylinder comprises a number of materials, such as a cylinder of membrane that encloses the cell, the intracellular fluid that fills the cylinder, and ion conductances in the membrane that allow electrically charged ions to pass into and out of the cell.

Our analog neurons are synthesized by combining modular circuits that emulate the various biophysical properties of neurons relevant to their electrophysiological behavior.[1] As in mathematical simulations, the electrical circuits that emulate the various properties are combined into compartments, and the compartments in turn are connected to each other as a tree that represents the neurons's morphology according to standard approximations (Koch & Segev, 1989). (See Figure 1.) Each compartment of the analog neuron contains a capacitor, which represents the capacitance of the patch of neuronal membrane associated with that compartment; a resistor, which represents the axial resistance of the cable process; and a set of voltage-, ion-, or ligand-sensitive conductances, each of which

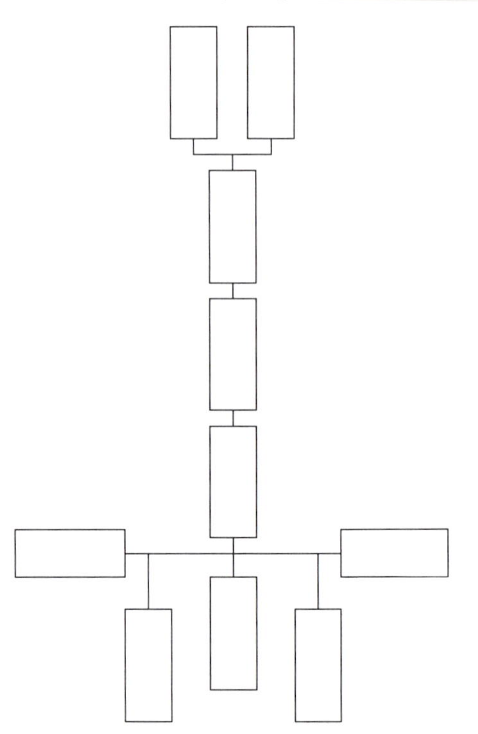

FIGURE 1 The morphology of a neuron modeled by constructing a tree of compartments, each of which is electronically equivalent to a portion of the dendritic arborization of the cell.

represents a population of ion-permeable channels that span the nerve cell membrane. As in mathematical cable models of neurons, the morphology of the cell may be approximated by any number of compartments to varying degrees of accuracy. A representative compartment is shown in Figure 2.

COMPARTMENTS

Passive Compartments

The foundation of the compartmental model is the passive, linear behavior that is attributed to the biophysics of the cell membrane and the intracellular fluid (Rall, 1977). The capacitance of the cell membrane is naturally emulated in the analog by the intrinsic capaci-

[1] A tutorial description of the well-known (Mead, 1989) primitive elements that are used in our circuit modules is provided in the last section of this chapter, *Basics*.

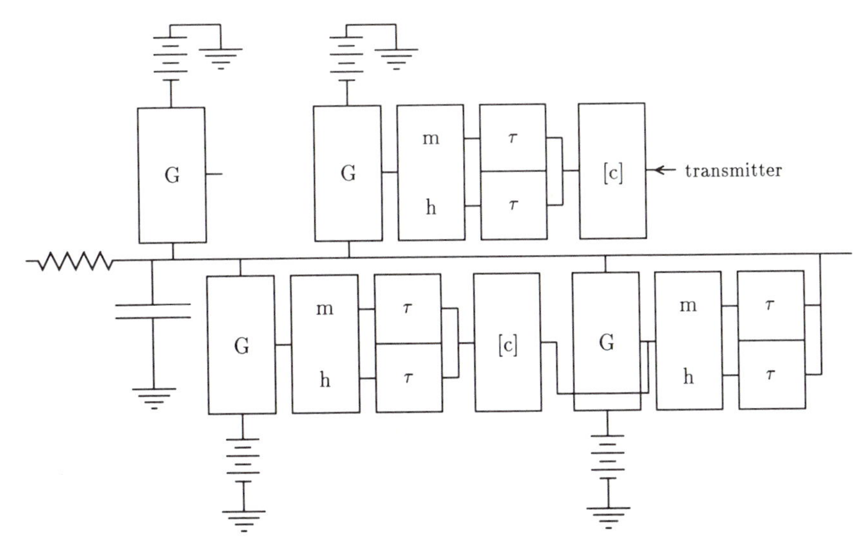

FIGURE 2 An example compartment comprising an axial resistor, a capacitor, a passive leak conductance and three types of active conductance, one voltage-sensitive, one ligand-(transmitter-) sensitive, and one that is sensitive to ion-concentration (usually calcium). The lower concentration, [c], element is shown connected to a voltage-sensitive conductance because the concentration, *c*, of the ion is affected by the voltage-regulated flow of that ion into the cell.

tances of the CMOS material. The intracellular fluid within each mathematically ideal compartment offers an axial resistance to the flow of current. This resistance establishes the electrical coupling between adjacent compartments. It is usually convenient to consider the resistance as an axial conductance, G_a. We have implemented this conductance using the *Hres* circuit developed by Mead (Mead, 1989). This is a resistance circuit whose value can be varied over many orders of magnitude by a control voltage. The Hres has a limited linear response range (about 100 mV), outside of which the current through the resistor saturates. Saturation of the axial resistance can limit the amount of synaptic current that can flow into the soma of the cell, and it is important to consider this effect when designing the neuron morphology. If many compartments are included, the linear range of the resistor may be large enough to encompass the voltage gradients that occur between the compartments of the neuronal model.

In addition to the axial conductance, each compartment also offers a leakage conductance, G_l, across its cylindrical membrane. Physiologically, the ion currents that flow across this average conductance drive the membrane voltage toward a potential, E_l, which is an aggregate of the potentials of the ions that leak across G_l. The leakage conductance is implemented by a wide-input range transconductance amplifier (see *Basics*, last section). This circuit also has a central linear range and saturating behavior, and the conductance in the linear range can be varied over orders of magnitude by a control voltage. The effects of the saturation of the transconductance amplifier are not significant because the leakage conductance forms a very tiny fraction of the total conductance of the neuron when the neuron is producing action potentials. The leakage conductance plays its most significant role in the limited voltage region in which the neuron is below threshold. The GA[2] and GLEAK circuits suffice to construct in CMOS VLSI approximations to the

[2] Subscripted notation is used to denote physical properties, such as G_a, and uppercase is used to denote the analogous circuit, such as GA.

traditional passive, linear cable models of dendrites (Rall, 1977). Cable models of dendrites are attractive because they are mathematically tractable. However, there is substantial experimental evidence that besides the action potential mechanisms of the soma, many kinds of neurons also have a variety of active conductances in their dendritic membranes (McCormick, 1990).

Active Compartments

Membrane conductances are said to be active when they are controlled by local conditions, such as electrical polarization of the cell membrane, or by the concentrations of various chemicals.[3] The interaction of active conductances, particularly the voltage-dependent conductances, give rise to the interesting nonlinear electrical behaviors of neurons.

Membrane conductance is the average of the conductances of individual ion channels (Hille, 1992). Particular types of channels conduct specific ions. In the case of voltage-sensitive channels, their ability to conduct depends on the voltage gradient across the membrane. A simplified view of this mechanism is shown in Figure 3. A charged channel floats in the membrane. When there is no gradient across the membrane, the preferred orientation of the channel is in the plane of the membrane, and so it does not conduct its preferred ions. When the membrane is suitably polarized, the charged channel is drawn across the membrane, and the preferred ions are able to flow down their electrochemical gradient. The fraction of channels that are conductive at any membrane polarization can be derived from the Boltzmann distribution (Hodgkin & Huxley, 1952; Hille, 1992). The open fraction is $O/(O + C) = 1/(1 + e^{-zVm/kT})$, which explains the commonly observed logistic relation between ion conductance and membrane potential (see Figure 3B).

The kinetics of individual openings and closings are determined by the intrinsic properties of the particular channel type. At the conductance level, these kinetics

A

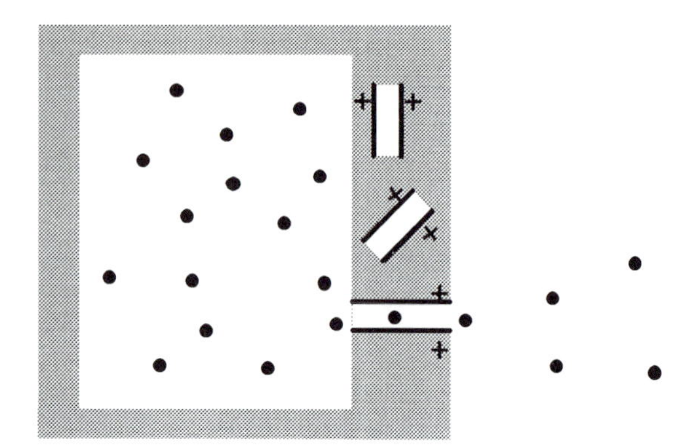

B

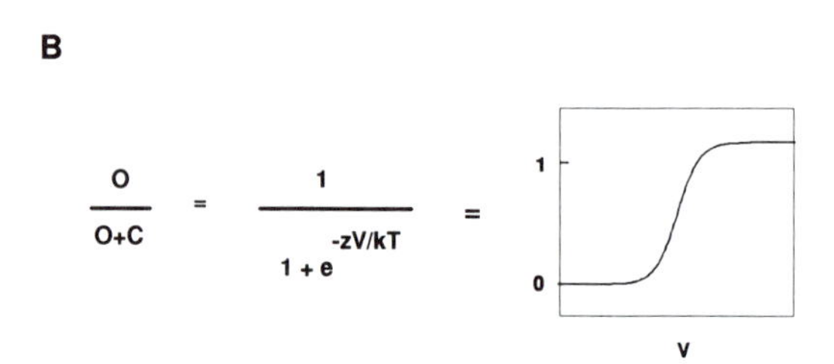

FIGURE 3 Boltzmann behavior of channels. (A) The neuronal membrane (gray) contains charged channels that are able to conduct specific charged ions (filled circles), provided that the membrane is sufficiently polarized to orient the channels across its thickness. (B) The expected fraction of open channels as a function of membrane polarization can be derived from the Boltzmann distribution. The relationship is logistic. kT, thermal energy per charge carrier; V, transmembrane voltage; z, channel charge.

can be viewed as the interaction of an activation and an inactivation process (m and h processes, respectively), each of which is governed by a time constant. The activation time constant governs the approach to the maximum conductance determined by the prevailing membrane voltage. The inactivation time constant, which is slower than the activation time constant, governs the relaxation of the conductance back toward the nonconductive state. The m and h processes combine to make the conductance change more or less transient. Some ion conductances have very long or infinite inactivation time constants and are said to be "noninactivating."

[3] Some membrane conductances are controlled by a combination of electrical and chemical factors.

The principles that govern ligand- or ion-sensitive conductances are similar to those described earlier. These common principles suggest a modular design strategy for their analog CMOS counterparts. The compartment of Figure 2 has three different kinds of active-conductance circuits, which will be described in detail in the following sections. Each of these circuits has a conducting element G whose magnitude is controlled by an analog voltage. In the case of the active channels, this analog voltage is computed by activation/inactivation circuits labeled (m h). The voltage output of the (m h) circuit represents the state of the conductance as the fraction of the population that is in the active or conducting state, and the fraction of the population that is in the inactive or nonconducting state. The kinetics of the m and h processes are represented by the low-pass elements labeled τ.

The state of each kind of conductance is a function of a different controlling variable, the membrane voltage, or a voltage computed by an element labeled [c], which represents the concentration of either a ligand or an ion. The ligand concentration circuit is used as a synapse. It calculates the time-varying transmitter concentration resulting from the activity of a presynaptic cell. The ionic concentration circuit usually represents calcium, which accumulates as a result of the opening of a calcium-permeable, voltage-sensitive conductance. This arrangement is indicated in Figure 2 by the membrane- and voltage-sensitive G coupling into the [c] element.

The conductance elements, activation elements, and concentration elements are described next. Each has several possible instantiations. The choice of which is appropriate depends on the properties of the particular conductance to be emulated.

Conductance Elements

The G element, illustrated in Figure 4, is the simplest circuit. In some cases, it comprises just a single tran-

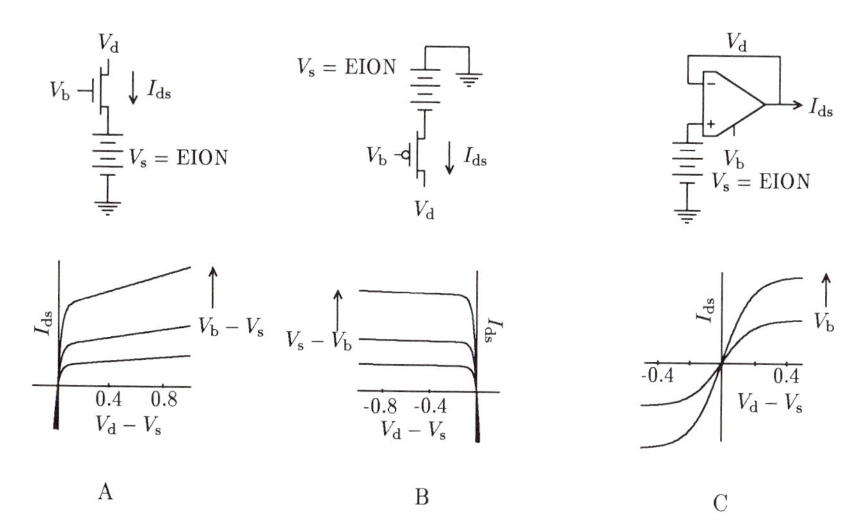

FIGURE 4 Inhibitory (A), excitatory (B), and shunting (C) conductance elements whose magnitude is modulated by an activation voltage, V_b. In each case, the drain current I_{ds} flows between the drain voltage V_d and the source voltage V_s. These voltages are analogous to the membrane voltage across the membrane capacitor (not shown) and the ionic potential, respectively. The current/voltage relation of each element is shown for two or three values of V_b. The degree of the dependence of the current on the voltage across the transistor depends on transistor-channel length. Longer transistors act more like current sources, producing a nearly constant current for large voltages across the transistor. These simulated current-voltage curves assume that the transistors are operating in the subthreshold operating regime.

sistor.[4] Three different types of conductance element are shown—excitatory, inhibitory, and shunting. In each case, the conductance element couples the membrane capacitor C_m to a battery that represents the potential of the ion(s) to which the channel is permeable. The membrane potential across the capacitor is the drain voltage V_d of the transistor, and the ion potential is the source voltage of the transistor, V_s. The drain current I_{ds} that flows between V_d and V_s is analogous to the transmembrane current. The magnitude of the current depends on the voltage gradient across the transistor and also its conductance. The magnitude of the conductance is modulated by the voltage V_b generated by the activation/inactivation circuit.

The excitatory and inhibitory conductances are just single transistors. This restricts their useful operating range in the neuronal circuits. If the voltage on C_m falls below the inhibitory potential, or rises above the excitatory potential, then the direction of current flow through the respective transistors reverses and its magnitude increases exponentially. Consequently, the single transistor conductance element should be used only for ions such as potassium, sodium, and calcium whose potentials are at or beyond the limits of the membrane's expected voltage operating range. For ions such as chloride, whose reversal potential is near the resting potential of the cell, a transconductance amplifier should be used.

The transconductance amplifier (see last section of this chapter, *Basics*) is used as a resistor when bidirectional current flow between the two terminals of the resistor is not required. For example, when the transconductance amplifier is used as an inhibitory shunting conductance, the V^+ input is tied to the reversal potential of the shunting ion, and the V^- input is tied to the output, which drives the membrane capacitor, C_m. In this way, the transconductance amplifier conducts current onto and off of the membrane capacitor symmetrically around the ionic potential. However, unlike a true resistor, this current does not actually flow into the V^+ terminal. Instead, it flows into the power supply. This is an operational detail which does not affect the behavior of the membrane voltage. The most important feature of the transamp in the shunting inhibition application is the linear portion of its current/voltage relation.[5] In this range, the transamp approximates a true conductance. The linear range can be extended by the addition of source-degeneration diodes to the base of the differential pair (see Figure 15, last section).

Kinetics

The kinetics of our circuit modules are very simply approximated by first-order, low-pass filters (see *Basics*, last section). Filters of higher order can be constructed by concatenating these first-order filters. A first-order, low-pass filter controls the kinetics of the active-conductance circuits (labeled τ in Figure 2). The τ circuit module is constructed with a transconductance amplifier and a capacitor. Together these elements function like a leaky resistance-capacitance (RC) integrator whose time constant can be varied over orders of magnitude by changing the transconductance of the amplifier.

Activation Elements

The key circuit of the ion conductance model is the activation/inactivation circuit. Two variants of this circuit are shown in Figure 5. Both implement the logistic relation between a control variable, such as membrane voltage, and membrane ion conductance. (See Figure 3). They do this by means of the differential pair, which naturally offers a logistic relationship between the current flowing though each of its limbs (see *Basics*, last section). The activation and inactivation halves of the circuit are such differential pairs. The threshold for activation and inactivation are set by the gate voltages to one side of the differential pair. The opposing gate voltage is typically a temporally

[4] See the last section of this chapter, *Basics*, for a brief description of the behavior of the transistor in the subthreshold operating regime.

[5] The linear range of such a modified transconductance amplifier is larger than the linear range of the voltage-controlled resistor (Hres) circuit.

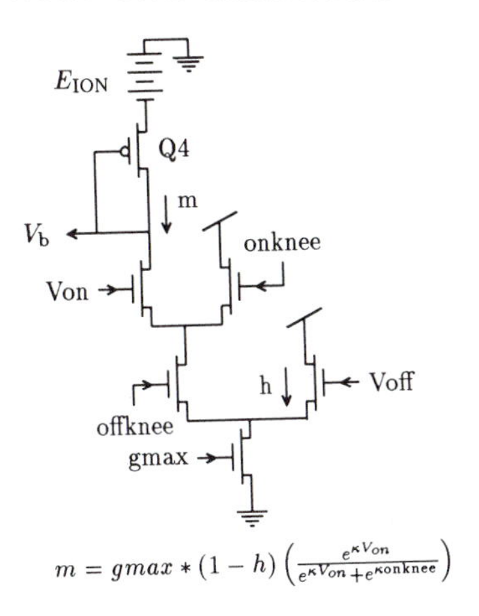

$$m = gmax * (1 - h)\left(\frac{e^{\kappa Von}}{e^{\kappa Von} + e^{\kappa onknee}}\right)$$

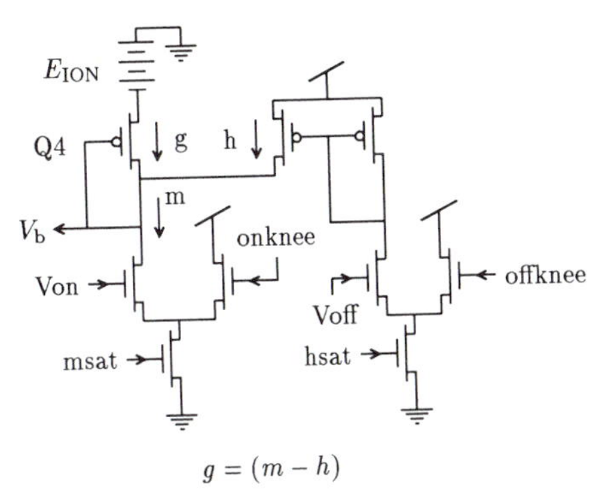

$$g = (m - h)$$

FIGURE 5 Activation/inactivation circuits. A. A multiplicative circuit. Activation m is multiplied by inactivation h. The diode-connected transistor, Q4, generates the final conductance control variable, V_b. V_b is generated in relation to the voltage on the source of transistor Q4. This voltage should be the same as that on the conductance element that will be controlled by V_b if a one-to-one scaling is to be maintained. B. A subtractive circuit. The total degree of activation g is the difference between activation m and inactivation h.

filtered version of the controlling voltage variable, such as the membrane voltage or a chemical concentration computed by a concentration module. The current through the m leg of the activation differential pair represents the degree of activation of the membrane conductance, whereas the current through the h leg of the inactivation differential pair represents the degree of inactivation. The activation variable is not normalized to 1, as in the traditional Hodgekin-Huxley formulation. Instead, the activation level is proportional to the conductance that is ultimately required by the cell. The *gmax* voltage sets the maximum conductance of the channel population represented by the activation/inactivation element.

In the circuit shown in Figure 5A, the degree of activation and the degree of inactivation are multiplied by placing the differential pairs in series. The current available for activation is steered by the inactivation differential pair. As inactivation proceeds, less of the *gmax* bias current flows through the activation differential pair. A multiplicative design is favored over the subtractive design shown in Figure 5B if the power constraints on the circuit are very tight. The activation current m is fed into a current-mirror diode (see *Basics*, last section), which generates the control signal V_b for the conductance module G (Figure 4).

In the subtractive design of the activation/inactivation circuit, the degree of inactivation h is subtracted from the degree of activation m to compute the final conductance. The h current of the inactivation transconductance amplifier and the m output current of the activation amplifier are summed into a current-mirror diode. The current-mirror diode half-wave rectifies the result of the subtraction, so that no matter what the values of m and h, the value of the conductance is never negative. The silicon neurons that we have fabricated to date all use a subtractive, rather than a multiplicitive activation/inactivation circuit. The sub tractive version of the activation/inactivation circuit has the advantage of one more free parameter than the multiplicative circuit, and that parameter can be used to supply a partial inactivation if m_{sat} is larger than h_{sat}.

Unlike the maximum conductances, which are voltage-regulated parameters, the steepnesses of the activation and inactivation functions are determined at the time of fabrication. The steepness of the function is controlled by the efficacy with which the gate voltages on the differential pair steer current between the two limbs of the pair. Current through a transistor cannot change more quickly than a factor of e every 25 mV voltage change on the gate, and typically changes a factor of e every 40 mV.[6] The current through a population of active channels changes by a factor of e every $25/n$ mV, where n is the gating charge of the channel. For this reason, the voltage range of the silicon neuron is scaled to be n (n is typically ≈ 10) times larger than the voltage range of a real neuron. By scaling the voltage range, the degree of activation of the channel population as a fraction of the neuron's operating range is conserved.

By modifying the circuits slightly, the steepnesses of the activation and inactivation functions can be changed. A shallow-pitched response can be obtained by adding source-degeneration diodes into the activation/inactivation differential pairs. The activation/inactivation circuit in Figure 5B, which has no source degeneration, has a moderately steep response because the tail of the current through the differential pair is long. If a sharper onset is required, then full transconductance amplifiers can be substituted for the differential pairs, as shown in Figure 8. The full transconductance amplifier subtracts the current going through the limb of the threshold from the limb controlled by the membrane voltage. The activation level may go negative, but the current-mirror diode performs a half-wave rectification. The use of a full transconductance amplifier generates a steep activation/inactivation function that begins abruptly when the membrane voltage crosses threshold.

[6] In the subthreshold transistor, the voltage on the gate controls the height of the energy barrier over which the charge carriers with Boltzmann-distributed energies must pass in order to flow across the channel of the transistor. The barrier height is changed electrostatically by the voltage on the CMOS gate, and cannot change more than the voltage on the gate itself. In the biological system, the situation is slightly different. Instead of the ions crossing the membrane, it is the gating charges of the channel that obey a Boltzmann distribution. If n gating charges are moved through the membrane per channel opening, then the voltage on the membrane need only change one nth as much as the voltage on the transistor gate for the same number of channels to open. The amount of current flowing across the nerve membrane is proportional to the number of open channels.

Concentration Elements

Two variants of the concentration circuit, [c], are shown in Figure 6. The concentration circuits are essentially leaky RC integrators, with time constants set by V_τ. Charges representing molecules of the species whose concentration is being calculated are gated onto the capacitor. The voltage on this capacitor represents

A

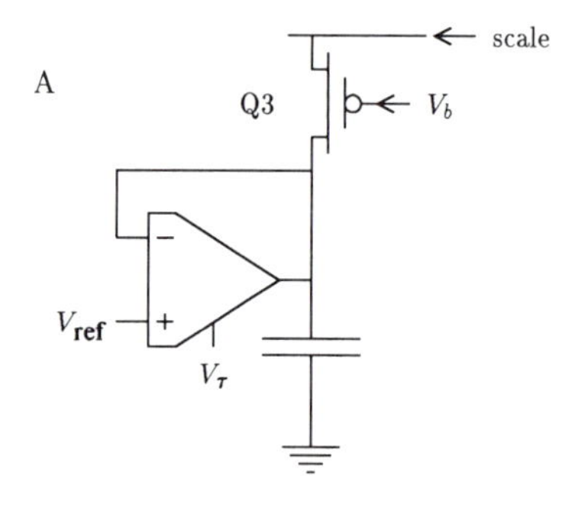

B

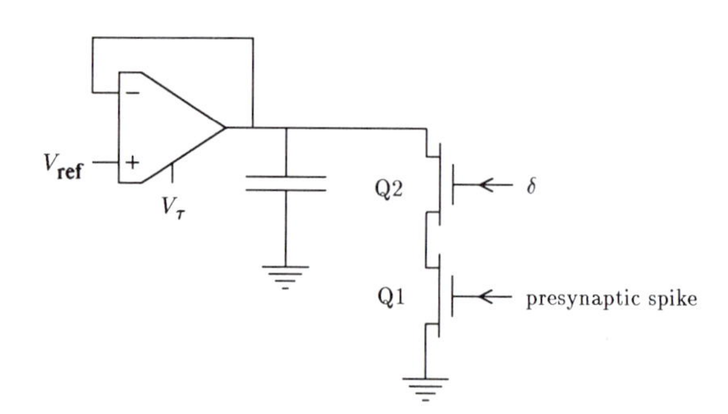

FIGURE 6 Concentration circuits. A. Intracellular ion concentration is calculated by a leaky integrator. Influx of ion is regulated by an activation/inactivation circuit, which provides the control voltage, V_b. The source of transistor Q3 is tied to a control voltage that allows the current representing ion flux into the compartment to be scaled independently of the associated electrical current flowing into the cell. B. This concentration circuit calculates the neurotransmitter concentration at the synapse. Transmitter accumulates on the capacitor at a rate dependent on δ when a presynaptic action potential is applied to Q1. Transmitter concentration decays via the action of the transconductance amplifier. The rate of decay is controlled by V_τ.

concentration. The concentration decays with the time constant of the leaky integrator. In the absence of input, the concentration decays to a resting level that is determined by the voltage applied to the positive input of the transconductance amplifier.

Concentration circuits are used to calculate the concentrations of neurotransmitter at a synapse, and also to calculate the calcium concentration inside the cell. To increase the transmitter concentration at a synapse, charge is gated off the concentration capacitor by presynaptic action potentials (transformed to logic pulses) applied to the gate of transistor Q1 (Figure 6). The amount of transmitter accumulated per action potential is determined by the duration of the logic pulse (assumed constant) and the voltage δ applied to the gate of transistor Q2. An additional conductance model, shown in Figure 2, represents the channels that are activated by the neurotransmitter. Thus, the strength of the synapse (the charge deposited on the membrane capacitor of the postsynaptic cell in response to a presynaptic input) is dependent on a number of factors: δ, the time constants of the low-pass filters, the maximum conductance of the activation/inactivation circuit, and the prevailing postsynaptic membrane voltage.

Calcium concentration within the cell is calculated by a concentration circuit. The source of calcium is a voltage-dependent activation/inactivation circuit (Figure 2), which controls flow of calcium into the neuron. The calcium influx has two effects: the current charges the membrane capacitance through a conductance, G, and the ions accumulate as a concentration of calcium via the action of the concentration circuit. The concentration circuit accumulates a charge proportional to the calcium charge flowing onto the membrane capacitor onto a separate capacitor that represents calcium ion concentration. The amount of charge entering the concentration circuit is scaled independently of the current onto the membrane capacitance, thereby maintaining independent control of the concentration. The scaling factor is applied to the source of transistor Q3, shown in Figure 6. The same circuit can be used to model the concentration of any ion type in the neuron model. The concentration can be modeled independently in each compartment, or the concentration capacitors in adjoining compartments may be resistively coupled to create a continuously distributed ion concentration. This coupling essentially creates another morphological neuron model, which, instead of calculating the electrical behavior of the neuron, computes its chemical state with respect to a particular ion.

THE ACTION POTENTIAL

The most basic feature of electrophysiology that these circuit modules must emulate is the generation of action potentials. The action potential was characterized in the squid giant axon by Hodgkin and Huxley (1952). They separated the current that flows during the action potential into two components, the sodium current and the potassium current. The action potential is triggered by a depolarization (increase) of the membrane voltage caused by an external current source. As the membrane voltage increases, the sodium current is activated and flows into the cell, further increasing the potential on the membrane capacitance. Eventually, the sodium current disappears, because of the delayed inactivation of the sodium channels. The potassium current is also activated by the membrane depolarization, but with a slower time constant. The potassium current flowing out of the cell discharges the membrane capacitance. The potassium current stops when the membrane voltage falls below the level for its activation. The repolarization of the membrane allows the sodium channel to deinactivate, so that it is ready to generate another action potential.

When the action potential cycle of conductance change is completed, the cell either returns to rest or begins to depolarize again if the external current is still present. At high spike rates, the frequency of action potential generation depends on the magnitude of the external current relative to the potassium current as a function of time. When the input current exceeds the potassium current, the membrane depolarizes and another action potential begins. Ultimately, the occurrence of action potentials is limited by the requirement

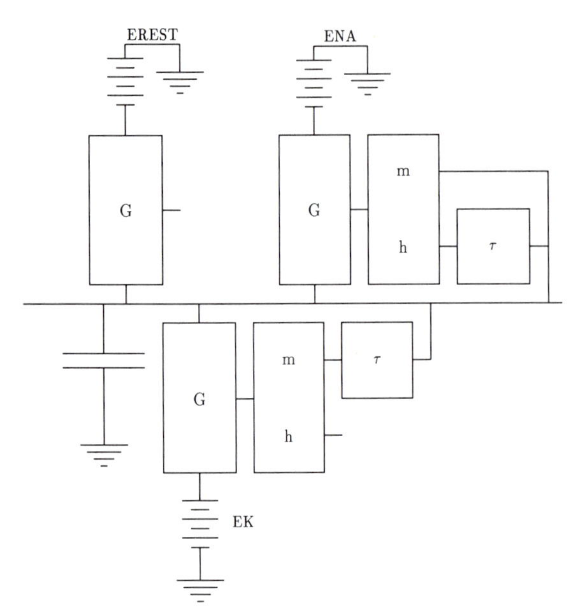

FIGURE 7 A simple silicon neuron that is able to generate action potentials. It comprises a capacitor, a leakage conductance, a sodium conductance, and a potassium conductance.

Circuits in Action

The action potential in the silicon neuron is generated the same way as in the biological neuron. A block diagram of a simple silicon neuron with membrane capacitor, leakage conductance, voltage-sensitive sodium conductance, and voltage-sensitive potassium conductance is shown in Figure 7. The detailed activation/inactivation circuits are shown in Figure 8 and Figure 9. The kinetics of opening and closing the sodium and potassium channels are controlled by the low-pass filters feeding the activation/inactivation circuits. The membrane voltage, and delayed membrane voltages controlling all of the activation/inactivation differential pairs, and the resulting currents are shown in Figure 10. The kinetics and scaled magnitudes of the currents giving rise to the action potential in the biological neuron are replicated by the concurrent dynamical interaction of the circuit modules.

that the membrane hyperpolarize long enough for the sodium channel to deinactivate. If the input current is large enough to depolarize permanently the membrane so that deinactivation cannot occur, the action potential generation mechanism fails.

SILICON NEURONS

Ultimately, these circuit modules are combined to create analogs of silicon neurons. For example, we have combined circuit modules emulating the various calcium currents, the calcium-dependent potassium current, the potassium A-current, and the sodium and

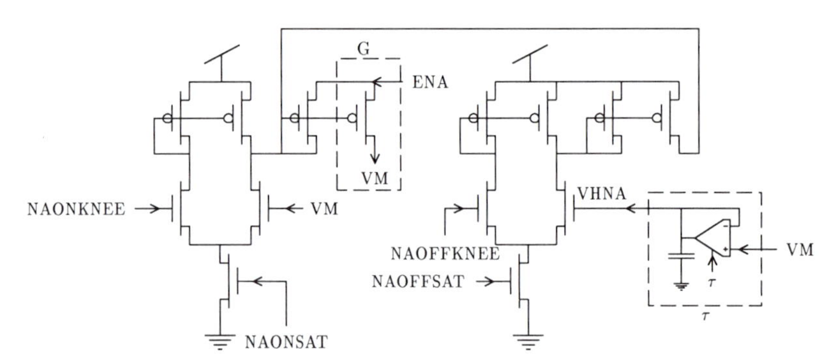

FIGURE 8 The sodium activation/inactivation circuit is composed of two transconductance amplifiers and a diode-connected transistor (see also Figure 5B). Box labeled G, membrane conductance element; box labeled τ, low-pass filter that controls the kinetics of inactivation.

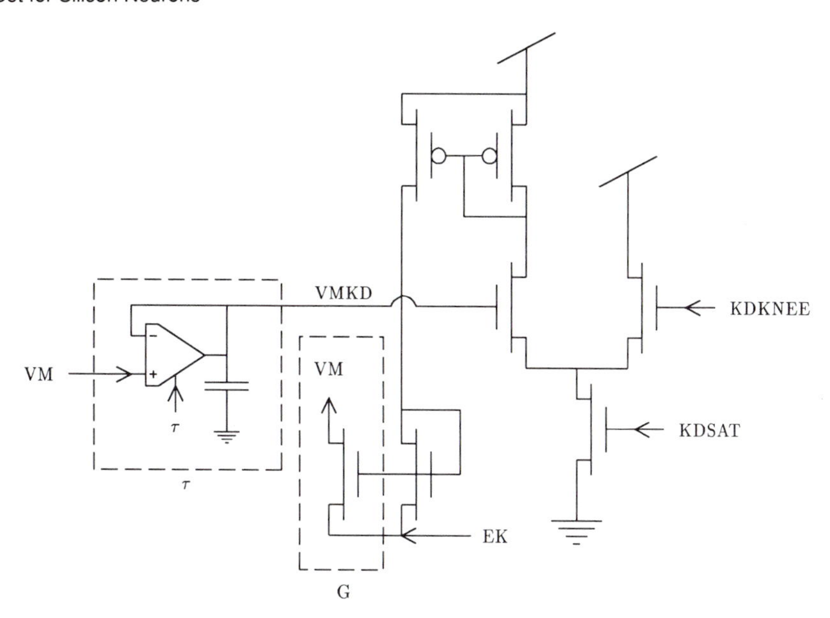

FIGURE 9 The potassium activation/inactivation circuit has only one differential pair, because this channel does not inactivate. Compare with Figure 8. Box labeled G, membrane conductance element; box labeled τ, low-pass filter that controls the kinetics of activation.

potassium spike currents and fabricated them on a MOSIS *tinychip*.[7] Such silicon neurons mimick the electrophysiology and temporal pattern of action-potential discharge of real cortical pyramidal cells, as shown in Figure 11.

VLSI designers have taken a wide range of approaches to the design of neurons for artificial neural networks. A survey of work in the field can be found Ramacher and Ruckert (1991) and IEEE (1992, 1993). Many of these designs use pulse-based encoding, similar to the action-potential output encoding of real neurons. However, even in these cases, the goal has been to reproduce a fixed input/output relation, rather than to understand how underlying biophysical mechanisms give rise to a range of input/output relations, as seen at different stages of adaptation or in different neuronal cell types. Designers take this approach in the interest of producing compact circuits. We have made many approximations in the design of our circuit modules for the same reason. For example, our conductances are saturating, rather than linear functions. Therefore, our silicon neurons would not perform like real neurons if they were clamped at voltages far away from the resting potential. The linear regions of our elements are arranged in such a way that the deviation from true neuronal behavior is minimized when the cell is operating in the physiological region. The trade-offs in design depend on familiarity with the critical aspects of the goal and with the idiosyncrasies of the implementing medium.

The significance of the modules described here is not in their detailed circuitry. Instead, it is in the correspondence between the parameters of the circuit modules and parameters describing the biophysical elements that underly neuronal electrophysiology. For example, the expression of a greater number of channels of a given type in the neuronal membrane can be approximated by increasing the maximum conductance of the activation/inactivation circuit. Different types of channels can be approximated by changing

[7] MOSIS tinychips are 2.22×2.25 mm^2. Our prototype silicon neurons were fabricated in the Orbit p-well double-poly 2μ process.

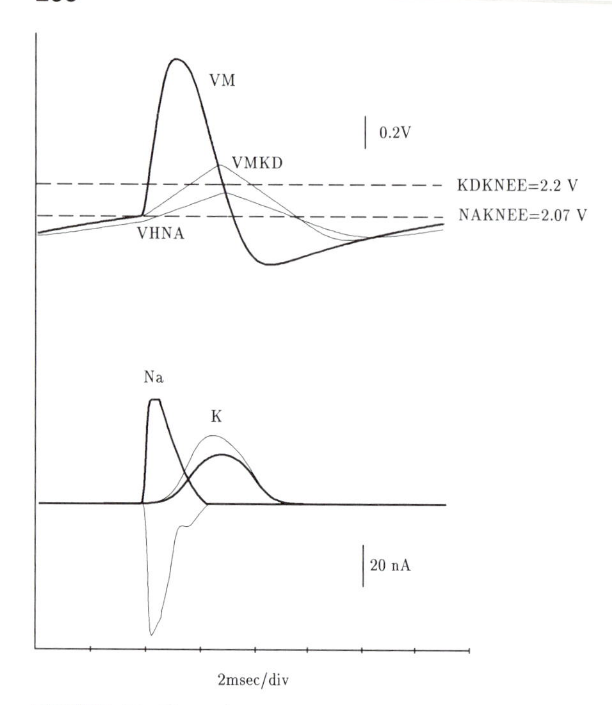

FIGURE 10 The action potential is generated by the dynamics of two channel modules that emulate the sodium channel and the delayed-rectifier potassium channel. The membrane voltage, VM, is plotted in thick line. The low-pass filtered versions of the membrane voltage that control sodium inactivation, VHNA, and potassium activation, VMKD, are plotted in thin line. The threshold level for sodium activation and inactivation were set to the same voltage, indicated by NAKNEE. The threshold for potassium activation is indicated by KDKNEE. The lower plot shows the conductances, plotted in thick line, of the sodium and potassium channels during the spike, as well as the resulting currents that flow onto the membrane capacitance. The notch in the sodium current relative to the sodium conductance is due to the diminished driving potential across the sodium conductance as the membrane voltage approaches the sodium reversal potential. These plots were obtained by circuit simulation, because not all of these variables were instrumented on fabricated chips.

the voltage thresholds and time constants of activation and inactivation. This correspondence to the underlying biophysics allows the same circuit to emulate different types of biological neurons. By changing the control voltages to the circuit modules, we are able to reconfigure the same circuit that emulated a regularly adapting cortical pyramidal cell to emulate (for example) a bursting cell (Figure 12). The range of reconfigurability is determined by the complexity of the circuit. Modules can be inactivated and therefore effectively removed from the total circuit, but no new modules can be added to the existing circuit. If additional properties are required, another silicon neuron with different morphology and different types of channels can be fabricated by using variations of the basic circuit modules.

As with any construction kit, much pleasure can be gained from contemplating future projects. In addition to constructing different neuronal types, we can use these circuit modules to emulate different modes of operation of a single type of neuron. For example, the actions of neuromodulators that change the efficacy of one or another channel population can be emulated through dynamical control of circuit parameters. By maintaining fidelity to underlying biophysical principles, we hope to capture the behavior of a broad range of neuronal types under a range of physiological conditions. This set of circuit modules will enable us to build artificial neurons that embody the diversity of neuronal function in the nervous system. Emulation of a range of biologically realistic neurons is the first step toward building large neuronal networks that operate in real time.

BASICS

This section briefly describes the basic functional elements that constitute the circuits of the silicon neuron. We include these descriptions for the benefit of neuroscientists who are interested in the operation of the silicon neuron, but who have little previous experience with CMOS circuits. A more detailed description and explanation of these circuit elements can be found in Carver Mead's book, Analog VLSI and Neural Systems (Mead, 1989). The input/output relations for the transistor, the current mirror, the differential pair, the transconductance amplifier, the low-pass filter, and the resistor are summarized here without explanation. The graphs in this section are simulated responses, rather than actual circuit measurements.

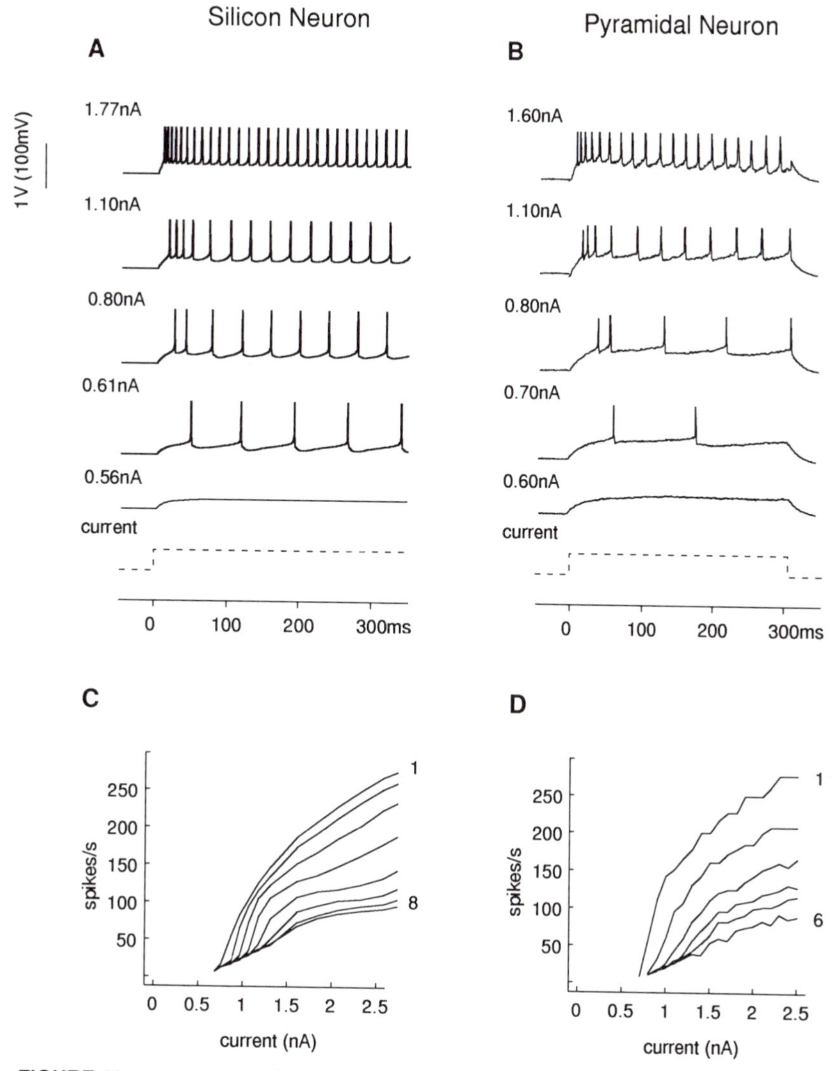

FIGURE 11 A comparison of the current-discharge curves of a silicon neuron and a pyramidal neuron. Temporal responses of a silicon neuron (A) and a cortical slice neuron (B) to steps of injected current. Current-discharge curves for the first eight interspike intervals of the silicon neuron (C) and for the first seven interspike intervals of the cortical slice neuron (D). (Tops of action potentials have been truncated.)

The Transistor

The transistor is an active, three-terminal device. Current flows between the source and the drain of the transistor under the modulatory control of the gate. In the complementary metal-oxide semiconductor (CMOS) technology, transistors come in two types, N-type and P-type. The current flows from a source of positive potential in a P-type transistor and into a

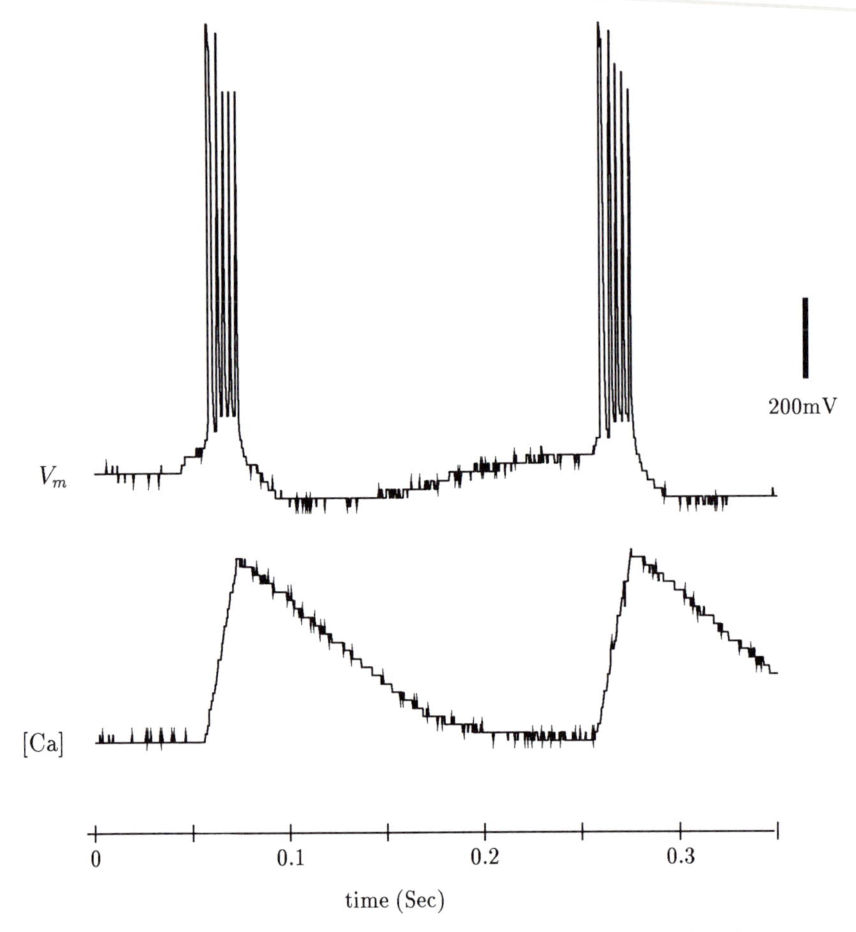

FIGURE 12 The silicon neuron emulates a bursting cell. The upper trace shows the silicon neuron producing bursts of action potentials in response to a constant input current. (Noise in this recording arises mainly from the digitizing oscilloscope.) The lower trace shows the voltage that represents the calcium concentration within the cell. The calcium enters the cell during the burst. Calcium opens a calcium-dependent potassium channel. The kinetics of this channel are delayed, so the calcium builds up in the cell before the potassium current begins to act. The cell hyperpolarizes after the burst until the calcium concentration decays and the cycle repeats itself.

source of negative potential in an N-type transistor, shown in Figure 13. In an N-type transistor operating in the subthreshold regime, the current through the channel between the drain and the source is an exponential function of the voltage difference between the gate and the source and a nonlinear saturating function of the voltage difference between the drain and the source. The equation for current flow between the drain and the source is

$$I_{\mathrm{DS}} = I_0 e^{\kappa V_{\mathrm{G}} - V_{\mathrm{S}}} \left(1 - e^{-(V_{\mathrm{D}} - V_{\mathrm{S}})} + \frac{(V_{\mathrm{D}} - V_{\mathrm{S}})}{V_0}\right) \tag{1}$$

where all voltages are in units of the thermal voltage,

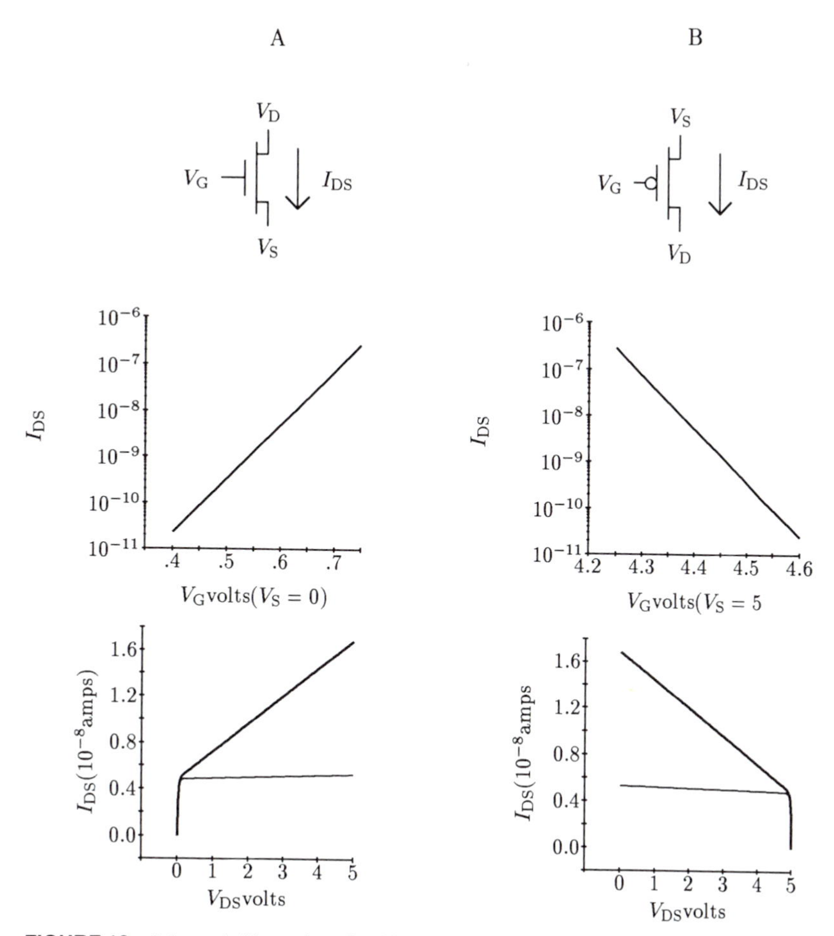

FIGURE 13 Schematic illustration of an N-type (A) and P-type (B) transistor and their associated transfer functions. The top graphs shows the drain-source current as a function of gate voltage, with the magnitude of the drain-source equal to 1 V. The lower graphs show the drain-source current as a function of drain-source voltage for a gate voltage of 0.6 V. The two curves are estimated for a transistor with channel length of a half micron whose Early voltage is about 2 V (thick line), and for a transistor with channel length of 3.5 microns whose Early voltage is about 50 V (thin line).

kT/q. The variable κ is a fabrication process-dependent parameter approximately equal to 0.7. It measures the efficacy with which charges placed on the gate change the Boltzmann energy barrier to current flow. This parameter is analogous to the gating charge of the channel through the nerve membrane, n, which we assume to have a value of 7. The variable V_0 is a geometry-dependent parameter called the Early voltage. It reflects the extent to which the transistor behaves as a current source. Transistors with long channels have large Early voltages of about 50 V, whereas transistors with short channels may have Early voltages as low as 2 V. Calculations of circuit transfer functions often assume that the transistor is in saturation, so that

$e^{-(V_D - V_S)} \approx 0$, and that the Early effect is negligible, either because the Early voltage is large or because the drain-source voltage is nearly constant. In this case, the transistor is an effective current source. The expression for the current through the transistor depends only on gate voltage if the source voltage is fixed:

$$I_{sat} = I_0 e^{\kappa V_G - V_S} \tag{2}$$

The Current Mirror

The current mirror is a two-transistor circuit that receives a current input and produces a current output that is inverted relative to the input current (Figure 14). The input transistor is a diode-connected transistor M1 that transforms an input current I_{in} into a voltage, V_b. The exponential dependence on gate voltage of the current through the diode-connected input transistor determines the gate-source voltage and thus also its drain-source voltage. If V_b is applied to another transistor M2 that is identical in every way (including the drain-source voltage) to M1, then the current out of M2 is the same as the current into M1. Changing the conditions of M2 relative to M1 allow the generated current to be modified. The equation for current flow through a transistor is used to estimate the scaling be-

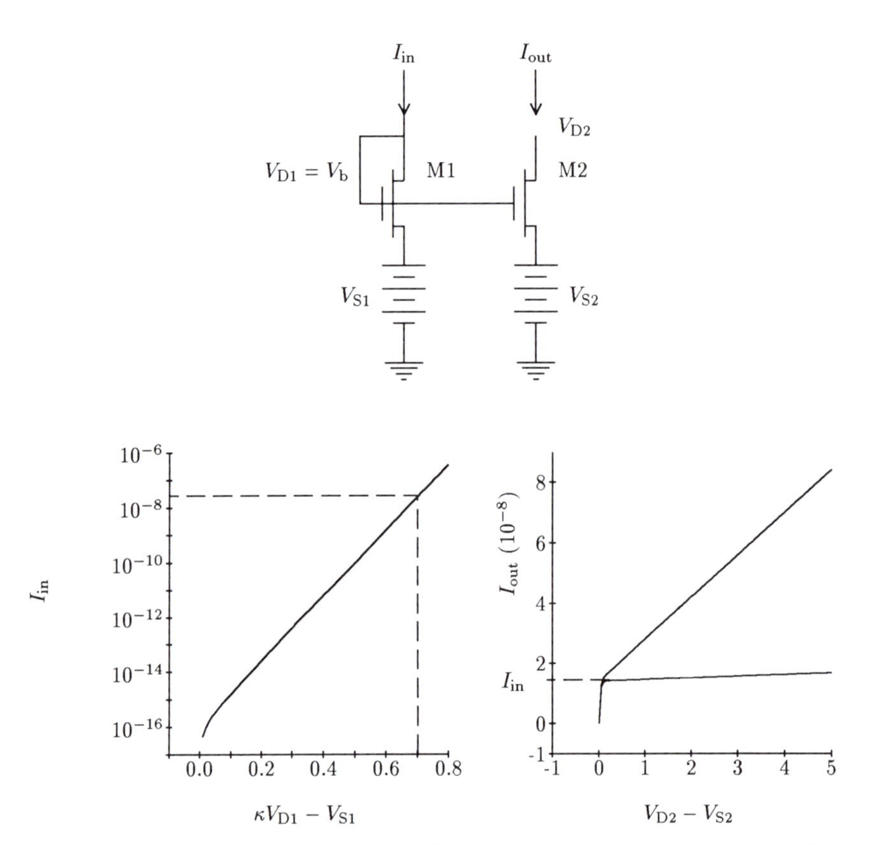

FIGURE 14 Current mirror and associated transfer functions. The input current is plotted on the left graph, relative to the gate-source voltage on the input transistor, M1. The output current is plotted on the right graph, for a particular input current, as a function of the drain-source voltage of the output transistor M2.

tween input current and output current. For example, if M2 is very short, the Early voltage of M2 is small and I_{out} will have a strong dependence on the drain-source voltage of M2. If the source potential of M2 is different from that of M1, then I_{out} will be multiplied by a constant factor, $e^{(V_{S1} - V_{S2})}$ in the subthreshold regime of transistor operation. Mirrors can be made either of N-fets or P-fets depending on the signs of the desired input and output currents.

The Differential Pair

The differential pair is a pair of transistors with a common source (Figure 15). A third *bias* transistor supplies current to the differential pair. The bias current limits the total current that can flow through the differential pair transistors. The voltage on the common source of the differential pair follows the larger of the input voltages so that the total current flowing through

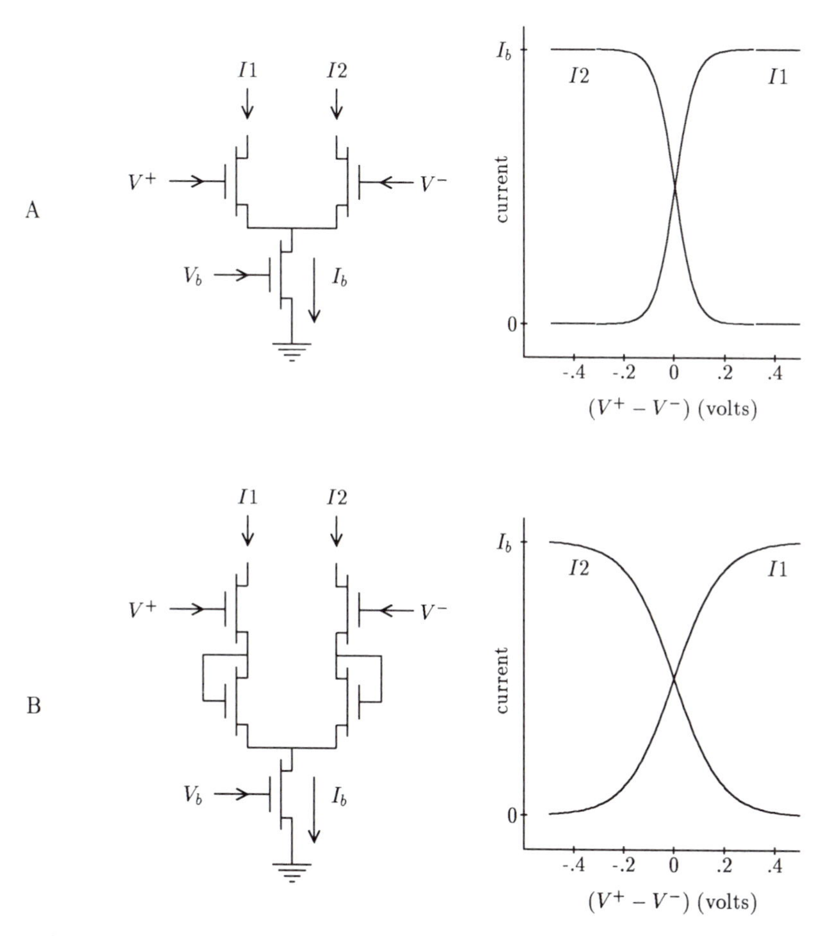

FIGURE 15 The differential pair. (A) The differential pair compares two input voltages, V^+ and V^- and splits a current, I_b, between the two branches, thereby generating currents I^+ and I^-. (B) The input-voltage range of the central linear portion of the current/voltage relation can be increased by adding diodes to the base of the differential pair.

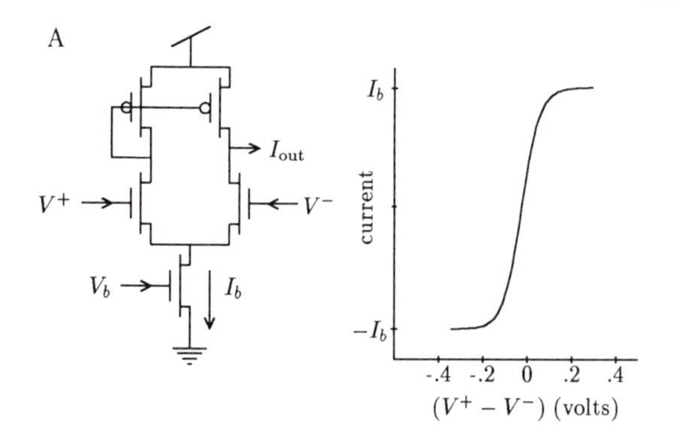

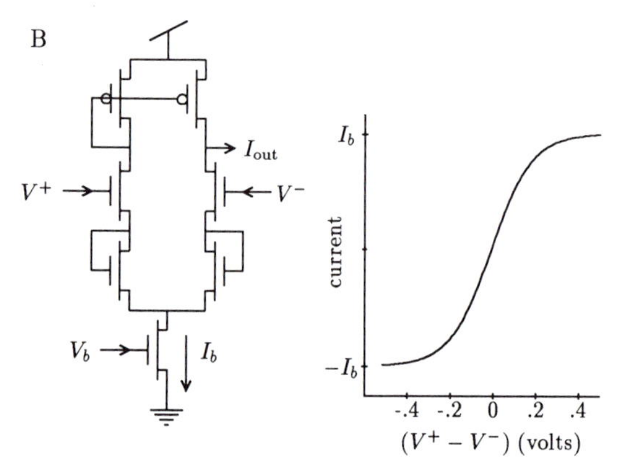

FIGURE 16 Schematic representation of the normal (A) and wide-input range (B) transconductance amplifiers and associated transfer functions.

the differential pair does not exceed the bias current. These constraints can be mathematically expressed, using the relationship for the current through a transistor in saturation (shown earlier). They are

$$I_1 = I_0 e^{\kappa V_1 - V} \tag{3}$$

$$I_2 = I_0 e^{\kappa V_2 - V} \tag{4}$$

$$I_b = I_1 + I_2 = I_0 e^{-V}(e^{\kappa V_1} + e^{\kappa V_2}) \tag{5}$$

These equations can be solved for the common source voltage V to yield expressions for the currents I_1 and I_2:

$$I_1 = I_b \frac{e^{\kappa V_1}}{e^{\kappa V_1} + e^{\kappa V_2}} \tag{6}$$

and

$$I_2 = I_b \frac{e^{\kappa V_2}}{e^{\kappa V_1} + e^{\kappa V_2}} \tag{7}$$

The voltage input range for which the output is linear can be extended using source-degeneration diodes (Watts *et al.*, 1992). The currents I_1 and I_2 in this case are given by

$$I_1 = I_b \frac{e^{\frac{\kappa^2}{\kappa+1} V_1}}{e^{\frac{\kappa^2}{\kappa+1} V_1} + e^{\frac{\kappa^2}{\kappa+1} V_2}} \tag{8}$$

and

$$I_2 = I_b \frac{e^{\frac{\kappa^2}{\kappa+1} V_2}}{e^{\frac{\kappa^2}{\kappa+1} V_1} + e^{\frac{\kappa^2}{\kappa+1} V_2}} \tag{9}$$

The Transconductance Amplifier

The transconductance amplifier is a differential pair whose branch currents, I^- and I^- have been subtracted by means of a current mirror (Figure 16). The output current I_{out} is the difference between I^+ and I^-. The equation for the output current of a normal transconductance amplifier is

$$I_{out} = I_b \tanh\left(\frac{\kappa(V_1 - V_2)}{2}\right) \tag{10}$$

The equation for the output current from a wide-input range transconductance amplifier is

$$I_{out} = I_b \tanh\left(\frac{\frac{\kappa^2}{\kappa+1}(V_1 - V_2)}{2}\right) \tag{11}$$

The Low-Pass Filter

The transconductance amplifier is connected to a capacitor C to form a low-pass filter (Figure 17). The amplifier, with its output tied to its negative input, behaves like a resistor from the point of view of the

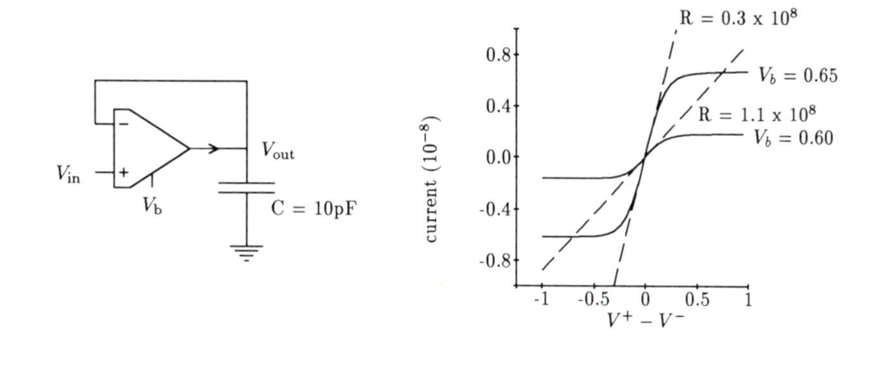

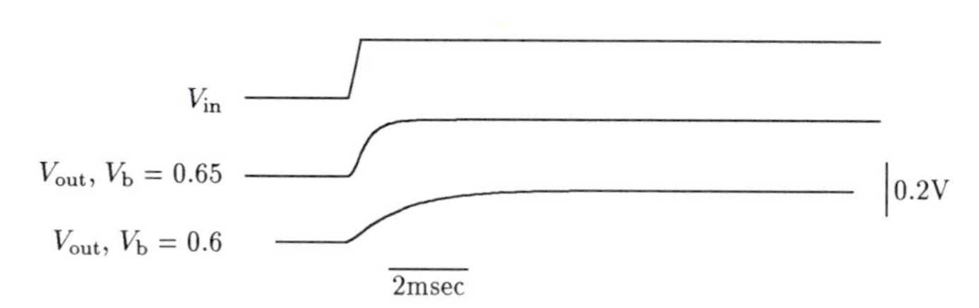

FIGURE 17 Schematic diagram of the low-pass filter and transfer function of the transconductance amplifier showing the region of linear resistance. The response of the filter to a step input is shown below for two different filter time constants.

capacitor. The resistance of the amplifier is determined by the voltage V_b. The time constant τ of the filter is the product of the resistance of the amplifier and the capacitance, $\tau = RC$. The equation for the time evolution of the output of the filter to an input step is

$$V_{out} = V_{in}(1 - e^{-t/\tau}) \tag{12}$$

The range of voltage over which the transconductance amplifier behaves linearly is roughly $\pm$ 0.25 V. Voltage steps larger than this range cause the amplifier to exceed its slew-rate limit, so the temporal response is slower than that predicted by a linear model.

A simple temporal differentiator measures the difference between an input and a low-pass filtered version of the input. This operation is performed by a transconductance amplifier whose positive input is connected to V_{in} and whose negative input is connected to V_{out}.

The Resistor

The resistor circuit and its current-to-voltage relation are shown in Figure 18. The resistor is linear in the center of its range. The magnitude of the resistance can be controlled by application of an analog potential V_g which generates a current I_b in the resistor bias circuit. The equation for the current through the resistor as a function of the voltage $(V_1 - V_2)$ is

$$I = I_b \tanh\left(\frac{V_1 - V_2}{2}\right) \tag{13}$$

In the linear range, the resistance is

$$R = \frac{2kT/q}{I_b} \tag{14}$$

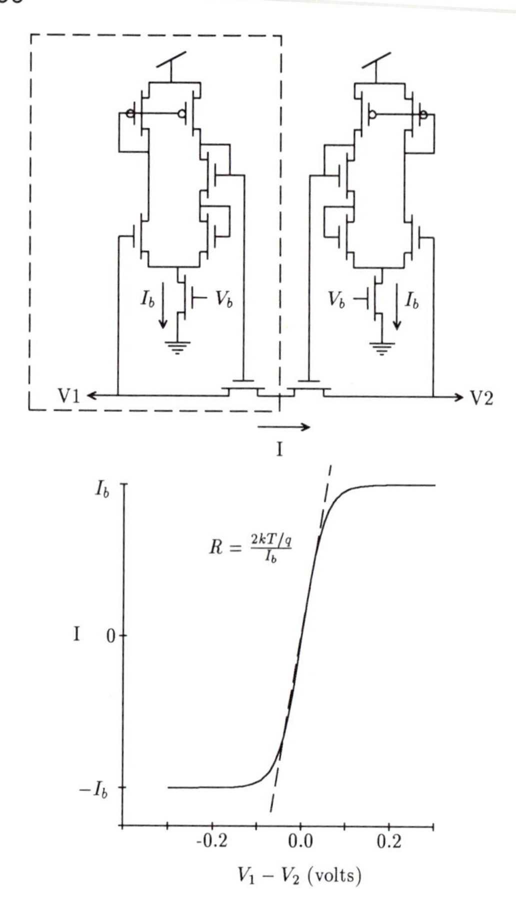

FIGURE 18 Circuit diagram and input/output relation of a voltage-controlled resistor.

References

Hille, B. (1992). *Ionic channels of excitable membranes.* Sunderland, MA: Sinauer.

Hodgkin, A. L., & Huxley, A. F. (1952). A quantitative description of membrane current and its application to conduction and excitation in nerve. *Journal of Physiology (London)*, **117**, 500–544.

IEEE Transactions on Neural Networks, **3**, May 1992.

IEEE Transactions on Neural Networks, **4**, July 1993.

Koch, C., & Segev, I. Eds. (1989). *Methods in neuronal modeling: From synapses to networks.* Cambridge, MA: MIT Press.

Mahowald, M., & Douglas, R. J. (1991). A silicon neuron. *Nature*, **354**, 515–518.

McCormick, D. A. (1990). Membrane properties and neurotransmitter actions. In G. Shepard (Ed.), *The synaptic organization of the brain* (3rd ed., pp. 220–243). New York: Oxford University Press.

Mead, C. A. (1989). *Analog VLSI and neural systems.* Reading, MA: Addison-Wesley.

Rall, W. (1977). Core conductor theory and cable properties of neurons. In E. R. Kandel, J. M. Brookhardt, & V. B. Mountcastle (Eds.), *Handbook of physiology: The nervous system* (Vol. 1, pp. 39–98). Baltimore, MD: Williams & Wilkins.

Ramacher, U., & Ruckert, U. (1991). *VLSI design of neural networks.* Boston: Kluwer.

Watts, L., Kerns, D. A., Lyon, R. F., & Mead, C. A. (1992). Improved implementation of the silicon cochlea. *IEEE Journal of Solid-State Circuits*, **27**, 692–700.

15

VLSI Implementation of Neural Networks

Federico Faggin and Carver Mead

INTRODUCTION

The recent resurgence of interest in neural networks (NNs) has resulted in the application of NNs to a variety of problem domains. The initial results show great promise, which in turn is motivating research in appropriate implementation technologies. Several approaches are now being explored; at one extreme of the spectrum is algorithmic research for software running on a conventional computer, whereas at the other extreme researchers are exploring radically new computational structures, such as optical computing. A more conservative approach relies on adapting the existing silicon-based CMOS VLSI (complementary metal oxide semiconductor, very large scale integration) technology to the unique needs of NN computation.

Which approach is appropriate is clearly dependent on the characteristics of the problems we wish to solve. For certain applications, a software implementation may be adequate; however, where speed, power dissipation, size, weight, and cost requirements are demanding, a solution using conventional computers (including advanced parallel architectures) may not be feasible. This drawback is particularly severe for the real-time solution of problems in vision, robotics, speech recognition, and so on.

In the course of this chapter, we shall provide evidence for the position that, both in advanced research and in the application of NNs, we require a new type of hardware, one that is tuned to the new paradigm and that offers a price to performance ratio several orders of magnitude better than that obtained with conventional hardware. We shall also explore the key issues and alternatives of NN hardware implementations using silicon-based VLSI technology. It will become apparent that NNs require and are inspiring the development of a new design style. This new design uses massively parallel computation with analog processors, has very low power dissipation, and has an inherently high degree of fault tolerance due to the use of adaptive elements, appropriate data representations and collective processing. We shall examine a few examples of this emerging design. At the end of the chapter, we shall offer our assessment of future prospects.

INFORMATION PROCESSING IN ANIMAL NEURAL SYSTEMS

It is insightful to view the nervous system as an organ that is designed to accomplish optimally a set of information processing tasks under a number of constraints, and to view evolution as a creative process that identifies and sorts out the best available strategies to perform the task. The information processing task of a nervous system, in this view, is to function as a survival-directed, real-time simulator both of the external environment and of the self as it interacts with the environment. The critical constraints that need to be satisfied include:

- Use only the existing biochemical machinery (certain small changes are allowed).
- Use the smallest possible information to specify the structure and the assembly process of the system.
- Use the smallest possible amount of tissue for the system's construction and use the smallest possible amount of metabolic energy for its operation.
- Construct the system to operate with the highest possible number of imperfect or even nonoperative components to achieve graceful degradation of performance with increasing component degradation and failure.

Some of the critical innovations that evolutionary dynamics has created which both accomplish the task and satisfy the constraints are the following:

- The overall architecture of the system is completely specified *a priori* and reflects the nature of the information processing task to be performed.
- The regularities (spatial and temporal correlations) of the sensory data are used to provide the additional information required to complete the specification of the system.
- A variety of adaptive elements are used to modify the hardware so that the system is capable of working with "components" of unknown and variable specifications, including faulty components, gradually completes (self-)construction while it operates continually as additional information is acquired, and stores information with several decaying time constants.
- The data representations used require minimum signaling frequency (to reduce the energetic requirement) and allow for fault tolerance. The critical information is carried by many units (distributed information). The hierarchical, contextual structure of natural objects is reflected in the representation.
- The system comprises massively parallel building blocks (to satisfy the computational requirement at minimum power), composed of regular structures (minimum specification) and connected mostly by local wiring (minimum size).
- The computation is analog (optimization of computational speed, power dissipation, and economy of hardware) and communications are digital (best signal to noise ratio).

NEUROMORPHIC INFORMATION PROCESSING STRUCTURES

If we wish to solve the same class of problems that nervous systems were designed to solve, then we must take seriously the approach that nature has evolved. Clearly, it is necessary to understand to what extent the individual biological innovations have been dictated by the intrinsic nature of the problem, and how much they are motivated by the unique constraints of the available biological hardware. Obviously, we too must use the available construction materials, and today the only ones are silicon-based. (It would be nice to have the other technology—70-year guaranteed, solar-powered, self-replicating computers! We could solve, once and for all, all our manufacturing problems!)

Except for the choice of materials, it appears that all constraints placed on animal systems also apply to a silicon-based implementation technology, although less stringently. Thus, we believe that the solutions adopted by nature should strongly influence our choice of architecture in implementations of NNs.

The Human Central Nervous System

The human central nervous system (CNS) is a three-dimensional structure composed of approximately

10^{15} interconnected components (synapses). It dissipates few watts, occupies approximately 10^{-3} m^3, is dominated by wiring (less than 1% of the volume of the CNS is made up of synapses), and can perform approximately 10^{16} operations per second. Advanced general-purpose computers are three-dimensional structures composed of approximately 10^8 interconnected components (transistors), or 10^{11} components if we include magnetic mass storage. They dissipate about 1000 W, occupy a volume of about 1 m^3, are dominated by packaging and wiring (less than 1% of the volume is in the silicon components), and can perform approximately 10^9 operations per second.

It is thus apparent that the CNS is a far more efficient computational structure than are state-of-the-art computers. We can quantify this statement by observing that all models of neural networks require at least one multiply and one add operation for each synaptic evaluation. We may suppose that the absolutely minimum computation carried out by the neuron is one operation for each nerve pulse arriving at a synapse. If we assume the average firing rate to be 10 spikes per second, then synapses will perform 10^{16} operations per second, as mentioned. It might be argued that the arrival of a nerve pulse at a synapse results in far less than one operation, because the dendritic tree is merely averaging a very large number of inputs. Recent experiments on the visual system of the fly (de Ruyter van Steveninck & Bialek, 1988) have yielded compelling evidence that individual synaptic inputs represent significant computation. Under conditions where visual input is sparse, the fly makes the decision to turn based on the arrival of only two or three spikes.

Our estimates of the number of neural interactions include only the computation initiated by spiking neurons, and completely ignore the inhibitory interaction of many nonspiking interneurons, as well as all the dendro-dendritic interactions of neurons, both of which operate by graded release of neurotransmitter. If these interactions and the computation required for synaptic plasticity are taken into account, the actual computation accomplished by the brain may be substantially larger than the foregoing numbers indicate.

We can view the energy consumed by a synaptic computation either in metabolic terms, or as the energy required to charge up a small patch of nerve membrane to a potential of approximately 100 mV. In either case, the energy dissipated per synaptic event is somewhat less than 10^{-15} J. Only a single application of this transition energy is required to achieve the equivalent of one or more "operations" in the computational sense. At an average firing rate of 10 spikes per second, the power dissipated by synaptic activity in the brain is a few watts. We can similarly estimate the power required for propagating neural signals in the axonal white matter to be less than 1 W.

In the digital CMOS VLSI technology with which modern computers are constructed, power is dissipated in charging and discharging the capacitance associated with the gates of individual transistors and with the wires that connect the gates to the sources and drains of other transistors that are driving the gates. The work required to charge the gate capacitance is absolutely necessary; that required to charge the wires is a function of the organization of the chip and of the required speed.

In 1990 technology, charging the gate of a minimum-sized transistor requires about 10^{-13} J. This energy is the elementary transition energy in our digital technology. To evaluate the potential of current digital practice, we consider a contemporary digital signal processing (DSP) chip containing about 10^5 transistors and operating at 10^7cycles per second. DSP chips represent today the best performance achievable with a single chip device. Such a chip dissipates about 1 W, and can accomplish about 10^7 operations per second (10 MOPS), or about one operation per cycle. This corresponds to 100 nJ per operation. One should note however, that the energy dissipation per operation at the system level is substantially higher than 100 nJ due to the additional external interface devices, memory, power supplies, and so on.

The necessary computation in such a chip can be accomplished by the operation of about 10^4 transistors in each cycle (approximately 10% of the transistors change state in any given clock cycle). Our best digital technology thus requires 10^4 times more transitions per operation than the CNS requires.

The energy dissipation required to do the necessary switching in the DSP chip is considerably less than the actual power dissipated. We multiply the required number of switching events per operation (10^4) by the switching energy of a minimum-sized transistor (10^{-13} J) and find that the necessary work is 10^{-9} J per operation. At 10^7 operations per second, the necessary power consumption is only 0.01 W of the 1 W actual power dissipation. The additional energy is dissipated in driving the massive amount of wire per transistor gate, and the use of transistors of larger than minimum size to drive these wires rapidly. Thus, there is a factor of about 100 in energy dissipation per operation that we could, in principle, gain if we reduced the capacitance of the wiring attributable to each transistor gate to the same order of magnitude as that attributable to a single minimum-sized transistor.

It has often been asserted that the brain is profligate in its use of wire. There are nerve bundles containing many millions of individual fibers that connect major areas of the brain. Collectively, these bundles form the white matter, and occupy about one-half of the brain's total volume. From the preceding computations, however, we can see that the brain is frugal in its use of communication resources. In terms of energy dissipated in communication relative to that used for essential synaptic computation, the CNS is about 100 times more effective than computers are. This factor is directly attributable to the brain's computational arrangements which minimize wire length per synapse. Although numerous axons connect distant regions of the brain, at the end of each axon there is a local axonal arborization, used to distribute axonal signals to several thousand synapses. The length of the axon attributable to each synapse is thus reduced by this fan-out factor. This sharing of wire, and the minimization of local dedicated wiring, is possible because most computations in the brain are confined to a local neighborhood. In computers, we achieve wiring sharing by using bus structures, but computer designs are still 100 times less efficient in this regard than brains are.

We can summarize these findings as follows. Present-day digital chips dissipate about 10^8 times as much energy per operation as do brains (10^{-7} versus 10^{-15} J per operation). Of this energy, a factor of 100 is attributable to the irreducible gating energy of the devices, a factor of 10^4 to the number of devices undergoing a transition in one operation, and a factor of 100 to the more efficient use of wire by brain architectures. We can reduce by a factor of about 100 the power dissipated per operation in the silicon technology by reducing power supply voltages and by shrinking device dimensions, until we reach the fundamental physical limits of the technology. We can improve the wiring factor by adopting more local connectivity patterns in dedicated VLSI architectures. There remains the factor of 10^4 or more in transitions per operation, which is a direct result of the analog computation at the synapse.

Unless we change the architecture of computers drastically, any additional increase in computational power will require a proportional increase in power dissipation. At 1 nJ per operation, 10^{16} operation per second would require 10 MW power dissipation! Similar calculations apply to the volume required per operation per second. If very high computational power is an inherent requirement for solving real-time real-world problems, then we have no hope of creating intelligent autonomous machines unless we change radically the architecture of computers or the technology we use to make those machines.

Important Characteristics of the Central Nervous System

Important characteristics of the CNS are the extremely large number of components and the high connectivity of these components—the natural consequence of the highly parallel organization. Let us consider how well the VLSI technology stacks up against biological hardware.

It appears feasible to be able to develop an architecture and a design methodology that allow a sufficient level of fault tolerance to make possible wafer-scale integration within the next 10 years. We shall say more about this possibility later. For the purpose of this analysis, chips the size of one wafer could be feasible in 5 to 8 years and commercially available in about 10 years.

The diameter of wafers in advanced production lines today is 6 in. By the turn of the century, wafer diameters of 10 to 12 in. could be in production. During the same period of time, minimum line widths, which presently are at about 1.0 to 1.2 μm for advanced production technology, will have improved to 0.3 to 0.4 μm. Thus, the level of integration will increase from 500 million components per wafer today (assuming an average area of 200 μm^2 per component) to 10 billion components per wafer on a 10 in. wafer, with 0.35 μm minimum line width.

If we extrapolate the chip size and complexity of digital hardware, by the turn of the century we may achieve chip sizes of 25 x 25 mm^2 with about 30 million components for logic devices and 128 million bits of memory (dynamic RAM).

We thus have a tremendous motivation to explore new data representations, architectures, and design methodologies that will permit us to achieve fault-tolerant structures and thus wafer-scale integration. Understanding and following the organizing principles used by the nervous system seems to be a sensible start.

Let us now examine how we can achieve a high level of interconnections with VLSI technology. There are two aspects that need to be considered: interconnections within the chip and interconnections between chips. There are limitations in both areas. Inside the chip the problem resides in the small number of levels of interconnections available (two to three now, maybe three to five by the turn of the century) and in the high real-estate cost of wiring. In practice, this means that buses of a few hundred to a few thousand wires can be implemented only if they are part of the active circuitry. Between chips, the limitation is the number of pins of a packaged chip, which is confined not to exceed a few hundred pins. By contrast, in the nervous system, a typical nerve bundle may contain tens of thousands of axons (equivalent to the wiring from subsystem to subsystem) and some bundles, such as the optic nerve, contain about 1 million axons. Within an individual functional unit, such as enthorhinal cortex or V1, the number of interconnections (equivalent to the interconnection buses inside a chip, and from chip to chip if the function must be partitioned in more than one chip) is many orders of magnitude larger.

This situation would be hopeless if it were not for the fact that we can operate silicon-based components and metal-based wires at much higher speed (thus dissipating much more power) than that at which biological components and wires operate.

We can therefore use the time dimension to overcome (partly) the lack of a third physical dimension, by time multiplexing onto a single wire information that would otherwise be carried by several wires. Such a technique, called time division multiplexing (TDM), was originally developed in telephony in order to reduce the number of wires or communication channels between switching stations. The technique takes advantage of the fact that, if the available channel bandwidth is much larger than the bandwidth of the individual signal to be transmitted, many signals can be sampled periodically and their measured values (the samples) can be sent down the channel in a prescribed order. The signals can then be reconstructed at the receiving end of the channel. For the original signal to be reconstructed without loss of information, it is sufficient to sample the signal at a rate of at least twice the maximum frequency component contained in the signal (Nyquist theorem); thus, a channel with bandwidth B can carry the information of N signals with $N = B/(2\Sigma f_i)$, where f_i is the maximum frequency component of signal i. It appears that the information bandwidth of axons is approximately 1 kHz, whereas an individual wire in today's technology can carry 20 to 50 million bits per second. If each sample value can be quantized into 16 bits of information, then one wire can carry about 1000 separate axonal signals.

This 1000 to 1 improvement in interconnection capability brings us much closer to the biological scale and removes what otherwise would have been an insurmountable obstacle. The future availability of wafer-scale integration will further alleviate the interconnection problem by relaxing even more the limitation on the number of interconnection levels within a chip (now a wafer) and those on the number of pins. In summary, it appears that the expected evolution of current technology will make possible in 10 years the construction of neuromorphic systems, containing as an upper limit about 1000 wafers. Each wafer could integrate approximately 10 billion components, for a total component count of 10^{13} components (this estimate includes both the number of components used for the synapse and the components used for supporting purposes). Typical dedicated, single-application systems will then contain 1 to 10 wafers. If the function of a synapse requires the equivalent of 10 components, then the upper limit is a system with 10^{12} synapses, dissipating about 10 kW (assuming 10 W of power dissipation per wafer) and occupying a volume of about 1 m^3. (The corresponding values for the human brain are 10^{15} synapses, 10 W, and 10^{-3} m^3.)

Although we still lack about three orders of magnitude compared to biological hardware for each of the important performance criteria (complexity, power dissipation, and volume), VLSI technology will allow us to explore this new horizon without necessarily having to develop a new technology based on a different physics.

HARDWARE IMPLEMENTATION OF NEURAL NETWORKS

In this section we shall consider some of the key implementation issues and we shall explore the possible hardware implementation approaches of NNs using VLSI technology.

Practical Applications of Neural Networks

To study or use NNs in practical applications, we must either simulate the network operation with a computer or physically build the network with specialized hardware. Conventional techniques use software simulation of the network dynamics on general purpose computers, scientific computers, supercomputers, or workstations provided with hardware accelerators. Hardware accelerators in the form of subsystems or boards that plug into workstations or other computers are widely available today. They can be used as floating-point, vector, or array processors, or may perform more specialized functions using multiple multiply–accumulate chips or multiple digital signal processor chips. These simulation techniques can provide computational speeds ranging from a few MOPS to up to a few hundred MOPS. They are widely known and shall not be considered any further here.

Before we turn our attention to higher performance, lesser known hardware approaches, we need to discuss one major issue that has profound consequences on the choice of the implementation method. This is the issue of dedicated versus general-purpose (programmable, reconfigurable) NN hardware. There are three major sources of variability in neural networks:

1. The specific equations describing the behavior of a single neuron.
2. The interconnection pattern of the various neurons constituting the network.
3. The characteristics and interconnections of the input–output lines to the network.

If we wish to specify conveniently, simulate, debug, and run in real-time a variety of NNs under the program control of a host system, then we need general purpose (GP) hardware. On the other hand, if all critical system specifications are known in advance and will not change after development, then we can use dedicated hardware.

As we shall see, the constraints on the design of GP hardware are much more severe than those on dedicated hardware. Furthermore, a completely programmable and reconfigurable solution may be impossible, since the nature of the input–output lines cannot always be known in advance. Thus custom-tailored hardware may be required, at least at the level of the input–output interface.

An even more serious challenge to designers of massively parallel GP NN hardware is the interconnection requirement. Interconnections present little problem if they are done in software, once sufficient memory is available. The speed requirement, however, cannot be met in most cases with this method.

The only solution, then, demands the use of a large number of hardware processing elements, connected physically by wires.

We should emphasize here that this interconnection problem is different from the one discussed previously. Here we are concerned not with realizing a large number of wires, but rather with implementing the variability in the pattern of interconnections that is a requirement for GP NN simulation. We avoid this problem completely in a dedicated NN, where the wiring pattern is known in advance, while the former problem still remains. One unfortunate consequence of this state of affairs is that GP NN hardware will always achieve a substantially lower level of functionality per chip than dedicated NN hardware.

Nonetheless, there is a strong need for highly parallel GP NN hardware, capable of achieving operating speed close to that possible with dedicated hardware, because in many real-time applications we must use real sensory and actuating signals before we attempt a dedicated hardware implementation. For example, adaptive learning must be done in real time if the data acquisition is controlled by the output of the learning system. In addition, we can envision many applications where the actual adaptive-learning process takes place in a laboratory environment, while production units are nonadaptive and simply duplicate, perhaps in read only memory (ROM), the synaptic weights learned in the laboratory.

It is therefore evident that we need GP NN hardware if we are to develop a workable approach to the design of cost-effective NN hardware.

If programmable and reconfigurable NN hardware proves to be impractical, then the only real time simulation method left is to develop a number of standard NN building blocks that can be physically interconnected to construct a breadboard of the network to be simulated. This scenario is similar to the current practice of breadboarding digital systems using arithmetic and logic units, programmable logic arrays, memory chips, and so on.

The development of such a set of standard building blocks might be desirable even if completely programmable and reconfigurable devices were to exist. Such building blocks would certainly offer a cost to performance ratio and functional density intermediate between dedicated and fully programmable hardware, and thus could be quite useful for field testing and for low-volume NN applications.

Analog versus Algorithmic Computation

The next major NN implementation issue centers around the degree to which an analog instead of an algorithmic computational approach is desirable; in addition, we must decide the degree to which analog instead of digital design techniques are to be favored.

Suppose we have a problem expressible as a differential equation, and that the solution of the differential equation in real-time constitutes the solution of the problem. Suppose further that we wish to find an automatic method for solving the equation. We can then approach the problem in two ways.

1. We can convert the differential equation into a difference equation, map it into a recursive procedure (an algorithm), and have a computer go through the procedure mechanically. If the computer provides its answers in time, we have solved the problem.
2. We can identify and construct a new physical system whose behavior is expressed exactly by the differential equation we wish to solve. We then let the physical system "behave." The observed behavior is exactly the solution to our problem. In other words, the physical system evolves in continuous time in a fashion analogous to the problem system. Here, too, it is necessary to be able to control the time constants of the analog system to match those of the problem system; otherwise, we may not be able to solve the problem in real time.

It is important to realize that the analog method of solving the problem does not necessarily imply an analog design technique. It is possible to design digital building blocks (processing elements) to perform elementary operations, such as addition, multiplication, and integration, and to connect them appropri-

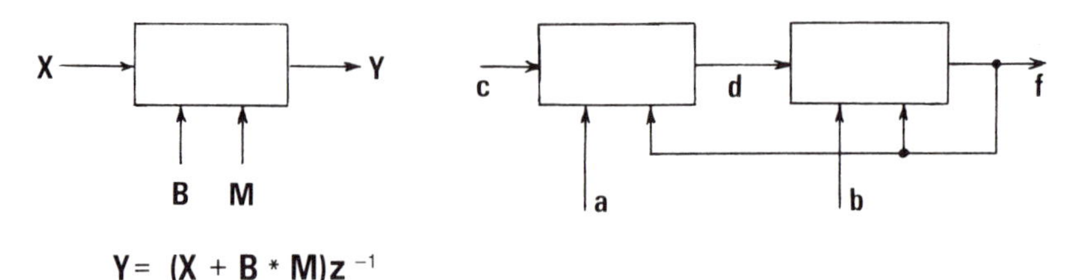

FIGURE 1 The analog computation approach to the solution of differential equations does not necessarily imply the use of analog design techniques as shown here. *Left*, digital building block performing the operation $Y = (X + B * M)z^{-1}$, that is the output, Y, is equal to the sum of the input X and the product $B * M$ delayed by one time step. With a number of these building blocks properly connected, we can solve any set of linear and nonlinear differential equations, as long as the nonlinearities can be approximated by polynomials (Mead & Wawrzynek, 1985). *Right*, illustration of how few building blocks can be interconnected to solve the differential equation $(d^2f/dt^2) - b(df/dt) - af - c = 0$. We can easily see how the solution is arrived at by noting that, based on the operation performed by the processor, $f = (d + b * f)\, z^{-1}$ and $d = (c + a * f)\, z^{-1}$. Solving for f we obtain, $f(z^2 - bz - a) = c$, which is exactly the difference equation we wish to solve.

ately so that a set of differential equations is solved in real time (see Figure 1). It is clear that, as the complexity of the problem increases, method 1 requires more memory and more time, whereas method 2 requires more processors and more interconnections. These interconnections, however, may be mostly local. The analog approach is germane to NNs since NNs could be viewed as a complex dynamic system governed by a set of nonlinear differential equations defining both the interconnections and the functional behavior of each element of the network.

Dedicated Neural Network Hardware

We shall restrict our discussion to the problem of building dedicated NN hardware, since dedicated hardware represents the lowest-cost solution, and cost considerations are the most important factor in determining the applicability of NN technology to practical problems.

We therefore assume here that (somehow) there exists a simulation and development environment enabling the design of dedicated hardware, where the design of the simulator hardware itself may be substantially different from that of the dedicated hardware. We now examine the three major approaches to the construction of massively parallel NN hardware; digital, conventional analog, and neuromorphic.

We shall consider a digital implementation where the function of each neuron is mapped into a separate special purpose processor with local memory. This approach offers a reasonable degree of parallelism, intermediate between one processor per synapse and one processor per neuron sheet.

The local memory contains the synaptic weights and acts as a synchronizing buffer between the input data stream and the processor operation. The operation of the processor is simple and can be controlled by an internal state machine. In this implementation, the synaptic weights can be stored in ROM if having the device continue to learn after deployment is undesirable.

The major problem to be solved for such a device would be the interconnections. However, judicious use of TDM (time division multiplexing, as previously discussed) coupled with the *a priori* knowledge of the interconnections allows us to solve this problem. The processor needs only to be fast enough to keep up with the rate of the input data flow, which in turn depends critically on the number and duration of the TDM slots. For example: assume that 64 time slots, each of 100 nsec duration, are used, that 24-bit fixed-point arithmetic is performed every 100 nsec by the processor and that there are 128 synapses, then 48 pins (24 x 128/64 = 48) are required to carry the input information. With today's technology, we can integrate two to four of these elements in a single chip.

If we need to be more precise or need to perform at a faster rate than 10 MOPS per processor, then we may be able to integrate only one neuron per chip. Conversely, if lower speed and precision is acceptable, then we can use a serial communication protocol, further reducing the number of pins and using less real estate for each processor, since fewer parallel operations are required. In the latter case we may be able to double the number of neurons per chip. In any event, for the near future, we can integrate less than 10 neurons per chip if we use fully digital techniques and dedicate one processor per neuron.

Fully Analog Implementations

With the exception of extremely simple systems, we can rule out the possibility of a fully analog implementation. As we discussed, we can solve the wiring problem only by using some kind of multiplexing—a digital technique. Furthermore, if long-distance communication is required, digital communication techniques perform much better than analog ones. Analog systems therefore will be hybrids of analog and digital techniques, computation can be done most profitably with analog techniques, whereas communication is best with digital techniques.

Long-Term Storage

If analog information needs to be stored for only a short time, say less than 1 sec, then a simple capacitor can be used. However, if we need a long time constant, then a capacitor is no longer adequate unless we periodically restore its charge. The restoration process is relatively simple if only binary information is to be stored (this is the technique used in the design of dynamic random access memories, or RAMs), but becomes increasingly complex as the number of states to be identified grows. At some point, the most reasonable solution is to store the information in binary form and then to perform a digital to analog conversion where needed. In fact, such a technique has already been used in experimental NN hardware (Raffel, Mann, Berger, Soares, & Gilbert, 1987).

The only physical mechanism available in conventional VLSI technology that has the property of retaining information for a very long time (hundreds of years) is storing a charge in a floating polysilicon gate. A floating gate is a piece of polysilicon that is not connected to anything and is completely surrounded by thermally grown oxide (SiO_2). Because the oxide is an extremely good insulator, any charge that found its way to the gate would stay there almost indefinitely (about 300 years at 50° C). If the floating gate is also the control electrode for a MOS transistor, we can sense the amount of charge stored in the floating gate by measuring the current flowing through the device. Thus, we have a method for reading the information.

To write information into the floating gate, we must get electrons to cross the oxide to and from the gate. There are two physical mechanisms that we can use to achieve this purpose: hot electron injection and tunneling (see Figure 2). Floating gate devices have been used extensively for the fabrication of electrically programmable read only memory (EPROM) and electrically erasable and programmable ROM (EEPROM). In these applications, however, only digital information is stored in such devices. In NN applications, we need to store analog information; therefore, it is important to develop structures for which the amount of charge that is injected to or extracted from the gate is precisely controllable and for which the injection process is reasonably replicable from device to device (Faggin & Lynch, 1986).

Floating gate devices can be used in the design of circuits that perform three critical functions essential for the practical realization of analog NN VLSI systems. The first is self-compensation: the ability of circuits to automatically and continuously compensate for the inherent lack of precision of their constituent components. The second function is learning. Learn-

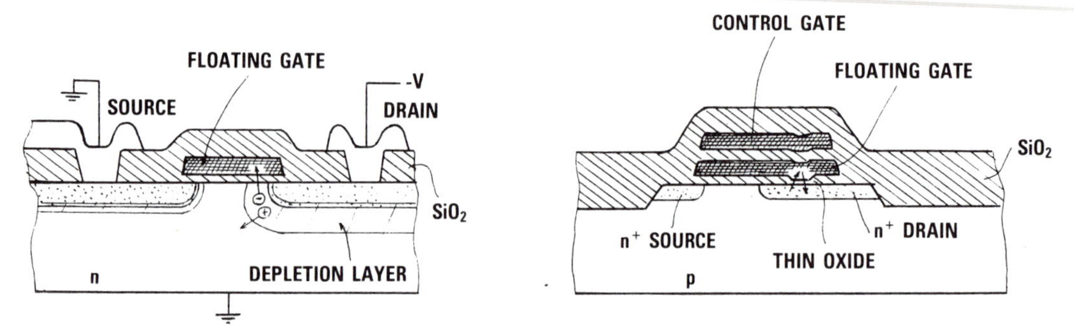

FIGURE 2 Two examples of the polysilicon floating-gate electrodes that can be used to store analog values in the form of an electric charge. *Left,* an MOS P-channel device with a floating polysilicon gate. Electrons can be injected into the gate if their energy exceeds the energy barrier between the single-crystal silicon Fermi level and the edge of the oxide conduction band. To generate electrons with high kinetic energy—called hot electrons—we apply to the drain junction a high voltage, sufficient to bring the junction into breakdown. In the breakdown condition, a small percentage of the carriers generated in the depletion region near the surface of the device are hot electrons. Hot electrons that have sufficient energy to overcome the oxide barrier (about 4 eV) will find their way to the floating gate, charging it negatively. In such a device, we can accomplish the removal of electrons from the floating gate by exposing the device to electromagnetic radiation of sufficient energy (in the ultraviolet spectrum) to carry electrons from the conduction band of the polysilicon floating gate to the conduction band of the oxide. *Right,* an N-channel device using tunneling to transport electrons to and from the floating gate. The device has a region of very thin oxide (50 to 100 Å) between the floating gate and the drain. If sufficiently high voltage (10 to 20 V) is applied between the floating gate and the drain, the electric field will be high enough for electrons to tunnel quantum-mechanically through the oxide barrier. Reversing the direction of the electric field will move electrons from the floating gate into the drain. Thus, this structure allows us to increase or decrease the electric charge stored in the floating gate solely by electrical means. The purpose of the control gate is to induce capacitively a voltage in the floating gate of the correct polarity to make possible the three operations of increasing, decreasing, and sensing the stored charge.

ing not only requires the ability to write, read, and permanently store an analog physical variable (electric charge in this case) but also the ability to effectively modify the charge in real time and in accordance with a variety of learning paradigms. The last requirement is to be capable of changing the topology of the network. Floating gate devices can also be effectively used for this purpose.

If obtaining the required level of reproducibility proves to be too difficult or impractical, we will need to explore other lesser known and less well characterized physical properties. It is our belief, however, that adequate floating gate structures can be fabricated for NN applications. We shall assume, for this discussion, that an appropriate analog storage element has been developed.

Traditional Analog Implementation

In the traditional analog implementation style, the critical circuit elements are the operational amplifier and certain passive components such as resistors and capacitors, the precision of which sets the precision of the circuit. If resistors of the appropriate value and precision cannot be fabricated readily, as is the case with a typical VLSI CMOS process technology, then switched capacitors can be used instead. If we wish to solve a set of linear differential equations, we can wire up the various components in a well known and relatively straightforward manner to achieve the solution. More difficult is the analysis of precision and stability. Depending on the specific design, such performance can be almost entirely determined by the components external to the operational amplifier or by the characteristics of the operational amplifier itself.

In any event, the evaluation of the precision and stability is highly circuit dependent and requires detailed knowledge of parameters such as voltage offsets, temperature drift, long-term drift, and parasitic elements, which may be difficult or even impossible to measure. Therefore, in general, we must use very conservative design approaches, sacrificing precision to guarantee stability.

Assume that we have a problem in which a set of analog input signals are given and a set of analog output signals are to be computed based on the solution of a set of differential equations. If we wish to compare the performance of a given analog implementation with that of a given algorithmic digital implementation, then we are confronted with a difficult problem for which no general rules are easily drawn. Only experience can help guide us through this analysis, and can help us to identify good solutions. We shall, however, examine one aspect of the problem that illustrates some of the key issues.

Suppose we need to solve a set of linear differential equations with a digital computer. The Z-transform theory provides a method to solve the equivalent set of difference equations and to guarantee that, if a stable solution exists, then the computer simulation will also converge to a stable solution with an error that can be evaluated based on the chosen precision (value quantization) and the duration of the time step. The situation is not so benign if we are dealing with a set of nonlinear differential equations. In this case, no general method exists to evaluate the numerical stability of the solution. Therefore we cannot determine *a priori* how many bits of precision are required and how small the time step should be.

We can visualize the problem in terms of the sampling theorem mentioned earlier. A time-sampled system can faithfully emulate a continuous-time system only as long as there are at least two sample points per cycle of the highest frequency present in any signal present in the system. For a linear system, we can limit the bandwidth of incoming signals to less than one-half of the sampling rate, and be certain that the sampling theorem is satisfied. The basis for our certainty is that a linear system cannot generate signals with any frequency components that are not already present in its input signals. With a nonlinear system, the behavior of the system is different in a fundamental way. The system itself can generate components at frequencies that are far higher than the frequency of components present in the input waveforms. These high frequencies alias against the sampling rate (set by the time step used by the digital system) and create many artifacts that have no counterpart in the dynamic system being simulated.

One commonly observed and particularly noxious artifact is the alternate-sample instability. Operating on data present at one sample time, the system generates a correction that, in the original system, would be at a frequency well within that allowed by the sampling theorem. Through nonlinearities in the system, this signal generates frequencies larger than one-half of the sample rate, which the system cannot distinguish from frequencies less than one-half the sample rate (aliasing). The phase characteristics of these signals, however, may be unrelated to those of the lower frequency. In particular, the phase of the aliased signals may be such that feedback in the system is positive for them when it is negative for their unaliased counterpart. Because the system cannot distinguish the aliased from the unaliased signals, it applies the correction appropriate for the unaliased one, which further amplifies the aliased one.

This kind of instability can occur in any time-sampled system, digital or analog. An additional source of instability arises from the quantization of values in a digital system. A signal changing at a slow rate generates an output that has high-frequency components due to the discrete transitions between quantized states. These components can alias against the sampling rate, further degrading the stability of the system. We emphasize again that all these instabilities can, and do, arise during the simulation of a perfectly well behaved continuous-time dynamical system.

Therefore, once we have established a given precision and a given time step, if we observe an instability we cannot distinguish whether such an instability is due to the nature of the problem itself or is an artifact of the simulation process. The only way to clarify the cause of the instability would be to repeat the computation with shorter time steps and perhaps higher precision. This may not be possible for a real-time application, if the sensory information is not stored anywhere but must flow through the system and produce an appropriate response in a single pass. An analog implementation operating in continuous time does not suffer from this limitation. Since nonlinearities appear to be essential ingredients in the kind of information processing done by NNs (any network that makes decisions cannot be entirely linear), this limitation of time-sampled systems may be rather serious. Indeed, limit cycles and chaos have commonly been observed in the computer simulation of networks that are perfectly well behaved when operated as continuous-time systems.

An additional attractive property of computation with continuous-valued analog circuits is that an analog implementation with even fairly low absolute precision remains responsive to much smaller changes of the variables than does a digital implementation having the same absolute precision. In other words, with an analog implementation we can easily achieve a much higher level of relative precision than of absolute precision. This capability is very important when dealing with real-world signals.

In conclusion, although designing NNs using the accumulated wealth of analog design experience is not a straightforward task, it appears nonetheless possible. With this design style we could achieve substantial improvements in the levels of integration and power dissipation compared to those of a fully digital implementation—perhaps a factor of 10 or so for the same speed. Real progress in this field, however, requires that we achieve a deeper understanding of how the nervous system manages to create such a robust, stable, and sensitive system using quite imprecise components. Only by adopting this design philosophy can we achieve an adequate level of integration to design truly sophisticated NN systems.

The Neuromorphic Approach

Using the neuromorphic approach, we can achieve significant improvements in computational hardware capability as compared with that obtainable with conventional analog and digital techniques. This novel methodology is the result of the pioneering work of Mead at California Institute of Technology and is the subject of his recent book (Mead, 1989). Although this design approach is still nascent, it is already providing encouraging results and could well be the hardware approach of choice for future NN implementation. This neuromorphic approach starts by identifying several structural levels in the nervous system, then attempts to capture these structures' organizing principles. At the lowest level, we identify the computational primitives of the nervous system and design silicon analogs by creatively harnessing the available physics of semiconductors. At the next level we study how various ensembles of primitives can be organized to perform complex computational tasks, such as sensory preprocessing, or elementary learning networks. At the next level we study workable architectures capable of solving practical problems, such as character recognition, or object identification, by combining structures of the preceding levels. A common thread in this research is the use of adaptive elements not just as the analogs of synapses, but also as elements that compensate for device-to-device and circuit-to-circuit variations, as well as allowing us to achieve a degree of hardware modification or reconfigurability. In addition, certain nonlinear device and circuit characteristics naturally present in the technology are considered highly desirable to facilitate the information processing task, instead of undesirable effects we must strive to eliminate. Finally, we consider it indispensable to develop circuits that display robustness and fault-tolerance through the use of adaptive elements and collective structures, that is, structures having the property that the information they represent is somewhat independent of the detailed behavior of the underlying hardware.

We shall discuss two examples of this emerging design style. First, we shall consider a CMOS transconductance amplifier operating in subthreshold (Figure 3). Its behavior can be described by the following equation:

$$I_{out} = I_b \tanh (V_1 - V_2) / 2 = I_b \tanh V_{in} / 2 \qquad (1)$$

The output current, I_{out}, is equal to the bias current,

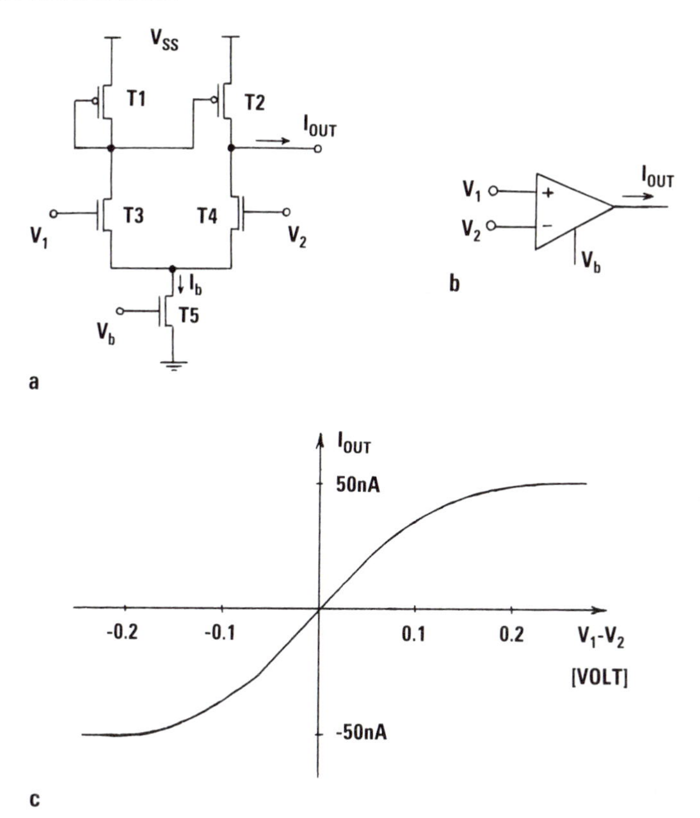

FIGURE 3 The transconductance amplifier operating in the subthreshold region. (a) Schematic of the amplifier. T1 and T2 are P-channel MOS transistors, all other transistors are N-channel. The bias current, I_b, is set by the gate voltage V_b on transistor T5. If V_b is less than the threshold voltage of T5, then I_b is an exponential function of V_b ($I_b = I_s e^{Vb}$) and the amplifier is said to operate in the subthreshold region; the output current is given by Equation 1. (b) Symbol commonly used to represent the transconductance amplifier. (c) Output current as a function of the input voltage difference. Notice that the linear region extends only a few KT/(qc); beyond that, the current rapidly saturates to + or - I_b.

I_b, multiplied by the hyperbolic tangent of one-half the voltage difference appearing at the inputs of the amplifier. In Equation 1 the voltage is measured in units of KT/(qc), where K is the Boltzmann constant, T is the absolute temperature, q is the electron charge, and c is a process-dependent constant. For a typical process, KT/(qc) is about 40 mV at room temperature. For small signals, the amplifier behaves like a transconductance with a value, G_m, proportional to the bias current:

$$G_m = \partial I_{\text{out}} / \partial V_{\text{in}} = I_b \, \text{cq} / (2\text{KT}) \tag{2}$$

For small signals, the output current is given by:

$$I_{\text{out}} = G_m (V_1 - V_2) \tag{3}$$

Suppose now that we have N transconductance amplifiers connected as shown in Figure 4. Each individual amplifier is connected as a follower, so-called because

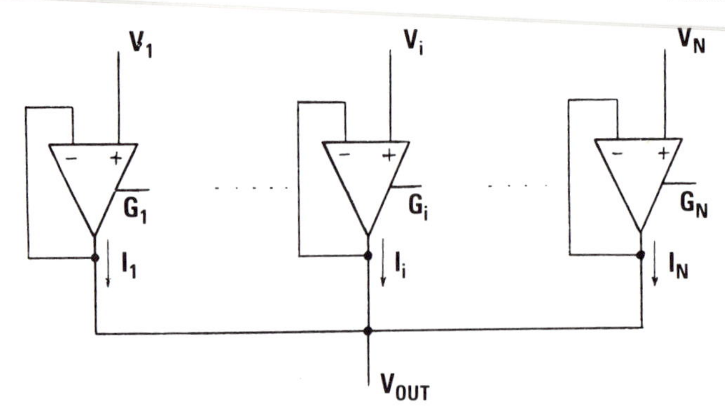

FIGURE 4 Schematic of the follower–aggregation circuit. Each amplifier contributes a current, I_i, proportional to the difference between its input voltage, V_i, and the computed output voltage, V_{out}. For a given (V_i - V_{out}), the current is proportional to the transconductance, G_i, of the amplifier.

the output voltage of an isolated follower is very close to (within a few KT / (qc) of) the input voltage.

If the ith amplifier has a transconductance G_i and an input voltage V_i, then application of Kirchoff's Law for the output node yields

$$\sum_{i=1}^{n} G_i \left(V_i - V_{out} \right) = 0 \qquad (4)$$

We can rearrange Expression 4 to give the output voltage:

$$V_{out} = \frac{\sum_{i=1}^{n} G_i V_i}{\sum_{i=1}^{n} G_i} \qquad (5)$$

This follower-aggregation circuit computes the weighted average of the input voltages V_1, V_2,..., V_n. Expression 5 is valid, however, only in the linear regime of each amplifier.

We have now just seen that a transconductance amplifier has a tanh transfer characteristic; that is, it has a very nonlinear behavior (a sigmoid) in which the output current is strictly limited.

Therefore, input voltages very different from V_{out} have no more impact on V_{out} than a voltage a few KT/(qc) different from V_{out} has. As long as most inputs are close to the average value, V_{out} will not follow the few inputs that are way off scale.

The follower-aggregation circuit provides one of the simplest examples of a robust circuit—a circuit whose output is not much affected by a few bad data points or by a few faulty components. This circuit can also serve to illustrate a principle that all robust structures must obey. We achieve robustness by combining two factors.

1. a large number of inputs carrying overlapping but not identical information
2. limitation by a suitable nonlinearity of the effect that each input can have on the output

The transconductance of each amplifier in the system can be viewed as the confidence assigned to the input of that amplifier, and therefore the influence that the output of that amplifier should have on the outcome of the collective computation. When data are sparse, as in the case of the fly visual system mentioned earlier, a single input with substantial confidence can dominate the outcome. What then becomes of the robustness of the computation if that single input is a bad mistake? A ubiquitous feature of brain function is adaptation. Any signal that has not changed in a long time cannot be carrying information, and hence cannot be assigned a high confidence. In the circuit of Figure 4, we would assign a confidence to a particular amplifier related to the rate of input voltage change to that amplifier. At the instant a signal went bad, that change could not be distinguished from real data. After

a

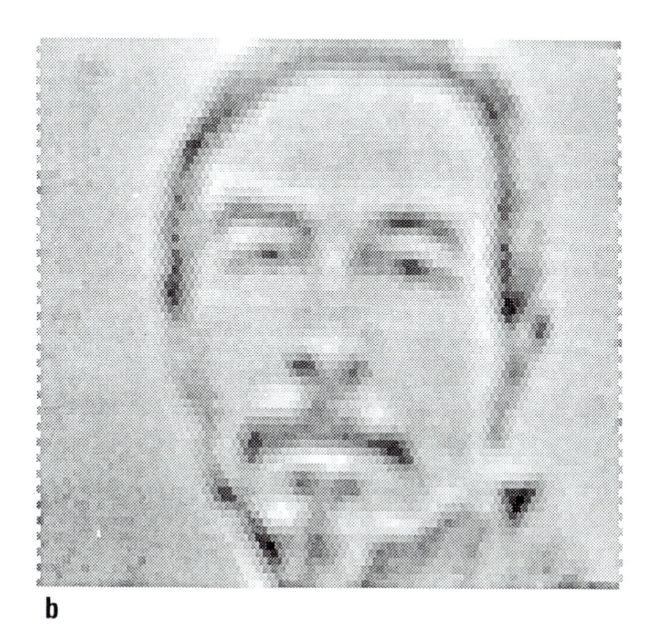

b

c

FIGURE 5 Results of sensing and processing of an image by RET 1B, a VLSI orientation-selective silicon retina. (a) Original image, a portion of which has been focused on the chip surface. Notice the poor quality of the input image (halftone). (b) Display of the chip output after the center-surround operation has been performed. The effect of this operation is to have the output image smoothed, edge-enhanced, and made relatively independent from the absolute level of illumination of the input image. (c) Result of the orientation computation done by the chip. The display shows, for each pixel, a vector the magnitude of which is proportional to the magnitude of the intensity gradient and the direction of which is orthogonal to that of the intensity gradient. This computation is done on the intensity of the center-surround processed image.

some time, however, the weighting of that signal would decay, leaving the computation to other inputs that are changing with time.

As an example of a more sophisticated design using the neuromorphic approach, Figure 5 shows the result of a silicon retina chip performing photodetection and image processing on the same chip (Allen, Mead, Faggin, & Gribble, 1988).

The image processing consists of first producing a voltage proportional to the logarithm of the light intensity at each pixel position. This operation is then followed by computation of the center–surround (equivalent to a convolution with a difference of Gaussians), followed by computation of the local orientation of the intensity gradient. At each pixel position, four (analog) values are computed: the light intensity corrected by the center–surround computation, and the projections on three axes of the local orientation of the intensity gradient.

This test chip has approximately 5000 pixels and an equivalent number of processors. It can also do computations at the rate of 1000 to 10,000 images per second, dissipating less than 100 mW. A conservative estimate of the computation required to achieve the same results using conventional computers would be 5×10^8 to 5×10^9 operations per second.

With this new design style, it appears possible to achieve substantial improvements over traditional analog design by more closely matching the silicon implementation to the unique requirements of NNs.

An essential aspect of neuromorphic computation is the use of adaptive structures capable of compensating for offsets and other component mismatches, including certain types of component failure. This capability can substantially improve the precision, robustness, and manufacturing yield of circuits fabricated using these techniques. Furthermore, if the same structures that are used for learning are also used for self-compensation, then, in the process of learning, one can also automatically adjust for component variations and thus achieve a precision at the system level substantially higher than the raw precision of the individual components. This is in sharp contrast with a sequential digital hardware implementation, where the system-level precision is generally substantially lower than the precision of its components (the arithmetic units) due to error accumulation during the recursive portion of the algorithm.

Using widely available 2-μm CMOS process technology and neuromorphic implementation, we can now routinely design VLSI components operating with 10^{-11} J/operation. This compares with 10^{-7} J/operation achievable with a digital DSP chip using the same technology. This four-orders-of-magnitude improvement in energy per operation is expected to decrease by a factor of ten, to three orders of magnitude, once the minimum line width is reduced to 0.35 μm by the turn of the century.

With 1990 state-of-the-art technology (1 μm), it is now feasible to integrate about 100,000 simple processors onto a single chip. Each processor is capable of performing one operation in approximately one μsec. The result is a VLSI neuromorphic chip with the computational power of 100 GOPS (10^{11} operations per second). Each processor is not limited to performing a multiply–add operation, but can just as easily compute more complex operations such as exponentials, sigmoids, logarithms, Gaussians, multiquadrics $(X^2 + C^2)^{1/2}$ and more.

Since the physics of the basic semiconductor devices do the work, the resultant function may only be an approximation of the true mathematical function. This fact, however, seldom seems to matter when dealing with the class of problems NNs are intended to solve. It should be noted that this impressive speed is the result of using massive parallelism with individual processors of modest performance.

While further progress is required, we believe that neuromorphic design will make possible the design of circuits with a high degree of fault tolerance. This capability, combined with the low power dissipation, will lead us naturally to wafer-scale integration. Wafer-scale integration will immediately increase by two orders of magnitude the feasible level of integration while significantly relaxing the wiring problem.

It is clearly with wafer-scale integration that NNs will come of age, but they are not there yet. We know where to look for inspiration and ideas, but we still have much work to do.

FUTURE PROSPECTS

Despite the progress made in NN research over the past few years, we have more to do before we can prove unconditionally that NNs are a viable and necessary alternative to conventional techniques to solve the problems of autonomous, intelligent machines.

Although researchers have obtained excellent results by applying NNs to simple problems (Lapedes & Faber, 1987; Mel, 1988; Tesauro & Sejnowski, 1988), it is not obvious that the same approach can work if the complexity of the problem is increased. In fact, early indications are that scaling may be a problem. This finding should not be surprising, given that the available artificial NN architectural paradigms (Hopfield networks, back propagation, and so on) are exceedingly crude and naive when compared with the deliberate and complex architectures and structures found in animal nervous systems.

A second area in which advances in our understanding are absolutely necessary is data representation. We must determine what are the general principles governing how information is mapped at the various levels of representation so that robust, intelligent behavior emerges. For example, what are the (possible many) encoding strategies used by spiking neurons? How are spatio-temporal patterns processed? In particular, how do we learn sequences of complex spatio-temporal patterns? How can we map effectively a multidimensional space into two dimensions such that the next level of information processing can make use of local neighborhood operations? The basic questions are numerous and addressing them will take time and a substantial amount of human and financial resources.

This is a special time, however, a time when the knowledge, the technology, and the need have reached a climactic phase. Hundreds of brilliant young minds have already sensed this unique time and have joined the research effort. Both government and industry are now financially supporting the NN research. Thus, all of the necessary ingredients of success are now present to properly develop this field.

Looking ahead with an application orientation and a hardware perspective, we expect this field to develop along the following lines. The first sensory chips for visual and auditory processing will become commercially available in a few years. Such devices will perform effectively the computationally intensive tasks of sensory preprocessing (from the receptor level up to various levels of feature extraction). For simple tasks, these devices could become front ends of conventional computers. For more complex tasks, however, it will be necessary to interpose adaptive-learning hardware between the sensory preprocessing unit and the conventional computer. Simple adaptive building blocks and a supporting system architecture general enough to enable the construction of a variety of learning networks will become available next, 3 to 5 years hence. The availability of sensory and learning chips will increase dramatically the ability to do research and develop applications for real-time problems. Early commercial applications using the new hardware will become available 5 to 7 years from now. In a few more years, we shall see widespread use of the NN technology.

The evolution of NNs could be similar to the one observed after the introduction of the first microprocessor in 1971; it was not until the introduction of the Intel 8080 in 1974 that sufficient computational resources were available for widespread applications to occur; two more years passed before we could safely say that microprocessors were here to stay.

We expect that a similar evolution will take place with NNs. The 8080 of NNs will be the first wafer-scale integration device to be commercially available, which we should see by the turn of the century. Much of the intervening work will comprise building the knowledge and technological foundations to make that device possible. So the real story of NN applications will begin in about 10 years. Between now and then, we will be mainly learning through experiment. Most of the applications that NN will make possible are unforeseeable now, just as in 1964 we could not have foreseen most of the microprocessor applications. In fact, if an omniscient visionary had predicted many of today's microprocessor applications then, we would surely have ridiculed that person mercilessly.

Truly revolutionary ideas have the power to produce mostly unexpected consequences and to alter in perceptible ways the way humans perceive themselves and relate to their environment. We believe that today's NNs contain the seed of the next revolution in information processing; that revolution will lead to the creation of intelligent, autonomous machines.

References

Allen, T., Mead, C., Faggin, F., & Gribble, G. (1988). Orientation selective VLSI retina. *Proceedings Visual Communications and Image Processing*, 1040–1046.

Faggin, F., & Lynch, G. (1986). Brain learning and recognition emulation circuitry and method of recognizing events. U.S. Patent No. 4802103. Filed June 3, 1986. Issued January 31, 1989.

de Ruyter van Steveninck, R. R., & Bialek, W. (1988). Real-time performance of a movement-sensitive neuron in the blowfly visual system: Coding and information transfer in short spike sequences. *Proceedings of the Royal Society of London, Series B* **234,** 379–414.

Lapedes, A., & Farber, R. (1987). Genetic database analysis with neural nets. *Proceedings of the IEEE Conference on Neural Information Processing Systems—Natural and Synthetic*, **77.**

Mead, C. (1989). *Analog VLSI and neural systems.* Reading, MA: Addison-Wesley.

Mead, C., & Wawrzynek, D. (1985). A new discipline for CMOS design: An architecture for sound synthesis. In *Chapel Hill Conference on Very Large Scale Integration* (pp. 87–104). Computer Science Press.

Mel, B. W. (1988). MURPHY: A robot that learns by doing. *Neural information processing systems, Denver, CO, 1987* (pp. 544–533). New York: American Institute of Physics.

Raffel, J., Mann, J., Berger, R., Soares, A., & Gilbert, S. (1987). A generic architecture for wafer-scale neuromorphic systems. *IEEE 1st International Conference on Neural Networks* **3,** 501–513.

Tesauro, G., & Sejnowski, T. (1988). A 'neural network' that learns to play backgammon. *Neural information processing systems, Denver, CO, 1987* (pp. 794–803). New York: American Institute of Physics.

16

Smart Vision Chips: An Overview

Christof Koch

INTRODUCTION

In the last 10 years, significant progress has been made in understanding the first steps in visual processing. A large number of well-studied algorithms exist that locate edges, compute disparities along these edges or over areas, estimate motion fields, and find discontinuities in depth, motion, color, and texture. Several key problems remain. One is the integration of information from different modalities; that is, how can disparity obtained from binocular stereo be combined with the depth information contained within the motion field (Barrow & Tenenbaum, 1981; Poggio, Gamble, & Little, 1988)? Fusion of information is expected to increase greatly the robustness and fault tolerance of current vision systems, as it is most likely the key toward fully understanding vision in biological systems.

A more immediate problem is the fact that a vision algorithm is very expensive in terms of computer cycles. A color CCD camera, sampling its visual field using a 1024 by 1024 pixel grid will generate approximately 100 million bytes of information each second that need to be transmitted and further processed. For simple image processing algorithms, such as edge detection or filtering, on the order of 10 to 100 elementary operations must be performed for each pixel in the image, requiring a computational throughput of about 10^8 to 10^9 operations per second. It is clear that even supercomputers will have difficulties with the computational load imposed by only moderately complex early vision algorithms. Furthermore, for most mass-market applications, computational speed is not the only constraint. Although supercomputers, such as the Connection Machine CM-5 or the Intel Delta Touchstone, can execute certain low-level algorithms in real time, it seems hardly practical or cost efficient to carry around a ton of silicon and steel to compute the distance to the next car or find the level of liquid in a bottle. Such potential widespread applications have driven the search for highly integrated systems that implement early vision algorithms in real time, as well as in a robust, low-power, and compact manner.

Animals, of course, devote a large fraction of their nervous system to vision. Approximately 270,000 out of 340,000 neurons in the 0.5-mg "heavy" brain of the common house fly *Musca domestica* are considered to be "visual" neurons (Strausfeld, 1975), whereas approximately 55% of the human cerebral cortex is given over to the computations underlying the perception of depth, color, motion, and object recognition. One way for technology to bypass the computational bottleneck is to likewise construct special-purpose digital vision hardware. Here, the output of a conventional video-camera is fed over a serial line into application-specific integrated circuits (ASICs) that convolve the image with various standardized operators for smoothing the image, detecting edges, and so on. This approach has been pursued using both metal-oxide semiconductor (MOS) as well as charge-coupled device (CCD) circuit fabrication technologies (Ruetz & Brodersen, 1987; Denayer, Vanzieleghem, & Jespers, 1988; Yamada, Hasegawa, Mori, Sakai, & Aono, 1988; Chiang & Chuang, 1991; Van der Wal & Burt, 1992; Kemeny et al., 1992). Here, the image intensities are digitized at each location in the camera and then passed through a single serial line into the digital processor. The two main components limiting the size of the image that can be processed in real time or the complexity of the image algorithm is (1) the clocking speed of the serial connection between the camera and the computer and (2) the processing power of the processor following the camera, almost always a digital microprocessor. Why not, however, integrate the image acquisition and the image processing steps (i.e., the camera and the computer) into a totally parallel system, bypassing the need for any serial bottleneck? Such an approach is known today as *focal plane (image) processing* (for a review, see Fossum, 1989; and Koch and Li, In press).

However, why should the image intensities be digitized if they could be processed within their original analog domain? Why not, in fact, exploit the physics of circuits to build very compact, analog special-purpose vision systems? In fact, such analog computers represent a time-honored method for computing (Karplus, 1958) and, for a time in the 1950s and 1960s, offered a clear alternative to the digital computer of the time. Such a *smart sensor paradigm*, in which as much as possible of the signal processing is incorporated into the sensor and its associated circuitry to reduce transmission bandwidth and subsequent stages of computation, is emerging as a viable approach for a large range of products in the commercial, industrial, and military market. Most of the papers cited in the following pages can be found in an edited reprint collection (Koch and Li, In press).

Analog Circuits for Vision: The Early Years

This idea was first explicitly raised by Horn at the Massachusetts Institute of Technology (1974), when he proposed the use of a two-dimensional hexagonal grid of resistances to solve for the inverse Laplace transform. This is the crucial operation in an algorithm for determining the lightness (in his case, the reflec-

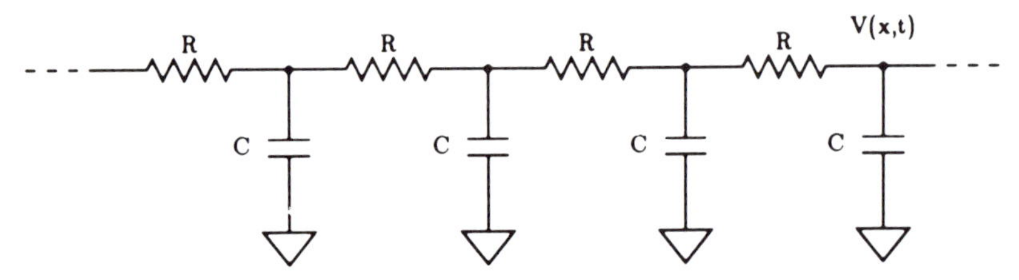

FIGURE 1 One-dimensional, lumped-element resistor/capacitor transmission line. The incoming light intensity is converted into the initial voltage distribution $V(x,0)$. The voltage $V(x,t)$ along the line is given by the convolution of $V(x,0)$ with a Gaussian of variance $\sigma^2 = t/(RC)$ (Knight, 1983). The voltage at time t then corresponds to the image intensity convolved with a Gaussian of variance σ^2. From Koch (1990).

tance) of objects from their image. An attempt to build an analog network for vision was undertaken by Knight (1983) for the problem of convolving images with the difference of two Gaussians, a good approximation of the Laplacian of a Gaussian filter of Marr and Hildreth (1984). The principal idea is to exploit the dynamic behavior of a resistor/capacitor transmission line, illustrated in Figure 1. In the limit that the grid becomes infinitely fine, the behavior of the system is governed by the diffusion equation:

$$RC\frac{\partial V(x,t)}{\partial t} = \frac{\partial^2 V(x,t)}{\partial x^2} \tag{1}$$

If the initial voltage distribution is $V(x,t = 0)$ and if the boundaries are infinitely far away, the solution voltage is given by the convolution of $V(x,0)$, with a progressively broader Gaussian distribution (Knight, 1983). Thus, a difference of two Gaussians can be computed by converting the incoming image into an initial voltage distribution, storing the resulting voltage distribution after a short time and subtracting it from the voltage distribution at a later time. A resistor/capacitor plane yields the same result in two dimensions. Practical difficulties prevented the successful implementation of this idea.

CCD Strategies

Charge-coupled device technology represents an alternative strategy to implement image sensing and image processing onto a single chip. Here, incoming light intensity is converted into a variable amount of charge trapped in potential "wells" at each pixel. The *electric charge* represents the analog variable that is shifted at each cycle. The charge, and therefore the associated image intensity at that pixel, can then be split, scaled, or delayed in various ways.

This trick was first exploited by researchers at Hughes Research Laboratories (Nudd, Nygaard, Thurmond, & Fouse, 1978) by using three one-dimensional CCD strips. They demonstrated blurring and simple edge enhancement algorithms operating at about 10 kHz with a 4-bit intensity resolution. Convolving images using a Gaussian smoothing operator was successfully tried by Sage at MIT's Lincoln Laboratory (Sage, 1984), based on an earlier idea of Knight's (1983). By using appropriate clocking signals, the original charge from the image can be divided by two and shifted into adjacent wells. A second step further divides and shifts the charges and so on (Figure 2), causing the charge in each pixel to spread out in a diffusive manner, accurately described by a binominal

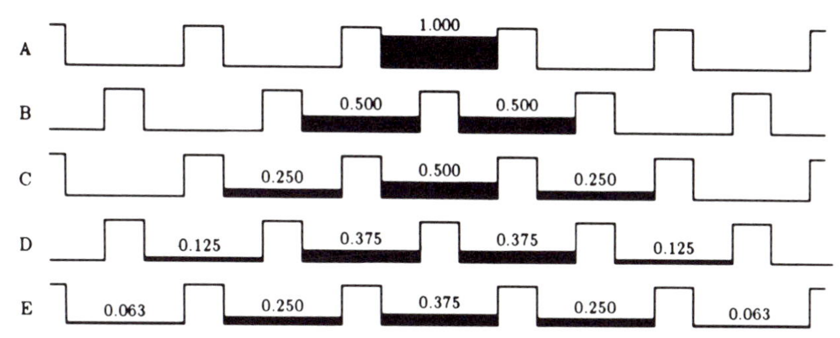

FIGURE 2 Schematic of a potential "well" CCD structure evolving over time. The initial charge across the one-dimensional array is proportional to the incoming light intensity. The charge packet shown in A is then shifted into the two adjacent wells by an appropriate clocking method. Because the total charge is conserved, the charge per well is halved (B). In subsequent cycles (C, D, and E), the charge is further divided and shifted, resulting in a binominal charge distribution. After several steps, the charge distribution is very similar to a Gaussian distribution. From Koch (1990).

convolution. This represents, after a few iterations, a good approximation to a Gaussian convolution. Sage extended this work to the two-dimensional domain (Sage & Lattes, 1987) by first effecting the convolution in the x and then in the y direction. Their 288×384 pixel device convolves images (using up to 40 mixing cycles) at 60 Hz.

Yang (1991) generalized this technique to permit the convolution of the incoming image with a difference-of-two-Gaussians operator, implementing the standard Laplacian of a Gaussian edge detection operator. Realistic circuit simulations predict that a 64×64 pixel CCD imager with only 15% additional chip area for the convolution circuitry should be able to perform edge detection at 1000 frames/sec. Charge-coupled devices can be packed extremely densely: Kodak introduced a 1.3 million-pixel, color CCD camera in 1989 and 4 million-pixel cameras are now becoming available (Khosla, 1992; for a CCD programmable image processor, see Chiang & Chuang, 1991). A crucial difference to analog complementary MOS (CMOS) implementations is that CCD circuits do not operate in a continuous-time mode but instead require a large number of different clock signals.

Parallel Resistive Networks for Machine Vision

Given the rapidly advancing technology for smart vision sensors, more researchers are turning toward the development of analog networks for implementing a great variety of different early vision tasks. Horn (1990) provides one of the best collections of such machine vision tasks and their associated circuits, summarizing the mathematical operations that can be performed within resistive networks for various filtering and edge detection schemes, computing the first and second moments of the brightness distribution and evaluating the optical flow for various specific situations, such as movement toward a plane, or rotational or translational motion. These principles were successfully applied in the center-of-mass chip (Standley, 1991) described below.

Analog VLSI and Neural Systems

The current leader in the field of analog sensory devices that include significant signal processing is Carver Mead at the California Institute of Technology (Mead, 1989). During the last ten years, he has developed a set of subcircuit types and design practices for implementing a variety of vision circuits using subthreshold analog CMOS Very Large Scale Integrated (VLSI) technology. His best known design is the *silicon retina* (Sivilotti, Mahowald, & Mead, 1987; Mead & Mahowald, 1988), a device that computes the spatial and temporal derivative of an image projected onto its phototransistor array. The version illustrated schematically in Figure 3A has two major components. The photoreceptor consists of a phototransistor that feeds current into a circuit element with an exponential current–voltage characteristic (Mead, 1985). The output voltage of the receptor V is logarithmic over four to five orders of magnitude of incoming light intensity, thus performing automatic gain control, analogous to the cone photoreceptors of the vertebrate retina (for an up-to-date account of such photoreceptors see Mann, 1991, and Delbrück & Mead, 1994; see also Koch and Li, 1994). This voltage is then fed into a 48×48 element hexagonal resistive layer with uniform resistance values R. The photoreceptor is linked to the grid by a conductance of value G. An amplifier senses the voltage difference across this conductance and thereby generates an output at each pixel proportional to the difference between the receptor output and the network potential. Formally, if the voltage at pixel i, j is V_{ij}, and the current being fed into the network at that location $I_{ij} = G(\tilde{V}_{ij} - V_{ij})$, the steady state is characterized by

$$6V_{ij} - V_{i-1,j+1} - V_{i+1j+1} - V_{ij+1} - V_{i-1j} - V_{ij-1} - V_{i+1j-1} = RI_{ij} \quad (2)$$

On inspection, this turns out to be the simplest possible discrete analog of the Laplacian differential operator ∇^2. In other words, given an infinitely fine grid and the voltage distribution $V(x,y)$, this circuit computes the current $I(x,y)$ via

$$5^2 V = RG(\tilde{V} - V) = RI \quad (3)$$

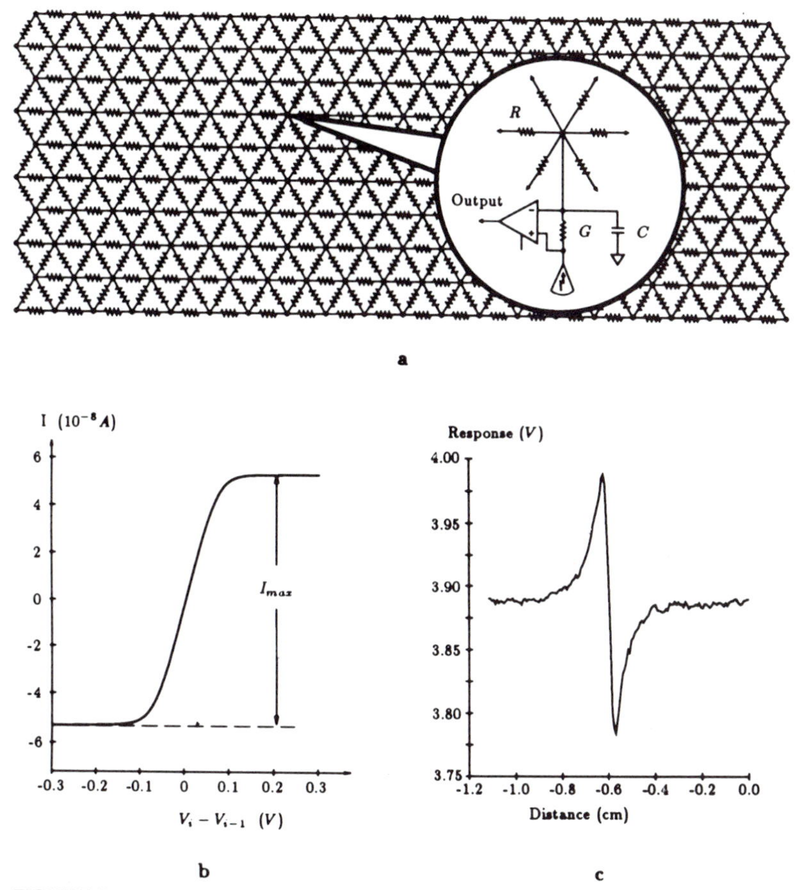

FIGURE 3 The "silicon" retina. a. Diagram of the hexagonal resistive network with an enlarged single element. A photoreceptor, whose output voltage is proportional to the logarithm of the image intensity, is coupled to the resistive grid. The output of the chip is proportional to the voltage difference between the photoreceptor and the grid. b. The current-voltage relationship for Mead's resistive element used in building the silicon retina. As long as the voltage gradient is less than ≈ 100 mV, the circuit acts like a linear resistive element. The output current saturates for larger voltages. The slope, and therefore the value of the resistance (determined by I_{max}), can be varied over five orders of magnitude. c. The experimentally measured voltage response of a 48 by 48 pixel version of the retina when a step intensity edge is moved past one pixel. This response is similar to the one expected by taking the second spatial derivative of the smoothed incoming light intensity. Adapted from Mead and Mahowald (1988).

The current I at each grid point—proportional to $\tilde{V} - V$ and sensed by the amplifier—then corresponds to a spatially high-passed filtered version of the logarithmic compressed image intensity. Operations akin to temporal differentiation can be achieved by adding capacitive elements (Sivilotti et al., 1987). The required resistive elements of this circuit are designed by exploiting the current-voltage relationship (Figure 3B) of a small 6 transistor circuit, *Hres*, instead of using the resistance of a special metallic process. As long as the voltage across the device is within its linear range (a couple of 100 mV), it behaves like a constant resistance whose value can be controlled over five orders of magnitude. The current saturates for larger voltage values, a nonlinearity with very desirable effects. The response of the silicon retina to a one-dimensional edge projected onto the photoreceptors is shown in Figure 3C. The voltage trajectory can be approximated well by the second spatial derivative of the smoothed and logarithmically compressed image irradiance. In two dimensions, the response is similar to that of convolving the compressed image with the difference-of-two-Gaussian edge detection operator (Marr & Hildreth, 1980).

Mead's principal motivation for this work comes from his desire to understand and emulate neurobiological circuits (Mead, 1989, 1990). He argues that the physical restrictions on the density of wires, the lack of very accurate circuit components due to local variations in the fabrication process, and the cost of communications imposed by the spatial layout of the electronic circuits are similar to the constraints imposed on biological circuits. Furthermore, the silicon medium provides both the computational neuroscience and engineering communities with tools to test theories under realistic, real-time conditions. To further the spread of this technology into the general academic community, all circuits are built via the U.S. government-sponsored silicon foundry MOSIS using their 2.0 μm process.

The Silicon Retina: Recent Developments

Two recent developments from the Mead laboratory include the *time-change retina* (Delbrück & Mead, 1994) and a *second-order retina* (Boahen & Andreou, 1992). In the former, a local and rapid adapting element (without the use of ultraviolet light) combined with an ingenious clocking scheme leads to a very clean 68 $\times$ 43 pixel retina, with very little apparent offsets and responding only to changes in the image, approximating a first-temporal derivative operation without any spatial filtering (Delbrück & Mead, 1994). The 90 $\times$ 92 pixel current-mode retina of Boahen and Andreou (1992) effectively implements a second-order resistive grid in which each pixel is connected not only with its adjacent neighbors but also with neighbors one pixel removed. As shown previously (Poggio, Voorhees, & Yuille, 1988), such a spatial filter is more stable and robust than first-order filtering.

A commercial application of an analog VLSI vision chip is the I-1000 chip developed by Mead and Faggin at Synaptics (Stix, 1992). This device, meant for reading the standardized printed bank-account information on personal checks, consists of a 400-pixel retina feeding into a 14-node (one node for each character in the check code) neural network. The network computes the closest match to the previously defined template within 1 μsec and outputs the matching node (plus a confidence level).

A potential problem in designing the type of networks discussed here is that unwanted oscillations can spontaneously arise when large populations of active elements are interconnected through a resistive grid. These oscillations can occur even when the individual elements are quite stable. Using methods from nonlinear circuit theory, Wyatt and Standley (1989) have shown how this flaw can be circumvented. They have proven that if each linear active element in isolation is designed to satisfy the experimentally testable Popov criterion from control theory (which guarantees that a related operator is positive real), then stability of the overall interconnected system is guaranteed. Furthermore, their stability proof is not invalidated by the presence of any unmodeled nonlinear resistors or capacitors at unknown locations in the grid, as commonly occurs in integrated circuits (Wyatt & Standley, 1989).

Learning Synaptic Weights and Offsets

A problem plaguing analog subthreshold circuits are random offsets, which vary from location to location and which are caused by fluctuations in the accuracy of the fabrication process as well as by dark currents. Such offsets, although usually not as problematic for digital circuits, can be very disruptive when operating in the subthreshold analog domain, given the exponential dependency of the drain current on the gate voltage (Mead, 1989). This is particularly true when computing spatial or temporal derivatives. Mead (1989; Mahowald, 1991) has developed a variant of the *floating gate technology* used for resetting programmable read-only memory cells (EPROM) by means of ultraviolet light. Whereas previously the chips were bombarded with ultraviolet (UV) radiation to erase memory, Glasser (1985) demonstrated how this technology could be used selectively to write a *0* or a *1* into the cell. Mead is the first to have applied this technique to the analog domain by building a local feedback circuit at every node of the retina (Figure 3) that senses the local current and attempts to keep it at or near zero by charging up a capacitor located between two layers of poly positioned above each node. Exposure to UV light excites electrons sufficiently to enable them to surmount the potential barrier at the silicon/silicon dioxide interface and to reach the polycrystalline floating gate. For all practical purposes, the charge on the floating gate will remain there forever.

To adapt the retina, a blank homogeneous image is projected for 10–20 min onto the chip in the presence of UV light. This effectively creates a "floating" battery at each pixel, inducing a current that counteracts the effect of the offset current at that particular location. It is even possible to demonstrate afterimage-like phenomena (Mahowald, 1991).

Recent variants of the "floating gate" technique use standard analog silicon structures to add or subtract charge from the floating gate in a controlled manner rather than the cumbersome usage of UV light. The charge-injection mechanisms employed are electron quantum mechanical tunneling and so-called "hot electrons," that is, electrons that have been greatly accelerated by a very high electric field. By combining these two techniques appropriately, electrons can be selectively removed from the polycrystalline silicon floating gate structure by using interpoly tunneling and can be added to the floating gate by using "hot electron injection." Applications for such circuits for learning local circuit parameters (such as the relative contrast over a particular neighborhood) or for conventional synaptic learning abound.

Regularization Theory and Analog Networks

Problems in vision are usually inverse problems; the two-dimensional intensity distribution on the retina or camera must be inverted to recover physical properties of the visible three-dimensional surfaces surrounding the viewer. More precisely, these problems are ill posed in that they either admit to no solution, to infinitely many solutions, or to a solution that does not depend continuously on the data. In general, additional constraints must be applied to arrive at a stable and unique solution. One common technique to achieve this "standard regularization" (Poggio, Torre, & Koch, 1985) is via minimization of a given "cost" functional (for earlier examples of this, see Grimson, 1981; Horn & Schunck, 1981; Terzopoulos, 1983; Hildreth, 1984; Nagel & Enkelmann, 1986). The first term in these functionals assesses how much the solution diverges from the measured data. The second term measures how closely the solution conforms to certain *a priori* expectations, for instance, that the final surface should be as smooth as possible. Let us briefly consider the problem of fitting a two-dimensional surface through a set of noisy and sparse depth measurements, a well-explored problem in computer vision (Grimson, 1981). Specifically, a set of sparse depth measurements, assumed to be corrupted by some noise process, are given on a two-dimensional lattice, d_{ij}. It is obvious that infinitely many surfaces, f_{ij} can be fitted through the sparse data set. One way to regularize this problem is to find the surface f that minimizes

$$\sum_{i,j}((f_{i+1j} - f_{ij})^2 + (f_{ij+1} - f_{ij})^2) + \lambda\sum_{i,j}(f_{ij} - d_{ij})^2 \qquad (4)$$

in which λ depends on the signal-to-noise ratio and the second sum only contains contributions from those locations i where data exists. Equation (4) represents the simplest possible functional, even though many alternatives exist (Grimson, 1981; Terzopoulos, 1983; Harris, 1987). This and all other quadratic regularized variational functionals of early vision can be solved within simple linear resistive networks by virtue of the fact that the electrical power dissipated in linear networks is quadratic in the current or voltage (Poggio et al., 1985; Poggio & Koch, 1985). The resistive network will then converge to its unique equilibrium state in which the dissipated power is at a minimum (subject to the source constraint). The static version of this statement is known as *Maxwell's Minimum Heat Theorem*.

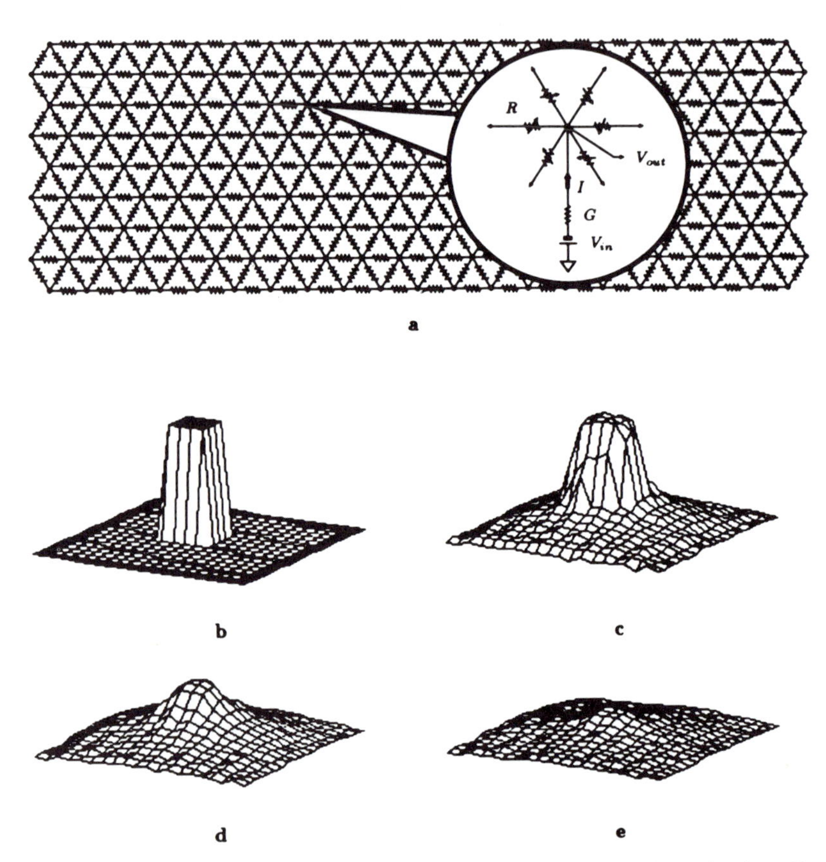

FIGURE 4 Surface interpolating network. a. At those locations where depth data is available, the values of the battery V_{in} and of the conductance G are set to their appropriate values. The horizontal resistances R are built using the nonlinear circuit of Figure 3B. The output is the voltage V_{out} at each location, corresponding to the reconstructed "smooth" surface. Experimental results from a 48 by 48 pixel chip are shown in the three lower figures. b. The input voltage V_{in}, corresponding to a flat, 2-pixel wide strip around the periphery and a central 4-pixel-wide tower (solid coloring). At these locations, the conductance G is set to a constant fixed value, whereas G is zero everywhere else. No data are present in the area between the bottom of the tower and the outside strip. c, d, and e show the output voltage for a high, medium, and low value of the transversal resistance R, respectively. If R is small enough, the resulting smoothing will flatten out the central tower. From Koch (1989) Seeing Chips: Analog VLSI circuits for computer vision. *Neural Computation* **1**, 184–200.

The performance of an experimental 48 × 48 analog CMOS *surface interpolation and smoothing chip* is illustrated in Figure 4 (Luo, Koch, & Mead, 1988). The steady state of the network in Figure 4A minimizes Equation (4) if the voltage V_{out} is identified with the discretized solution surface f_{ij}, the battery V_{in} with the data d_{ij}, and the product of the variable conductance-to-ground G_{ij} and the constant horizontal resistance R with λ (see Figure 4). The power minimized by this circuit is then formally equivalent to the functional of Equation (4). For an infinitely fine grid and a voltage source $E(x,y)$, the surface interpolation chip computes the voltage distribution $V(x,y)$ according to the modified Poisson equation (see also Horn, 1974):

$$\nabla^2 V + RGV = RGE \tag{5}$$

with either a "zero voltage along the boundary" or a "no current across the boundary" boundary condition. If RG is a constant across the grid, this equation is sometimes known as the Helmholtz equation. Note that the difference to Equation (3) lies in the choice of observable current I versus voltage V.

Another working circuit exploiting the properties of resistive grids is the *zero-crossing chip* of Bair and Koch (1991; Figure 5). This device uses on-chip photoreceptors and the natural filtering properties of resistive networks to approximate the difference-of-Gaussians operator proposed by Marr and Hildreth (1980). The chip localizes the thresholded zero-crossings associated with the difference of two exponential weighting functions along a one-dimensional, 64-pixel retina in real time. The output of the chip, a 64-bit word that codes for edges between each of the 64 pixels, has been fed into a conventional programmable microprocessor at a rate of 300 Hz to evaluate the apparent speed of cars on a highway (Koch et al., 1991, and Horiuchi et al., 1992).

Many of the problems in early vision, such as detecting edges, computing optical flow, estimating depth from two images, and so on, have a similar architecture, with resistive connections among neighboring nodes implementing the constraint that objects in the real world tend to be smooth and continuous. For instance, an algorithm for estimating depth from two images—exploiting an analogy between binocular stereo and optical flow—can be solved within a linear resistive network nearly identical with the one shown in Figure 4A (Chhabra & Grogan, 1988). This method uses the gradient of the image brightness to compute

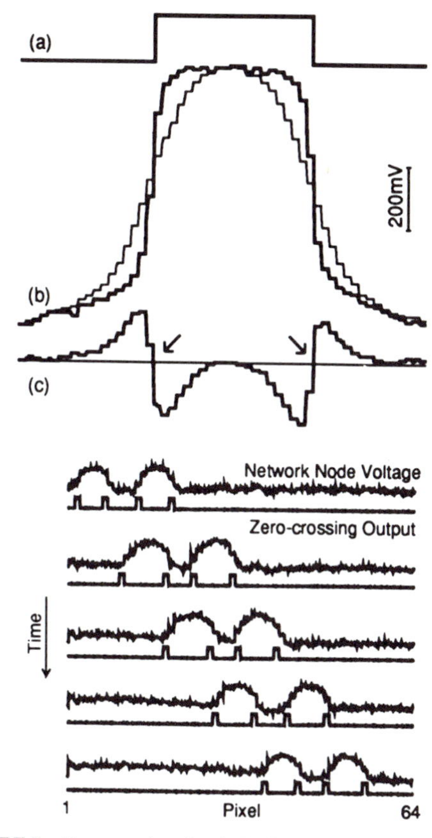

FIGURE 5 Zero-crossing circuit. Performance of a 64-pixel resistive grid chip that approximates thresholded zero-crossings of the difference-of-Gaussians (Bair & Koch, 1991). A. Input light intensity. B. The voltage trace from the two resistive networks. The difference between the two voltage traces is shown in C. If the spatial derivative of this signal at a zero-crossing is high enough, the chip outputs a digital 1, corresponding to an edge at that location (the two arrows). The positions of the zero-crossings (in each lower row) associated with two fingers waving approximately 1 m in front of the lens of the chip is shown at the bottom (in addition to the voltages from one of the two resistive networks in the upper rows). From Bair and Koch (1991).

depth directly. It should be emphasized, however, that in some cases, such as computing the optical flow, the resistances in the network depend on data and may even be negative, making hardware implementation difficult.

Discontinuities

However, the most interesting locations in any scene are arguably those locations at which some feature changes abruptly, for instance, the two-dimensional optical flow at the boundary between a moving figure and the stationary background, or the color across the sharp boundaries in a painting. Geman and Geman (1984; see also Blake & Zisserman, 1987) introduced the powerful concept of a binary line process at location i, j, $l_{i,j}$, which explicitly codes for the absence ($l_{ij} = 0$) or presence ($l_{ij} = 1$) of a discontinuity. Additional constraints, such that discontinuities should occur along continuous contours (as they do, in general, in the real world) or that they rarely intersect, can be incorporated into their theory, which is based on a statistical estimation technique (see also Marroquin, Mitter, & Poggio, 1987). In the case of surface interpolation and smoothing, maximizing the *a posteriori* estimate of the solution can be shown to be equivalent to minimizing:

$$\sum_{i,j}(1-l^h_{ij})(f_{i+1j}-f_{ij})^2+\sum_{i,j}(1-l^v_{ij})(f_{ij+1}-f_{ij})^2 +\lambda\sum_{i,j}(f_{ij}-d_{ij})^2+\alpha\sum_{i,j}V(l_{ij}) \qquad (6)$$

where l^h_{ij} and l^v_{ij} are the horizontal and vertical depth discontinuities, λ depends on the signal-to-noise ratio, α is a fixed parameter, and V is a potential function containing a number of terms penalizing or encouraging specific configurations of line processes (e.g., that they should occur along continuous contours). In the case of surface interpolation in a single dimension $V(l_i) = l_i$. In that case, the line process l_i between i and $i + 1$ will, in general, be set to 1 if the "cost" for smoothing, that is $(f_{i+1} - f_i)^2$, is larger than the parameter α; otherwise, $l_i = 0$. In two dimensions, this potential function can be considerably more complex (e.g., Marroquin et al., 1987; Hutchinson, Koch, Luo, & Mead, 1988). Discontinuities greatly improve the performance of early vision processes, because they allow algorithms to smooth over unreliable or sparse data as well as account for boundaries between figures and ground. They have also been used to demarcate boundaries in the intensity, color, depth, motion, and texture domains (Geman & Geman, 1984; Blake & Zisserman, 1987; Marroquin et al., 1987; Hutchinson et al., 1988; Poggio et al., 1988; Harris, Koch, Staats, & Luo, 1990). Figure 6 demonstrates their advantages in the case of computing the optical flow field induced by moving objects.

Line discontinuities can be implemented in at least two different ways. In a *hybrid* implementation, each line process is represented by a simple binary switch that can be implemented using a single pass transistors (Koch, Marroquin, & Yuille, 1986). When the switch (representing l^h_{ij}) is open, no current flows across the connection between the two adjacent nodes i, j and $i + 1, j$, no matter what the voltage gradient is between the two locations. If the switch is closed, that is, $l^h_{ij} = 1$, the normal interaction between the values at points i, j and $i + 1$, j occurs. The network operates by switching between two distinct modes. In the analog cycle, the network settles into the state of least power dissipation, *given a fixed distribution of switches*, whereas the switches are set to the state minimizing Equation (6) during the digital phase.

A more elegant and completely analog implementation combines both smoothing as well as the concept of a line process in a single, two-terminal nonlinear circuit device, termed *resistive fuse* (Harris, Koch, & Luo, 1990; Harris, Koch, Staats, & Luo, 1990). If the voltage gradient across the resistive fuse—corresponding to the difference in depth between two neighboring values of the surface—is small, for example, less then $\alpha^{1/2}$ to stay within the previous example of Equation (6), then the algorithm assumes that both points lie on the same surface but are corrupted by noise. In that case, the device acts as a resistance, conducting a current proportional to the voltage gradient. If, however, the voltage across the device is above a threshold (here, $\alpha^{1/2}$), the algorithm assumes

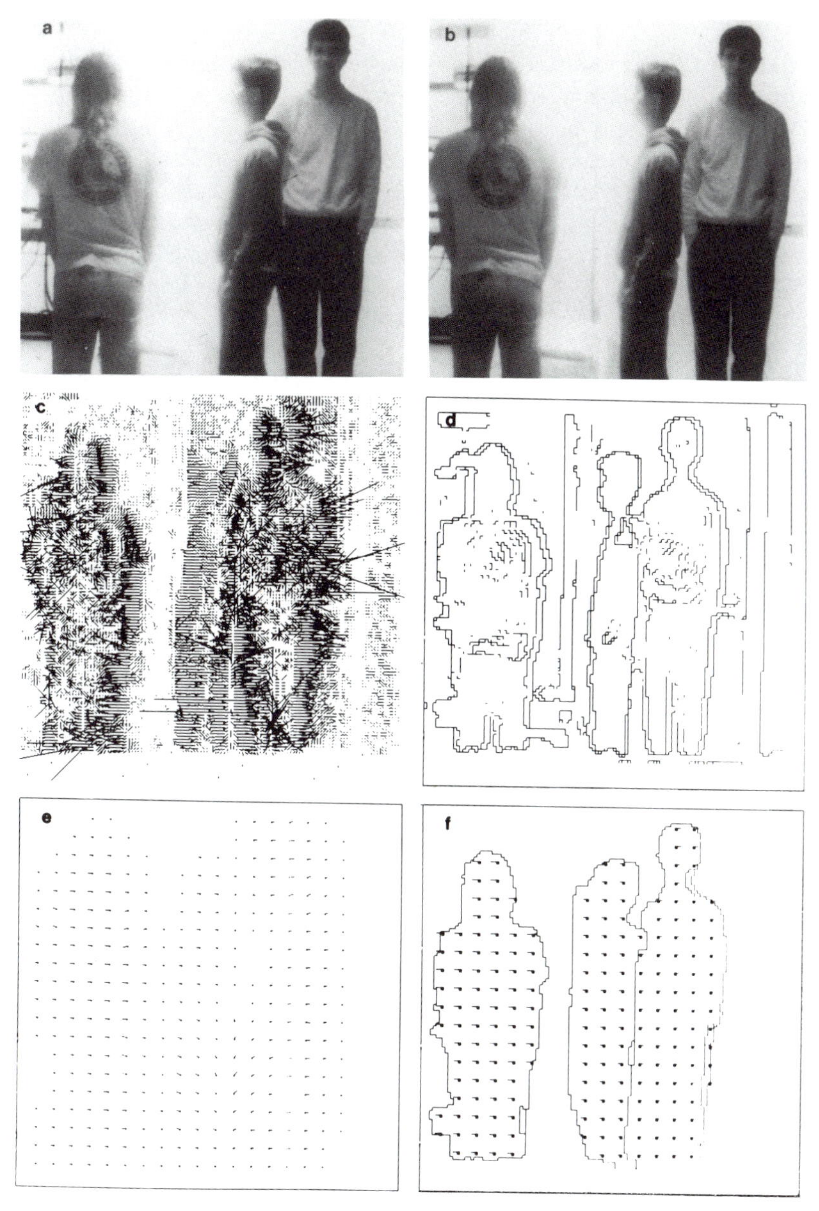

FIGURE 6 Optical flow with motion discontinuities. A and B show two images captured by a video camera. The two people on the left move toward the left, whereas the person on the right moves to the right. C. The initial local velocity data before smoothing. D. The thresholded zero-crossings of the Laplacian of a Gaussian of both images. E. Smooth and subsampled optical flow obtained by solving the associated convex variational functional (Horn & Schunck, 1981). Two undifferentiated "blobs" move to the left and one moves to the right. F. The subsampled optical flow computed with an analog continuous form of the "resistive fuse." To visualize the behavior of the fuses, we indicate their state with a solid line if the voltage difference across them exceeds some threshold value. The final optical flow indicates three moving people and was obtained by solving for the steady state of the associated nonlinear resistive network. From Harris, Koch, Staats, and Luo (1990).

that the two points lie on two different surfaces and no smoothing occurs; that is, the device open-circuits and no current flows. Figure 7 shows the measured I–V relationship for an analog version of such a resistive fuse (Harris, Koch, & Luo, 1990). Because usually no *a priori* information exists as to the optimal value of α, it is best to adopt a deterministic annealing method for varying α, closely related to continuation methods for minimizing the associated nonconvex variational functional of Equation (6) (Ortega & Rheinboldt, 1970). The notion of minimizing power in linear networks implementing quadratic regularization algorithms must be replaced by the more general notion of minimizing the total co-content J for these networks, where $J = \int_0^V f(V')dV'$ for a resistor defined by $I = f(V)$ (Harris, Koch, Luo, & Wyatt, 1989). Figure 8 illustrates the robust performance of an experimental test circuit.

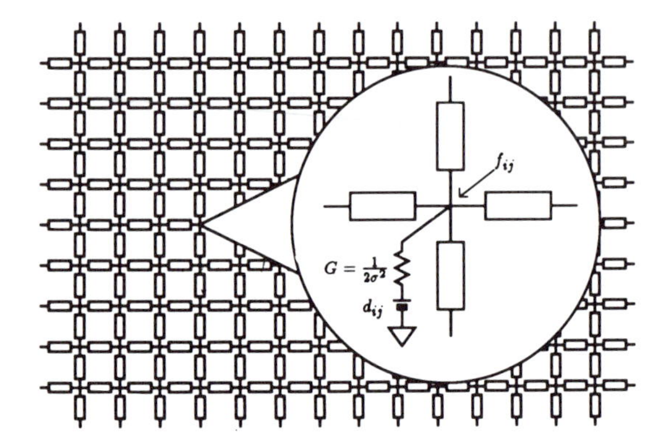

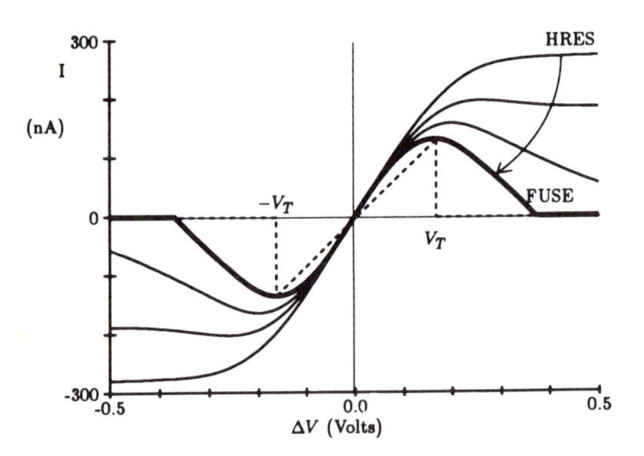

FIGURE 7 A. Diagram of a 20 × 20 pixel piecewise smoothing chip. A rectangular mesh of resistive fuse elements (shown as rectangles) provides the smoothing and segmentation ability of the network. The data are given as battery values d_{ij}, Equation (6), and the conductance G depends on the variance of the Gaussian noise corrupting the data. If no data is available, $G = 0$ at that location. The output is the voltage f_{ij} at each node. The steady state of the circuit corresponds to one of the local minima of the nonconvex variational functional of Equation (6). B. The I–V curve can be continuously varied from an hyperbolic tangent (HRES; Mead, 1989) to that of an analog fuse (FUSE). The curve of a binary fuse is shown in a dashed line. For voltages less than V_T across the fuse, the circuit acts as a resistor with fixed conductance. Above V_T, the current is either abruptly set to zero (binary fuse) or smoothly goes to zero (analog fuse). From Harris, Koch, and Luo (1990).

The flow field shown in Figure 6F was evaluated by having a conventional digital computer simulate the appropriate resistive network with fuses—in this case, two networks of the type shown in Figure 7 and corresponding to the x and y components of the optical flow field—interconnected with linear resistances. The flow field—induced by the time-varying image intensity $I(x,y,t)$—is regularized using a first-order smoothness constraint (Horn & Schunck, 1981). The amount of smoothing is governed by the constant conductance value R of the upper and lower horizontal grids. Figure 8 shows experimental data from an analog VLSI chip that implements Equation (6), performing piecewise smooth-surface interpolation.

Toward Object-Based Vision Chips

Early vision algorithms typically can be classified as either pixel-based or image-based. Pixel-based algorithms, such as edge detection and optical flow, produce a dense output on every grid point. Two vision chips that produce a few outputs by integrating information from the entire image (using resistive grids) are the *center-of-mass chip* of DeWeerth (1992; Figure 9) and Standley's *orientation chip* (1991; Wyatt et al., 1992). Both circuits exploit computational properties of a resistive grid (Horn, 1990); the first one by computing the "center of mass" of the incoming two-dimensional light distribution (centroid) and the second one by computing both the first as well as the second moment of the incoming light distribution. In the case of a single bright location in the image, the DeWeerth circuit (1992) outputs two values, the x and y positions of the centroid, computed over the 160 by 160 retina (Figure 9), whereas Standley's chip estimates the position and the orientation of a single elongated object to within ± 2 deg.

More recently, two experimental circuits assigned a single number to *one figure* in the input array (and not across the entire image as in the previous circuits). The *dynamic wire chip* (Liu & Harris, 1992) provides an avenue for object-based processing. The dynamic wire model provides dedicated lines of communication among groups of pixels that share a common property. The methodology consists of first configuring the switches in a two-dimensional network for the groups of pixels that are connected to each other, and second, utilizing the resulting dynamic connections for computation. These tasks can be accomplished by using a

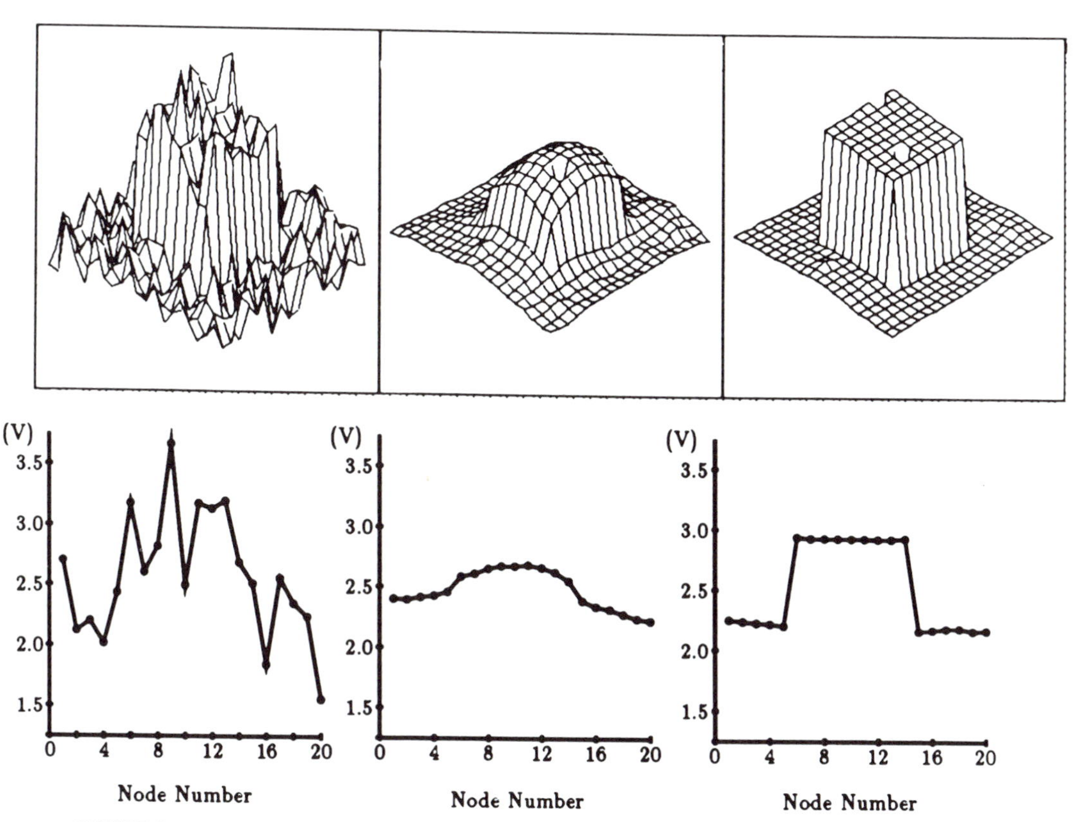

FIGURE 8 Experimental data from the chip described in Figure 7. A central tower with superimposed Gaussian noise is used as input data. Different from the case illustrated in Figure 4, data exists everywhere. The input is corrupted with noise, whose variance is set to 20% of the difference between the tower and the background (*upper left*). The voltage output with the fuse configured as a saturating resistor (*middle left*). The voltage output when the *I–V* curve of the fuse has been changed from the saturating resistor to that of the analog fuse (following the arrow in Figure 7B), while the conductance has been increased (*upper right*). The same behavior along a horizontal slice across the chip for an increased level of noise corresponding to 40% of the height of the central tower is illustrated in the lower panels. Notice the "bad" pixel in the middle of the tower (in the upper panels), whose effect is localized to a single element. From Harris, Koch, and Luo (1990).

network of resistors and switches. In one prototype circuit, this concept is implemented to evaluate the contour length associated with arbitrary figures projected onto the input array (Liu & Harris, 1992).

The *figure-ground chip* (Luo, Koch, & Mathur, 1992) performs figure–ground segregation of a scene, labeling all the points inside a designated figure by one voltage value and all other pixels outside this object using a different voltage value (Figure 10). The circuit, implemented using a network of resistances and switches, is very robust in the sense that it can deal with incomplete boundaries, as demonstrated in a 48 by 48 test circuit (Luo et al., 1992). The next generation of this circuit will be able to label multiple figures in a scene in parallel with different voltage values.

Analog Chips versus Digital Computers

As we have seen, all of the preceding circuits exploit the physics of the system to perform operations useful from a computational point of view. Thus, the time-varying voltage or charge distribution at some time in the networks of Figures 1, 2, or 3 corresponds to the solution, in this case, convolution of the image intensity with a difference-of-a-Gaussian. In the networks derived from standard or nonstandard regularization theory, the stationary voltage distribution corresponds to the interpolated surface (Figure 4), the location of an edge (Figure 5), the brightest point (Figure 9), or to the optical flow (Figure 6). These quantities are governed by Kirchhoff's laws instead of being symbolically computed via execution of software in a digital computer. Furthermore, the architecture of the analog resistive circuits reflects the nature of the underlying computational task, for instance, smoothing, whereas digital computers—being Turing universal—do not. This difference has interesting parallels for structure–function relationships in neurobiology. One of the advantages of analog circuits is that their operating mode is optimally suited to analog sensory data because they do not suffer from temporal aliasing problems. Furthermore, their robustness to imprecisions or errors in the hardware, their processing speed and low power consumption (Mead's retina requires less than 1 mW, most of which is used in the photo-conversion stage), and their small size make analog smart sensors very attractive for tele-robotic applications, remote exploration of planetary surfaces, and a host of industrial applications where their power hungry, heat producing, and bulky digital cousins are unable to compete.

The two principal drawbacks of analog VLSI circuits are their lack of flexibility and their imprecision. The preceding circuits are all hardwired to perform specific tasks, unlike digital computers which can be programmed to approximate any logical or numerical operation. Only parameters associated with this algorithm, for instance the smoothness in the case of Figures 4, 5, and 6, can be varied. Thus, digital computers appear vastly preferable for developing and evaluating new algorithms; analog implementations should only be attempted after such initial exploration of algorithms. Furthermore, although 12- and even 16-bit analog-to-digital converters are commercially available, it seems unlikely that the precision of analog vision circuits will exceed 7- to 8-bits in the next years. However, for a number of important tasks, such as navigation or tracking, the incoming intensity data are

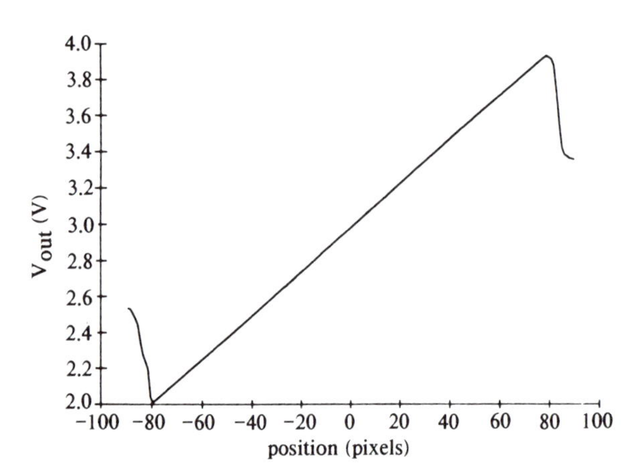

FIGURE 9 Measured performance of the 160 × 160 pixel center-of-mass chip (DeWeerth, 1992). Exploiting the weighting properties of a resistive grid (Horn, 1989), the circuit computes the centroid ("center of mass") of the brightness distribution across the entire focal plane. Shown here is the output voltage of the chip along one axis for an image consisting of a single bright area. Notice the linearity of the response within the range of the sensor (2–4 Volt). From DeWeerth (1992).

rarely more accurate than 1%. Moreover, it seems unlikely that the individual circuit components in biological systems, neurons, process data with more than, at most, 100 levels of resolution, that is, between 6 and 7 bits.

THE FUTURE

What is the shape of things to come? The current generation of vision chips can perform simple spatial and temporal filtering using array sizes of less than 100 by

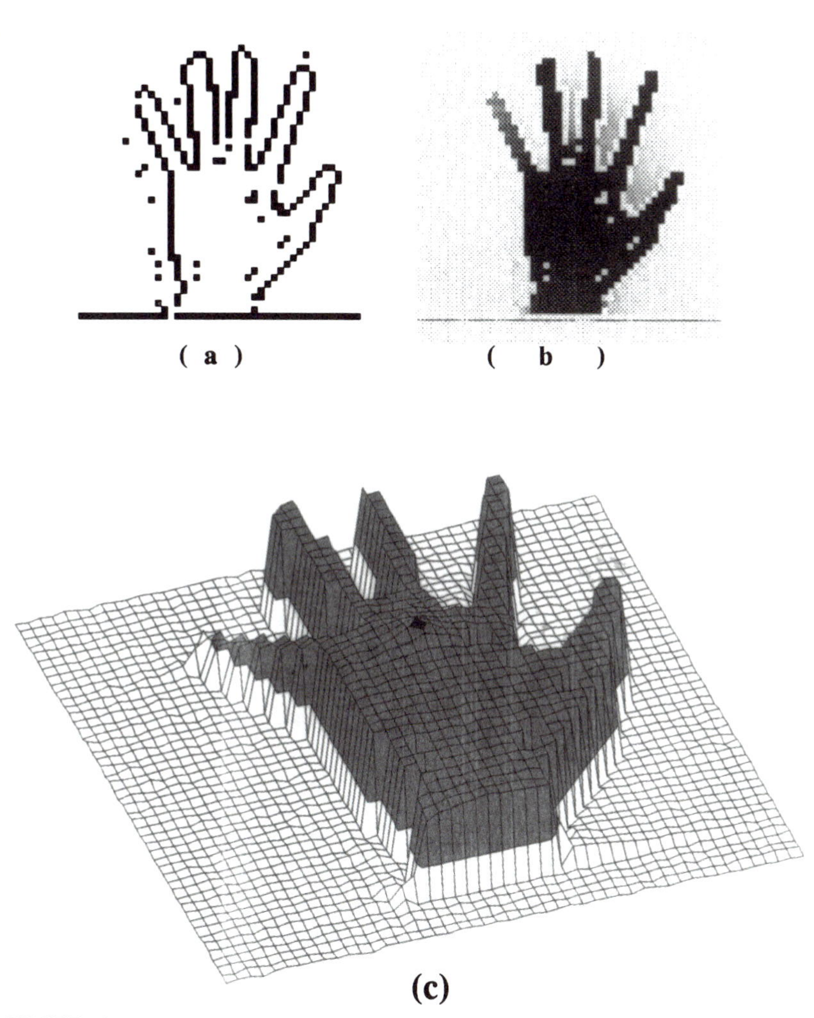

FIGURE 10 Measured performance of the 48 × 48 pixel Figure–Ground resistive grid chip, labeling all of the points in a possibly incomplete and broken boundary by one voltage value and all of the outside points by another. The output of the chip in the presence of a binary edge map scanned in from off-chip (A) are shown in B. The shape of the hand with its five fingers is apparent. Notice the voltage decrement along the little finger resulting from a very incomplete boundary here. If the output were to be thresholded at the level indicated by the voltage-amplitude plot in C, all points inside the hand would be labeled as belonging to the hand. From Luo et al. (1992).

100 pixels. Steady progress at the circuit fabrication level will no doubt lead to much larger array sizes with more accurate components. For instance, the 0.8 μm CMOS design rules available now will boost array size to well beyond 300 × 300 pixels. Although it is important that chips achieving this performance be built, smart vision sensors will only see large-scale industrial or commercial applications when more functionality is moved onto chip. Let me describe three promising approaches.

Estimating Motion

Many researchers have attempted to build vision chips that can reliably estimate the optical flow from the time-varying intensity values. So far, these attempts have failed when using purely analog chips (but has lead to some success using one-dimensional, analog preprocessing chips feeding into a conventional microprocessor; Koch et al., 1991). The computational reason underlying this failure is the ill-posed or ill-conditioned nature of the optical flow problem, making it a computation prone to failure in the presence of even moderately inaccurate hardware components (Poggio, Yang, & Torre, 1989).

However, for a variety of applications, the ability to estimate rapidly the displacement field across the entire retina is crucial. For instance, stand-alone chips with onboard photoreceptor arrays could predict the time to collision with an approaching object, avoid obstacles, estimate observer heading, and a variety of other tasks crucial to moving vehicles such as cars. It is therefore important that a number of independent attempts to design and build two-dimensional optical flow chips be tried (using different algorithms and circuit implementations; for the currently most promising approach using pulse-coded circuits, see Sarpeshkar, Bair, & Koch, 1993).

Neuromorphic Systems

The development of smart vision chips has been partly driven by Mead's belief (Mead, 1989) that building working systems mimicking parts of the nervous systems, such as the retina or cochlea, will lead both to a greater understanding of the function of these circuits and to new commercial products. This methodology has received a significant boost by the fabrication of the analog *silicon neuron* of Mahowald and Douglas (1991). By combining neurophysiological principles with silicon engineering, they managed to emulate efficiently the ionic currents that cause nerve impulses and control their spiking discharge. These "silicon neurons" are quite distinct (and substantially more complex) from silicon implementations of threshold neurons for artificial neural networks (for a review of this exciting subfield, see Mahowald, Douglas, & LeMoncheck, 1992).

Several research groups are now attempting to design and build an entire primitive visual system on one or several analog vision chips, including two silicon retinae with associated motors mimicking eye muscles, a tectum, the eye motor nuclei in the brain stem, and several cortical areas. Such a research program has as its primary goal understanding the elements necessary for a visual system to acquire images actively and move about in a constantly changing environment. *Building* such a system working in real time in a real environment will force designers to confront directly a number of issues that are usually disregarded in computer simulations. These modules will be located on different chips, communicating with each other using an event-driven, digital pulse code (Deiss, 1994).

Adaptation and Learning

To process visual information reliably in the real world, vision circuits must be able to operate over seven to eight orders of magnitude of light intensity (from moonlit to bright sunlight at the beach). Furthermore, given the limited accuracy of analog MOS fabrication technology, it is crucial that all circuits center their operating point on the mean value of the relevant variable (determined by its recent history). This is a trick nature appears to use at every stage, from the retina to high up in cortex. Smart analog vision chips should have the ability to adapt rapidly to recent input and only signal abrupt changes away from the operating point. The third-generation design of CMOS photoreceptors (the *hysteretic photoreceptor*) of Delbrück and Mead (1994) accomplishes this in an elegant manner.

A large application field would open up if vision chips could combine not only image acquisition and processing, but also if the same chips could learn to extract features from scenes and store them or to compare features from the currently viewed scene with stored ones. Within a conventional neural network context, this learning would be achieved by modifiable synapses between neurons. The most promising technology for implementing such synapses appears to be a variant of the floating-gate technology (Mead, 1988; Sin, 1992). This would also allow vision chips to optimize dynamically their performance by learning certain parameters, such as the amount of smoothing (λ in Equations (4)–(6)).

Several groups are working on analog MOS chips for implementing Gaussian *radial basis functions*. They are the key for "tabular lookup" learning procedures. Poggio and Girosi (1990) and Poggio and Edelman (1990) have shown how such an hypersurface reconstruction scheme can be used in object and face recognition and for very fast computer graphics applications.

By the time this book is published, we will see the use of analog vision chips proliferate for particular types of applications. Some of these chips will be based on principles that we learned from neurobiological circuits. It is personally very exciting and rewarding to participate in this transition from theory and neurobiology to real-life products.

Acknowledgments

The author thanks Tomothy Horiuchi for a careful reading of this manuscript. The design and fabrication of analog vision chips in our laboratory has been supported by the Office of Naval Research, the National Science Foundation, and by Rockwell International. We acknowledge MOSIS for quickly, reliably, and inexpensively fabricating our chips.

References

Bair, W., & Koch, C. (1991). An analog VLSI chip for finding edges from zero-crossings. In D. Touretzky (Ed.), *Neural Information Processing Systems* (pp. 399–405). San Mateo, CA: Morgan Kaufmann.

Barrow, H. G., & Tenenbaum, J. M. (1981). Computational Vision. *Proceedings of IEEE*, **69**, 572–595.

Blake, A., & Zisserman, A. (1987). *Visual reconstruction*. Cambridge, MA: MIT Press.

Boahen, K., & Andreou, A. (1992). A contrast sensitive silicon retina with reciprocal synapses. In J. E. Moody, S. J. Hanson, & R. P. Lippmann (Eds.), Advances in Neural Information Processing Systems 4 (pp. 764–772). San Mateo, CA: Morgan Kaufmann.

Chhabra, A. K., & Grogan, T. A. (1988). Estimating depth from stereo: Variational methods and network implementation. In *IEEE International Conference on Neural Networks*, San Diego.

Chiang, A. M., & Chuang, M. L. (1991). A CCD programmable image processor and its neural network applications. *IEEE Journal on Solid-State Circuits*, **26**, 1894–1901.

Delbrück, T., & Mead, C. A. (1994). Analog VLSI phototransduction by continuous, adaptive, logarithmic photoreceptor circuits. Computation and Neural Systems Program Memo, California Institute of Technology, May, 1994.

Deimel, P. P., Müller, G., & Baumeister, K. (1987). Transient photoresponse of amorphous silicon p-i-n diodes. *Mathematical Research Society Symposium Proceedings*, **95**, 639–644.

Deiss, R. S. (In press). Temporal binding in analog VLSI. World Congress on Neural Networks, San Diego, CA, 1994.

Denayer, T., Vanzieleghem, E., & Jespers, P. G. A. (1988). A class of multiprocessors for real-time image and multidimensional signal processing. *IEEE Journal on Solid State Circuits*, **23**, 630–638.

DeWeerth, S. (1992). Analog VLSI circuits for stimulus localization and centroid computation. *International Journal of Computer Vision*, **8**, 191–202.

Fossum, E. R. (1989). Architectures for focal plane image processing. *Optical Engineering*, **28**, 866–871.

Geman, S., & Geman, D. (1984). Stochastic relaxation, Gibbs distribution and the Bayesian restoration of images. *IEEE Transactions on Pattern Analysis and Machine Intelligence*, **6**, 721–741.

Glasser, L. A. (1985). A UV write-enabled PROM. In 1985 Chapel Hill Conference on VLSI (pp. 61–65). Rockville, MD: Computer Science Press.

Grimson, W. E. L. (1981). From images to surfaces. Cambridge, MA: MIT Press.

Harris, J. G. (1987). A new approach to surface reconstruction: The coupled depth/slope model. *Proceedings of International Conference on Computer IEEE 1. Vision*, pp. 277–283, London.

Harris, J. G., Koch, C., & Luo, J. (1990). A two-dimensional analog VLSI circuit for detecting discontinuities in early vision. *Science*, **248**, 1209–1211.

Harris, J. G., Koch, C., Luo, J., & Wyatt, J. (1989). Resistive fuses: Analog hardware for detecting discontinuities in early vision. In C. Mead & M. Ismail (Eds.), *Analog VLSI implementations of neural systems* (pp. 27–56). Norwell, MA: Kluwer.

Harris, J. G., Koch, C., Staats, E., & Luo, J. (1990). Analog hardware for detecting discontinuities in early vision. *International Journal of Computer Vision*, **4**, 211–223.

Hildreth, E. C. (1984). *The measurement of visual motion*. Cambridge, MA: MIT Press.

Horiuchi, T., Bair, W., Bishofberger, B., Lazzaro, J., Moore, A., & Koch, C. (1992). Computing motion using analog VLSI vision chips: An experimental comparison among different approaches. *International Journal of Computer Vision*, **8**, 203–216.

Horn, B. K. P. (1974). Determining lightness from an image. *Computer Graphics and Image Processing*, **3**, 277–299.

Horn, B. K. P., & Schunck, B. G. (1981). Determining optical flow. *Artificial Intelligence*, **17**, 185–203.

Horn, B. K. P. (1990). Parallel analog networks for machine vision. In P. H. Winston and S. A. Shellard (Eds.), Artificial Intelligence at MIT: Expanding Frontiers (Vol. 2, pp. 437–471). Cambridge, MA: MIT Press.

Hutchinson, J., Koch, C., Luo, J., & Mead, C. (1988). Computing motion using analog and binary resistive networks. *IEEE Computers*, **21**, 52–63.

Karplus, W. J. (1958). *Analog simulation: Solution of field problems*. New York: McGraw-Hill.

Kemeny, S. E., Torbey, H. H., Meadows, H. E., Bredthauer, R. A., La Shell, M. A., & Fossum, E. R. (1992). CCD Focal-plane image recognization processors for lossless image compression. *IEEE Journal on Solid-State Circuits*, **27**, 398–405.

Khosla, R. P. (1992). From photons to bits. *Physics Today*, **45**, 42–49.

Knight, T. (1983). *Design of an integrated optical sensor with on-chip preprocessing*. Ph.D. thesis, Dept. of Electrical Engineering and Computer Science, MIT, Cambridge, MA.

Koch, C. (1990). Resistive networks from computer vision: A tutorial. In S. F. Zornetzer, J. C. Davis, & C. Lau (Eds.) *An Introduction to Neural and Electronic Networks* (pp. 293–305). New York: Academic Press.

Koch, C., & Li, H., Eds. (1994). Vision Chips: Implementing Vision Algorithms with Analog VLSI Circuits. Los Alamitos, CA: IEEE CS Press.

Koch, C., Moore, A., Bair, W., Horiuchi, T., Bishofberger, B., & Lazzaro, J. (1991). Computing motion using analog VLSI vision chips: An experimental comparison among four approaches. *Proceedings of the IEEE Workshop on Visual Motion*, pp. 312–324.

Koch, C., Marroquin, J., & Yuille, A. (1986). Analog "neuronal" networks in early vision. *Proceedings of the National Academy of Sciences* (USA), **83**, 4263–4267.

Liu, S.-C., & Harris, J. (1992). Dynamic wires: An analog VLSI model for object-based processing. *International Journal of Computer Vision*, **8**, 231–239.

Luo, J., Koch, C., & Mathur, B. (1992). Figure-ground segregation using an analog VLSI chip. *IEEE Micro*, **12**, 46–57.

Luo, J., Koch, C., & Mead, C. (1988). An experimental subthreshold, analog CMOS two-dimensional surface interpolation circuit. *Neural Information and Processing Systems Conference*, Denver, November.

Mahowald, M. (1991). Silicon retina with adaptive photoreceptors. *Visual Information Processing: From Neurons to Chips, SPIE Proceedings*, **1473**, 52–58.

Mahowald, M., & Douglas, R. (1991). A silicon neuron. *Nature*, **354**, 515–518.

Mahowald, M. A., Douglas, R. J., LeMoncheck, J. E., & Mead, C. A. (1992). An introduction to silicon neural analogs. *Seminars on Neuroscience*, **4**, 83–92.

Mann, J. (1991). Implementing early visual processing in analog VLSI: Light adaptation. *Visual Information Processing: From Neurons to Chips, SPIE Proceedings*, **1473**, 128–136.

Marr, D., & Hildreth, E. C. (1980). Theory of edge detection. *Proceedings of the Royal Society of London B*, **207**, 187–217.

Marroquin, J., Mitter, S., & Poggio, T. (1987). Probabilistic solution of ill-posed problems in computational vision. *Journal of the American Statistics Association*, **82**, 76–89.

Mead, C. A. (1985). A sensitive electronic photoreceptor. *1985 Chapel Hill Conference on Very Large Scale Integration*, pp. 463–471.

Mead, C. A. (1989). *Analog VLSI and neural systems*. Reading, MA: Addison-Wesley.

Mead, C. A. (1990). Neuromorphic electronic systems. *Proc. IEEE* **78**: 1629–1636.

Mead, C. A., & Mahowald, M. A. (1988). A silicon model of early visual processing. *Neural Networks*, **1**, 91–97.

Nagel, H.-H., & Enkelmann, W. (1986). An investigation of smoothness constraint for the estimation of displacement vector fields from image sequences. *IEEE Pattern Analysis and Machine Intelligence*, **8**, 565–593.

Nudd, G. R., Nygaard, P. A., Thurmond, G. D., & Fouse, S. D. (1978). A charge-coupled device image processor for smart sensor applications. *SPIE Image Understanding Systems & Industrial Applications*, **155**, 15–22.

Ortega, J. M., & Rheinboldt, W. C. (1970). *Iterative solution of nonlinear equations in several variables*. New York: Academic Press.

Poggio, T., & Edelman, S. (1990). A network that learns to recognize 3D objects. *Nature*, **343**, 263–265.

Poggio, T., Gamble, E. B., & Little, J. J. (1988). Parallel integration of vision modules. *Science*, **242**, 436–440.

Poggio, T., & Girosi, F. (1990). Theory of networks for learning. *Science*, **247**, 978–981.

Poggio, T., & Koch, C. (1985). Ill-posed problems in early vision: From computational theory to analogue networks. *Proceedings of the Royal Society of London B*, **226**, 303–323.

Poggio, T., Torre, V., & Koch, C. (1985). Computational vision and regularization theory. *Nature*, **317**, 314–319.

Poggio, T., Voorhees, H., & Yuille, A. (1988). A regularized solution to edge detection. *Journal of Complexity*, **4**, 106–123.

Poggio, T., Yang, W., & Torre, V. (1989). Optical flow: Computational properties and networks, biological and analog. In R. Durbin, C. Miall, & G. Mitchison (Eds.), The Computing Neuron (pp. 355–370). Wokingham, UK: Addison-Wesley.

Ruetz, P. A., & Brodersen, R. W. (1987). Architectures and design techniques for real-time image processing IC's. *IEEE Journal on Solid-State Circuits*, **22**, 233–250.

Sage, J. P. (1984). Gaussian convolution of images stored in a charge-coupled device. *Quarterly Technical Report* (No. 1, p. 53–59). Lexington, MA: MIT Lincoln Laboratory.

Sage, J. P., & Lattes, A. L. (1987). A high-speed two-dimensional CCD Gaussian image convolver. *Quarterly Technical Report*, (No. 1). Lexington, MA: MIT Lincoln Laboratory.

Sarpeshkar, R., Bair, W., & Koch, C. (1993). Visual motion computation in analog VLSI using pulses. In J. E. Moody, S. J. Hanson, & R. P. Lippmann (Eds.), In Advances in Neural Information Processing Systems 5, (pp. 781–788). San Mateo, CA: Morgan Kaufmann.

Sin, C. K. (1992). EEPROM as an analog storage device with particular application in neural networks. *IEEE Transactions on Electrical Devices*, **39**, 1410–1424.

Sivilotti, M. A., Mahowald, M. A., & Mead, C. A. (1987). Real-time visual computation using analog CMOS processing arrays. In *1987 Stanford Conference on VLSI* (pp. 295–312). Cambridge, MA: MIT Press.

Standley, D. L. (1991). An object position and orientation IC with embedded images. *IEEE Journal on Solid-State Circuits*, **26**, 1853–1859.

Stix, G. (1992). Check it out: A retina on a chip eyeballs bad paper. *Scientific American*, **267**(2), 119–120.

Strausfeld, N. (1975). *Atlas of an insect brain*. Heidelberg: Springer.

Terzopoulos, D. (1983). Multilevel computational processes for visual surface reconstruction. *Computer Graphics and Image Processing*, **24**, 52–96.

Van der Wal, G. S., & Burt, P. J. (1992). A VLSI pyramid chip for multiresolution image analysis. *International Journal on Computer Vision*, **8**, 177–189.

Wyatt, J. L., Keast, C., Seidel, M., Standley, D., Horn, B., Knight, T., Sodini, C., Lee, H.-S., & Poggio, T. (1992). Analog VLSI systems for image acquisition and fast early vision processing. *International Journal on Computer Vision*, **8**, 217–230.

Wyatt, J. L., & Standley, D. L. (1989). Criteria for robust stability in a class of lateral inhibition networks coupled through resistive grids. *Neural Computation*, **1**, 58–67.

Yamada, H., Hasegawa, K., Mori, T., Sakai, H., & Aono, K. (1988). A microprogrammable real-time image processor. *IEEE Journal on Solid-State Circuits*, **23**, 216–223.

Yang, W. (1991). Analog CCD processors for image filtering. *Visual Information Processing: From Neurons to Chips, SPIE Proceedings*, **1473**, 114–127.

17

A Digital VLSI Architecture for Real-World Applications

Dan Hammerstrom

INTRODUCTION

As the other chapters of this book show, the neural network model has significant advantages over traditional models for certain applications. It has also expanded our understanding of biological neural networks by providing a theoretical foundation and a set of functional models.

Neural network simulation remains a computationally intensive activity, however. The underlying computations—generally multiply-accumulates—are simple but numerous. For example, in a simple artificial neural network (ANN) model, most nodes are connected to most other nodes, leading to $O(N^2)$ *connections*.[1] A network with 100,000 nodes, modest by biological standards, would therefore have about 10 billion connections, with a multiply-accumulate operation needed for each connection. If a state-of-the-art workstation can simulate roughly 10 million connections per second, then one pass through the network takes 1000 sec (about 20 min). This data rate is much too slow for real-time process control or speech recognition, which must update several times a second. Clearly, we have a problem.

This performance bottleneck is worse if each connection requires more complex computations, for instance, for incremental learning algorithms or for more realistic biological simulations. Eliminating this computational barrier has led to much research into building custom Very Large Scale Integration (VLSI) silicon chips optimized for ANNs. Such chips might perform ANN simulations hundreds to thousands of times faster than workstations or personal computers—for about the same cost.

The research into VLSI chips for neural network and pattern recognition applications is based on the premise that optimizing the chip architecture to the computational characteristics of the problem lets the designer create a silicon device offering a big improvement in performance/cost or "operations per dollar." In silicon design, the cost of a chip is primarily determined by its two-dimensional area. Smaller chips

[1]The "order of" $O(F(n))$ notation means that the quantity represented by O is approximate for the function F within a multiplication or division by n.

are cheaper chips. Within a chip, the cost of an operation is roughly determined by the silicon area needed to implement it. Furthermore, speed and cost usually have an inverse relationship: faster chips are generally bigger chips.

The silicon designer's goal is to increase the number of operations per unit area of silicon, called *functional density*, in turn, increasing operations per dollar. An advantage of ANN, pattern recognition, and image processing algorithms is that they employ simple, low-precision operations requiring little silicon area. As a result, chips designed for ANN emulation can have a higher functional density than traditional chips such as microprocessors. The motive for developing specialized chips, whether analog or digital, is this potential to improve performance, reduce cost, or both.

The designer of specialized silicon faces many other choices and trade-offs. One of the most important is flexibility versus speed. At the "specialized" end of the flexibility spectrum, the designer gives up versatility for speed to make a fast chip dedicated to one task. At the "general purpose" end, the sacrifice is reversed, yielding a slower, but programmable device. The choice is difficult because both traits are desirable. Real-world neural network applications ultimately need chips across the entire spectrum.

This chapter reviews one such architecture, CNAPS[2] (Connected Network of Adaptive ProcessorS), developed by Adaptive Solutions, Inc. This architecture was designed for ANN simulation, image processing, and pattern recognition. To be useful in these related contexts, it occupies a point near the "general purpose" end of the flexibility spectrum. We believe that, for its intended markets, the CNAPS architecture has the right combination of speed and flexibility. One reason for writing this chapter is to provide a retrospective on the CNAPS architecture after several years' experience developing software and applications for it.

The chapter has three major sections, each framed in terms of the capabilities needed in the CNAPS computer's target markets. The first section presents an overview of the CNAPS architecture and offers a rationale for its major design decisions. It also summarizes the architecture's limitations and describes aspects that, in hindsight, its designers might have done differently. The section ends with a brief discussion of the software developed for the machine so far.

The second section briefly reviews applications developed for CNAPS at this writing.[3] The applications discussed are simple image processing, automatic target recognition, a simulation of the Lynch/Granger Pyriform Model, and Kanji OCR. Finally, to offer a broader perspective of real-world ANN usage, the third section reviews non-CNAPS applications, specifically, examples of process control and financial analysis.

THE CNAPS ARCHITECTURE

The CNAPS architecture consists of an array of processors controlled by a sequencer, both implemented as a chip set developed by Adaptive Solutions, Inc. The sequencer is a one-chip device called the CNAPS Sequencer Chip (CSC). The processor array is also a one-chip device, available with either 64 or 16 processors per chip (the CNAPS-1064 or CNAPS-1016). The CSC can control up to eight 1064s or 1016s, which act like one large device.

These chips usually sit on a printed circuit board that plugs into a host computer, also called the control processor (CP). The CNAPS board acts as a coprocessor within the host. Under the coprocessor model, the host sends data and programs to the board, which runs until done, then interrupts the host to indicate completion. This style of operation is called "run to completion semantics." Another possible model is to use the CNAPS board as a stand-alone device to process data continuously.

[2] Trademark Adaptive Solutions, Inc.

[3] Because ANNs are becoming a key technology, many customers consider their use of ANNs to be proprietary information. Many applications are not yet public knowledge.

The CNAPS Architecture

Basic Structure

CNAPS is a single instruction, multiple data stream (SIMD) architecture. A SIMD computer has one instruction sequencing/control unit and many processor nodes (PNs). In CNAPS, the PNs are connected in a one-dimensional array (Figure 1) in which each PN can "talk" only to its right or left neighbors. The sequencer broadcasts each instruction plus input data to all PNs, which execute the same instruction at each clock. The PNs transmit output data to the sequencer, with several arbitration modes controlling access to the output bus.

As Figure 2 suggests, each PN has a local memory,[4] a multiplier, an adder/subtracter, a shifter/logic unit, a register file,[5] and a memory addressing unit. The entire PN uses fixed-point, two's complement arithmetic, and the precision is 16 bits, with some exceptions. The PN memory can handle 8- or 16-bit reads or writes. The multiplier produces a 24-bit output; an 8×16 or 8×8 multiply takes one clock, and a 16×16 multiply takes two clocks. The adder can switch between 16- or 32-bit modes. The input and output buses are 8 bits wide, and a 16-bit word can be assembled (or disassembled) from two bytes in two clocks.

A PN has several additional features (Hammerstrom, 1990, 1991) including a function that finds the PN with the largest or smallest values (useful for winner-take-all and best-match operations), various precision and memory control features, and OutBus arbitration. These features are too detailed to discuss fully here.

The CSC sequencer (Figure 3) performs program sequencing for the PN array and has private access to a program memory. The CSC also performs input/output (I/O) processing for the array, writing input data to the array and reading output data from it. To move data to and from CP memory, the CSC has a 32-bit bus, called the AdaptBus, on the CP side. The CSC also has a direct input port and a direct output port used to connect the CSC directly to I/O devices for higher-bandwidth data movement.

[4]Currently 4 KB per PN.

[5]Currently 32, 16-bit registers.

Neural Network Example

The CNAPS architecture can run many ANN and non-ANN algorithms. Many SIMD techniques are the same in both contexts, so an ANN can serve as a general example of mapping an algorithm to the array. Specifically, the example here shows how the PN array simulates a layer in an ANN.

Start by assuming a two-layered network (Figure 4) in which—for simplicity—each node in each layer maps to one PN. PN_i thus simulates the node n_{ij}, where i is the node index in the layer and j is the layer index. Layers are simulated in a time-multiplexed manner. All layer 1 nodes thus execute as a block, then all layer 2 nodes, and so on. Finally, assume that layer 1 has already calculated its various $n_{i,1}$ outputs.

The goal at this point is to calculate the outputs for layer 2. To achieve this, all layer 1 PNs simultaneously load their output values into a special output buffer and begin arbitration for the output bus. In this case, the arbitration mode lets each PN transmit its output in sequence. In one clock, the content of PN_0's buffer is placed on the output bus and goes through the sequencer[6] to the input bus. From the input bus, the value is broadcast to all PNs (this out-to-in loopback feature is a key to implementing layered structures efficiently). Each PN then multiplies node $n_{0,1}$'s output with a locally stored weight, $w_{0,1}$.

On the next clock, node $n_{1,1}$'s output is broadcast to all PNs, and so on for the remaining layer 1 output values. After N clocks, all outputs have been broadcast, and the inner product computation is complete. All PNs then use the accumulated value's most significant 8 bits to look up an 8-bit nonlinear output value in a 256-item table stored in each PN's local memory. This process—calculating a weighted sum, then passing

[6]This operation actually takes several clocks and must be pipelined. These details are eliminated here for clarity.

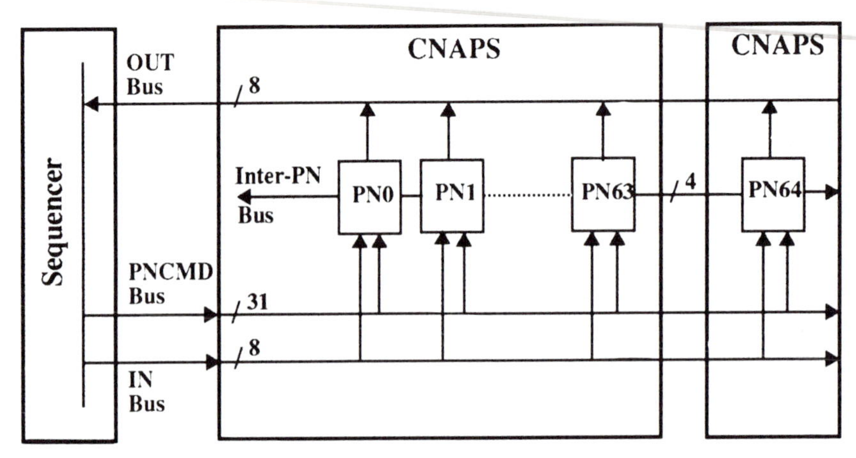

FIGURE 1 The basic CNAPS architecture. CNAPS is a single instruction, multiple data (SIMD) architecture that uses broadcast input, one-dimensional interprocessor communication, and a single shared output bus.

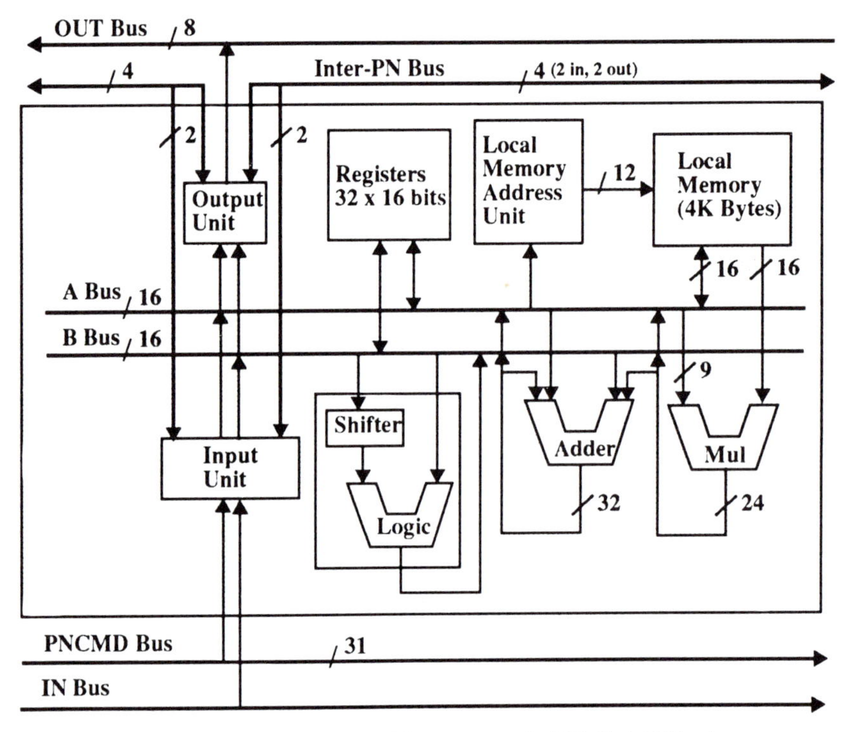

FIGURE 2 The internal structure of a CNAPS processor node (PN). Each PN has its own storage and arithmetic capabilities. Storage consists of 4096 bytes. Arithmetic operations include multiply, accumulate, logic, and shift. All units are interconnected by two 16-bit buses.

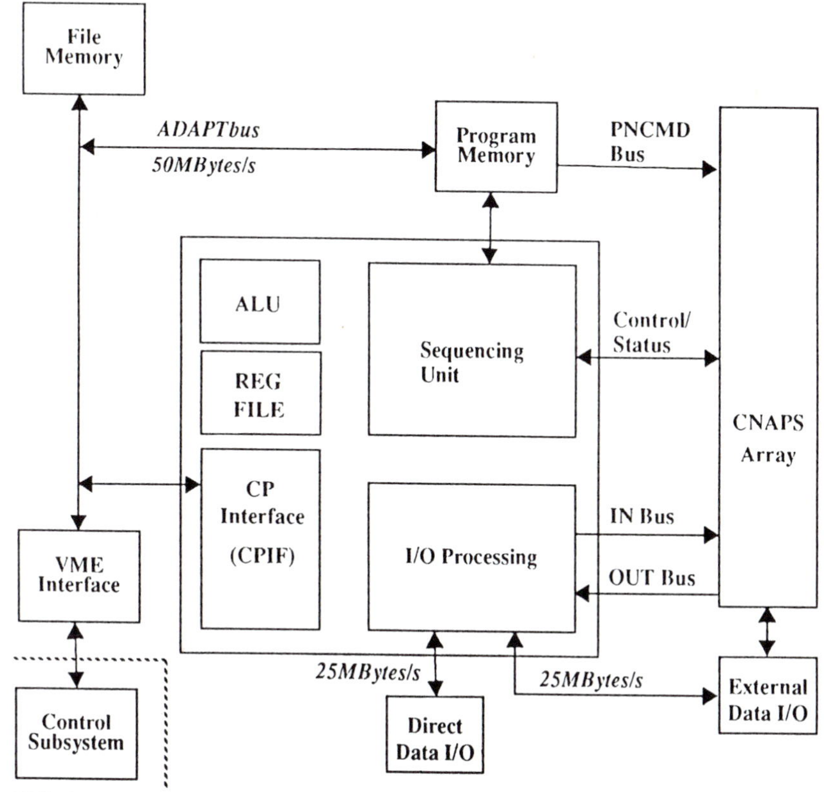

FIGURE 3 The CNAPS sequencer chip (CSC) internal structure. The CSC accesses an external program store, which contains both CSC and CNAPS PN array instructions. PN array instructions are broadcast to all PNs. CSC instructions control sequencing and all array input and output.

it through a function stored in a table—is performed for each output on each layer. The last layer transmits its output values through the CSC to an output buffer in the CP memory.

The multiply-accumulate pipeline can compute a connection in each clock. The example network has four nodes and uses only four clocks for its 16 connections. For even greater efficiency, other operations can be performed in the same clock as the multiply-accumulate. The separate memory address unit, for instance, can compute the next weight's address at the same time as the connection computation; and the local memory allows the weight to be fetched without delay.

An array of 256 PNs can compute $256^2 = 65536$ connections in 256 clocks. At a 25-MHz clock frequency, this equals 6.4 billion connections per second (back-propagation feed-forward) and over 1 billion connection updates per second (back-propagation learning). An array of 64 PNs (one CNAPS-1064 chip), for example, can store and train the entire NetTalk (Sejnowski & Rosenberg, 1986) network in about 7 sec.

Physical Implementation

The CNAPS PN array has been implemented in two chips, one with 64 PNs (the CNAPS-1064; Griffin

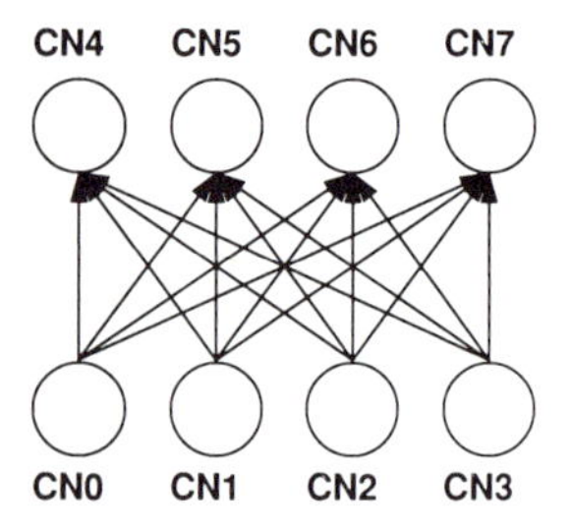

Broadcast by PN0 of CN0's output to CN4, 5, 6, 7 takes 1 clock

N^2 connections in N clocks

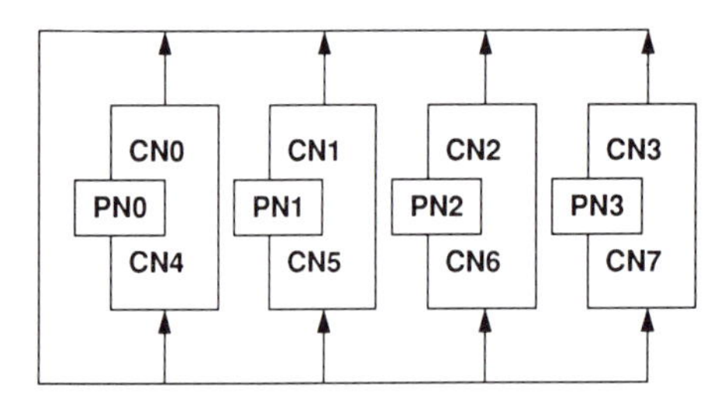

FIGURE 4 A simple two-layered neural network. In this example, each PN emulates two network nodes. PNs emulate the first layer, computing one connection each clock. Then, they sequentially place node output on the OutBus while emulating, in parallel, the second layer.

et al., 1990 Figure 5) and the other with 16 PNs (the CNAPS-1016). Both chips are implemented in a 0.8 micron CMOS process. The 64-PN chip is a full custom design and is approximately 26 mm on a side and has more than 14 million transistors, making it one of the largest processor chips ever made. The simple computational model makes possible a small, simple PN, in turn permitting the use of redundancy to improve semiconductor yield for such a device.

The CSC is implemented using a gate array technology, using a 100,000-gate die and is about 10 mm on a side.

The next section reviews the various design decisions and the reasons for making them. Some of the features described are unique to CNAPS; others apply to any digital signal processor chip.

Major Design Decisions

When designing the CNAPS architecture, a key question was where it should sit relative to other computing devices in cost and capabilities. In computer design, flexibility and performance are almost always inversely related. We wanted CNAPS to be flexible enough to execute a broad family of ANN algorithms as well as other related pattern recognition and preprocessing algorithms. Yet, we wanted it to have much higher performance than state-of-the-art workstations and—at the same time—lower cost for its functions.

Figure 6 shows where we are targeting CNAPS. The vertical dimension plots each architecture by its flexibility. Flexibility is difficult to quantify, because it involves not only the range of algorithms that an architecture can execute, but also the complexity of the problems it can solve. (Greater complexity typically requires a larger range of operations.) As a result, this graph is subjective and provided only as an illustration.

The horizontal dimension plots each architecture by its performance/cost—or operations per second per dollar. The values are expressed in a log scale due to the orders-of-magnitude difference between traditional microprocessors at the low end and highly custom, analog chips at the high end. Note the *technology barrier*, defined by practical limits of current semiconductor manufacturing. No one can build past the barrier: you can do only so much with a transistor; you can put only so many of them on a chip; and you can run them only so fast.

For pattern recognition, we placed the CNAPS architecture in the middle, between the specialized analog chips and the general-purpose microprocessors. We wanted it to be programmable enough to solve many real-world problems, and yet have a performance/cost about 100 times faster than the highest performance RISC processors. The CNAPS applications discussed later show that we have provided sufficient flexibility to solve complex problems.

In determining the degree of function required, we must solve all or most of a targeted problem. This need results from *Amdahl's law*, which states that system

FIGURE 5 The CNAPS PN array chip. There are 64 PNs with memory on each die. The PN array chip is one of the largest processor chips ever made. It consists of 14 million transistors and is over 26 mm on a side. PN redundancy, there are 16 spare PNs, is used to guarantee high yields.

performance depends mainly on the slowest component. This law can be formalized as follows:

$$S = \frac{1}{(op_f * s_f) + (op_h * s_h)} \tag{1}$$

where S is the total system speed-up, op_f is the fraction of total operations in the part of the computation run on the fast chip, s_f is the speedup the chip provides, op_h is the fraction of total operations run on the host computer without acceleration. Hence, as op_f or s_f get large, S approaches $1/op_h$. Unfortunately, op_f needs to be close to one before any real system-level improvement occurs, as shown in the following example.

Suppose there are two such support chips to choose from: the first can run 80% of the computation with 20× improvement on that 80%; the second can run only 20%, but runs that 20% 1000× faster. By Amdahl's law, the first chip speeds up the system by more than 400%, whereas the second—and seemingly faster—chip speeds up the system by only 20%. So Amdahl tells us that flexibility is often better than raw performance, especially if that performance results

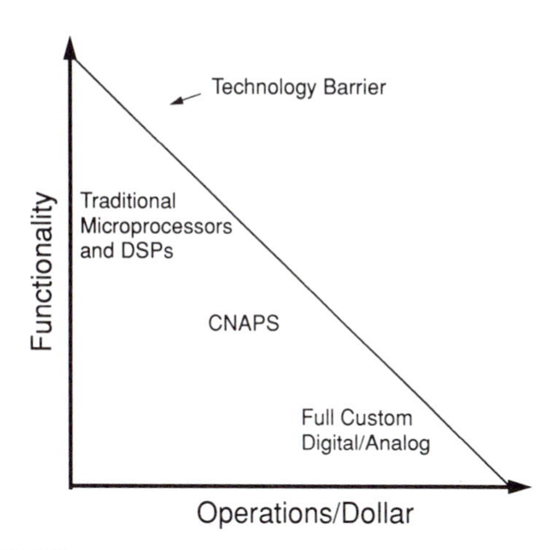

FIGURE 6 Though subjective, this graph gives a rough indication of the CNAPS market positioning. The vertical dimension measures the range of functionality of an architecture; the horizontal dimension measures the performance/cost in operations per second per dollar. The philosophy behind CNAPS is that by restricting functionality to pattern recognition, image processing, and neural network emulation, a larger performance/cost is possible than with traditional machines (parallel or sequential).

from limiting the range of operations performed by the device.

Digital

Much effort has been dedicated to building analog VLSI chips for ANNs. Analog chips have great appeal, partly because they follow biological models more closely than digital chips. Analog chips also can achieve higher functional density. Excellent papers reporting research in this area include Mead (1989), Akers, Haghighi, and Rao (1990), Graf, Jackel, and Hubbard (1988), Holler, Tam, Castro, and Benson (1989), and Alspector (1991). Also, see Morgan (1990) for a good summary of digital neural network emulation.

Analog ANN implementations have been primarily academic or industrial research projects, however. Only a few have found their way into the real world as commercial products: getting an analog device to work in a laboratory is one thing; making it work over a wide range of voltages, temperatures, and user capabilities is another. In general, analog chips require much more stringent operating conditions than digital chips. They are also more difficult to design and, after implementation, less flexible.

The semiconductor industry is heavily oriented toward digital chips. Analog chips represent only a minor part of the total output, reinforcing their secondary position. There are, of course, successful analog parts, and there always will be, because some applications require analog's higher functional density to achieve their cost and performance constraints, and those applications can tolerate analog's limited flexibility. Likewise, there will be successful products using analog ANN chips. Analog parts will probably be used in simple applications, or as a part of a larger system in more complex applications.

This prediction follows primarily from the limited flexibility of analog chips. They typically implement one algorithm, hardwired into the chip. A hardwired algorithm is fine if it is truly stable and it is all you need. The field of ANN applications is still new, however, so most complex implementations are still actively evolving—even at the algorithm level. An analog device cannot easily follow such changes. A digital, programmable device can change algorithms by changing software.

Our major goal was to produce a commercial product that would be flexible enough and provide sufficient precision to cover a broad range of complex problems. This goal dictated a digital design, because digital could offer accurate precision and much more flexibility than a typical CMOS analog implementation. Digital also offered excellent performance and the advantages of a standardized technology.

Limited, Fixed-Point Precision

In both analog and digital domains, an important decision is choosing the arithmetic precision required. In analog, precision affects design complexity and the amount of compensation circuitry required. In digital, it effects the number of wires available as well as the size and complexity of memory, buses, and arithmetic units. Precision also affects the power dissipation.

In the digital domain, a related decision involves floating-point versus fixed-point representation. Floating-point numbers (Figure 7) consist of an exponent (usually 8 bits representing base 2 or base 16) and a mantissa (usually 24 bits). The exponent is set so that the mantissa is always normalized; that is, the most significant "1" of the data is in the most significant position of the field. Adding two floating-point numbers involves shifting at least one of the operands to get the same exponent. Multiplying two floating-point

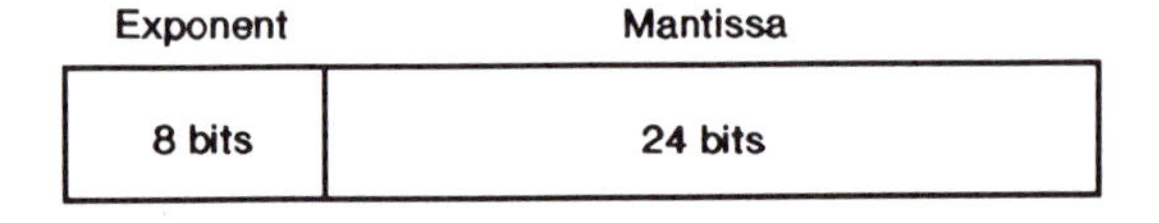

FIGURE 7 A floating point number. A single-precision, IEEE compatible floating point configuration is shown. The high order 8 bits constitute the exponent; the remaining 24 bits, the mantissa or "fractional" part. Floating point numbers are usually normalized so that the mantissa has a 1 in the most significant position.

numbers involves separate arithmetic on both exponents and mantissas. Both operations require postnormalizing shifts after the arithmetic operations.

Floating point has several advantages. The primary advantage is dynamic range, which results from the separate exponent. Another is precision, due to the 24-bit mantissas. The disadvantage to floating point is its cost in silicon area. Much circuitry is required to keep track of both exponents and mantissas and to perform pre- and postoperation shifting of the mantissa. This circuitry is particularly complicated if high speed is required.

Fixed-point numbers consist of a numeral (usually 16 to 32 bits) and a radix point (in base 2, the binary point). In fixed point, the programmer chooses the position of the radix point. This position is typically fixed for the calculation, although it is possible to change the radix point under software control by explicitly shifting operands. For many applications needing only limited dynamic range and precision, fixed point is sufficient. It is also much cheaper than floating point because it requires less silicon area.

After choosing a digital signal representation for CNAPS, the next question was how to represent the numbers. Biological neurons are known to use relatively low precision and to have a limited dynamic range. These characteristics strongly suggest that a digital computer for emulating ANN structures should be able to employ limited precision fixed-point arithmetic. This conjecture in turn suggests an opportunity to simplify significantly the arithmetic units and to provide greater computational density. Fixed-point arithmetic also places the design near the desired point on the flexibility versus performance/cost curve (Figure 6).

To confirm the supposition that fixed point is adequate, we performed extensive simulations. We found that for the target applications, 8- or 16-bit fixed-point precision was sufficient (Baker & Hammerstrom, 1989). Other researchers have since reached the same conclusion (Hoehfeld and Fahlman, 1992; Shoemaker, Carlin, & Shimabukuro, 1990). In keeping with experimental results, we used a general 16-bit resolution inside the PN. One exception was to use a 32-bit adder to provide additional head-room for repeated multiply-accumulates. Another was to use 8-bit input and output data buses, because most computations involve 8-bit data and 8- or 16-bit weights, and because busing external to the PN is expensive in silicon area.

SIMD

The next major decision was how to control the PNs. A computer can have one or more instruction streams and one or more data streams. Most computers are single instruction, single data (SISD) computers. These have one control unit and one processor unit, usually combined on one chip (a microprocessor). The control unit fetches instructions from program memory and decodes them. It then sends data operations such as add, subtract, or multiply to the processing unit. Sequencing operations, such as branch, are executed by the control unit itself. The SISD computers are serial, not parallel.

Two major families of parallel computer architectures have evolved: multiple instruction, multiple data (MIMD) and single instruction, multiple data (SIMD). MIMD computers have many processing units, each of which has its own control unit. Each control/processing unit can operate in parallel, executing many instructions at once. Because the processors operate independently, MIMD is the most powerful and flexible parallel architecture. The independent, asynchronous processors also make MIMD the most difficult to use, requiring complex processor synchronization.

The SIMD computers have many processors but only one instruction stream. All processors receive the same instruction at the same time, but each acts on its own slice of the data. SIMD computers thus have an array of processors and can perform an operation on a block of data in one step. SIMD computing is often called "data parallel" computing, because it applies one control thread to multiple local data elements, executing one instruction at each clock.

SIMD computation is perfect for vector and matrix arithmetic. Because of Amdahl's law, however, SIMD is cost-effective only if most operations are matrix or vector operations. For general-purpose computing, this is not the case. Consequently, SIMD machines are poor general-purpose computers and rarer than SISD or even MIMD computers. Our target domain is not general-purpose computing, however. For ANNs and other image and signal processing algorithms, the dominant calculations are vector or matrix operations. SIMD fits this domain perfectly.

The SIMD architecture is a good choice for practical reasons, too. One advantage is cost: SIMD is much cheaper than MIMD, because there is only one control unit for the entire array of processors. Another is that SIMD is easier to program than MIMD, because all processors do the same thing at the same time. Likewise, it is easier to develop computer languages for SIMD, because it is relatively easy to develop parallel data structures where the data are operated on simultaneously. Figure 8 shows a simple CNAPS-C program that multiplies a vector times a matrix. Normally, vector matrix multiply takes n^2 operations. By placing each column of the matrix on each PN, it takes n operations on n processors.

In sum, SIMD was better than MIMD for CNAPS because it fit the problem domain, was much more economical, and easier to program.

```
# define N 20
# define K 30
typedef scaled 8 8 arithType;
domain Krows
      {arithType sourceMatrix[N];
      arithType resultVector;} dimK[K];


main()
{ int n;
      [domain dimK].{
      resultVector = 0;
      for (n=0; n < N; n++)
      resultVector += sourceMatrix[n] * getchar();
      }
}
```

FIGURE 8 A CNAPS-C program to do a simple vector–matrix multiply. The "data-parallel" programming is evident here. Within the loop, it is assumed because of the domain declaration that there are multiple copies of each matrix element, one on each PN. The program takes N loop iterations, which would require N^2 on a sequential machine.

Broadcast Interconnect

The next decision concerned how to interconnect the PNs for data transfer, both within the array and outside it. Computer architects have developed several interconnect structures for connecting processors in multiprocessor systems. Because CNAPS is a SIMD machine, we were interested only in synchronous structures.

The two families of interconnect structures are *local* and *global*. Local interconnect attaches only neighboring PNs. The most common local scheme is NEWS (North-East-West-South, Figure 9). In NEWS, the PNs are laid out in a two-dimensional array, and each PN is connected to its four nearest neighbors. A one-

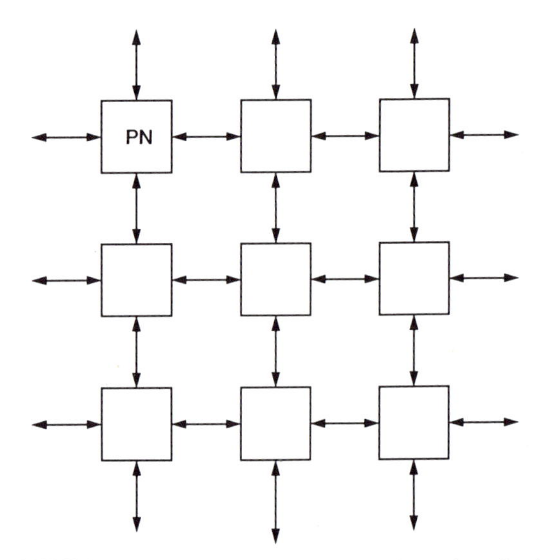

FIGURE 9 A two-dimensional PN layout. This configuration is often called a "NEWS" network, because each PN connects to its north, east, west, and south neighbor. These networks provide more flexible intercommunication than a one-dimensional network, but are more expensive to implement in VLSI and difficult to make work when redundant PNs are used.

dimensional variation connects each PN only to its left and right neighbors.

Global interconnect permits any PN to talk to any other PN, not just to its immediate neighbors. There are several possible configurations with different levels of performance/cost. At one end of the scale, *crossbar* interconnect is versatile because it permits random point-to-point communications, but expensive [the cost is $O(n^2)$, where n is the number of PNs]. At the other end, *broadcast* interconnect is cheaper but less flexible. Here, one bus connects all PNs, so any one PN can talk to any other (or set of others) in one clock. On the other hand, it takes n clocks for all PNs to have a turn. The cost is $O(1)$. In between crossbar and broadcast are other configurations that can emulate a cross-bar in $O(\log n)$ clocks and have cost $O(n \log n)$.

Choosing an interconnect structure interacted with other design choices. We decided against using a *systolic* computing style, in which operands, intermediate results, or both flow down a row of PNs using only local interconnect. Systolic arrays are harder to program. They are also occasionally inefficient because of the clocks needed to fill or empty the pipeline—peak efficiency occurs only when all PNs see all operands. Choosing a systolic array would have permitted us to use local interconnect, saving cost. Deciding against it forced us to provide some form of global interconnect.

Choosing "global" leads to the next choice: what type? The basic computations in our target applications required "one-to-many" or "many-to-many" communication almost exclusively. We therefore decided to use a broadcast bus, which uses only one clock for one-to-many communication. In the many-to-many case, n PNs can talk to all n PNs in n clocks. Broadcast interconnect thus allows n^2 connections in n clocks. Such $O(n^2)$ total connectivity occurs often in our applications. An example is a back-propagation network in which all nodes in one layer connect to all nodes in the next.

Another advantage is that broadcast interconnection is synchronous and fits the synchronous SIMD structure quite well. We were able to use a "slotted" protocol, in which each connection occurs at a known time on the bus. Because the time is known, there is no need to send an address with each data element, saving wires, clocks, or both. Also, the weight address unit can "remember" the slot number and use it to address the weight associated with the connection. A single broadcast bus is simple, economical to implement, and efficient for the application domain. In fact, if every PN always communicates with every other PN, then broadcast offers the best possible performance/cost.

Broadcast interconnection has several drawbacks. One problem is its inefficiency for some point-to-point communication patterns, in which one PN talks with one other PN anywhere in the array. An example of such a pattern is the "perfect shuffle" used by the fast Fourier transform (FFT; Figure 10). This pattern takes n clocks on the CNAPS broadcast bus and is too slow to be effective. Consequently, CNAPS implements the compute-intensive discrete Fourier transform (DFT) instead of the communication-intensive FFT. The DFT requires $O(n^2)$ operations; the FFT, $O(n \log n)$. If $n \approx p$,

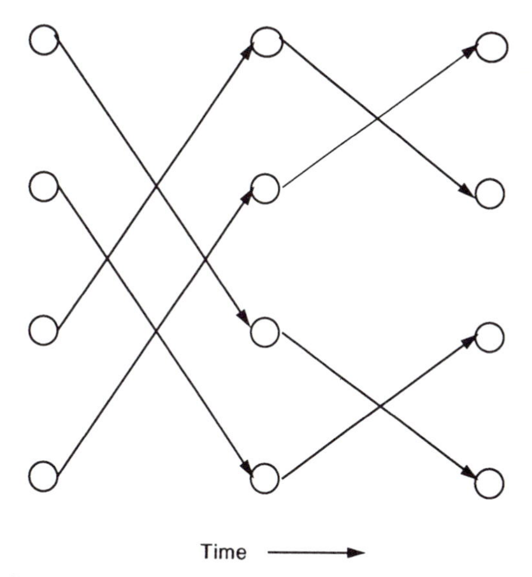

FIGURE 10 The intercommunication pattern of a fast Fourier transform (FFT) . A butterfly intercommunication pattern for four nodes. This pattern is difficult for CNAPS to do in less than N clocks (where N is the number of nodes) with broadcast intercommunication.

where p is the number of PNs, then CNAPS can perform a DFT in $O(n)$ clocks. If $n > p$, then performance can approach the $O(n\log n)$ of a sequential processor.

Another problem involves computation localized in a portion of an input vector, where each PN operates on a different (possibly overlapping) subset of the elements. Here, all PNs must wait for all inputs to be broadcast before any computation can begin. A common example of this situation is the limited receptive field structure, often found in image classification and character recognition networks. The convolution operation, also common in image processing, uses similar localized computation. The convolution can proceed rapidly after some portion of the image has been input into each PN, because each PN operates independently on its subset of the image.

When these subfields overlap (such as is in convolution), a PN must communicate with its neighbors. To improve performance for such cases, we added a one-dimensional inter-PN pathway, connecting each PN to its right and left neighbors. (One dimension was chosen over two to allow processor redundancy, discussed later). The CNAPS array therefore has both global (broadcast) and local (inter-PN) interconnection. An example of using the inter-PN pathway might be image processing, where a column of each image is allocated to each PN. The inter-PN pathway permits efficient communication between columns—and, consequently, efficient computation for most image-processing algorithms.

A final problem is sparse random interconnect, where each node connects to some random subset of other nodes. Broadcast, from the viewpoint of the *connected* PNs, is in this case efficient. Nonetheless, when a slotted protocol is used, many PNs are idle because they lack weights connected to the current input and do not need the data being broadcast. Sparse interconnect affects all aspects of the architecture, not just data communication. To improve efficiency for sparsely connected networks, the CNAPS PN offers a special memory technique called virtual zero, which saves memory locations that would otherwise be filled with zeros by not loading zeros into memory for unused connections. The Virtual Zero technique does not help the idle PN problem, however. Full efficiency with sparse interconnect requires a much more complex architecture, including more individualized control per PN, more complex memory-referencing capabilities, and so on, and is beyond the scope of this chapter.

On-Chip Memory

One of the most difficult decisions was whether to place the local memory on-chip inside the PN or off-chip. Both approaches have advantages and drawbacks—it was a complex decision with no obvious right answer and little opportunity for compromise.

The major advantage of off-chip memory is that it allows essentially unlimited memory per PN. Placing memory inside the PN, in contrast, limits the available memory because memory takes significant silicon area. Increasing PN size also limits the number of PNs. Another advantage to off-chip memory is that it allows the use of relatively low-cost commercial memory chips. On-chip memory, in contrast, increases the cost per bit—even if the memory employs a commercial memory cell.

The major advantage of on-chip memory is that it allows much higher bandwidth for memory access. To see that bandwidth is a crucial factor, consider the following analysis. Recall that each PN has its own data arithmetic units, therefore each PN requires a unique memory data stream. The CNAPS-1064 has 64 PNs, each potentially requiring up to 2 bytes per clock. At 25 MHz, that is 25M * 64 * 2 = 3.2 billion bytes/sec. Attaining 3.2 billion bytes/sec from off-chip memory is difficult and expensive because of the limits on the number of pins per chip and the data rate per pin.[7] An option would be to reduce the number of PNs per chip, eroding the benefit of maximum parallelism.

Another advantage to on-chip memory is that each PN can address different locations in memory in each

[7] For most implementations, the bit rate per pin is roughly equal to the clock rate, which can vary anywhere from 25 to 200 MHz. There are some special interface protocols which now allow up to 500 Mbits/sec per pin.

clock. Systems with off-chip memory, in contrast, typically require all PNs to address the same location for each memory reference to reduce the number of external output pins for memory addressing. With a shared address only a single set of address pins is required for an entire PN array. Allowing each PN to have unique memory addresses, requires a set of address pins for each PN, which is expensive. Yet, having each PN address its own local memory improves versatility and speed, because table lookup, string operations, and other kinds of "indirect" reference are possible.

Another advantage is that the total system is simpler. On-chip memory makes it possible to create a complete system with little more than one sequencer chip, one PN array chip, and some external RAM or ROM for the sequencer program. (Program memory needs less bandwidth than PN memory because SIMD machines access it serially, one instruction per clock.)

It is possible to place a cache in each PN, then use off-chip memory as a backing store, which attempts to gain the benefits of both on-chip and off-chip memory by using aspects of both designs. Our simulations on this point verified what most people who work in ANNs already suspected: Caching is ineffective for ANNs because of the nonlocality of the memory reference streams. Caches are effective if the processor repeatedly accesses a small set of memory locations, called a *working set*. Pattern recognition and signal processing programs rarely exhibit that kind of behavior; instead, they reference long, sequential vector arrays.

Separate PN memory addressing also reduces the benefit of caching. Unless all PNs refer to the same address, some PNs can have a cache miss and others not. If the probability of a cache miss is 10% per PN, then a 256-PN array will most likely have a cache miss every clock. But because of the synchronous SIMD control, all PNs must wait for the one or more PNs that miss the cache. This behavior renders the cache useless. A MIMD structure overcomes the problem, but increases system complexity and cost.

As this discussion suggests, local PN memory is a complex topic with no easy answers. Primarily because of the bandwidth needs and because we had access to a commercial density static RAM CMOS process, we decided to implement PN memory on chip, inside the PN. Each PN has 4 KB of static RAM in the current 1064 and 1016 chips.

CNAPS is the only architecture for ANN applications we are aware of that uses on-chip memory. Several designs have been proposed that use off-chip mem-ory. The CNS system being developed at Berkeley (Wawrzyneck, Asanovic, & Morgan, 1993), for instance, restricts the number of PNs to 16 per chip. It also uses a special high-speed PN-to-memory bus to achieve the necessary bandwidth. Another system, developed by Ramacher at Siemens (Ramacher et al., 1993) uses a special systolic pipeline that reduces the number of fetches required by forcing each memory fetch to be used several times. This organization is efficient at doing inner products, but has restricted flexibility. HNC has also created a SIMD array called the SNAP (Means & Lisenbee, 1991). It uses floating-point arithmetic, reducing the number of PNs on a chip to only four—in turn, reducing the bandwidth requirements.

The major problem with on-chip memory is its limited memory capacity. Although this limitation does restrict CNAPS applications somewhat, it has not been a major problem. With early applications, the performance/cost advantages of on-chip memory have been more important than the memory capacity limits.

Redundancy for Yield Improvement

During the manufacture of integrated circuits, small defects and other anomalies occur, causing some circuits to malfunction. These defects have a more or less random distribution on a silicon wafer. The larger the chip, the greater the probability that at least one defect will occur there during manufacturing. The number of good chips per wafer is called the *yield*. As chips get larger, fewer chips fit on a wafer and more have defects, therefore, yield drops off rapidly with size. Because wafer costs are fixed, cost per chip is directly related to the number of good chips per wafer. The result is that bigger chips cost more. On the other hand, bigger chips do more, and their ability to fit more function into a smaller system makes big chips worth more. Semiconductor engineers are constantly pushing the limits to maximize both function and yield at the same time.

One way to build larger chips and maximize yield is to use *redundancy*, where many copies of a circuit are built into the chip. After fabrication, defective circuits are switched out and replaced with a good copy. Memory designers have used redundancy for years; where extra memory words are fabricated on the chip and substituted for defective words. With redundancy, some defects can be tolerated and still yield a fully functional chip.

One advantage of building ANN silicon is that each PN can be simple and small. In the CNAPS processor array chip, the PNs are small enough to be effective as "units of redundancy." By fabricating spare PNs, we can significantly improve yield and reduce cost per PN. The 1064 has 80 PNs (in an 8×10 array), and the 1016 has 20 (4×5). Even with a relatively high defect density, the probability of at least 64 out of 80 (or 16 out of 20) PNs being fully functional is close to 1.0. CNAPS is the first commercial processor to make extensive use of such redundancy to reduce costs. Without redundancy, the processor array chips would have been smaller and less cost-effective. We estimate a CNAPS implementation using redundancy has about a two-times performance/cost advantage over one lacking redundancy.

Redundancy also influenced the decision to use limited-precision, fixed-point arithmetic. Our analyses showed that floating-point PNs would have been too large to leverage redundancy; hence, floating point would have been even more expensive than just the size difference (normally about a factor of four) indicates. Redundancy also influenced the decision to use one-dimensional inter-PN interconnect. One-dimensional interconnect makes it relatively easy to implement PN redundancy, because any 64 of the 80 PNs can be used. Two-dimensional interconnect complicates redundancy and was not essential for our applications. We chose one-dimensional interconnect, because it was adequate for our applications and does not impact the PN redundancy mechanisms.

Limitations

In retrospect, we are satisfied with the decisions made in designing the CNAPS architecture. We have no regrets about the major decisions such as the choices of digital, SIMD, limited fixed point, broadcast interconnect, and on-chip memory.

The architecture does have a few minor bottlenecks that will be alleviated in future versions. For example, the 8-bit input/output buses should be 16-bit. In line with that, a true one-clock 16×16 multiply is needed, as well as better support for rounding. And future versions will have higher frequencies and more on-chip memory. The one-dimensional inter-PN bus is 2 bits, it should be 16 bits. Despite these few limitations, the architecture has been successfully applied to several applications with excellent performance.

Product Realization and Software

Adaptive Solutions has created a complete development software package for CNAPS. It includes a library of important ANN algorithms and a C compiler with a library of commonly used functions.[8] Several board products are now available and sold to customers to use for ANN emulation, image and signal processing, and pattern recognition applications.

CNAPS APPLICATIONS

This section reviews several CNAPS applications. Because of the nature of this book its focus is on ANN applications, although CNAPS has also been used for non-ANN applications such as image processing. Some applications mix ANN and non-ANN techniques. For example, an application could preprocess and enhance an image via standard imaging algo-

[8] CNAPS-C is a data parallel version of the standard C language.

rithms, then use an ANN classifier on segments of the image, keeping all data inside the CNAPS array for all operations.[9] A discussion of the full range of CNAPS's capabilities is beyond the scope of this paper. For a detailed discussion of CNAPS in signal processing, see Skinner, 1994.

Back-Propagation

The most popular ANN algorithm is back-propagation (BP; Rumelhart & McClellan, 1986). Although it requires large computational resources during training, BP has several advantages that make it a valuable algorithm:

- it is reasonably generic, meaning that one network model (emulation program) can be applied to a wide range of applications with little or no modification;
- its nonlinear, multilayer architecture lets it solve complex problems;
- it is relatively easy to use and understand; and
- several commercial software vendors have excellent BP implementations.

It is estimated that more than 90% of the ANN applications in use today use BP or some variant of it. We therefore felt that it was important for CNAPS to execute BP efficiently. This section briefly discusses the general implementation of BP on CNAPS. For more detail, see McCartor (1991).

There are two CNAPS implementations of BP, a single-precision version (BP16) and a double-precision version (BP32). BP16 uses unsigned 8-bit input and output values and signed 16-bit weights. The activation function is a traditional sigmoid, implemented by table lookup. BP32 uses signed 16-bit input and output values and signed 32-bit weights. The activation function is a hyperbolic tangent implemented by table lookup for the upper 8 bits and by linear extrapolation for the lower 8 bits. All values are fixed point. We have found that BP16 is sufficient for all classification problems. BP16 has also been sufficient for most curve-fitting problems, such as function prediction, which have more stringent accuracy requirements. In those cases in which BP16 does not have the accuracy of floating point, BP32 is as accurate as floating point in all cases studied so far. The rest of this section focuses on the BP16 algorithm. It does not discuss the techniques involved in dealing with limited precision on CNAPS.

Back-propagation has two phases. The first is feed-forward operation, in which the network passes data without updating weights. The second is error back-propagation and weight update during training. Each phase will be discussed separately. This discussion assumes that the reader already has a working understanding of BP.

Back-Propagation: Feed-Forward Phase

Assume a simple CNAPS system with four PNs and a BP network with five inputs, four hidden nodes, and two output nodes (34 total connections, counting a separate bias parameter for each node; Figure 11).

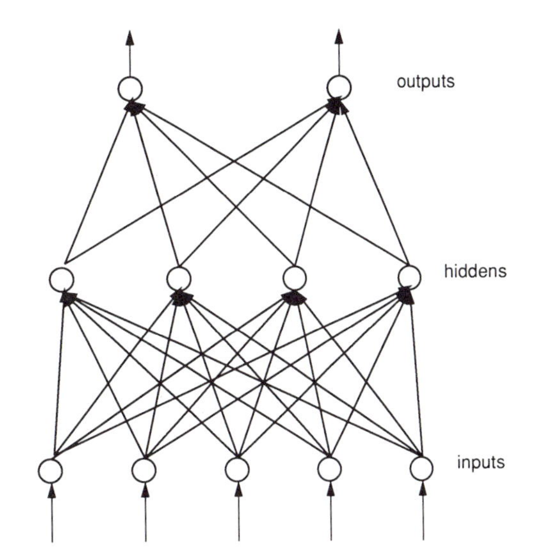

FIGURE 11 A back-propagation network with five inputs, four hidden nodes, and two output nodes.

[9] To change algorithms, the CSC need only branch to a different section of a program.

Allocate nodes 0 and 4 to PN0, nodes 1 and 5 to PN1, node 2 to PN2, and node 3 to PN3. When a node is allocated to a PN, the local memory of that PN is loaded with the weight values for each of the node's connections and with the lookup table for the sigmoid function. If learning is to be performed, then each connection requires a 2-byte weight plus 2 bytes to sum the weight deltas, and a 2-byte transpose weight (discussed below). This network then requires 204 bytes for connection information. Using momentum—ignored here for simplicity—would require more bytes per connection.

Each input vector contains five elements. The weight index notation is wA_{BC}, where A is the layer index, in our example, 0 for input, 1 for hidden, and 2 for output. B indexes the node in the layer, and C indexes the weight in the node. To start the emulation process, each element of the input vector is read from an external file by the CSC and broadcast over the Inbus to all four PNs. PN0 performs the multiply $v_0 * w1_{00}$; PN1, $v_0 * w1_{10}$; and so on. This happens in one clock. In the next clock, v_1 is broadcast, PN0 computes $v_1 * w1_{01}$; PN1, $v_1 * w1_{11}$; and so on. Meanwhile, the previous clock's products are sent to the adder, which initially contains zero.

All of the hidden-layer products have been generated after five clocks. One more clock is required to add the last product to the accumulating sum (ignoring the bias terms here for simplicity). Next, all PNs extract the most significant byte out of the product and use it as an address into the lookup table to get the sigmoid output. The read value then is put into the output buffer, and the PNs are ready to compute the output node outputs.

The next step is computing the output-layer node values (nodes 4 and 5). In the first clock, PN0 transmits its output (node 0's output) onto the output bus. This value goes through the CSC and comes out on the Inbus, where it is broadcast to all PNs. Although only PN0 and PN1 are used, all PNs compute values (PN2 and PN3 compute dummy values). PN0 and PN1 compute $n_0 * w2_{00}$ and $n_0 * w2_{01}$. In the next clock, node 1's value is broadcast and $n_1 * w2_{10}$ and $n_1 * w2_{11}$ are computed, and so on. After four clocks, PN0 and PN1 have computed all products. One more clock is needed for the last addition; then, a sigmoid table lookup is performed. Finally, the node 4 and 5 outputs are transmitted sequentially on the Outbus, and the CSC writes them into a file.

Let a *connection clock* be the time it takes to compute one connection. For standard BP, a connection requires a multiply-accumulate plus, depending on the architecture, a memory fetch of the next weight, the computation of that weight's address, and so on. For the CNAPS PN, a connection clock takes one cycle. On a commercial microprocessor chip, a connection clock can require one or more cycles, because many commercial chips cannot simultaneously execute all operations required to compute a connection clock: weight fetch, weight address increment, input element fetch, multiply, and accumulate. These operations can take up to 10 clocks on many microprocessors. Much of this overhead is memory fetching, because many state-of-the-art microprocessors are making more use of several levels of intermediate data caching. However, as discussed previously, ANNs are notorious cache busters, so many memory and input element fetches can take several clocks each.

Simulating a three-layer BP network with N_I inputs, N_H nodes in the hidden layer, and N_O nodes in the output layer will require $(N_I * N_H) * (N_H * N_O) + N_O$ connection clocks for nonlearning, feed-forward operation on a single-processor system. On CNAPS, assuming there are more PNs than hidden or output nodes, the same network will require $N_I + N_H + N_O$ connection clocks. For example, assume that $N_I = 256$, $N_H = 128$, and $N_O = 64$. For a single processor system, the total is 73,792 connection clocks; for CNAPS, 448. If a workstation takes about four cycles on average per connection, which is typical, to compute a connection, then CNAPS is about 600× faster on this network.

Back-Propagation: Learning Phase

The second and more complex aspect of BP learning is computing the weight delta for each connection. A detailed discussion of this computation and its CNAPS implementation is beyond the scope of this chapter, so only a brief overview is given here. The computation is more or less the same as a sequential implementation. The basic learning operation in BP is to compute an error signal for each node. The error signal is proportional to that node's contribution to the output error (the difference between the target output vector and the actual output error). From the error signal, a node can then compute how to update its weights. At the output layer, the error signal is the difference between the feed-forward output vector and the target output vector for that training vector. The output nodes can compute their error signals in parallel.

The next step is to compute the delta for each output node's input weight (the hidden-to-output weights). This computation can be done in parallel, with each node computing, sequentially, the deltas for all weights of the output node on this PN. If a batching algorithm is used, then the deltas are added to a data element associated with each weight. After several weight updates have been computed, the weights are updated according to an accumulated delta.

The next step is to compute the error signals for the hidden-layer nodes, which requires a multiply-accumulate of the output-node error signals through the output-node weights. Unfortunately, the output-layer weights are in the wrong place (on the output PNs) for computing the hidden-layer errors; that is, the hidden nodes need weights that are scattered among the output PNs, which can best be represented as a transpose of the weight matrix for that layer. In other words, a row of the forward weight matrix is allocated to each PN. When propagating the error back to the hidden layer, the inner product uses the column of the same matrix which is spread across PNs. A transpose of the weight matrix makes these columns into rows and allows efficient matrix-vector operations. A transpose operation is slow on CNAPS, taking $O(N^3)$ operations. The easiest solution was to maintain two weight matrices for each layer, the feed-forward version and a transposed version for the error back-propagation. This requires twice the weight memory for each hidden node, but permits error propagation to be parallel, not serial. Although the new weight value need only be computed once, it must be written to two places. This duplicate transpose weight matrix is required only if learning is to be performed.

After the hidden-layer error signals have been computed, the weight delta computation can proceed exactly as previously described. If more than one hidden layer is used, then the entire process is repeated for the second hidden layer. The input layer does not require the error signal.

For nonbatched weight update, in which the weights are updated after the presentation of each vector, the learning overhead requires about five times more cycles than feed-forward execution. A 256-PN (four-chip) system with all PNs busy can update about one billion connections per second, almost one thousand times faster than a Sparc2 workstation. A BP network that takes an hour on a Sparc2 takes only a few seconds on CNAPS.

Simple Image Processing

One major goal of CNAPS was flexibility because, by Amdahl's law, the more the problem can be parallelized the better; therefore, other parallelizable, but non-ANN, parts of the problem should also be moved to CNAPS where possible. Many imaging applications, including OCR programs, require image processing before turning the ANN classifier loose on the data. A common image-processing operation is convolution by spatial filtering.

Using spatial (pixel) filters to enhance an image requires more complex computations than simple pixel operations require. Convolution, for example, is a common operation performed during feature extraction to filter noise or define edges. Here, a kernel, an M by M dimensional matrix, is convolved over an image. In the following equation, for instance, the local kernel k is convolved over an N by N image a to produce a filtered N by N image b:

$$b_{i,j} = \sum_{p,q} k_{p,q} a_{i-p,j-q} \tag{2}$$

$$(i \leq i,j \leq N)(1 \leq p,q \leq M)$$

Typical convolution kernels are Gaussian, differences-of-Gaussian, and Laplacian filters. Because of their inherent parallelism, convolution algorithms can be easily mapped to the CNAPS architecture. The image to be filtered is divided into regions of "tiles," and each region is then subdivided into columns of pixel data. The CNAPS array processes the image one row at a time. Pixels from adjacent columns are transferred between neighboring PNs through the inter-PN bus. A series of $(M - 1)/2$ transfers in each direction is made so that each PN can store all the image data needed for the local calculation. Once the PN has in local memory all the pixels in the "support" for the convolution being computed, the kernel, k, is broadcast simultaneously to all PNs. This kernel can come from external data memory, or be sequentially from M PNs. The actual computation is just our familiar inner-product.

Because of the parallel structure of this algorithm, all PNs can calculate the convolution kernel at the same time, convolving all pixels in one row simultaneously. Using different kernels, this convolution process can be carried out several times, each time with a different type of spatial filtering performed on the image.

For a 512×512 image and 512 PNs (one column allocated per PN), a 3×3 kernel can be convolved over all pixels in 1.6 msec, assuming the image is already loaded. A 7×7 kernel requires 9.6 msec.

Naval Air Warfare Center

At the Naval Air Warfare Center (NAWC) at China Lake, California, ANN technology has been aimed at air-launched tactical missiles. Processing sensor information on board these missiles demands a computational density (operations per second per cubic inch) far above most commercial applications. Tactical missiles typically have several high-data-rate sensors, each with its own separate requirements for high-speed processing. The separate data must then be fused, and the physical operation of the missile controlled. All this must be done under millisecond or microsecond time constraints and in a volume of a few cubic inches. Available power is measured in tens of watts. Such immense demands have driven NAWC researchers toward ANN technology.

For some time (1986 to 1991), many believed that analog hardware was the only way to achieve the required computational density. The emergence of wafer scale, parallel digital processing (exemplified by the CNAPS chip) has changed that assessment, however. With this chip, we have crossed the threshold at which digital hardware—with all its attendant flexibility advantages—has the computational density needed to be useful in the tactical missile environment. Analog VLSI may still be the only way to overcome some of the most acute time-critical processing problems on board the missile, for example, at the front end of an image-processing system. A hybrid system combining the best of both types of chips may easily turn out to be the best solution.

Researchers at NAWC have worked with several versions of the CNAPS system. They have easily implemented cortico-morphic computational structures on this system—structures that were difficult or impossible under the analog constraints of previous systems. They have also worked with Adaptive Solutions to design and implement a multiple-controller CNAPS system (a multiple SIMD architecture or MSIMD) with high-speed, data-transfer paths between the subsystems, and they are completing the design and fabrication of a real-time system interfaced to actual missile hardware. The current iteration will be of the SIMD form, but the follow-on will have the new MSIMD structure.

Because of the nature of the work at NAWC, specific results cannot be discussed here. Some general ideas merit mention, however. Standard image-processing techniques typically only deal with spatial detail, examining a single frame of the image in discrete time. One advantage to the cortico-morphic techniques developed by NAWC is that they incorporate the temporal aspects of the signal into the classification process. In target tracking and recognition applications, temporal information is at least as important as spatial information. The cortico-morphic processing paradigm, as implemented on the CNAPS architecture, allows sequential processing of patches of data in real time, similar to the processing in the vertebrate retina and cortex.

One important near-term application of this computational structure is in the area of adaptive, nonuniformity compensation for staring focal plane arrays. It appears also that this structure will allow the implementation of three-dimensional wavelet transforms where the third dimension is time.

Lynch/Granger Pyriform Implementation

Researchers Gary Lynch and Richard Granger (Granger et al., this volume) at the University of California, Irvine, have produced an ANN model based on their studies of the pyriform cortex of the rat. The algorithm contains features abstracted from actual biological operations, and has been implemented on the CNAPS parallel computer (Means & Hammerstrom, 1991).

The algorithm contains both parallel and serial elements, and lends itself well to execution on CNAPS. Clusters of competing neurons, called *patches* or *subnets*, hierarchically classify inputs by first competing for the greatest activation within each patch, then subtracting the most prominent features from the input as it proceeds down the lateral olfactory tract (LOT, the primary input channel) to subsequent patches. Patch activation and competition occur in parallel in the CNAPS implementation. A renormalization function analogous to the automatic gain control performed in pyriform cortex also occurs in parallel across competing PNs in the CNAPS array.

Transmission of LOT input from patch to patch is an inherently serial element of the pyriform model, so opportunities for parallel execution for this part of the model are few. Nevertheless, overall speedups for execution on CNAPS (compared to execution on a serial machine) of 50 to 200 times are possible, depending on network dimensions.

Refinements of the pyriform model and applications of it to diverse pattern recognition applications continue.

Sharp Kanji

Another application that has successfully used ANNs and the CNAPS system is a *Kanji* optical character recognition (OCR) system developed by the Sharp Corporation of Japan. In OCR, a page of printed text is scanned to produce a bit pattern of the entire image. The OCR program's task is to convert the bit pattern of each character into a computer representation of the character. In the United States and Europe, the most common representation of Latin characters is the 8-bit ASCII code. In Japan, because of their unique writing system, it is the 16-bit JIS code.

The OCR system requires a complex set of image recognition operations. Many companies have found that ANNs are effective for OCR because ANNs are powerful classifiers. Many commercial OCR companies, such as Caere, Calera, Expervision, and Mimetics, use ANN classifiers as a part of their software.

Japanese OCR is much more difficult than English OCR because Japanese has a larger character set. Written Japanese has two basic alphabets. The first is *Kanji*, or pictorial characters borrowed from China. Japanese has tens of thousands of *Kanji* characters, although it is possible to manage reasonably well with about 3500 characters. Sharp chose these basic *Kanji* characters for their recognizer.

The second alphabet is *Kana*, composed of two phonetic alphabets (*hiragana* and *katakana*) having 53 characters each. Typical written Japanese mixes *Kanji* and *Kana*. Written Japanese also employs arabic numerals and Latin characters also found in business and newspaper writing. A commercial OCR system must be able to identify all four types of characters. To add further complexity, any character can appear in several different fonts.

Japanese keyboards are difficult to use, so a much smaller proportion of business documentation than one sees in the United States and other western countries is in a computer readable form. This difficulty creates a great demand for the ability to read accurately printed Japanese text and to convert it to the corresponding JIS code automatically. Unfortunately, because of the large alphabet, computer recognition of written Japanese is a daunting task. At the time this chapter is being written, the commercial market consists of slow (10–50 characters/sec), expensive (tens of thousands of dollars), and marginally accurate (96%) systems. Providing high speed and accuracy for a reasonable price would be a quantum leap in capability in the current market.

Sharp Corporation and Mitsubishi Electric Corporation have both built prototype Japanese recognition systems based on the CNAPS architecture. Both systems recognize a total of about 4000 characters in 15 or more different fonts at accuracies of more than 99% and speeds of several hundred characters per second. These applications have not yet been released as commercial products, but both companies have announced intentions to do so.

Sharp's system uses a hierarchical three-layer network (Hammerstrom, 1993; Togawa, Ueda, Aramaki, & Tanaka, 1991; Figures 12 and 13). Each layer is based on the Kohonen's Learning Vector Quantization (LVQ), a Bayesian approximation algorithm that shifts the node boundaries to maximize the number of correct classifications. In Sharp's system, unlike backpropagation, each hidden-layer node represents a character class, and some classes are assigned to several nodes. Ambiguous characters pass to the next layer. When any layer unambiguously classifies a character, it has been identified, and the system moves on to the next character.

The first two levels take as input a 16 × 16 pixel image (256 elements) (Figure 12). With some exceptions, these layers classify the character into multiple subcategories. The third level has a separate network per subcategory (Figure 13). It uses a high-resolution 32 × 32 pixel image (1024 elements), focusing on the subareas of the image known to have the greatest differences among characters belonging to the subcategory. These subareas of the image are trained to tolerate reasonable spatial shifting without sacrificing accuracy. Such shift tolerance is essential because of the differences among fonts and shifting during scanning.

Sharp's engineers clustered 3303 characters into 893 subcategories containing similar characters. The use of subcategories let Sharp build and train several small networks instead of one large network. Each small network took its input from several local receptive fields designed to look for particular features. The locations of these fields were chosen automatically during training to maximize discriminative information. The target features are applied to several positions within each receptive field, enhancing the shift tolerance of the field.

On a database of scanned characters that included more than 26 fonts, Sharp reported an accuracy of 99.92% on the 13 fonts used for training and 99.01% accuracy on characters on the 13 fonts used for testing. These results show the generalization capabilities of this network.

NON-CNAPS APPLICATIONS

This section discusses two applications that do not use CNAPS (although they could easily use the CNAPS BP implementation).

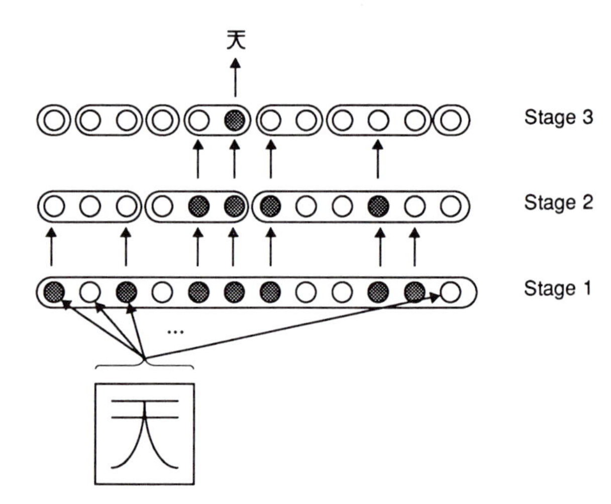

FIGURE 12 A schematicized version of the three-layer LVQ network that Sharp uses in their *Kanji* OCR system. The character is presented as a 16 × 16 or 256-element system. Some characters are recognized immediately; others are merely grouped with similar characters.

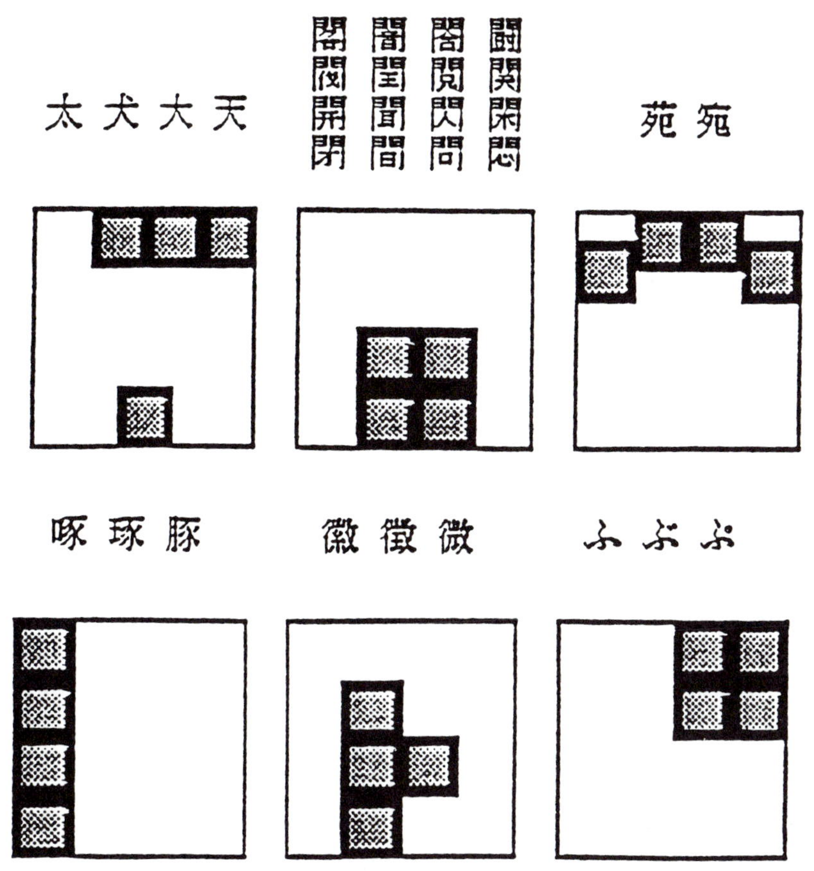

FIGURE 13 Distinguishing members of a group by focusing on a group-specific subfield. Here, a more detailed 32 × 32 image is used (Togawa et al., 1991).

Nippon Steel

ANNs are starting to make a difference in process control for manufacturing. In many commercial environments, controlling a complex process can be beyond the best adaptive control systems or rule-based expert systems. One reason for this is that many natural processes are strongly nonlinear. Most adaptive control theory, on the other hand, assumes linearity. Furthermore, many processes are so complex that there is no concise mathematical description of the process, just large amounts of data.

Working with such data is the province of ANNs, because they have been shown to extract, from data alone, accurate descriptions of highly complex, nonlinear processes. After the network describes the process, it can be used to help control it. Another technique is to use two networks, where one models the process to be controlled and the other the inverse control model. An *inverse network* takes as input the desired state and returns the control values that place the process in that state.

There are many examples of using ANNs for industrial process control. This section describes an application in the steel industry, developed jointly by Fujitsu Ltd., Kawasaki, and Nippon Steel, Kitakyushu-shi, Japan. The technique is more effective than any previous technique and has reduced costs by several million dollars a year.

This system controls a steel production process called *continuous casting*. In this process, molten steel is poured into one end of a special mold, where the molded surface hardens into a solid shell around the molten center. Then, the partially cooled steel is pulled out the other end of the mold. Everything works fine unless the solid shell breaks, spilling molten steel and halting the process. This "breakout" appears to be caused by abnormal temperature gradients in the mold, which develop when the shell tears inside the mold. The tear propagates down the mold toward a second opening. When the tear reaches the open end, a breakout occurs. Because a tear allows molten metal to touch the surface of the mold, an incipient breakout is a moving hot spot on the mold. Such tears can be spotted by strategically placing temperature sensing devices on the mold. Unfortunately, temperature fluctuation on the mold makes it difficult to find the hot spot associated with a tear. Fujitsu and Nippon Steel developed an ANN application that recognizes breakout almost perfectly. It has two sets of networks: the first set looks for certain hot spot shapes; the second, for motion. Both were developed using the back-propagation algorithm.

The first type of network is trained to find a particular temperature rise and fall between the input and output of the mold. Each sensor is sampled 10 times, providing 10 time-shifted inputs for each network forward pass. These networks identify potential breakout profiles. The second type of network is trained on adjacent pairs of mold input sensors. These data are sampled and shifted in six steps, providing six time-shifted inputs to each network. The output indicates whether adjacent sensors detect the breakout temperature profile. The final output is passed to the process-control software which, if breakout conditions are signalled, slows the rate of steel flow out of the mold.

Training was done on data from 34 events including nine breakouts. Testing was on another 27 events including two breakouts. The system worked perfectly, detecting breakouts 6.5 sec earlier than a previous control system developed at considerable expense. The new system has been in actual operation at Nippon Steel's Yawata works and has been almost 100% accurate.

Financial Analysis

ANNs can do nonlinear curve fitting on the basis of data points used to train the networks. This characteristic can be used to model natural or synthetic processes and then to control them by predicting future values or states. Manufacturing processes such as the steel manufacturing described earlier are excellent examples of such processes. Financial decisions also can benefit from modeling complex, nonlinear processes to predict future values.

Financial commodities markets—for example bonds, stocks, and currency exchange—can be viewed as complex processes. Granted, these processes are noisy and highly nonlinear. Making a profit by predicting currency exchange rates or the price of a stock does not require perfect accuracy, however. Accounting for all of the statistical variance is unneeded. What is needed is only doing better than other people or systems.

Researchers in mathematical modeling of financial transactions are finding that ANN models are powerful estimators of these processes. Their results are so good that most practitioners have become secretive about their work. It is therefore difficult to get accurate information about how much research is being done in this area, or about the quality of results. One academic group publishing some results is affiliated with the London Business School and University College London, where Professor A. N. Refenes (1993) has established the NeuroForecasting Centre. The Centre has attracted more than £1.2 million in funding from the British Department of Trade and Industry, CitiCorp, Barclays-BZW, the Mars Corp., and several pension funds.

Under Professor Refenes's direction, several ANN-based financial decision systems have been created for computer-assisted trading in foreign exchange, stock and bond valuation, commodity price prediction, and global capital markets. These systems have shown better performance than traditional automatic systems. One network, trained to select trading strategies, earned an average annual profit of 18%. A traditional system earned only 12.3%.

As with all ANN systems, the more you know about the environment you are modeling, the simpler the network, and the better it will perform. One system developed at the NeuroForecasting Centre models international bond markets to predict when capital should be allocated between bonds and cash. The system models seven countries, with one network for each (Figure 14). Each network predicts the bond returns for that country one month ahead. All seven predictions for each month are then presented to a software-based portfolio management system. This system allocates capital to the markets with the best predicted results—simultaneously minimizing risk.

Each country network was trained with historical bond market data for that country between the years 1974 and 1988. The inputs are four to eight parameters, such as oil prices, interest rates, precious metal prices, and so on. Network output is the bond return for the next month. According to Refenes, this system returned 125% between 1989 and 1992; a more conventional system earned only 34%. This improvement represents a significant return in the financial domain. This system has actually been used to trade a real investment of $10 million, earning 2.4% above a standard benchmark in November and December of that year.

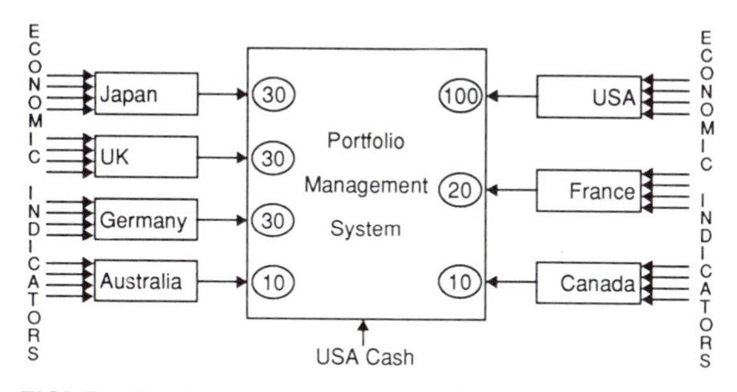

FIGURE 14 A simple neural network-based financial analyzer. This network consists of seven simple subnetworks, each trained to predict bond futures in its respective market. An allocation expert system is used to allocate a fixed amount of cash to each market (Refenes, 1993).

CONCLUSION

This chapter has given only a brief view into the CNAPS product and into the decisions made during its design. It has also briefly examined some real applications that use this product. The reader should have a better idea about why the various design decisions were made during this process and the final outcome of this effort. The CNAPS system has achieved its goals in speed and performance and, as discussed, is finding its way into real world applications.

Acknowledgments

I would like to acknowledge, first and foremost, Adaptive Solutions' investors for their foresight and patience in financing the development of the CNAPS system. They are the unsung heros of this entire effort. I would also like to acknowledge the following people and their contributions to the chapter: Dr. Dave Andes of the Naval Air Warfare Center, and Eric Means and Steve Pendleton of Adaptive Solutions.

The Office of Naval Research funded the development of the implementation of the Lynch/Granger model on the CNAPS system through Contracts No. N00014 88 K 0329 and No. N00014-90-J-1349.

References

Akers, L. A., Haghighi, S., & Rao, A. (1990). VLSI implementations of sensory processing systems. In *Proceedings of the Neural Networks for Sensory and Motor Systems (NSMS) Workshop*, March.

Alspector, J. (1991). Experimental evaluation of learning in a neural microsystem. In *Advances in Neural Information Processing Systems III*. San Mateo, CA: Morgan Kaufman.

Baker, T., & Hammerstrom, D. (1989). Characterization of artificial neural network algorithms. In *1989 International IEEE Symposium on Circuits and Systems*, pp. 78–81, September.

Griffin, M., et al. (1990). An 11 million transistor digital neural network execution engine. In *The IEEE International Solid State Circuits Conference*.

Graf, H. P., Jackel, L. D., & Hubbard, W. E. (1988). VLSI implementation of a neural network model. *IEEE Computer*, **21**(3), 41–49.

Hammerstrom, D. (1990). A VLSI architecture for high-performance, low-cost, on-chip learning. In *Proceedings of the IJCNN*.

Hammerstrom, D. (1991). A highly parallel digital architecture for neural network emulation. In J. G. Delgado-Frias & W. R. Moore

(Eds.), *VLSI for artificial intelligence and neural networks*. New York: Plenum Press.

Hammerstrom, D. (1993a). Neural networks at work. In *IEEE Spectrum*, pp. 26–322, June.

Hammerstrom, D. (1993b). Working with neural networks. In *IEEE Spectrum*, pp. 46–53, July.

Hoehfeld, M., & Fahlman, S. E. (1992). Learning with limited numerical precision using the cascade-correlation algorithm. *IEEE Transactions on Neural Networks*, **3**(4), July.

Holler, M., Tam, S., Castro, H., & Benson, R. (1989). An electrically trainable artificial neural network (eTANN) with 10,240 floating gate synapses. In *Proceedings of the IJCNN*.

McCartor, H. (1991). Back-propagation implementations on the adaptive solutions neurocomputer chip. In *Advances in neural information processing systems II*, Denver, CO. San Mateo, CA: Morgan Kaufman.

Mead, C. (1989). Analog VLSI and Neural Systems. New York: Addison-Wesley.

Means, E., & Hammerstrom, D. (1991). Piriform model execution on a neurocomputer. In *Proceedings of the IJCNN*.

Means, R. W., & Lisenbee, L. (1991). Extensible linear floating point simd neurocomputer array processor. In *Proceedings of the IJCNN*.

Morgan, N. Ed. (1990). *Artificial neural networks: Electronic implementations*. Computer Society Press Technology Series and Computer Society Press of the IEEE, Washington, DC.

Ramacher, U. (1993). Multiprocessor and memory architecture of the neurocomputer synapse-1. In *Proceedings of the World Congress of Neural Networks*.

Refenes, A. N. (1993). Financial modeling using neural networks. *Commercial applications of parallel computing*. UNICOM.

Rumelhart, D., & McClellan, J. (1986). *Parallel distributed processing*. New York: MIT Press.

Sejnowski, T., & Rosenberg, C. (1986). NetTalk: A parallel network that learns to read aloud. *Technical Report JHU/EECS-86/01*, The Johns Hopkins University Electrical Engineering and Computer Science Department.

Shoemaker, P. A., Carlin, M. J., & Shimabukuro, R. L. (1990). Back-propagation learning with coarse quantization of weight updates. In *Proceedings of the IJCNN*.

Skinner, T., (1994). Harness multiprocessing power for DSP systems. In *Electronic Design*, February 7.

Togawa, F., Ueda, T., Aramaki, T., & Tanaka, A. (1991). Receptive field neural networks with shift tolerant capability for *kanji* character recognition. In *Proceedings of the International Joint Conference on Neural Networks*, June.

Wawrzyneck, J., Asanovic, K., & Morgan, N. (1993). The design of a neuro-microprocessor. *IEEE Transactions on Neural Networks*, **4**(3), May.

18

Synthetic Neural Systems in the 1990s

Lex A. Akers, David K. Ferry, and Robert O. Grondin

INTRODUCTION

Motivation for Electronic Implementation of Neural Systems

The 1990s will see an explosive growth of systems incorporating neural networks. Hardware implementations of neuromorphic systems are mandatory for many real-time applications such as vision and speech recognition, robotics, and numerous other interactive control and signal processing applications. Software simulation, by its very nature of instruction steps and clock cycles, is much slower than the propagation delay through a circuit. Also, the highly parallel system of equations used to simulate the neural models are usually solved on serial computers, further slowing their execution time when compared with direct hardware implementation. For example, computer run times for simulation of training have been reported in terms of days and weeks (J. Wang & L. Clark, personal communication). The progression to supercomputers will not significantly reduce this time because the host computer's architecture is so ill-suited to the problem being simulated. On the other hand, hardware implementations can be devised that incorporate the speed of both physically generating the system equations and utilizing parallel architectures. Compared with simulations, processing speed enhancements of over three orders of magnitude have been reported for custom hardware systems (Mackie, Graf, & Schwartz, 1988). The speed advantage of the hardware approach will continue to expand as the systems are scaled to the number of processing elements found in biological systems.

Why is speed so important in neural network processing? A biological system can respond, for example, to visual information within a few hundred milliseconds. This response time has obvious advantages for survival, and allows adaptive behavior in a dynamic environment. For interacting with humans, large-scale synthetic neural systems (SNS) must also respond in similar time constants to be useful. Therefore, only hardware-implemented systems offer the speed, as well as size and cost, necessary to make SNS widely available and interactive.

Approaches for Electronic Implementations

Hardware implementations span a wide range of approaches. At one end of the spectrum are hardware accelerators that enhance the processing speed of the host computer. The hardware for this method is easy to install and use, relatively inexpensive, and offers two to three orders of magnitude of enhanced processing speed. The other end of the spectrum is the design and use of custom Very Large Scale Integrated (VLSI) chips. This approach contributes an additional three to four orders of magnitude of enhanced processing speed over accelerator boards. These VLSI systems have low power consumption, are small in size, and are inexpensive in large volumes. Only the VLSI chips offer the potential of matching the processing speed, high density, and low power consumption of complex biological systems.

Two disparate groups of workers are presently engaged in VLSI chip implementations of neural networks. The first is committed to electronic-based implementations of neural networks, using standard or custom VLSI chips. The driving force for this group is to implement the neural system equations, not to design novel integrated circuits. The electronic systems are only a vehicle to obtain the desired system behavior and through-put. The second group desires to build fault-tolerant, adaptive VLSI chips, and is much less concerned with whether the design rigorously duplicates the neural models. This group is composed of VLSI designers who realize the opportunities of obtaining biological behavior in hardware. They see numerous problems for VLSI designers on the horizon, and recognize that nature has found solutions to many of these problems. As will be discussed later, VLSI technology currently lets us have the capability of fabricating systems that border on biological complexity. The central problem in the construction of a VLSI neural network is that the design constraints of VLSI differ from those of biology (Walker & Akers, 1988a). In particular, the high fan-in/fan-out of biology imposes connectivity requirements such that the VLSI chip implementation of a highly interconnected neural network of just a few thousand neurons would require a level of connectivity which exceeds the current or even projected interconnection density of Ultra Large Scale Integrated (ULSI) systems (Ferry, Akers, & Greeneich, 1988). Fortunately, highly layered, limited interconnect networks can be designed that are functionally similar to highly connected systems (Akers, Walker, Ferry, & Grondin, 1988).

Coupling in Large Scale Systems

There is another reason for studying SNS. In conventional descriptions of VLSI circuits, each switching device is assumed to behave in the same manner within the total system as it does when it is isolated. The full function of the system or integrated circuit (IC) is determined solely by the interconnections used to join the individual devices together. A different function can only be assigned to the system by redesigning the interconnections—a practical impossibility for most systems, but in practice accomplished in some programmable systolic arrays. The conventional clear separation of device design from system design thus depends on being able to isolate each individual device from the environment of the other devices, except for the planned effects occurring through the interconnection network. This simplification is likely to be seriously in error for submicron configured systems, in which the devices are packed much closer together. As a result of the much closer packing, the isolation of one device from another will be difficult to achieve. Instead, one must begin to think about methods of using the interaction between devices as a tool to accomplish distributed information processing within the VLSI chip. This approach is exactly the approach used by neural networks.

Several methods of modeling interdevice coupling at the chip level are possible. One of the most promising involves modeling the chip as a cellular array and employing cellular automata to describe its behavior (Preston & Duff, 1984). In a cellular array, each element is bidirectionally connected to all of its nearest neighbors. The "state" or status of each element therefore depends on the state of its neighbors. When an input stimulus is applied to the array, the states of all the affected elements evolve in a highly coupled fashion until a stable array state is reached. The response of a cellular automaton to a given input can be predicted, and cellular arrays using nearest-neighbor coupling have been shown to behave as an associative memory, similar to neural networks.The basic structural difference between cellular automata and neural networks is that neural networks are heavily interconnected over long distances. Obviously, these long interconnects pose serious problems for the layout on a chip. A much more appropriate basic architecture would be one that is based on interconnections only between nearest or second-nearest neighbors.

Applications of Neural Networks to Intelligent Processing

In semiconductor processing, there is a move to introduce automation of each of the various processing modules. Efforts at total automation, in the sense of introducing intelligent control, have met with failure because of the overall complexity of the process and the lack of adequate process models for each step in the process. Efforts at introducing intelligent processing at the module level have similarly met with limited success, but for somewhat different reasons. In general, approaches have been centered on the idea of developing a computer-based, intelligent (or "smart") system. As with the development of expert systems in other areas of manufacturing, this approach requires the joint efforts of a processing expert and a computer artificial intelligence (AI) expert, as the approach requires the development of a computer-based set of "rules" for carrying out the processing. Along the way, an adequate process model for comparing the process parameters *in situ* is needed as well. This approach has not worked well in the semiconductor industry for the simple reason that processing modules are relatively rapidly evolving systems. Because of the relatively long development time required for expert systems, it is the experience of the semiconductor industry that by the time the expert system is developed and operational, the particular process module is obsolete.

To incorporate effectively intelligent manufacturing concepts within the semiconductor processing arena, it is necessary to break the limitation of the long development time required by expert systems. The synthetic neural network is optimally positioned to do just this. First, the SNS is a learning system, which does not require the AI expert, but may be trained by the processing module engineers because the SNS learns from comparison of actual output data with desired output data for a given input reference set. Second, its learning may be an evolutionary process, because it updates the knowledge database stored within its effective content-addressable memory. Thus, as the process evolves, the knowledge database is evolving at the same time because it is continually learning new data. Third, its learning time is quite short when compared with the development time of an expert system. Finally, the computational time is considerably shorter than that required to conduct a tree-based data search to find the appropriate rules in the expert system. Thus, the SNS promises to provide a more powerful, and useful, intelligent processor for the module than can be achieved by an expert system.

The inherent ability to segment the SNS into layers is also important, because different data sets and model parameters may be introduced as input to the SNS at different levels. Thus, the SNS has the intrinsic ability to assign different importance weighting to different input sets. The trend to build SNS within natural VLSI architectures promises to enhance the advantages of this approach over that of the expert system. Rather than adding necessary memory for storing the rule base of the expert system (and which must be searched in an ever-increasing, time-consuming fashion), the SNS can be added as a processing board within a small computer environment. This speeds up the overall process by introducing effective parallelism into the intelligent processing environment, exactly the opposite effect of expanding the database in the expert system.

Summary

With the motivation for the electronic implementation in VLSI of neural systems articulated, we will next describe the opportunities and constraints imposed by VLSI technology. Electronic neural systems must conform to the limitations of the fabricating technology. This discussion, as well as considering input and output pin restrictions, leads to the conclusion that highly layered, limited interconnect architectures are ideally suited for VLSI implementation. A training methodology for these architectures will also be described and two analog–digital circuits will be discussed. Last, our vision of the future of electronic implementation will be explored.

VLSI TECHNOLOGY

A brief historical review of VLSI technology is useful for two reasons. First, an understanding of the evolution of the technology will illustrate why at this phase of its growth the opportunities for implementing large scale complex neural systems exist. During this discussion, it will become apparent why VLSI designers are excited about the prospects of acquiring various biological characteristics in their systems. Second, the constraints of VLSI technology will be described. These constraints have direct impact on the development of synthetic neural chips. New neural system architectures will need to be developed to allow the production of VLSI chips with large numbers of neurons. For the reader already familiar with this technology, this section may be skipped.

The Semiconductor Revolution

The so-called semiconductor revolution has really just been an evolution. Fabrication techniques have steadily improved and design rules have shrunk, all in a very fast but predictable pattern. The real semiconductor revolution occurred around 1960 when Noyce and Kilby independently developed techniques to interconnect a few semiconductor devices on a die. This was later termed Small Scale Integration (SSI) and resulted in the implementation of basic digital gates such as NAND and NOR. A reduction of the size of components on a die, allowing more to be included, resulted in Medium Scale Integration (MSI), which in turn allowed counters, clocks, and flip-flops to be implemented on one die. At this point, the tremendous advantage of scaling became apparent. The cost of the die is only loosely tied to the number of components on the die. Hence, continued scaling could offer powerful systems on a die at a much lower cost than conventional multichip systems (Ferry et al., 1988). The technology quickly jumped to Large Scale Integration (LSI) and was used to make the very popular handheld calculators. The continuing reduction in component size, along with process improvements, allowed the development of VLSI and the making of 64K DRAMs. We are currently at a ULSI level of integration with over 16 million devices on a die. A look at the short history of this technology shows that the number of components on a die has doubled approximately every 2 years, resulting in an exponential growth. Figure 1 illustrates this growth for the last 25 years. This exponential growth has continued, despite many predictions of its imminent falloff. We presently have in production microprocessor chips with more than 8 million devices and memory chips with more than 16 million devices, and chips having more than 64 million devices have been built and are being prepared for introduction into production by 1995.

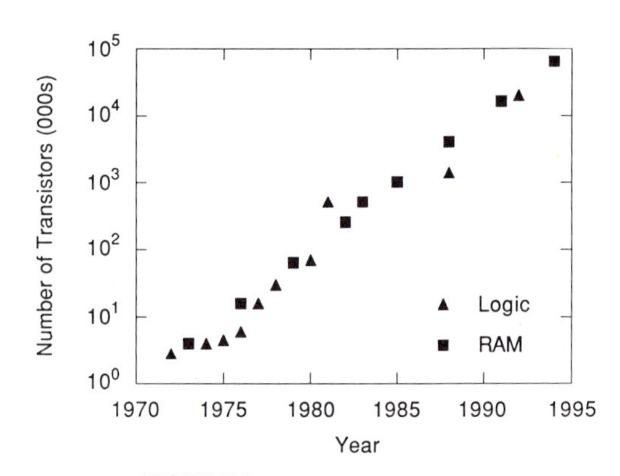

FIGURE 1 Device count per year.

Opportunities with VLSI

This increase in devices has allowed the direct electronic implementation of complex systems on a chip. The increase in device count has occurred for three reasons. (1) As discussed earlier, the size of each device has been reduced by improvements in fabrication and lithographic methods. (2) The chip area has increased. Figure 2 illustrates this increase in chip area (Myers, Yu, & House, 1986). In practice, chip size has gone from being barely visible to an inch on a side. (3) Circuit cleverness has reduced the number of components needed in a design. For example, the storage cell for a memory traditionally required six transistors. Innovative circuit design techniques have reduced this requirement to only one transistor, a tremendous saving of space when multiplied by millions. To demonstrate this, Figure 3 shows the space occupied by 1 million bits (1 MB; Asai, 1986). Notice in Figure 3 that the density of neurons in the human brain is located near the 16 MB chip.

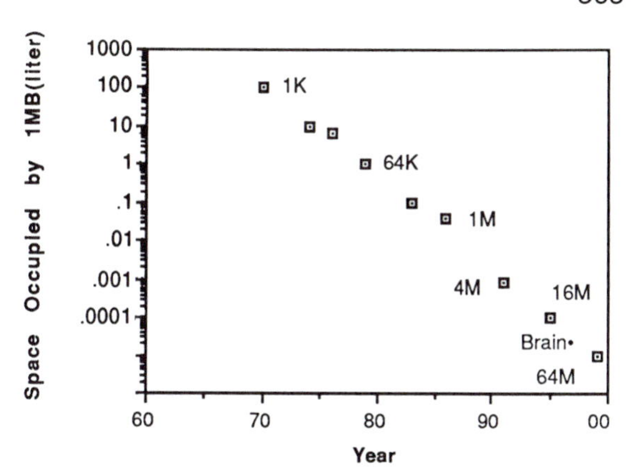

FIGURE 3 Space required to store 1 million bits.

VLSI Constraints

As device density has increased and cost per device has decreased exponentially, problems in the technology have also arisen. Prediction of critical problems is necessary to allow time for alternatives or solutions to be developed. In ULSI, several major problems are predicted. First, yields must be high—reduced yield means a vastly increased cost per chip. If yield cannot be maintained and improved, the cost of a chip increases; and instead of a few dollars per chip, costs of several thousands of dollars per chip are quite likely in low-yield production. Next, design time is a serious problem. The next generation of a product has to be available less than 2 years after initial chip introduction. To reduce the length of design time, standard cells and interactive graphics systems have been introduced. Still, as shown in Figure 4, the cost of a design has rapidly increased over the years. As systems are made larger, they still have to be tested. The testing time is directly related to the number of states in a system, which rapidly increases as the number of devices is increased. Hence, to test completely many of today's systems requires a time in excess of 10^{10} years, which is clearly impossible. Hence, designs are only partially simulated and tested. One additional limitation in ULSI is that the number of levels of intercon-

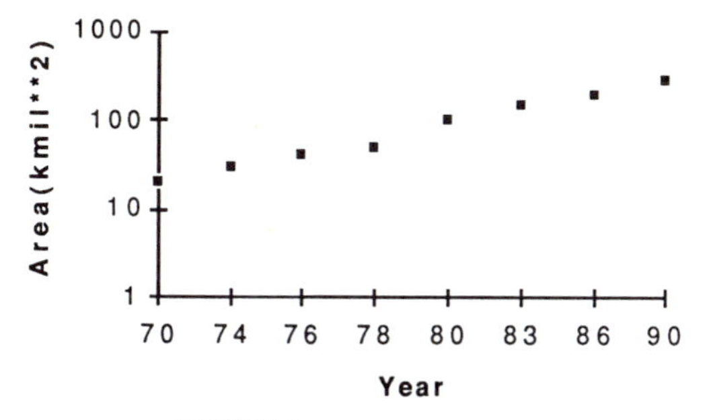

FIGURE 2 Chip area per year.

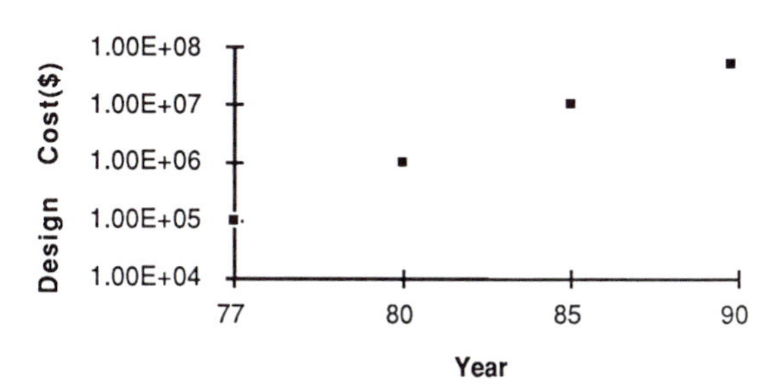

FIGURE 4 Cost of a design.

nects has grown only slowly in the last 33 years. Levels of interconnects have increased only from two to five. This very slow increase in interconnection density is not expected to change. This has important consequences for hardware implementations of SNS and will be further discussed later.

The introduction of fault tolerance in VLSI and ULSI systems is a means of achieving both a specified system reliability and acceptable production yields. In memory circuits, fault tolerant techniques such as including spare rows and columns have been widely used. For microprocessors and controllers, traditional approaches such as incorporating error detection and correction codes, and allowing self-test and recovery, have been adopted. Also, the use of redundant sections of the circuit, with removal of the faulty sections at circuit test, offers additional fault-tolerant operation. New techniques that use the special properties of a function block, such as an arithmetic logic unit (ALU), to provide error detection and correction have been tried, but this approach does not easily extend to other functional blocks. Even if functional level fault tolerance is achieved, the coordination of the different approaches to achieve system level fault-tolerant operation is difficult. These techniques also require additional die area and must be designed into the circuit. What is needed is an architecture that incorporates fault-tolerant behavior as a natural characteristic of the circuit. Neural inspired architectures offer this property.

Neural inspired architectures do not store data in fixed, specified locations, but rather in a distributed fashion, as a small variation in the impedances of interconnections spread over a large area. The loss of an interconnection or processing element in most cases will have no effect on system behavior. This natural robustness to individual device and interconnect failure provides a graceful degradation of system performance as the number of failed devices increases. Instead of the system exhibiting catastrophic failure, one observes reduced associations and accuracy. This natural robustness potentially offers high chip yields without the need for error detection, correcting codes, and spare sections of the circuit.

Coherency in VLSI

What we really need, if we are to make a quantum jump forward in the capabilities of microchips, is to develop a coherent, parallel type of processing that provides robustness and which is not sensitive to failure of a few individual gates. The problem with using arrays of devices, highly integrated within a chip and coupled to each other, is not one of making the arrays, but is one of introducing the hierarchical control structure necessary to implement fully the various necessary system or computer algorithms. In other words, how are the interactions between the devices orchestrated so as to map a desired architecture onto the array itself? We have suggested that these arrays could be considered as local cellular automata (Ferry, Grondin, & Porod, 1987), but this does not alleviate the problem of global control, which must change the local computational rules in order to implement a general algorithm. Huberman and Hogg (1984) have studied the nature of attractors on finite sets in the context of iterative arrays, and have shown in a simple example how several inputs can be mapped onto the same output. The ability to change the function during processing allowed them to demonstrate adaptive behavior in which dynamical associations are made between different inputs that initially produced sharply distinct outputs. However, these remain only the initial small steps toward the required design theory to map algorithms into network architecture.

LIMITED INTERCONNECT ARCHITECTURES

From the previous discussion of VLSI technology, it was shown that highly interconnected architectures are not well supported. As the number of synthetic neurons are scaled to large numbers, fully connected architectures will require significantly larger numbers of input/output (I/O) pins. Large, fully connected SNS could require tens of thousands of I/Os, whereas packages with hundreds of I/Os are difficult to build. In this section, we discuss how the proper choice of the architecture can reduce this explosive growth in I/O. We also demonstrate that highly layered, limited interconnect architectures can be functionally similar to fully connected architectures. We also discuss the training of highly layered, limited interconnect architectures.

Rent's Rule and Information Flow

It is often suggested that large scale integration of semiconductor devices, such as the integration of large scale neural networks, will eventually entail a significantly large number of interconnection pin-outs at the periphery of (or distributed throughout) the chip. This fact has led to a number of studies which invariably give the results in terms of a relationship that has become known as *Rent's rule*. Empirically, this relationship gives the number of pins for a given size module in terms of a power law dependent on the number of gates (or the effective number of functional blocks) in the module. The validity of this relationship is based on a number of studies of possible interconnections of present gate arrays, or of experiments actually carried out on master slice-type chips.

The considerations that lead to ideas such as Rent's rule arise from topographical details of the implementation of a given architecture in an integrated system. In the earliest form, the problem is one of partitioning the system graph into appropriate modules, each of which would probably be a separate chip in early configurations. The partitioning problem is one of assigning the required logic blocks in such a manner so that the total numbers of pins and interconnection wires can be simultaneously minimized. In this context, a logic block is some arbitrary (and usually not specified) primitive function. These blocks are then interconnected by the system graph, called a *block graph* or *net*. In early machines, these modules would be composed of at most a few blocks. The choice of assigning blocks and nodes into modules is the partitioning problem. Because each possible assignment of blocks into modules specifies the necessary module–module interconnections, the total number of pins and interconnections is then dependent on the details of the partition selected. In fact, however, the beauty of Rent's rule is that there is (perhaps) a general rule that buries the details and concentrates on the more universal aspects. Rent's rule has been formulated as

$$P = KG^{p} \tag{1}$$

where P is the number of pins, G is the number of gates (or blocks in the early work), K is the number of pins per block, and p is a general exponent. Values of the parameters K and p have been found for early IBM machines, for RCA machines, and for others. One particularly useful study was carried out by Landman and Russo (1970) and another by Chiba (1975). They considered graphs of 670–12,700 logic circuits and blocks ranging from a single NOR gate to 30 circuits. Rent's rule was also found to be obeyed, and the parameters varied over the range 3–5 for K and 0.6–0.7 for p. Although the requirement on pins and interconnections is significant for predicting growth in future machines, Rent's rule is also significant in that the exponent p is very important in predicting what the average interconnection length within the chip will be. Chips with an exponent $p > 0.5$, such as fully connected neural networks, generally are found to have long interconnection lengths as a rule, whereas chips with an exponent $p < 0.5$ have an interconnection length (in terms of circuit pitches) that is independent of the overall number of gates.

It has generally been the experience in VLSI that the number of pins is much smaller than expected in Rent's rule, at least in terms of the exponents found in gate arrays. This trend has been studied for the reported number of pins for microprocessor chips and for functional chips (other than gate arrays) of both GaAs and Si. In addition to microprocessors, these chips include memories, multipliers, and multiplexers. From this study, the best fit to the pin requirements of highly integrated circuits is found to be (Ferry, 1985)

$$P = 7G^{0.21} \tag{2}$$

which differs appreciably from the earlier forms. These circuits are *functionally partitioned* circuits. The far fewer number of pins required for such highly integrated circuitry suggests that future pin-intensive circuits, such as supercomputers, will have to be designed to be functionally partitioned. Synthetic neural systems will also have to be designed with this style of architecture.

The pin-to-gate ratio can be interpreted directly in terms of surface-to-volume considerations. Consider a system of functional cells, which are laid on an d-dimensional grid ($d = 1, 2, 3$). Those cells that lie on the interior of the system are gates and those that lie on the periphery are pins (this ignores the scaling difference that arises from the fact that pin pads are larger than gate areas). If those cells are all of the same characteristic length, the edge of a cube in $d = 3$, the sides of a square in $d = 2$, and a unit length in $d = 1$, we can easily establish a Rent's rule for each case. For example, in a square array ($d = 2$) of cells, with N^2 blocks, there are $4N - 4$ edge blocks, and $N^2 - 4(N - 1)$ interior blocks, and for large N,

$$P = 4G^{1/2} \tag{3}$$

Similarly, for $d = 3$,

$$P = 6G^{2/3} \tag{4}$$

and for $d = 1$,

$$P = 2G^0 \tag{5}$$

These considerations suggest that we interpret Rent's rule as giving us the information flow dimension of the integrated circuit chip. This is certainly in keeping with the concept in microprocessors, which are generally von Neumann architectures incorporating a one-dimensional information flow. Another aspect of this is that we are trying to planarize the interconnections. If we have a dimension $d > 2$, as in fully connected neural networks, we cannot really use a planar graph, and the chip will have a much larger number of wire crossings because of the long average interconnect length. This is clearly a detrimental factor in the design of large scale neural networks, and stresses the need for short, local interconnects.

As mentioned earlier, the exponent p also affects the average length of an interconnection on the chip. If $p > 0.5$, the average interconnection length is a function of the number of pins and gates on the chip. This means that any device must drive, *on the average*, a longer wire as device dimensions are scaled downward. On the other hand, if $p < 0.5$, the average interconnection length is independent of the number of gates, and is fixed at approximately 2.8 circuit pitches (Ferry, 1985). This has an extra benefit. If the average interconnection length scales with the device dimensions, as in this latter case, then the speed and power improvements that arise from downward scaling device size will appear directly as enhancements in system clock speed. On the other hand, if the average interconnection length increases with the number of gates, then there is a loss of improvement from size reductions, and the system clock speed cannot be increased proportionately with the downsizing of individual devices. This is probably why improvements in clock speed in large computers has not kept pace with the improvements in microprocessor-based workstations.

A Computational Model

Can a network with a reduced interconnection scheme perform in a functionally similar manner to a neural network? By *functionally similar*, we mean more than just compute. The network must also have similar relaxation behavior, fault tolerance, robustness, and, as a precondition, must be trainable. We approach the general idea that neural networks and limited interconnect architectures, like cellular automata, are isomorphic by discussing general computation itself. Consider an array of N logic elements, which are supposed to represent all of the central processor and part of the memory. We can describe the instantaneous state of the system by one of two possible descriptors: (1) a vector of length N whose elements are 0 or 1, corresponding to the state of the appropriate logic element, or (2) a vector of length 2^N whose elements are all 0 except for a single element whose value is 1. This latter description is the one we shall choose, and the single entry identifies which of the possible 2^N combinations is the proper state of the system. This is analogous to the pure-state description of quantum mechanics in which the density matrix has a single entry corresponding to the pure state. We call this state vector $\mathbf{X}$, and its value at time step n of the system's evolution is $\mathbf{X}_n$. As the system is clocked, $\mathbf{X}_n$ evolves according to the iteration

$$\mathbf{X}_{n+1} = \mathbf{M}\mathbf{X}_n \tag{6}$$

where **M** is the state transition matrix.

The formulation of Equation (6) is not completely abstract, however. This formulation is exactly that described by Peretto and Niez (1986) in their investigation of stochastic dynamics of neural networks. In particular, we want to relax our requirement that **X** have only a single entry, but describe it as a mixed-state representation in which each entry is the probability of that state existing. We then immediately recognize that the elements of **M** are transition probabilities (Huberman & Hogg, 1984), and we can then use **M** for discussing the statistical mechanics of the temporal evolution of neural networks. More importantly, each element of **M** is the transition probability between state i and state j that results from the switching of a single logic element. Thus, the structure and entries of **M** allow us to formulate (in principle) a network architecture that achieves the desired **M**.

It is also important to note that the form of the entries is

$$M_{ji} = W(j|i) = \frac{1}{2N}[1 - f(x)] \tag{7}$$

for the networks of interest. For neural networks, $f(x)$ is usually taken as a simple sigmoidal function which has a smooth, monotonic transition between levels. In general, functions like $\arctan(x)$ and $U_o(x)$ (the Heavyside unit step function) are functions of this class. In limited interconnect neural architectures, $f(x)$ can also be sigmoidal, and for other reduced-interconnection architectures, such as cellular array networks, $f(x)$ is not a sigmoidal function, but is a multivalued function that represents a truth table mapping between input and output, a mod_2-type function.

The simplified model that is customarily used for fully and limited interconnect neural systems and cellular arrays is

$$x_j = \sum_i C_{ji} y_i - \phi_j \tag{8}$$

where y_i is the value of the state i (presumably 0 or 1, but it could also be a continuous variable in analog systems), C_{ji} is the interconnection strength from the output of state i to the input of state j, and ϕ_j is a threshold value for the state j. The key factor in Equation (8) is that the entry in **M** can be changed either by changing the interconnection weights C_{ji} or by changing the threshold ϕ_j. Thus, the primary difference between fully and limited interconnected neural networks, and cellular arrays, lies in the neighborhood chosen for the interconnection sum in Equation (8) and in the function $f(x)$. In this regard, the networks are formally equivalent, and both are representable by the iteration

$$y_j^{t+1} = f\left[\sum_R C_{ji} y_i^t - \phi_j + I_j^t\right] \tag{9}$$

where R is a neighborhood, and we have added a bias I_j which can represent a learning function. If R is complete and f is sigmoidal, we have a fully connected neural network. If R is reduced from full interconnection and the network has multiple layers, we have a limited connect neural network. Finally, if R is local and f is a digital truth table, we have a cellular automata network.

It is important to note that in the neural network hierarchy, fully connected systems are the most general type of network, with many possible subsets. Several important types have been described by Hinton and Anderson (1981). Of particular interest are the layered networks, in which information propagates from an input layer through several buried layers to an output layer. These networks are only fully connected between layers. Learning can be incorporated readily using the popular backward-propagation algorithm.

Training Limited Interconnect Architectures

Highly layered, limited interconnected architectures are especially well suited for VLSI implementations. Fortunately, highly layered, limited interconnect networks can be formed that are functionally equivalent to highly connected systems (Akers et al., 1988). Our objective is to design highly layered, limited interconnect synthetic neural architectures and to develop training algorithms for systems made from these chips.

Several algorithms for training networks of binary threshold elements have been proposed. The topic of network synthesis for threshold elements goes back to the 1960s when it was believed that the way to obtain more powerful computers was to increase the power of the processing elements. All of the methods considered during this period were for the synthesis of complex Boolean, that is, linear functions. Methods in this section are for multilayered networks implementing nonlinear functions. The earliest method is the training algorithm for Widrow's ADELINE, which employed a single-layer perceptron followed by a single second-layer node which employed a simple function such as majority voting (Widrow, 1962). The single layer is essentially a linear filter and can therefore be adapted by using any linear filter algorithm, including least mean squares (LMS) learning. Widrow's latest method for multilayered binary networks is to employ a method he calls *least disturbance*, which basically involves a search through network space, modifying only those connections which most contribute to the reduction of output error (Widrow & Winter, 1988).

Back-propagation is a powerful algorithm for adapting layered networks, but this method typically requires a continuous, differentiable activation function, such as a sigmoid, in order to determine the error gradient at every node in the network. This problem may be circumvented by determining a "desired" state for every node in the network. Le Cun (1985) described a simple procedure for determining the desired states, working backward from the desired responses at the output layer. With desired states determined, any learning algorithm for single-layer systems can be employed.

In backward error propagation (BEP), the output error is defined as

$$E = \frac{1}{2} \sum_{C} \sum_{J} (y_j - d_j)^2 \tag{10}$$

where C is an index over the number of input–output pairs, J is the index of the output units, y_j is the actual output value, and d_j is the desired output value at each output unit. The output of each processing element is a nonlinear function of its input, x_j,

$$y_j = \frac{1}{1 + \exp(-x_j)} \tag{11}$$

where the input x_j is a weighted sum of the outputs of units in the previous layer. The weighted transmittance value between the previous output y_i and the next input x_j is w_{ij}. x_j is expressed as

$$x_j = \sum_i y_i w_{ij} \tag{12}$$

The use of a differentiable node-transfer function allows the use of the chain rule to determine the change in the output error resulting from each transmittance value change. The procedure for adaptation using the Delta rule involves applying input vectors at the input layer and accumulating the difference between the actual and desired output vectors for the entire example set. The change for each transmittance value as a function of total output error is

$$\Delta w(t) = -\epsilon \frac{\partial E}{\partial w(t)} + \alpha \Delta w(t-1) \tag{13}$$

where t is incremented by 1 for each sweep through the whole set of input–output cases, ϵ is a proportionality constant that determines the magnitude of transmittance adjustment at each step, and α is a momentum factor that determines the contribution of the previous gradients to the transmittance change. This procedure is repeated until the total accumulated error falls below some minimum value.

Because the networks described earlier employ binary threshold elements, additional considerations are necessary. For this type of network, desired values for the states of the internal processing elements as well as the output nodes must be defined. A procedure has been described (Plaut, Nowlan, & Hinton, 1986) in which internal desired states are calculated as

$$\begin{aligned} d_i &= 1 \text{ if } \sum_j w_{ij}(2d_j - 1) > \mu_i \\ \text{else } d_i &= 0 \end{aligned} \tag{14}$$

where j is the index over the succeeding layer of processing elements and i is the index of the layer under consideration. Here, μ is an arbitrary threshold for the generation of the backward-propagated desired state. Plaut has modified this procedure to generate a "criticality" term for each desired state associated with each hidden unit. This ratio is a numerical representation of the "confidence" associated with the desired state selected for each hidden unit and is used to modify the weight update. The transmittance change for any connection then becomes

$$\Delta w_{ji}(t) = \epsilon(d_j - y_j)\, c_j y_i + \alpha \Delta w_{ij}(t-1) \qquad (15)$$

where c is the criticality of any node in a given layer, defined by

$$c_i = \frac{\text{abs}\left[\sum_j w_{ij}\,(2d_j - 1)\, c_j\right]}{\sum_j \text{abs}\,[(2d_j - 1)\, c_j]} \qquad (16)$$

In this way, desired states which have a high criticality will exert the greatest influence on the modification of a given connection.

Each weight in this type of network develops in a way that not only converges on the training set but also constrains all units to the saturated end points of their activation functions. In addition, because all units are forced to the end points, each unit ends up acting as a simple threshold element implementing a Boolean function. For this reason, binary threshold elements may be used to replace the continuous sigmoids in feed-forward operation after training has completed.

Additional considerations are required for sparsely connected networks adapted with BEP (Walker & Akers, 1988b). Efficient internal network representations are necessary for generalization of a given problem in a perceptron network. A general rule for accurate generalization in perceptrons with full interconnection between layers is that they must contain a minimum number of hidden units necessary to encode the invariant properties of the input training set. In a similar fashion, sparsely connected perceptrons that generalize have a minimum number of active internal units. Learning algorithms for these networks must not only develop internal abstractions with a minimum number of processing elements, but must also route required signals to the proper locations within the network. Low fan-in and the use of random initial weight values serves to reduce the probability that a signal necessary for a given hidden unit to generate a required abstraction will be routed in. For this reason, the generation of a "high" desired state within a hidden unit must be made artificially difficult. In this way, the majority of connections will be eliminated from the forward signal path unless they are needed. Fortunately, the presence of a positive threshold value in each processing element means that each node is hard to turn on and easy to turn off. Another necessary *a priori* consideration is that the inputs must be arranged in a way that ensures that they can be connected to all of the correct output nodes with the number of layers provided.

Sparsely connected perceptrons are generally more prone to local minima during training. Fully connected perceptrons—providing more signal paths and a richer, more redundant set of internally formed abstractions—more easily find a near-optimal solution (in the LMS sense) in weight space. Because gradient descent is basically a shortcut search method through weight space, greater redundancy and overlapping of internal representations of information improve the probability of convergence to a near-optimal solution of the training set. It is well known, however, that networks which have an overspecified layer of hidden units converge quickly, but generalize untrained cases poorly.

Sparse networks provide less redundancy and have a more difficult time forming the necessary internal abstractions, because routing of signals to the correct hidden unit to form a needed abstraction now becomes important. An advantage, however, is that because fewer points in weight space are available to enhance convergence with the given training set, the probability is higher that the given weight set has captured the invariant properties of the desired network function.

The following example shows how this algorithm finds a local minimum solution to the XOR problem. Figure 5 shows the "prewired" routing through the sparse array and the necessary Boolean logic functions formed by the unique weight patterns at each node. Figure 6 shows the weight pattern formed using back-propagation of desired states. Interesting is the similarity and location of the Boolean functions formed within the network. Figure 7 is the classical solution to the XOR problem.

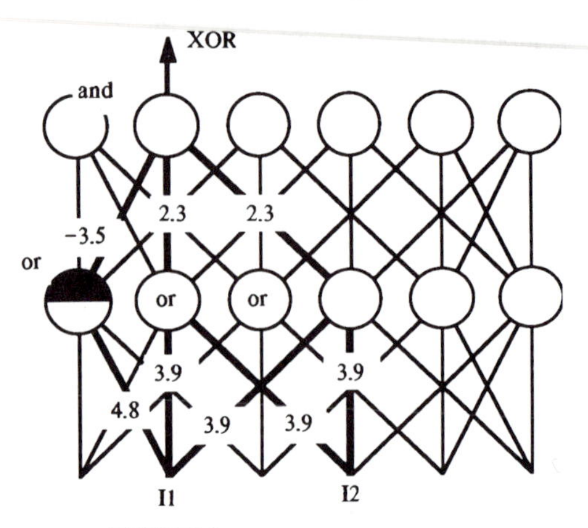

FIGURE 6 Adapted XOR network.

ELECTRONIC IMPLEMENTATIONS

A team at Bell Laboratories (Graf et al., 1986) have designed, fabricated, and tested SNS that are composed of 22 and 256 neurons. Both designs used resistors for the interconnection elements that could not be changed once they were fabricated. For a system to be able to respond to a changing environment, the interconnection values must change with time. The Bell Labs group (Graf et al., 1986), Caltech group (Sivilotti, Emerling, & Mead, 1986), and Arizona State Group (Akers et al., 1989) have designed chips with programmable interconnection values.

Analog–Digital Implementation

Synthetic neural systems are limited by VLSI constraints, and not neural constraints. A principal VLSI constraint is cost, and the cost of a chip is directly related to its die area. By exploiting the natural functions available with analog circuits, such as summation, less die area is consumed than with a digital implementation of the same function. Of course, ana-

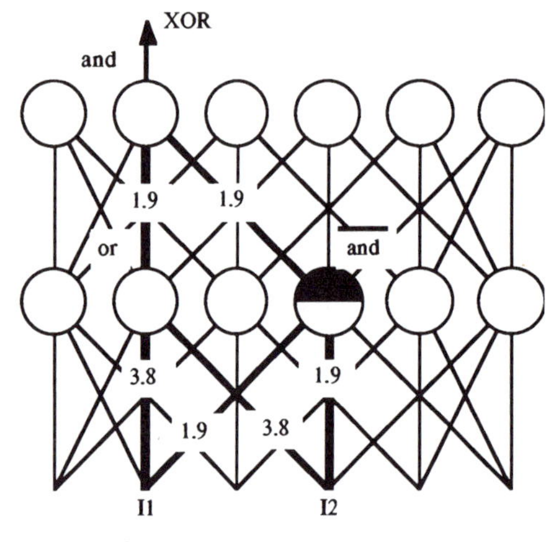

FIGURE 5 Prewired XOR algorithm.

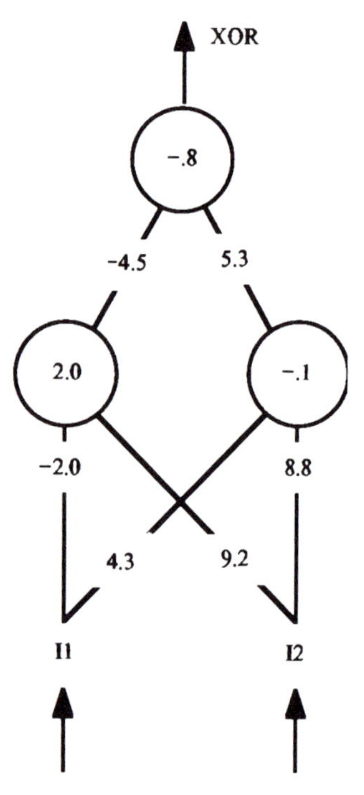

FIGURE 7 PDP XOR algorithm.

log VLSI designs have additional problems not found in digital systems. For example, power dissipation can be a serious problem for large chips. Whereas digital circuits are relatively insensitive to varying device parameters, in most cases, analog circuits are very sensitive to these variations. Analog neural-like circuits must be designed not to rely on the absolute behavior of each individual transistor, but rather on the cooperative behavior of large numbers of devices.

To take advantage of the compactness of analog circuits, and the device process tolerance associated with digital circuits, a hybrid analog–digital circuit has been developed (Akers et al., 1989; Hasler, 1988). This cell is designed for use in limited-interconnect architectures, so tens of thousands of neurons can be fabricated on a die. Figure 8 is the circuit diagram of the limited interconnect analog neural cell, and Figure 9 is the timing diagram. The operation of the cell is as follows. Weights, W1, W2, and W3 are stored dynamically on the gates of transistors T1, T2, and T3. Notice only PMOS transistors (T18, T19, and T20) are used to pass and isolate the weights instead of transmission gates. This is allowable because only weights above the device threshold are important, and hence a degraded low state voltage has no effect on circuit performance. For inputs of 5 V, the drain parasitic capacitors of T7, T8, and T9 are charged by current flowing through the pass transistors T4, T5, and T6 to a voltage equal to the weight minus the device threshold voltage. For inputs of 0 V, the pass transistors will allow the capacitors to discharge. Although exact multiplication of the input and the weight is not done, shifting of the circuit's logical threshold voltage and modifying the training algorithm compensates for this behavior. In fact, one of the very useful characteristics of neural networks is that an exact generation of products is not necessary. Once the storage capacitor in each branch is charged, clock Φ1 is turned off to isolate the signal from the input. Turning on Φ2 allows the signals to be analog summed and compared to the logical threshold of the first inverter. The first inverter needs to be of minimum size to allow acceptable charge transfer. The circuit on top of the PMOS pull-up device allows the logical threshold, and hence the neural threshold, to be set at a voltage lower than $.5V_{dd}$. The output inverter restores the output voltage level and drives the next stage. Because this circuit uses only positive weights, a shunting transistor T11 is used to provide inhibition. Arrays made with this cell can perform as a complete logic family. Figure 10 shows a circuit simulation of the cell.

A system formed by replicating the analog cell operates in the following manner. To keep the number of I/O pins at reasonable numbers as the system is expanded, we share the input, weights, and output pins.

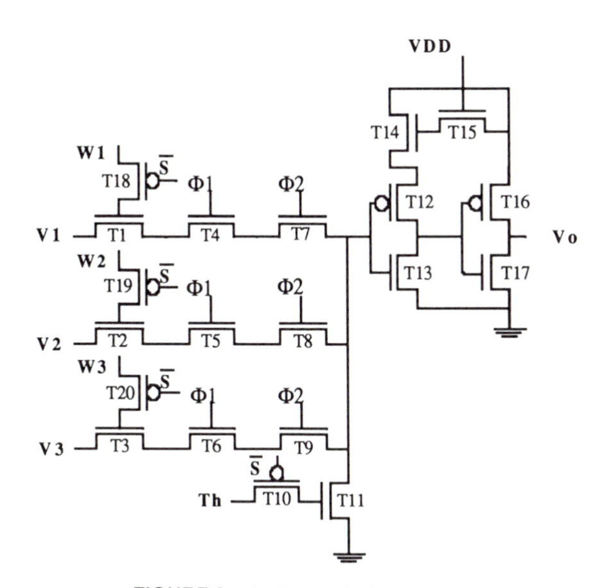

FIGURE 8 Analog synthetic neural cell.

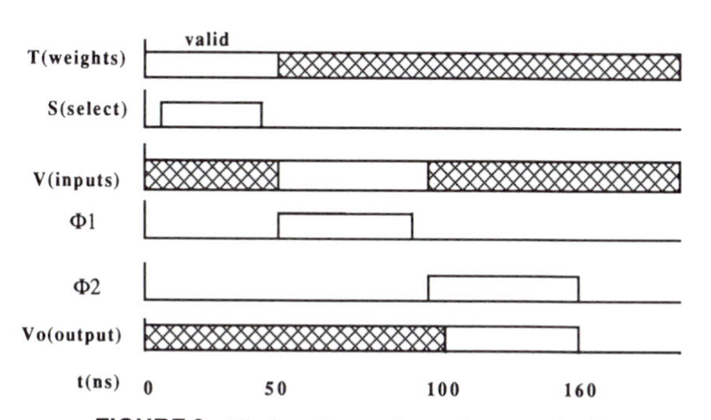

FIGURE 9 Timing diagram for analog neural cell.

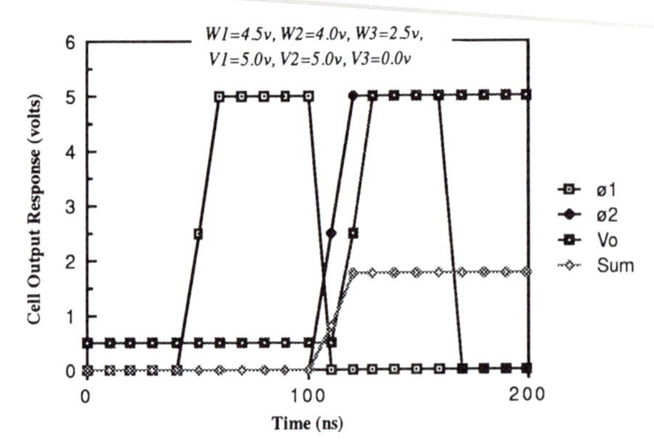

FIGURE 10 Circuit simulation of cell.

For the implementation of 512 neurons in an array 32 wide by 16 long, 32 lines are used for the weights and I/O. The four unique weights are multiplexed to the single line running to each cell. Hence, a row is selected for a write, then weights for each cell are written. This continues for four cycles after which all weights for the row addressed have been written. The next row can then be selected for weight loading. The whole array can be loaded in approximately 10 μsec, an order of magnitude faster than the discharge time constant for an individual gate. Figure 11 shows the block diagram for the chip. For efficient data flow through the layers, the clock lines are interchanged in every other row. After the weights have been loaded, the inputs are loaded, and the output vector ripples through the layers.

Analog-Pulse Implementation with On-Chip Learning

The real challenge for the 1990s is synthetic neural chips with on-chip learning. The circuits discussed so far have learning calculated off-chip. For real-time control and processing, this is not practical. The learning algorithm must be an integral part of the neural circuit. The next design discussed has an unsupervised learning algorithm implemented as an integral part of the neural circuit. This processing node acts as a preprocessing stage to reduce the dimension of its inputs while generating efficient representations for additional layers of processing elements.

Pulses are used for interprocessor communication. Pulse-coded information allows both robust information transfer and space efficient computation (Murray, 1989). Regeneration and cross-talk problems associated with analog communication are greatly reduced by using digital signals. Inputs and outputs of a processing node are represented by a sequence of pulses. Analog inputs have values proportional to the duration of the pulse divided by the period between pulses.

Figure 12 shows a block diagram of the complete analog input–output processing system. First, a set of tapped delay lines and level-sensitive neurons expand the dimensionality of the analog signal. Tapped delay lines time-embed input signals, converting time information into a spatial representation required by the feed-forward transfer function of the processing nodes. Level-sensitive neurons convert amplitude information to spatial representation. This is required if the individual processing nodes in the dimensional reduction layer have insufficient accuracy to code the analog signal. The next layer combines multiple inputs to reduce the dimensionality of the first layer. It must also provide an efficient representation for the output layer. The output layer maps the internal representation to the correct output value. A layer of neurons using a supervised training algorithm performs the final output-mapping task.

An adaptive system must continuously adjust to the nonstationary statistics provided by the environment. To maintain an optimal representation for the output layer, the dimensional reduction unit must continuously adapt through weight modification. Several researchers have shown that Hebbian weight modification rules can form efficient representations of the input statistics when used in conjunction with a method to force each processing node to extract a different feature from the inputs (Oja, 1982; Leen, Rudnick, & Hammerstrom, 1989; Sanger, 1989). For this continuous-time VLSI implementation, the magnitude of the weights must be controlled. The weight modification rule proposed by Oja uses negative feedback from the outputs to control the magnitude of the

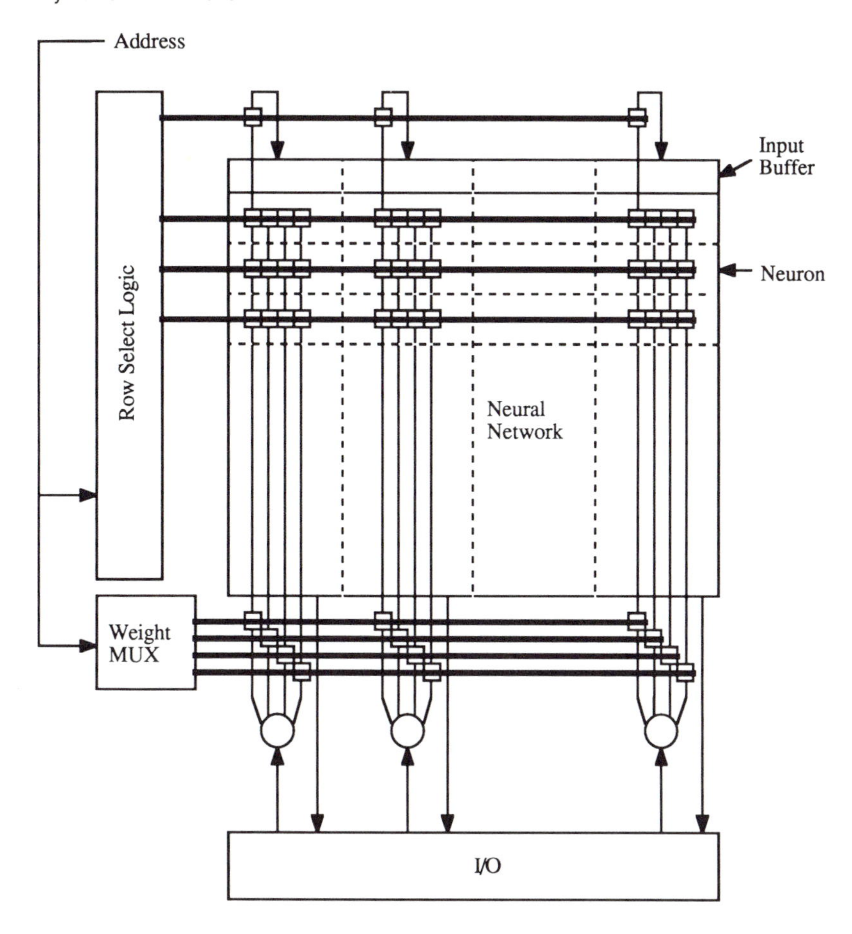

FIGURE 11 Chip architecture

weights (Oja, 1982). This design utilizes the concept of negative output feedback to control the weight magnitude. This rule combined with a variable lateral inhibition mechanism forces each processing node to extract a different feature from the inputs while learning.

To test the feasibility of a VLSI internal layer based on this weight modification rule, two integrated circuit

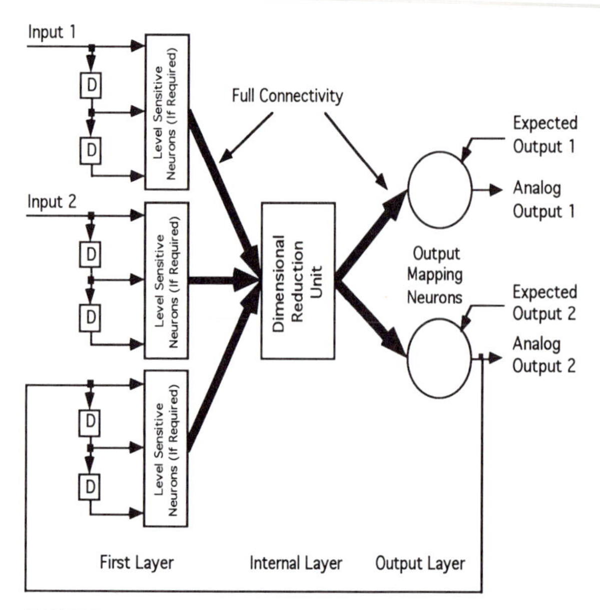

FIGURE 12 A complete nonlinear input-output mapping system.

chips were designed, fabricated, and tested (Donald & Akers, 1993). The first chip contained one processing node connected so that the state of individual function blocks could be monitored and controlled. The second chip contained an array of four processing nodes. In the array, each processing node shared nine inputs connected to adaptable synapses and two inputs connected to nonadapting synapses. The nonadapting synapses provide two-quadrant multiplication, whereas the adaptable synapses provide only first-quadrant multiplication. Adaptable synapses also contain additional circuitry for weight modification. The adaptable synapse measures 105 μm × 168 μm, including interconnects, using a 2-μm CMOS MOSIS process with two layers of polysilicon and two layers of metal. The output layer was not implemented on this chip. It is planned to have the output layer implemented on a separate chip. Because digital pulses communicate analog values between chips, EMI-type noise will have little effect on the accuracy.

The feed-forward function consists of a sum of products between an input vector and weight vector. This provides a measure of the match between the weight vectors and input. Additionally, some type of output nonlinearity, typically a sigmoidal function, splits the output probability density into two groups, indicating a decision. In this system, the dimensional reduction layer performs a linear sum-of-products computation. Nonlinearities producing a decision are provided by the next processing layer.

Figure 13 shows the circuit used for the sum-of-products function, which includes one synapse and the output function. Summation of the product of the weight and input occurs by injecting a current proportional to the synaptic weight into a summing node for the duration of each input pulse. The summing node receives the current produced by a complete row of synapses producing summation through Kirchhoff's current law. An integrating capacitor (C1) converts the current pulses to a voltage change at the summing node. The neuron output function compares the voltage on the summing node to a threshold voltage, V_{thresh}, and produces an active high digital output if the threshold voltage is exceeded. The output increases the summing node voltage by positive feedback through C2. Each output function block requires one C1-C2 set. When implemented in a two layer polysilicon process, C1 and C2 have areas of 9792 μm^2 and 2496 μm^2, respectively. C1 is approximately 5 pF, and when combined with C2 provides an increase of 1 V in the output switching voltage when the output goes high.

Input to the synapse circuit, Figure 13, consists of a pulsed active high signal IN. The input pulse enables the transmission gate formed by Q1 to transmit the voltage present on the drain-connected transistor Q2 to the gate of transistor Q3. When the input is low, Q4 shorts the gate of Q3 to Vdd to reduce leakage into the summing node through Q3. This structure forms a simple current mirror that can be switched on for the duration of the input pulse. The current delivered by the current mirror depends on the voltage on C3, the weight capacitor.

When active, the pulse output from the output function block enables a current sink to remove charge from C1. The output turns off when the voltage on C1 falls below the threshold voltage. Transistor Q5a enables the current sink formed by Q5b. This action pro-

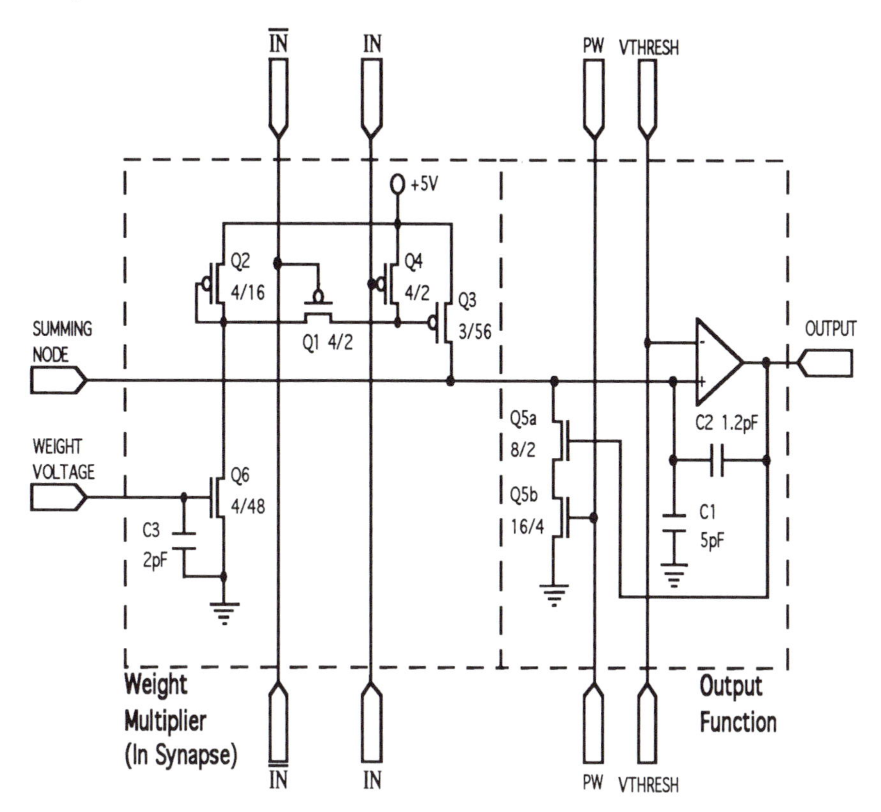

FIGURE 13 Single synapse and summation circuitry.

duces a series of pulses with the average period between the pulses equal to the sum of the products of the weights and inputs. A differential comparator generates the threshold function. The comparator allows us to control externally the threshold voltage, which allows more control of the operation of the circuit. For electrical tests, the current through Q5b was adjusted so pulse widths of 250 ns were obtained.

The transistor sizes in the synapse current mirror, the size of C1, and the hysteresis of the output determine the maximum output frequency of the processing node. A nominal pulse width of 250 ns and a maximum frequency of 1 MHz were chosen for this design.

A modified, unsupervised learning algorithm, using concepts developed by Oja, is implemented to determine the change in the weight to best represent the input statistics. For a single synapse element, Oja's weight modification rule is

$$w_i(t + \tau) = w_i(t) + \lambda y(t)(x_i(t) - y(t)w_i(t)) \quad (17)$$

where, $x_i(t)$ is the input, $w_i(t)$ is the weight, $y(t)$ is the output from the linear sum-of-products function, and λ is the learning rate or weight increment size.

One of the main advantages of Oja's weight modification rule is the use of feedback from the neuron output to normalize the weights. This has several significant advantages in a VLSI hardware implementation. First, if the system is stable, the weight equilibrium is independent of the weight increment size (λ). Second, negative feedback reduces the effects of nonlinearities in the weight-to-output transfer function resulting from the implementation. In this implementation, we take advantage of the insensitivity to the forward weight transfer function to simplify circuit design and allow continuous adaptation.

Figure 14 shows the synapse weight modification circuit. Basically, the weight modification circuit operates as follows. The weight update is calculated in two stages. The argument $(x - yw)$ is calculated by using pulses that enable current sources to charge and discharge a capacitor. The charging is proportional to x, and the discharging is proportional to yw. Then, a switch capacitor system produces the multiplication of the output y with the argument $(x - yw)$ (Madani, Garda, Belhaire, & Devos, 1991).

The modification circuit generates a weight update using three phases. The first phase sets the capacitor C5 to a predetermined voltage (LRNLEV). During the second phase, the statistical sampling phase, input pulses charge capacitor C5 at a rate that depends on $V_{PreGain}$ using the current source formed by Q7. Simultaneously, output pulses, when active, enable a current sink that discharges C5 at a rate that depends on the weight capacitor voltage V_{C3} and the output. If Q7 and Q10 are operating in saturation with gain parameters β_7 and β_{10}, the voltage on C5 at time t after the reset phase is

$$V_{C5}(t) = \frac{1}{C5}\left(\mathrm{fl}_7 x(V_{cc} - V_{PreGain} - |V_{tp}|)^2 - \mathrm{fl}_{10}\, y(V_{C3} - |V_{tn}| - V_{LrnLev})^2\right)t \tag{18}$$

where x is the active fraction of $\overline{PRE}$, and y is the active fraction of POST during the statistical sampling

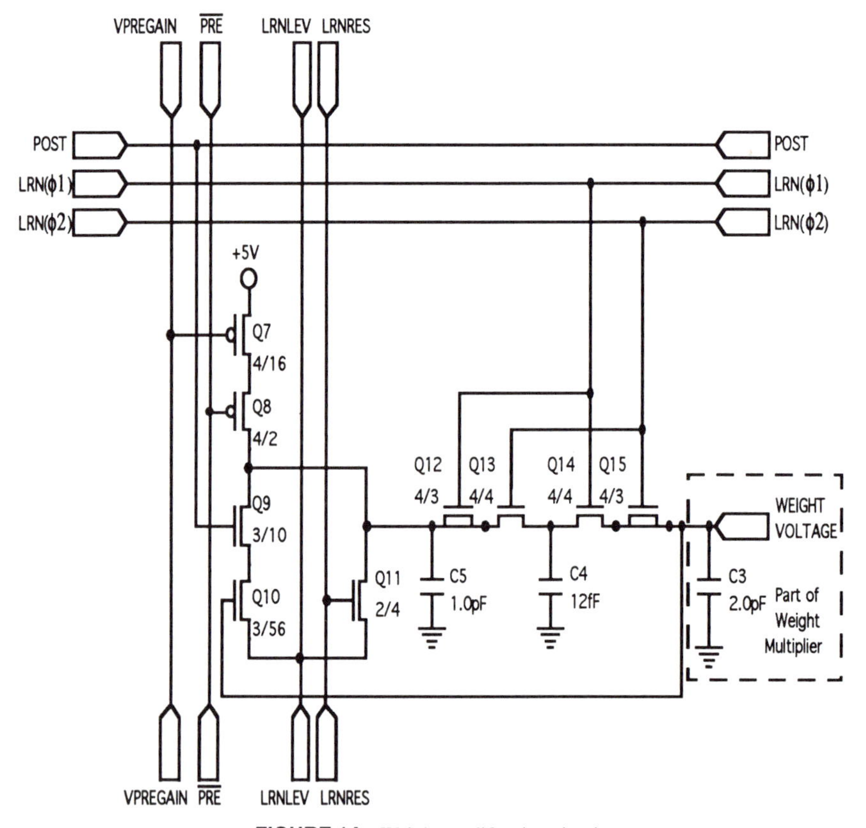

FIGURE 14 Weight modification circuit.

phase (SAMPEN active). The signal SAMPEN is active during the statistical sampling phase and is used to gate $\overline{PRE}$ and POST. By controlling the period of the statistical sampling phase, (t), we can control the final voltage on C5.

The current mirror for the weight multiplication function supplies current into the summing node that depends on the voltage on C3 and is proportional to the current through the drain of Q6. The gate of Q10 also connects to the weight storage capacitor C3, which generates a drain current proportional to the drain current of Q6. Assuming that the initial voltage at $t = 0$ was zero, the voltage on C5 at the end of the statistical sampling period is

$$V_{C5} = \nu(d_{PreGain}x - yw) \tag{19}$$

where ν is a proportionality constant with units of voltage, and $d_{PreGain}$ is an external gain constant that depends on $V_{PreGain}$.

A switched capacitor is used to calculate the final multiplication. During the third phase, the learning phase, each output pulse transfers charge from the statistical sampling capacitor C5 to the weight capacitor C3. The size of charge transfer must be as small as possible so that the time constant of the weight filter is as long as possible (λ small). Because y is proportional to output pulse frequency, when the output operates in the linear region, the output can be used to set the frequency of the switched capacitor system. During the learning phase, an output pulse generates a set of nonoverlapping signals LRN(ϕ1) and LRN(ϕ2). These signals enable switches formed by Q13 and Q14 first to move charge from C5 to C4 and then finally C3. This incrementally modifies the voltage on C3 for each output pulse. The ratio between C3 and C4, and the difference between the weight voltage V_{C3} and the voltage on C5 sets the size of the weight increment.

Figure 15 shows experimental measurements of the weight counter. In this test, the sampling capacitor was set to 4.8 V, and the weight capacitor was set to 1.3 V. Learning was then enabled and each processing element output pulse incremented the weight capacitor. In this test, the increment size was approximately 6 mV per output pulse.

When the weight is large, the weight modification used in the circuit can be derived as

$$\begin{aligned} w_i(t + \tau_2) = w_i(t + \tau_1) + \lambda y(t + \tau_1) \\ \{(d_{PreGain}x_i(t) - y(t)w_i(t)) - \delta w(t + \tau_1)\} \end{aligned} \tag{20}$$

where δ is a proportionality constant relating the voltage on C3 and the weight. Note the implementation of the training rule departs slightly from Oja's.

An array of neurons using a modified Hebbian weight modification rule requires a mechanism to force each neuron to extract a different feature from the inputs (Oja, 1982; Leen et al., 1989; Sanger, 1989). Sanger (1989) incorporates a Gram-Schmidt orthogonalization mechanism in the synapse to force the weights to be orthogonal. Leen et al. (1989) adds a matrix of weights between outputs that uses an inverse Hebbian modification rule to force the outputs to be uncorrelated. In this system, we use a variable lateral inhibition mechanism to force the outputs to be uncorrelated during learning and allow overlap between outputs during recall. During learning, this inhibition mechanism acts as a winner-take-all circuit. During recall, the inhibition mechanism allows several outputs to become active if the presented pattern lies between trained classes.

Pulse computation allows a simplified inhibition circuit. The lateral inhibition circuit uses a global wire-or signal and a gate circuit to allow the neuron with the greatest sum of products to inhibit all the other neurons in the array. This inhibition occurs only during the high output pulse of the active neuron. When the output pulse is high, it resets all nonfiring neuron-summing node voltages to the inhibition voltage (V_{inh}). After the active neuron's pulse occurs, synapses attached to each neuron charge the summing node capacitor at a rate that depends on the sum of products between the weights and inputs. The time for the summing node capacitor to charge to the threshold and cause the neuron to fire depends not only on the sum-of-products, but also on the initial voltage of the summing node. Note that the active neuron also resets its summing node voltage below the threshold to $V_{thresh} - V_{\Delta th}$ after it fires. The relation between the initial volt-

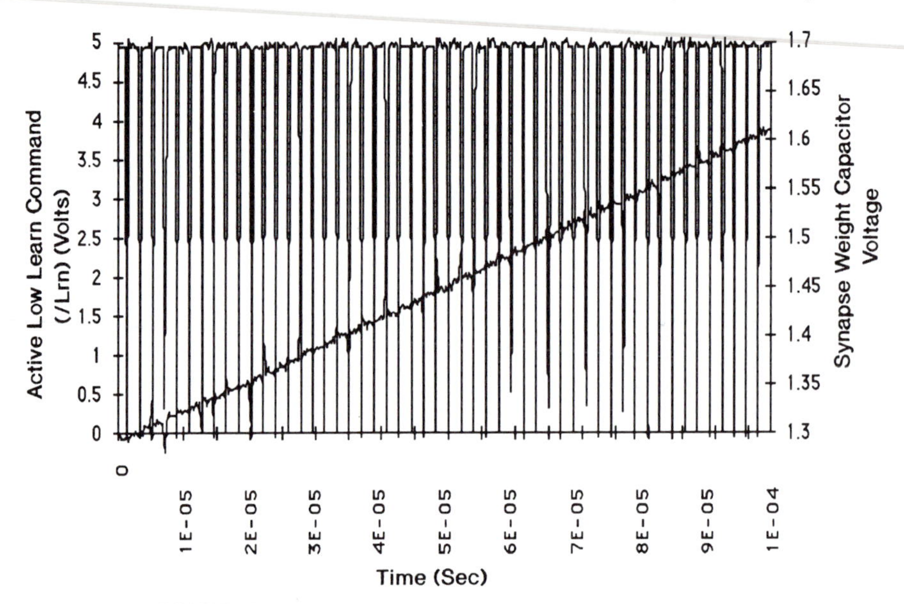

FIGURE 15 Measurements of weight voltage during weight increment.

age of the inhibited and the active neurons determines how much greater the sum-of-products must be for an inactive neuron to become active. If the inhibited summing node voltage of a neuron is less than the reset voltage of the active neuron, the inhibited neuron must have its sum-of-products greater than those of the active neuron for it to become active. Conversely, if the neuron's summing node voltage is inhibited to a value greater than the reset voltage, this neuron can become active if its sum-of-products is slightly less than the active neuron.

By varying the inhibition voltage, the sensitivity to changes in the input can be controlled. Normal operation of an array of neurons would first set the inhibition voltage higher than the reset voltage, allowing neurons with close pattern matches to become active. The inhibition voltage would then be reduced, resulting in only one active neuron for the duration of the learning phase. Reducing the value of the inhibition voltage allows the network to reduce its sensitivity to changes in the input feature during synapse modification in a real-time control system. For the tests presented here, we fixed the inhibition voltage to a value slightly lower than the reset voltage by using an external voltage source.

Figure 16 shows the circuit used for the lateral inhibition mechanism. One transistor, Q18, per chip acts as a load transistor for the inhibition line. Each neuron output block ties to the single inhibition line with a pull-up transistor Q17. This produces a wire-or computation that activates Q16 to set the summing node to the inhibition voltage, INHLEV, in all neurons except the currently firing neuron. When the neuron fires, transistor Q19 disconnects the gate of the inhibition transistor Q16 from the inhibition line, and Q20 shorts the gate to ground to prevent spurious triggering of Q16. Just after the gate of Q16 is disconnected from the summing line, Q17 pulls the inhibition line active. The size of Q22 contributes to this delay. The inhibition line, INH, becoming active causes the inhibition transistor Q16 to connect the summing node to INHLEV on all neurons except the neuron(s) that has an active output pulse while the inhibition line is active.

To demonstrate how the network processes information, two sine waves with adjustable phase shift were provided to the network. Before being applied to the network, the sine waves were converted to pulses. Outputs from the pulse generators were applied to the four-neuron classifier chip. Inputs and outputs from the chip were recorded during weight modification. Depending on the phase shift between the two inputs, one or two outputs became active and indicated specific features at the inputs.

Eigenvectors of the input correlation matrix provide the best basis vectors for the representation of the input statistics (Sanger, 1989). The correlation matrix is a statistical measure used to determine the degree of similarity between elements. In the simple case, the matrix elements are just the time average of the two inputs, for example, X1 * X1, X1 * X2, X2 * X1, and X2 * X2. The cross-correlation matrix of two sine waves offset by half of the amplitude of the sine wave have eigenvectors $\{w_1, w_2\}$ of {{0.707,0.707}, and {0.707,0.707}}. Because outputs from the system provide the only method of information exchange between nodes, the outputs should have activations that correspond to the eigenvectors. We expect one output to be maximum when the two inputs have the largest product, and one output to be maximum when one input is maximum and the other input is minimum. Note that the system can only have positive weights, inputs, and outputs. The system converges on the weights of {{0.707,0.707}, and {1,0}}. This produces one output to be maximum when the two inputs have the largest product, and one output to be maximum when one input is maximum and the other input is minimum.

Figures 17, 18, and 19 show results from experimental tests using three phase shifts. For the experimental test described, we only needed and therefore used two neurons, two inputs and four weights. We intend to use other configurations for future experiments. The first graph shows activation of one out-

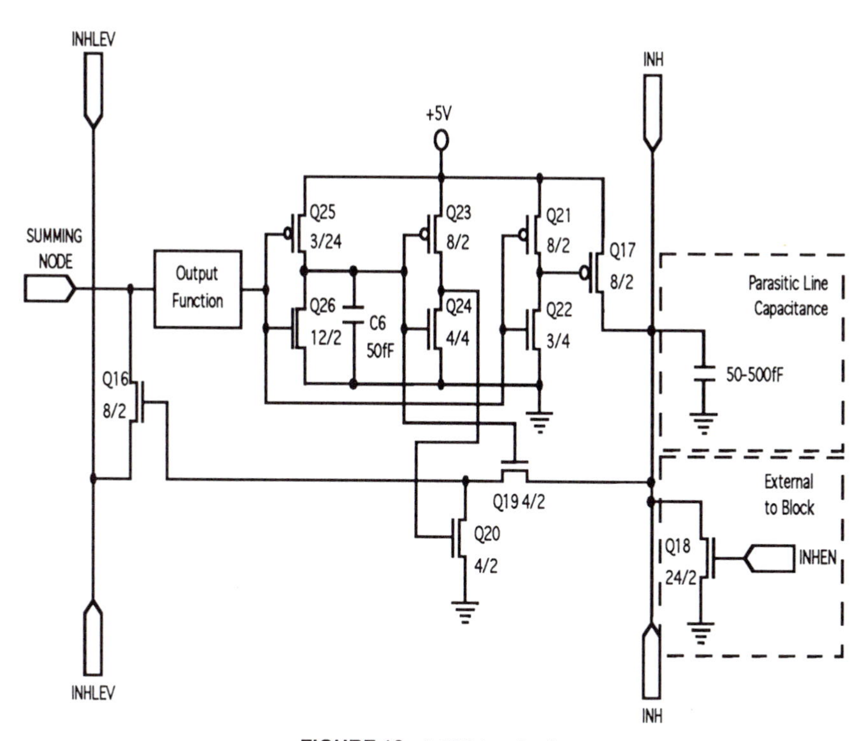

FIGURE 16 Inhibition circuit.

put when the inputs were phase shifted by 36 deg (Figure 17). For a phase shift of 36 deg, the eigenvalue for the second eigenvector is too small for the network to extract, resulting in only one feature recognized by the network. At a 72-deg phase shift, the network can detect two features in the input as shown in Figure 18. The theoretical transition point between two activations is 0.3 msec and Figure 18 shows a measured transition point of 0.225 msec. Finally, Figure 19 shows the case where the inputs have a 108-deg phase shift. In this case, two outputs are active, indicating that two features were detected.

Alternate Architectures

Synthetic neural system chips are direct attacks on the basic problem of ULSI, which, simply put, is What do you do with a billion devices? There are three key

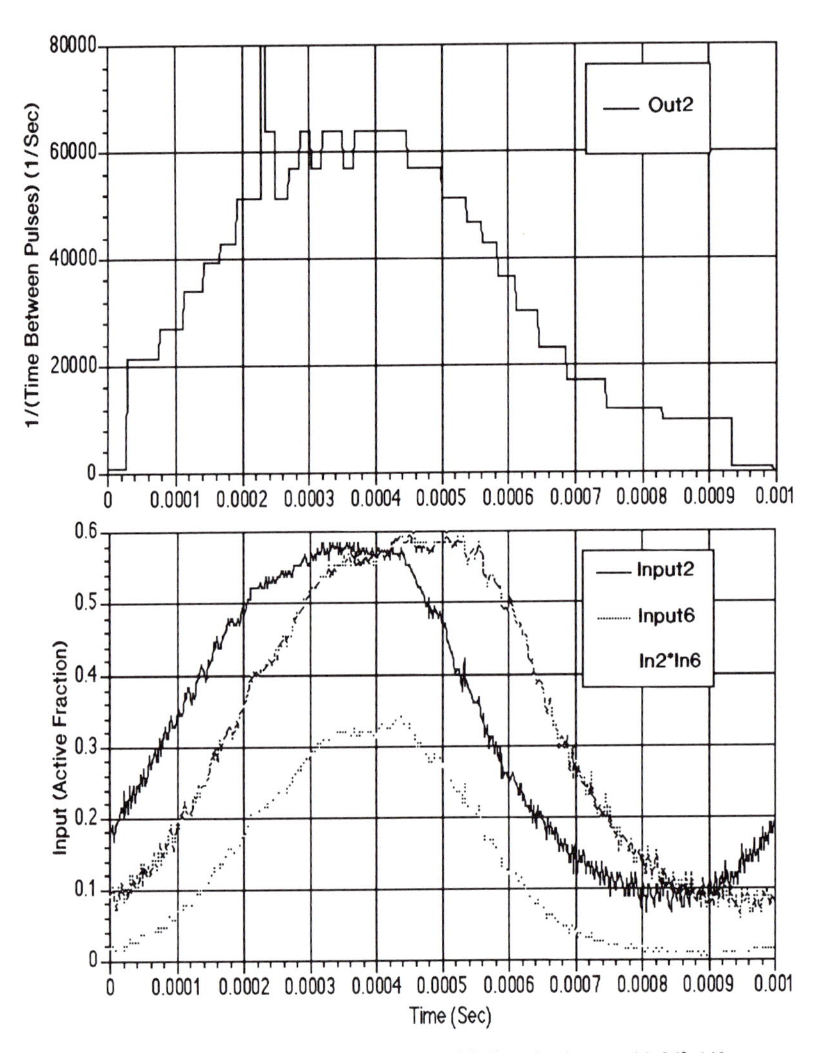

FIGURE 17 Measured outputs and inputs while learning inputs with 36° shift.

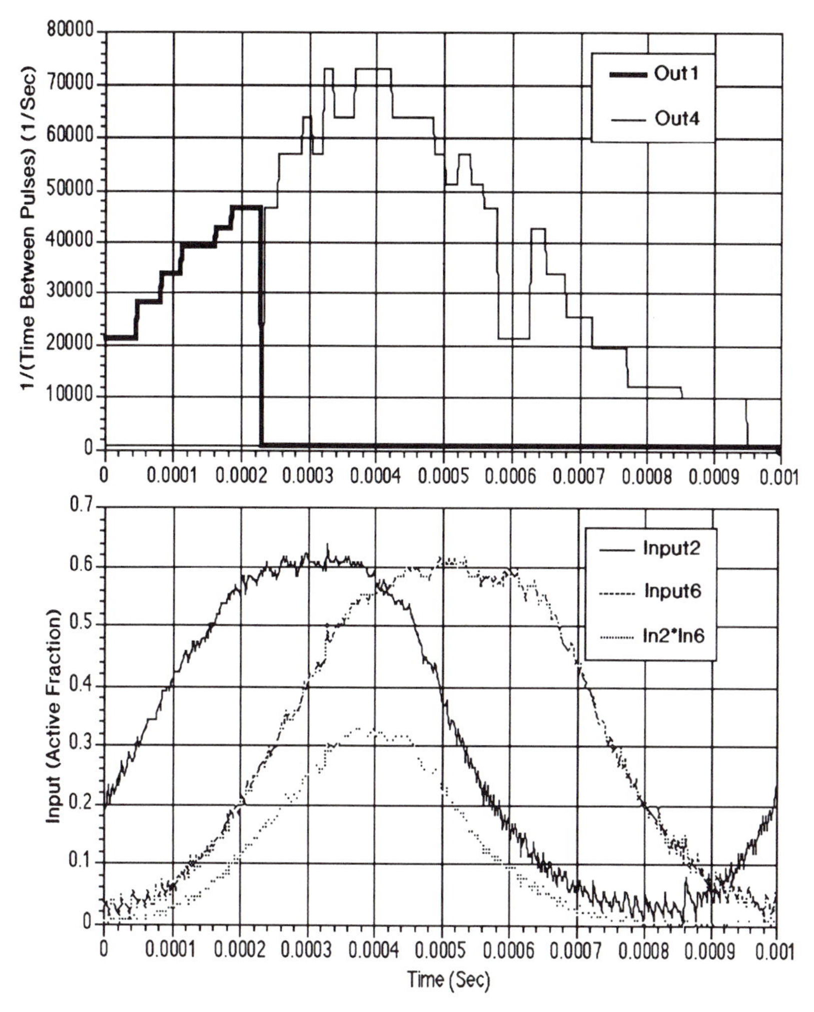

FIGURE 18 Measured outputs and inputs while learning inputs with 72° shift.

features that underlie much of the potential success of SNS. These are the use of large numbers of processing elements (PEs), some form of concurrent processing, and some form of adaptation. It is important to remember, however, that neural networks are not the only systems possible that implement these three key features, and we have shown that similarities exist between disparate approaches (Ferry et al., 1987). One such architecture, a pipelined content addressable memory (CAM), will be discussed here, and a comparison between it and neural networks helps to answer the following question: What are the advantages and disadvantages of SNS when compared with alternative architectures?

The basic structure of this architecture is a pipeline, each stage of which contains a word of memory and some comparison logic. Data flows into the top of the pipeline, and at the bottom, we recover this data along with the address of the pipeline stage that contained in its memory the closest match to the input and a mea-

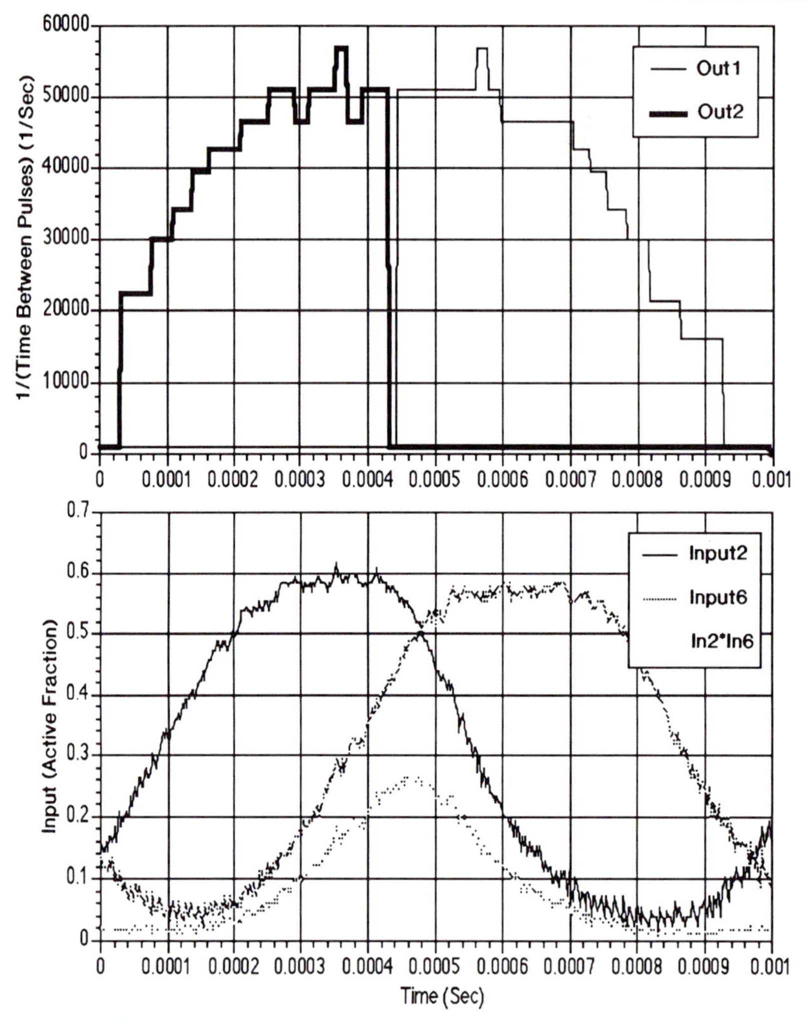

FIGURE 19 Measured outputs and inputs while learning inputs with 108° shift.

sure of the closeness of this match. Obviously, the comparison logic of each stage compares the input data with the stored data, determines the closeness of this match, and compares this closeness with that of the previous best match. The address and closeness of the best of these two is then fed along with the input data to the next stage of the pipeline. A single pipeline has been fabricated and has passed some crude functional testing (Clark & Grondin, 1987).

At this point we can already begin to compare this scheme with an SNS. First, an SNS usually destroys any input data unless special care is taken, whereas here this data is naturally recovered. Second, an SNS functioning as an associative memory gives no measure of the closeness or quality of the match between the input "key" and the remembered data, whereas here this closeness is supplied. In an SNS, interference can occur between two memories and it is often true that when the system learns something new, previous knowledge is degraded somewhat. No such interference or degradation occurs in the pipelined CAM. This interference between individual memories makes it difficult to predict the actual capacity of an SNS, whereas for the pipelined CAM, the capacity is easily determined from knowledge of the number of pipelines, the length of these pipelines, and the word length. Whereas the SNS is resistant to noise degradation of the input data, so is the CAM and for largely the same reason. Because the output is chosen by a "best-match" strategy, noise degradation of the input introduces errors in the CAM only when it has the effect of corrupting the input in a fashion that causes it to falsely appear to be a different portion of the learned corpus. Such effectively false inputs would fool the SNS as well. For that matter, it is difficult to envision any memory that is insensitive to having the input effectively constitute a lie. Although an SNS is insensitive to various "soft" errors in the storage process, the error-resistance of the CAM in this regard could be improved quite easily by utilizing the error-correcting codes commonly seen in modern memory systems (Chen & Hsiao, 1984). At this point, it would appear that the SNS is an inferior architecture, but we will now consider an example that shows that although an SNS may in fact be an inferior associative memory, there are other applications in which it should out perform systems built around pipelined CAMs.

One of the more notable successes of neural networks is NETtalk (Sejnowski & Rosenberg, 1986). This is a software simulation in which a neural network was taught to transliterate from an English text input to a phonemic output. The same task has been performed by a pipelined CAM-based system (Clark & Grondin, 1987). Indeed, a pipelined CAM is quite effective in this task and can rapidly learn to reproduce a large corpus with 100% accuracy. The best result by NETtalk was 98% learned. The trade-off between the two systems roughly can be described as giving the learning speed advantage to the pipelined CAM while giving an advantage to the neural network in terms of its ability to respond to novel input, words that were not included in the learning corpus. For example, after a large number of words have been taught to the neural network, it will tend to transliterate the word "hurling" properly even if it has never seen this word and was taught very similarly spelled words such as "hurting" as part of the learning corpus. The pipelined CAM, however, under the same circumstances would transliterate the word "hurling" as "hurting" (assuming that this is the closest match from the learning corpus) and would therefore make an error. However, Wang and Clark (1988) have demonstrated that the network may not actually be generalizing. Although it is true that neural network does better than the CAM at novel input that differs only slightly from the training inputs, this situation reverses for novel input that is more dissimilar to the training inputs. The network does not follow the rules as might be expected, but instead responds with gibberish. This indicates an incomplete mapping from input to output, not incomplete general rules.

This comparison clearly brings out the connectionist basis of a neural network. In a neural network, each connection is determined by interaction between individual pieces of knowledge. The resulting interference between individual facts results in the network's "forgetting" old information when new information is added. However, a constructive interference can also occur and this constructive interference (which underlies many of the advantages of SNS) is a very difficult one to emulate with other approaches such as a CAM. A central issue is whether or not we can alleviate the difficulties posed by this interference without throwing away the advantages as well.

Automata Theory, Hierarchy, and Neural Networks

The preceding assertion that a layered, feed-forward network trained by back-propagation does not learn how to perform rule-based activities is not one which is universally accepted. Some, for example, claim to have trained such networks to perform arithmetic operations. Examination of such claims, however, reveals that although arithmetic was used to generate the training set, the network generally could not be used to handle longer strings of numerals. The network, in fact, does not follow an algorithmic approach, but instead has essentially memorized a table and cannot handle the corresponding general problem. We argue that there is an important difference between implementation of algorithmic behavior and memorization of a mapping of input to output. In the slightly more theoretical terminology of automata theory, there is an important difference between finite state machines and combinatorial logic. A simple feed-forward network with no memory does the latter, but not the former. Long-term goals therefore include introducing both memory and feedback into the SNS.

There is an obvious splitting of problems into those which involve subsymbolic behaviors, and those which involve symbolic manipulation. Perception problems are excellent examples of problems that are dominated by subsymbolic operations, because sensory input is not symbolic in nature. For example, when we look at a flower, it is photons and not symbols that are incident on the retina. A subsequent recognition of a flower by the neural system is a result of a subsymbolic manipulation which turned nonsymbolic input into symbolic output. Such recognition problems are ones which conventional devices do not address effectively and we see great potential for SNS in this area. Here, however, we will address the other problem type, the area of symbolic manipulation.

Once we decide to focus on symbolic manipulation, the general theory of automata becomes applicable. An automata is a formal way of representing a system that receives symbolic input, has a set of possible internal states, and emits symbolic output determined by the input and the internal state. A particular example that we have explored (Shiue & Grondin, 1989) is a nondeterministic finite automaton. This class has importance because it has been shown that there exists such an automata for any regular grammar, and therefore a wide variety of rule-based systems can be mapped into such a system. The connections between automata and neural networks have been explored before (Shiue & Grondin, 1987) (Kemke, 1989), but the general role that they may play with regard to the implementation of an SNS on silicon has not been explained.

The key concept in this explanation is hierarchy. Complicated systems consisting of millions of transistors are designed in an hierarchical fashion. One does not start at one corner of the chip and begin to wire transistors together. Instead, one starts by describing the functionality of the chip as a whole. Then, the chip area is subdivided into functional blocks, each of which will consist of many transistors. These blocks are described functionally. In other words, at this point, the key issue is not the production of a detailed design of the interior circuitry of the block, but instead the specification of the functions it performs. Each block is then subdivided into a set of smaller blocks, which again are functionally described. After several repetitions, one finally reaches the transistor level. Note that the hierarchy here is not a hierarchy of networks with lower level networks providing input to higher level networks. Instead, it is a hierarchy of increasingly detailed descriptions of a single network. The resulting ability to suppress temporally unwanted, although eventually necessary, detail is a key to successful design of complex systems.

Automata provide a basis for hierarchical approaches to neural network and learning system design. For example, a different approach can be adopted to the construction of a system that learns rules. In the automata approach, one can implement algorithms that perform grammatical inference (Fu & Booth, 1975; Radhakrishnan & Nagaraja, 1987), thus allowing the grammar to be explicitly determined before the network is laid out. The corresponding automata is then mapped into SNS. The network architecture then clearly reflects the rule-based learned. This is an obvious contrast with the back-propagation approach in which the network is laid out first, and then the problem is studied. In the automata approach, the learning process is shifted to a higher hierarchical level and in the learning algorithm itself no detailed knowledge of the actual connection weights is used. The automata approach does not avoid the generalization problem, however. In a back-propagation approach, the generalization difficulties arise from the simple fact that there may be many functions that perfectly produce the examples of the learning corpus and therefore one cannot be sure after training that the network has selected the right mapping. In the context of grammatical inference, the problem is that there are many possible languages and therefore one cannot be sure that the right grammar has been inferred. This type of difficulty is inherent in the problem itself and is not a result of any particular approach.

We are presently searching for learning algorithms that sit someplace between the two levels just mentioned. Our goal is to apply them to a given chip implementation of an SNS in which there is only a limited ability to reconfigure the network. In particular, the connectivity of the network that is the potential existence of a connection path between two elements is permanently fixed, but the actual connection weight is adaptable. The actual example that we are studying is a layered, feed-forward network built from the hybrid analog–digital cell discussed earlier. As the output of this cell is binary, layers beyond the first layer have binary input. Each cell on such a layer has three binary input lines and one binary output line. In our hierarchical approach, we functionally represent such a cell by an eight-row truth table. As there are two possible results for each of these eight combinations of three binary inputs, there are 256 possible truth tables. Our approach differs from the grammatical inference approach in that the structure of the network will not be dictated by an inferred grammar. On the other hand, a learning algorithm operating at this hierarchical level differs from the more conventional neural network algorithms in that a straightforward gradient descent approach is difficult to implement because we are no longer representing the cell by a differentiable function. The learning algorithms that we are exploring are evolutionary algorithms in which we mutate a cell from an initial truth table to a different truth table. We then either retain or reject the mutation on the basis of improved network accuracy. A feature worth noting is that we can actually choose mutations that do not change the cell's response to certain inputs if we feel that the network's performance for such input is satisfactory. At this intermediate hierarchical level, we do not care what the connection weights are. We only care about the functionality of the cell. The weights will be selected later once the desired functionality is discovered.

THE FUTURE: THE ULTIMATE INTEGRATED CIRCUIT

To end this chapter, we discuss our vision of the ultimate integrated circuit. A combination of wafer scale integration and ultra large scale integration will culminate in a "chip" containing several billion transistors. The central dilemmas are, first, that it is unlikely that all of these devices will work; second, that we will never be able to test this system; and last, that we do not know what it is that we will do with all of these devices. We would suggest that techniques that answer the first two of these dilemmas will almost certainly lead to an answer to the last.

The chip must be based on a regular array of regularly connected elements, or it is unlikely that we will ever be able to complete the initial design. These elements and connections, however, will be plastic. We will be able to reconfigure the system locally to solve any problems posed by a poorly functioning chip. We can even envision systems in which we handle problems such as the tendency of threshold voltages to vary as one moves across a wafer by allowing various regions of the chip to "learn" that the voltages of the next region are a little low. In short, our chip may contain regional dialects and accents. This type of adaptability and insensitivity to parameter fluctuations will be essential if our yield is to be greater than zero. What is sometimes not mentioned is that they also play a role in long-term reliability as well, where one is concerned about temporal variations in device parameters through effects such as oxide charging.

While obviously a significant amount of on-chip testing will occur, even then the chip will not be fully testable. In a certain sense, however, the worry about testability is false. Our goal is to build systems which when confronted with novel, unanticipated situations, will usually function in a reasonable fashion. For such a chip, intended to brave new worlds, the time will come when we will have to stop testing it and let it leave the nest. Our challenge therefore is to learn how we can build systems capable of being "parented" rather than quality-assured.

The neural system of an insect may contain on the order of a million neurons. Although a neuron is much more complicated than a transistor, our ultimate chip has several orders of magnitude more transistors, and it seems to not be totally unrealistic to anticipate achieving similar complexity. Insects such as a bee are capable of flight control, fine motor control, genuine navigation, simple pattern recognition, simple communication (the famous bee dance), scheduling the time of day when it is optimal to visit certain locations, developing a map of their surroundings, and a large number of tasks that no robot today can imitate. Attaining similar behaviors as these depends critically on developing adaptability, learning capabilities, and a large measure of self-testing. These, whether obtained through an SNS, a pipelined CAM, or some reconfigurable automata, are where the frontiers of ULSI will be found.

Acknowledgments

We acknowledge the central role of our students James Donald, Larry Clark, Jennifer Wang, Mark Walker, L. C. Shieu, Mary Snyder, Paul Hasler, W. Fu, Ataru Shimodaire, Wen Looi, C. C. Goh, Siamack Haghighi, Jerome Brandt, and Arun Rao in formulating many of these ideas and correcting our mistakes.

References

Akers, L. A., Walker, M., Ferry, D. K., & Grondin, R. O. (1988). Limited interconnectivity in synthetic neural systems. In R. Eckmiller & C. von der Malsburg (Eds.), *Neural computers* (pp. 407–416). Berlin: Springer-Verlag.

Akers, L. A., Walker, M., Ferry, D. K., & Grondin, R. O. (1989). A limited interconnect highly layered synthetic neural architecture. In J. Delgado & W. Moore (Eds.), *VLSI for Artificial Intelligence* (pp. 407–416). Norwell, MA: Kluwer Academic.

Asai, S. (1986). Semiconductor memory trends. *Proceedings of the IEEE*, **74**, 1623.

Chen, C., & Hsiao, M. (1984). Error-correcting codes for semiconductor memory applications. *IBM Journal of Research and Development*, **28**, 124.

Chiba, T. (1975). Impact of the LSI on high-speed computer packaging. *IEEE Transactions on Computers*, **C-27**, 319.

Clark, L., & Grondin, R. (1987). Comparison of a pipelined "best match" content addressable memory with neural networks. *Proceedings of the IEEE Neural Network Conference, San Diego*, **III**, 411.

Donald, J., & Akers, L. (1993). An Adaptive Neural Processing Node. *IEEE Neural Networks*. Vol. 4, No. 3, p. 413.

Ferry, D. K. (1985). Interconnect length and VLSI. *IEEE Circuits and Devices*, **1**, 39.

Ferry, D. K., Akers, L. A., & Greeneich, E. (1988). *Ultra large scale integrated microelectronics*. New York: Prentice Hall.

Ferry, D., Grondin, R., & Porod, W. (1987). Interconnections, dissipation and computation. *VLSI Electronics*, **15**, 451.

Fu, K., & Booth, T. (1975). Grammatical interence: Introduction and survey—Part 1. *IEEE Transactions on Systems, Man, and Cybernetics*, **SMC-5**, 95.

Graf, G., Jackel, L., Howard, R., Howard, B., Stranghn, B., Denker, J., Hubbard, W., & Hasler, P. (1988). A synthetic neural cell. *WESCON*.

Hinton, G., & Anderson, J. (1981). *Parallel models of associative memory*. Hillsdale, NJ: Erlbaum.

Huberman, B., & Hogg, T. (1984). Adaptation and self repair in parallel computing structures. *Physical Review Letters*, **52**, 1048.

Kemke, C. (1989). Representing neural networks models by finite automata. In L. Personnaz and G. Dreyfus (Eds.), *Neural networks from models to applications* (pp. 372–379). Paris-IDSET.

Landman, B. S., & Russo, R. L. (1970). On a pin verse block relationship for partitions of logic graphs. *IEEE Transactions on Computers*, **C-20**, 1469.

Le Cun, Y. (1985). Learning process in an asymmetric threshold network. In E. Bienenstock (Ed.), *Disordered systems and biological organization*. New York: Springer-Verlag.

Leen, T., Rudnick, M., & Hammerstrom, D. (1989). Hebbian feature discovery improves classifier efficiency. *Proceedings of the International Joint Conference on Neural Networks*, Washington, DC.

Mackie, S., Graf, H., & Schwartz, D. (1988). Implementations of neural networks models in silicon. *In* R. Eckmiller & C von der Malsberg (Eds.), *Neural Computers* (pp. 467). Berlin: Springer-Verlag.

Madani, K., Garda, P., Belhaire, E., & Devos, F. (1991). Two analog counters for neural network implementation. *IEEE Journal of Solid-State Circuits*, **26** (7).

Murray, A. F., Corso, D. D., & Tarassenko (1991). Pulse-stream VLSI neural networks mixing analog and digital techniques. *IEEE Trans. Neural Networks*, Vol. 2, No. 2, p. 193–204.

Myers, G., Yu, A., & House, D. (1986). Microprocessor technology trends. *Proceedings of the IEEE*, **74**, 1605.

Oja, E. (1982). A simplified neuron model as a principal component analyzer. *Journal of Mathematical Biology*, **15**, 267–273.

Peretto, P., & Niez, J. (1986). Long-term memory storage capacity of multiconnected neural networks. *Biological Cybernetics*, **54**, 53.

Plaut, D., Nowlan, S., & Hinton, G. (1986). Experiments on learning by back propagation. In *Computer Science Technical Report*. Carnegie-Mellon.

Preston, K., & Duff, M. (1984). *Modern cellular automata*. New York: Plenum Press.

Radhakrishnan, V., and Nagaraja, G. (1987). Inference of regular grammar via skeletons. *IEEE Transactions on Systems, Man, and Cybernetics*, **SMC-17**, 982.

Sanger, T. (1989). An optimality principle for unsupervised learning. In David Touretzky, (ed.), *Advances in Neural Information Processing Systems 1* (pp. 11–19). San Mateo, CA: Morgan Kaufmann.

Sejnowski, T., & Rosenberg, C. (1986). NETtalk: A parallel network that learns to read aloud. (Tech. Rep. no. 86/101). Baltimore, MD: Johns Hopkins University.

Shiue, L. C., & Grondin, R. (1987). On designing fuzzy learning neural-automata. *Proceedings IEEE Neural Network Conference San Diego*, **II**, 299.

Shiue, L. C., & Grondin, R. (1989). An automata approach to reconfigurable learning network design. In L. Personnaz & G. Dreyfus (Eds.), *Neural networks from models to applications* (pp. 380–388). Paris-IDSET.

Sivilotti, M., Emerling, M., & Mead, C. (1986). VLSI architectures for implementations of neural networks. *AIP Conference Proceedings*, **151**, 408.

Walker, M., & Akers, L. A. (1988a). A neuromorphic approach to adaptive digital circuitry. *Proceedings of the IEEE Conference on Computers and Communication*, 19.

Walker, M., & Akers, L. A. (1988b). Training of limited-interconnect feedforward neural arrays. *Abstracts of the First Annual INNS Meeting*, 372.

Wang, J., & Clark, L. (1988). Personal communication.

Widrow, B. (1962). Generalization and information storage in networks of adaline neurons. In M. Yovits, G. Jacobi, & G. Goldstein (Eds.), *Self-organizing systems* (pp. 435–461). New York: Spartan.

Widrow, B., & Winter, R. (1988). Neural nets for adaptive filtering and adaptive pattern recognition. *IEEE Computer*, **21**, 25.

IV

Computational and Mathematical Considerations

19

Covariance Storage in the Hippocampus

Terrence J. Sejnowski and Patric K. Stanton

INTRODUCTION

This chapter is primarily concerned with the use of modeling techniques to uncover principles of brain function. This is a different but related enterprise from the practical problems of building machines that solve engineering problems. These two goals, however, are not incompatible. An advance in our understanding of how the brain works is likely to provide new designs for massively parallel computers; the technology being developed could, in turn, be used to help model the brain. However, even within the domain of computational neuroscience there are a number of modeling approaches that should be distinguished (Sejnowski, Koch, & Churchland, 1988).

Realistic Brain Models

One modeling strategy consists of using very large scale simulations that attempt to incorporate as much of the cellular detail as is available (Koch & Segev, 1989). We call these realistic brain models. While this approach to simulation can be very useful, the realism of the model is both a strength and a weakness. As the model is made increasingly realistic by adding more variables and parameters, the danger is that the simulation ends up as poorly understood as the nervous system itself. Equally worrisome, since we do not yet know all the cellular details, important features may be inadvertently left out, thus invalidating the results. Finally, realistic simulations are highly computation intensive. Present constraints limit simulations to tiny nervous systems or small components of more complex systems. Only recently has sufficient computer power been available to go beyond the simplest models. Realistic models require a substantial empirical database; it is all too easy to make a complex model fit a limited subset of the data.

Simplifying Brain Models

Because even the most successful realistic brain models may fail to reveal the function of the tissue, computational neuroscience needs to develop simplifying models that capture important principles. Textbook examples in physics that admit exact solutions are typically unrealistic, but they are valuable because they illustrate physical principles. Minimal models that reproduce the essential properties of physical systems, such as phase transitions, are even more valuable. The study of simplifying models of the brain can provide a conceptual framework for isolating the basic computational problems and understanding the computational constraints that govern the design of the nervous system. Simplifying models are essential but are also dangerously seductive; a model can become an end in itself and lose touch with nature.

In this chapter we present a set of theoretical ideas and experimental results about plasticity in the nervous system. We will show that a combination of modeling and experimentation can help overcome some of the limitations inherent in the complexity of the brain. In the next few sections we present several simplifying neural network models that were developed for modeling associative memory. Later in this chapter we describe a novel form of Hebbian synaptic plasticity in the mammalian hippocampus that confirms predictions made from covariance models of associative memory. There is already some evidence to indicate that similar forms of plasticity are also found in the cerebral cortex.

ASSOCIATIVE MEMORY

In 1949 Donald Hebb published *The Organization of Behavior* in which he introduced several hypotheses about the neural substrates of learning and memory, including the Hebb learning rule or Hebb synapse. The Hebb rule and variations on it have also served as the starting point for the study of information storage in simplifying models (Sejnowski, 1981; Kohonen, 1984; McClelland & Rumelhart, 1986; Rumelhart & McClelland, 1986; Sejnowski & Tesauro, 1989). Many types of networks have been studied—networks with random connectivity, networks with layers, networks with feedback between layers, and a wide variety of local patterns of connectivity. Even the simplest network model has complexities that are difficult to analyze.

The Hebb synapse, or Hebb rule, has been used to signify a wide variety of ideas and mechanisms, so it would be worthwhile to start by examining what Hebb (1949) actually proposed. "When an axon of cell A is near enough to excite cell B or repeatedly or persistently takes part in firing it, some growth process or metabolic change takes place in one or both cells such that A's efficiency, as one of the cells firing B, is increased."

This verbal description is a general statement about the factors and conditions that could be important for changing synaptic strengths. There are two ways to sharpen the statement. First, we can formalize the verbal description as a quantitative equation, and second, we can specify in greater physiological detail what is meant by having one cell "excite" another.

Consider first a neuron A, with average firing rate $V_A(t)$, that projects to neuron B, with average firing rate $V_B(t)$. The synaptic connection from A to B has a strength value W_{BA}, which determines the degree to which activity in A is capable of exciting B. In linear models, the average postsynaptic depolarization of B due to A is taken to be the product of the firing rate V_A times the synaptic strength value W_{BA}. In other models the relationship between the inputs and outputs could be nonlinear. The statement of Hebb above states that the strength of the synapse W_{BA} should be modified in some way which is dependent on both activity in A and activity in B. The most general expression which captures this notion is

$$\Delta W_{BA}(t) = F\ (V_A(t), V_B(t)) \tag{1}$$

which states that the change in the synaptic strength W_{BA} at any given time is some as yet unspecified function F of both the presynaptic and the postsynaptic firing rates. Given this general form of the assumed learning rule, it is then necessary to choose a particular form for the function $F(V_A, V_B)$. The most straightforward interpretation of what Hebb said is a simple product

$$\Delta W_{BA}(t) = \varepsilon\ V_A(t)\ V_B(t) \tag{2}$$

where ε is a numerical constant usually taken to be small. However, there are many other choices possible for the function $F(V_A, V_B)$. The choice depends on the particular architecture and the problem at hand. Equation 2 might be appropriate for a simple associative memory task, but for other tasks one would need different forms of the function $F(V_A, V_B)$. For example, in classical conditioning, the precise timing relationships of the presynaptic and postsynaptic signals are important, and plasticity must then depend on the rate of change of firing, or on the "trace" of the firing rate, that is, weighted average over previous times, rather than simply depending on the current instantaneous firing rate (Tesauro, 1986; Klopf, 1988).

Probably the most important and most thoroughly explored use of the Hebb rule is in the formation of associations between one stimulus or pattern of activity and another. The Hebb rule is appealing for this use because it provides a way of forming global associations between macroscopic patterns of activity in assemblies of neurons using only the local information available at individual synapses. It is important to keep in mind that much more complex rules are possible, such as a function $F(V_A, V_B, V_C)$, which depends additionally on a third neuron C in a heterosynaptic fashion.

The earliest models of associative memory were based on network models in which the output of a model neuron was assumed to be proportional to a linear sum of its inputs, each weighted by a synaptic strength. Thus,

$$V_B(t) = \sum_{A=1}^{N} W_{BA} V_A(t) \tag{3}$$

where V_B are the firing rates of a group of M output cells, and V_A are the firing rates of a group of N input cells, and W_{BA} is the synaptic strength between input cell A and output cell B. Note that A and B are being used here as indices to represent one output of a group of cells.

The transformation between patterns of activity on the input vectors to patterns of activity on the output vectors is determined by the synaptic weight matrix, W_{BA}. How should this matrix be chosen if the goal of the network is to associate a particular output vector with a particular input vector? The earliest suggestions were all based on the Hebb rule (Steinbuch, 1961; Longuet-Higgins, 1968; Anderson, 1970; Kohonen, 1970). It is easy to verify by direct substitution of Equation 2 into Equation 3 that the increment in the output is proportional to the desired vector and the strength of the learning rate, ε, can be adjusted to scale the outputs to the desired values.

More than one association can be stored in the same matrix, as long as the input vectors are not too similar to each other. This is accomplished by using Equation 2 for each input–output pair. This model of associative storage is simple and has several attractive features. First, the learning occurs in only one trial; second, the information is distributed over many synapses, so that recall is relatively immune to noise or damage; and third, input patterns similar to stored inputs will give output similar to the stored outputs, a form of generalization. This model also has some strong limitations, such as interference between information associated with similar input vectors. Nonlinear models can overcome some of these limitations (Kohonen, 1984; Hopfield & Tank, 1986); however, the learning algorithms used in these models are similar to those presented here for simpler linear models.

THE COVARIANCE RULE

One problem with any synaptic modification rule that can only increase the strength of a synapse is that the synaptic strength will eventually saturate at its maximum value. The weights can be reduced by nonspecific decay, but the stored information will also decay and be lost at the same rate. Another approach is to renormalize the total synaptic weight of the entire terminal field from a single neuron to a constant value (von der Malsburg, 1973). This could be accomplished, for example, by a mechanism for heterosynaptic depression, in which the persistent firing of neuron A, which increased the strength of the synapse to neuron B, would depress the strengths of all other synapses on neuron B.

Alternatively, a more flexible learning rule could be used that decreased the strength of a plastic synapse as specifically as the Hebb rule increased it. The covariance rule is an example of a variation on the Hebb rule that solves the problem of dynamical range and saturation at a plastic synapse, while permitting differential modifications contingent upon the statistics of the presynaptic and postsynaptic activities (Sejnowski, 1977a,b). According to this rule, the change in strength of a plastic synapse should be proportional to the covariance between the presynaptic firing and postsynaptic firing

$$\Delta W_{BA}(t) = \varepsilon\,(V_B(t) - \langle V_B \rangle)\,(V_A(t) - \langle V_A \rangle) \qquad (4)$$

where $\langle V_B \rangle$ are the average firing rates of the output neurons, averaged over a longer time interval than V_B, and $\langle V_A \rangle$ are the average firing rates of the input neurons (Chauvet, 1986). Thus, the strength of the synapse should increase if the firing of the presynaptic and postsynaptic elements are positively correlated, decrease if they are negatively correlated, and remain unchanged if they are uncorrelated.

The covariance rule is a special case of the form of the Hebb rule in Equation 1. It differs from the simple Hebb rule in Equation 2 by the addition of an extra term. If we take the time average of Equation 4 we can rewrite it in the form

$$\langle \Delta W_{BA}(t) \rangle = \varepsilon\,(\langle V_B(t)\, V_A(t) \rangle - \langle V_B \rangle \langle V_A \rangle) \qquad (5)$$

Both terms on the right hand side have the same form as the simple Hebb synapse in Equation 2. Thus, the covariance learning algorithm can be realized by applying the Hebb rule relative to a "threshold" that varies with the product of the time-averaged presynaptic and postsynaptic activity levels. The time scale for taking the average activity levels must be longer than that for the synaptic plasticity. The effect of the threshold is to ensure that no change in synaptic strength should occur if the average correlation between the presynaptic and postsynaptic activities is at chance level.

The covariance form of the Hebb rule in Equation 5 has the important advantage that the strength of the synapse can be used throughout its dynamical range. Thus, the strength of a synapse that was near its maximum value could be selectively decreased if the activity of the presynaptic terminal and the postsynaptic neuron were negatively correlated. This would occur if the postsynaptic neuron was active while the presynaptic terminal was inactive, or *vice versa*.

SYNAPTIC PLASTICITY IN THE HIPPOCAMPUS

Until recently, preparations for studying long term plasticity at synapses were not available, so it was difficult to test Hebb's hypothesis. Phenomena such as posttetanic potentiation last only a few minutes. Seventeen years ago Bliss and Lømo (1973) identified a long lasting enhancement of synaptic strength in the mammalian hippocampus, now called long term potentiation (LTP). When input fibers to pyramidal cells were stimulated at a high frequency, in the 30–100 Hz range, synaptic strengths remained elevated for many hours. This effect was homosynaptic, since the potentiated synapses were the same ones that were stimulated.

Experiments designed to test the involvement of the postsynaptic cell in the generation of LTP were reported by Kelso, Ganong, and Brown (1986), Malinow and Miller (1986), and Gustafsson et al. (1987). They stimulated the presynaptic terminals with a high frequency tetanus while simultaneously injecting current into a postsynaptic cell with an intracellular microelectrode. They reported that pairing the stimulus with a depolarization produced LTP, but pairing with hyperpolarization blocked the induction of LTP. This is consistent with a Hebbian mechanism.

In his description of the conditions for plasticity, Hebb specified the excitation of the postsynaptic cell leading to its firing an action potential. In the hippocampus, LTP can be induced even when action potentials in the postsynaptic cell are blocked (Kelso, et al., 1986). Evidently, it is enough for the postysnaptic cell to be strongly depolarized at the same time that the presynaptic terminal is stimulated. Although this does not change the spirit of the Hebb rule, the details make a difference with regard to dendritic processing. Neighboring synapses on a dendrite could communi-

TABLE 1 Summary of the Combinations of Presynaptic Activity and Levels of Postsynaptic Potentials that Lead to Different Forms of Synaptic Plasticity in the Hippocampus

	Postsynaptic activity	
Presynaptic activity	Hyperpolarization	Depolarization
Low	—	Heterosynaptic depression
High	Homosynaptic depression (LTD)	Hebbian potentiation (LTP)

cate through depolarization without involving more distant synapses. An influence that spread through somatic action potentials would involve many more synapses, but a mechanism that was based on subthreshold depolarization would allow each branch of dendrite to act semi-independently in locally regulating synaptic plasticity (Finkel & Edelman, 1985). Thus, the focus that Hebb placed on the cell as the processing unit may have to shift to a finer level, perhaps to the level of individual dendritic branches (Shepherd et al., 1985). Of course, the ionic currents generated in the dendrites sum in the cell body to produce an output which is typically encoded as trains of action potentials.

If depolarization of the postsynaptic cell together with presynaptic activity is sufficient to produce LTP, then a weak presynaptic stimulus, which by itself is not strong enough to produce LTP, should be potentiated when paired with the strong stimulation of another separate pathway. This, in fact, happens and has been called associative LTP because of the cooperativity between inputs (McNaughton, Douglas, & Goddard, 1978; Levy & Steward, 1979, 1983; Barrionuevo & Brown, 1983). Thus, correlations *between* neighboring synapses could be detected with this mechanism and information in the correlations could be stored through associative LTP of the relevant synapses.

Table 1 is a summary of the possible conditions for plasticity based on coincidence or noncoincidence of presynaptic and postsynaptic activity. The Hebbian condition occurs when they are both active. There is some evidence for synaptic depression of the sort that would be needed for the covariance rule in Equation 5. When one set of inputs to an area is inactive (Dunwiddie & Lynch, 1977; Lynch, Dunwiddie, & Gribkoff, 1977, Levy & Steward, 1979) or weakly active (Levy & Steward, 1983) during the stimulation of a strong input, the strengths of the inactive synapses are depressed. This is a heterosynaptic form of depression and does not depend on the pattern of weak input activity. Also, the duration of the depression is typically not as long lasting as LTP. A candidate mechanism for long term depression (LTD) should have roughly the same strength and duration as LTP itself.

We recently searched for conditions under which the stimulation of a hippocampal pathway, rather than its inactivity, could produce either long term depression or potentiation of synaptic strengths, depending on the pattern of stimulation (Stanton & Sejnowski, 1989). The stimulus paradigm that we used (Figure 1) is based on the finding that high frequency bursts of stimuli at 5 Hz are optimal in eliciting LTP (Larson & Lynch, 1986). This is close to the 5–6 Hz theta rhythm normally recorded in the hippocampus during some behaviors associated with learning. A strong bursting stimulus was applied to the Schaffer collaterals and a weak low frequency stimulus was applied to a separate subicular input on the opposite side of the recording site; each shock of the weak input was either superimposed on the middle of each burst of the strong input (IN PHASE) or occurred symmetrically between the bursts (OUT OF PHASE).

Extracellular evoked field potentials were recorded from the apical dendritic and somatic layers of CA1 pyramidal cells. The weak stimulus train was first applied alone and did not itself induce long lasting

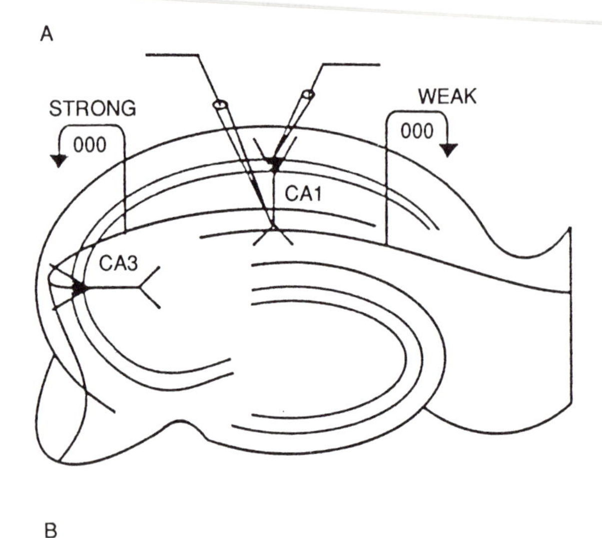

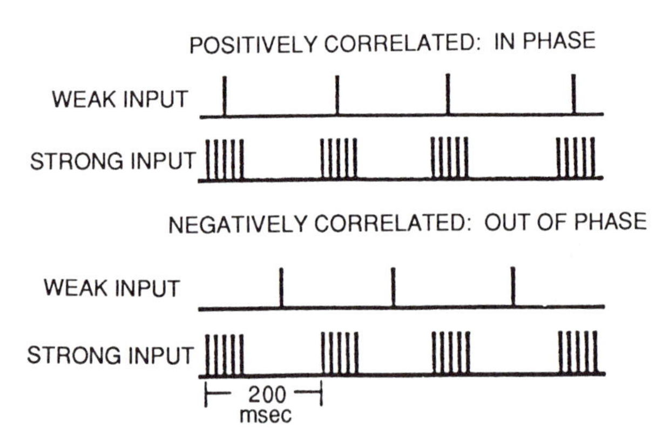

FIGURE 1 Hippocampal slice preparation and associative stimulus paradigms. A. Schematic diagram of the *in vitro* hippocampal slice showing recording sites in the CA1 pyramidal cell somatic (stratum pyramidale) and dendritic (stratum radiatum) layers, and stimulus sites activating Schaffer collateral (STRONG) and commissural (WEAK) afferents. B. Schematic diagram of stimulus paradigms used. Strong input stimuli (STRONG INPUT) were four trains of 100 Hz bursts. Each burst had five stimuli and the interburst interval was 200 msec. Each train lasted 2 sec and had a total of 50 stimuli. Weak input stimuli (WEAK INPUT) were four trains of shocks at 5 Hz frequency, each train lasting for 2 sec. When these inputs were IN PHASE, the weak single shocks were superimposed on the middle of each burst of the strong input, as shown. When the weak input was OUT OF PHASE, the single shocks were placed symmetrically between the bursts.

changes in synaptic strength. The strong site was then stimulated alone, which elicited homosynaptic LTP of the strong pathway while not significantly altering the amplitude of responses to the weak input. When weak and strong inputs were activated in phase, there was an associative LTP of the weak synapses (Figure 2A). Both the synaptic excitatory postsynaptic potential and population action potential were significantly enhanced for at least 60–180 min following stimulation.

In contrast, when weak and strong inputs were applied out of phase, we observed an associative long term depression of the weak input synapses (Figure 2B). There was a marked reduction in the population spike with smaller decreases in the EPSP. Note that the stimulus patterns applied to each input were identical in these two experiments, and only the relative phase of the weak and strong stimuli was altered. With these stimulus patterns, synaptic strength could be repeatedly enhanced and depressed in a single slice (Figure 2C).

The simultaneous depolarization of the postsynaptic membrane and activation of glutamate receptors of the *N*-methyl-D-aspartate (NMDA) subtype appears to be necessary for LTP induction (Collingridge, Kehl, & McLennan, 1983; Harris, Ganong, & Cotman, 1984; Wigstrom & Gustafsson, 1984). The spread of current from strong to weak synapses in the dendritic tree paired with glutamate release from the weak input could account for the ability of a strong pathway to associatively potentiate a weak one (Barrionuevo & Brown, 1983). Consistent with this hypothesis, we find that the NMDA receptor antagonist 2-amino-5-phosphonovaleric acid (AP5, 10 μM) blocks the induction of associative LTP in CA1 pyramidal neurons. In contrast, the application of AP5 to the bathing solution at this same concentration had no significant effect on associative LTD. Thus, the induction of depression seems to involve mechanisms different from potentiation.

In further experiments, intracellular recordings from CA1 pyramidal neurons were made using standard techniques. Induction of associative LTP (Figure 3, WEAK S+W IN PHASE) produced an increase in

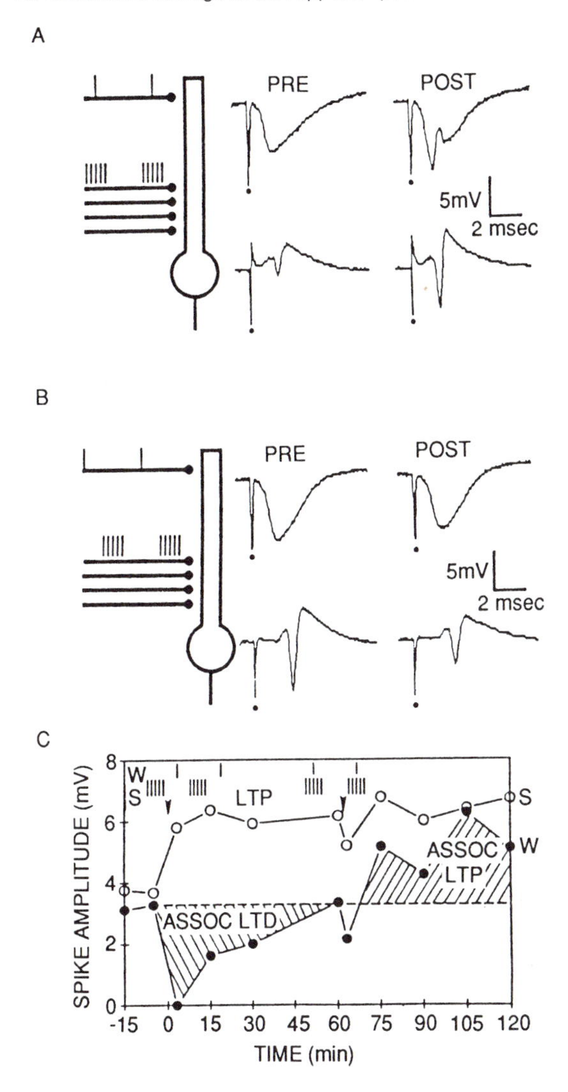

FIGURE 2 Illustration of associative long-term potentiation (LTP;A) and associative long-term depression (LTD;B) using extracellular recordings. A. Associative LTP of evoked excitatory postsynaptic potentials (EPSPs) and population action potential responses in the weak input. Test responses are shown before (Pre) and 30 min after (Post) application of weak stimuli in phase with the coactive strong input. B. Associative LTD of evoked EPSPs and population spike responses in the weak input. Test responses are shown before (Pre) and 30 min after (Post) application of weak stimuli out of phase with the coactive strong input. C. Time course of the changes in population spike amplitude observed at each input for a typical experiment. Test responses from the strong input (S, open circles) show that the high-frequency bursts (5 pulses/100 Hz, 200 msec interburst interval as in Figure 1) elicited synapse-specific LTP independent of other input activity. Test responses from the weak input (W, filled circles) show that stimulation of the weak pathway out of phase with the strong one produced associative LTD (Assoc LTD) of this input. Associative LTP (Assoc LTP) of the same pathway was then elicited following in-phase stimulation. Amplitude and duration of associative LTD or LTP could be increased by stimulating input pathways with more trains of shocks.

amplitude of the excitatory postsynaptic potential (EPSP) and a lowered action potential threshold in the weak pathway, as reported previously (Barrionuevo & Brown, 1983). Conversely, the induction of associative LTD (Figure 3, WEAK S+W OUT OF PHASE) was accompanied by a long lasting reduction of EPSP elicited by weak input stimulation.

A weak stimulus that is out of phase with a strong stimulus arrives when the postsynaptic neuron is hyperpolarized as a consequence of inhibitory postsynaptic potentials and afterhyperpolarization from mechanisms intrinsic to pyramidal neurons. This suggests that postysnaptic hyperpolarization coupled with presynaptic activation may trigger LTD. To test this hypothesis, we injected current with intracelluar microelectrodes to hyperpolarize or depolarize the cell while stimulating a synaptic input at low frequency. Pairing the injection of depolarizing current with weak input stimulation led to LTP of those synapses (Figure 4A, STIM), while a control input inactive during the stimulation did not change (CONTROL), as reported previously (Kelso, et al., 1986; Malinow & Miller, 1986; Gustafsson et al., 1987). Conversely, prolonged hyperpolarizing current injection paired with the same weak stimuli led to induction of LTD in the stimulated pathway (Figure 4B, STIM), but not in the unstimulated pathway (CONTROL). The application of either depolarizing current, hyperpolarizing current, or the weak 5 Hz synaptic stimulation alone did not induce long term alterations in synaptic strengths. Thus, hyperpolarization and simultaneous presynaptic activity is sufficient for the induction of LTD in CA1 pyramidal neurons.

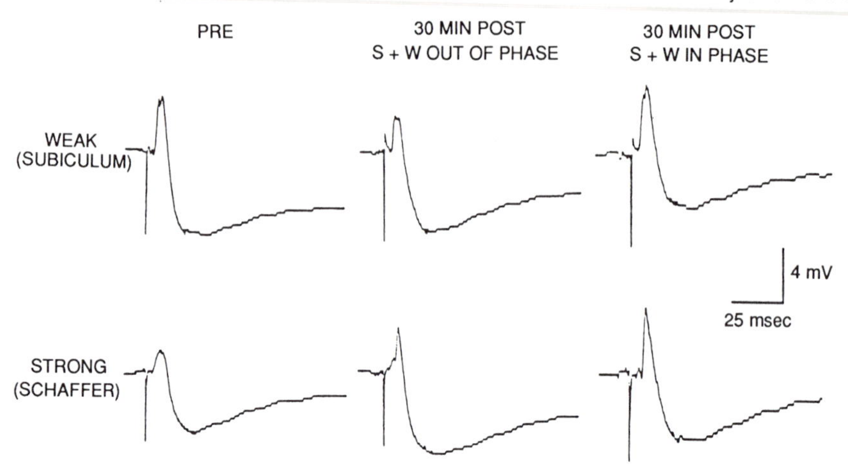

FIGURE 3 Demonstration of associative LTP and LTD using intracellular recordings from a CA1 pyramidal neuron. Intracellular EPSPs prior to repetitive stimulation (Pre), 30 min after out of phase stimulation (S+W OUT OF PHASE), and 30 min after subsequent in phase stimuli (S+W IN PHASE). The strong input (Schaffer collateral side, lower traces) exhibited LTP of the evoked EPSP independent of weak input activity. Out of phase stimulation of the weak (subicular side, upper traces) pathway produced a marked persistent reduction in EPSP amplitude. In the same cell, subsequent in phase stimuli resulted in associative LTP of the weak input that reversed the LTD and enhanced amplitude of the EPSP past the original baseline.

The properties of associative LTD described here make it a good candidate for the covariance learning rule outlined in the last section. Since there is a large variation in the strength and duration of both LTP and LTD in different slices, we designed a stimulus pattern to compare them in the same slice at the same time. The stimulus combined both the weak input shocks superimposed with the bursts and between the bursts, so that on average there was no net covariance between weak and strong inputs. This stimulus produced no net change in synaptic strength using extracellular recording techniques, as predicted by the covariance rule in Equation 4. Thus, the associative LTP and LTD mechanisms appear to be balanced.

SYNAPTIC PLASTICITY IN THE VISUAL CORTEX

Neurons in the visual cortex of cats and monkeys respond preferentially to oriented bars and edges. A network model for the development of neuronal selectivity incorporating the Hebbian form of plasticity was proposed by Bienenstock, Cooper, and Munro (1982). The BCM model requires patterned visual inputs and in this respect is similar to an earlier proposal by von der Malsburg (1973). The BCM algorithm for synaptic modification is a special case of the general Hebb rule in Equation 1,

$$\Delta W_{BA} = \phi(V_B, < V_B >) V_A \quad (6)$$

where the function $\phi(V_B, <V_B>)$ is shown in Figure 5. The synapse is strengthened when the average postsynaptic activity exceeds a threshold, and is weakened when the activity falls below the threshold level. Furthermore, the threshold varies according to the average postsynaptic activity

$$\theta = < V_B >^2 \quad (7)$$

Bienenstock et al. (1982) show that this choice has desirable stability properties and allows neurons to become selectively sensitive to common features in input patterns.

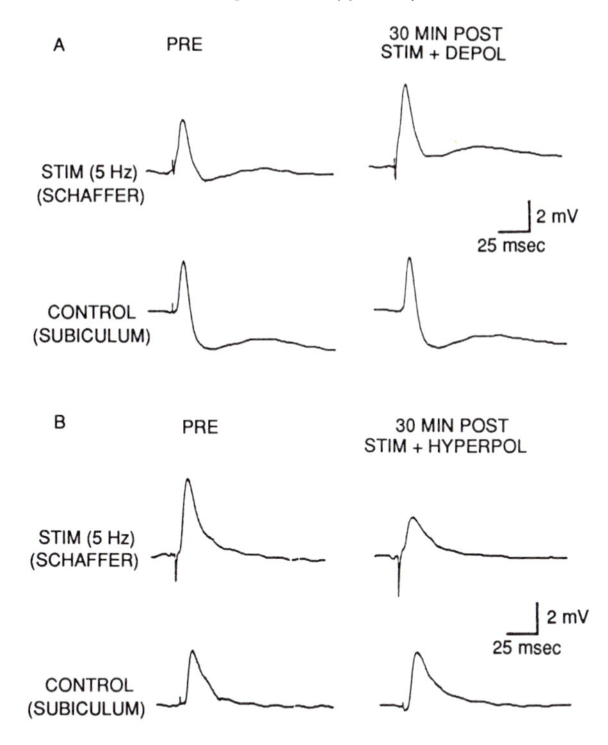

FIGURE 4 Pairing of postsynaptic hyperpolarization with stimulation of synapses of CA1 hippocampal pyramidal neurons produces LTD specific to the activated pathway, while pairing of postsynaptic depolarization with synaptic stimulation produces synapse-specific LTP. A. Intracellular invoked EPSPs are shown at stimulated (STIM) and unstimulated (CONTROL) pathway synapses before (Pre) and 30 min after (Post) pairing a 20 mV depolarization (constant current, +2.0 nA) with 5 Hz synaptic stimulation. The stimulated pathway exhibited associative LTP of the EPSP while the control, unstimulated input showed no change in synaptic strength. (RPM = -65 mV; R_N = 35 MΩ) B. Intracellular EPSPs are shown evoked at stimulated and control pathway synapses before (Pre) and 30 min after (Post) pairing a 20 mV hyperpolarization (constant current, -1.0 nA) with 5 Hz synaptic stimulation. The input (STIM) activated during the hyperpolarization showed associative LTD of synaptic evoked EPSPs while synaptic strength of the silent input (CONTROL) was unaltered.

The covariance form of the Hebb synapse has been used by Linsker (1986) to model the formation of receptive fields in the early stages of visual processing. The model is a layered network having limited connectivity between layers and uses the learning rule in Equation 5. As the learning proceeds, the units in the lower layers of the network develop on-center and off-center receptive fields that resemble the receptive fields of ganglion cells in the retina, and elongated receptive fields develop in the upper layers of the network that resemble simple receptive fields found in visual cortex. This model demonstrates that some of the properties of sensory neurons could arise spontaneously during development by specifying the general pattern of connectivity and a few parameters to control the synaptic plasticity. One surprising aspect of the model is that regular receptive fields develop even though only spontaneous activity is present at the sensory receptors.

The visual response properties of neurons in the visual cortex of cats and monkeys are plastic during the first few months of postnatal life, and can be permanently modified by visual experience (Wiesel & Hubel, 1965; Sherman & Spear, 1982). Normally, most cortical neurons respond to visual stimuli from either eye. Following visual deprivation of one eye by eyelid suture during the critical period, the ocular preference of neurons in primary visual cortex shifts toward the nondeprived eye. In another type of experiment, a misalignment of the two eyes during the critical period produces neurons that respond to only one eye and, as a consequence, binocular depth perception is impaired. These and many other experiments have led to testable hypotheses for the mechanisms underlying synaptic plasticity during the critical period (Bear, Cooper, & Ebner, 1987).

Singer (1987) has suggested that the voltage-dependent entry of calcium into spines and the dendrites of postsynaptic cells may trigger the molecular changes required for synaptic modification in visual cortex. This hypothesis is being tested at a molecular level using a combined pharmacological and physiological technique. NMDA receptor antagonists infused into visual cortex block the shift in ocular dominance normally associated with monocular deprivation (Kleinschmidt, Bear, & Singer, 1986). The NMDA receptor is a candidate mechanism for triggering synaptic modification because it allows calcium to enter a cell only if the neurotransmitter binds to the receptor while the

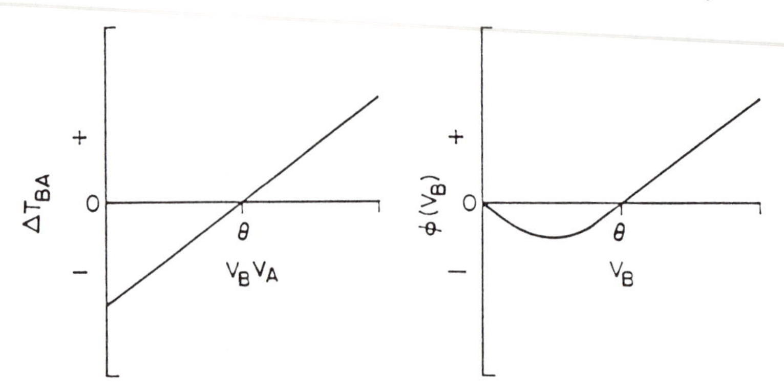

FIGURE 5 *Left*, change in synaptic strength ΔT_{BA} as a function of the average correlation $\langle V_A V_B \rangle$ between the presynaptic and postsynaptic activity levels, as indicated in Equation 5. The threshold θ is given by $\langle V_B \rangle \langle V_A \rangle$. *Right*, the postsynaptic factor $\phi(V_B, \langle V_B \rangle)$ in the BCM learning algorithm in Equation 6, where the threshold $\theta = \langle V_B \rangle^2$ (Bienenstock, Cooper, & Munro, 1982).

postsynaptic membrane is strongly depolarized. In a sense, the NMDA receptor is a "Hebb molecule" since it is only activated when there is a conjunction of presynaptic and postsynaptic activity. The NMDA receptor also is critically involved in the induction of LTP in the hippocampus (Collingridge et al., 1983; Harris et al., 1984; Wigstrom & Gustafsson, 1984).

Stent (1973) has suggested that the effects of monocular deprivation could be explained if the synaptic weight were to decrease when the synapse is inactive and the postsynaptic cell is active. This is similar to the condition that leads to heterosynaptic depression in the hippocampus (Lynch et al., 1977; Levy & Steward, 1979, 1983). The evidence for synaptic depression found in visual cortex during the critical period would correspond to the upper right corner of Table 1, in which presynaptic activity is absent but postsynaptic activity is normal.

The conditions that correspond to the lower left corner of Table 1 have only recently been tested in the visual cortex. Chronic hyperpolarization of neurons was produced by infusion of muscimol, a GABA agonist, while one of the eyes was sutured shut during the critical period. When the eye was opened, neurons in visual cortex near the site of infusion could be driven only by the closed eye, in contrast to neurons more distant from the infusion site, which could only be driven from the open eye (Reiter & Stryker, 1987). One interpretation of these results is that presynaptic transmitter release onto hyperpolarized cells leads to a long term depression of the input synapses arising from the lateral geniculate nucleus, but that inactive terminals are not affected. These conditions are similar to those reported here that lead to associative LTD in the hippocampus (Stanton & Sejnowski, 1989). Thus, the Hebbian mechanisms found in the hippocampus are likely to be found elsewhere in the central nervous system.

The mechanisms for plasticity in the cerebral cortex during development may be related to mechanisms responsible for synaptic plasticity in the adult. The evidence so far favors the general form of Hebbian plasticity in Equation 1. However, the details of how this plasticity is regulated on short and long time scales may be quite different during development and in the adult. Recently, it has been shown that the receptive field properties of cells in cat visual cortex can be altered even in the adult by visual experience paired with ionophoretic excitation or depression of cellular activity (Fregnac, Schulz, Thorpe, & Bienenstock, 1988; Greuel, Luhmann, & Singer, 1988). These results are consistent with the presence of Hebbian covariance mechanisms, though the complexity of visual cortex prevents a direct interpretation of these results at the level of identified synapses.

CONCLUSIONS

The experiments on associative synaptic depression in the hippocampus summarized in this chapter were based on ideas that first arose in the context of modeling some aspects of memory. These simplifying models leave out most of the biological details of real neurons and do not even refer to specific brain areas. What such models provide is a general framework for thinking about the complex relationships that could exist between signals in neural circuits like those found in the hippocampus. Simplifying models suggest possible experiments and help with interpreting their outcomes. In our experiments, the choice of the stimulus parameters was inspired by the prediction made by the covariance model of memory that anticorrelation was the critical condition for synaptic depression (Sejnowski, 1977b).

However, the covariance model did not provide the details of the stimulus paradigm, but only the general conditions. The choice of 100 Hz for the burst rate and 5 Hz for the burst repetition rate was determined by properties of the hippocampus. What the model did provide was the idea that synaptic depression comparable in magnitude and duration to LTP is likely to be found in the hippocampus, and the general properties that would characterize its occurrence. The covariance model pointed to anticorrelation as the key variable. The model only narrowed the range of possibilities. Two forms of synaptic depression have in fact been found in the hippocampus in different areas under different conditions. These are likely to have different functions. The homosynaptic form of LTD that we described in this chapter has many of the characteristics needed to balance LTP.

LTP and LTD are candidate mechanisms for long term information storage in neuronal networks. If the strength of synapses in the hippocampus can be enhanced or depressed repeatedly, then these coupled mechanisms could also provide a means for implementing a "working memory." This type of memory could be used to temporarily store the information needed to accomplish a task. For example, the strengths of some synapses could be incremented by LTP and these strengths maintained for an indefinite interval. When this information was no longer needed, the synaptic strengths could be selectively decreased with LTD. It is known that the long term memories of facts and events are stored not in the hippocampus, but in the cerebral cortex, so it will be of particular interest to determine whether mechanisms similar to LTP and LTD are found there as well.

More needs to be known about the timing relationships for LTP and LTD, and also about the spatial integration of information within dendritic trees. Realistic models can help with sorting out these relationships, but only if enough data can be obtained to fully constrain the models. Here is an example of how two different types of models, both simplifying and realistic, can each contribute, on different levels, to the solution of difficult problems in storing and retrieving information in neural populations.

Acknowledgments

This research was supported by grants from the National Science Foundation and the Office of Naval Research to T.J.S.

References

Anderson, J. A. (1970). Two models for memory organization using interacting traces. *Mathematical Biosciences*, **8**, 137–160.

Barrionuevo, G., & Brown, T. H. (1983). Associative long-term potentiation in hippocampal slices. *Proceedings of the National Academy of Sciences of the United States of America*, **80**, 7347–7351.

Bear, M., Cooper, L. N., & Ebner, F. F. (1987). A physiological basis for theory of synapse modification. *Science*, **237**, 42–48.

Bienenstock, E. L., Cooper, L. N., & Munro, P. W. (1982). Theory for the development of neuron selectivity: Orientation specificity and binocular interaction in visual cortex. *Journal of Neuroscience*, **2**, 32–48.

Bliss, T. V. P., & Lømo, T. (1973). Long-lasting potentiation of synaptic transmission in the dentate area of the anaesthetized rabbit following stimulation of the perforant path. *Journal of Physiology,* **232**, 334–356.

Chauvet, G. (1986). Habituation rules for theory of the cerebellar cortex. *Biological Cybernetics*, **55**, 201–209.

Collingridge, G. L., Kehl, S. L., & McLennan, H. (1983). Excitatory amino acids in synaptic transmission in the Schaffer collateral–commissural pathway of the rat hippocampus. *Journal of Physiology*, London, **334**, 33–46.

Dunwiddie, T., & Lynch, G. (1977). Long-term potentiation and depression of synaptic responses in the rat hippocampus: Localization and frequency dependence. *Journal of Physiology*, **276**, 353–367.

Finkel, L. H., & Edelman, G. M. (1985). Interaction of synaptic modification rules within populations of neurons. *Proceedings of the National Academy of Sciences of the United States of America*, **82**, 1291–1295.

Fregnac, Y., Schulz, D., Thorpe, S., & Bienenstock, E. (1988). A cellular analogue of visual cortical plasticity. *Nature*, **333**, 367–370.

Greuel, J. M., Luhmann, H. J., & Singer, W. (1988). Pharmacological induction of use-dependent receptive field modifications in the visual cortex. *Science*, **242**, 74–77.

Gustafsson, B., Wigstrom, H., Abraham, W. C., & Huang, Y. Y. (1987). Long-term potentiation in the hippocampus using depolarizing current pulses as the conditioning stimulus to single volley synaptic potentials. *Journal of Neuroscience*, **7**, 774–780.

Harris, E. W., Ganong, A. H., & Cotman, C. W. (1984). Long-term potentiation in the hippocampus involves activation of N-methyl-D-aspartate receptors. *Brain Research*, **323**, 132–137.

Hebb, D. O. (1949). *The organization of behavior*. New York: Wiley.

Hopfield, J. J., & Tank, D. W. (1986). Computing with neural circuits. *Science*, **233**, 625–633.

Kelso, S. R., Ganong, A. H., & Brown, T. H. (1986). Hebbian synapses in hippocampus. *Proceedings of the National Academy of Sciences of the United States of America*, **83**, 5326–5330.

Kleinschmidt, A., Bear, M. F., & Singer, W. (1986). *Neuroscience Letters Supplement*, **26**, S58.

Klopf, A. H. (1988). A neuronal model of classical conditioning. *Psychobiology*, **16**, 85–125.

Koch, C., & Segev, I. (1989). *Methods in neuronal modeling: From synapse to networks.* Cambridge, MA: MIT Press.

Kohonen, T. (1970). Correlation matrix memories. *IEEE Transactions on Computers*, **C-21,** 353–359.

Kohonen, T. (1984). *Self-organization and associative memory.* New York: Springer-Verlag.

Larson, J., & Lynch, G. (1986). Induction of synaptic potentiation in hippocampus by patterned stimulation involves two events. *Science*, **232**, 985–987.

Levy, W. B., & Steward, O. (1979). Synapses as associative memory elements in the hippocampal formation. *Brain Research*, **175**, 233–245.

Levy, W. B., & Steward, O. (1983). Temporal contiguity requirements for long-term associative potentiation/depression in the hippocampus. *Neuroscience*, **8**, 791–797.

Linsker, R. (1986). From basic network principles to neural architecture: Emergence of orientation columns. *Proceedings of the National Academy of Sciences of the United Statesof America*, **83**, 8779–8783.

Longuet-Higgins, H. C. (1968). Holographic model of temporal recall. *Nature*, **217**, 104–107.

Lynch, G. S., Dunwiddie, T. V., & Gribkoff, V. K. (1977). Heterosynaptic depression: A post-synaptic correlate of long-term potentiation. *Nature*, **266**, 737–739.

Malinow, R., & Miller, J. P. (1986). Postsynaptic hyperpolarization during conditioning reversibly blocks induction of long-term potentiation. *Nature*, **320**, 529–531.

McClelland, J. L., & Rumelhart, D. E. (Eds.) (1986). *Parallel distributed processing* (Vol. 2. *Psychological and biological models*). Cambridge, MA: MIT Press.

McNaughton, B. L., Douglas, R. M., & Goddard, G. V. (1978). Synaptic enhancement in fascia dentata: Cooperativity among coactive afferents. *Brain Research*, **157**, 277–293.

Reiter, H. O., & Stryker, M. P. (1987). A novel expression of plasticity in kitten visual cortex in the absence of postsynaptic activity. *Society for Neuroscience Abstracts*, **13**, 1241.

Rumelhart, D. E., & McClelland, J. L. (Eds.) (1986). *Parallel distributed processing* (Vol. 1. *Foundations*). Cambridge, MA: MIT Press.

Sejnowski, T. J. (1977a). Statistical constraints on synaptic plasticity. *Journal of Mathematical Biology*, **69**, 385–389.

Sejnowski, T. J. (1977b). Storing covariance with nonlinearly interacting neurons. *Journal of Mathematical Biology*, **4**, 303–321.

Sejnowski, T. J. (1981). Skeleton filters in the brain. In G. E. Hinton & J. A. Anderson (Eds.), *Parallel models of associative memory.* Hillsdale, NJ: Erlbaum.

Sejnowski, T. J., Koch, C., & Churchland, P. S. (1988). Computational neuroscience. *Science*, **241**, 1299–1306.

Sejnowski, T. J., & Tesauro, G. (1989). Building network learning algorithms from Hebbian synapses. In J. L. McGaugh & N. M. Weinberger (Eds.), *Brain organization and memory: Cells, systems, and circuits.* New York: Oxford Univ. Press.

Shepherd, G. M., Brayton, R. K., Miller, J. P., Segev, I., Rinzel, J., & Rall, W. (1985). Signal enhancement in distal cortical dendrites by means of interactions between active dendritic spines. *Proceedings of the National Academy of Sciences of the United States of America*, **82**, 2192–2195.

Sherman, S. M., & Spear, P. D. (1982). Organization of visual pathways in normal and visually deprived cats. *Physiological Reviews*, **62**, 738–855.

Singer, W. (1987). Activity-dependent self-organization of synaptic connections as a substrate of learning. In J. P. Changeux & M.

Konishi (Eds.), *The neural and molecular bases of learning*. New York: Wiley.

Stanton, P. K., & Sejnowski, T. J. (1989). Associative long-term depression in the hippocampus induced by Hebbian covariance. *Nature*, **339,** 215–218.

Steinbuch, K. (1961). Die lernmatrix. *Kybernetik*, **1**, 36–45.

Stent, G. S. (1973). A physiological mechanism for Hebb's postulate of learning. *Proceedings of the National Academy of Sciences of the United States of America*, **70**, 997–1001.

Tesauro, G. (1986). Simple neural models of classical conditioning. *Biological Cybernetics*, **55**, 187–200.

von der Malsburg, C. (1973). Self-organization of orientation selective cells in the striate cortex. *Kybernetik*, **14**, 85–100.

Wiesel, T. N., & Hubel, D. H. (1965). Comparison of the effects of unilateral and bilateral eye closure on cortical unit responses in kittens. *Journal of Neurophysiology*, **28**, 1029–1040.

Wigstrom, H., & Gustafsson, B. (1984). A possible correlate of the postsynaptic condition for long-lasting potentiation in the guinea pig hippocampus *in vitro*. *Neuroscience Letters*, **44**, 327–332.

20

Computation of Motion by Real Neurons

Norberto M. Grzywacz and Tomaso Poggio

INTRODUCTION

Properties and limitations of neurons and synapses are crucial in determining the algorithms used by the brain to perform specific computational tasks. In this chapter we consider the problem of computing motion. We first review the present understanding of the computational problem of extracting the optical flow from the time-dependent image. We identify two stages in solving this problem: obtaining the *initial measurements* and *regularizing* those initial measurements. The first stage consists of choosing a specific definition of the optical flow. In biological terms, a definition of the optical flow is equivalent to an elementary *motion detector*. We argue that there are several definitions, or detectors, that are very similar in terms of the desired task, that of computing qualitative and robust properties of the motion field. The thesis of this chapter is that the biologically *correct* detector should be sought among the ones that are most plausible in terms of neural implementation. To give examples of what such detectors may be, this chapter focuses on the mechanisms of direction selectivity in the retina. Direction selectivity arises from the interaction between an excitatory and an inhibitory pathway. The two main candidates for the mechanism underlying this interaction, the *threshold model* and the shunting *inhibition model*, and their variants are discussed. The chapter summarizes new experimental evidence in favor of the shunting model. Also, we review the spatial and temporal properties of the excitatory and inhibitory pathways. Finally, the chapter discusses some of the computations that the relevant biological mechanisms are well suited to implement.

The computational level of analysis of a complex task may provide useful constraints about the algorithms that a system—artificial or biological—could use. Constraints imposed by the neurons (in the case of biological systems) are, however, at least as important (see Poggio, 1983). We need to characterize properties and limitations of the biophysical mechanisms underlying information processing in the brain in order to connect a computational analysis with neural circuitry. Neural network models that do not take into account the relevant and known properties of membranes and cells may well turn out to be completely irrelevant for understanding the brain. In this chapter we will consider one specific example, the computation of motion. We first define the computational task. (In the following we will use some of the material from Poggio, Yang, & Torre, 1988.)

VELOCITY FIELD AND OPTICAL FLOW

It is possible to associate a 3-D velocity field to moving objects. This 3-D velocity field is generally unobservable in the image plane (consider a white rotating sphere) for any mechanism or algorithm that has access only to the time dependent brightness distribution in the image plane. If features in the image were identifiable that correspond to locations on the surface of objects, the 2-D velocity field could be computed from image data. The 2-D velocity field is the projection of the true 3-D velocity field. (One can show that the correct 3-D velocity field can be computed from its 2-D projection if the latter is available exactly and without noise.) It is possible, however, to define an "optical flow" from the time dependent intensities in the image in such a way that it will be as close as possible to the "true" 2-D velocity field in most circumstances.

Fennema and Thompson (1979), Horn and Schunck (1981), and others have defined such optical flow in terms of the "image constraint equation"

$$\frac{dE}{dt} = 0 \tag{1a}$$

where E is the brightness of the image. This equation defines an optical flow $\mathbf{u}^{\mathrm{FT}}$ as (by the chain rule of derivatives)

$$\nabla E \bullet \mathbf{u}^{\mathbf{FT}} + \frac{\partial E}{\partial t} = 0 \tag{1b}$$

Although Horn and Schunck (1981) make clear that this optical flow is generally different from the true velocity field, several papers seem to assume that this is not the case. The flow defined by Equation 1 has often been referred to in the literature as *the* optical flow, although it is not based on physical constraints that are general enough in our 3-D world and is therefore somewhat arbitrary. However, Definition 1 does satisfy the requirements of simplicity and of providing the true velocity field for a special case, that is, translation in the image plane. Equation 1 is only one of the infinitely many optical flows, that is, motion detectors, that one can define.

This discussion suggests therefore that there are several possible definitions of optical flow and not just one. In particular, *optical flows are observable fields,* **u**, *that are usually close to the true velocity field,* **v**, *at least qualitatively.* The choice of the appropriate definition to use will depend on its closeness to the motion field (for the given task) but also on ease and robustness of computation for a given implementation. This is especially important from the biological point of view. Properties of the biophysical mechanisms play a critical role in determining the type of detectors that are feasible.

TWO STEPS IN THE COMPUTATION OF OPTICAL FLOW

Whatever definition of the optical flow is used there must be another step in order to obtain a satisfactory field; regularization over space–time is needed, possibly in a way to preserve discontinuities in the field. The reasons for the regularization step are that the initial data provided by the local detectors (we call *detectors* the mechanism implementing the local *definitions* of optical flow) may be (1) noisy, (2) sparse, and (3) nonunique. Condition (1) is always true in practice and is therefore the main motivation for regularization. Depending on the definition and on the image data the initial motion data may be sparse and even nonunique as in the case of Equation 1.

Thus the computation of the optical flow requires two conceptually separate stages: (1) local evaluation of the chosen definition of optical flow via a corresponding motion detector and (2) regularization of these initial motion data. In this chapter we will analyze the first stage by listing several possible detectors, which are equivalent to definitions of the optical flow, characterizing some of their properties and discussing them from the point of view of biologically plausible implementations. We will then concentrate on two models and their variants that are the most likely to explain directional selectivity of retinal ganglion cells. This chapter does not discuss the second stage of the computation of motion, the regularization step (for formulations of the second step, see Horn & Schunck, 1981; Hildreth, 1984; Yuille & Grzywacz, 1988; for a recent overview of the situation see Poggio, Yang, & Torre, 1988).

THE APERTURE PROBLEM

Marr and Ullman (1981) recognized a problem in the initial computation of motion. If a straight contour is observed through a small aperture, only the component of velocity normal to the contour can be measured—the tangential component is invisible. They called this the "aperture problem." The problem is intrinsic to the situation and is valid for any mechanism or detector or definition of the optical flow that has access to data through an aperture. Notice, however, that the problem may disappear if the contour is not straight, or if other distinctive features are available within the aperture. We call this the *weak* aperture problem (Figure 1).

The optical flow defined by Equation 1 suffers, of course, from the same problem. In fact, the problem for this definition is considerably worse: only the component of $\mathbf{v}$ that is tangential to the intensity gradient is defined by Equation 1 for any intensity distribution, not only for straight edges. This is a very pathological case of the aperture problem, which holds for this particular definition of optical flow. We call this the *strong* aperture problem (Figure 1).

It should be clear that the weak aperture problem cannot be avoided. Any definition of the optical flow and therefore any physical detector will suffer from it. The strong form of the aperture problem, however, exists only for special definitions of the optical flow, as we will see later. In practice, the weak form is not a serious problem; straight segments without additional features are rare and isolated (see the next section).

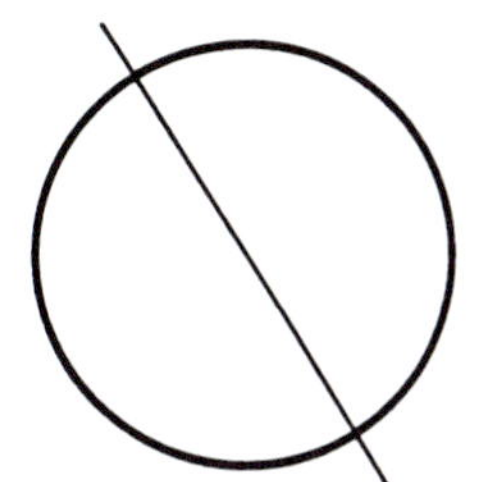
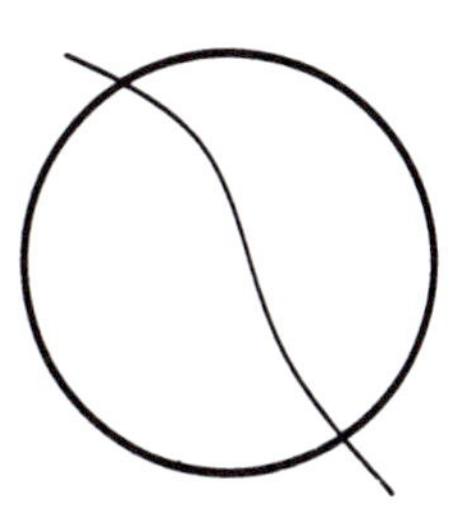

FIGURE 1 Any motion detector, that is, any definition of optical flow, fails to recover the motion component along the straight line edge as seen through an aperture, since only the motion component perpendicular to the straight line is physically observable. This is the weak aperture problem (*left*) and cannot be overcome by any physical mechanism or definition of optical flow. If a motion detector can only recover one component of motion for a curved edge, it is said to suffer from the strong aperture problem (*right*) since both components of motion are physically observable through an aperture when the image is not simply a straight edge.

DEFINITIONS OF THE OPTICAL FLOW

We consider here three main definitions.

(1) The Fennema–Thompson optical flow. This is the same definition used by Horn and Schunck (1981) and is given by Equation 1, that is

$$\frac{\mathrm{d}E}{\mathrm{d}t} = 0 \tag{1a}$$

which defines an optical flow $\mathbf{u}^{\mathbf{FT}}$ as:

$$\nabla E \bullet \mathbf{u}^{\mathbf{FT}} + \frac{\partial E}{\partial t} = 0 \tag{1b}$$

(2) The definition proposed by Girosi, Verri, and Torre (1989) and given by

$$\frac{\mathrm{d}}{\mathrm{d}t}\nabla E = 0 \tag{2a}$$

which defines a new optical flow $\mathbf{u}^{\mathbf{GVT}}$ as

$$\nabla \frac{\partial E}{\partial x} \bullet \mathbf{u}^{\mathbf{GVT}} + \frac{\partial^2 E}{\partial x \partial t} = 0$$

$$\nabla \frac{\partial E}{\partial y} \bullet \mathbf{u}^{\mathbf{GVT}} + \frac{\partial^2 E}{\partial y \partial t} = 0 \qquad (2b)$$

(3) Correlation models, mainly associated with the name of Reichardt (Hassenstein & Reichardt, 1956), some of which are shown in Figure 2, define a class of optical flows—and, in an equivalent way, they define a class of motion detectors—that may be a first order approximation to the discrimination of motion in insects (Hassenstein & Reichardt, 1956) and humans (van Santen & Sperling, 1984).

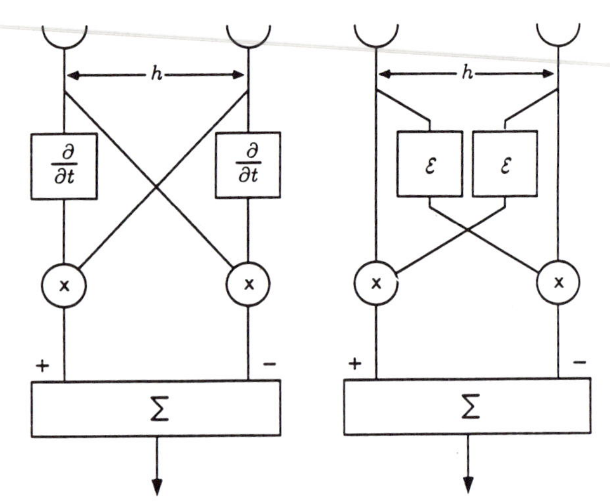

FIGURE 2 Two types of Reichardt's correlation detectors. The detectors compute the product of the time signals measured by neighboring photosensors after they have been filtered in different ways. The time average of the output is direction selective in the sense that inversion of the direction of motion will give a different average output. In the case of the detectors shown here, the output for motion in one direction is exactly the opposite of the output for motion in the opposite direction (because of their antisymmetry). (*Left*) The first model contains a high-pass filter in the direct channels. (*Right*) The second model contains a delay in the cross channels; the delay can be replaced by a low-pass filter.

The Fennema–Thompson Optical Flow

Verri and Poggio (1987) showed in detail that Equation 1 does not in general provide the correct velocity field. They proved that $\mathbf{u}^{\mathrm{FT}}$ is the physically correct 2-D projection of the 3-D velocity field only for translations (in 3-D) of a Lambertian surface. For rotations, for instance, Equation 1 provides $\mathbf{u}^{\mathrm{FT}}$ which is different from the true $\mathbf{v}$. If the surface reflectance contains a specular component, $\mathbf{u}^{\mathrm{FT}}$ is never the true velocity. Verri and Poggio provide formulae for the discrepancy. They argue that, because of this situation, several definitions of optical flow could be considered, as long as they are usually close to the correct velocity. There is no reason for ascribing privileged status to Equation 1.

Equation 1 suffers from the strong aperture problem since it is a single scalar equation while the field it defines is a two-dimensional vector field. Thus the initial data represented by $\mathbf{u}^{\mathrm{FT}}$ are nonunique and require regularization, in the form suggested by Horn and Schunck and others.

The Girosi–Verri–Torre Flow

A new definition is given by Equation 2, that is

$$\frac{\mathrm{d}}{\mathrm{d}t} \nabla E = 0 \qquad (2a)$$

which defines a new optical flow $\mathbf{u}^{\mathrm{GVT}}$ as

$$\nabla \frac{\partial E}{\partial x} \bullet \mathbf{u}^{\mathbf{GVT}} + \frac{\partial^2 E}{\partial x \partial t} = 0$$

$$\nabla \frac{\partial E}{\partial y} \bullet \mathbf{u}^{\mathbf{GVT}} + \frac{\partial^2 E}{\partial y \partial t} = 0 \qquad (2b)$$

This definition has, of course, as any other realizable definitions, the weak aperture problem (for a straight line the $det[Hessian\, E]$ is zero and only the component tangential to the intensity gradient is recoverable), but usually provides the two components of $\mathbf{u}^{\mathrm{GVT}}$, not just one, and thus avoids the strong aperture problem.

This new optical flow is identical to the true velocity field for translation in the image plane, and is otherwise different (an exact calculation similar to that of Verri and Poggio has yet to be carried out). The definition seems to be in practice a good approximation to the velocity field. Other definitions with the same flavor as Equation 2 have been proposed.

The Correlation Model

Figure 2*a* shows a detector of the correlation type. Figure 2*b* shows another one. Reichardt, Egelhaaf and Schlögl (1989) show that a 2-D network of correlation detectors will have an output response $\mathbf{u}^{\mathbf{R}}$ given by

$$\mathbf{u_x}^{\mathbf{R}} = \mathrm{h}\left(\frac{\partial E}{\partial x}\frac{\partial E}{\partial t} - E\frac{\partial^2 E}{\partial x \partial t}\right)$$
$$\mathbf{u_y}^{\mathbf{R}} = \mathrm{h}\left(\frac{\partial E}{\partial y}\frac{\partial E}{\partial t} - E\frac{\partial^2 E}{\partial y \partial t}\right) \tag{3}$$

in the continuous approximation limit (h is constant). If $E(x,y,t) = E(x + w_x t, y + w_y t)$ then $\mathbf{u}^{\mathbf{R}}$ is undetermined only when the $det[Hessian(\log \mathrm{E})]$ is zero (in particular for straight lines).

This means that in the most interesting pathological case, which is $E(x,y) = E(x + \mathrm{k}y)$, both Equations 2 and 3 are underdetermined. In particular, both definitions suffer from the weak aperture problem but not from the strong one. For this very special case of rigid translation in the image plane, the 2-D continuous approximation to the correlation model is essentially equivalent to the definition given by Equation 2 in the following sense. They both fail for 1-D patterns but they can both compute the correct velocity otherwise (provided that appropriate derivatives of E are available). Thus for the simple case of translation in the image plane (corresponding to parallel translation of a Lambertian surface in space under orthographic projection), the measurements provided by the correlation network plus Equation 3 are equivalent to the definition of optical flow given by Equation 2. Under the same condition, the "old" definition, Equation 1, provides only one correct component of the flow. Under more general conditions, all three definitions are expected to provide values that are not equal to the true 2-D velocity field and are different from each other. All three definitions suffer, of course, from the weak form of the aperture problem (Figure 1). In practice, this is not a problem for Equations 2 and 3 since the ($det[Hessian\ E]$ close to zero), which leads to noisy estimates of optical flow, is a serious problem likely to occur on a set of points (in the image), which is not of measure zero. Some form of subsequent regularization is one way to correct for this.

In general, one may define the *class of correlation models* as those smooth input–output time-invariant systems (or functionals) that have a Volterra expansion with nonzero second order and possibly first order terms but zero higher order terms. This definition is justified by the fact that all the characteristic properties of the average responses of all studied correlation models are captured by the second order systems defined earlier and that all studied correlation models are second order systems (Poggio & Reichardt, 1973; Geiger & Poggio, 1975).

The general form for a second order system is

$$y = h_0 + \sum_{j=1}^{2} \int \cdots \int u_j(x_1, x_j, t - \tau_1, t - \tau_j) \prod_{r=1}^{j} f(x_r, \tau_r) dx_1 dx_j d\tau_1 d\tau_j, \tag{4}$$

where $f(x,\tau)$ represents the time-varying image, u contains the spatio-temporal filtering and the nonlinear operations, and $h0$ is constant (see Equation 5 of Geiger & Poggio, 1975). One can prove that *any model that consists of a finite number of inputs (corresponding to the photoreceptors) followed by linear operations and quadratic no-memory nonlinearities (in any combination and order) is a special case of Equation 4.*

From this definition, it follows that there are several versions of the correlation model, for instance:

- The F model (low-pass filters instead of delays—Thorson, 1966)
- The F–H model (low-pass filters and high-pass filters—Reichardt, 1969)
- The modified Reichardt model by van Santen and Sperling (1984)

- The energy model of Adelson and Bergen (1985; see also Heeger, 1987; and Grzywacz and Yuille, 1990)
- The model of Watson and Ahumada (1985), if provided with a quadratic nonlinearity at its output (a nonlinearity is needed in any true motion detector; Poggio & Reichardt, 1973).

Models such as the three introduced earlier, and the shunting and threshold models, to be introduced, can be represented by a functional power series expansion of the Volterra type. (The shunting model is in fact an analytic functional, that is, the corresponding functional power series exists and converges everywhere, Poggio & Torre, 1978.) Equation 4 is a truncated Volterra series and, as such, may be considered an approximation of those definitions. Under general circumstances, terms of order higher than the second may not be negligible and the approximation will correspondingly be quantitatively poor.

MODELS OF MOTION DETECTION: A CRITICISM

All of the preceding models are, strictly speaking, nonbiological: The brain does not have, to the best of our present knowledge, multipliers, Fourier analyzers, and the like. At best, neurons may approximate one of these models. These models have been very useful in extracting the relevant properties of biological motion detection and in driving the experimental work. Now, with the amount of available data on the psychophysics and the physiology of motion perception and our increasing knowledge of the biophysics of neurons, the time seems ripe to consider directly models that are biologically plausible.

DIRECTIONAL SELECTIVITY IN THE RETINA

In the remainder of this chapter, we focus on retinal cells that are selective to the direction of visual motion. This section provides some background and presents open questions on the mechanisms of directional selectivity.

Figure 3 summarizes the working hypothesis on how retinal directional selectivity arises. It is due to the interaction of an excitatory and an inhibitory pathway (Barlow & Levick, 1965). Both pathways flow from the photoreceptors through bipolar and amacrine cells towards the directionally selective ganglion cells. There is evidence that the transmitter of the excitatory amacrine synapse is acetylcholine (Ariel & Daw, 1982; Masland, Mills, & Cassidy, 1984; Ariel & Adolph, 1985) and of the inhibitory synapse is GABA (Caldwell, Daw, & Wyatt, 1978; Ariel & Daw, 1982; Ariel & Adolph, 1985). However, it is still unclear whether the site of action of the inhibitory synapse is pre- (Masland et al., 1984; DeVoe et al., 1989; Dowling, 1987; Vaney, 1990; Borg-Graham & Grzywacz, 1992) or postsynaptic (Werblin, 1970; Marchiafava, 1979; Dowling, 1987) to the dendritic tree of the ganglion cells. [In some cells, these pathways are replicated for both light ON and light OFF (Famiglietti, Kaneko, & Tachibana, 1977; Famiglietti, 1983a; Amthor, Oyster, & Takahashi, 1984; Amthor, Takahashi, & Oyster, 1989; Amthor & Grzywacz, 1993; Grzywacz & Amthor, 1993).]

Directional selectivity arises from the spatial separation of the receptive fields of the excitatory and inhibitory pathways and from their different time courses. Particularly, some data suggest that the inhibitory pathway is slower than the excitatory pathway (Barlow & Levick, 1965; Wyatt & Daw, 1975). In this case, for motions in the null direction, the slower rise of inhibition is compensated by its earlier activation. On the other hand, for the preferred direction, the inhibition arrives too late to the interaction site to be effective. We deal with the spatio-temporal properties of the excitation and inhibition later.

Another important property of retinal directionally selective cells is that small regions of their receptive field are directionally selective, which suggests the existence of directionally selective receptive field subunits (Barlow & Levick, 1965; Grzywacz, Amthor, & Merwine, 1994). In 1978, Torre and Poggio argued that the subunits impose a theoretical problem if

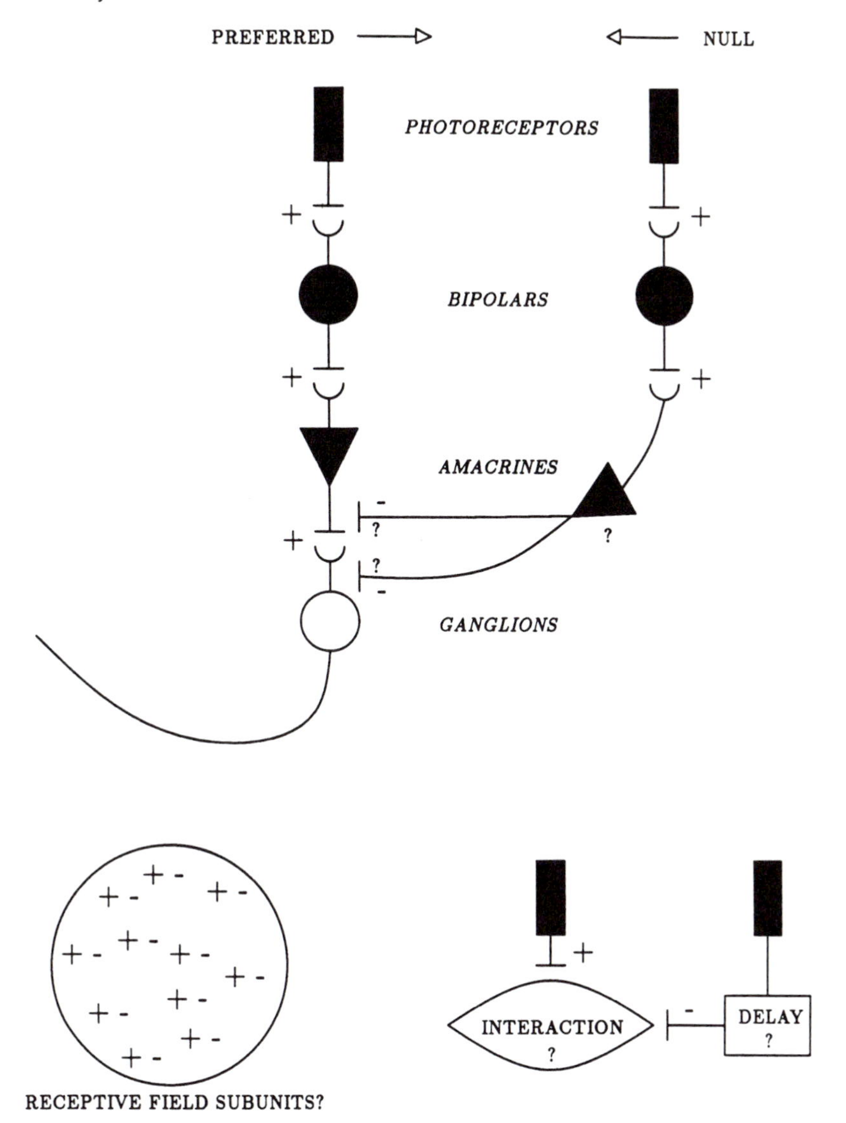

FIGURE 3 State of the art and open questions in the study of directional selectivity in the retina. This selectivity originates from the interaction between an excitatory and an inhibitory pathway. The current evidence suggests that both pathways pass through amacrine cells. Directionality arises from the spatial segregation of excitation and inhibition and from the difference in their time courses. (The label "Delay" does not mean fixed delay, but that the inhibition is slow; see text on Spatiotemporal Properties.) Some open questions include: What amacrine cells are involved in the process? What is the interaction between excitation and inhibition? Where does the interaction occur? How are the different time courses produced? What is the origin of the directionally selective subunit?

directional selectivity is to arise in the ganglion cells, as the evidence at the time suggested. They pointed out that the problem would especially arise if the inhibitory synapse acted only through hyperpolarization. Although such an inhibition would kill the response for null direction motions, it would also do so for the preferred direction. This is because the hyperpolarization would propagate throughout the dendritic tree (Figure 4).

To explain the locality of directional selectivity, Torre and Poggio (1978) proposed a nonhyperpolarizing inhibitory synapse, whose action is to shunt excitatory currents in specific branches of the dendritic tree.

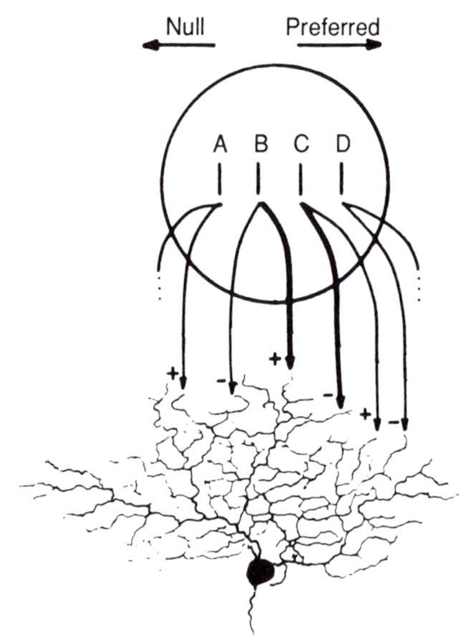

FIGURE 4 The problem of generating directionally selective subunits in ganglion cells. The figure displays a directionally selective ganglion cell stained with horseradish peroxidase (Amthor, Oyster, & Takahashi, 1984) and its hypothetical inputs from the cell's receptive field. Imagine that the inhibitory inputs are hyperpolarizing synapses and that the only existing nonlinearity is in the spike generation in the soma (the black blob). For the null direction, the excitatory inputs are properly inhibited (for example, the input from C arrives to the cell simultaneously with that from B). However, there is also inhibition for the preferred direction (for example, A inhibits B). Thus, if subunits arise in ganglion cells, the underlying nonlinearity must occur at isolated dendritic branches.

A hyperpolarizing model is also possible, if a threshold operation confines the excitatory–inhibitory interactions to preganglionic cells or synapses (Barlow & Levick, 1965; Torre & Poggio, 1978; Grzywacz, & Koch, 1987) or to dendrites with active currents as suggested, for instance, by Torre and Poggio (1978). (For evidence of such currents see Llinas & Nicholson, 1971; Miller & Dacheux, 1976; Llinas & Sugimori, 1980.) In the next two sections, we discuss the threshold and shunting interactions in more detail.

THE THRESHOLD MODEL

In this section, we discuss how the threshold model may generate directional selectivity and how the retina may implement this model.

Figure 5*A* illustrates an electrical version of this model. We assume cell isopotentiality, neglect temporal effects of the membrane capacitance, and set the resting potential to zero. The model has excitatory, g_e, and inhibitory, g_i, conductances, which are much smaller than the membrane's leak conductance, g_{leak}, and has excitatory, E_e, and inhibitory, E_i, batteries. The model assumes that the inhibitory conductance is activated by a synapse, which in turn is activated by the stimulation of the receptive field with slit *A* (Figure 5*A*). Similarly, stimulation with slit *B* activates the excitatory conductance. The response of this model is approximately

$$R = Thr\left(\frac{E_e g_e + E_i g_i}{g_{leak}}\right) \tag{5}$$

where $Thr(x) = R_0 \int D(x)dx$, where $D(x)$ is a generalized density probability function and $R_0 > 0$ (Thr rises bounded by 0 and R_0). The mean of $D(x)$ is the threshold of Thr.

The threshold model works as follows (Figure 5*B*): *In the preferred direction, inhibition is negligible and the excitatory conductance may raise the membrane potential above threshold. In the null direction, inhibition prevents threshold from being reached.*

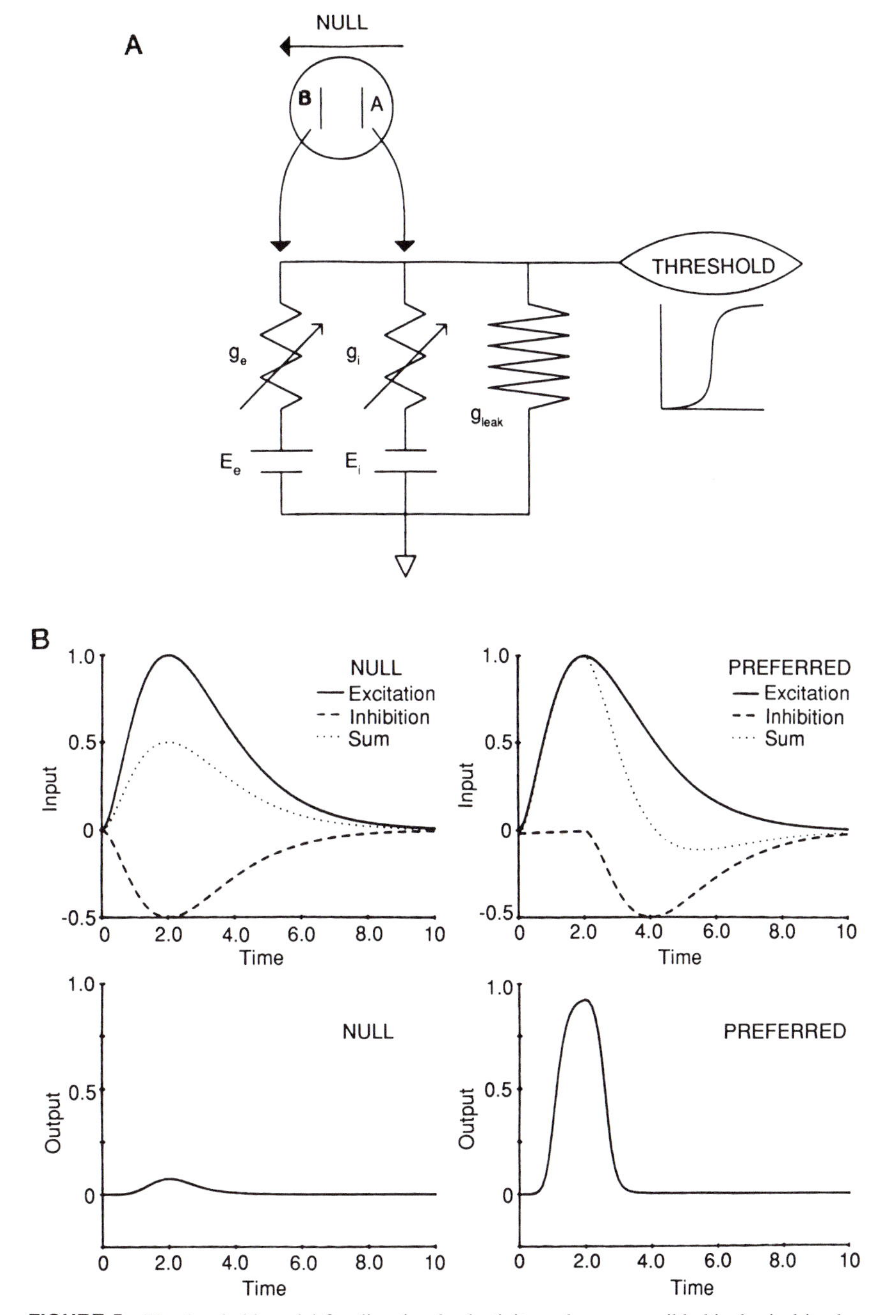

FIGURE 5 The threshold model for directional selectivity and some possible biophysical implementations in the retina. (A) Slits *A* and *B* in the receptive field activate hyperpolarizing and depolarizing synapses respectively, and the resulting potential is thresholded. (B) For the null, but not for the preferred direction the depolarization and hyperpolarization are simultaneous and the output (thresholded sum) is small. (C) Dendritic spikes may be a threshold mechanism for the subunits. (D) Also, possibly, one entire pre-ganglionic cell (an amacrine?) may underlie each given subunit. (E) Finally, the threshold of the synaptic release from starburst amacrines (this one recovered by Masland, Mills, & Cassidy, 1984) could generate subunits, because their outputs (heavy arrows) are distal in the dendritic tree.

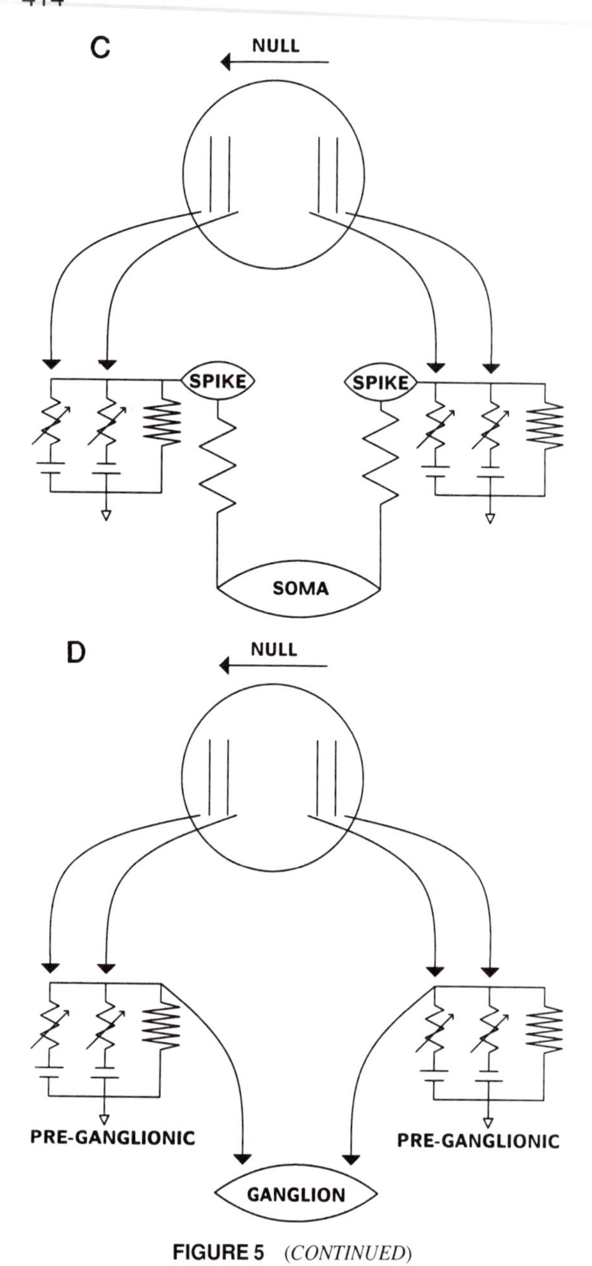

FIGURE 5 (*CONTINUED*)

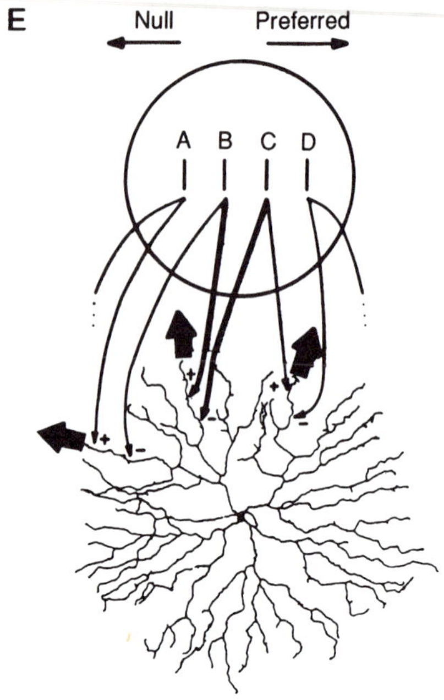

FIGURE 5 (*CONTINUED*)

What are the neural mechanisms that may implement the threshold operation required in the threshold model (Grzywacz & Koch, 1987)?

One form of neural threshold occurs in the generation of spikes (Rushton, 1937; Hodgkin & Huxley, 1952). The only retinal cells known to spike are ganglion cells and certain amacrine cells (Dowling & Werblin, 1969; Miller & Dacheux, 1976; Miller, 1979). If directional selectivity arises in the former, then somatic spikes cannot account for the directionally selective subunits (see the caption of Figure 4).

Although somatic spikes in ganglion cells cannot account for the locality of directional selectivity (Barlow & Levick, 1965), dendritic spikes may (Figure 5*C*). Such spikes have been reported in other areas of the brain (Llinas and Nicholson, 1971; Llinas and Sugimori, 1980) and even in some amacrine cells (Miller & Dacheux, 1976; Miller, 1979). However, there is no evidence yet of dendritic spikes in ganglion cells, though they have multiple active currents (Lasater & Witkovsky, 1990). In any case, it is still unclear how well a dendritic spike model for directional selectivity could explain the available data.

Another example of neural threshold appears in the relationship between pre- and postsynaptic potentials in synapses (Katz & Miledi, 1967; Kusano, 1970). Such a mechanism may implement the directionally selective subunits through the use of one preganglionic cell for each subunit (Figure 5*D*). But also, amacrine-ganglion dendro-dendritic synapses (Famiglietti & Kolb, 1976) may implement the subunits, since these synapses are distal in the dendritic tree (Figure 5*E*). In fact, the complex morphology of the starburst cells (Figure 5*E*, Famiglietti, 1983b, 1991; Miller & Bloomfield, 1983; Masland et al., 1984) is consistent with such a possibility.

SHUNTING INHIBITION

This section shows how shunting inhibition may work to generate directional selectivity. Figures 6*A* and 6*B* illustrate how a nonhyperpolarizing synapse can shunt excitatory currents in one branch of a cell's dendritic tree to inhibit its responses. We display an electrical model of such cell's branch again assuming isopotentiality, neglecting membrane capacitance, and setting the resting potential to zero. In this example, the inhibitory conductance is near the soma while the excitatory conductance is more distal. The excitatory conductance has a correspondent battery that injects, upon activation, a positive current into the cell. For the preferred direction, the signal from *B* to g_e arrives before the signal from *A* to g_i and the current flows towards the soma. However, null direction motions activate g_e and g_i simultaneously and this current escapes through g_i before reaching the soma.

Shunting inhibition may account for the locality of directional selectivity (Figure 6*C*; Barlow & Levick, 1965). This is because the shunting inhibition of a given dendritic branch barely affects other branches. In fact, if a shunting synapse acts upon a given branch, this synapse will not stand on the path to the soma of an excitatory current generated on another branch (Figure 6*A*). This on-the-path condition for optimal shunting inhibition has been demonstrated through computer simulations (Rall, 1964, 1967) and mathematically (Koch, Poggio, & Torre, 1983).

Next, we present a simple mathematical equation that indicates how shunting inhibition works. This equation is based on the model illustrated in Figure 6*D*, which assumes the cell to be isopotential. (Also, the model neglects the temporal effects of the membrane capacitance and assumes the excitatory conductance, g_e, to be small compared with the membrane leaking conductance, g_{leak}.) The equation that describes the behavior of the model is approximately

$$V = \frac{E_e g_e}{g_{leak} + g_i} \tag{6}$$

where V is the membrane's potential. *Thus, the effect of a shunting synapse is a division-like inhibition of the cells' response.*

Finally, we would like to point out that shunting inhibition may act on ganglion cells or on preganglionic cells to generate directional selectivity. Evidence for shunting inhibition in direction selective retinal ganglion cells stimulated with moving stimuli has been provided in the turtle and frog (Marchiafava, 1979; Watanabe & Murakami, 1984).

NEURAL AND COMPUTATIONAL INTERACTIONS

In this section, we compare some properties of the shunting and threshold mechanisms for directional selectivity with the form of nonlinear mechanism most often used by computational theories of optical flow: the second-order nonlinearity. This is important, since models based on such nonlinearity appear to account for motion discrimination in insects (Hassenstein & Reichardt, 1956) and humans (van Santen & Sperling, 1984; Adelson & Bergen, 1985). Thus, one wonders whether this nonlinearity may approximate the behavior of motion sensitive neurons.

It has been shown (Thorson, 1966; Poggio & Torre, 1978; Torre & Poggio, 1978) that, if the shunting conductance is small, then a second-order nonlinearity can approximate the interaction between excitation and inhibition. To see this in a simple case, expand

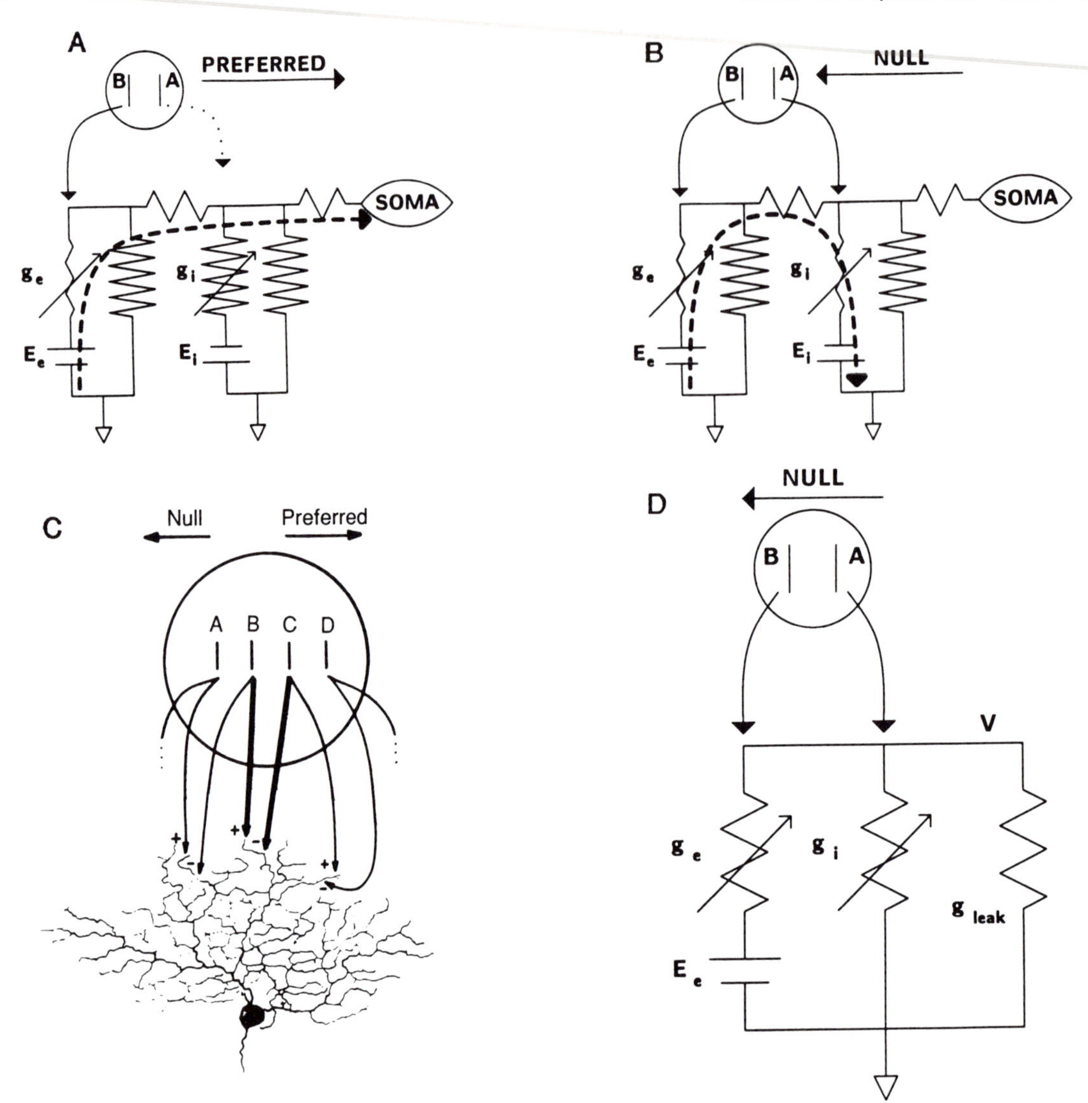

FIGURE 6 The shunting inhibition model for directional selectivity and its implications. (A) For motions in the preferred direction, the inhibition arrives late to the interaction site from the receptive field (dotted arrow) and the excitatory current (dashed bold arrow) flows toward the soma. (B) On the other hand, for the null direction, the inhibitory conductance is large (small resistance) and it shunts away the excitatory current. (C) Shunting inhibition may mediate the subunits, since its action on the dendritic tree is local. To be most effective, the shunting synapse must be on the path between the inhibited excitatory synapse and the soma. (D) The lumped version of the shunting model from which Equation 1 was developed.

Equation 6 in a Taylor series in g_i and neglect the high-order terms:

$$V \approx \frac{E_e}{g^2_{leak}} (g_e\, g_{leak} - g_e g_i) \quad (7)$$

Such an approximation is also possible for the threshold model, if *Thr* in Equation 5 is smooth (Grzywacz & Koch, 1987). A derivation in terms of Volterra series for a more realistic approximation of the Hodgkin-Huxley model was given by Torre and Poggio (1977).

However, one expects that in real neurons these approximations may be quantitatively very poor for two reasons: (1) because the gain of the inhibitory process is large (see next section), and thus even weak stimuli lead to large inhibitions, and (2) because nonlinearities not involved in directional selectivity (for example, somatic spikes) may contaminate an otherwise second-order behavior.

In fact, Grzywacz and Koch (1987) found these approximations to be poor in a quantitative comparison of the shunting and threshold models with second-order models. To achieve this goal, they analyzed properties of second-order systems that were previously proposed for the optomotor behavior of insects (Poggio & Reichardt, 1973; Geiger & Poggio, 1975). They showed that the biophysical models deviate significantly from second-order behavior for conductance modulations as low as 30%.

Furthermore, the responses of retinal directionally selective cells are not well approximated by a second-order model (Grzywacz, Amthor, & Mistler, 1990). For example, Figure 7 shows that these cells do not possess a superposition property characteristic of second-order models. This property is as follows. The average response to a moving spatially periodic stimulus is equal to the sum of the average responses to the stimulus' Fourier components presented in isolation (Poggio & Reichardt, 1973; Geiger & Poggio, 1975). More precisely, let y and y_i be the average responses to $\sum_j I_j \cos(j2\pi\lambda(x - vt + \phi_j))$ and $I_i \cos(i2\pi\lambda(x - vt))$ respectively, where I_j and ϕ_j are constants, λ is the spatial frequency of the stimulus, and v its velocity, then this property states:

$$y = \sum_{j=0}^{\infty} y_j \quad (8)$$

In conclusion, *pure second-order nonlinearities, which are the key mechanism in some models for optical flow, are poor quantitative approximations to the nonlinearities of ON–OFF directionally selective ganglion cells of the rabbit retina.*

How does the apparent second-order motion discrimination in insects and humans originate? The answer is not known, but we raise four possible explanations. The first is that, in insects and in mammalian cortices, the cellular mechanisms for motion discrimination may be different from the retinal mechanisms. However, given the visual system's nonlinearities, it would be surprising that a single cell could behave as

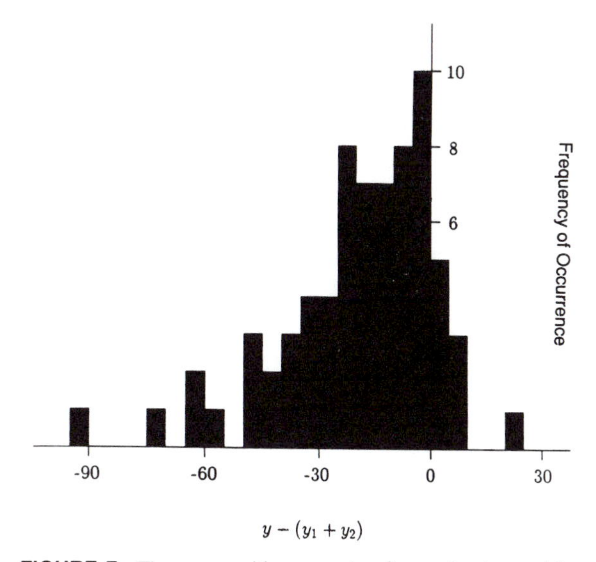

FIGURE 7 The superposition property of second-order models does not hold for ON–OFF directionally selective ganglion cells in the rabbit retina. The stimuli were two low-contrast sinusoidal gratings drifting in the preferred direction presented singly and in phase-locked pairs. The computer accumulated spike histograms over multiple cycles for each presentation, and then over multiple presentations. The figure is the histogram of values of the subtraction of the sum of the integrated responses to the isolated gratings from the integrated response to the phase-locked pairs. This distribution is negatively biased, which contrasts with the unbiased prediction by second-order models. (This figure is from Grzywacz, Amthor, & Mistler, 1990.)

a pure second-order model (but see Emerson, Citron, Vaughn, & Klein, 1987). A second explanation is that the resolution of the behavioral methods used with insects and humans is not sufficient to detect nonlinearities of higher order. Another possibility is that a form of averaging could take place across large ensembles of cells and cancel out some of the individual high order nonlinearities. Finally, noise, together with averaging intrinsic to the whole system and to the experimental techniques, could partially "linearize" the system, and thus reduce its high order nonlinearities (Poggio, 1975).

EVIDENCE FOR SHUNTING INHIBITION

This section presents evidence that a shunting inhibition mechanism underlies directional selectivity in the retina. The material presented here is a short sketch of the work of Amthor and Grzywacz (1991, 1993).

The synaptic model illustrated in Figure 8*A* provides the rationale for the experiments described in this section. For null direction apparent motion, the investigators determined the effect of varying the contrast of the first, inhibitory, slit on responses elicited by the second, excitatory, slit also presented at various contrasts (Figure 9*A*). The expected form of this interaction is different for a shunting versus a threshold mechanism. If the directionally selective inhibition is based on a hyperpolarizing (subtractive) mechanism followed by a threshold, the effect of the inhibitory slit should be to shift the response versus contrast function of the excitatory slit downward, but preserve its general shape (Figures 8*C* and *E*). However, if the directionally selective inhibition is based on a shunting mechanism, whose effect is to divide the excitatory current by the inhibitory conductance, there will be a marked change in the vertical scale of the response versus contrast function and, thus, changes in this function's general shape (Figures 8*B* and *D*).

To follow this rationale, Amthor and Grzywacz (1991, 1993) recorded extracellular responses to two slits flashed on and off within the receptive field centers of ON-OFF directionally selective ganglion cells in the rabbit retina such that apparent motion in the null (Figure 9*A*) or preferred direction was simulated.

The investigators measured the effects of inhibition on the response versus contrast functions using repeated trials of a number of excitatory and inhibitory contrasts (Figure 9*A*). They found the best sum-of-least-squares fit for the mean response at all excitatory and inhibitory contrasts using four general models that contained both hyperpolarizing and shunting components. Figure 8*A* illustrates one of the four models compared with the data. The features that distinguish it from the other three models are a large resistance between the sites of excitation and inhibition, and the nonsaturation of the inhibition. It turns out that the conclusions presented here do not depend on which of these models is used. Figure 9*B* shows the data for one cell and the best fit of the model.

In virtually every characteristic, the effect of the inhibiting slit is consistent with shunting, but not hyperpolarizing inhibition. As found by best fitting to the data, the model parameter that corresponds to the inhibitory battery is not significantly different than zero. On the other hand, the parameter that corresponds to the shunting conductance is significantly positive. The membrane resistance decreased by a factor of 2.8 from the noninhibition situation for the inhibitory contrast of 30% (*a high gain inhibition*).

A direct illustration of these conclusions is shown in Figure 9*C*, which compensates for response saturation and thus linearizes the curves (compare with Figures 8*B–E*). The intercept with the ordinate is insignificantly different than zero, which again suggests a negligible hyperpolarization. In addition, the slope decreases with the inhibitory contrasts—a direct evidence for shunting inhibition.

SPATIOTEMPORAL PROPERTIES

To understand completely the properties of directionally selective cells, one must also study the time course and spatial distribution of the underlying processes. In this section, we review recent results on these issues and discuss possible mechanisms for the time course

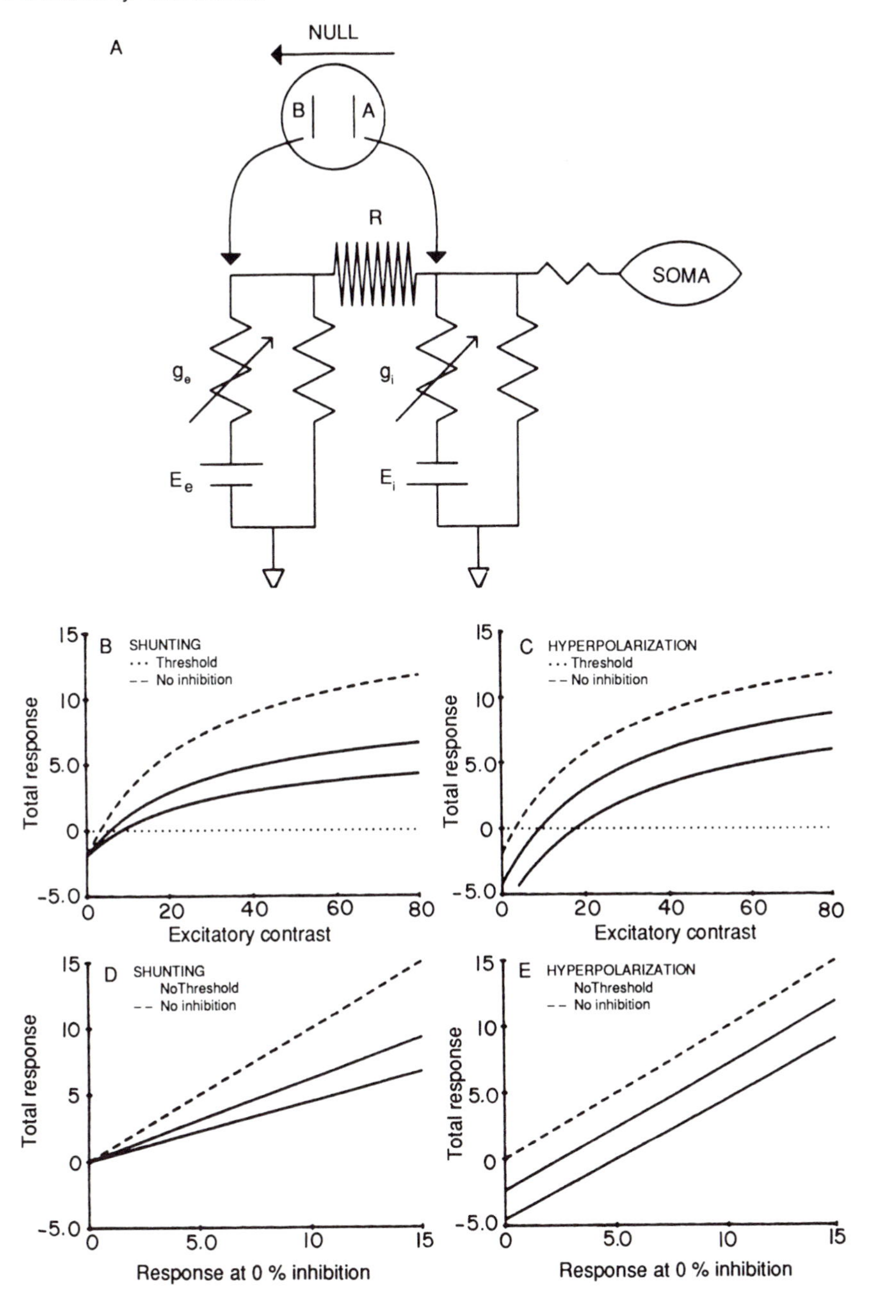

FIGURE 8 Theoretical differences between biophysical implementations of the shunting and threshold models. A. A synaptic model for directional selectivity, whose inhibition may act through hyperpolarization (mediated by the battery E_i) followed by threshold and by shunting (mediated by the conductance g_i). B,C. The response of the cell as a function of excitatory contrast parametric on the inhibitory contrast (increasing inhibition leads to lower curves). D,E. Linearization of the curves in B and C through the use of the zero inhibition curve as abscissa. Note that the shunting model changes the ordinate scale of the curves, but that the hyperpolarizing threshold model shifts the curves in parallel.

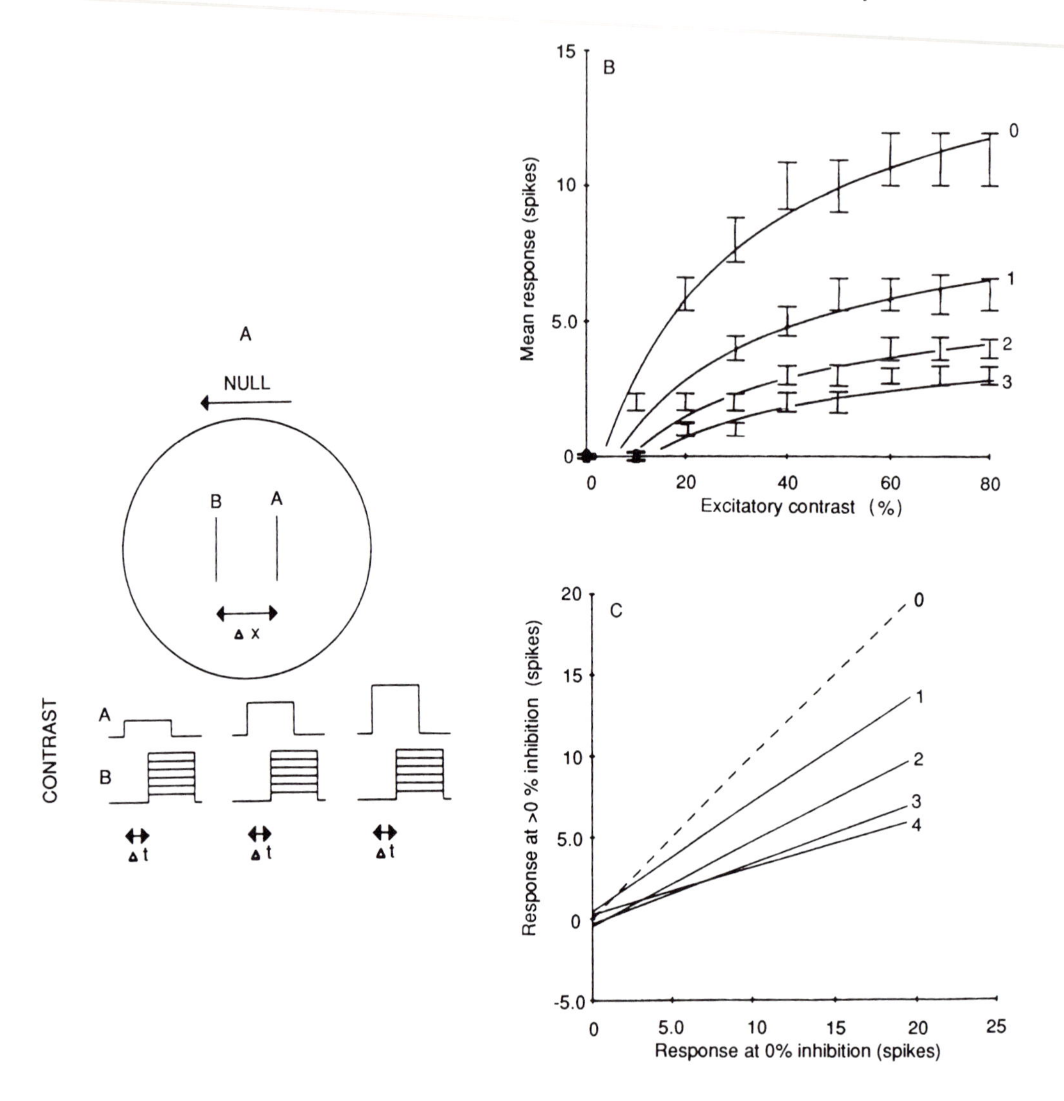

FIGURE 9 Evidence that shunting inhibition mediates directional selectivity in retinal ganglion cells. (A) The stimuli controlled independently the excitatory and inhibitory synapses by controlling slit contrasts in two-slit apparent motions. (B) The response to the second slit as a function of its contrast parametric on the first slit's contrast and the best fit of the Figure 8 model (solid lines). Inhibition for 0, 0%; 1, 10%; 2, 20%; 3, 30%; cell E380c3.1. (C) Linear regression lines with the zero-inhibition curve as abscissa (see Figures 8*D* and *E*). As predicted by the shunting inhibition model (Figures 8*B* and *D*), but not by the hyperpolarizing threshold model (Figures 8*C* and *E*), the curves change their ordinate scale without parallel shifts. Inhibition for 0, 0%; 1, 10%; 2, 20%; 3, 30%; 4, 40%; cells E375c2.1 and E380c3.1.

of inhibition. Amthor and Grzywacz (1993) studied the inhibition's time course by varying the delay between slits undergoing an apparent motion (the slits were left on for long periods of time; Figure 10*A*). They measured the integrated response to the second slit with and without the first slit. Figure 10*B* plots these responses as function of the delay.

The inhibition appears to be sustained, as indicated by the fall in the response to the second slit even 1 sec after the first slit onset, for null direction motions. However, the inhibition's impulse response lasts only 50–200 msec (Wyatt & Daw, 1975; Amthor & Grzywacz, 1993). (If the first slit remains on, the inhibition falls after 4 sec, not shown, possibly due to adaptation.) In contrast to the inhibition, the excitation appears to be transient with a response that lasts no more than 200 msec after the onset of the slit.

What is the source of the observed transience of excitation? The first cells in the retinal pathway to exhibit full transience are the amacrine cells (Dowling & Werblin, 1969; Kaneko, 1971; Naka & Ohtsuka, 1975). A proposed mechanism for this transience is a putative negative feedback from amacrines to bipolar cells (Werblin et al., 1988), which is hinted by anatomical data (Dowling, 1968, 1979; Burkhardt, 1972; Richter & Ullman, 1982). Another proposed mechanism depends on active membrane properties of the amacrine cells (Barnes & Werblin, 1986). Presynaptic depression (Takeuchi, 1958; Mallart & Martin, 1968; Rosenthal, 1969) or desensitization of postsynaptic receptors (Katz & Thesleff, 1957; Magleby & Pallota, 1981) are possible though unlikely mediators of the transience, since the time constant of these processes is never faster than 5 sec (Magleby, 1987).

The sustained inhibition may be mediated by sustained amacrine cells (Kaneko, 1971; Toyoda, Hashimoto, & Ohtsu, 1973; Sakuranaga & Naka, 1985a,b). The responses of the sustained amacrine cells resemble those of the distal neurons in the retina, and especially those of horizontal and bipolar cells (Sakuranga & Naka, 1985a,b).

Amthor and Grzywacz (1993) also found that inhibition around any point of the receptive field is spatially asymmetric, contiguous to the point, and may extend by more than half the field. To reach this conclusion, they performed an experiment similar to the one just described, but which varied the slit's distance (Figure 11). These results are different than those of Barlow and Levick (1965), who found shorter extents for the inhibition. However, the results agree with Wyatt and Daw's findings (1975). These results challenge the inhibition locality as the mechanism for directionally selective subunits. Rather the subunits probably arise from the locality of interaction between excitation and inhibition possibly due to shunting inhibition (Torre & Poggio, 1978), which allows the inhibition's asymmetry and contiguity (Wyatt & Daw, 1975; Koch et al., 1983; Amthor & Grzywacz, 1993).

Finally we point out that, for the preferred direction, one observes a small but significant facilitation (Figure 10B; Barlow & Levick, 1965; Grzywacz & Amthor, 1993). It is not known whether this facilitation is directionally selective or whether it is overpowered in the null direction by the inhibition.

COMPUTATIONS THAT NEURONS PERFORM

Because we have seen in the previous sections that neurons do not implement strictly the mathematical operations of some of the models, the following question arises. What are the properties of the computations that real neurons perform?

Retinal directionally selective cells may be relatively invariant to shape, are very sensitive to small contrasts, and separate the computations of motion direction and speed. In the rest of this section, we expand on this claim.

Similar to second-order models, the shunting inhibition model is relatively invariant to the input's Fourier components phase, and thus invariant to the input's shape. This was shown in Grzywacz and Koch's (1987) calculations.

Since inhibition has high gain, one expects that directional selectivity is sensitive to contrast, that is, cells may be directionally selective even for very low contrasts. Support for this hypothesis appears in the analysis by Grzywacz and Amthor of a mathematical

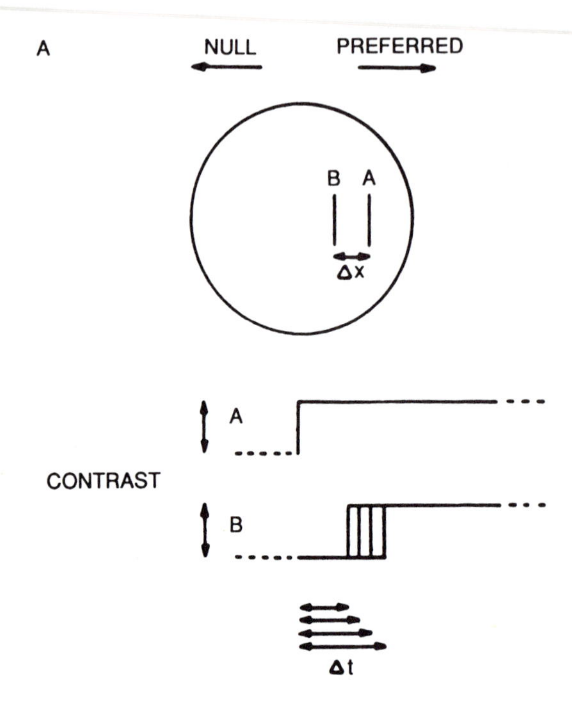
A
NULL
PREFERRED
B A
Δx
A
CONTRAST
B
Δt

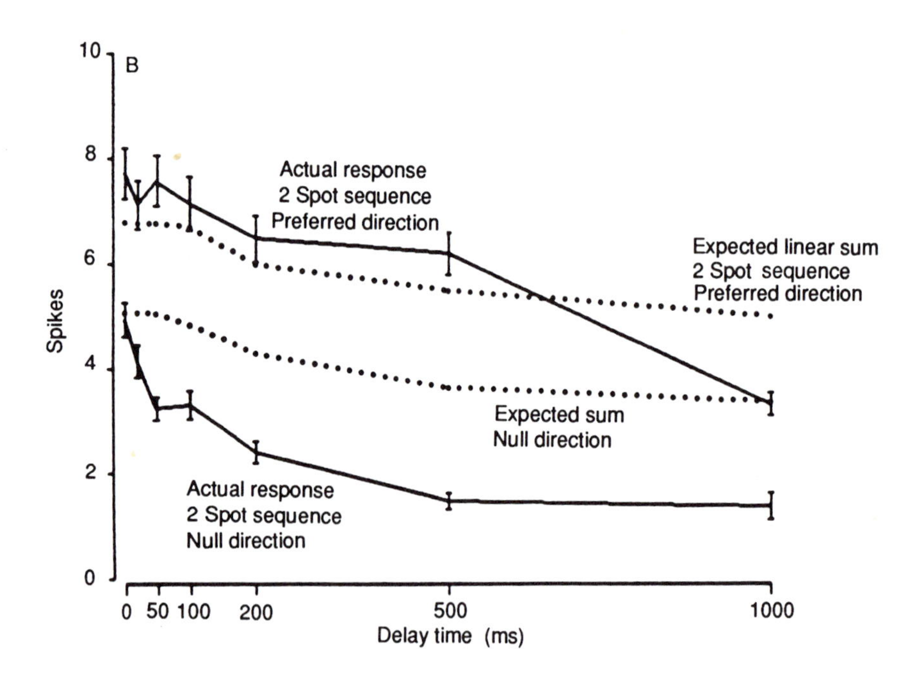
B
Spikes
Actual response
2 Spot sequence
Preferred direction
Expected linear sum
2 Spot sequence
Preferred direction
Expected sum
Null direction
Actual response
2 Spot sequence
Null direction
0
2
4
6
8
10
0 50 100 200 500 1000
Delay time (ms)

model for directional selectivity (1989). Thus, it is not surprising that retinal directionally selective responses were recorded with 1% contrasts (Grzywacz et al., 1990).

Finally, we argue that the spatio-temporal properties of excitation and inhibition allow independent computations of motion's speed and direction. This has been shown elsewhere in the analysis of a mathematical model (Grzywacz & Amthor, 1989), but is only sketched here.

Figure 12*A* presents the argument by comparing two limiting cases: (1) a transient and spatially shifted inhibition and (2) a sustained and contiguous inhibition (the case supported by data). In the first case, the inhibitory action is highly dependent on speed. In the second, this action is robust against speed because, for sufficiently low speeds, the inhibition waits until the stimulus arrives. Thus, it is expected in the second case, but not in the first, that the computation of motion direction is relatively invariant to speed. A similar property is well known from work on some correlation models (Reichardt, 1969).

Amthor and Grzywacz (1988) confirmed this prediction for retinal directionally selective cells in the rabbit by measuring a directional selectivity index as a function of speed. The index was insensitive to speed for speeds that ranged more than two decades (Figure 12*B* lower panel). Nevertheless, cells may still be speed selective (or more precisely, selective for temporal frequencies), since their absolute responses are tuned to speed (temporal frequency; Figure 12*B*, upper panel). This tuning is probably due to the band pass characteristics (transience) of the amacrine cells that excite the directionally selective ganglion cells.

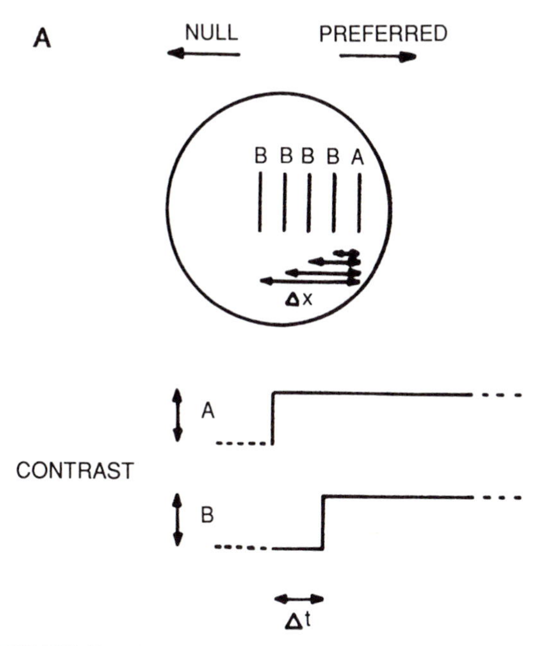

FIGURE 11 The spatial distribution of inhibition around points of the receptive field. (A) The experiment was identical to the one in Figure 10, except that this time the distance between the slits also varied. (B) Null direction response versus delay after first spot. The circle represents the receptive field center and the numbered rectangles are the positions at which the slits appeared. (The spatial scale of rectangle positions in the circle is correct, that is, for example, slits 2 and 5 were more than half the receptive size away from each other.) Note that the inhibition spreads for long distances in the receptive field and that it is contiguous to the inhibited point. (Cell E386c1.1, 50 trials, 20% mod. 1 sec flash.)

DISCUSSION

The problem of how neurons compute motion is not yet fully solved. Mechanisms different from the ones

FIGURE 10 The time course of the inhibition following a long lasting light increment. (A) The experimenters varied the delay between slits on a two-slit apparent motion, keeping contrast and distance constant. (B) Plot of the integrated response to the second slit as a function of delay (with the spikes due to the first slit subtracted) when presented in the apparent motion sequence (solid lines), or in isolation (dotted lines). Note that, for the null direction, it takes about 200 msec for the inhibition to rise, but that after that it is sustained. (Also note the small facilitation for short delays for preferred direction motions. Cell E385c1.1, 83 trials, 10% contrast, squareware mod. From Amthor & Grzywacz, 1988.)

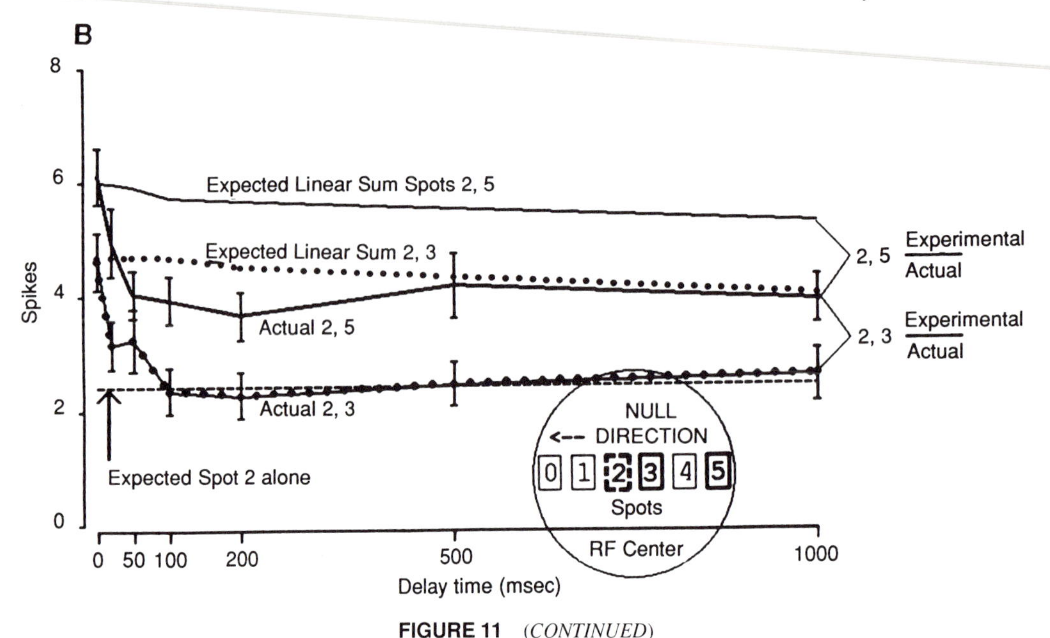

FIGURE 11 (*CONTINUED*)

we have emphasized (and, correspondingly, different definitions of the optical flow) may be used in other parts of the brain, such as the visual cortex. The mechanism of shunting inhibition, though it seems, according to the recent evidence we have reviewed here, to be the one used by the retina, is not yet established beyond any doubt. From the present knowledge, however, we can draw a number of interesting conclusions about how neurons compute, some general and some more specific. The main lesson is that there may be several somewhat different formulations of a computational task and that the biologically *correct* choice is determined, among other reasons, by the constraints imposed by the biological mechanisms. A second potentially important point is that the threshold mechanism of spikes or synaptic transmissions is not the only mechanism used by neurons to perform elementary computations. The mechanism of directional selectivity in the retina seems to be shunting inhibition and not a threshold. Also, it is possible that the basic unit of computation is not just a neuron but also a piece of dendritic membrane. Thus, most neural network models may be of little biological relevance, since they are built upon a model of the neuron that is too simple—just a threshold unit. The properties of a network depend strongly on its basic elements and operations, and not only on the overall computation to be performed. Therefore, to understand the computations of the brain it is essential, we claim, to understand first the basic biophysical mechanisms of information processing before developing complex networks.

A more specific point regards the *validity* of the biological models of motion detection. It is certainly good to develop and use nonphysiological models, such as the correlation model, as long as they are used within their limits. For instance, the correlation model does a good job of capturing important properties of many motion detection schemes, because in a certain mathematical sense, it is the simplest motion detector and, under certain conditions, it is an approximation of a large class of (smooth) motion detectors. (The Volterra expansion formalism and its conditions of validity make this argument rigorous; Poggio & Reichardt, 1980.) First-order models of this type are

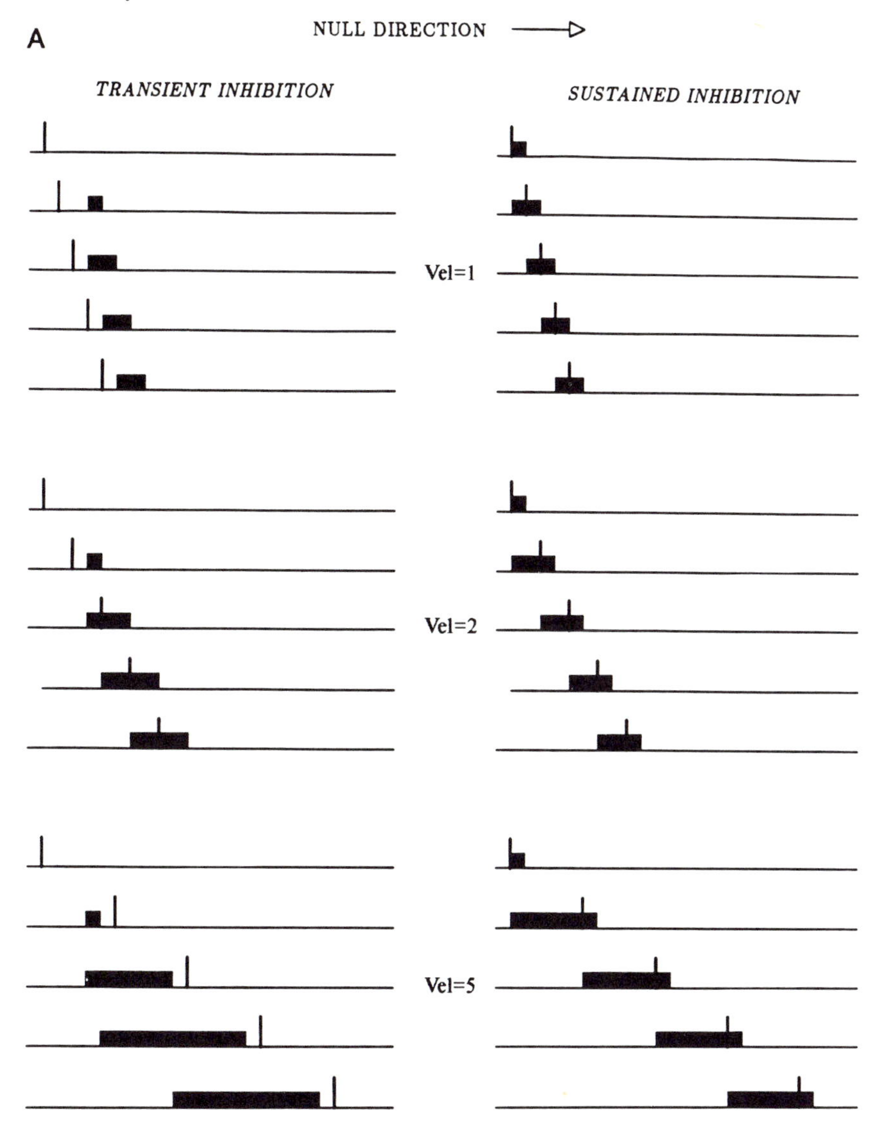

FIGURE 12 The computation of speed and direction by directionally selective cells. (A) In this diagram, the abscissa is space, horizontal lines are time snapshots, vertical lines indicate slit position, and rectangles indicate inhibited positions. Two limiting situations make different predictions. Transient and separated inhibitions work for a narrow speed range, but sustained and contiguous inhibitions (Figures 10 and 11) always work for sufficiently low speeds. (B) Responses to drifting sinusoidal gratings of retinal directionally selective cells conform with the latter, since their selectivity is relatively speed invariant. (*lower panel*, ordinate: $y_{ds} = (y_p - y_n)/(y_p + y_n)$; y_p and y_n are average responses in preferred and null directions, respectively.) However, the cells' responses to the preferred direction are tuned to speed (*upper panel*, •, Amthor & Grzywacz; °, Wyatt & Dow). In this experiment, the spatial frequency was 0.5 cycles/degree and the contrast was 40.6%. (This figure is from Amthor & Grzywacz, 1988.)

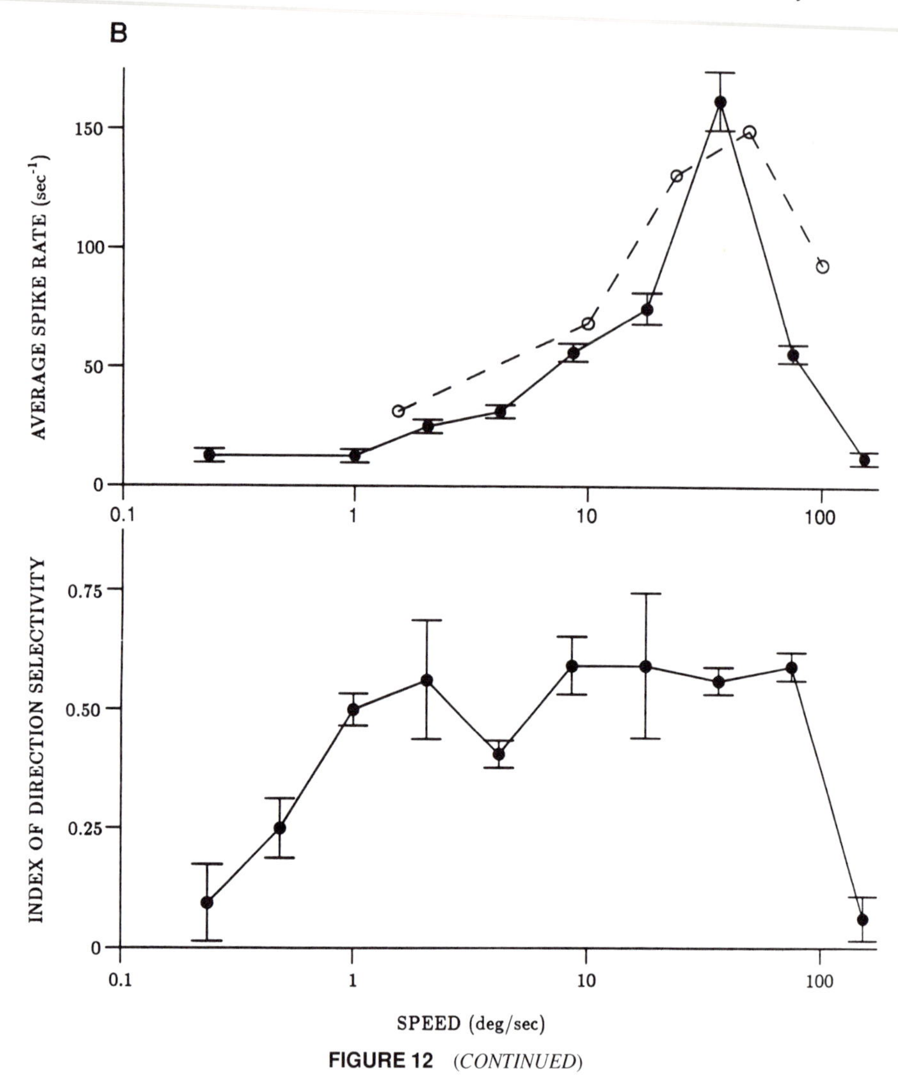

FIGURE 12 *(CONTINUED)*

useful to summarize data and plan new experiments. However, they are only approximations and are not faithful to the physiological constraints. In fact, we argue that neurons do not share certain important properties of second-order models, for example, the superposition property (Figure 7), and may have important properties that these models do not have (see the section on Computations that Neurons Perform). Thus, whenever sufficiently detailed data are available, these approximations should be abandoned and replaced by biologically detailed (and correct) models, such as the shunting inhibition model. In the future, it may be necessary to develop even finer and more faithful models to take directly into account morphological, physiological, and pharmacological data.

References

Adelson, E. H., & Bergen, J. R. (1985). Spatio-temporal energy models for the perception of motion. *Journal of the Optical Society of America A*, **2**, 284–299.

Amthor, F. R., & Grzywacz, N. M. (1988). The time course of inhibition and the velocity independence of direction selectivity in the rabbit retina. *Investigative Ophthalmology & Visual Science*, **29**, 225.

Amthor, F. R., & Grzywacz, N. M. (1991). Nonlinearity of the inhibition underlying retinal directional selectivity. *Visual Neuroscience*, **6**, 95–104.

Amthor, F. R., & Grzywacz, N. M. (1993). Inhibition in directionally selective ganglion cells of the rabbit retina. *Journal of Neurophysiology*, **69**, 2174–2187.

Amthor, F. R., Oyster, C. W., & Takahashi, E. S. (1984). Morphology of ON-OFF direction-selective ganglion cells in the rabbit retina. *Brain Research*, **298**, 187–190.

Amthor, F. R., Takahashi, E. S., & Oyster, C. W. (1989). Morphologies of rabbit retina ganglion cells with complex receptive fields. *Journal of Comparative Neurology*, **280**, 97–121.

Ariel, M., & Adolph, A. R. (1985). Neurotransmitter inputs to directionally sensitive turtle retinal ganglion cells. *Journal of Neurophysiology*, **54**, 1123–1143.

Ariel, M., & Daw, N. W. (1982). Pharmacological analysis of directionally sensitive rabbit retinal ganglion cells. *Journal of Physiology*, **324**, 161–185.

Barlow, H. B., & Levick, W. R. (1965). The mechanism of directionally selective units in rabbit's retina. *Journal of Physiology*, **178**, 477–504.

Barnes, S., & Werblin, F. (1986). Gated currents generate single spike activity in amacrine cells of the tiger salamander retina. *Proceedings of the National Academy of Sciences of the United States of America*, **83**, 1509–1512.

Borg-Graham, L. J., & Grzywacz, N. M. (1992). A model for the directional selectivity circuit in the retina: Transformation by neurons singly and in concert. In T. McKenna, J. Davis, & S. F. Zornetzer (Eds.), Simple Neuron Computation (pp. 347–375). San Diego, CA: Academic Press.

Burkhardt, D. A. (1972). Effects of picrotoxin and strychnine upon electrical activity in the proximal retina. *Brain Research*, **43**, 246–249.

Caldwell, J. H., Daw, N. W., & Wyatt, H. J. (1978). Effects of picrotoxin and strychnine on rabbit retinal ganglion cells: Lateral interactions for cells with more complex receptive fields. *Journal of Physiology*, **276**, 277–298.

DeVoe, R. D., Carras, P. L., Criswell, M. H., & Guy, R. B. (1989). Not by ganglion cells alone: directional selectivity is widespread in identified cells of the turtle retina. In R. Weiler & N. N. Osborne (Eds.), Neurobiology of the Inner Retina, NATO ASI Series, (Vol. H31, pp. 235–246) Springer Verlag, Berlin, Germany.

Dowling, J. E. (1968). Synaptic organization of the frog retina: an electron microscopic analysis comparing the retinas of frogs and primates. *Proceedings of the Royal Society of London, Series B*, **170**, 205–228.

Dowling, J. E. (1979). Information processing by local circuits: the vertebrate retina as a model system. In F. O. Schmitt & F. G. Worden (Eds.), *The neurosciences* (4th Study Program. Cambridge, MA: MIT Press.

Dowling, J. E. (1987). *The retina: An approachable part of the brain.* Cambridge, MA: Harvard Univ. Press.

Dowling, J. E., & Werblin, F. S. (1969). Organization of retina of the mudpuppy, *Necturus maculosus*. *Journal of Neurophysiology*, **32**, 315–338.

Emerson, R. C., Citron, M, C., Vaughn, W. J., & Klein, S. A. (1987). Nonlinear directionally sensitive subunits in complex cells of cat striate cortex. *Journal of Neurophysiology*, **58**, 33–64.

Famiglietti, E. V. (1983a). ON and OFF pathways through amacrine cells in mammalian retina: The synaptic connection of starburst amacrine cells. *Vision Research*, **23**, 1265–1279.

Famiglietti, E. V. (1983b). Starburst amacrine cells and cholinergic neurons: Mirror-symmetric ON and OFF amacrine cells of rabbit retina. *Brain Research*, **261**, 138–144.

Famiglietti, E. V. (1991). Synaptic organization of starburst amacrine cells in rabbit retina: Analysis of serial thin sections by electron microscopy and graphic reconstruction. *Journal of Comparative Neurology*, **309**, 40–70.

Famiglietti, E. V., Kaneko, A., & Tachibana, M. (1977). Neuronal architecture of ON and OFF pathways to ganglion cells in carp retina. *Science*, **198**, 1267–1268.

Famiglietti, E. V., & Kolb, H. (1976). Structural basis for ON and OFF-center responses in retinal ganglion cells. *Science*, **194**, 193–195.

Fennema, C. I., & Thompson, W. B. (1979). *Computer Graphics and Image Processing*, **9**, 301–315.

Geiger, G., & Poggio, T. (1975). The orientation of flies towards visual patterns: On search for the underlying functional interactions. *Biological Cybernetics*, **19**, 39–54.

Girosi, F., Verri, A., & Torre, V. (1989). Constrains on the computation of optical flow. *IEEE Workshop on Visual Motion*, **116**, 124.

Grzywacz, N. M., & Amthor, F. R. (1988). What are the directionally selective subunits of rabbit retinal ganglion cells? *Society for Neuroscience Abstracts*, **14**, 603.

Grzywacz, N. M., & Amthor, F. R. (1989). A computationally robust anatomical model for retinal directional selectivity. In D. S. Touretzky (Ed.), *Advances in neural information processing systems*, (Vol. **1**, pp. 477–484) Morgan Kaufman.

Grzywacz, N. M., & Amthor, F. R. (1993). Facilitation in directionally selective ganglion cells of the rabbit retina. *Journal of Neurophysiology*, **69**, 2188–2199.

Grzywacz, N. M., Amthor, F. R., & Merwine, D. K. (In Press). Directional hyperactivity in ganglion cells of the rabbit retina. *Visual Neuroscience*.

Grzywacz, N. M., Amthor, F. R., & Mistler, L. A. (1990). Applicability of quadratic and threshold models to motion discrimination in the rabbit retina. *Biological Cybernetics*, **64**, 41–49.

Grzywacz, N. M., & Koch, C. (1987). Functional properties of models for direction selectivity in the retina. *Synapse*, **1**, 417–434.

Grzywacz, N. M., & Yuille, A. L. (1990). A model for the estimate of local image velocity by cells in the visual cortex. *Proceedings of the Royal Society of London, B*, **239**, 129–161.

Hassenstein, B., & Reichardt, W. E. (1956). Systemtheoretische Analyse der Zeit-, Reihenfolgen und Vorzeichenauswertung bei der Bewegungsperzeption des Russelkäfers *Chlorophanus. Zeitschrift für Naturforschung, Teil B*, **11**, 513–524.

Heeger, D. J. (1987). A model for extraction of image flow. *Journal of the Optical Society of America A*, **4**, 1455–1471.

Hildreth, E. C. (1984). *The measurement of visual motion*. Cambridge, MA: MIT Press.

Hodgkin, A. L., & Huxley, A. F. (1952). A quantitative description of membrane current and its application to conduction and excitation in nerve. *Journal of Physiology*, **117**, 500–544.

Horn, B. K. P., & Schunck, B. G. (1981). Determining optical flow. *Artificial Intelligence*, **17**, 185–203.

Kaneko, A. (1971). Physiological studies of single retinal cells and their morphological identification. *Vision Research, Supplement*, **3**, 17–26.

Katz, B., & Miledi, R. (1967). A study of synaptic transmission in the absence of nerve impulses. *Journal of Physiology*, **192**, 406–436.

Katz, B., & Thesleff, S. (1957). A study of the "desensitization" produced by acetylcholine at the motor end-plate. *Journal of Physiology*, **138**, 63–80.

Koch, C., Poggio, T., & Torre, V. (1983). Retinal ganglion cells: a functional interpretation of dendritic morphology. *Proceedings of the National Academy of Sciences of the Uited States of America*, **80**, 2799–2802.

Kusano, K. (1970). Influence of ionic environment on the relationship between pre and postsynaptic potentials. *Journal of Neurobiology*, **1**, 435–457.

Lasater, E. M., & Witkovsky, P. (1990). Membrane currents of spiking cells isolated from turtle retina. *Journal of Comparative Physiology*, **167**, 11–21.

Llinas, R., & Nicholson, C. (1971). Electrophysiological properties of dendrites and somata in alligator Purkinje cells. *Journal of Neurophysiology*, **34**, 532–551.

Llinas, R., & Sugimori, M. (1980). Electrophysiological properties of in vitro Purkinje cell somata in mammalian cerebellar slices. *Journal of Physiology*, **305**, 171–195.

Magleby, K. L. (1987). Short-term changes in synaptic efficacy. In G. M. Edelman, W. E. Gall, & W. M. Cowan (Eds.), *Synaptic function*. New York: Wiley.

Magleby, K. L., & Pallota, B. S. (1981). A study of desensitization of acetylcholine receptors using nerve-released transmitter in the frog. *Journal of Physiology*, **316**, 225–250.

Mallart, A., & Martin, A. R. (1968). The relation between quantum content and facilitation at the neuromuscular junction of the frog. *Journal of Physiology*, **196**, 593–604.

Marchiafava, P. L. (1979). The responses of retinal ganglion cells to stationary and moving visual stimuli. *Vision Research*, **19**, 1203–1211.

Marr, D., & Ullman, S. (1981). Directional selectivity and its use in early visual processing. *Proceedings of the Royal Society of London, Series B*, **211**, 151–180.

Masland, R. H., Mills, W., & Cassidy, C. (1984). The functions of acetylcholine in the rabbit retina. *Proceedings of the Royal Society, London*, **223** (B), 121–139.

Miller, R. F. (1979). The neuronal basis of ganglion-cell receptive-field organization and the physiology of amacrine cells. In F. O. Schmitt & F. G. Worden (Eds.), *The neurosciences (4th Study Program)*. Cambridge, MA: MIT Press.

Miller, R. F., & Bloomfield, S. A. (1983). Electroanatomy of an unique amacrine cell in the rabbit retina. *Proceedings of the National Academy of Sciences of the United States of America*, **80**, 3069–3073.

Miller, R. F., & Dacheux, R. F. (1976). Dendritic and somatic spikes in mudpuppy amacrine cells: Identification and TTX sensitivity. *Brain Research*, **104**, 157–162.

Naka, K. I., & Ohtsuka, T. (1975). Morphological and functional identifications of catfish retinal neurons. II. Morphological identification. *Journal of Neurophysiology*, **38**, 72–91.

Poggio, T. (1975). Stochastic linearization, central limit theorem and linearity in (nervous) 'black-boxes.' *Atti del III Congresso di Cibernetica e Biofisica*, 349-358.

Poggio, T. (1983). Visual algorithms. In O. J. Braddick & A. C. Sleigh (Eds.), *Physical and biological processing of images*. New York: Springer-Verlag.

Poggio, T., & Reichardt, W. E. (1973). Consideration on models of movement detection. *Kybernetik*, **13**, 223–227.

Poggio, T., & Reichardt, W. (1980). On the representation of multi-input systems: Computational properties of polynomial algorithms. *Biological Cybernetics*, **37**, 167–186.

Poggio, T., & Torre, V. (1978). A new approach to synaptic interactions. In R. Heim & G. Palm (Eds.), *Approaches to complex systems*. Berlin: Springer-Verlag.

Poggio, T., Yang, W., & Torre, V. (1988). Optical flow: Computational properties and networks, biological analog. In R. Durbin, C. Miall, & G. Mitchison (Eds.), *The computing neuron*. Reading, PA: Addison-Wesley.

Rall, W. (1964). Theoretical significance of dendritic trees for neuronal input-output relations. In R. F. Reiss (Ed.), *Neural theory of modeling*. Stanford, CA: Stanford University Press.

Rall, W. (1967). Distinguishing theoretical synaptic potentials computed for different soma-dendritic distributions of synaptic input. *Journal of Neurophysiology*, **30**, 1138–1168.

Reichardt, W. (1969). Movement perception in insects. In W. Reichardt (Ed.), *Processing of optical data by organisms and machines*. London: Academic Press.

Reichardt, W., Egelhaaf, M., & Schlögl, R. W. (1988). Movement detectors provide sufficient information for local computation of 2-D velocity field. *Die Naturwissenschaften*, **75**, 313–315.

Richter, R., & Ullman, S. (1982). A model for the temporal organization of X and Y type receptive fields in the primate retina. *Biological Cybernetics*, **43**, 127–145.

Rosenthal, J. (1969). Post-tetanic potentiation at the neuromuscular junction of the frog. *Journal of Physiology*, **203**, 121–133.

Rushton, W. A. H. (1937). Initiation of the propagated disturbance. *Proceedings of the Royal Society of London, Series B*, **124**, 210.

Sakuranaga, M., & Naka, K. I. (1985a). Signal transmission in the catfish retina. II. Transmission to type-N cell. *Journal of Neurophysiology*, **53**, 390–410.

Sakuranaga, M., & Naka, K. I. (1985b). Signal transmission in the catfish retina. II. Transmission to type-N cell. *Journal of Neurophysiology*, **53**, 411–428.

Takeuchi, A. (1958). The long-lasting depression in neuromuscular transmission of frog. *Journal of Physiology*, **8**, 102–113.

Thorson, J. (1966). Small-signal analysis of visual reflex in the locust. II. Frequency dependence. *Kybernetik*, **3**, 52–66.

Torre, V., & Poggio, T. (1977). A Volterra representation of some neuron models. *Biological Cybernetics*, **27**, 113–124.

Torre, V., & Poggio, T. (1978). A synaptic mechanism possibly underlying directional selectivity to motion. *Proceedings of the Royal Society of London, Series B*, **202**, 409–416.

Toyoda, J., Hashimoto, H., & Ohtsu, K. (1973). Bipolar-amacrine transmission in the carp retina. *Vision Research*, **13**, 295–307.

van Santen, J. P. H., & Sperling, G. (1984). A temporal covariance model of motion perception. *Journal of the Optical Society of America A*, **1**, 451–473.

Vaney, D. I. (1990). The mosaic of amacrine cells in the mammalian retina. In N. N. Osborne, & G. Chader (Eds.), *Progress in Retinal Research* (Vol. 9, pp. 49–100). Elmsford, NY: Pergamon Press.

Verri, A., & Poggio, T. (1987). Against quantitative optical flow. *Proceedings of the First International Conference on Computer Vision*, 171–180.

Watanabe, S. I., & Murakami, M. (1984). Synaptic mechanisms of directional selectivity in ganglion cells of frog retina as revealed by intracellular recordings. *Japanese Journal of Physiology*, **34**, 497–511.

Watson, A. B., & Ahumada, A. J. (1985). Model of human visual-motion sensing. *Journal of the Optical Society of America A*, **2**, 322–342.

Werblin, F. S. (1970). Response of retinal cells to moving spots: Intracellular recording in *Necturus maculosus*. *Journal of Neurophysiology*, **32**, 339–355.

Werblin, F., Maguire, G., Lukasiewicz, P., Eliasof, S., & Wu, S. M. (1988). Neural interactions mediating the detection of motion in the retina of the tiger salamander. *Visual Neuroscience*, **1**, 317–329.

Wyatt, H. J., & Daw, N. W. (1975). Directionally sensitive ganglion cells in the rabbit retina: Specificity for stimulus direction, size, and speed. *Journal of Neurophysiology*, **38**, 613–626.

Yuille, A. L., & Grzywacz, N. M. (1988). A computational theory for the perception of coherent visual motion. *Nature*, **333**, 71–74.

21

Brain Style Computation: Learning and Generalization

David E. Rumelhart

The last several years have witnessed a remarkable growth in interest in the study of brain style computation. This effort has variously been characterized as the study of neural networks, connectionist architectures, parallel distributed processing systems, neuromorphic computation, artificial neural systems, and other names as well. The common theme to all these efforts has been an interest in looking at the brain as a model of a parallel computational device very different from that of a traditional serial computer. The strategy has been to develop simplified mathematical models of brainlike systems, and then to study these models to understand how various computation problems can be solved by such devices. The work has attracted scientists from a number of disciplines. It has attracted neuroscientists who are interested in making models of the neural circuitry found in specific areas of the brains of various animals, physicists who see analogies between the dynamical behavior of brainlike systems and the kinds of nonlinear dynamical systems familiar in physics, computer engineers who are interested in fabricating brainlike computers, workers in artificial intelligence who are interested in building machines with the intelligence of biological organisms, psychologists who are interested in the mechanisms of human information processing, mathematicians who are interested in the mathematics of such brain style systems, philosophers who are interested in how such systems change our view of the nature of mind and its relationship to brain, and many others. The wealth of talent and the breadth of interest has made the area a magnet for bright young students.

Although proposals differ in the detail, the most common models consider the neuron as the basic processing unit. A typical model neuron in illustrated in Figure 1. Each such processing unit is characterized by an activity level (representing of the state polarization of a neuron), an output value (representing the firing rate of the neuron), a set of input connections (representing synapses on the cell and its dendrite), a bias value (representing an internal resting level of polarization of the neuron), and a set of output connections (representing a neuron's axonal projections). Each of these aspects of the unit are represented mathematically by real numbers. Thus, each connection has an associated weight (synaptic strength) which deter-

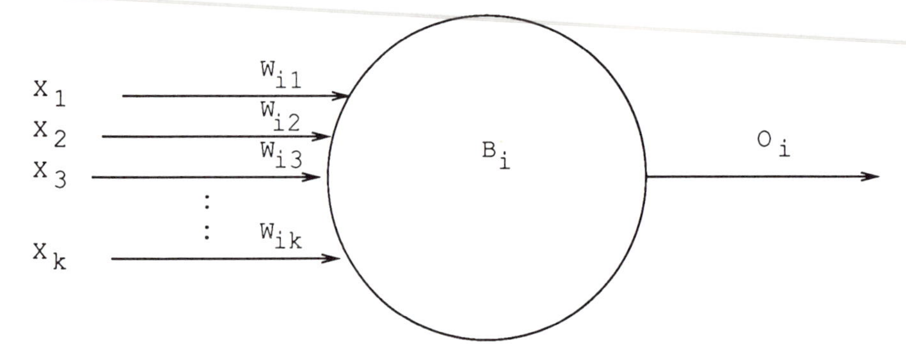

FIGURE 1 A typical model neuron.

mines the effect of the incoming input on the activation level of the unit. The weights may be positive (excitatory) or negative (inhibitory). Frequently, the input lines are assumed to sum linearly yielding an activation value for unit i at time t, given by

$$\alpha_i(t) = \sum_j w_{ij} o_j(t) + \beta_i$$

where w_{ij} is the strength of the connection from unit j to unit i, β_i is the unit's bias value and o_j is the output value of unit j. Note that the effect that the output of a particular unit has on the activity of another unit is jointly determined by its output level and the strength (and sign) of its connection to that unit. If the sign is negative, it lowers the activation, if the sign is positive it raises the activation. The magnitude of the output and the strength of the connection determines the amount of the affect. The output of such a unit is normally a nonlinear function of its activation value. A typical choice of such a function is a sigmoid. The logistic

$$o_i(t) = \frac{1}{1 + e^{\alpha_i(t)/T}}$$

illustrated in Figure 2, will be employed in the examples illustrated below. The parameter of the logistic T, yields functions of differing slopes. As T approaches zero, the logistic becomes a simple logical threshold function which takes on the value of 1 if the activity level is positive, and 0 otherwise.

A brain style computational device consists of a large richly interconnected network of such units. In real brains there are tens of billions of such units and tens of trillions of such connections. Such a network is a general computing device. The function it computes is determined by the pattern of connections. Thus, the configuration of connections is the analog of a program. The goal is to understand the kinds of algorithms that are naturally implemented by such networks.

It is clear enough that the kinds of units outlined earlier could be used to implement conventional logic gates. Even a cursory comparison of neurons with conventional hardware makes it clear that the design constraints on computation in brains differ radically from those in silicon. Perhaps the most striking comparison is in terms of processing speed. Whereas components in conventional machines operate in the range of a few nanoseconds, neurons operate in the range of a few milliseconds—a factor of about one million. This means that human processes that take less than a second can involve only one hundred or so time steps. Since most of the processes psychologists have studied—perception, memory retrieval, speech processing, sentence comprehension—take less than a second, it makes sense to impose what Feldman (1985) calls the "100-step program" constraint. That is, plausible explanations for these mental phenomena must not require more than about one hundred elementary sequential operations. Given that the processes we seek to characterize are often quite complex and may

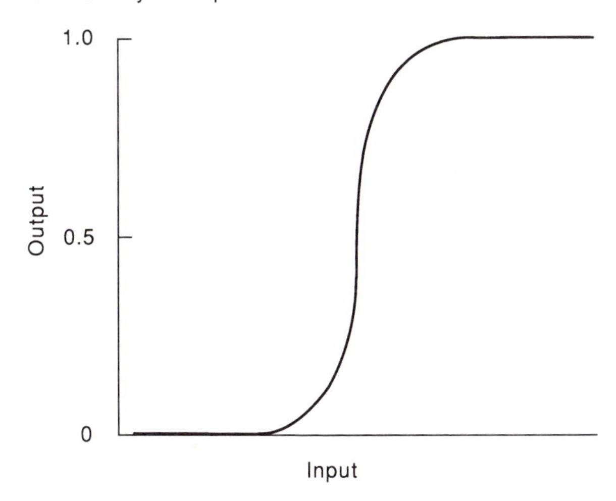

FIGURE 2 The logistic activation function $o_i = f(a_i) = 1/(1 + e^{-a_i})$ where $a_i = \sum_j w_{ij} o_j + \beta_i$.

involve consideration of large numbers of simultaneous constraints, our algorithms must involve considerable parallelism. Thus, although a serial computer could be created out of the kinds of components represented by our units, such an implementation would surely violate the 100-step program constraint for any but the simplest processes. Some might argue that although parallelism is obviously present in much of human information processing, this fact alone need not greatly modify our world view. This is unlikely. The speed of components is a critical design constraint. Although the brain has slow components, it has very many of them. The human brain contains billions of such processing elements. Rather than organize computation with many, many serial steps, as we do with modern computers whose steps are very fast, the brain must deploy many, many processing elements cooperatively and in parallel to carry out its activities. These design characteristics, among others, lead to a general organization of computing which is very large relative to the fan-in and fan-out conventional processing elements. There are thousands or even tens of thousands of connections on the dendrites of a single neuron. In mammalian cortex, the activity of a given cell is rarely enough to cause activity in the postsynaptic neuron. Rather, it is the cooperative activity of numbers of input units which determine the output of a given cell. Input from a given unit is rarely either necessary or sufficient. This suggests a model of operation that is rather more statistically based than logically based.

Although there has been a good deal of activity recently, the study of brain style computation has its roots over 50 years ago in the work of McCulloch and Pitts (1943) and slightly later in Hebb's famous *The Organization of Behavior* (1949). The early work in artificial intelligence was torn between those who believed that truly intelligent systems could best be built on computers modeled after brains (Selfridge, 1955; Widrow & Hoff, 1960; Rosenblatt, 1962) and those like Newell, Shaw, and Simon (1958), McCarthy (1959), and Minsky and Pappert (1969) who believed that the intelligence was fundamentally symbol processing of the kind readily modeled on the von Neumann computer. For a variety of reasons, the symbol processing approach became the dominant theme in artificial intelligence. The reasons for this were both positive and negative. On the one hand, the stored program digital computer became the standard of the computer industry. Such computers were easy to design and easy to program. The symbol processing, logic based approach to AI is well suited for such an architecture. On the other hand, the fundamentally parallel brain style systems, such as Rosenblatt's Perceptron system, were not well suited to implementation on serial computers. Moreover, the Perceptron turned out to be rather more limited than first expected (Minsky & Pappert, 1969) and this discouraged both scientists and funding agencies. Although work continued throughout the 1970s by a number of workers including Amari, Anderson, Arbib, Fukishima, Grossberg, Kohonen, and others, and although a number of important results were obtained during this period, the work received relatively little attention.

The 1980s showed a rebirth in interest. There seem to be at least five reasons for this. Three of the reasons are essentially pragmatic and two theoretical. First, on the more pragmatic side:

- Today's computers are much faster than those of the 1950s and 1960s. It is thus possible to use conventional computers to simulate and experiment with much larger and more interesting networks than ever before.
- Everyone believes that the future for faster computers must be in parallel computation. Unfortunately, there is no generally accepted paradigm for parallel computation. It is generally easier to build parallel computers than to find algorithms efficient for them. There is a hope that algorithms which prove efficient and effective on brain style computers may prove a useful general paradigm for parallel computation.
- The basic empirical tools of neuroscience are expanding and we are learning more and more about how the neuron functions and how neurons communicate to one another, but little is known about how to go from this information about specific neurons to a theoretical account of how large networks of such neurons might function. It is hoped that the theoretical tools developed in the study of brain style computational systems will allow for the modeling of real neural networks.

In addition to these three reasons for interest, there are two theoretical results which have now been developed and appreciated.

- The first of these results is due to Hopfield (1982) and provides the mathematical foundation for understanding the dynamics of an important class of networks. These ideas have been extended and applied by Cohen and Grossberg (1983), Hinton and Sejnowski (1985), Smolensky (1986), and a number of others to provide us with a useful mathematical understanding of how networks such as these might be configured to solve important optimization problems.
- The second result is an extension of the work of Rosenblatt and Widrow and Hoff, to deal with learning in complex networks and thereby provide an answer to one of the most severe criticisms of the original perceptron work.

The rest of this chapter provides an explication of these two results along with some examples.

PARALLEL COMPUTATION WITH NETWORKS OF SIMPLE UNITS

Imagine a network of units connected to all of the rest. Imagine that the input data on which the computation is to rest is fed to the system by applying external excitation or inhibition to certain of the units (the input units). Once these units are externally activated, the dynamics of the network takes over. Each unit excites or inhibits those units to which it is directly connected (depending on whether it is positively or negatively connected to the neighbor). These units in turn excite or inhibit their neighbors. Certain networks will eventually reach a stable point so that the activation value of each unit is constant and the network reaches an equilibrium point. If the system is working properly, the final state of the units (the output units) constitutes the result of the computation. This method of computing, settling or relaxing to a solution, contrasts sharply with the more conventional use of branching, subroutine calls, and the like to organize the computation and thereby calculate the result. In the case of the network, we want the solution to emerge from the interaction among the set of units in the network. There is no executive guiding the computation. All the knowledge must be in the set of interconnections. How can we organize a network so that it will behave properly? Hopfield (1982) provided an important insight when he observed that purely local operations within networks of the kind we have been discussing can, under certain conditions, be seen to be optimizing a global measure of performance. In particular, in the class of symmetric, asynchronous networks the network can be shown to be locally optimizing the following measure of the system's energy,

$$E(\mathrm{t}) = -\left[\sum_{i,j} w_{ij} o_i(t) o_j(t) + \sum_i (I_i + \beta_i) o_i(t)\right]$$

where I_i is the magnitude of the external input to unit i. The conditions for such an optimization can be easily understood by considering a binary output unit which takes on value 1 if its activation is positive and 0 otherwise. In this case, we can show that on each update of each unit the global energy is either decreased or stays the same. To see this it is useful to compute the derivative of the energy function with respect to the output of a given unit. The derivative of E with respect to any unit, i, in the system is given by

$$\frac{\partial E(t)}{\partial o_i(t)} = -\left[\sum_j w_{ij} o_i(t) + \mathrm{I}_i + \beta\right] = -\alpha_i(t)$$

Now since the derivative is just the negation of the activation, the unit will take on its maximum value whenever the derivative of the energy with respect to the output of the unit is negative and will take on its minimum value whenever the derivative is positive. This means that in either case the energy will be reduced if the unit is changed or stay the same if it was not. Thus, since the energy either decreases or stays the same, this process of letting the network assign output value to its units will eventually reach a minimal energy point in which no changes remain that will further reduce the energy in the system. Such a stable point is the end of the computation and the values assigned to the output units are considered the final result of the computation. There are two important issues associated with this sort of computational paradigm. First, we must assign values to the connections so that the minimal energy states correspond to the desired solutions of the problem and second, this procedure only assures that the system will find a local minimum in the energy landscape when what we often want is to find the global minimum.

Hinton and Sejnowski (1983) have addressed both of these issues. It is useful, at the start, to think of units as representing hypotheses about the problem to be solved and connections between pairs of units as representing constraints among the hypotheses. If the truth of one hypothesis provides positive evidence for the truth of another, they should be positively connected. On the other hand, if the truth of one hypothesis provides negative evidence for the truth of another, they should be negatively connected. The magnitude of the connections are determined by the strength of the evidence. Zero connections correspond to independence between the hypotheses. In this case, finding the pattern of values which minimizes the energy function can be equivalent to finding the set of hypotheses which are maximally consistent with the input data and with one another.

Figure 3 illustrates a simple example. The figure illustrates the so-called graph cutting problem. The problem is to partition the nodes of a random graph into two equal classes while maximizing the number of links within a class and minimizing the number of links between classes. This is a very difficult problem which, in general, requires a number of steps which increase exponentially with the size of the problem. It is, however, rather easy to develop a network capable of finding very accurate approximate solutions to the problem. For simplicity, we assign a unit to each node in the graph. Each node stands for the hypothesis that the node in question is a member of class A. For simplicity we choose the value of 1 to designate that the hypothesis is true and -1 to designate that the hypothesis if false (i.e., the node is to be in class B). We use unit weights to correspond to the links of the graph.

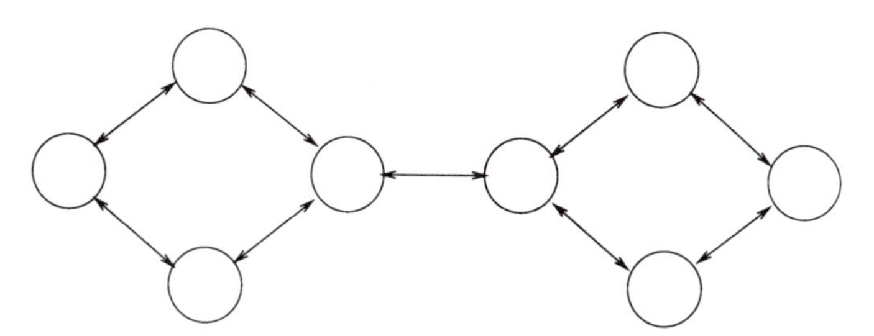

FIGURE 3 The graph partitioning problem.

Whenever two nodes are connected in the graph, we have a connection of value 1 between the corresponding units in the network. Note that we can write down the difference between the number of links within a class and the number of links between classes as

$$D = \frac{1}{2} \sum_{i,j} w_{ij} o_i o_j$$

This follows because whenever a link connects a pair of nodes in the same class we add one to D and whenever we find a link connecting nodes in a different class we subtract one from D (the factor of 1/2 keeps us from counting links twice). The similarity of D to the energy function is obvious. Unfortunately, finding the maximum value of D (or minimizing $-D$) doesn't quite solve our problem because D can be trivially maximized by putting all of the nodes in one of the classes. We must add the constraint that the two partitions be of the same magnitude. We can add an additional term by noting that the algebraic sum of the output values will be zero whenever the number of nodes in each class is equal. Thus, minimizing the following term will ensure an equal number of nodes in each class:

$$M = (\sum_i o_i)^2$$

Finally, the desired solution occurs when we maximize D and minimize M. From D and M we can construct a final energy function that we want to minimize:

$$E = -D + M = -\frac{1}{2} \sum w_{ij} o_i o_j + \lambda (\sum_i o_i)^2$$

where λ is a number weighting the relative importance of the two terms in the energy function. Taking the derivative yields the appropriate output function for each unit in the network:

$$\frac{\partial E}{\partial o_i} = -\frac{1}{2} \sum w_{ij} o_j + 2\lambda \sum o_i$$

Renormalizing to get rid of the constants and letting the new constant weighting the second term be α, we find that letting

$$\alpha_i = \sum w_{ij} o_j - \alpha \sum o_i = \sum (w_{ij} - \alpha)\, o_j$$

yields $w_{ij} = 1$ wherever there is a link between the corresponding nodes, and 0 otherwise. This is equivalent to a network with links of negative α between all units without a link and $1\text{-}\alpha$ for all units with links. Peterson and Anderson (1987) have shown very promising results for this sort of network solving the graph partitioning problem.

Although the constraint satisfaction interpretation given earlier represents a useful heuristic for assigning values to connections, it is not always easy to find a set of values which weighs the various terms of the energy function appropriately. In this case we need to find methods whereby the network can learn the weight values required for the task. I will turn to a discussion of learning shortly, but first, we consider the issue of local minima.

Local versus Global Minima In his original work, Hopfield (1982) was building a content addressable memory and he wanted each memory to be a minimum in the network. The idea was that the network would be started at a region "close" to one of the minima in the energy landscape and then would fall into that minimum, thereby retrieving the closest stored memory. In this case, global minima are not important. For the sorts of applications discussed in this chapter, however, the problem of local minima can be severe.

Once the problem has been cast as an energy minimization problem and the analogy with physics has been noted, the solution to the problem of local goodness maxima can be solved in essentially the same way that flaws are dealt with in crystal formation. One standard method involves annealing. Annealing is a process whereby a material is heated and then cooled very slowly. The idea is that, as the material is heated, the bonds among the atoms weaken and the atoms can reorient relatively freely. They are in a state of high energy. As the material is cooled, the bonds begin to strengthen, and as the cooling continues, the bonds eventually become sufficiently strong that the material freezes. If we want to minimize the occurrence of flaws in the material, we must cool slowly enough so that the effects of one particular coalition of atoms has time to propagate from neighbor to neighbor throughout the whole material before the material freezes. The cooling must be especially slow as the freezing temperature is approached. During this period the bonds are quite strong so that the clusters will hold together, but they are not so strong that atoms in one cluster might not change state to line up with those in an adjacent cluster, even if it means moving into a momentarily more energetic state. In this way annealing can move a material toward a global energy minimum.

The solution then is to add an annealing-like process to the network models and have them employ a kind of *simulated annealing*. The basic idea is to add a global parameter analogous to temperature in physical systems and therefore called temperature. This parameter should act to decrease the strength of connections at the start and then change to strengthen them as the network is settling. Moreover, the system should exhibit some random behavior so that, instead of always moving downhill in energy space, when the temperature is high it will sometimes move uphill. This will allow the system to "step out of" local minima that are not very deep and explore other parts of the energy landscape to find the global minimum. This is just what Hinton and Sejnowski (1983) proposed in the Boltzmann machine, what Geman and Geman (1984) have proposed in the *Gibbs sampler*, and what Smolensky (1986) has proposed in harmony theory.

The essential update rule employed in all of these models is probabilistic and is given by the logistic function illustrated earlier. This differs from the Hopfield model in two important ways. First, these materials are *stochastic*. That is, the update rule specifies only a probability that the units will take on one or the other of their values. This means that the system need not necessarily go downhill in energy—it can move uphill as well. Second, the behavior of the systems depends on a global parameter, *temperature*, which can start out high and be reduced during the settling phase. These characteristics allow these systems to implement a simulated annealing process. These changes are important because of mathematical results which, in effect, guarantee that the system will end up in a global minimum if annealed slowly enough.

LEARNING BY EXAMPLE

The problem of learning in these brain style networks is the problem of finding a set of connection strengths that allow the network to carry out the desired computation. In this section I will focus on only one of the many forms of learning and only one of the learning paradigms that have been proposed and studied. In particular, I will discuss error correction learning as applied to the problem of pattern association. The situation I have in mind is illustrated in Figure 4. There is a set of input units which are connected, through a set of so-called hidden units, to a set of output units. In the general case, there may be any number and configuration of hidden units and connections among the units. [For simplicity, I will restrict discussion here to the case of feedforward networks in which the activity of a given unit cannot influence, even directly, its own inputs. A more complete account in which recurrent networks as well as feedforward cases are dealt with is given in Rumelhart, Hinton, and Williams (1986).] The network is provided with a set of example input–output pairs (a training set) and is to modify its connections so as to approximate the function from which the input/output pairs have been drawn. The networks are then tested for ability to generalize. The error correction learning procedure is simple enough in conception. The procedure is as follows. During training an input is put into the network and it flows through the network generating a set of values on the output units. Then, the actual output is compared with the desired target and a match is computed. If the output and target match, no change is made to the net. However, if the output differs from the target then a change must be made to some of the connections. The problem is to determine which connections in that entire network were at fault for the error; this is called the *credit assignment* (or perhaps better the blame assignment) problem.

Although Rosenblatt's (1962) perceptron convergence procedure and Widrow and Hoff's (1960) LMS (least mean square) learning procedure and its variants have been around for some time, these learning procedures are limited to one-layer networks involving only

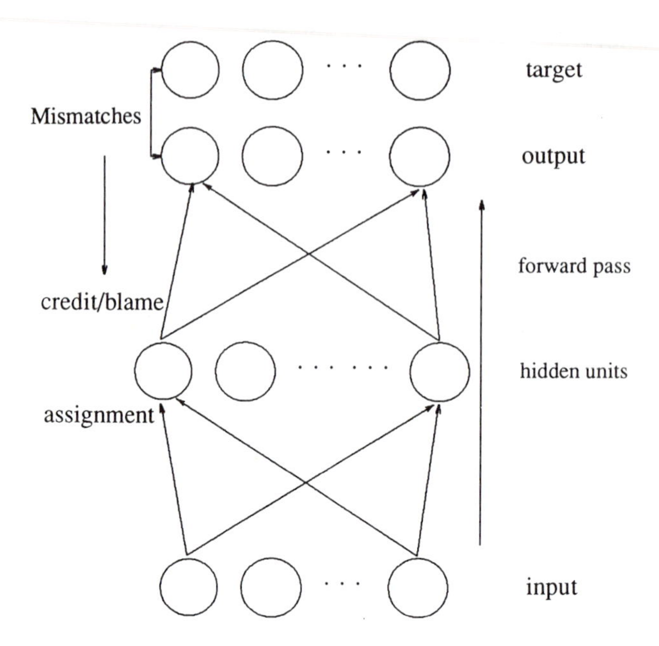

FIGURE 4 A multilayer network. In this case the information coming to the input units is recoded into an internal representation and the outputs are generated by the internal representation rather than by the original pattern. Input patterns can always be encoded, if there are enough hidden units, in a form so that the appropriate output pattern can be generated from any input pattern.

input and output units. There are no hidden units in these cases and no ability for the network to create its own internal representation. The coding provided by the external world has to suffice. Nevertheless, these networks have proved useful in a wide variety of applications. Perhaps the essential character of such networks is that they map similar input patterns to similar output patterns. This is what allows these networks to make reasonable generalizations and perform reasonably on patterns that have never before been presented. The similarity of patterns in a connectionist system is determined by their overlap. The overlap in such networks is determined outside the learning system itself by whatever produces the patterns.

The constraint that similar input patterns lead to similar outputs can lead to an inability of the system to learn certain mappings from input to output. Whenever the representation provided by the outside world is such that the similarity structure of the input and output patterns is very different, a network without internal representations (i.e., a network without hidden units) will be unable to perform the necessary mappings.

Minsky and Pappert (1969) have provided a careful analysis of conditions under which such systems are capable of carrying out the required mappings. They show that in many interesting cases, networks of this kind are incapable of solving the problems. On the other hand, as Minsky and Pappert also pointed out, if there is a layer of simple perceptron-like hidden units, as shown in Figure 5, with which the original input pattern can be augmented, there is always a recoding (i.e., an internal representation) of the input patterns in the hidden units in which the similarity of the patterns among the hidden units can support any required map-

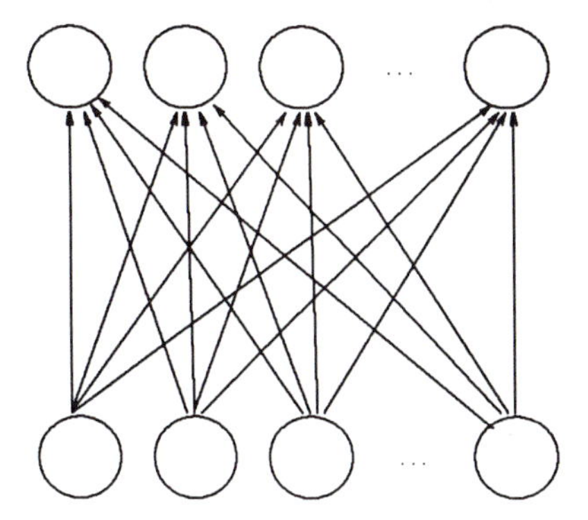

FIGURE 5 Three-layer network with hidden units. Similar inputs yield similar outputs.

ping from the input to the output units. Thus, if we have the right connections from the input units to a sufficiently large set of hidden units, we can always find a representation that will perform any mapping from input to output through these hidden units.

The existence of networks such as this illustrates the potential power of hidden units and internal representations. The problem, as noted by Minsky and Pappert, is that, whereas there is a very simple guaranteed learning rule for all problems that can be solved without hidden units (the perceptron convergence procedure, or the variation due originally to Widrow & Hoff, 1960), there has been no equally powerful rule for learning in multi layer networks.

It is clear that if we hope to use these connectionist networks for general computational purposes, we must have a learning scheme capable of learning its own internal representations. We have developed a generalization of the perceptron and LMS learning procedures, which we call back propagation, which allows the system to learn to compute arbitrary functions. The constraints inherent in networks without self-modifying internal representations are no longer applicable. The 1980s have led to the development of a rather simple yet powerful solution to this problem (cf. Rumelhart et al., 1986). The basic idea is to define a measure of the overall performance of the system and then to derive a learning rule which is designed to optimize that performance. The weights and biases can then be determined (and blame assigned) by a gradient descent procedure. The basic idea is that each unit is assigned a measure of blame proportional to the gradient of a measure of overall performance with respect to the activity of that unit. A unit for which small changes in activity yield large changes in performance is thus assigned more blame than a unit whose activity has little effect on the performance measure.

In order to carry out this method we must begin by defining a performance measure and then finding a computational method for determining the gradient which is well suited to networks of brainlike elements. In this case, we define the performance of the system as

$$E = \sum_{p,i} (t_{ip} - o_{ip})^2$$

where i indexes the output units, p indexes the I/O pairs to be learned, t_{ip} indicates the target for a particular output unit on a particular pattern, o_{ip} indicates the actual output for that unit on the pattern, and E, the performance measure, is the total error of the system. The goal, then, is to find a principle of learning which is both generally consistent with the principles of brain style computation (i.e., involves only local computation and simple processing units connected through excitatory and inhibitory connections) and minimizes this error function.

As it turns out, there is a simple recursive method of computing the relevant gradient and thereby implementing the gradient descent learning procedure in brain style networks. The basic procedure is a two stage process. First, an input is applied to the network. Then, after the system has processed for some time, certain of the units of the network are informed of the values they ought to have at this time. If they have attained the desired values, the weights are unchanged. If they differ from the target values, then the weights are changed according to the difference between the actual value the units have attained and the target for those units. This difference is essentially the derivative of the error function with respect to the output units, that is, it represents the "blame" assigned to the output units. This derivative is both the basis for changing the connections to the output units and for computing the derivative of the error with respect to the activity of those units impinging on the output units. Each such unit can compute its own blame by taking the weighted sum of all of the derivatives of the units to which it projects multiplied by the strength of the weights connecting these units to the output elements. Then, based on this signal, the weights projecting into these "second layer" units are modified after which the "blame" is similarly passed back another layer. This process continues until the signal reaches the input units or until it has been passed back for a fixed number of times. Then a new input pattern is presented and the process repeats. Although the procedure may sound difficult, it is actually quite simple and easy to implement within these nets.

No further attempt to define the details of the learning procedure will be offered here. The Rumelhart et al. (1986) paper should be consulted for further details. The important point for our purposes here is that back propagation learning is a procedure for discovering an internal representation in which patterns which are functionally similar (i.e., must be responded to in the same way) are represented by similar patterns. This is derived by defining a global cost function which is minimized through a local, iterative, gradient descent procedure well suited to brainlike networks. Although we have no direct evidence that back propagation is employed in actual brains, we do know that it is a relatively powerful mechanism which could be implemented in neural tissue. (Note, some observers assume that, since the back flow of information cannot go through the axons, this learning procedure cannot be implemented in a real brain. There is, of course, no need for the error correction information to be sent down the very same wires which carried the forward flow of processing. The information would most likely be transmitted through back projections as normal activity levels.) Given the difficult learning tasks with which brains must deal it seems likely that evolution has discovered a process at least as complex as gradient descent learning. We turn now to an example which illustrates the kinds of representations that are discovered by such a learning procedure and how these representations facilitate generalization to new cases.

Although most applications of adaptive networks of this sort have been to the domain of perceptual processing, the nature of what is learned and the nature of the generalizations can often be better understood with examples which are more symbolic in nature. It is easier to see the "logic" of the generalization. Therefore, I have chosen to consider how a network might store the information illustrated in Figure 6. The figure illustrates a set of symbolic information about the relationship among categories and their properties. The data structure, called a semantic network, summarizes a number of facts. It summarizes, for example, the fact that a canary can sing. It also summarizes the fact that a canary is a bird and that a canary can fly. This later fact follows from the fact that properties of higher level nodes in a semantic network can be inherited by lower level nodes. Thus, for example, a rose can grow because a rose is a flower, a flower is a plant, a plant is a living thing, and a living thing can grow. Knowledge structures such as these are common place in many AI knowledge based systems. One of the central features of such a network is that new information can easily be added and many new inferences readily made. For example, suppose that we add to the system that a "sparrow is a bird." In this case, we can directly "infer" that a sparrow can fly, that a sparrow has wings, that a sparrow has feathers, that a sparrow is an animal, that a sparrow is a living thing, that a sparrow can grow, etc. All of these inferences follow from the addition of one fact to the data base. If we were to propose that a connectionist account is appropriate here, it would be nice if such inferences were "automatically" made by the connectionist system as well. In the following paragraphs I show how to construct just such a network which, when taught a new fact about sparrows, makes the appropriate generalizations.

Consider the connectionist network illustrated in Figure 7. In this case the lower units, labeled nodes and relations, are the input units. There is one "node" unit for each node in the semantic network. That is, one for canary, one for robin, one for bird, animal, liv-

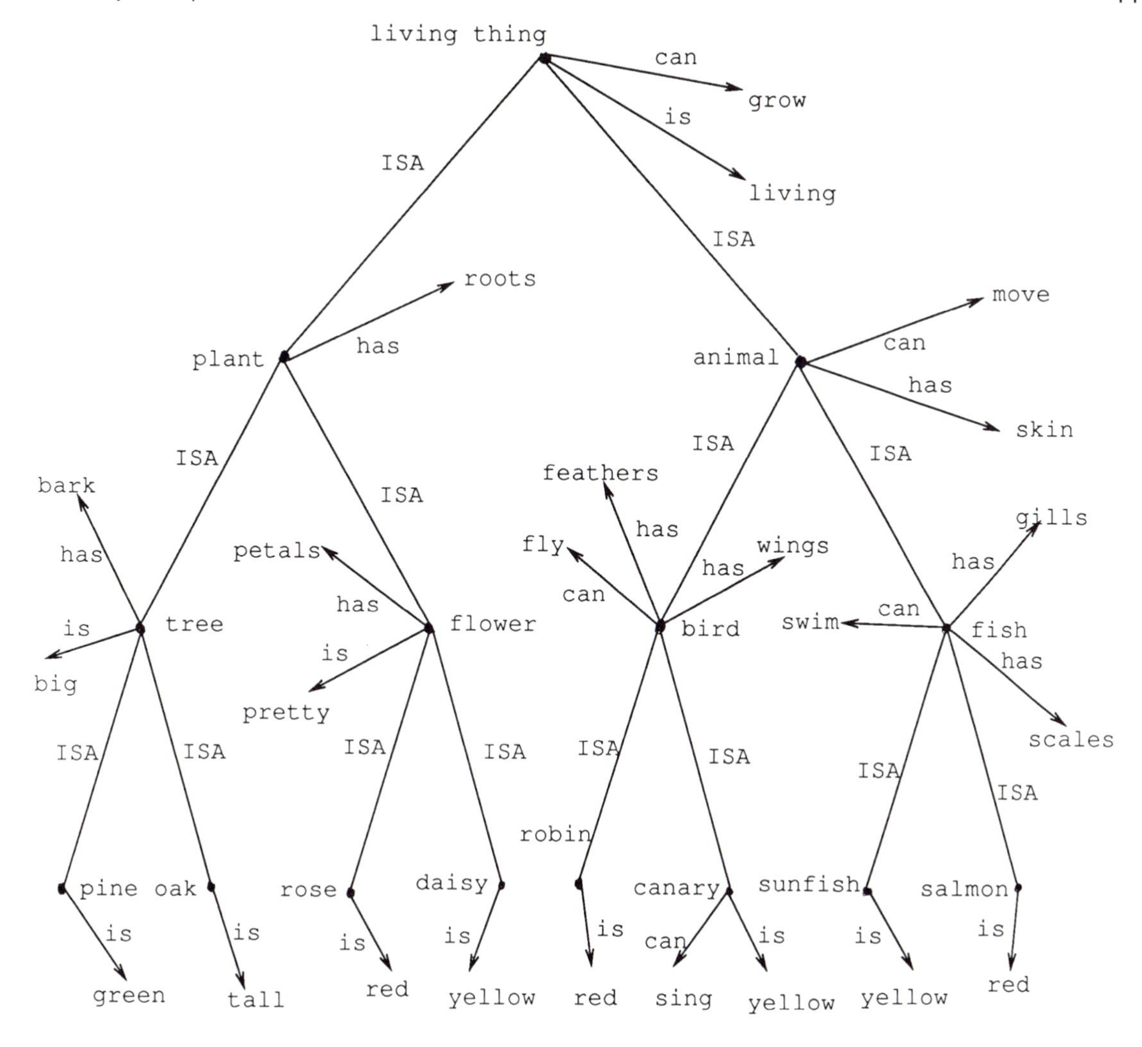

FIGURE 6 A semantic network showing the relationships among various plants and animals.

ing thing, pine, oak, daisy, flower, tree, etc. Similarly, there is one "relation" unit for each of the four arc labels—isa, has, is, and can. The output units are shown across the top of the figure. They are divided into four groups corresponding to each arc type. There is an output unit for each action any of the plants or animals can do. For example, there is one for sing, one for grow, one for fly, swim, move, etc. Similarly, there is an output unit for each "property" that a plant or animal in the semantic network can have. Thus, there is one for wings, feathers, gills, etc. Similarly, there is an output for everything that something "isa" and one for everything that something in the semantic network "is." The connectionist network is trained by presenting it with facts randomly chosen from the semantic network. For example, it is presented with the input "canary can" by turning on the units for "canary" and "can" in the input layer and training it to turn on the

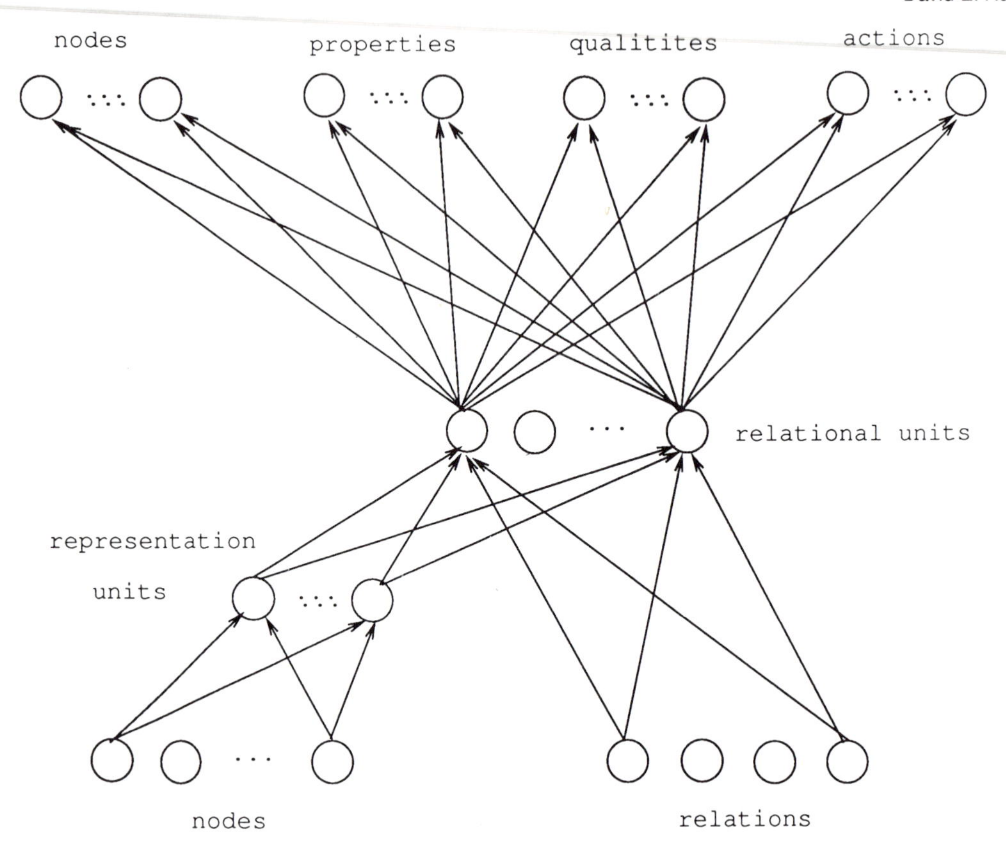

FIGURE 7 A connectionist network for storing symbolic information.

units for "sing," "fly," "move," and "grow" in the output layer. If it fails to turn on the correct units in the output, the back propagation learning algorithm is employed to change the connections in the network so that next time this particular query is presented the network will do a better job of turning on the correct output units. This process is repeated for all of the facts implicit in the semantic network until, for any such input given to the connectionist network, the system will produce the correct answer as an output.

It will be noted from the previous figure that the connectionist network contains two groups of "hidden" unit layers. The first layer, labeled as the representational units, receive inputs only from the "node" inputs. The second layer of hidden units, labeled the "relational" units, receive inputs from both the "representational" hidden units and from the four units representing arc labels. These relational units then project to all of the output units. The rationale for this configuration of hidden units is that the representational hidden units receive inputs only from the nodes and that therefore they should come to represent the terms (such as canary, robin, etc.). Moreover, we should find that the representation for "pine" and "oak" should be more similar to one another than either is to "rose" or "daisy" and these representations should be more similar to one another than any of them are to "canary," "robin," etc. Thus, by looking at the patterns over the representational units when various units are turned on we can study the representations developed by the

TABLE 1 The Representations Developed for the 15 Input Concepts

oak – . . – – +			
	tree – + – . – +		
pine – – – – – +		plant – – – + – –	
rose – – – – – –			
	flower – + – . – –		
daisy – + – – – –			
canary + + – + + +			
			living thing – + – . – +
	bird . – – + + –		
robin + – – + + +			
		animal + + . + – –	
salmon + + + – + +			
	fish + + + + + –		
sunfish + + + + + +			

network for these concepts without contamination from the specific label type being queried. The "relational" units that the distributed representations develop at the "representational" level are to be combined with the information from the arc type to turn on the relevant output units.

We find that, after presenting the facts of the semantic network to the connectionist network a few hundred times, the network has learned to produce the right answer for each of the inputs in the training set. At this point it is useful to examine the kinds of representations of the input actually developed at the representational units. Recall that the point of developing a hidden unit layer is to develop a representation of the input in which stimuli which require similar responses have similar representations, no matter how similar or dissimilar the actual input patterns are. In this case, the input pattern for "canary" has nothing in common with the input pattern for "robin" (indeed, since the input pattern for each concept involved turning on a single unit in the input layer, all input patterns were orthogonal to one another). One way of seeing the representational patterns associated with an input concept is to look at the patterns of activity over the representational units arising for each input pattern. Table 1 shows the patterns. The table is laid out as a tree on its side. On the left are the leaves of the tree and on the right is the root of the tree. Associated with each concept is a sequence of six plus, minus, or dot symbols. There are six because in this particular run of the experiment six representational units were used. The plus means that the corresponding representational unit was on, the minus means that it was off, and the dot means that it was near a 0.5 level of activation. The first thing to notice is that one of the units, in this case the first unit, indicates whether the input is plant or animal. It is off for plants, on for animals. Notice also that for plants the sixth unit indicates whether it was a "tree" or a "flower," while for animals the third unit indicates "bird" versus "fish." Why did the system choose to allocate units in this way? The reason is simply that in order to solve the problem it had to represent patterns to which it must respond similarly in a similar fashion and that these are the most important dimensions of similarity for this data set. The remaining units are a bit harder to identify, but they apparently encode the more idiosyncratic information about the concepts (such as the color, the size, whether they can sing or not, etc.). The example shown here is just the result of one of many such experiments and the results are always qualitatively the same—one unit for plant versus animal, one for each grouping within plant and animal, and the remaining units encoding the remaining facts about the concepts. Figure 8 shows essentially the same results in a more geometric form. This figure shows a cluster analysis of the representational vectors. We observe that the major break is between plants and animals, that the next level of

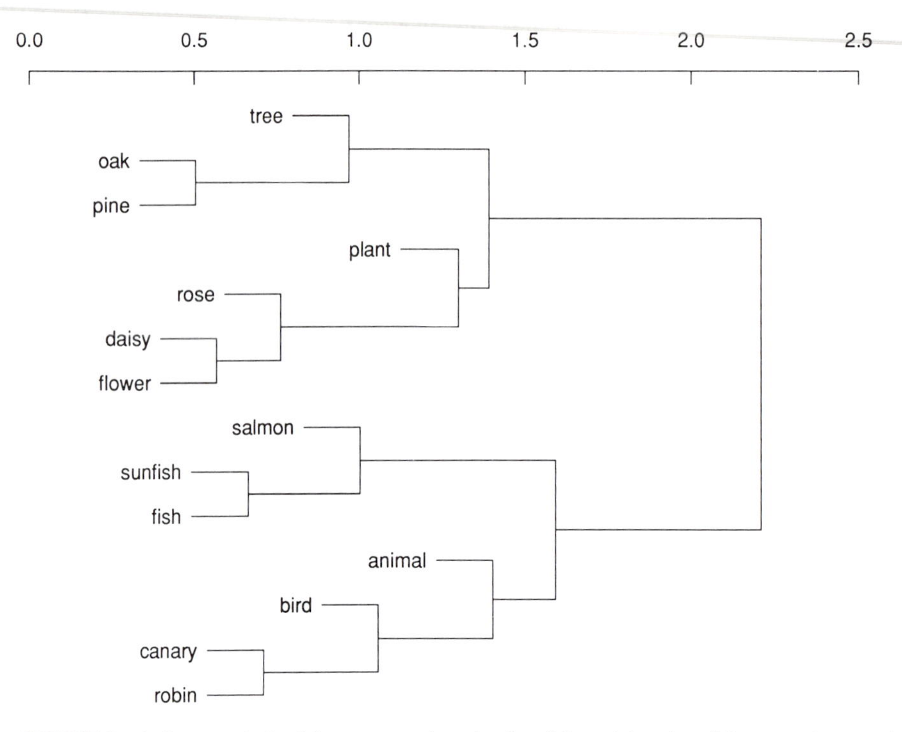

FIGURE 8 A cluster analysis of the representations developed through learning of the semantic network.

break is between trees and flowers on the one side, and fish and animals on the other side.

The next question revolves around the issue of generalization. In this case we have shown that the network has been able to store all of the desired information and that it has done so by building internal representations which reflect the conceptual structure implicit in the data being stored. What of new information? If it is taught about a new item, a sparrow, and told that a sparrow is a bird, will it know that a sparrow can fly, has feathers, is an animal etc.? In order to test this I presented the network with a new pattern it had never seen before, namely, the pattern in which a (previously unused) node for sparrow was turned on and the arc label "isa" was turned on and the system was taught to turn on the unit for "bird." During this training period all connections were "frozen" except that from the sparrow unit to the representational units. In other words, the learning of "sparrow is a bird" amounts to finding a set of connections from "sparrow" to the representational units so that they (the representational units) will turn on "bird." Thus, the representation for sparrow must be very much like that for other birds and, indeed, the network discovered a representation that was very much like a weakened version of "robin" for its representation of "sparrow." Thus, when I test to see what a "sparrow can" do, it responded with "fly," "move," and "grow" just as it does for "robin." Similarly, it responded that a sparrow has feathers, wings, and other bird properties. Figure 9 shows the placement of "sparrow" in the cluster analysis.

If we understand how the process works, it appears that the generalization process is simple and natural, as it is, of course. It depends, however, for its success on a kind of learning procedure in which computation is based on the similarity of representations and a kind of learning procedure whose mode of learning is discov-

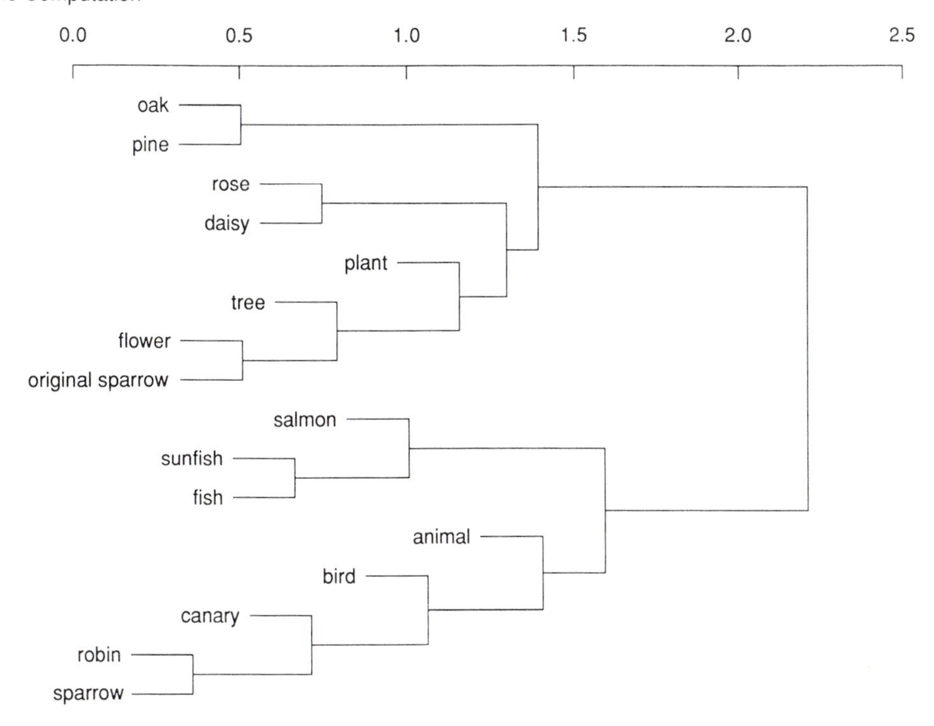

FIGURE 9 A cluster analysis showing the placement of the concept "sparrow" both before (original sparrow) and after learning that a "sparrow is a bird." Note that originally sparrow, because of the random starting weights, is located in the midst of some of the plants. Learning that the representation of sparrow is very similar to that of "robin" leads them to cluster together in the diagram.

ering representations which satisfy these constraints. Note, for example, that no simple statistical analysis of the input patterns will solve this problem. The system responds that a sparrow can fly despite the fact that it has never seen that combination of inputs before. Moreover, it has only seen sparrow in the context of "isa bird" and, although it had seen "can" before, "can" itself is only very weakly correlated with "fly". Most of the time when "can" has been on, "fly" has been off. It is crucial for the generalization to work that the network be able to form "categories" at the representational level which can themselves be associated (in the context of "can" with "fly"). On the usual semantic network account, inheritance is a basic form of inference. In this case, the equivalent to inheritance comes about through the similarity of derived representations.

To summarize the section of learning from example, I have shown how a simple learning mechanism, which could easily be constructed out of the kinds of connections and computational materials found in real brains, can be used to encode information in brainlike networks in such a way that new representations are formed and important generalizations are naturally made. The question of whether this particular learning mechanism actually underlies learning in real brains is unknown, but given the kind of machinery available in the brain and the apparent power of this kind of learning procedure, it would seem reasonable to suppose that this or some similar mechanism leading to the same kinds of results is very likely to have evolved. An open question is, of course, "if the brain implements this sort of learning mechanism, how is it implemented?"

CONCLUSION

I have tried to sketch the basic computational ideas that have fueled the spurt of interest and activity in the study of brain style computation. Our understanding of the computational properties of such networks is still in its infancy. There is much to learn. As we learn more from neuroscience, as we learn more from the mathematical analysis of the computational properties of these networks, and as we learn more from attempts to build more realistic "neural-inspired" models of human cognition we are building a database of knowledge from which, I expect, will come the neurobiological, psychological, and computational theories and models of the future.

References

Cohen, M. A., & Grossberg, S. (1983). Absolute stability of global pattern formation and parallel memory storage by competitive networks. *IEEE Transactions on Systems, Man, and Cybernetics*, **13,** 815–826.

Feldman, J. A. (1985). Connectionist models and their applications: Introduction. *Cognitive Science*, **9,** 1–2.

Geman, S., & Geman, D. (1984). Stochastic relaxation, Gibbs distributions, and the Bayesian restoration of images. *IEEE Transactions on Pattern Analysis and Machine Intelligence*, **6**, 721–741.

Grossberg, S. (1976). Adaptive pattern classification and universal recoding: Part I. Parallel development and coding of neural feature detectors. *Biological Cybernetics*, **23**, 121–134.

Hebb, D. O. (1949). *The organization of behavior*. New York: Wiley.

Hinton, G. E., & Sejnowski, T. J. (1983). Optimal perceptual inference. *Proceedings of the IEEE Computer Science Conference on Computer Vision and Pattern Recognition*, 448–453.

Hinton, G. E., & Sejnowski, T. J. (1986). Learning and relearning in Boltzmann machines. In D. E. Rumelhart & J. L. McClelland (Eds.) *Parallel distributed processing* (Vol. 1. *Foundations*). Cambridge, MA: MIT Press.

Hopfield, J. J. (1982). Neural networks and physical systems with emergent collective computational abilities. *Proceedings of the National Academy of Sciences of the United States of America*, **79**, 2554–2558.

Jordan, M. I. (1986). Attractor dynamics and parallelism in a connectionist sequential machine. In *Proceedings of the Eighth Annual Meeting of the Cognitive Science Society*. Hillsdale, NJ: Erlbaum.

McCarthy, J. (1959). Comments. In *Mechanisms of thought processes: Proceedings of a symposium held at the National Physical Laboratory, November 1958. (Vol. I)*. London: Her Majesty's Stationery Office.

McClelland, J. L., & Rumelhart, D. E. (1986). *Parallel distributed processing (Vol. 2. Psychological and biological models)*. Cambridge, MA: MIT Press.

McCulloch, W. S., & Pitts, W. (1943). A logical calculus of the ideas immanent in nervous activity. *Bulletin of Mathematical Biophysics*, **5**, 115–133.

Minsky, M., & Pappert, S. (1969). *Perceptrons*. Cambridge, MA: MIT Press.

Newell, A., Shaw, J. C., & Simon, H. (1958). Elements of a theory of human problem solving. *Psychological Review*, **51**, 239–243.

Peterson, C., & Anderson, J. R. (1987). A mean field theory learning algorithm for neural networks. *Complex Systems*, **1**(5), 995–1019.

Reddy, D. R., Erman, L. D., Fennell, R. D., & Neeley, R. B. (1973). The hearsay speech understanding system: An example of the recognition process. *Proceedings of the International Conference on Artificial Intelligence*, 185–194.

Rosenblatt, F. (1962). *Principles of neurodynamics*. New York: Spartan.

Rumelhart, D. E. (1988). *Generalization and the learning of minimal networks by back propagation*. In preparation.

Rumelhart, D. E., Hinton, G. E., & Williams, R. J. (1986). Learning internal representations by error propagation. In D. E. Rumelhart & J. L. McClelland (Eds.), *Parallel distributed processing* (Vol. 1. *Foundations*). Cambridge, MA: MIT Press.

Rumelhart, D. E. & McClelland, J. L. (Eds.) (1986). *Parallel distributed processing* (Vol. 1. *Foundations*). Cambridge, MA: MIT Press.

Selfridge, O. G. (1955). Pattern recognition in modern computers. *Proceedings of the Western Joint Computer Conference*.

Smolensky, P. (1986). Information processing in dynamical systems: Foundations of harmony theory. In D. E. Rumelhart & J. L. McClelland (Eds.), *Parallel distributed processing* (Vol. 1. *Foundations*). Cambridge, MA: MIT Press.

Widrow, G., & Hoff, M. E. (1960). Adaptive switching circuits. *Institute of Radio Engineers, Western Electronic Show and Convention, Convention Record, Part 4*, 96–104.

22

Network Self-Organization in the Ontogenesis of the Mammalian Visual System

Christoph von der Malsburg

GENERAL INTRODUCTION

This chapter discusses the process of network self-organization, which is fundamental to the organization of the brain. The process takes place on several temporal scales: the ontogenetic/learning time scale of hours, days, and years; and probably also the functional time scale of fractions of a second to minutes. Several concepts and tools of network self-organization are introduced here. I start with a discussion of self-organization in general and subsequently demonstrate the essential mechanisms, with the help of examples taken from the ontogenesis of the visual system. Although a particular mathematical formulation has been chosen for this discussion, I do not intend to put it forward as a general basis for network self-organization. A canonical mathematical formulation of self-organization has yet to be developed.

Self-Organization

One often speaks of some structural trait of an organism as being "genetically determined." This seems to imply that the genes contain a blueprint describing the organism in full detail. However, all the stages of brain organization (not just evolution) more or less strongly involve an element of self-organization: an element of creativity. It has often been emphasized that the genes cannot, in any naive sense, contain the full information necessary to describe the brain. Cerebral cortex alone contains on the order of 10^{14} synapses. Forgetting considerations of genome size, one can hardly imagine how ontogeny could select the correct wiring diagram out of all of the alternatives if all were equally likely. Besides, judging from the variability of the vertebrate brain structure, the precision of the ontogenetic process is not sufficient to specify individual connections.

The conclusion one must draw is that ontogeny makes use of self-organization, that is, of general rules to generate neural structure and of principles of error correction. Above all, ontogenesis can only produce structures with a high degree of regularity, for example, homogeneity, repetitivity, or continuity. Knowing the mechanism of ontogeny is of extreme importance: one cannot understand the function of the brain without knowing its structure, and one cannot know the structure of the brain without knowing the principles of its ontogenesis.

Abstract Scheme of Organization

There are well-studied paradigms of pattern formation, especially in physics, physical chemistry, and astronomy: convection, crystallization (or more generally, phase transitions), reaction–diffusion systems (the emergence of spatial and temporal chemical patterns, e.g., in the Zhabotinski-Belusov reaction), and star and galaxy formation. I will attempt to give here a general description of the basic mechanisms of organization by using the important example of convective pattern formation, the so-called Bénard problem.

Organization takes place in systems consisting of a large number of interacting elements, such as atoms in a liquid or crystal or small subvolumes of liquid in convection currents, in a reaction–diffusion system or in an evolving star system, or, in the application that is of interest here, synapses in nerve networks. Initially, self-organizing systems are in a relatively undifferentiated state: atoms move randomly and all subvolumes of the liquid are in the same state of motion or have the same chemical composition. Then, some small, typically random deviations from that state arise; for example, some convective fluid motion sets in. To stress the random nature of typical small deviations, they are called *fluctuations*.

In the prime example, the Bénard phenomenon, a flat vessel is filled with liquid and its bottom is homogeneously heated. As long as the temperature gradient is below a certain threshold, heat is conducted from the lower to the upper surface without bulk movement of the liquid. However, above that threshold, the warmer, lighter liquid near the bottom rises and cooler liquid from the top flows down. Under homogeneous conditions, this flow pattern is very regular and has the form of hexagons or rolls.

From this and many other organizing systems the following three principles may be abstracted:

1. *Fluctuations self-amplify.* This self-amplification is analogous to reproduction in Darwinian evolution. In the Bénard system, fluctuations are created by thermal motion. If a small column of liquid moves upward, more warm liquid is drawn in from the bottom, the column becomes less dense, and its upward movement is accelerated. Downward movement accelerates analogously.

2. *Limitation of resources leads to competition among fluctuations and to the selection of the most vigorously growing (the ''fittest'') at the expense of the others.* In the Bénard system, upward movement in one place requires downward movement in other places. The columns with the least density will win and rise.

3. *Fluctuations cooperate. The presence of a fluctuation can enhance the fitness of some of the others, in spite of the overall competition in the field. (In many systems the ''fitness'' of a fluctuation is identical with the degree of cooperation with other fluctuations.)* The liquid near a column of rising liquid is dragged up by viscosity.

The identification of these three principles with features of a concrete system is sometimes ambiguous. In the Bénard system, competition in terms of upward movement might also be seen as cooperation between upward movement occurring in one place and downward movement occurring in another place. Whole coherent patterns of movement, again, compete as long as there is local contradiction between them: liquid cannot move up and down at the same place.

A fundamental and very important observation about organizing systems is the fact that global order can arise from local interactions. Many originally random local fluctuations can coalesce into a globally ordered pattern of deviations from the original state. The intermolecular forces acting within a volume of liquid are of extremely short range, yet the patterns of convective movement they give rise to may be coherent and ordered on a large scale. This fact will be one of extreme importance to the brain, in which local interactions between neighboring cellular elements create states of global order, ultimately leading to coherent behavior.

The stage for the organization of a pattern is set by the forces between elements and by initial and boundary conditions. In the Bénard system, these forces are the hydrodynamic interactions, gravity, thermal conduction, and expansion. Boundary conditions are set by temperatures at the upper and lower boundary and by the form of the vessel. In the nervous system, the stage for the generation of connection patterns is ultimately set by prespecified rules for the interaction of cellular processes and signals, and by the environment. Because nerve cells are connected by long axons, there is an important and exciting difference between the nervous system and most other examples studied so far. Neural interactions are not necessarily topologically arranged; connected cells are "neighbors" although they may be located at different ends of the brain. This gives rise to genuinely new phenomena. Some of the ordered structures within the nervous system may not "look" ordered to our eye, which relies essentially on spatial continuity. However, in the concrete cases considered here, ordinary space will still play a dominant role.

An organizing system may contain a symmetry such that there are several equivalent organized patterns. These compete with each other during organization. In the Bénard system, if set up in a circular pan, any organized pattern could be rotated around the center of the pan by an arbitrary angle to obtain another valid pattern. One of these has to be spontaneously selected during pattern formation, a process that is called *spontaneous symmetry breaking*. When the boundary or initial conditions are slightly deformed, so that the original symmetry is destroyed, one organized pattern is favored. In general, self-organizing systems react very sensitively to symmetry-breaking influences.

GENERAL NEURAL NETWORK ORGANIZATION

Two types of variables are relevant to network organization: signals and interconnections. Signals are the action potentials that are propagated down the axonal trees of neurons. Connections control neural interactions and are characterized by weight variables. These measure the size of the effect exerted on the postsynaptic membrane by arriving nervous impulses. Correspondingly, organization takes place on two levels: activity and connectivity.

On the ontogenetic time scale, one is interested mainly in network self-organization, which has the following general form. Assume that previous processes have already set up a primitive network. This network, together with input signals, creates activity patterns, and these activity patterns in turn modify connections by synaptic plasticity. The feedback loop between changes in synaptic strengths and changes in activity patterns must be positive, so that coherent deviations from the undifferentiated state self-amplify, conforming to the first of the principles previously formulated. The process is constrained by the requirement that modifications in a synaptic connection have to be based on locally available signals. These are the presynaptic signals, the postsynaptic signal, and possibly modulatory signals that are broadcast by central structures. The postsynaptic signal could be a local dendritic signal or the outgoing axonal signal.

The requirements of self-reinforcement and locality suffice to specify the mechanism of synaptic plasticity in excitatory synapses: A strong synapse leads to coincidences of pre- and postsynaptic signals which, in turn, increase the strength of the synapse. Hebb (1949) gave this formulation:

> When an axon of cell A is near enough to excite cell B and repeatedly or persistently takes part in firing it, some growth process or metabolic change takes place in one or both cells such that A's efficiency, as one of the cells firing B, is increased (p. 62).

This rule is referred to as "Hebbian plasticity." The corresponding rule for inhibitory synapses would have a synapse strengthened if it was successful in inhibiting the postsynaptic element. At present, however, most authors consider inhibition as a rigid service system that does not take part in network self-organization.

Hebb's rule corresponds to the "self-reproduction" of the general scheme of organization. To stabilize the system, some competition for limited "resources" has to be introduced. Most likely, there is a mechanism of isostasy, by which each cell keeps the temporal average of its activity (taken over the span of some hours) constant. As a consequence, the increase in strength in some synapses must be compensated for by a decrease in others. Only the more successful synapses can grow; the less successful ones weaken and eventually disappear. For technical reasons, some models discuss a simpler competition rule for synapses, in which the sum of the synaptic weights of all synapses converging on a cell is kept constant. (Although this rule leads to certain functional deficits and is probably not realistic, I employ it here for its technical convenience.) Synaptic plasticity, constrained by competition, implements organizing principles 1 and 2.

One synapse on its own cannot efficiently produce favorable events. For that, it needs the cooperation of other synapses that converge onto the same postsynaptic neuron and that carry coincident signals. This implements the third organizing principle. In order for such coincidences to occur consistently, there must be a causal connection between presynaptic cells. Synaptic plasticity is the means by which the nervous system detects such causal connections. Coincidences may result from excitatory links between presynaptic neurons. They may, however, also be caused by simultaneous stimulation of sensory cells, in which case they point to the existence of causal connections in the external world.

The rules of cooperation and competition act on a local scale. The phenomenon of self-organization is the emergence of globally ordered states, as discussed in context with the emergence of global convection patterns in the Bénard phenomenon. The term *global order* is used for configurations that bring the local rules into a state of optimal mutual consistency with each other. The fact that the external world takes part in the game leads to the adaptation of the nervous system to it.

The rules for the adjustment of synaptic weights that have been introduced are able to produce ordered connection patterns. However, they do not necessarily organize the nervous system for optimal biological utility. For this, two types of controls are necessary: (1) genetic control of boundary conditions and interaction rules to favor certain useful connection patterns; and (2) control by central structures that are able to evaluate the degree of biological desirability of activity states. If a state proves to be useful, a gating signal is sent to all of the brain, or to an appropriate part of it, to authorize synaptic plasticity. That state is thereby stabilized, and the likelihood for its future appearance is increased.

Central control as the *only* criterion for growth or decay of synapses is not sufficient. Assume our nervous system evaluates the usefulness of its state once per second. It then could create less than 3×10^9 bits of information in our lifetime, for that is about the maximum number of seconds given to us. This certainly is not sufficient to regulate the strengths of all of the 10^{14} synapses of our cerebral cortex. On the other hand, this amount of information may be sufficient to select from among the relatively small universe of ordered connectivity patterns that can be created by rules of local cooperation and competition under predetermined constraints.

Having briefly introduced relevant principles, I now present a more detailed discussion of a few paradigmatic cases of network organization, stressing the application to the visual system. Retinotopy, ocularity, and orientation specificity are important examples which have been studied intensively both theoretically and experimentally over several decades.

ESTABLISHMENT OF TOPOLOGICAL MAPS

Biological Background

At some stage in the development of the vertebrate embryo, the fibers of retinal ganglion cells grow out through the eye stalk toward the brain and establish retinotopic connections there. Neighboring cells in retina contact neighboring positions in the target structure. The most intensively studied case is that of retino-tectal connections in amphibia and fish. The interesting question is how fibers find their correct target positions. This problem of retinotopy has long been recognized as an important paradigm of ontogenetic brain organization in general. There are many topological fiber projections between thalamic nuclei and cortical areas, and also between and within cortical areas. Retinotopy is one of those few biological phenomena studied with enough experimental intensity to allow their theoretical issues to be settled.

For a long time, the most puzzling aspect of the retinotopy problem was a mixture of rigid genetic determination on the one hand and plasticity on the other. The orientation of the projection is reliably prespecified, for instance, nasal retina reliably connecting to caudal tectum. On the other hand, magnification and position of the map are adjusted with flexibility so that all of the existing retinal tissue maps to all of the existing tectum, even if the sizes of these structures vary under some physiological conditions or after experimental manipulation. This property is called *systems matching*. For instance, because growth in retina and in tectum are disparate—retina grows in a concentric fashion, tectum (in some species) from front to back—the already existing fiber projections shift in an ordered fashion to achieve systems matching. The picture is further complicated by evidence that fibers are able to follow tectal tissue that has been grafted to a new position within tectum (for a review of various experiments, see Fraser, 1985), seemingly proving the existence of rigid addresses in tectal cells.

The apparent contradictions are resolved by, and all types of experiments are consistent with, the assumption of the following three mechanisms:

A. There is a mechanism to guide fibers to tectum.

B. There is a mechanism to position fiber terminals within tectum. This mechanism is responsible for rigid constraints on the mapping. The nature of the mechanism is not yet known, but there is the widespread conviction that it is based on chemical marker gradients in retina and tectum.

C. There is a fiber-sorting mechanism that improves the precision of the mapping over that attained by mechanism B alone, and which is activity-dependent (Harris, 1980). This mechanism is able to account for systems matching in the retino-tectal system (for review, see Schmidt & Tiemann, 1985).

In spite of many ongoing controversies within the retino-tectal community, the preceding statements have gained wide acceptance. The division of labor between mechanisms B and C, and the nature of C, were first formulated by Willshaw and von der Malsburg (1976). In the present context, I concentrate on the fiber-sorting mechanism C, because it deals with activity-dependent network organization and can be generalized to other interesting cases, in particular to processes in the cerebral cortex.

Basic to the fiber-sorting mechanism is the fact that the spontaneous activity of neighboring retinal ganglion cells is correlated because of excitatory connections in retina (Mastronarde, 1983; Meister, Wong, Baylor, & Shatz, 1991). These correlations carry complete information about neighborhood relationships within retina, without coding for retinal position directly. They are used for fiber sorting in the following way: Retinal fibers establish tentative contacts on tectum. Through these contacts they impress their activity patterns on tectal cells. Because of excitatory links within tectum, activity in neighboring tectal cells is correlated, just as in retina. Retino–tectal contacts undergo a selective growth process, in which successful contacts grow in strength and less successful ones decay. Success is measured by the degree to which a contact is able to induce correlations between pre- and postsynaptic signals, and it depends on the number and strength of other fibers linking the same retinal and tectal locations. Competition among contacts is introduced by the inability of a tectal cell to receive more than a certain number of contacts and by the inability of retinal fibers to support more than a given number of contacts ("conservation of axonal arbor"). If all goes well, only fibers that have a maximal number of neighboring fibers (neighboring in the sense of tectum and retina) survive the competition. Moreover, there are no retinal spots that project to more than one spot of tectum, and there are no tectal spots that receive fibers from more than one spot of retina. This is a fully retinotopic mapping, which is distinguished by maximally conforming to the preceding constraints.

Formal Description

I concentrate here mainly on the sorting mechanism C, which illustrates network organization best. The temporal development of retino–tectal connections is described by a differential equation of the type:

$$\dot{W}_i = W_i F_i - W_i \sum_j W_j F_j / N \quad (1)$$

Here, W_i is the strength of synapse i, $\dot{W}_i$ is its rate of change, F_i is a rate coefficient, and N is the total number of synapses competing with each other. Assume that $\Sigma_i W_i / N = 1$ (otherwise it would quickly converge to 1). The quantity $\bar{F} = \Sigma_j W_j F_j / N$ can then be interpreted as the weighted mean of the coefficients F_i. Accordingly, Equation (1) can be written

$$\dot{W}_i = W_i (F_i - \bar{F}) \quad (2)$$

making it evident that the synapses that will grow are those with a growth coefficient above average, whereas the others diminish. As a consequence, the weighted average $\bar{F}$ grows, and more and more synapses fall below threshold, the system finally settling into the state in which only the synapses with maximal $F_i = \bar{F}$ survive.

According to Equation (1) or (2), nonexistent synapses ($W_i = 0$) cannot grow. To represent the formation of new synapses, a constant rate α of synapse formation is added to the growth term $W_i F_i$ and its average in the competitive term, turning the set of differential equations into

$$\dot{W}_i = \alpha + W_i F_i - W_i \sum_j (\alpha + W_j F_j) / N \quad (3)$$

or, equivalently,

$$\dot{W}_i = \alpha (1 - W_i) + W_i (F_i - \bar{F}) \quad (4)$$

In this differential equation, α acts as a convenient control parameter. The α-dependent term pulls the synaptic strengths toward the value 1, whereas the other term destabilizes that value. With positive α, more than just the synapses with maximal F_i can survive.

Retino–tectal connections have to be labeled by two indices, ρ and τ, for a retinal and a tectal cell, respectively, replacing W_i by $W_{\tau\rho}$. Conservation of axonal and dendritic arbor is implemented in the differential equation by a tendency for the averaged weights for all of the presynaptic and postsynaptic competitors of a given synapse ($\tau\rho$) to converge to 1. This average is

$$B_{\tau\rho}(X) = (\sum_{\tau'} X_{\tau'\rho} + \sum_{\rho'} X_{\tau\rho'})/(2N) \quad (5)$$

where X stands for any array and N is the number of neurons both in retina and in tectum. With these changes, one arrives at the set of differential equations that are used for describing the self-organization of retino–tectal connections:

$$W_{\tau\rho} = \alpha + W_{\tau\rho} F_{\tau\rho} - W_{\tau\rho} B_{\tau\rho} (\alpha + WF) \quad (6)$$

For the growth coefficients $F_{\tau\rho}$, see Equation (13). The first two terms on the right-hand side describe growth of synapse ($\tau\rho$), the third term describes its competition with all synapses to the same tectal cell τ and from the same retinal cell ρ, B being the average over the growth terms of all of those other synapses (for a more thorough explanation, see Häussler & von der Malsburg, 1983). In analogy to Equation (4), this equation can be rewritten as

$$W_{\tau\rho} = \alpha(1 - W_{\tau\rho}) + W_{\tau\rho}(F_{\tau\rho} - B_{\tau\rho}(WF)) \quad (7)$$

The parameter α regulates the rate of formation of new synapses, arriving at the tectum by virtue of mechanism A. Positioning mechanism B is easily implemented here by starting the calculation with a more or less precisely ordered retinotopic mapping as the initial condition for $W_{\tau\rho}$.

Before Equation (6) or (7) can be used, the dependency of $F_{\tau\rho}$ on cellular signals and the dependency of signals on network structure have to be defined. The general Hebbian idea is that $F_{\tau\rho}$ is proportional to an appropriate measure of correlation between signals in retinal cell ρ and tectal cell τ. Activity arises spontaneously in retina and is propagated through W to the tectum. Suppose $f^r(t)$ is the normalized deviation of spontaneous activity in ganglion cell r from its temporal average, the only thing that needs to be known about f^r is that its correlation for different cells is

$$\langle f^r(t)\, f^{r'}(t)\rangle = \delta_{rr'} \tag{8}$$

The total activity in retinal cell ρ is

$$c^R_\rho(t) = \sum_r D^R_{\rho r} f^r(t) \tag{9}$$

Here, $D^R_{\rho r}$ describes the propagation of activity from cell r to cell ρ in retina. It is assumed to be a smooth and monotonically falling function of the distance $|\rho - r|$ (the latter tacitly implying periodic boundary conditions). Linear signal propagation is assumed for simplicity. For the interpretation of Equation (9), it is immaterial whether $D^R_{\rho r}$ describes forward propagation of activity from an earlier generation of cells, or whether it describes feedback propagation within the same layer of cells. In the latter case, Equation (9) would be the result of some rapid iteration of signal exchange. In either case, the important property of $c^R_\rho(t)$ is the form of its autocorrelation function $\langle c^R_\rho{}' c^R_\rho\rangle$. Given Equations (8) and (9), this has the form:

$$\begin{aligned}\langle c^R_{\rho'} c^R_\rho\rangle &= \sum_{rr'} D^R_{\rho' r'} D^R_{\rho r} \langle f^{r'}(t)\, f^r(t)\rangle = \\ &\sum_r D^R_{\rho' r} D^R_{\rho r} = \bar{D}^R_{\rho'\rho}\end{aligned} \tag{10}$$

The last equality sign defines a function $\bar{D}$ that again depends only on $|\rho' - \rho|$ and which is just a bit broader than $D^R_{/\rho - r/}$.

The total input signal afferent to tectal cell τ is

$$I_\tau(t) = \sum_{\rho'} W_{\tau\rho'} c^R_{\rho'}(t) \tag{11}$$

After propagation within tectum with the propagation kernel D^T, tectal activity is

$$c^T_\tau(t) = \sum_{\tau'\rho'} D^T_{\tau\tau'} W_{\tau'\rho'} c^R_{\rho'} \tag{12}$$

Now it is time to calculate the covariance of the signals on the presynaptic and postsynaptic sides of synapse (τρ), which will serve as growth coefficient in Equation (6) or (7):

$$\begin{aligned}F_{\tau\rho} = \langle c^T_\tau(t) c^R_\rho(t)\rangle &= \sum_{\tau'\rho'} D^T_{\tau\tau'} W_{\tau'\rho'} \langle c^R_{\rho'} c^R_\rho\rangle = \\ &\sum_{\tau'\rho'} D^T_{\tau\tau'} W_{\tau'\rho'} \bar{D}^R_{\rho'\rho}\end{aligned} \tag{13}$$

With the previous assumptions about the retinal signal propagator and with similar assumptions about the tectal one, this is just a low-pass filtered version of the connectivity matrix W. For a given synapse (τρ), the growth coefficient $F_{\tau\rho}$ is a weighted sum of the strengths of neighboring synapses, that is, the strengths of connection from points neighboring ρ in retina to points neighboring τ in tectum.

The differential Equation (6) together with Equation (13) implements the features of the sorting mechanism C as described in the previous section. A simulation is shown in Figure 1. Like all self-organizing systems, the fiber-sorting mechanism described here has the capability of creating global order from local interactions. All such systems, however, are in danger of getting caught in local minima, which are of less-than-perfect global order. In the case given here, such minima are mappings that are only piecewise retinotopic—the different pieces do not fit together in orientation or position. Once such a partially disorganized mapping is created, it is impossible for fibers near the borders of domains to collect all their retinotopic neighbors, because some of them would have to tunnel through foreign territory. The danger of getting trapped in local minima is reduced by the positioning mechanism B, which already creates some retinotopic order in the initial state of the sorting mechanism. However, because the positioning mechanism (if properly discussed) has the same problem of avoiding local minima, and because this is a problem of general importance for self-organizing systems, I will discuss here some of the factors that are important to avoid local minima.

The issue is conveniently studied with the help of stability analysis. This has been carried through for the case of one-dimensional retina and one-dimensional tectum in Häussler & von der Malsburg (1983), of which a qualitative outline is given here. (The two-dimensional case will be treated in a forthcoming pa-

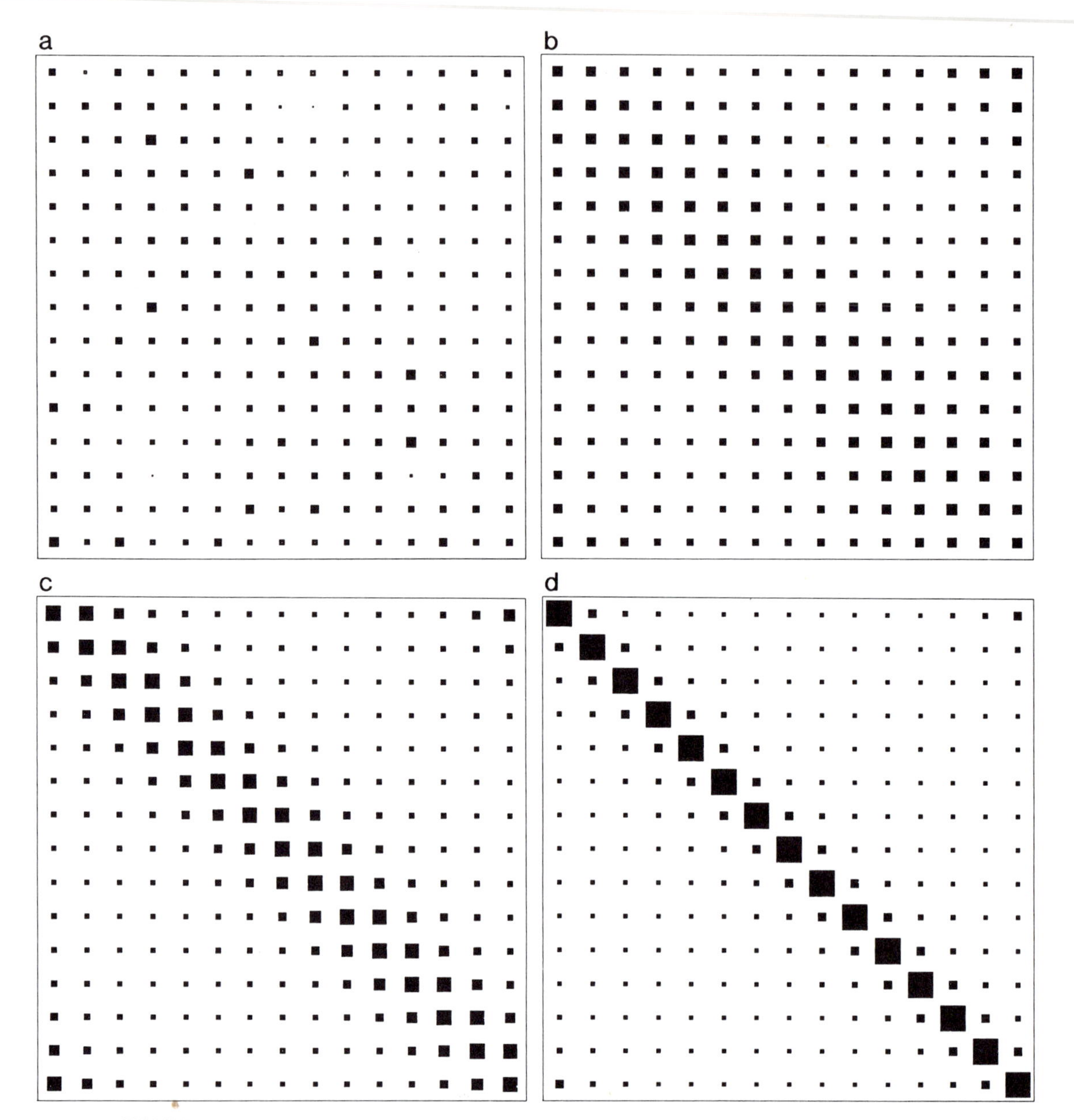

FIGURE 1 Development of a retinotopic mapping: simulation of Equation (6), one-dimensional retina and tectum with wraparound boundary conditions. The retinal coordinate runs horizontally, the tectal vertically. The area of the small squares indicates the strengths $W_{\tau\rho}$ of synapses. The temporal sequence is left to right, top to bottom. In the initial state, the symmetry between the two orientations of the mapping is slightly broken to speed up the process. During the simulation, the control parameter α is slowly reduced.

homogeneous solution $W_{\tau\rho} = W_0 = 1$. Consider the equivalent form of Equation (7). It has two terms. The first of them stabilizes W_0, with a strength that is proportional to α, whereas the second term is destabilizing, having the general form of (1). Stability analysis now proceeds by rewriting Equation (7) in terms of the deviation $V_{\tau\rho} = W_{\tau\rho} - W_0$ from the stationary state and linearizing the equation in $V_{\tau\rho}$, striking out all terms of higher order (there are actually terms up to order 3). This approximation is accurate for small $V_{\tau\rho}$.

The resulting set of N^2 linear equations (N being the number of cells in retina and in tectum) has N^2 independent solutions, which I will call *modes*. Each mode grows or decays exponentially with its own characteristic rate constant, called its *eigenvalue*. The modes have a simple form if wraparound boundary conditions are used for retina and tectum. In that case, they are harmonic waves, that is, a product of a sinusoidal function of ρ and a sinusoidal function of τ. Modes differ in spatial frequency along each of the two coordinates, starting from zero frequency (corresponding to a constant function).

It is now important to know the spectrum of eigenvalues of the modes. In this case, it turns out that the spectrum contains $-\alpha$ as an additive constant common to all eigenvalues. Thus, α is a very convenient control parameter with which the spectrum can be shifted up and down. It further turns out that the smallest of all eigenvalues is that of the constant mode ($V = 1$), followed by all modes that are constant along either ρ or τ. This is due to the damping effect of the B-operator in Equation (7). The shape of the spectrum of all the other modes, which vary along both ρ and τ, is strongly determined by the shape of the low-pass filter $F_{\tau\rho}$, Equation (13). The higher the spatial frequency of a mode, the stronger the damping effect exerted by that filter, and the lower the eigenvalue of that mode. Hence, the modes with the highest eigenvalue are those with the lowest nonzero spatial frequency, corresponding to just one cycle, both along retina and tectum. There are two such modes (not counting versions shifted along retina or tectum), which correspond to two broad diagonals of different orientation in the (ρ, τ) matrix V. By appropriate choice of α, the spectrum can be shifted such that only the eigenvalues of these two diagonals are positive and all other modes have negative eigenvalues. Some further analysis (Häussler & von der Malsburg, 1983) shows that the nonlinear interactions between the two diagonal modes are of a competitive nature, such that one of them will eventually win and the other will die out. This amounts to spontaneous symmetry breaking between the two possible orientations of the retino–tectal mapping.

Once one of the globally ordered diagonal modes has won, one can safely lower α to permit modes with higher frequency to grow. Due to nonlinear interactions, the broad diagonal will excite those higher modes that are parallel to it and which have the same phase. All these modes conspire to sharpen the retino–tectal mapping, until, with $\alpha = 0$, a very precise and globally ordered retinotopic mapping is established. For a simulation see Figure 1.

The original proposal of a sorting mechanism based on signal correlations, together with simulations of the two-dimensional case is in Willshaw & von der Malsburg (1976). A version based on chemical marker induction instead of on electrical signals, together with a didactic analogy ("tea trade model") is contained in von der Malsburg and Willshaw (1977). A version with more biological detail in the rules for the making and breaking of connections and with extensive simulations of experimental results is given in Willshaw and von der Malsburg (1979). A critical comparison with other theories is given in von der Malsburg and Willshaw (1981). An algorithmic caricature of the mechanism is known as Kohonen learning (Kohonen, 1982). The formulation given here, using linear signal propagation, is not an efficient basis for simulations. According to it, the postsynaptic signals c^T_τ are a very undifferentiated and flat function of position τ, being the result of repeated low-pass filtering of the original retinal spontaneous activity $f^r(t)$; see Equations (9) and (12). Much faster convergence is achieved by employing nonlinear sigmoid input–output functions in retina and tectum, creating broad localized blobs of strong activity on a silent background.

OCULARITY DOMAINS

There are several cases in which fiber systems originating in the two eyes innervate common target structures, for example, lateral geniculate body, optic tectum, or layer IV of visual cortex. This is the case in animals with binocular vision (and in frogs after certain experimental manipulations; Constantine-Paton & Law, 1978). Both fiber systems are organized in a retinotopic fashion and overlap so that corresponding points in the two eyes innervate the same small region of the target structure. Ocularity domains are formed as a small local deviation from this pattern by the segregation of the two types of fibers over a small distance. These domains can have the form of patches, tufts, stripes, or layers.

It has first been shown in von der Malsburg (1979) that the fiber-sorting mechanism described earlier can account for the formation of ocularity stripes if the system has two input sheets and if the fibers coming from the two sheets can be distinguished on the basis of their signals (Figure 2). Although the treatment in von der Malsburg (1979) makes explicit use of signals, the description here will go as far as possible on the basis of just their pair correlations. A straightforward

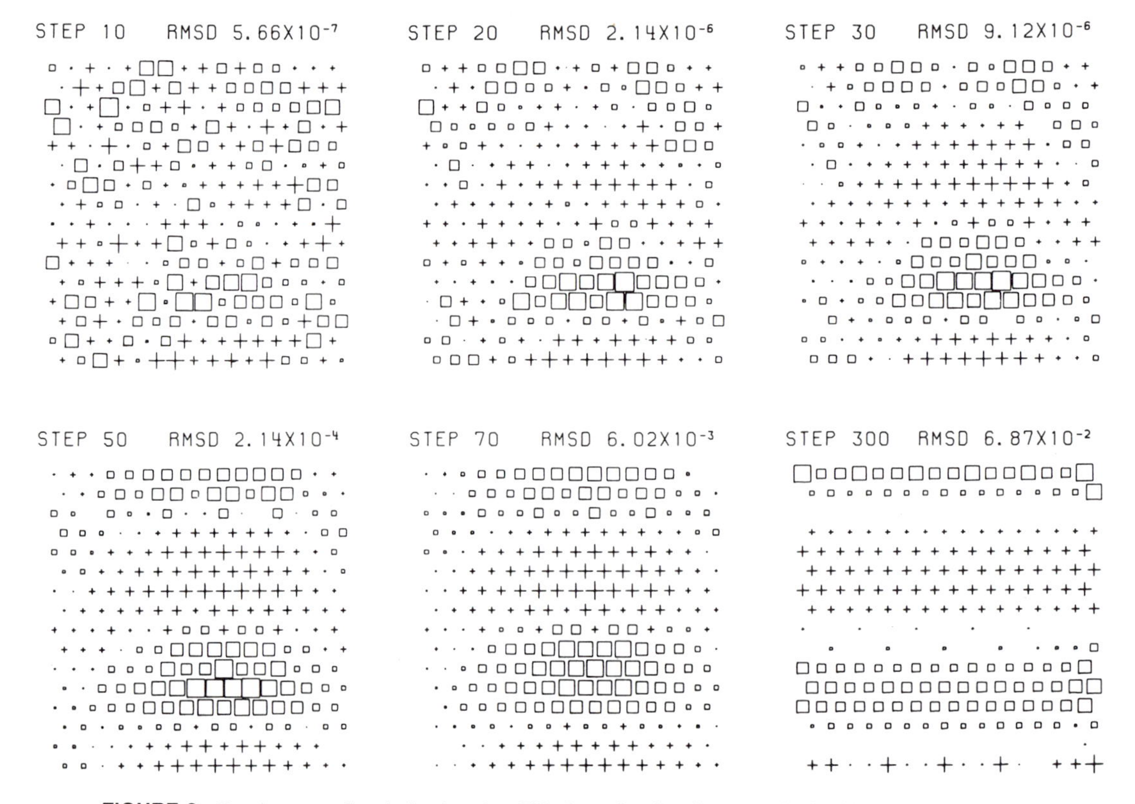

FIGURE 2 Development of ocularity domains. This figure is taken from von der Malsburg (1979). A two-dimensional patch of cortex receives input from two corresponding patches of retina. Relative size and sign of the left–right difference of innervation strength is represented by crosses and squares, indicating preponderance of one ocularity or the other. Within each figure, the size of symbols is normalized to the same maximal size. The root mean square of the left–right difference in innervation strength is given above each figure.

generalization of the Hebbian correlation Equation (13) for the case with two retinae is

$$F^{\gamma}_{\tau\rho} = \langle c^{T}_{\tau}(t)c^{\gamma}_{\rho}(t)\rangle = \sum_{\tau'\rho'\gamma'} D^{T}_{\tau\tau'} W^{\gamma'}_{\tau'\rho'} \langle c^{\gamma'}_{\rho'} c^{\gamma}_{\rho}\rangle \quad (14)$$

where $\gamma \in \{l, r\}$ labels the two retinae of ocularity l and r. This is to be inserted in Equation (7). It is natural to assume that for signals from the same retina correlations are as discussed before, whereas for signals from different retinae correlations are zero or even negative. This correlation is to be inserted into Equation (7). The B-operator generalizes to

$$B^{\gamma}_{\tau\rho}(X) = (\sum_{\tau'} X^{\gamma}_{\tau'\rho} + \frac{1}{2}\sum_{\rho'\gamma'} X^{\gamma'}_{\tau\rho'})/(2N) \quad (15)$$

If Equations (14) and (15) where just inserted into Equation (7) and the simulation started with a near to homogeneous (all-to-all) mapping and large control parameter α, a bad instability would result. The two retinae would just carve up the postsynaptic space into a few large chunks and would develop independent and separate retinotopic mappings within those chunks. Some measure has to be taken to limit the spatial scale over which the fibers of different ocularity actually segregate from each other to account for the observation that the two projections form a common retinotopic mapping that is only locally broken up into ocularity domains. Ocularity domain formation is actually a process that takes place very late in ontogeny, long after the establishment of retinotopic mappings (and also very long after the establishment of orientation columns; see next section). One way to model this is to use the following assumption. Early in ontogenesis, the fiber systems of the two eyes conspire to form a single retinotopic mapping. Only after this has been set up does segregation start. One possible way to model this sequence of events is to assume that the interocular correlation $\langle c^{l}_{\rho} c^{r}_{\rho'}$ is (for small $|\rho - \rho'|$) positive at first, and that it diminishes to small or negative values only after a fairly precise retinotopic mapping has formed. (Positive correlations for fibers of different ocularity afferent to cortex could be set up by mutual excitation within the lateral geniculate body.) By that time the formation of new fibers has ceased ($\alpha = 0$) and the appropriate version of Equation (7) now is

$$\dot{W}^{\gamma}_{\tau\rho} = W^{\gamma}_{\tau\rho}(F^{\gamma}_{\tau\rho} - B^{\gamma}_{\tau\rho}(WF)) \quad (16)$$

Now, growth of fibers is restricted to nonzero connections and all changes must take place within the small projection areas of retinal fibers that have developed in the early stage. It is necessary to assume that the term $D^{T}_{\tau\tau'}$ has shrunk to a range comparable to the size of the ocularity domains in this stage of development. (This is a natural assumption if that term is actually not mediated by the postsynaptic cells but rather as a consequence of axonal sprouting behavior.)

Equations of the type (14) through (16) have first been formulated in Miller, Keller, and Stryker (1989), where the formation of ocularity stripes is also demonstrated in simulations. Their formulation deviates from the one here in two details. Their version of Equation (15) contains only the sum running over ρ and γ, and they freeze the explicit W terms on the right-hand side of Equation (16) to a constant receptive field $A_{\tau\rho}$.

ORIENTATION DOMAINS

This review of network self-organization so far has concentrated on cases where synaptic growth can be conditioned on merely binary correlations and where these correlations can be expressed as a simple function of synaptic strengths, as in Equation (13). In many cases, however, the underlying phenomenon by its very nature comprises higher order correlations and the nonlinear dependence of signals on synaptic strengths is essential. A case in point is the ontogenetic development of orientation sensitivity in vertebrate visual cortex. In this section, I describe a model that accounts for the following experimental data, derived from cat or primate (similar facts hold for many other species):

1. Most neurons in visual cortex are orientation sensitive. They best respond to the presence of light

bars or edges within their receptive fields when these stimuli are oriented within a restricted range around an optimal orientation.

2. Orientation selectivity is distributed over the cortical surface in a continuous fashion, with occasional interruptions. Iso-orientation domains, comprising cells selective for one orientation, are arranged in the form of irregular ripples.

3. Going from neuron to neuron in cortex, the position of receptive fields are subject to a large positional scatter. Similarity of orientation in two neurons is not contingent on overlap between their receptive fields (Hubel & Wiesel, 1974a,b).

4. This organization is already present, although in immature form, when the animal first opens its eyes, so that visual experience cannot be held responsible for its formation (Hubel & Wiesel, 1974c).

Before setting out to formulate a model for these facts, it is necessary to discuss in slightly more formal terms the adult structure that is to be developed ontogenetically. I will focus on one point of retina (or the geniculate body) and the extended patch of cortex that is innervated by it. Of interest are the points in time when a barlike stimulus hits the point in retina. These stimuli are classified according to their orientation in retinal coordinates. They form a one-parameter family, the parameter being the orientation θ of the stimulus (counted cyclically from 0 to 180 deg). In response to a stimulus, a subset of (somewhat less than one half of) the neurons in the cortical patch is excited to fire. This subset, being part of an "orientation domain," is spatially organized, as evidenced with the help of the deoxyglucose method (Singer, 1981; Hubel, Wiesel, & Stryker, 1977; Humphrey, Skeen, & Norton, 1980) or with the help of optical recording (Blasdel & Salama, 1986). Orientation domains have the general appearance of ripples of sand in flowing water, and they are more or less ordered, depending on the species studied. There has been much discussion as to whether orientation domains are to be idealized as ringlike or spokelike centered on nonoriented spots, or as waves. For the present purposes, this discussion is of minor importance, but to have something simple in mind, domains may be idealized as regular plane waves, perhaps best exemplified in the tree shrew (Humphrey et al., 1980). When the orientation θ of the retinal stimulus is continuously rotated, the orientation domains in cortex shift or deform continuously too. By the time the retinal stimulus has turned through 180 deg, cortical activity again takes the shape of the first pattern encountered. Thus, the activity patterns in cortex also form a one-parameter family which can be labeled by the parameter θ counted cyclically from 0 to 180 deg.

When an extended oriented visual stimulus is applied to the retina, a larger cortical patch is activated, which expresses the shape of the stimulus by its shape, and which expresses the orientation of the stimulus locally by the ripple pattern of the orientation domains within the patch. When the stimulus moves without changing its orientation, the envelope shifts around, making the same member of the orientation pattern family visible in different parts of cortex. On the other hand, when the orientation of the stimulus is rotated, the cortical orientation domains shift through the different members of the family.

What is the detailed neurophysiological mechanism by which a cortical neuron decides whether to fire or not in response to an afferent stimulus? In this discussion, I distinguish three mechanisms that correspond to three sets of fibers. The first is the afferent receptive field aRF (or "classical receptive field"), the second is the intracortical receptive field iRF (or "nonclassical receptive field"), and the third is the formation of activity patterns in cortex by local activity feedback. The relative importance of these three mechanisms is difficult to determine experimentally, although it is a main point of discussion in the theoretical literature.

When discussing the ontogeny of orientation domains, two periods need to be distinguished—early ontogenesis, taking place before eye opening, and late ontogenesis, which has the benefit of patterned vision. The assumption underlying most models for the ontogenesis of orientation specificity is that the dominant mechanism is the aRF, cortical pattern formation playing a modulating influence. Let me refer to these models as aRF-based models. I will argue that they are difficult to reconcile with experimental fact. The model presented here, in contrast, places the main burden of orientation specificity in the immature cortex on cortical pattern formation, the iRF exerting a weak but important modulating influence. I will refer to this model as iRF-based.

In aRF models, the orientation specificity of a cortical neuron is the consequence of a spatial arrangement of retinal sensitivity within the confines of the afferent receptive field. An arrangement of excitatory subfields elongated along a certain orientation predisposes the neuron to respond preferentially to stimuli of that orientation. In aRF models, this arrangement is created during early ontogenesis. Linsker (1986) first proposed an ontogenetic mechanism based on signals of the afferent fibers having an isotropic correlation function, analogous to $c^R_{\rho'} c^R_{\rho}$. This correlation is supposed to be positive over short distance and negative over longer distance, both fitting within the confines of the aRF. Coupled with a nonlinearity in the synaptic plasticity rule, these correlations lead to spontaneous breaking of the circular symmetry of the receptive field and thus create orientation specificity. Miller (1994) has simplified Linsker's formulation, giving it a form very similar to the one I have used for the formation of retinotopy and ocularity domains presented earlier. The formation of elongated ON and OFF regions within the receptive field of a cortical neuron, according to Miller's model, is analogous to the formation of ocularity stripes. The great attraction of Linsker's idea lies in the fact that it does not need any oriented stimulus within retina to produce oriented aRFs. However, for the Linsker mechanism to work, the profile of the correlation function has to fit the size of the receptive field within close tolerance. This is difficult to accommodate if cortical cells vary in receptive field size, which is actually the case (for cat, aRF size varies according to Albus, 1975a, by a factor of 7–9 at all eccentricities; for monkey, it varies according to Schiller, Finlay, & Volman, 1975, by factors of 3–4).

In the Linsker model, both in his version and in Miller's, orientation domains are formed in the following way. Two neighboring cortical cells have overlapping receptive fields and have excitatory coupling. Because of the coupling, the cells are simultaneously active with high probability, and because of the receptive field overlap, they see virtually the same symmetry breaking noise patterns. Thus they develop the same or similar orientation preference. Miller has shown in simulations (Miller, 1994) that, given a regular array of receptive field positions (each one retinotopically shifted by a small fraction of receptive field diameter with respect to its neighbors), orientation specificity is indeed organized continuously (around occasional point defects, just as observed in many species).

Unfortunately, the Linsker model for early ontogenesis cannot deal with the experimental fact, listed earlier, that the receptive fields of cortical cells have a large retinotopic scatter that nevertheless does not disturb the very regular progression of orientation sensitivity. There are many cases in which neighboring neurons have completely nonoverlapping receptive fields and yet have very similar optimal orientation.

I am now going to propose a new model for the early ontogenesis of orientation specificity. It is of the iRF type and relies on the existence of running waves of spontaneous activity in retina, such as those that have been found experimentally (Meister et al., 1991). These waves are projected to cortex by the retinotopic mapping from retina via the lateral geniculate body. Within the cortical region activated by the wave, a rapid process of self-organization takes place by which the plexus of local connections creates one of the patterns of a one-parameter family of ripple patterns. Cortical neurons (or at least a subset of them) are assumed to receive input connections from other cortical neurons within an area that is large compared with the (projection of the) scatter in receptive field positions. Such connections have been described at least for the adult (Gilbert, Hirsch, & Wiesel, 1990). These connections form the iRF of those neurons. The model supposes that the connections of the iRF are plastic in the Hebbian sense.

Consider a particular patch of cortex (see Figure 3). For the sake of the present discussion, call the ripple patterns within the patch "micropatterns," and use the term "macropattern" for the grosser activity pattern in the cortical region surrounding the patch at the time the patch is swept by a retinal wave. The micropatterns form a one-parameter family, the parameter having the topology of the circle. (Again, the simplest ideal version of the micropatterns is straight waves, the parameter being the phase of the wave.) The macropatterns are straight swaths of activity that run through the cortical patch. Also the macropatterns form a circular one-parameter family, the parameter being the orientation of the wave.

All the reorganization that is required during early ontogenesis of orientation specificity is the establishment of a reliable mapping between the two families of patterns. According to the model, this happens simply in the following way. When the projection of a retinal wave crosses the particular cortical patch, one of the micropatterns is activated (this choice is initially random). The neurons active in the micropattern (or at least some of them) receive intracortical signals that come from the macropattern corresponding to the orientation of the retinal wave. The synapses of these active fibers are strengthened in a Hebbian fashion, giving some of the neurons active in the micropattern an elongated iRF. The next time a wave of similar orientation crosses the patch, the same micropattern will be selected, some of its neurons being preactivated

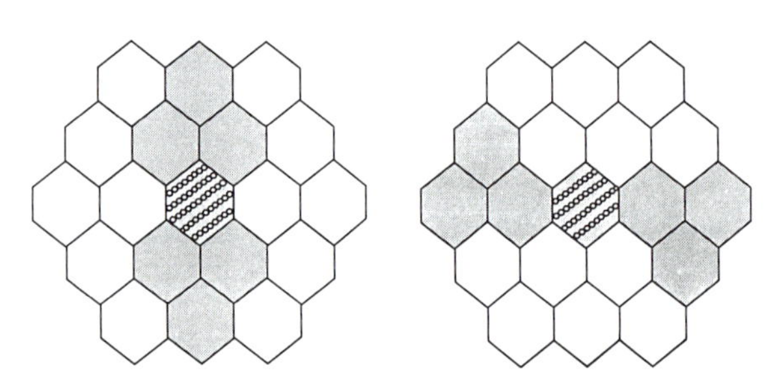

FIGURE 3 Model for the development of orientation domains. Both the macropatch (a patch of layer A, large hexagon composed of smaller hexagons) and the corresponding micropatch (patch of layer I, central hexagon) are shown. Each of the two versions shows the network when a retinal wave (indicated by shading) just passes over the center of the macropatch, producing a micropattern within the micropatch (rows of circles). Waves of different orientation produce micropatterns of different phase.

through their iRF. Overlapping macropatterns will activate overlapping micropatterns, these have similar compound iRFs because of the many neurons that are common to them. The developing mapping from the macropatterns to the micropatterns will therefore be continuous. If one could now record from individual cells in cortex, one would find that they responded selectively to retinal waves within a small range of orientations. This corresponds to orientation specificity of the iRF type.

Although cortical cells are now orientation specific, there is still nothing in their own aRF that corresponds to that specificity. However, as soon as the eyes open, cells receive precisely patterned input constellations that have been produced by bars or edges of light passing over their receptive fields. Moreover, because of their iRFs, cells only respond to a narrow range of orientations. With the help of Hebbian plasticity in the afferent synapses, the oriented patterns of activity can now be turned into precisely structured aRFs that are, for instance, elongated along the orientation of the stimulus. This structuring of aRFs cannot be done in the early ontogenetic period because the retinal activity waves are much too broad for this purpose.

Now to formulate the model in mathematical terms. Group the neurons of cortex into two layers and designate their activity as $c^A(t)$ and $c^I(t)$. The A-layer corresponds best to layer IV of cortex and is controlled directly by the afferent input from retina via the geniculate body. The I-layer corresponds best to layers above and below layer IV. Consider a large patch ("macropatch") of the A-layer, sampled at a low spatial density, designating individual points in it by the index a (which refers to a two-dimensional coordinate). In the I-layer, consider a small patch ("micropatch") centered on the macropatch and sampled at a much higher density to allow the description of micropatterns. A possible sample scheme is indicated in Figure 3. The activity in the A-layer is dominated by retinal input. Assume for it the form of straight waves of activity. Of interest are the moments when a wave just crosses the central micropatch. These bars form a family of macropatterns. The synaptic connection strengths between a sample a within the A-layer and a sample i within the I-layer is given by W_{ia}. When an activity wave passes over the retinal region projecting to the cortical micropatch, its cells receive an input I_i through their direct receptive fields. Simultaneously, they receive intracortical input from a bar-shaped region in the macropatch. The dynamics of activity within the micro-patch is described by

$$\dot{c}_i^I = -\alpha c_i^I + (K * S(c^I))_i + \sum_a W_{ia} S(c_a^A) + I_i \quad (17)$$

Here, α is a decay constant (not to be confused with the control parameter used previously), the star designates a convolution, and S is the sigmoid input–output function

$$S(c) = \frac{1}{1 - e^{-\beta(c - c_o)}} \quad (18)$$

with steepness parameter β and threshold c_o. $K_{i-i'}$ is a convolution kernel describing the fiber plexus within the I-layer. It has the general form

$$K(x,y) = qG^{\xi\eta}(x,y) - pG^{\xi'\eta'}(x,y) \quad (19)$$

where x is one component of the distance vector $i - i'$, y the other, p and q are constants, and the Gaussian G has the form

$$G^{\xi\eta}(x,y) = e^{-\frac{x^2}{2\xi^2} - \frac{y^2}{2\eta^2}} \quad (20)$$

It is assumed that ξ', η' are larger than their unprimed counterparts by some factor, say 2, and that ξ and η differ slightly to model a natural local anisotropy in the local fiber plexus (which may, of course, vary in orientation from region to region in cortex).

The activity of the micropatch, described by Equation (17), is zero when there is no input activity. However, when a wave is passing over its retinal projection region, both the direct input I_i and the indirect intracortical input $\sum_a W_{ia} S(c_a^A)$ spring to life and push c^I to positive values. Now, assume first that the iRFs W_{ia} are unstructured and that they give a flat though slightly noisy intracortical input signal, independently of the spatial profile of c_a^A. When the activity c^I in the micropatch reaches a point with critical steepness $S'(c^I)$, the activity pattern c^I branches away from homogeneity and a micropattern is created. When this critical steepness is reached just near the inflection point of S, then the growing pattern will be a standing wave (rather than a set of blobs). The orientation of that wave is determined by the orientation of the anisotropy in the kernel Equation (19), and its phase is selected by the noise in the input.

At this point, when both c^I and c^A are positive, the iRFs W_{ia} are modified by Hebbian plasticity:

$$\dot{W}_{ia} = hS(c_i^I)\{S(c_{a'}^A) - W_{ia} \sum_{a'} S(c_{a'}^A)\} \quad (21)$$

where h is a time constant. The first term corresponds to Hebbian growth in response to coincident activity on the presynaptic and postsynaptic sides of the connection ia, whereas the second term makes sure that $\sum_{a'} W_{ia'}$, the sum of synaptic strengths converging on a position i, is kept constant at 1. Due to Equation (21), simultaneously occurring macropatterns and micropatterns will be associated with each other. More specifically, an A-wave of a particular *orientation* will become associated with an I-wave of a particular *phase*. When a wave with the same orientation traverses the projection area of the micropatch again, it will create a wave in A and a patterned input into I that will select the micropattern with the same phase as before. In this way, a reliable association between macropatterns and micropatterns is developed. Overlapping macropatterns, that is, waves with similar orientation, will associate with overlapping micropatterns, that is, with waves of similar phase. Correspondingly, optimal orientation will vary smoothly within the I-layer: it will be constant along the wave crests of the micro-patterns in I and will progress linearly in the direction at a right angle to the crests. It is not necessary that the iRFs of subregions within layer I be strongly modified. In fact, it would suffice if only a small minority of neurons had appropriate intracortical connections at all. It is only necessary that a wave crossing the macropatch create enough modulation in the input pattern to the micropatch reliably to break the symmetry between micropatterns.

A system very similar to the one described in Equations (17) through (21) has been simulated in Malsburg (1973); the difference between the two models lies mainly in the interpretation of the two layers. The A-layer was then regarded as a patch of retina, and only the present I-layer was modeled for cortex. The simulations of Malsburg (1973) showed that the form of receptive fields is adapted to the stimuli and that neighboring cells are likely to specialize to neighboring orientations, validating the preceding claims.

The model proposed here accounts for the early ontogenesis of orientation specificity in the presence of large random scatter in the retinotopic arrangement of afferent fibers. In cat (Albus, 1975) and in monkey (Hubel & Wiesel, 1974a) this scatter is at least as large as the average diameter of receptive fields. This makes it impossible for aRF theories of early ontogenesis to account for the regular progression of optimal orientation that is actually observed. The aRFs are simply too small to be structured by the crude waves of retinal spontaneous activity. The difference is made here by employing the iRFs of cortical cells, which are much larger in terms of retinal coordinates.

The formulation of network self-organization given in Equations (17) through (21) differs in a significant way from the earlier formulations that were employed as models for the establishment of retinotopy and of ocularity domains. Both retinotopy and ocularity are phenomena that can be expressed in terms of pair correlations of signals. These in turn can be expressed linearly in terms of the plastic synapses. The signal variables could in this way be eliminated from the equation for synaptic plasticity. Orientation, on the other hand, cannot be expressed in terms of pair correlations (these would be isotropic for an ensemble of patterns of all orientations). Consequently, one would have to work with correlations of higher order, which, with linear signal dynamics, would contain higher order products of the connectivity variables. Autonomous differential equations for synaptic weights would accordingly become rather unwieldy. This is why I take here the more general approach of using activity variables directly for describing the ontogenesis of orientation domains.

Acknowledgments

This work was supported by grants from the Human Frontier Science Program, the Bundesministerium Für Forschung und Technologie (413-5839-01 IN 101 B/9) and AFOSR (F 49620-93-1-0109). I thank J.-M. Fellous for a critical reading of the manuscript and for his help with the preparation of my figures.

References

Albus, K. (1975). A quantitative study of the projection area of the central and the paracentral visual field in area 17 of the cat. I. The precision of the topography. *Experimental Brain Research*, **24**, 159–179.

Blasdel, G. G., & Salama, G. (1986). Voltage-sensitive dyes reveal a modular organization in monkey striate cortex. *Nature*, **321**, 579–585.

Constantine-Paton, M., & Law, M. I. (1978). Eye specific termination bands in tecta of three-eyed frogs. *Science*, **202**, 639–641.

Fraser, S. E. (1985). Cell interactions involved in neuronal patterning: An experimental and theoretical approach. In G. M. Edelman, W. E. Gall, & W. M. Cowan (Eds.), *Molecular bases of neural development*. Neurosciences Research Foundation.

Gilbert, C. D., Hirsch, J. A., & Wiesel, T. N. (1990). Lateral interactions in visual cortex. *Cold Spring Harbor Symposium on Quantitative Biology*, **LV**, 663–667.

Harris, W. A. (1980). The effects of eliminating impulse activity on the development of the retinotectal projection in salamanders. *Journal of Comparative Neurology*, **194**, 303–317.

Häussler, A. F., & von der Malsburg, C. (1983). Development of retinotopic projections. An analytical treatment. *Journal of Theoretical Neurobiology*, **2**, 47–73.

Hebb, D. O. (1949). *The organization of behavior*. New York: Wiley.

Hubel, D. H., & Wiesel, T. N. (1974a). Ordered arrangement of orientation columns in monkeys lacking visual experience. *Journal of Comparative Neurology*, **158**, 307–318.

Hubel, D. H., & Wiesel, T. N. (1974b). Sequence regularity and geometry of orientation columns in monkey striate cortex. *Journal of Comparative Neurology*, **158**, 267–294.

Hubel, D. H., & Wiesel, T. N. (1974c). Uniformity of monkey striate cortex: A parallel relationship between field size, scatter and magnification factor. *Journal of Comparative Neurology*, **158**, 295–306.

Hubel, D. H., Wiesel, T. N., & Stryker, M. P. (1977). Orientation columns in monkey visual cortex demonstrated by the 2-deoxyglucose autoradiographic technique. *Nature*, **269**, 328–330.

Humphrey, A. L., Skeen, L. C., & Norton, T. T. (1980). Topographic organization of the orientation column system in the striate cortex of the tree shrew (*Tupaia glis*): II. Deoxyglucose mapping. *Journal of Comparative Neurology*, **192**, 549–566.

Kohonen, T. (1982). Self-organized formation of topologically correct feature maps. *Biological Cybernetics*, **43**, 59–69.

Linsker, R. (1986). From basic network principles to neural architecture: Emergence of orientation columns. *Proceedings of the National Academy of Sciences* (*USA*), **83**, 8779–8783.

Mastronarde, D. N. (1983). Correlated firing of cat retinal ganglion cells: I. Spontaneously active inputs to X- and Y-cells. *Journal of Neurophysiology*, **49**, 303–324.

Meister, M., Wong, R. O. L., Baylor, D. A., & Shatz, C. J. (1991). Synchronous bursts of action potentials in ganglion cells of the developing mammalian retina. *Science*, **252**, 939–943.

Miller, K. D. (1994). A model for the development of simple cell receptive fields and the ordered arrangement of orientation columns through activity-dependent competition between ON- and OFF-center inputs. *Journal of Neuroscience*, **14**, 409–441.

Miller, K. D., Keller, J. B., & Stryker, M. P. (1989). Ocular dominance column development: Analysis and simulation. *Science*, **245**, 605–615.

Schiller, P. H., Finlay, B. L., & Volman, S. F. (1975). Quantitative studies of single-cell properties in monkey striate cortex: I. Spatiotemporal organization of receptive fields. *Journal of Neurophysiology*, **39**, 1288–1319.

Schmidt, J. T., & Tieman, S. B. (1985). Eye-specific segregation of optic afferents in mammals, fish, and frogs: The role of activity. *Cellular and Molecular Neurobiology*, **5**, 5–34.

Singer, W. (1981). Topographic organization of orientation columns in the cat visual cortex. A deoxyglucose study. *Experimental Brain Research*, **44**, 431–436.

von der Malsburg, C. (1973). Self-organization of orientation sensitive cells in the striate cortex. *Kybernetik*, **14**, 85–100.

von der Malsburg, C. (1979). Development of ocularity domains and growth behaviour of axon terminals. *Biological Cybernetics*, **32**, 49–62.

von der Malsburg, C., & Willshaw, D. J. (1977). How to label nerve cells so that they can interconnect in an ordered fashion. *Proceedings of the National Academy of Sciences of the United States of America*, **74**, 5176–5178.

von der Malsburg, C., & Willshaw, D. J. (1981). Cooperativity and brain organization. *Trends in NeuroSciences*, **April**, 80–83.

Willshaw, D. J., & von der Malsburg, C. (1976). How patterned neural connections can be set up by self-organization. *Proceedings of the Royal Society of London* , **194**, 431–445.

Willshaw, D. J., & von der Malsburg, C. (1979). A marker induction mechanism for the establishment of ordered neural mappings: Its application to the retinotectal problem. *Philosophical Transactions of the Royal Society of London, Series* B, **287**, 203–243.

23

A Neural Network Architecture for Autonomous Learning, Recognition, and Prediction in a Nonstationary World

Gail A. Carpenter and Stephen Grossberg

INTRODUCTION

In a constantly changing world, humans are adapted to alternate routinely between attending to familiar objects and testing hypotheses about novel ones. We can rapidly learn to recognize and name novel objects without unselectively disrupting our memories of familiar ones. We can notice fine details that differentiate nearly identical objects and generalize across broad classes of dissimilar objects. This chapter describes a class of self-organizing neural network architectures—called ARTMAP—that are capable of fast, yet stable, on-line recognition learning, hypothesis testing, and naming in response to an arbitrary stream of input patterns (Carpenter, Grossberg, Markuzon, Reynolds, & Rosen, 1992; Carpenter, Grossberg, & Reynolds, 1991). The intrinsic stability of ARTMAP allows the system to learn incrementally for an unlimited period of time. System stability properties can be traced to the structure of its learned memories, which encode clusters of attended features into its recognition categories, rather than slow averages of category inputs. The level of detail in the learned attentional focus is determined moment-by-moment, depending on predictive success: an error caused by overgeneralization automatically focuses attention on additional input details, enough of which are learned in a new recognition category so that the predictive error will not be repeated.

An ARTMAP system creates an evolving map between a variable number of learned categories that compress one feature space (e.g., visual features) to learned categories of another feature space (e.g., auditory features). Input vectors can be either binary or analog. Computational properties of the networks enable them to perform significantly better in benchmark studies than alternative machine learning, genetic algorithm, or neural network models. Some of the critical problems that challenge and constrain any such autonomous learning system will be illustrated next. Design principles that work together to solve these problems are also outlined. These principles are realized in the ARTMAP architecture, which is specified as an algorithm. Finally, ARTMAP dynamics are illustrated by means of a series of benchmark simulations.

Reprinted and revised, with permission, from *Neural and Synergetic Computers*, H. Haken (ed.) Copyright © 1989 by Springer-Verlag.

CRITICAL PROBLEMS TO BE SOLVED BY AN AUTONOMOUS LEARNING SYSTEM

ARTMAP performance success is based on a set of design principles that are derived from an analysis of learning by an autonomous agent in a nonstationary environment (Table 1). Realization of these principles enables a self-organizing ARTMAP system to learn, categorize, and make predictions about a changing world, as follows.

Rare Events

A successful autonomous agent must be able to learn about rare events that have important consequences, even if these rare events are similar to a surrounding cloud of frequent events that have different consequences. *Fast learning* is needed to pick up a rare event on the fly. For example, a rare medical condition could be either a unique case or the harbinger of a new epidemic. A slightly different chemical assay could either be a routine variation or predict the biological activity of a new drug. Many feed-forward neural network systems, such as back-propagation, require a form of slow learning that tends to average over similar event occurrences. ARTMAP can rapidly group or single out events, depending on their predictive outcomes.

Large Nonstationary Databases

Rare events typically occur in a nonstationary environment whose event statistics may change rapidly and unexpectedly through time. Individual events may also occur with variable probabilities and durations, and arbitrarily large numbers of events may need to be

TABLE 1 Autonomous Learning and Control in a Nonstationary World

An ARTMAP system can reconcile conflicting requirements and autonomously learn about:

RARE EVENTS
—requires FAST learning

LARGE NONSTATIONARY DATABASES
—requires STABLE learning

MORPHOLOGICALLY VARIABLE EVENTS
—requires MULTIPLE SCALES of generalization (fine/coarse)

MANY-TO-ONE AND ONE-TO-MANY RELATIONSHIPS
—requires categorization and naming for expert knowledge

To realize these properties, ARTMAP systems:

PAY ATTENTION
—ignore masses of irrelevant data
TEST HYPOTHESES
—discover predictive constraints hidden in data streams
CHOOSE BEST ANSWERS
—quickly select globally optimal solution at any stage of learning
CALIBRATE CONFIDENCE
—measure on-line how well a hypothesis matches the data
DISCOVER RULES
—identify transparent if-then relations at each learning stage
SCALE
—preserve all desirable properties in arbitrarily large problems

processed. Each of these factors tends to destabilize the learning process within feed-forward algorithms. New learning in such systems tends to wash away unselectively the memory traces of old, but still useful, knowledge. By using such an algorithm, for example, learning new faces could erase the memory of a parent's face. More generally, learning a new type of expertise could erase the memory of previous expert knowledge. ARTMAP contains a *self-stabilizing memory* that permits accumulating knowledge to be stored reliably in response to arbitrarily many events in a nonstationary environment under incremental learning conditions. Learning may continue until the system's full memory capacity, which can be chosen arbitrarily large, is exhausted.

Morphologically Variable Types of Events

In many environments, some information, including rulelike inferences, is coarsely defined whereas other information is precisely characterized. Otherwise expressed, the morphological variability of the data may change through time. For example, we may recognize one photograph as an animal and see a similar one as a picture of our own pet. Under autonomous learning conditions, a system typically has to adjust constantly how coarse the generalization, or compression, of particular types of data should be. Multiple scales of generalization, from fine to coarse, need to be available on an as-needed basis. ARTMAP automatically adjusts its scale of generalization to match the morphological variability of the data, based on predictive success. The network embodies a Minimax Learning Rule that conjointly minimizes predictive error and maximizes generalization by using only the information that is locally available under incremental learning conditions in a nonstationary environment. This property has been used to suggest how the inferotemporal cortex can learn to recognize both fine and coarse information about the world (Carpenter & Grossberg, 1993), as demonstrated by neurophysiological experiments of Desimone (1992), Harries and Perrett (1991), Miller, Li and Desimone (1991), Mishkin (1982), and Spitzer, Desimone and Moran (1988), among others.

Many-to-One and One-to-Many Relationships

In ARTMAP learning, many-to-one code compression occurs in two stages: categorization and naming. For example, during categorization of printed letter fonts, many similar exemplars of the same printed letter may establish a single recognition category, or compressed representation (Figure 1). Different printed letter fonts or written exemplars of the letter may establish additional categories. Each of these categories carries out a many-to-one map of exemplar into category. During naming, all of the categories that represent the same letter may be associatively mapped into the letter name, or prediction. Compressed many-to-one maps are thus constructed from both unsupervised (categorization) and supervised (naming) learning.

Conversely, one-to-many learning is also used to build up expert knowledge about an object or event. A single visual image of a particular animal may, for example, lead to learning that predicts animal, dog, beagle, and my dog Rover (Figure 2). A computerized

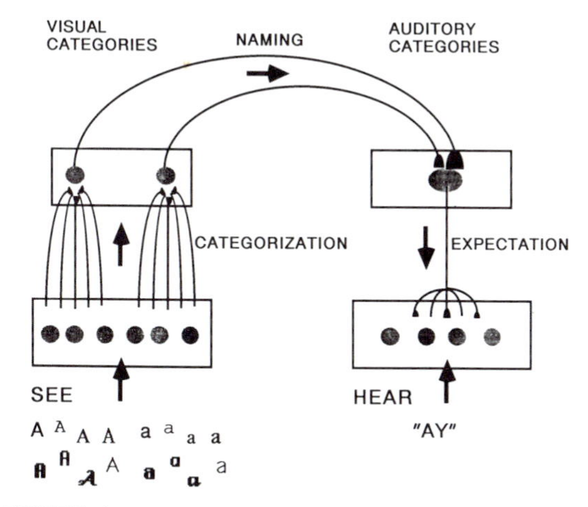

FIGURE 1 Many-to-one learning combines categorization of many exemplars into one category, and labeling of many categories with the same name.

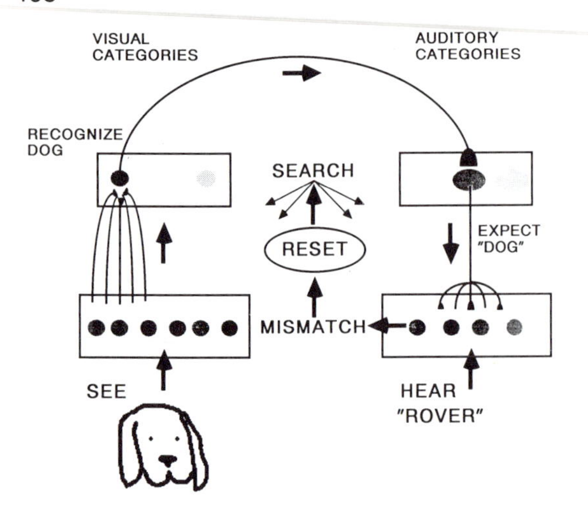

FIGURE 2 One-to-many learning enables one input vector to be associated with many output vectors. If the system predicts an output that is disconfirmed at any given stage of learning, the predictive error drives a memory search for a new category to associate with the new prediction without degrading its previous knowledge about the input vector.

record of a patient's medical checkup may lead to a series of predictions about the patient's health.

In feed-forward networks, the attempt to learn more than one prediction about a single input leads to unselective forgetting of previously learned predictions, for the same reason that these algorithms become unstable in response to nonstationary data. In particular, error-based learning systems, including multilayer perceptrons such as back-propagation (Rosenblatt, 1958; Rumelhart, Hinton, & Williams, 1986; Werbos, 1974), find it difficult, if not impossible, to solve the critical problems just described.

ARTMAP DESIGN PRINCIPLES

ARTMAP systems solve the critical design problems because they implement a qualitatively different set of heuristics from error-based learning systems, as follows (Table 1).

Paying Attention

ARTMAP learns top-down expectations (also called prototypes, primes, or queries) that allow the system to ignore masses of irrelevant distributed data. These queries "test the hypothesis" that is embodied by a recognition category, or symbol, as they suppress features not in the prototypical attentional focus. When one object is recognized as "dog" versus "Rover," distinct top-down expectations focus attention on distinct feature clusters. ARTMAP therefore embodies properties of intentionality. A large mismatch between a bottom-up input vector and a top-down expectation (Figure 2) can drive an adaptive memory search that carries out hypothesis testing for a better category, as described next.

Hypothesis Testing and Match-Based Learning

ARTMAP actively searches for recognition categories, or hypotheses, whose top-down expectations provide an acceptable match to bottom-up data. The top-down expectation learns a prototype that focuses attention upon that cluster of input features that it deems relevant. If no available category, or hypothesis, provides a good enough match, then selection and learning of a new category and top-down expectation is automatically initiated. When the search discovers a category that provides an acceptable match, the system locks into an attentive resonance through which the distributed input and its symbolic category are bound together. During this resonantly bound state, the input exemplar refines the adaptive weights of the category on the basis of any new information in the attentional focus. Thus ARTMAP carries out match-based learning, rather than error-based learning, because a category modifies its previous learning only if its top-down expectation matches the input vector well enough to risk changing its defining characteristics. Otherwise, hypothesis testing selects a new category on which to base learning of a novel event, thereby preserving information in both the new and the old categories.

Choosing Globally Best Symbolic Answer

In many learning algorithms, as learning proceeds, local minima or less-than-optimal solutions are selected. In ARTMAP, at any stage of learning, an input exemplar first selects the category whose top-down expectation provides the globally best match. This top-down expectation thereby acts as a prototype for the class of all the input exemplars that its category represents. After learning self-stabilizes, every input directly selects the globally best matching category without any search. This category symbolically represents all the inputs that share the same prototype. Before learning self-stabilizes, familiar events gain direct access to the globally best category without any search, even if they are interspersed with unfamiliar events that drive hypothesis testing for better matching categories. A lesion in the *orienting subsystem* that mediates the hypothesis testing or memory search process leads to a memory disorder that strikingly resembles clinical properties of medial temporal amnesia in humans and monkeys after lesions of the hippocampal formation (Carpenter & Grossberg, 1993). These and related data properties provide support for the hypothesis that the hippocampal formation carries out an orienting subsystem function as one of its several functional roles.

Learned Prototypes and Exemplars

The learned prototype represents the cluster of input features that the category deems relevant on the basis of its past experience. The prototype represents the features to which the category "pays attention." In cognitive psychology, an input pattern is called an *exemplar*. A fundamental issue in cognitive psychology concerns whether the brain learns prototypes or exemplars. Some argue that the brain learns prototypes, or abstract types of knowledge, such as being able to recognize that a particular object is a face or an animal. Others have argued that the brain learns individual exemplars, or concrete types of knowledge, such as being able to recognize a particular face or a particular animal. Recently, it has been increasingly realized that some sort of hybrid system is needed that can acquire both types of knowledge (Smith, 1990). ARTMAP is such a hybrid system. It uses the Minimax Learning Rule to control how abstract or concrete—how fuzzy—a category becomes in order to conjointly minimize predictive error and maximize predictive generalization.

Calibrate Confidence

A confidence measure, called *vigilance*, calibrates how well an exemplar must match the prototype that it selects. Otherwise expressed, vigilance measures how well the chosen hypothesis must match the data. If vigilance is low, even poorly matching exemplars can then be incorporated into one category, so compression and generalization by that category are high. The symbol here is more abstract. If vigilance is high, then even good matches may be rejected, and hypothesis testing may be initiated to select a new category. In this case, few exemplars activate the same category, so compression and generalization are low. In the limit of very high vigilance, prototype learning reduces to exemplar learning.

The Minimax Learning Rule is realized by adjusting the vigilance parameter in response to a predictive error. Vigilance is first low, to maximize compression. When a predictive error occurs, vigilance is increased just enough to initiate hypothesis testing to discover a better category, or hypothesis, with which to match the data. In this way, a minimum amount of generalization is sacrificed to correct the error. This process is called *match tracking* because vigilance tracks the degree of match between exemplar and prototype in response to a predictive error.

IF-THEN Rule Discovery

At any stage of learning, a user can translate the learned weights of an ARTMAP system into a set of IF-THEN rules that completely characterize the decisions of the system. These rules evolve as ARTMAP is exposed to new inputs. Suppose, for example, that n visual categories are associated with the auditory prediction "AY." Backtrack from prediction "AY" along the associative pathways whose adaptive weights have learned to connect the *n* visual categories to this prediction (Figure 1). Each of these categories codes a "reason" for predicting "AY." The prototype of each category embodies the set of features, or constraints, whose binding together constitutes that category's "reason." The IF-THEN rule takes the form: IF some of the features of any of these *n* categories are found bound together, within the fuzzy constraints that would lead to selection of that category, THEN the prediction "AY" holds. Keeping in mind that ARTMAPs carry out hypothesis testing and memory search to discover these rules, we can see that ARTMAPs are a type of self-organizing production system (Laird, Newell, & Rosenbloom, 1987) that evolves adaptively from individual input–output experiences, as in case-based reasoning.

IF-THEN rules of ARTMAP can be extracted from the system at any stage of the learning process. This property is particularly important in applications such as medical diagnosis from a large database of patient records, when doctors may want to study the rules by which the system reaches its diagnostic decisions. Some of these rules may already be familiar to the doctors; others may represent novel constraint combinations (symptoms, tests, treatments, etc.) which the doctors could then evaluate for their possible medical significance. This property also sheds light on how humans believe that brains somehow realize rulelike behavior although brain anatomy is not algorithmically structured in a traditional sense. The Minimax Learning Rule determines how abstract these rules will become in response to any prescribed environment. Typical databases generate a mixture of a few broad rules, with few constraints and many exemplars, plus a set of more highly specified special cases (Carpenter, Grossberg, & Reynolds, 1991).

Table 2 summarizes some medical and other benchmark studies that compare the performance of ARTMAP with alternative recognition and prediction models. Three of these benchmarks are summarized here. These and other benchmarks are described elsewhere in greater detail (Carpenter, Grossberg, &

TABLE 2 Benchmark Studies

Database benchmark:
MACHINE LEARNING (90–95% correct)
ARTMAP (100% correct)
Training set an order of magnitude smaller.
Medical database:
STATISTICAL METHOD (60% correct)
ARTMAP (91% correct)
Incremental improvement.
Transparent "rules" from critical feature clusters.
Letter recognition database:
GENETIC ALGORITHM (82% correct)
ARTMAP (96% correct)
Database benchmarks:
BACK-PROPAGATION (10,000–20,000 training epochs)
ARTMAP (1–5 epochs)
Used in applications where other algorithms fail, e.g., Boeing CAD Group Technology (T. Caudell et al.)
Part design reuse and inventory compression.
Need fast stable learning and search of a huge (16 million) and continually growing nonstationary parts inventory.

Iizuka, 1992; Carpenter, Grossberg, Markuzon, Reynolds, & Rosen, 1992; Carpenter, Grossberg, & Reynolds, 1991).

Properties Scale

One of the most serious deficiencies of many algorithms is that their desirable properties tend to break down as small toy problems are generalized to large-scale problems. In contrast, all of the desirable properties of ARTMAP scale to arbitrarily large problems. Recall, however, that ARTMAP solves a particular class of problems, not all problems of learning or intelligence. The categorization and inference problems that ARTMAP does handle well are, however, core problems in many intelligent systems, and include technology bottlenecks for many alternative approaches.

ARTMAP ARCHITECTURE

Each ARTMAP system includes a pair of ART modules (ART_a and ART_b), as in Figure 3. During supervised learning, ART_a receives a stream $\{\mathbf{a}^{(p)}\}$ of input patterns and ART_b receives a stream $\{\mathbf{b}^{(p)}\}$ of input patterns, where $\mathbf{b}^{(p)}$ is the correct prediction given $\mathbf{a}^{(p)}$. These modules are linked by an associative learning network and an internal controller that ensures autonomous system operation in real time. The controller is designed to create the minimal number of ART_a recognition categories, or "hidden units," needed to meet accuracy criteria. As previously noted, this is accomplished by realizing a Minimax Learning Rule that conjointly minimizes predictive error and maximizes predictive generalization. This scheme automatically links predictive success to category size on a trial-by-trial basis using only local operations. It works by

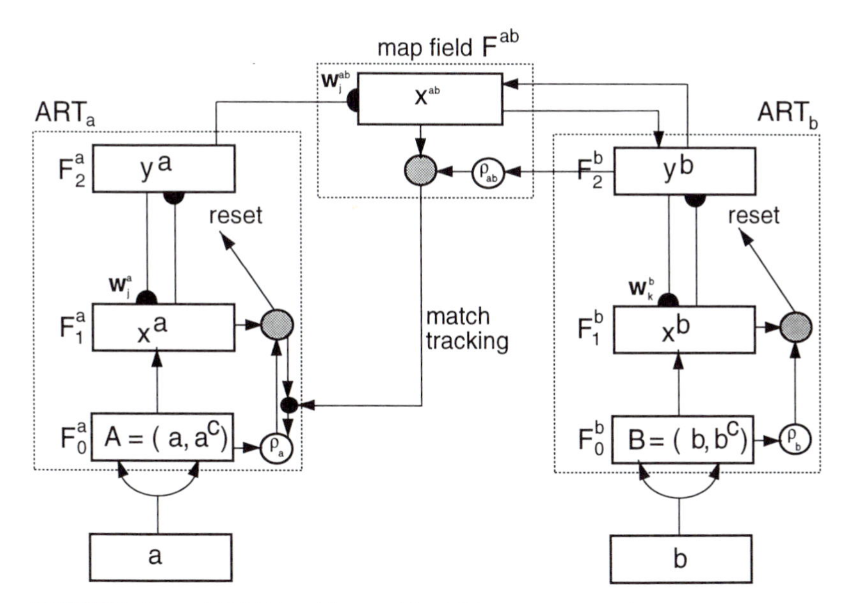

FIGURE 3 Fuzzy ARTMAP architecture. The ART_a complement coding preprocessor transforms the M_a vector $\mathbf{a}$ into the $2M_a$ vector $\mathbf{A} = (\mathbf{a},\mathbf{a}^c)$ at the ART_a field F_0^a. $\mathbf{A}$ is the input vector to the ART_a field F_1^a. Similarly, the input to F_1^b is the $2M_b$ vector $(\mathbf{b},\mathbf{b}^c)$. When a prediction by ART_a is disconfirmed at ART_b, inhibition of map field activation induces the match tracking process. Match tracking raises the ART_a vigilance (ρ_a) to just above the F_1^a-to-F_0^a match ratio $|\mathbf{x}^a|/|\mathbf{A}|$. This triggers an ART_a search which leads to activation of either an ART_a category that correctly predicts $\mathbf{b}$ or to a previously uncommitted ART_a category node.

increasing the vigilance parameter ρ_a of ART_a by the minimal amount needed to correct a predictive error at ART_b (Figure 4).

Parameter ρ_a calibrates the minimum confidence that ART_a must have in a recognition category, or hypothesis, that is activated by an input $\mathbf{a}^{(p)}$ in order for ART_a to accept that category, rather than search for a better one through an automatically controlled process of hypothesis testing. Lower values of ρ_a enable larger categories to form. These lower ρ_a values lead to broader generalization and higher code compression. A predictive failure at ART_b increases the minimal confidence ρ_a by the least amount needed to trigger hypothesis testing at ART_a by using a mechanism called *match tracking*. Match tracking sacrifices the minimum amount of generalization necessary to correct the predictive error. It increases the criterion confidence just enough to trigger hypothesis testing, and hypothesis testing leads to the selection of a new ART_a category, which focuses attention on a new cluster of $\mathbf{a}^{(p)}$ input features that is better able to predict $\mathbf{b}^{(p)}$. From the combination of match tracking and fast learning, a single ARTMAP system can learn a different prediction for a rare event than it learns for a cloud of similar frequent events in which it is embedded.

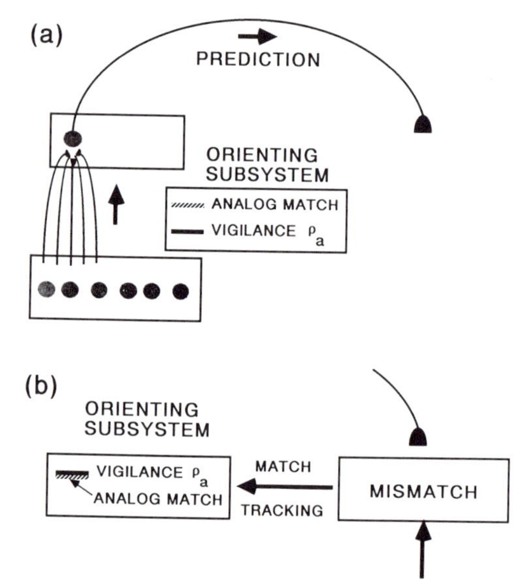

FIGURE 4 Match tracking: (a) A prediction is made by ART_a when the baseline vigilance ρ_a is less than the analog match value. (b) A predictive error at ART_b increases the baseline vigilance value of ART_a until it just exceeds the analog match value, and thereby triggers hypothesis testing that searches for a more predictive bundle of features to which to attend.

An ARTMAP simulation algorithm will now be summarized. When input components are binary, the ARTMAP system is constructed from ART 1 component modules (Carpenter & Grossberg, 1987). ART 1 dynamics are simulated via a series of operations that include binary intersection ($\cap$). When input components are analog (real-valued), the set-theoretic intersection operator can be replaced by a fuzzy intersection ($\wedge$), or component-wise minimum (Zadeh, 1965). Binary ART 1 then is thereby transformed into fuzzy ART (Carpenter, Grossberg, & Rosen, 1991), and binary ARTMAP (Carpenter, Grossberg, & Reynolds, 1991) becomes fuzzy ARTMAP (Carpenter, Grossberg, Markuzon, Reynolds, & Rosen, 1992). Algorithms for the more general systems, fuzzy ART and fuzzy ARTMAP, will now be specified.

FUZZY ART ALGORITHM

ART Field Activity Vectors

Each ART system includes a field F_0 of nodes that represent a current input vector; a field F_1 that receives both bottom-up input from F_0 and top-down input from a field F_2 that represents the active code or category. The F_0 activity vector is denoted $\mathbf{I} = (I_1, \ldots, I_M)$, with each component I_i in the interval [0,1], $i = 1, \ldots, M$. The F_1 activity vector is denoted $\mathbf{x} = (x_1, \ldots, x_M)$ and the F_2 activity vector is denoted $\mathbf{y} = (y_1, \ldots, y_N)$. The number of nodes in each field is arbitrary.

Weight Vector

Associated with each F_2 category node $j (j = 1, \ldots, N)$ is a vector $\mathbf{w}_j \equiv (w_{j1}, \ldots, w_{jM})$ of adaptive weights, or long-term memory (LTM) traces. Initially

$$w_{j1}(0) = \ldots = w_{jM}(0) = 1 \tag{1}$$

Then each category is said to be *uncommitted*. After a category is selected for coding it becomes *committed*. As shown, each LTM trace w_{ji} is monotone nonincreasing through time and hence converges to a limit. The fuzzy ART weight vector $\mathbf{w}_j$ subsumes both the bottom-up and top-down weight vectors of the ART 1 neural network.

Parameters

Fuzzy ART dynamics are determined by a choice parameter $\alpha > 0$; a learning rate parameter $\beta \in [0,1]$; and a vigilance parameter $\rho \in [0,1]$.

Category Choice

For each input $\mathbf{I}$ and F_2 node j, the *choice function* T_j is defined by

$$T_j(\mathbf{I}) = \frac{|\mathbf{I} \wedge w_j|}{\alpha + |w_j|} \tag{2}$$

where the fuzzy AND operator $\wedge$ is defined by

$$(\mathbf{p} \wedge \mathbf{q})_i \equiv \min(p_i, q_i) \tag{3}$$

and where the city-block norm $|\cdot|$ is defined by

$$|\mathbf{p}| \equiv \sum_{i=1}^{M} |p_i| \tag{4}$$

for any M-dimensional vectors $\mathbf{p}$ and $\mathbf{q}$. For notational simplicity, $T_j(\mathbf{I})$ in Equation (2) is often written as T_j when the input $\mathbf{I}$ is fixed.

The system is said to make a *category choice* when at most one F_2 node can become active at a given time. The category choice is indexed by J, where

$$T_J = \max\{T_j : j = 1 \ldots N\} \tag{5}$$

If more than one T_j is maximal, the category j with the smallest index is chosen. In particular, nodes become committed in order $j = 1, 2, 3, \ldots$. When the Jth category is chosen, $y_J = 1$; and $y_j = 0$ for $j \neq J$. In a choice system, the F_1 activity vector $\mathbf{x}$ obeys the equation

$$\mathbf{x} = \begin{cases} \mathbf{I} & \text{if } F_2 \text{ is inactive} \\ \mathbf{I} \wedge \mathbf{w}_J & \text{if the } J^{th}\ F_2 \text{ node is chosen} \end{cases} \tag{6}$$

Resonance or Reset

Resonance occurs if the *match function* $|\mathbf{I} \wedge \mathbf{w}_J|/|\mathbf{I}|$ of the chosen category meets the vigilance criterion:

$$\frac{|\mathbf{I} \wedge \mathbf{w}_J|}{|\mathbf{I}|} \geq \rho \tag{7}$$

That is, by Equation (6), when the Jth category is chosen, resonance occurs if

$$|\mathbf{x}| = |\mathbf{I} \wedge \mathbf{w}_J| \geq \rho |\mathbf{I}| \tag{8}$$

Learning then ensues, as defined next. *Mismatch reset* occurs if

$$\frac{|\mathbf{I} \wedge \mathbf{w}_J|}{|\mathbf{I}|} < \rho \tag{9}$$

that is, if

$$|\mathbf{x}| = |\mathbf{I} \wedge \mathbf{w}_J| < \rho |\mathbf{I}| \tag{10}$$

Then the value of the choice function T_J is set to 0 for the duration of the input presentation to prevent the persistent selection of the same category during search. A new index J is then chosen by Equation (5). The search process continues until the chosen J satisfies Equation (7).

Learning

Once the search ends, the weight vector $\mathbf{w}_J$ is updated according to the equation

$$\mathbf{w}_J^{(\text{new})} = \beta(\mathbf{I} \wedge \mathbf{w}_J^{(\text{old})}) + (1 - \beta)\mathbf{w}_J^{(\text{old})} \tag{11}$$

Fast learning corresponds to setting $\beta = 1$. The learning law used in the EACH system of Salzberg (1990) is equivalent to Equation (11) in the fast-learn limit with the complement coding option described next.

Fast-Commit Slow-Recode Option

For efficient coding of noisy input sets, it is useful to set $\beta = 1$ when J is an uncommitted node, and then to take $\beta < 1$ after the category is committed. Then, $\mathbf{w}_J^{(\mathrm{new})} = \mathbf{I}$ the first time category J becomes active. Moore (1989) introduced the learning law Equation (11), with fast commitment and slow recoding, to investigate a variety of generalized ART 1 models. Some of these models are similar to fuzzy ART, but none includes the complement coding option. Moore described a category proliferation problem that can occur in some analog ART systems when a large number of inputs erode the norm of weight vectors. Complement coding solves this problem.

Input Normalization/Complement Coding Option

Proliferation of categories is avoided in fuzzy ART if inputs are normalized. *Complement coding* is a normalization rule that preserves amplitude information. Complement coding represents both the ON-response and the OFF-response to an input vector $\mathbf{a}$. To define this operation in its simplest form, let $\mathbf{a}$ itself represent the ON-response. The complement of $\mathbf{a}$, denoted by $\mathbf{a}^c$, represents the OFF-response, where

$$a_i^c \equiv 1 - a_i \tag{12}$$

The complement coded input $\mathbf{I}$ to the field F_1 is the $2M$-dimensional vector

$$\mathbf{I} = (\mathbf{a}, \mathbf{a}^c) \equiv (a_1, \ldots, a_M, a_1^c, \ldots, a_M^c) \tag{13}$$

Note that

$$\begin{aligned} |\mathbf{I}| &= |(\mathbf{a}, \mathbf{a}^c)| \\ &= \sum_{i=1}^{M} a_i + (M - \sum_{i=1}^{M} a_i) \\ &= M \end{aligned} \tag{14}$$

so inputs preprocessed into complement coding form are automatically normalized. Where complement coding is used, the initial condition (1) is replaced by

$$w_{j1}(0) = \ldots = w_{j,2M}(0) = 1 \tag{15}$$

FUZZY ARTMAP ALGORITHM

The fuzzy ARTMAP system incorporates two fuzzy ART modules (ART_a and ART_b) that are linked together via an inter-ART module (F^{ab}) called a *map field*. The map field is used to form predictive associations between categories and to realize the *match tracking rule* whereby the vigilance parameter of ART_a increases in response to a predictive mismatch at ART_b. The interactions mediated by the map field F^{ab} may be operationally characterized as follows.

ART_a and ART_b

Inputs to ART_a and ART_b are in the complement code form: for ART_a, $\mathbf{I} = \mathbf{A} = (\mathbf{a},\mathbf{a}^c)$; for ART_b, $\mathbf{I} = \mathbf{B} = (\mathbf{b},\mathbf{b}^c)$ (Figure 3). Variables in ART_a or ART_b are designated by subscripts or superscripts a or b. For ART_a, let $\mathbf{x}^a \equiv (x_1^a \ldots x_{2Ma}^a)$ denote the F_1^a output vector; let $\mathbf{y}^a \equiv (y_1^a \ldots y_{Na}^a)$ denote the F_2^a output vector; and let $\mathbf{w}_j^a \equiv (w_{j1}^a, w_{j2}^a, \ldots, w_{j,2Ma}^a)$ denote the jth ART_a weight vector. For ART_b, let $\mathbf{x}^b \equiv (x_1^b \ldots x_{2Mb}^b)$ denote the F_1^b output vector; let $\mathbf{y}^b \equiv (y_1^b \ldots y_{Nb}^b)$ denote the F_2^b output vector; and let $\mathbf{w}_k^b \equiv (w_{k1}^b, w_{k2}^b, \ldots, w_{k,2Mb}^b)$ denote the kth ART_b weight vector. For the map field, let $x^{ab} \equiv (x_1^{ab}, \ldots, x_{Nb}^{ab})$ denote the F^{ab} output vector, and let $\mathbf{w}_j^{ab} \equiv (w_{j1}^{ab}, \ldots, w_{jNb}^{ab})$ denote the weight vector from the jth F_2^a node to F^{ab}. Vectors $\mathbf{x}^a$, $\mathbf{y}^a$, $\mathbf{x}^b$, $\mathbf{y}^b$, and x^{ab} are set to $\mathbf{0}$ between input presentations.

Map Field Activation

The map field F^{ab} is activated whenever one of the ART_a or ART_b categories is active. If node J of F_2^a is chosen, then its weights $\mathbf{w}_J^{ab}$ activate F^{ab}. If node K in F_2^b is active, then the node K in F^{ab} is activated by one-to-one pathways between F_2^b and F^{ab}. If both ART_a and ART_b are active, then F^{ab} becomes active only if ART_a predicts the same category as ART_b via the weights $\mathbf{w}_J^{ab}$. The F^{ab} output vector $\mathbf{x}^{ab}$ obeys

$$\mathbf{x}^{ab} = \begin{cases} \mathbf{y}^b \wedge \mathbf{w}_J^{ab} & \text{if the } J^{th}\ F_2^a \text{ node is active and } F_2^b \text{ is active} \\ \mathbf{w}_J^{ab} & \text{if the } J^{th}\ F_2^a \text{ node is active and } F_2^b \text{ is inactive} \\ \mathbf{y}^b & \text{if } F_2^a \text{ is inactive and } F_2 \text{ is active} \\ \mathbf{0} & \text{if } F_2^a \text{ is inactive and } F_2^b \text{ is inactive} \end{cases} \quad (16)$$

By Equation (16), $\mathbf{x}^{ab} = \mathbf{y}^b \wedge \mathbf{w}_J^{ab} = \mathbf{0}$ if the prediction $\mathbf{w}_J^{ab}$ is disconfirmed by $\mathbf{y}^b$. Such a mismatch event triggers an ART_a search for a better category, as follows.

Match Tracking

At the start of each input presentation, the ART_a vigilance parameter ρ_a equals a baseline vigilance $\overline{\rho_a}$. The map field vigilance parameter is ρ_{ab}. A predictive mismatch is detected when

$$|\mathbf{x}^{ab}| < \rho_{ab}|\mathbf{y}^b| \quad (17)$$

Then, ρ_a is increased until it is slightly larger than $|\mathbf{A} \wedge \mathbf{w}_J^a|\ |\mathbf{A}|^{-1}$, where $\mathbf{A}$ is the input to F_1^a, in complement coding form. After match tracking,

$$|\mathbf{x}^a| = |\mathbf{A} \wedge \mathbf{w}_J^a| < \rho_a|\mathbf{A}| \quad (18)$$

where J is the index of the active F_2^a node, as in Equation (10). When this occurs, ART_a search leads either to activation of another F_2^a node J with

$$|\mathbf{x}^a| = |\mathbf{A} \wedge \mathbf{w}_J^a| \geq \rho_a|\mathbf{A}| \quad (19)$$

and

$$|\mathbf{x}^{ab}| = |\mathbf{y}^b \wedge \mathbf{w}_J^{ab}| \geq \rho_{ab}|\mathbf{y}^b| \quad (20)$$

or, if no such node exists, to the shutdown of F_2^a for the remainder of the input presentation.

Map Field Learning

Learning rules determine how the map field weights w_{jk}^{ab} change through time, as follows. Weights w_{jk}^{ab} in $F_2^a \rightarrow F^{ab}$ paths initially satisfy

$$w_{jk}^{ab}(0) = 1 \quad (21)$$

During resonance with the ART_a category J active, $\mathbf{w}_J^{ab}$ approaches the map field vector $\mathbf{x}^{ab}$. With fast learning, once J learns to predict the ART_b category K that association is permanent; that is, $w_{JK}^{ab} = 1$ for all time.

THE GEOMETRY OF FUZZY ART

Fuzzy ARTMAP dynamics will be illustrated here by a benchmark simulation problem, circle-in-the-square. The low dimensions of this problem ($M_a = 2$, $N_a = 1$) allow the evolving category structure to be illustrated graphically. To do this, a geometric interpretation of fuzzy ART will now be outlined. For definiteness, let the input set consist of two-dimensional vectors $\mathbf{a}$ preprocessed into the four-dimensional complement coding form. Thus

$$\mathbf{I} = (\mathbf{a}, \mathbf{a}^c) = (a_1, a_2, 1 - a_1, 1 - a_2) \quad (22)$$

In this case, each category j has a geometric representation as a rectangle R_j, as follows. Following Equation (22), the weight vector $\mathbf{w}_j$ can be written in complement coding form:

$$\mathbf{w}_j = (\mathbf{u}_j, \mathbf{v}_j^c) \quad (23)$$

where $\mathbf{u}_j$ and $\mathbf{v}_j$ are two-dimensional vectors. Let vector $\mathbf{u}_j$ define one corner of a rectangle R_j and let $\mathbf{v}_j$ define another corner of R_j (Figure 5A). The size of R_j is defined to be

$$|R_j| \equiv |\mathbf{v}_j - \mathbf{u}_j| \quad (24)$$

which is equal to the height plus the width of R_j in Figure 5A.

In a fast-learn fuzzy ART system, with $\beta = 1$ in Equation (11), $\mathbf{w}_J^{(new)} = \mathbf{I} = (\mathbf{a},\mathbf{a}^c)$ when J is an uncommitted node. The corners of $R_J^{(new)}$ are then given by $\mathbf{u}_J = \mathbf{a}$ and $\mathbf{v}_J = (\mathbf{a}^c)^c = \mathbf{a}$. Hence $R_J^{(new)}$ is just the point $\mathbf{a}$. Learning increases the size of each R_j. In fact, the size of R_j grows as the size of $\mathbf{w}_j$ shrinks during learning. The maximum size of R_j is limited by the size of the vigilance parameter, with $|R_j| \leq 2(1 - \rho)$. During each fast-learning trial, R_J expands to $R_J \oplus \mathbf{a}$, the minimum rectangle containing R_J and $\mathbf{a}$ (Figure 5B). The corners of $R_J \oplus \mathbf{a}$ are given by $\mathbf{a} \wedge \mathbf{u}_J$ and $\mathbf{a} \vee \mathbf{v}_J$, where the fuzzy AND (intersection) operator $\wedge$ is defined by Equation (3), and the fuzzy OR (union) operator $\vee$ is defined by

$$(\mathbf{p} \vee \mathbf{q})_i \equiv \max(p_i, q_i) \tag{25}$$

(Zadeh, 1965). Hence, by Equation (24), the size of $R_J \oplus \mathbf{a}$ is given by

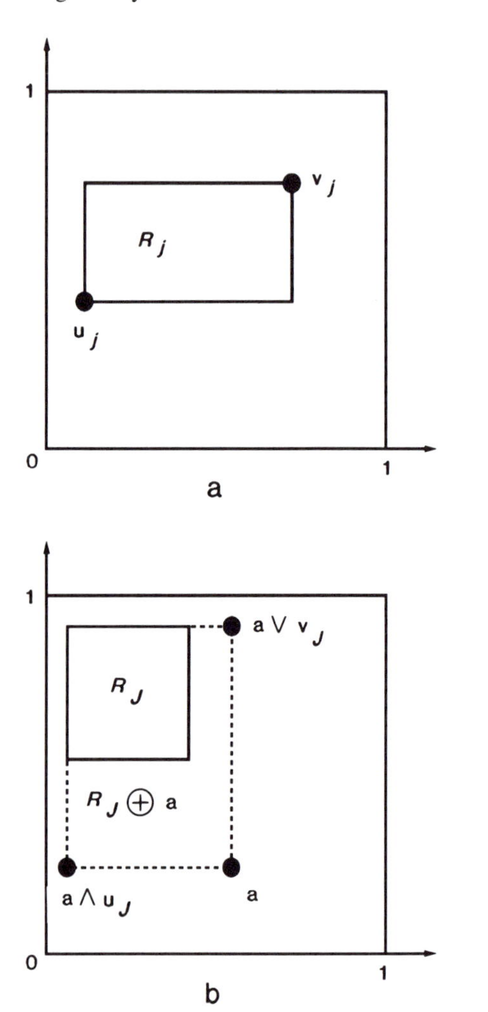

FIGURE 5 Fuzzy ART weight representation. (a) In complement coding form with $M = 2$, each weight vector $\mathbf{w}_j$ has a geometric interpretation as a rectangle R_j with corners $(\mathbf{u}_j, \mathbf{v}_j)$. (b) During fast learning, R_J expands to $R_J \oplus \mathbf{a}$, the smallest rectangle that includes R_J and $\mathbf{a}$, provided that $|R_J \oplus \mathbf{a}| \leq 2(1 - \rho)$.

$$|R_J \oplus \mathbf{a}| = |(\mathbf{a} \vee \mathbf{v}_J) - (\mathbf{a} \wedge \mathbf{u}_J)| \tag{26}$$

However, reset leads to another category choice if $|R_J \oplus \mathbf{a}|$ is too large. In summary, with fast learning, each R_j equals the smallest rectangle that encloses all vectors $\mathbf{a}$ that have chosen category j, under the constraint that $|R_j| \leq 2(1 - \rho)$.

SIMULATION: CIRCLE-IN-THE-SQUARE

The circle-in-the square problem requires a system to identify which points of a square lie inside and which lie outside a circle whose area equals half that of the square. This task was specified as a benchmark problem for system performance evaluation in the DARPA Artificial Neural Network Technology (ANNT) Program (Wilensky, 1990). Wilensky examined the performance of 2-n-1 back-propagation systems on this problem. He studied systems in which the number (n) of hidden units ranged from 5 to 100, and the corresponding number of weights ranged from 21 to 401. Training sets ranged in size from 150 to 14,000. To avoid overfitting, training was stopped when accuracy on the training set reached 90%. This criterion level was reached most quickly (5000 epochs) in systems with 20 to 40 hidden units. In this condition, approximately 90% of test set points, as well as training set points, were correctly classified.

Fuzzy ARTMAP performance on this task after one training epoch is illustrated in Figures 6 and 7. As training set size increased from 100 exemplars (Figure 6A) to 100,000 exemplars (Figure 6D) the rate of correct test set predictions increased from 88.6% to 98.0%, whereas the number of ART_a category nodes increased from 12 to 121. Each category node j required four learned weights $\mathbf{w}_j^a$ in ART_a plus one map field weight $\mathbf{w}_j^{ab}$ to record whether category j predicts that a point lies inside or outside the circle. Thus, for example, 1-epoch training on 100 exemplars used 60 weights to achieve 88.6% test set accuracy. Figure 7 shows the ART_a category rectangles R_j^a established in each simulation of Figure 6. Initially, large R_j^a estimated large areas as belonging to one or the other

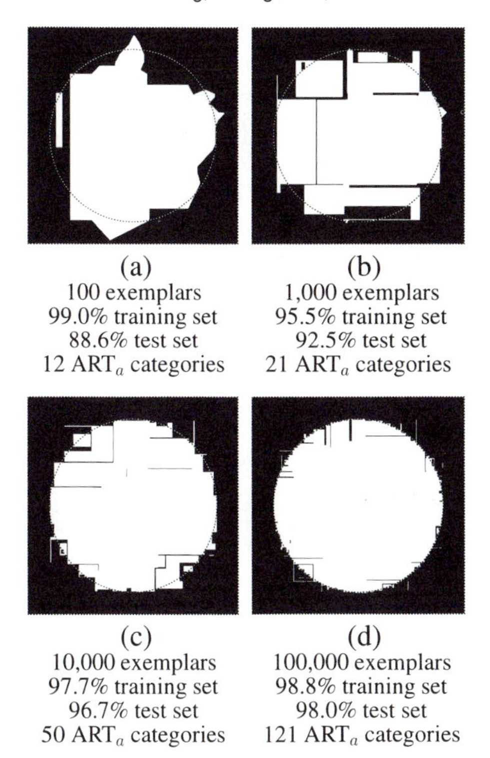

FIGURE 6 Circle-in-the-square test set response patterns after one epoch of fuzzy ARTMAP training on (a) 100, (b) 1,000, (c) 10,000, and (d) 100,000 randomly chosen training set points. Test set points in white areas are predicted to lie inside the circle and points in black areas are predicted to lie outside the circle. The test set error rate decreases, approximately inversely to the number of ART$_a$ categories, as the training set size increases.

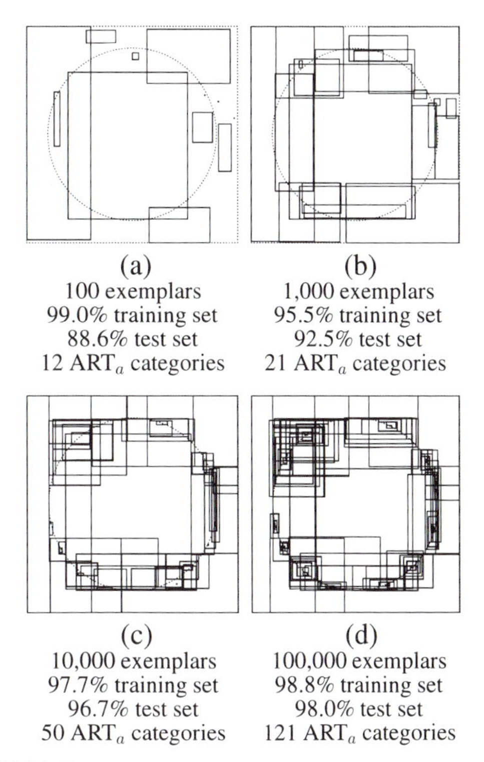

FIGURE 7 Fuzzy ARTMAP category rectangles R_j^a for the circle-in-the-square simulations of Figure 6. Small rectangles are created near the map discontinuities as the error rate drops toward 0.

category plus three point rectangles created near the decision boundary to correct errors (Figure 7A). Additional R_j^a improved accuracy, especially near the boundary of the circle (Figure 7D). The map can be made arbitrarily accurate provided the number of ART$_a$ nodes is allowed to increase as needed. As in Figure 5, each rectangle R_j^a corresponds to the four-dimensional weight vector $\mathbf{w}_j^a = (\mathbf{u}_j^a, (\mathbf{v}_j^a)^c)$, were $\mathbf{u}_j^a$ and $\mathbf{v}_j^a$ are plotted as the lower-left and upper-right corners of R_j^a, respectively.

Figure 8 depicts the response patterns of fuzzy ARTMAP on another series of circle-in-the-square simulations. The simulations used the same training sets as in Figure 6, but with each training set input presented for as many epochs as were needed to achieve 100% predictive accuracy on the training set. In each case, test set predictive accuracy increased, as did the number of ART$_a$ category nodes. For example, with 10,000 exemplars, 1-epoch training used 50 ART$_a$ nodes to give 96.7% test set accuracy (Figure 6C). The same training set, after 6 epochs, used 89 ART$_a$ nodes to give 98.3% test set accuracy (Figure 8C).

Figure 6 showed a test set error rate that is reduced from 11.4% to 2.0% as training set size increases from 100 to 100,000 in 1-epoch simulations. Figure 8 showed how a test set error rate can be further reduced if exemplars are presented for as many epochs as necessary to reach 100% accuracy on the training set. An ARTMAP *voting strategy* provides a third way to eliminate test set errors. The voting strategy assumes a fixed set of training exemplars, with the input ordering randomly assembled before each individual simulation. After the simulation, the prediction of each test set item is recorded. Voting selects the outcome predicted by the largest number of individual simulations. In case of a tie, one outcome is selected at random. The number of votes cast for a given outcome provides a measure of predictive confidence at each test set point. Given a limited training set, voting across a few simulations can improve predictive accuracy by a factor that is comparable to the improvement that could be attained by an order of magnitude more training set inputs, as shown in the following example.

A fixed set of 1000 randomly chosen exemplars was presented to a fuzzy ARTMAP system on five independent 1-epoch circle-in-the-square simulations. After each simulation, inside/outside predictions were recorded on a 1000-item test set. Accuracy on individual simulations ranged from 85.9% to 93.4%, averaging 90.5%, and the system used from 15 to 23 ART_a nodes. Voting by the five simulations improved test set accuracy to 93.9% (Figure 9C). In other words, test set

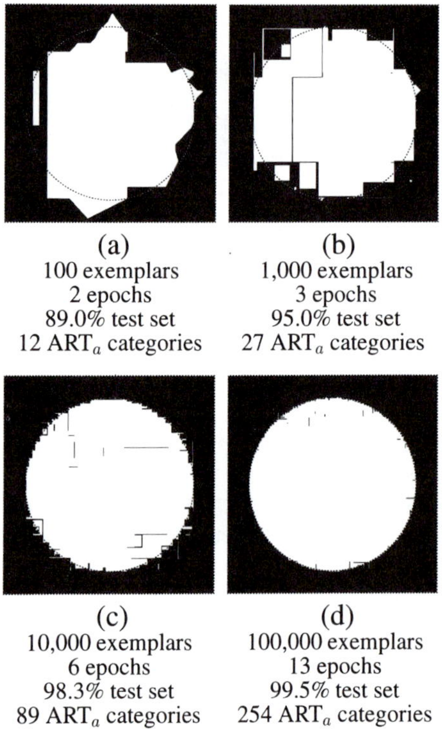

(a)
100 exemplars
2 epochs
89.0% test set
12 ART_a categories

(b)
1,000 exemplars
3 epochs
95.0% test set
27 ART_a categories

(c)
10,000 exemplars
6 epochs
98.3% test set
89 ART_a categories

(d)
100,000 exemplars
13 epochs
99.5% test set
254 ART_a categories

FIGURE 8 Circle-in-the-square test set response patterns with exemplars repeatedly presented until the system achieved 100% correct prediction on (a) 100, (b) 1,000, (c) 10,000, and (d) 100,000 training set points. Training sets were the same as those used for Figures 6 and 7. Training to 100% accuracy required (a) 2 epochs, (b) 3 epochs, (c) 6 epochs, and (d) 13 epochs. Additional training epochs decreased test set error rates but created additional ART_a categories, compared with the one-epoch simulation in Figure 6.

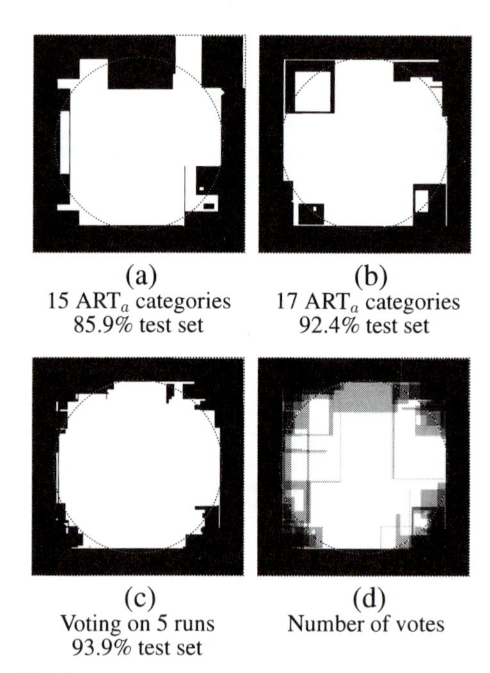

FIGURE 9 Circle-in-the-square response patterns for a fixed 1000-item training set. (a) Test set responses after training on inputs presented in random order. After one epoch that used 15 ART_a nodes, test set prediction rate was 85.9%, the worst of five runs. (b) Test set responses after training on inputs presented in a different random order. After one epoch that used 17 ART_a nodes, test set prediction rate was 92.4%, the best of five runs. (c) Voting strategy applied to five individual simulations. Test set prediction rate was 93.9%. (d) Cumulative test set response pattern of five 1-epoch simulations. Gray scale intensity increases with the number of votes cast for a point's being outside the circle.

errors were reduced from an average individual rate of 9.5% to a voting rate of 6.1%. Figure 9D indicates the number of votes cast for each test set point, and hence reflects variations in predictive confidence across different regions. Voting by more than five simulations maintained an error rate between 5.8% and 6.1%. This limit on further improvement by voting appears to be caused by random gaps in the fixed 1000-item training set. By comparison, a 10-fold increase in the size of the training set reduced the error by an amount similar to that achieved by five-simulation voting. For example, in Figure 6B, 1-epoch training on 1000 items yielded a test set error rate of 7.5%, whereas increasing the size of the training set to 10,000 reduced the test set error rate to 3.3% (Figure 6C).

In the circle-in-the square simulations, $M_a = 2$ and ART_a inputs **a** were randomly chosen points in the unit square. Each F_1^a input **A** had the form

$$\mathbf{A} = (a_1, a_2, 1 - a_1, 1 - a_2) \tag{27}$$

and $|\mathbf{A}| = 2$. For ART_b, $M_b = 1$. The ART_b input **b** was given by

FIGURE 10 Illustrative letter fonts used by Frey and Slate (1991).

$$\mathbf{b} = \begin{cases} (1) & \text{if } \mathbf{a} \text{ is inside the circle} \\ (0) & \text{otherwise} \end{cases} \tag{28}$$

In complement coding form, the $F_0^b \to F_1^b$ input **B** is given by

$$\mathbf{B} = \begin{cases} (1,0) & \text{if } \mathbf{a} \text{ is inside the circle} \\ (0,1) & \text{otherwise} \end{cases} \tag{29}$$

The fuzzy ARTMAP simulations used fast learning, defined by Equation (11) with $\beta = 1$; the choice parameter $\alpha \cong 0$ (the conservative limit) for both ART_a and ART_b; and the baseline vigilance parameter $\overline{\rho_a} = 0$. The vigilance parameters ρ_{ab} and ρ_b can be set to any value between 0 and 1 without affecting fast-learn results. In each simulation, the system was trained on the specified number of exemplars, then tested on 1000 or more points.

TWO ANALOG ARTMAP BENCHMARK STUDIES: LETTER AND WRITTEN DIGIT RECOGNITION

As summarized in Table 2, fuzzy ARTMAP has been benchmarked against a variety of machine learning, neural network, and genetic algorithms with considerable success. An illustrative study used a benchmark machine learning task that Frey and Slate (1991) developed and described as a "difficult categorization problem" (p. 161). The task requires a system to identify an input exemplar as one of 26 capital letters A–Z. The database was derived from 20,000 unique black-and-white pixel images. The difficulty of the task results from the wide variety of letter types represented: the twenty "fonts represent five different stroke styles (simplex, duplex, complex, and Gothic) and six different letter styles (block, script, italic, English, Italian, and German)" (p. 162). In addition, each image was randomly distorted, leaving many of the characters misshapen (Figure 10). Sixteen numerical feature attributes were then obtained from each character image, and each attribute value was scaled to a range of 0 to 15. The resulting Letter Image Recognition file is archived in the UCI Repository of Machine Learning Databases and Domain Theories (ml_repository@ics.uci.edu).

Frey and Slate used this database to test performance of a family of classifiers based on Holland's genetic algorithms (Holland, 1980). The training set consisted of 16,000 exemplars, with the remaining 4,000 exemplars used for testing. Genetic algorithm classifiers having different input representations, weight update and rule creation schemes, and system parameters were systematically compared. Training was carried out for five epochs, plus a sixth "verification" pass during which no new rules were created but a large number of unsatisfactory rules were discarded. In Frey and Slate's comparative study, these systems had correct prediction rates that ranged from 24.5% to 80.8% on the 4000-item test set. The best performance (80.8%) was obtained using an integer input representation, a reward sharing weight update, an exemplar method of rule creation, and a parameter setting that allowed an unused or erroneous rule to stay in the system for a long time before being discarded. After training, the optimal case that had 80.8% performance rate ended with 1,302 rules and eight attributes per rule, plus more than 35,000 more rules that were discarded during verification. (For purposes of comparison, a rule is somewhat analogous to an ART_a category in ARTMAP, and the number of attributes per rule is analogous to the size of ART_a category weight vectors.) Building on the results of their comparative study, Frey and Slate investigated two types of alternative algorithms, namely an accuracy-utility bidding system, that had slightly improved performance (81.6%) in the best case, and an exemplar/hybrid rule creation scheme that further improved performance, to a maximum of 82.7%, but that required the creation of over 100,000 rules before the verification step.

Fuzzy ARTMAP had an error rate on the letter recognition task that was consistently less than one third of that of the three best Frey-Slate genetic algorithm classifiers just described. In particular, after one to five epochs, individual fuzzy ARTMAP systems had a robust prediction rate of 90–94% on the 4000-item test set. The ARTMAP voting strategy consistently eliminated 25–43% of the errors, giving a robust prediction rate of 92–96%. Moreover fuzzy ARTMAP simulations each created fewer than 1070 ART_a categories, compared with the 1040–1302 final rules of the three genetic classifiers with the best performance rates. Most fuzzy ARTMAP learning occurred on the first epoch, with test set performance on systems trained for one epoch typically over 97% that of systems exposed to inputs for five epochs.

Rapid learning was also found in a benchmark study of written digit recognition, in which the correct prediction rate on the test set after one epoch reached over 99% of its best performance (Carpenter, Grossberg, & Iizuka, 1992). In this study, fuzzy ARTMAP was tested along with back-propagation and a self-organizing feature map. Voting yielded fuzzy ARTMAP average performance rates on the test set of 97.4% after an average number of 4.6 training epochs. Back-propagation achieved its best average performance rates of 96% after 100 training epochs. Self-organizing feature maps achieved a best level of 96.5%, again, after many training epochs.

In summary, on a variety of benchmarks, fuzzy ARTMAP has demonstrated much faster learning and better performance compared with alternative machine learning, genetic, or neural network algorithms. In addition, fuzzy ARTMAP can be used in applications where many other adaptive pattern recognition algorithms cannot perform well. These are the classes of applications where very large nonstationary databases need to be organized rapidly into stable variable-compression categories under real-time autonomous learning conditions.

CONCLUDING REMARKS

Fuzzy ARTMAP is one of a rapidly growing family of attentive self-organizing learning hypothesis testing and prediction systems that have evolved from the biological theory of cognitive information processing, of which ART forms an important part (Carpenter & Grossberg, 1991). ART modules have found their way into such diverse applications as the control of mobile robots, a Macintosh system that adapts to user behavior, diagnostic monitoring systems for nuclear plants, learning and search of airplane part inventories, medical diagnosis, three-dimensional visual object recognition, musical analysis, seismic recognition, sonar recognition, and laser radar recognition (Baloch & Waxman, 1991; Caudell, Smith, Johnson, Wunsch, & Escobedo, 1991; Gjerdingen, 1990; Goodman et al., 1992a,b; Keyvan, Durg, & Rabelo, 1993; Johnson, 1993; Seibert & Waxman, 1991). All of these applications exploit the ability of ART systems to learn to classify rapidly large databases in a stable fashion, to calibrate their confidence in a classification, and to focus attention on those featural groupings that they deem to be important on the basis of their past experience. We anticipate that the growing family of supervised ARTMAP systems will find an even broader range of applications because of their ability to adapt the number, shape, and scale of their category boundaries to meet the on-line demands of large nonstationary databases.

Acknowledgments

G. A. C. was supported in part by ARPA (N00014-92-J-4015), the National Science Foundation (NSF IRI-94-01659), and the Office of Naval Research (ONR N00014-91-J-4100). S. G. was supported in part by the Air Force Office of Scientific Research (AFOSR F49620-92-J-0225), ARPA (ONR N00014-92-J-4015), and the Office of Naval Research (ONR N00014-91-J-4100). The authors thank Cynthia Bradford, Robin Locke, and Diana Meyers for their valuable assistance in the preparation of the manuscript.

References

Baloch, A. J., & Waxman, A. M. (1991). Visual learning, adaptive expectations, and learning behavioral conditioning of the mobile robot MAVIN. *Neural Networks*, **4**, 271–302.

Carpenter, G. A., & Grossberg, S. (1987). A massively parallel architecture for a self-organizing neural pattern recognition machine. *Computer Vision, Graphics, and Image Processing*, **37**, 54–115. Reprinted in G. A. Carpenter, & S. Grossberg (Eds.), *Pattern recognition by self-organizing neural networks*. Cambridge, MA: MIT Press, 1991.

Carpenter, G. A., & Grossberg, S., Eds. (1991). *Pattern recognition by self-organizing neural networks*. Cambridge, MA: MIT Press.

Carpenter, G. A., & Grossberg, S. (1993). Normal and amnesic learning, recognition, and memory by a neural model of cortico-hippocampal interactions. *Trends in Neurosciences*, **16**, 131–137.

Carpenter, G. A., Grossberg, S., & Iizuka, K. (1992). Comparative performance measures of fuzzy ARTMAP, learned vector quantization, and back propagation for handwritten character recognition. *Proceedings of the International Joint Conference on Neural Networks, Baltimore*, **I**, 794–799. Piscataway, NJ: IEEE Service Center.

Carpenter, G. A., Grossberg, S., Markuzon, N., Reynolds, J. H., & Rosen, D. B. (1992). Fuzzy ARTMAP: A neural network architecture for incremental supervised learning of analog multidimensional maps. *IEEE Transactions on Neural Networks*, **3**, 698–713.

Carpenter, G. A., Grossberg, S., & Reynolds, J. H. (1991). ARTMAP: Supervised real-time learning and classification of nonstationary data by a self-organizing neural network. *Neural Networks*, **4**, 565–588. Technical Report CAS/CNS-TR-91-001. Boston, MA: Boston University. Reprinted in G. A. Carpenter, & S. Grossberg (Eds.), *Pattern recognition by self-organizing neural networks*. Cambridge, MA: MIT Press, 1991.

Carpenter, G. A., Grossberg, S., & Rosen, D. B. (1991). Fuzzy ART: Fast stable learning and categorization of analog patterns by an adaptive resonance system. *Neural Networks*, **4**, 759–771. Technical Report CAS/CNS-TR-91-015. Boston, MA: Boston University.

Caudell, T., Smith, S., Johnson, C., Wunsch, D., & Escobedo, R. (1991). An industrial application of neural networks to reusable design. *Adaptive neural systems*, Technical Report BCS-CS-ACS-91-001, Seattle, WA: The Boeing Company, pp. 185–190.

Desimone, R. (1992). Neural circuits for visual attention in the primate brain. In G. A. Carpenter & S. Grossberg (Eds.), *Neural networks for vision and image processing* (pp. 343–364). Cambridge, MA: MIT Press.

Frey, P. W., & Slate, D. J. (1991). Letter recognition using Holland-style adaptive classifiers. *Machine Learning*, **6**, 161–182.

Gjerdingen, R. O. (1990). Categorization of musical patterns by self-organizing neuronlike networks. *Music Perception*, **7**, 339–370.

Goodman, P. H., Kaburlasos, V. G., Egbert, D. D., Carpenter, G. A., Grossberg, S., Reynolds, J. H., Hammermeister, K., Marshall, G., & Grover, F. (1992a). Fuzzy ARTMAP neural network prediction of heart surgery mortality. *Proceedings of the Wang Institute Research Conference: Neural Networks for Learning, Recognition, and Control*, p. 48. Boston, MA: Boston University.

Goodman, P. H., Kaburlasos, V. G., Egbert, D. D., Carpenter, G. A., Grossberg, S., Reynolds, J. H., Rosen, D. B., & Hartz, A. J. (1992b). Fuzzy ARTMAP neural network compared to linear discriminant analysis prediction of the length of hospital stay in patients with pneumonia. *Proceedings of the IEEE International Conference on Systems, Man, and Cybernetics* (Chicago), pp. 748–753. New York: IEEE Press.

Harries, M. H., & Perrett, D. I. (1991). Visual processing of faces in temporal cortex: Physiological evidence for a modular organization and possible anatomical correlates. *Journal of Cognitive Neuroscience*, **3**, 9–24.

Holland, J. H. (1980). Adaptive algorithms for discovering and using general patterns in growing knowledge bases. *International Journal of Policy Analysis and Information Systems*, **4**, 217–240.

Keyvan, S., Durg, A., & Rabelo, L. C. (1993). Application of artificial neural networks for development of diagnostic monitoring system in nuclear plants. *American Nuclear Society Conference Proceedings*, April 18–21.

Johnson, C. (1993). Agent learns user's behavior. *Electrical Engineering Times*, June 28, 43–46.

Laird, J. E., Newell, A., & Rosenbloom, P. S. (1987). SOAR: An architecture for general intelligence. *Artificial Intelligence*, **33**, 1–64.

Miller, E. K., Li, L., & Desimone, R. (1991). A neural mechanism for working and recognition memory in inferior temporal cortex. *Science*, **254**, 1377–1379.

Mishkin, M. (1982). A memory system in the monkey. *Philosophical Transactions of the Royal Society of London Series B*, **298**, 85–95.

Moore, B. (1989). ART 1 and pattern clustering. In D. Touretzky, G. Hinton, & T. Sejnowski (Eds.), *Proceedings of the 1988 Connectionist Models Summer School* (pp. 174–185). San Mateo, CA: Morgan Kaufmann.

Rosenblatt, F. (1958). The perceptron: A probabilistic model for information storage and organization in the brain. *Psychological Review*, **65**, 386–408. Reprinted in J. A. Anderson & E. Rosenfeld (Eds.), *Neurocomputing: Foundations of research* (pp. 18–27). Cambridge, MA: MIT Press, 1988.

Rumelhart, D. E., Hinton, G., & Williams, R. (1986). Learning internal representations by error propagation. In D. E. Rumelhart & J. L. McClelland (Eds.), *Parallel distributed processing*. Cambridge, MA: MIT Press.

Salzberg, S. L. (1990). *Learning with nested generalized exemplars*. Hingham, MA: Kluwer.

Seibert, M., & Waxman, A. M. (1991). Learning and recognizing 3D objects from multiple views in a neural system. In H. Wechsler (Ed.), *Neural networks for perception* (Vol. 1). New York: Academic Press.

Smith, E. E. (1990). In D. O. Osherson & E. E. Smith (Eds.), *An invitation to cognitive science*. Cambridge, MA: MIT Press.

Spitzer, H., Desimone, R., & Moran, J. (1988). Increased attention enhances both behavioral and neuronal performance. *Science*, **240**, 338–340.

Werbos, P. (1974). *Beyond regression: New tools for prediction and analysis in the behavioral sciences*. Ph.D. thesis, Harvard University, Cambridge, MA.

Wilensky, G. (1990). Analysis of neural network issues: Scaling, enhanced nodal processing, comparison with standard classification. *DARPA Neural Network Program Review*, October 29–30.

Zadeh, L. (1965). Fuzzy sets. *Information Control*, **8**, 338–353.

Author Index

Numbers in italics refer to the pages on which the complete references are cited.

Subject Index